Rational Readings on Environmental Concerns

Edited by

Jay H. Lehr

VNR VAN NOSTRAND REINHOLD
New York

Copyright © 1992 by Van Nostrand Reinhold
Except for entries carrying © permissions

Library of Congress Catalog Card Number 92-3240
ISBN 0-442-01146-6

Manufactured in the United States of America

Published by Van Nostrand Reinhold
115 Fifth Avenue
New York, New York 10003

Chapman and Hall
2–6 Boundary Row
London, SE1 8HN, England

Thomas Nelson Australia
102 Dodds Street
South Melbourne 3205
Victoria, Australia

Nelson Canada
1120 Birchmount Road
Scarborough, Ontario M1K 5G4, Canada

16 15 14 13 12 11 10 9 8 7 6 5 4 3 2 1

Library of Congress Cataloging-in-Publication Data

Rational readings on environmental concerns / edited by Jay H. Lehr.
    p. cm.
    Includes bibliographical references.
    ISBN 0-442-01146-6
    1. Pollution—Environmental aspects. 2. Environmental protection.
3. Environmental policy. 4. Ecology. I. Lehr, Jay H., 1936–
TD176.7.R38 1992
363.7—dc20   92-3240
                                                             CIP
                                                              r92

The Editor

Dr. Jay Lehr is world renowned as a water scientist and is frequently quoted as a leading authority on ground water hydrology and hydrogeology in America. He served as Executive Director of the Association of Ground Water Scientists and Engineers for the last 25 years.

Dr. Lehr graduated from Princeton University in 1957 with a degree in geological engineering. Following a tour of duty in the U.S. Navy's Civil Engineering Corp, he received the nation's first Ph.D. in ground water hydrology from the University of Arizona in 1962.

A pioneer of much of the research currently being done throughout the world on ground water topics, Dr. Lehr's promotion of the need for increased data regarding underground resources has fostered the development of significant work in both public and private sectors. He has been involved in research topics such as Law, Regulations and Institutions for the Control of Ground Water Pollution, Impact of Abandoned Wells on Ground Water, Technology of Subsurface Wastewater Injection, Domestic Water Conditioning and Treatment, Well Maintenance and Rehabilitation, Groundwater Geothermal Heat Pump Applications, and Groundwater Rehabilitation and Aquifer Restoration Techniques.

He has assisted more than a dozen foreign governments in establishing ground water management programs, water well construction regulations, artificial recharge programs, and irrigation technology. Dr. Lehr is the author of ten published books including *Water Well Technology* and *Domestic Water Treatment*. His latest book on ground water is *Design and Construction of Water Wells: A Guide for Engineers*. Dr. Lehr is also a physical fitness expert and has published a popular text on that subject entitled *Fit Firm & 50*.

Dr. Lehr has written over 300 scientific articles for such journals as *Science*, *Journal of Soil and Water Conservation*, *Journal of Geological Education*, *Journal: Water Pollution Control Federation*, *Journal of Environmental Health*, *Water and Sewage Works*, Journal of *Ground Water*, and *Ground Water Monitoring Review*. He was editor/publisher of *Ground Water, Ground Water Monitoring Review,* and *Water Well Journal* for more than 25 years. Dr. Lehr now directs an organization dedicated to continuing education for environmental professionals.

Contents

Biotechnology

Cancer/Carcinogenesis

DDT

Dioxin

Electromagnetic Fields

Environmental Economics

Environmentalism: What's Real, What's Not

Greenhouse/Global Warming

Landfills

Media Coverage

Medicine

Nutrition

Ozone

Population

Radiation and Nuclear Energy

Radon

Recycling

Reverse Effects

Preface

The purpose of *Rational Readings on Environmental Concerns* is to evaluate environmental issues which arose initially through scientific investigation and were later promoted by advocacy groups, legislative initiatives, and public debate. Contributions to this book come from scientists held in the highest regard by their peers. The discussion of each environmental issue gives special attention to the uncertainties of science and the economic trade-offs that must be made in a democratic society.

We recognize the great diversity in the various public interest environmental organizations which comprise the environmental movement, together with the plethora of governmental and international agencies established to work on environmental issues. Some are respectful of science and optimistic about technology, while others are not. Our contemporary media and system of government, unfortunately, pay most attention to sensationalized scenarios predicting gloom-and-doom which result in much government decision-making which is not based on science and hard facts but rather allegations, prejudice and fear. Such environmental decisions are too often made without regard to comparative risks and priorities and thus can impose enormous economic damage without any comparable social benefit and, at times, even negatively impact both human health and the environment.

Environmental policies must be based on a process of careful, dispassionate, scientific inquiry and civil and democratic discourse which recognizes the genuine dilemmas society faces.

Fifty-one scientists have contributed 81 papers on 29 subjects to this text which is intended to shed light rather than heat on environmental concerns. We hope this will allow those in decision-making positions, in both the public and private sectors, members of the press, environmental activists and the public at large, to better evaluate these issues toward improving society's management of its environment in the future.*

Jay H. Lehr, Ph.D.

*NOTE: The Editor acknowledges the assistance of Peter Samuel and Greentrack International, a Washington, DC based reporting service in the preparation of this Preface.

Acknowledgments

The Editor would like to acknowledge the numerous contributions made by Priscilla Frasher who coordinated the entire communication effort between me and the many scientists contributing to this anthology. Her good nature and excellent organizational skills forged a brilliant group of authors into an enthusiastic team effort.

Tremendous thanks are also accorded to Merry Pryor and Louise Hambel. Their typesetting skills were noted by all the authors whose galley corrections were minimal, and for their patience and forbearance with the Editor during his extensive editing and abridgment efforts which were assisted by Anita Stanley. Additionally, Merry was responsible for articulately creating the interesting and balanced biographical summaries which preceded each article.

Rational Readings on Environmental Concerns

Introduction

Nature's Value
Michael M. Gemmell

Ecology's Ancestry
Michael M. Gemmell and Jay H. Lehr

Environmentalism and the Assault on Reason
Richard F. Sanford

Nature's Value

Michael M. Gemmell

Michael Gemmell is the editor and publisher of *The Free-Market Environmentalist*. He was previously affiliated with Dames & Moore as a project manager; and with Woodward-Clyde Consultants as an assistant project hydrologist.

Mr. Gemmell received his B.S. in geology from the University of California; and his M.S. in geology from California State University. He has contributed to *Ground Water Monitoring Review* and is a regular contributor to *The Free-Market Environmentalist*.

The driving force behind the environmental movement is the idea that nature is a value regardless of its utility to man. Its wonders are to be preserved rather than modified. Modification is destruction according to the environmentalist creed. The more extreme elements of the environmental movement such as the group *Earth First* say this explicitly. To one degree or another the environmental mainstream believes that nature's value is separate from its relationship to human well being.

What is the appeal of environmentalism? Why does it persist when its leading proponents such as Paul Ehrlich and Stephen Schneider keep prophesying disaster after disaster that never occur?

In my judgment, its appeal has two components: the manipulation of fear in regards to the dynamics of capitalism, and the offering of an allegedly positive alternative to relieve that fear. This the same technique used by intellectuals to attack capitalism for the last 200 years. However, earlier intellectuals addressed the social aspects of capitalism, while the environmentalists concentrate on the dynamics of capitalism as related to man's transformation of nature.

Capitalism is a political-economic system that is dynamic by nature. Its implementation led to the transformation of society and nature by breaking down social class barriers and by the use of technology to create industrial civilization. The transformation of both society and nature is inseparable for the flourishing of capitalism.

Sustaining the dynamics of capitalism requires an understanding and acceptance of change. It requires a consistent philosophic perspective showing that the inherent changes brought about by capitalism represent an overall improvement in the quality of life. Unfortunately, from its very inception its defenders have not displayed that perspective. For example, a prominent thinker of the Enlightenment, John Locke, combined elements of collectivism with individualism in his moral theories.

In the social realm, the contradictions of the Enlightenment thinkers were exploited by Rousseau, Marx, and others. They manipulated people's fear of change by asserting that workers would become slaves under capitalism. They offered a seductive vision of "utopia" based on communal solidarity.

After 200 years of trying to implement this vision its intellectual bankruptcy has finally become apparent. The realization that the individual must take responsibility for himself and not look to the state is being revived around the world. However, the defense for the other half of capitalism—the transformation of nature—has not fully emerged. Consequently, the environmental movement has moved into the void.

The environmentalists have employed the same technique as earlier intellectuals of manipulating people's fears of the dynamic changes inherent in capitalism, this time in regard to nature. In the 1970s it was the fear of global cooling, whereas now it is global warming, the ozone layer, etc. Their allegedly positive alternative to capitalism is the view that man can live in harmony with nature if he does not alter it with industrialization.

But their alleged love of nature is a smokescreen for their underlying fear (not to mention loathing) of the dynamics of capitalism. Their motives have been revealed as evidenced by their almost total lack of concern over Saddam Hussein's torching of Kuwait's oil fields. This action was the worst manmade environmental disaster in human history. (Even so, indications are that the effects are localized and the environment will recover. It remains to be seen whether the people of Kuwait will.) Yet, the reaction of the environmental community was lukewarm, at best. On the other hand, Exxon, in 1989, accidentally ran a tanker aground, lost millions of dollars, was convicted of a criminal offense, and still had the environmental community on its back like a pack of vultures. It is not an accident that an ardent anticapitalist, Tom Hayden, was the driving force behind the most coercive piece of environmental legislation ever proposed, California's "Big Green" initiative.

The environmentalists have promoted the idea that nature is benevolent to the point that it has become religious dogma. They have become religious zealots with nature as their god. They refuse to acknowledge that an advanced technological society is on balance safer than a non-technological society, and they continually twist scientific data to support that view. Freedom from religious or ideological intolerance is a fundamental right in the U.S., and the environmental mainstream is blatantly guilty of violating it.

The defense of capitalism needs a philosophic perspective that integrates environmental values into a capitalistic context. However, for this to occur, the standard of value has to be the enhancement of human life.

Nature is not inherently benevolent; yet it is inherently under-standable. It is the inherent logic of natural laws that allow us not just to utilize nature but to improve upon it. It is through knowledge that a genuine reverence for nature can develop. It is this knowledge that allows us to deal with nature with confidence rather than with the chronic fear propagated by the environmentalists.

Man's mistakes pale next to the destruction caused by natural forces. The Amoco Cadiz spill of 1978 was six times the size of the 1989 Exxon Valdez, yet within two years the French Brittany coast had recovered with the exception of some marsh areas which took several years longer (Political Economy Research Center News Release, April 1991). On a wider perspective, the divisions between different periods of the geologic time scale have been delineated based on mass extinctions. For example, at least twice in geologic history, at the end of the Paleozoic Era and the Cretaceous period (225 and 65 million years ago, respectively), up to 75% of the species on earth died out.

Even with these catastrophic changes the ecosystems of the world recovered and evolved. But Life does not end at the level of an ecosystem just as human life does not end at the level of a society. The ecosystem in the natural world and societies in the human world are terms describing the interactions of living organisms. They are not themselves entities capable of making life-sustaining decisions.

When individual human life is held as the standard, a pro-human form of environmentalism will naturally develop. No one cares more about nature than those who deal with it on a continuing basis for their livelihood. I have witnessed this phenomenon while attending the Redwood Region Logging Conference in the Pacific Northwest. The loggers of the Northwest care about the health of the forest. They know its rhythms why it changes, and how to utilize it for human well being. They know that short-sighted harvesting techniques lead to the depletion of valuable resources. People like the loggers in the northwest truly deserve the dual term industrialist/environmentalist.

Should we expect the new industrialist/environmentalists to be perfect "custodians" of the environment? Of course not. But it's a good bet that none of their mistakes will wipe out 75% of the species on the planet. Furthermore, through the application of science technology and a philosophic perspective that holds human life as the standard of value their mistakes will be insignificant compared to their benefits to both mankind and nature.

Reprinted, with permission, from
the *The Free-Market Environmentalist*,
Vol. 1, No. 3, August 1991.

Ecology's Ancestry

Michael M. Gemmell and Jay H. Lehr, Ph.D.

Please see biographical sketch for Mr. Gemmell on page 3.

Please see biographical sketch for Dr. Lehr on page v.

Western philosophers devolve individual rights

The current level of environmental awareness has its origins in the ecology movement, which began in the middle of the 19th century. It developed when the ideas of the Age of Reason were coming under attack.

The fundamental ideas set forth in the Age of Reason were that Man's rational faculty was his means of dealing with reality and that a rational faculty was the property of the individual, not society. One example of these ideas in action was the creation of the United States Constitution and its Bill of Rights that explicitly protected the rights of the individual.

By the end of the 18th century, Western philosophers began attacking the fundamental ideas of the Age of Reason. The most effective attack was mounted by the German philosopher Immanuel Kant. He contended that reality was subjective rather than objective. His alleged proof was that the senses distort reality rather than perceive it.

The German philosopher George Hegel carried the attack on objective reality even further. According to him, reality is inherently contradictory and true reality consists of the "identity of opposites." Hegel's alleged proof involved the use of a "dialectic" logic (combinations of opposites to form a higher truth).

These attacks on objective reality and Man's means of perceiving it (senses—logic—rational faculty) were used as a foundation for the attacks on the dominant ideas of ethics. Kant and Hegel contended, in essence, that what appeared to be true (i.e., humans were individual beings) did not represent the higher truth. That higher truth being, individuals were appendages of society. In the "Philosophy of Right," Hegel states: "A single person, I need hardly say, is something subordinate, and as such he must dedicate himself to the ethical whole. Hence if the state claims life, the individual must surrender it."

Into this framework of ideas, ecology was born.

6

In The Beginning

The origin of the ecologic world view

Ecology as an identifiable philosophy began with the German zoologist Ernst Haeckel (Ecology in the 20th century, A. Bramwell, 1989). Haeckel coined the term "ecology" in his 1866 book *Generelle Morphologie,* in which he defined ecology as the science of the relationships between organisms and their environment.

It is crucial to recognize that ecology was, from the very beginning, primarily a philosophic rather than a scientific area of study. Science is the systematic study, observation, experimentation, and development of theoretical explanations to explain some aspect of the natural world. Philosophy defines the fundamental nature of living things and their relationship to the physical world. By analogy, philosophy is the soil of the forest; science consists of the trees within it. Scientific study proceeds from a philosophic framework.

Ernst Haeckel's scientific work involved the study of microscopic forms of life. From this, he made sweeping assertions about the nature of life and Man's place in it. Like most 19th and 20th century ecologists, he attempted to influence politics with his ideas.

An understanding of the nature of ecology is necessary to grasp its uniquely destructive power. Although it is philosophical, it has a scientific veneer that has allowed it to intellectually disarm the public.

The Founding Principles

Ecology is based on two false principles

Two fundamental principles form the core of ecologic philosophy: holistic biology and resource economics. Holistic biology is the doctrine that all life, and even nonlife, is part of a larger collective organism. This organism has been called "Mother Nature," "Planet Earth," and "Gaia," to name a few. Ecology's conception of resource economics asserts that resources available for life should be conserved owing to the inevitable long-term loss of energy due to Man's activities. Those are the principles. The question is, are they true?

Holistic Biology

The most fundamental problem inherent in holistic biology is its lack of distinction between the living and the nonliving. Living organisms composed of matter use energy to sustain life. Inanimate matter, on the other hand, releases or absorbs energy in response to external forces. In other words, living entities exhibit goal-oriented behavior, while non-living entities do not. As an example, a child running down a hill to fetch water from a lake differs from a rock rolling down that same hill on its way to splashing into the lake.

In addition to obliterating the distinction between the living and nonliving, holistic biology falsely asserts that biological systems (eco-systems) act as collective organisms. The best argument put forth to support this theory references Darwin's survival of the fittest theory. This theory has been misapplied to situations in which a number of different organisms react too changes in a given ecosystem (e.g., drought, fire, etc.) Because a number of organisms are reacting to changes in an ecosystem, it is claimed, this represents collectivized behavior orchestrated by "Mother Nature."

The appearance of directed behavior by "Mother Nature" results from the inhabitants scrambling to adjust to altered conditions. Loss in plant biomass, due to a natural catastrophe, means animals that consume plants will have to compete harder to obtain a smaller amount of food. This effect, in turn, ripples up the food chain to affect other animals. In this scenario, a number of organisms are reacting to changed conditions to preserve their individual lives. It is not the orchestration of events of a larger collective organism.

Resource Economics

Ecology's resource economics, asserts that resources used are resources irretrievably lost. A review of energy transfer disproves this theory.

The earth is continually bathed in energy via the sun. The earth rotates so that areas receive energy during the day and radiate energy into space during the night. Energy that is not daily radiated back into space is incorporated into biologic organisms (photosynthesis for instance) or into various physical systems (weather cycles, etc.). Energy is continually in transit, but is not lost in either physical or biological systems.

Neither is it lost as a result of Man's influence. The only difference is by virtue of his rational faculty, Man has the ability to consciously redirect the flow of energy in physical and biological systems. Redirection of energy does not represent a loss of energy.

The use of fossil fuel resources for energy illustrates this principle. Fossil fuels have been used as energy sources for several centuries. If their use represents an irretrievable loss, then the cost of these resources, should increase with time. In fact, however, the opposite is true. Rather, costs of these resources relative to per capita income, when adjusted for inflation, have been declining for decades (Population Matters: People, Resources, Environment & Immigration, J. Simon, 1990).

Simply and truthfully put, energy is a renewable resource as are other material resources. Energy is contained in all forms of matter, as Einstein proved. His famous equation, $E=MC^2$, simply means energy and matter are different forms of one another. Each time Man unlocks an energy source, it is used to enhance his life and then, via one route or another, is transferred to the earth's physical or biological systems. It is not lost. The same is true for material resources. They change form, but they are not truly destroyed through Man's use.

Natural law decrees that matter can be converted to energy and vice versa. Man, with his rational faculty, discovered how this could be accomplished more efficiently over time by identifying better methods of extracting known resources and discovering new resources. From wood burning fire, to wood burning stoves, to coal, to oil, to nuclear fission, and someday to nuclear fusion, each step of the way has unleashed more efficient ways to use energy. That is why the more efficiently energy is used, the higher the standard of living that can be maintained.

Founding Principles To Historic Folly

Militant German ecologists pursue their world vision

It was inevitable that the ecologists would advocate force in bringing their ecologic ideals to life. False ideas cannot be implemented through rational discourse and the policies proceeding therefrom. In 1905, Haeckel became president of the Monist League, which advocated the implementation of the ecologic ideal by force. By the 1920s the "Back to the Land" movement and its anti-technologic ideal were in full force and capitalism was being actively attacked by ecologists. Capitalism was seen as being individualistic, not oriented toward nature, wasteful of resources, and nonholistic.

Coercive laws and other restrictive regulations are the first steps taken when people attempt to implement false ideas. These controls breed further controls. However, the ideas breeding these controls, or at least their destructive effects, will eventually become apparent to individuals within a society. With time, more and more people will react by trying to influence the political process. Political pull then becomes the dominant force directing the movement of a culture. This was the atmosphere of the 1920s and 1930s in Germany, and ecologic ideas were a major contributor to it.

In the 1930s the ecologists "Green Revolution" reached full flower in Germany. Their rallying cry, "Blood and Soil," was an ecological slogan indicative of the ecologists' holistic vision of biological roots to the soil. In the political sphere, ecologists lobbied successfully, for antivivisection laws (prevention of animal research for medicinal purposes), creation of nature reserves, implementation of organic farming (known as biodynamic farming), and the redistribution of large land holdings to the German peasants (Back-to-the-Land movement).

These laws became the policies of a political party that incorporated a major portion of the ecologists political agenda. This party also believed in the "Blood and Soil" ethic, and was known as the National Socialist Party. Its leader was Adolf Hitler.

Adolf Hitlers become possible when a culture is unable to resolve the contradictions in its philosophic framework. People eventually become confused, disoriented, and vulnerable to someone like Hitler, who promises them answers to all their problems.

From Kant to Hegel to Haeckel to Nazi Germany. One country's ideas and one country's actions. This can hardly be termed as a coincidence.

Kant preached self-sacrifice for one's fellow man; Hegel preached self-sacrifice for the "organic whole," and Haeckel preached self-sacrifice for "Mother Nature." Kant's ethics were tailormade for a totalitarian state, although he didn't explicitly advocate it. Hegel did. Haeckel did. The leading ecologists of the 1930s did. And they got it.

An ideology built on a false foundation of ideas is by its nature unstable. The German people had been inculcated with the concept of "unity" for more than a century. They believed it, but the rest of the world remained to be unified. Ecologic ideals were eventually seen as an impediment to the technological development needed for Germany's war effort. As a result, organic farming efforts were harassed by the SS and the Back-to-the-Land Movement faded during World War II.

With the end of World War II, Nazism and associated philosophic ideas such as ecology went into hibernation in Europe. However, German philosophy had not reached the climax of its influence in the United States of America.

Ecology's Resurrection

The ecology movement takes hold in the U.S.
The 1960s saw the climax of the influence of German philosophy in the United States and, not coincidentally, resurgence of the ecology movement.

The 60s were a poignant example of the power of philosophic ideas. With the end of World War II, the modern socialist ideas fueled by the ideas of Kant, Hegel, and others lost their appeal as a moral force. Intellectuals began to see what living for "society," "mankind," etc. meant in practice. This confusion led to skepticism toward any philosophic perspective. (The Ominous Parallells, L. Peikoff, 1982)

Student activists were explicitly antibusiness, antiprofit, antifamily, antireason, and antiestablishment. Tom Hayden accurately summarized their outlook in 1968: "First we will make the revolution, and then we will find out what for."

It is literally impossible to be skeptical toward everything. To do so is to completely paralyze thought and the human life that depends on it. Some allegedly positive values slipped in: the free speech movement, the peace movement, and the nature movement, also known as the ecology movement.

The free speech movement was thinly veiled nihilism. A small group of leftist radicals on the U.C. Berkeley campus led by Mario Savio demanded the right to say whatever they thought, wherever and whenever they thought, and accused anyone who opposed them of restricting their free speech. The free speech movement eventually died. Its proponents failed to realize that flag burning and other nihilistic displays would not endear them to the American people.

The peace movement was somewhat more successful. Largely through this movement's influence, the U.S. withdrew from the Vietnam War. The essence of this movement was that force should not be used, even in resisting force initiated by others. Its proponents ignored the moral difference between those that initiate force and those that resist force initiated by others. This movement continues to affect the thinking of U.S. leaders today.

However, the ecology movement was the undisputed winner in the movements started in the 1960s. It had one of America's most treasured values defending it—science. Or to be precise, a small number of scientists claiming to represent science. They were, in essence, the 20th century version of Ernst Haeckel, philosophers masquerading behind a veneer of science.

Ecology's 20th Century Mother

19th century falsehoods are resurrected

Rachel Carson began the resurrection of the ecology movement with the publishing of *Silent Spring* in 1962. She identified some legitimate problems involved in the use of pesticides. However, she didn't stop there. She made sweeping statements about how the profit motive, capitalism, and industrial society were to blame for the problems caused by pesticide use. Her partly scientific approach, combined with her emotionally-charged prose, brought her book to a wide audience.

The widespread readership of Carson's book prompted further attention. In 1968, Paul Ehrlich published *The Population Bomb: Population Control or Race to Oblivion?* in which he asserted that widespread death and famine would unfold worldwide if population and the associated industrial technology that was consuming enormous amounts of energy were allowed to go unchecked. In 1969, Senator Edmund Muskie hosted senate hearings focusing on environmental problems, during which biologist Barry Commoner claimed that to save the environment, a change in the political and social system, including changing the U.S. constitution, would be necessary. Uncontrolled technology was once again the culprit.

Probably the most prominent scientists to join the fray were Rene Dubos and George Wald. Dubos was a highly respected microbiologist at Rockefeller University. George Wald won a Nobel Prize for Physiology and Medicine in 1968. Dubos's message was that Western society, which was based on science, technology, and economic growth, was careening toward the death of the human race. The "unity" of life was very fragile and exhibited an "egg-shell delicacy." Wald contended that unless society was reorganized, he could not see "how the human race will get itself much past the year 2000."

These sentiments vividly illustrate the resurrection of the ideas of the 19th century ecologists. Holistic biology references such as death of the entire human race, unity of the planet, and an "egg shell delicacy" are clearly evident. References to ecology's resource economics such as uncontrolled technology, economic growth, and excessive population overconsuming resources come through unmistakably.

Philosophic Corruption To Scientific Lunacy

Studies at MIT fuel the ecology movement
When bad ideas are resurrected, destructive actions based on those ideas will inevitably follow. Two studies performed by MIT at the height of the environmental concern of the late 1960s bear this out.

The first study was published in 1970 and was the "Report of the Study of Critical Environmental Problems" (SCEP). The study assessed the impact of Man's activities on the earth. The report's findings stated explicitly:

- That critically needed scientific, technical, economic, industrial, and social data were fragmentary, contradictory, and in some cases unavailable.

- Few projections existed for rates of growth of various industrial sectors, relevant domestic and agricultural activities of Man and associated energy demands.

In fact, discussing specific results such as global warming, effects of oil spills on marine life, and discharge of industrial effluents into the ocean, the consistent response of this study were that not enough data were available. The statements concerning their understanding of global issues are particularly revealing:

- "CO_2 seems to have been increasing throughout the world at about 0.2 percent per year . . . but the length of record is too short to place much emphasis on the deviations from a linear trend."

- "The area of greatest uncertainty in connection with the effects of particles on the heat balance of the atmosphere is our current lack of knowledge of their optical properties in scattering or absorbing radiation from the sun or the earth."

- "Research is most needed in providing a closer specification of the present state of the planet and in developing a more complete understanding of the mechanisms of interaction between atmosphere, ocean and ecosystem."

Simply put, the authors of the study had little understanding of how the planetary systems worked. Nevertheless, the study rampantly speculated on the disastrous events that might occur if ill-understood chemical reactions reacted with ill-understood planetary systems, etc. etc.

In 1972, the "The Limits to Growth" study was published. It concluded that the human race had perhaps only 100 years left to live if its course remained unchanged. The study recommended an immediate end to economic growth. Not surprisingly, considering the prestige of MIT's name combined with these startling recommendations, more than three million volumes of the study were sold worldwide.

Although this volume in sales indicates some outstanding merchandising skills, it doesn't necessarily follow that great scientific methodology was inherent to the product. And, in fact, these "scientists" forgot to consider a small factor—the intelligence and resourcefulness of minds other than their own.

The study took present rates of growth and consumption and projected them over time while assuming that energy resources would remain static. They ignored or were unaware of the following:

- People could discover new resources.
- People could improve technology and utilize known resources more effectively.
- Population growth rates do not continue to grow exponentially, indefinitely. When industrialization occurs, birth rates fall.

These studies were ridiculed by Noble prize winning economist Gunnar Myrdal, among others. Unfortunately, the scattered public statements were not enough to counteract the immense circulation of volumes of the study. (Source for facts on "SCEP" and "Limits to Growth—The Apocalyptics, E. Efron, 1984, Chapter 1.)

These studies acted as the catalyst to implement ecologic philosophy as public policy. In 1970, the U.S. Environmental Protection Agency (EPA) was created, and laws were rapidly promulgated concerning pesticides, air and water pollution, toxic substances, use of resources, and a host of other issues concerning the "environment."

In many cases, these regulations responded to legitimate problems. Air and water pollution were occurring, and the misuse of toxic chemicals had been clearly documented. However, the sweeping scope of the laws addressing these issues indicated that the actual problems were seen not as instances of isolated ignorance or misuse, but as mere examples of a much larger problem. The legislation was based on ecology's fundamentally flawed principles of holistic biology and resource economics.

The Cure

Expanding individual property rights offers a free-market solution

The philosophic roots of capitalism are presented in Ayn Rand's 1967 book, *Capitalism the Unknown Ideal,* and are the foundation for free-market solutions to environmental issues. The fundamental principle that must be applied to environmental issues is individual rights. In most environmental contexts, this principle is specifically, individual property rights.

All human beings are individuals. We are not part of some collective organism called Humanity, Society, Mother Earth, etc. Because we are individuals we must be able to function as individuals, which means we must have the right to own property. Property rights also properly extend to those resources we obtain by our efforts. This includes land, air, water, minerals, oil, etc.

By expanding the principles of property rights to include air, water, and other mediums, the legitimate problems we face can be addressed without coercive methods.

Air Property Rights

Pollution occurs when property rights for a given medium are not properly specified. Air as a common pool resource requires a different application of property rights than land, but the principle is still valid.

More than one equitable, objective method is available for assigning property rights to air in a basin. Assigning effluent rights on the basis of the size of land holdings is offered here as one method. (Non-landowners would rent air effluent rights as with other property rights.)

As an example, assume a basin has an area of 1,000 acres and a carrying capacity of 1,000 units/day of a certain effluent (the equivalent of 1 unit/day/acre). If two landowners in the basin owned 0.5 and 5.0 acres respectively, under this method they would be entitled to emission permits for 0.5 and 5.0 units of effluent per day, respectively. These permits would apply to stationery and non-stationery (automobiles, etc.) sources of pollution.

If one landowner generates less effluent than his permits allow, he could sell some of his permits to another landowner in the basin. The application of property rights allows all parties to pursue their own interests while simultaneously being protected from potentially destructive actions, such as impacts from pollutants, by others.

Such an approach would also foster technological innovation to reduce effluents further. When effluent reduction translates to lower costs, via selling of effluent permits, higher profits will follow.

The Future

About the authors, Free-Market Environmentalists

As ones who have spent a significant portion of their lives in the outdoors; we know the beauty and value of the natural environment. And as ones who believe in material prosperity, we know the value of the industrial environment. It is our hope that this essay has shown you, we can have both in the same world.

Environmental quality and economic prosperity are achievable goals.

Reprinted from
The Free-Market Environmentalist
December 1990

Environmentalism and the Assault on Reason

Richard F. Sanford, Ph.D.

U.S. Geological Survey, M.S. 905, Denver Federal Center, Denver, CO 80225. (This paper was prepared by the author as a private individual with no contribution from the U.S. government or U.S. Geological Survey.)

Dr. Richard F. Sanford is a research geologist whose specialities include hydrology, economic geology, geochemistry, mineralogy, and statistics. He has been employed at the U.S. Geological Survey since 1978 and has authored more than 40 publications. His research has covered a broad spectrum of subjects from the evolution of the lunar crust to the formation of asbestos, uranium and precious-metal deposits.

Dr. Sanford has studied and lectured extensively on environmentalism. He has focused on a wide variety of scientific issues including global warming, acid rain, asbestos, and DDT. His understanding of the science is uniquely complemented by his knowledge of the philosophical basis of environmentalism.

He received a B.A., and an M.A. in geology from the Johns Hopkins University, 1973; and a Ph.D. in geology from Harvard University in 1978.

The papers in this volume echo a common theme, that the "proof" of man's destruction of the environment is consistently flawed. As shown in reviews here and elsewhere, the scientific method is being abused and ignored. The errors are not random, however, but are systematically biased toward attempting to prove the guilt of man in the alleged destruction of the planet. Objective science is disappearing and is being replaced by the pursuit of a philosophical agenda. Environmentalism today is a major symptom of a general crisis of objectivity. This paper examines the forms by which science is abused and attributes the thoroughness of this abuse to a concerted rejection of reason by philosophers, influential scientists and society.

A review of the major environmental issues shows that virtually every major error of scientific induction has been committed in the campaign to prove that man is destroying the planet. Scientific induction, the basis of the scientific method, consists of three principal functions: discovery of what exists, identification of how things are related, and determination of how things affect other things. Errors of induction are thus violations of these three principles. In a phrase, the three types of errors can be summarized as violations of existence, classification, and causality. First, and most blatant, is simply ignoring facts, typically when they do not support a hypothesis. Second, things can be misdefined, misclassified, or otherwise misidentified. Phenomena are defined or classified based on desired results rather than on essential characteristics. Third, errors involving cause and effect include, among others, attributing an effect to one cause when several may be possible and focusing on selected consequences of an action while ignoring other possible consequences.

Errors of existence typically involve ignoring known facts and are the most obvious type to recognize by an informed individual. Nonetheless, they are major and pervasive. Two conspicuous examples are the banning of DDT and the issue of acid rain. After seven months of testimony by 125 expert witnesses during 1971–1972, the evidence showed that DDT was not a major threat to animals or man, and that the benefits to man from pest reduction and malaria control far outweighed any risks. This conclusion was reached by the court-appointed hearings examiner as well as independent scientists. Nevertheless, DDT was banned essentially by one man, then-administrator of the Environmental Protection Agency, William Ruckelshaus, who had not attended any part of the hearings and had not even read the transcripts of the proceedings.[1]

More recently, the National Acid Precipitation Assessment Program (NAPAP) completed a 10-year study of the effects of acid rain in the United States at a cost of $600 million.[2] It is the most comprehensive and authoritative study of the subject ever performed. Nevertheless, during debate on the 1990 Clean Air amendment, the US Congress held no hearings to review the NAPAP study, and passed the amendment even though the sulfur reduction provision would cost in the $10's of billions per year[3] and probably would have no effect on the few acid lakes that exist or on the small acreage of forest actually damaged by acid rain.[4] In both instances, DDT and acid rain, a huge body of knowledge available to the decision-makers was simply ignored.

Lest the reader conclude that only politicians ignore known facts, the error is also widespread among scientists. A recent review article in *Science* claimed that the ozone "hole" was discovered in 1985,[5] when in fact the ozone "hole" was in existence when the first measurements were taken in 1956.[6] A recent work on wetlands by an environmental engineer and a botanist[7] makes no mention of the fact that malaria infects some 250 to 500 million and kills some 2.5 to 5 million people, mostly children, each year, and that wetlands are breeding grounds for malaria-carrying mosquitos.[8] In fact the only mention of mosquitos is a disparaging remark about the damage of mosquito-abating drainage *to the swamps*. Discussions allegedly proving massive extinction of species, subspecies and populations, never mention the genetic diversity *created* by man. For example, in 4000 years, man has created over 6500 varieties of apple alone.[9] Genetic engineering promises even more genetic diversity, but research in this area is vigorously opposed by environmental groups.

A more subtle, but more widespread error of existence than ignoring the facts is the omission of important qualifiers and uncertainties. A paper on climate modeling by a climatologist at NASA contained the following qualifications, "The experiments conducted here indicate that many uncertainties still exist . . . The lack of oceanic dynamic feedback in all models is a possibly important limitation . . . The medium grid runs produce too much rainfall . . . There is no guarantee that an increase in resolution will . . . make the . . . current climate simulation more accurate . . . Our true uncertainty in climate sensitivity is probably of the order of 100% . . . Until model differences are resolved, various basic questions about the climate of the doubled CO_2 world will remain uncertain . . . The problems . . . are formidable."[10] This is a remarkable criticism of global climate models by one of the climate modelers himself. However, when interviewed on television, the same scientist compared global warming to the attack on Pearl Harbor.[11]

Numerous uncertainties affect the measurement of the earth's mean surface temperature, yet global warming is often presented as a proven fact, as illustrated by James Hansen's widely publicized statement that he was 99% sure that the greenhouse effect had arrived.[12] However, careful studies show that most of the land surface measurements of increasing temperature have been affected by a systematic bias known as urban warming, which creates the illusion of global warming.[13]

While some environmentalists ignore known facts, others manufacture them. In 1978, Joseph Califano, then Secretary of Health and Human Services, asserted that asbestos was killing 67,000 people per year.[14] Sir Richard Doll, considered the dean of British epidemiologists, replied that Califano's estimate was for "political rather than scientific purposes,"[15] that is, fictitious. The actual number of deaths from asbestos is close to 500 per year.[16] However, the impression that asbestos is a deadly agent in any amount has persisted, and regulations based on that premise could cost some $50–150 billion in the United States alone.[17]

Errors of classification, including definition, typically involve grouping by nonessential characteristics and result in categories that are either too broad or too narrow. Such misclassification introduces an element of the arbitrary and subjective that corrupts subsequent research and policy. A particularly insidious definition is that of "wetlands." Wetlands, according to a widely accepted definition, is land "where the water table is usually at or near the surface" and includes land having certain types of plants or soil.[18] This definition is so broad that land that never has standing water nor is even moist at the surface is classified as wetland. Given the tremendous power of the US government to ban human use of wetlands, this definition provides rationalization for massive violations of private property rights and the lowering of future living standards for United States citizens.

A common error of misclassification is grouping together items having very different properties. For example, there are six different varieties of asbestos, but the EPA lumps them all together, despite the fact that the most common variety is much less harmful than the other five.[19] The result is to increase greatly the estimated risk from asbestos. In another example of invalid grouping, scientists seeking to prove that DDT caused thinning of egg shells and extinction of species lumped different populations and subspecies of brown pelican.[20] Such errors are widespread, and virtually always man is found guilty.

The converse of inappropriate grouping is inappropriate subdividing. An example is to treat subspecies and populations as species and thus effectively create more endangered "species" than there are in fact. The Mt. Graham red squirrel and the Florida panther are examples of such subspecies. Of the "species" currently listed under the Endangered Species Act, approximately 40 percent are subspecies and 10 percent are populations; only about half are actually species. Further, because the Endangered Species Act discourages hybridization, it is actually having the effect of reducing the viability of some of these populations.[21]

Errors of cause and effect are widespread. Multiple possible causes are routinely ignored. It is commonly assumed that if the earth's surface temperature is increasing, as claimed by proponents of the global warming theory, then the cause must be man's production of CO_2. However, other possible causes of warming, if it exists, are generally ignored or not publicized. For example, the northern hemisphere has been warming from natural causes since the Little Ice Age, which culminated in the 17th century.[22] Based on this natural trend, continued warming in the 20th century should be expected. For another example, DDT was supposed to have caused egg-shell thinning and declines in bird populations, according to environmentalist biologists. However, some of the many alternative likely causes were mercury contamination and disruption of breeding colonies by the biologists themselves.[23] This example also illustrates the common error of citing a statistical correlation as proof of a causal relationship, in the absence of a mechanism for such a relationship.

The destruction of the Amazon rain forest is commonly blamed on capitalistic man, specifically McDonald's Corporation, for buying beef. The commonly ignored cause is the socialist government of Brazil, which spent $2.5 billion between 1965 and 1983 to subsidize cutting of rainforest in order to rescue the country from irresponsible economic policies that also caused runaway inflation at a rate of one million percent over five years.[24]

When a causal relationship can be shown, such as one between a toxic substance and its health effects, inappropriate assumptions about the nature of the relationship are common. For example, the assumption of a linear dose-response relationship is widespread and ignores the presence of a threshold dose. Such errors have lead to very expensive but useless "solutions" to harmless levels of pesticides, asbestos, and dioxin.[25]

Ignoring multiple possible effects is as common as ignoring multiple possible causes. Commonly, remedial actions mandated by government agencies fail to have any desirable effect; the major effect is to impoverish the companies involved. For example, past attempts to clean up oil spills now appear to have been more harmful than simply allowing the oil to degrade by natural processes.[26] Nevertheless, oil companies have been and continue to be compelled to spend billions on cleanup operations. Restoring polluted aquifers to their original condition is another policy forced upon companies, however evidence shows that much aquifer pollution cannot be eliminated, but only contained, at best.[27]

In addition to damaging financial consequences, some environmental "solutions" cause major loss of human life. Banning mildly harmful chemicals often encourages or mandates the use of more dangerous substitutes. The attitude of indifference to human consequences is exemplified by a statement from one of the leading scientists in the fight to ban DDT, Dr. Charles Wurster.

> A reporter asked Dr. Wurster whether or not the ban on the use of DDT wouldn't encourage the use of the very toxic materials, Parathion, Azedrin and Methylparathion, the organo-pbosphates, [and] nerve gas derivatives. And he said "Probably". The reporter then asked him if these organophosphates didn't have a long record of killing people. And Dr. Wurster, reflecting the views of a number of other scientists, said "So what? People are the cause of all the problems; we have too many of them; we need to get rid of some of them; and this is as good a way as any."[28]

A further effect of the ban on DDT has been the resurgence of malaria, which was close to being eradicated in 1972, but today claims some 5 million human lives *per year*.[29] If such loss of human life is disregarded, then it is not surprising that massive financial losses are also ignored. The solution for virtually every environmental problem involves huge costs, typically in the $10's of billions per year. Some environmentalists simply ignore these consequences; others recognize and applaud them, based on the view that both human population and technology must be reduced at all cost. Such a view is expressed by David M. Graber, a biologist for the National Park Service, "Human happiness, and certainly human fecundity, are not as important as a wild and healthy planet . . . We have become a plague upon ourselves and upon the Earth . . . Until such time as *Homo sapiens* should decide to rejoin nature, some of us can only hope for the right virus to come along."[30]

As shown by these examples and others throughout this volume, the scientific method is being systematically abused under the guise of protecting the environment. The requirements of existence, classification, and causality are routinely ignored. The errors are too fundamental and the consequences too grave for these errors to be innocent. The consistent abuse of the scientific method indicates a wholesale rejection of objectivity. Further, there is a systematic bias which consistently attempts to incriminate man. His alleged crime is global destruction, and his punishment consists of limitations on his life-sustaining technology, industry, and freedom. The implicit premise is that man, by nature, is evil. Although explicit admissions, like those of Wurster and Graber, are rare, the facts give ample testimony to the widespread underlying premises.

Occasional published statements are explicit evidence that the abuse of facts is a conscious act on the part of environmentalists, including scientists and journalists, two groups who ought to have the highest regard for objectivity. Aldo Leopold, widely considered the father of modern environmentalism, admitted that he and other environmentalists appeal to man's self-interest when actually their goal is to preserve nature for its own sake, in spite of man's interests. "Of the 22,000 higher plants and animals native to Wisconsin, it is doubtful whether more than 5 percent can be sold, fed, eaten, or otherwise put to economic use. Yet these creatures are members of the biotic community, and . . . they are entitled to continuance. When one of these non-economic categories is threatened, and if we happen to love it, we invent *subterfuges* to give it economic importance"[31] (emphasis mine). Stephen Schneider, a climatologist and leading proponent of the global warming theory says, "To capture the public imagination . . . we have to offer up some scary scenarios, make simplified dramatic statements and little mention of any doubts one might have."[32] The following dialogue took place at a meeting of prominent journalists,

Charles Alexander declared, "As the science editor at Time, I would freely admit that on this issue [the environment] we have crossed the boundary from news reporting to advocacy." . . . NBC correspondent Andrea Mitchel told the audience that "clearly the networks have made the decision now, where you'd call it advocacy." . . . Washington Post editor Benjamin Bradlee chimed in, saying "I don't think there's any danger in doing what you suggest [taking an advocacy approach]. There's a minor danger in saying it because as soon as you say "To hell with the news. I'm no longer interested in news, I'm interested in causes," you've got a whole kooky constituency to respond to, which you can waste a lot of time on."[33]

Further, "Environmental reporter Dianne Dumanski of the *Boston Globe,* widely regarded as one of the best papers covering environmental issues in the country . . . readily admits to being progressive in her politics . . . '*There's no such thing as objective reporting*' she asserted" (emphasis mine).[34] Because admission of guilt is a rare phenomenon, one has to assume that many more scientists and journalists who are advocates of environmentalism have abandoned objectivity but succeed in maintaining appearances.

We have seen how the inductive method is violated by many scientists and ignored by environmentalist journalists and politicians. More than anything else, the systematic errors are a consequence of the dominant philosophical trend of late 20th century society: the rejection of reason. Although most of society pays lip service to reason, in fact, reason has been severely crippled and is near death. The campaign against reason takes many forms. It varies from the relegation of reason to a severely limited realm of empirical or "hard" facts to the wholesale embrace of intuition and mysticism over reason.

The problems start with the definition of reason itself. To Aristotle, reason is the process by which man gains knowledge of reality. This view depends on two basic premises, that objective reality exists, and that conscious beings, through their senses, gain an accurate awareness of reality. Knowledge is obtained by validating the content of one's mind according to the facts of reality. Truth, then, *corresponds* to reality, and reason is the process of gaining knowledge of reality.

This Aristotelian view has been thoroughly rejected among professional philosophers since Immanuel Kant. Kant postulated a fundamental metaphysical dichotomy between "true reality" and the "world of appearance." The mind, he said, cannot gain an accurate representation of reality but instead *creates* an artificial world, which is the one we conceive. Despite the theory's incredible complexity and its myriad interpretations and modifications by subsequent philosophers, three principles are universally accepted by modern philosophers since Kant. "True reality" is unknowable. Reason is limited. Value-judgements (or normative questions) are a matter of intuition (or faith). All three principles entail the rejection of reason and the embrace of scepticism, the doctrine that knowledge is impossible. Kant summarizes these principles in one uncharacteristically clear statement, "I have therefore found it necessary to deny *knowledge,* in order to make room for faith."[35]

After Kant, reason was no longer a means of grasping an independent reality and of ensuring objectivity. Instead, it has been regarded as a tool for distorting reality and as a means of imposing subjective desires upon it. This Kantian view of reason accounts for the widespread violation of the inductive method, the assault on reason, and the corresponding rise of environmentalism today.

Today's intellectual and scientific trends are explained by two principal ideas that derive from the Kantian views just mentioned. First, truth is redefined to mean, not *correspondence* to reality, but *coherence* among ideas. This is necessary given the premise that the mind cannot know reality. Reason then becomes, not a method of validating knowledge according to reality, but merely a means of checking that ideas are "meaningful," "internally consistent," "verifiable," or "falsifiable."

Second, the view that normative judgments belong to the world of intuition has lead to rampant nonobjectivity on issues of ethics and public policy that spills over into scientific issues.

Because of the Kantian dichotomy between the mind and reality, the dominant school in the philosophy of science does not seek truth but studies how people use language. "Reason" is reduced to analyzing sentences for "meaningfulness." Correct theories are not "right," but "verifiable" or "falsifiable." The dominant school in the philosophy of science today is logical positivism or logical empiricism. The basic premise of Ludwig Wittgenstein, leader of this school, is that the senses do not provide accurate information about things as they are.[36] "Prior to Wittgenstein, most philosophers had regarded philosophy as the pursuit of truth. Wittgenstein regarded it as a kind of therapy." That "therapy" consists of rejecting the "mental illness" of seeking truth from a knowable, objective reality and instead attempting to "dissolve" theoretical problems about the nature of existence.[37]

Much of logical positivism is a futile attempt to resolve the irresolvable conflict between the supposed impossibility of gaining knowledge and the obvious fact that man must, can and does gain knowledge that corresponds to reality. Thus, much of the philosophy of science is simply irrelevant to real science. As observed by P.R. Durbin, "A philosophical introduction to science . . . was always said to have a bearing on modern science, and many schools made it compulsory for science majors. Yet students for the most part . . . failed to see even the remotest connection with science."[38] These students correctly recognized what the philosophers of science refused to consider a problem, that modern philosophy of science had severed its own connection to reality.

Because of its false conception of reality and truth, the practical result of logical positivism and other mutations of Kantianism is that reason and science focus on theory rather than reality. Words take the place of facts. The mind focuses on its own content, or that of other minds, but not on the facts of reality. Ideas become primary, and the facts become subservient to them. This tendency is manifested today in the concerted attempt to justify the environmentally "correct" view despite inconvenient data that "cloud,"[39] specific issues. Repeatedly, reason, in the sense of validating ideas according to the facts of reality, is rejected.

One manifestation of the mind-reality split is the tendency to resolve issues, not by reference to reality (since correspondence is out), but by appealing to authority or consensus (since coherence is in). For example, when the head of the EPA recently declared that the ozone layer is thinning faster than previously thought,[40] his assertion was accepted uncritically and broadcast worldwide, before the data and results appeared in a published journal and were subjected to critical review by scientists. The implicit assumption is that we cannot know true reality anyway, so the best we can do is make our ideas "cohere" with those of a recognized authority. Appealing to consensus is another invalid substitute for reason and goes so far as polls of the general public by scientific organizations on questions such as global warming.[41]

Science today, therefore, is a mixture of two influences. True science, on one hand, implicitly relies on the Aristotelian concept of an objective reality knowable by reason. Although it lacks any intellectual support from professional philosophers, this remnant of Aristotealianism is the basis, *in fact,* for the valid science that does take place. On the other hand, the influence of Kantian philosophy is to place human consciousness (intuition) over the facts of reality. And this leads to the abuse of science that characterizes modern environmentalism.

To understand how reason is relegated to near irrelevancy, consider how scientists accept and respond to the Kantian split between matters of fact and matters of value. Scientists generally (not just environmentalist scientists), in the belief that all value-judgements are at root subjective, avoid taking any "advocacy" position. Their motivation is partly to preserve their reputation for objectivity, partly to avoid any possible embarrassment resulting from speaking on unfamiliar subjects outside their field, and partly because modern philosophy has become nonsensical and irrelevant. These views lead to the position that other people, with training in social sciences, philosophy, or theology should make the important decisions that guide society. Consequently, many scientists provide data for one small part of a much larger program and tend not to question the overall objectives or methodology of the program or their field as a whole or the direction of society. Such decisions are then left to those with power and influence. Implicitly the scientist can resort to the "Nuremberg defense," that is, "I was only doing my job." Thus, many scientists voluntarily limit themselves to a narrow realm in which reason rules while giving up the direction, guidance, and ultimate use of scientific investigation to the few who do concern themselves with basic goals. Thus they support the environmental agenda not by design but by default.

In the entire area of metaphysical and ethical questions and increasingly in the fields of science, the dominant substitute for reason is intuition (faith, feelings, revelation, or emotions) as the ultimate source of knowledge. Although Kant and subsequent philosophers fervently claim to be advocates of reason, they have redefined the concept of reason so that it has no connection to reality, and what they refer to as reason is, in fact, intuition.

The methodology of starting with a conclusion one feels is correct and then finding evidence to support it is an expression of the view that intuition is epistemologically primary. That is, the final "court of appeals" on any question is not the facts of reality, but one's intuitive feelings. Rationalization is another term for this approach.

Overt appeals to intuition are rare among scientists, but are common among intellectuals and philosophers. Even if reason is not explicitly rejected by those who hold intuition as primary, the two approaches are mutually exclusive, and the embrace of intuition as a basic means of obtaining knowledge necessitates the rejection of reason.

Intuition is the basic approach of environmentalist philosophers, intellectuals and activists, for example Arne Naess, the "ecosopher" who coined the term "deep ecology" begins a recent book, appropriately, with the section "Beginning with intuitions." His starting point is a feeling of "our world in crisis."[42]

Paul W. Taylor, professor of philosophy at City University of New York, has "the biocentric outlook" as the starting point of his philosophy. The biocentric outlook consists of four mystical views, the first of which is "humans are members of the Earth's Community of Life."[43] Despite pretensions of rationality, expressed in mind-numbing terminology and convoluted definitions, his starting point is his intuitive feelings about man and nature.

Environmental activists also start with intuitive feelings, "green rage," for example. " 'I pretty much feel,' says Rick Bailey, an Earth First! activist from Oregon, 'that the biological . . . and ecological foundation of this planet is under siege.' "[44] Although much is made of differences between mainstream environmentalist philosophers and radical activists, they share the same basic epistemology of intuition. The philosophers tend to construct elaborate systems based on their arbitrary premises; the activists reject systems on principle. "Most radical environmentalists look at systematic philosophy as the problem."[45] Philosophers and activists both start with intuition and both reject reason. The activists, in fact, are more consistent, for they dispense with the elaborate philosophical edifice, which they correctly perceive as a hypocritical rationalization, a "subterfuge," to use Leopold's term.

Just as intuition is the basis for the environmentalist epistemology, the basic premise in the environmentalist view of man is that he is inherently evil. Both mainstream environmental philosophers and radical activists share this view. The activist says "our culture is lethal to the ecology."[46] The philosopher says, the demise of the human species "would be no loss to other species, nor would it adversely affect the natural environment. On the contrary, other living things would be much benefited . . . Given the total, absolute, and final disappearance of *Homo sapiens*, then, not only would the Earth's Community of Life continue to exist but . . . the ending of the human epoch on Earth would most likely be greeted with a hearty 'Good riddance!'"[47] In practice, the main difference between philosophers and activists is that the activists attempt to carry out their principles in action, while the philosophers provide the moral and intellectual rationalizations they require. However, in both cases, man is seen as destructive, unnatural, and evil.

The symbiotic relationship between the "mainstream" environmental movement and the radicals is recognized by both parties. "'When I helped to found Earth First!,' writes Howie Wolke, 'I thought . . . we would make the Sierra Club look moderate by taking positions most people would consider ridiculous.'" And from the other side: "the national environmental leadership is in some instances perversely happy that Earth First! exists. 'Frankly, it makes us look moderate,' says Robert Hattoy, Southern California representative of the Sierra Club'"[48] The symbiotic relationship among philosophers, mainstream environmentalists and activists is based on a shared rejection of reason and condemnation of man. The same symbiotic relationship involves environmentalist scientists, as exemplified by Wurster and Graber, quoted above. For them, science is a facade to dignify preconceived opinions about the condition of the planet and the role of human beings.

In contrast to the type of scientist who attempts to preserve his objectivity by avoiding "advocacy" positions in public, the activist scientist leaps into moral controversies and public debate. Although, he too holds the view that moral judgements are necessarily subjective, he recognizes that man and society require moral guidance. And he gives it, whether it is objective or not. This results in a conflict of objectivity between the scientist as scientist and as human being. "As scientists, we are ethically bound to the scientific method . . . On the other hand, we are not just scientists but human beings as well," says Schneider, who goes on to say that his subjective belief that he is morally right overrides the uncertainties he has as an objective scientist and leads him to "offer up scary scenarios" etc.,[49] in other words to act as an environmental advocate rather than an objective scientist.

A valid alternative to the idea that value-judgements are necessarily subjective is the view that value-judgements can be objective, if they are based on the basic nature or identity of man. The basic fact of man's identity is that, if he is to survive, he must live according to the objective requirements of man's life. According to philosopher, Ayn Rand, "In answer to those philosophers who claim that no relation can be established between ultimate ends or values and the facts of reality, let me stress that the fact that living entities exist and function necessitates the existence of values and of an ultimate value which for any given living entity is its own life. Thus the validation of value judgements is to be achieved by reference to the facts of reality. The fact that a living entity *is*, determines what it *ought* to do."[50] Thus man's life provides an objective standard of value. Because man, in contrast to other living beings, *makes* virtually everything required for his survival, and because *production* requires the constant exercise of his reason, reason is then the basic means of survival and the fundamental mortal virtue for man. Thus, it is possible for man to be objective as a scientist and moral as a human being without conflict.

As shown, despite innumerable variations in approach and goals, the environmental movement shares a common rejection of reason, either actively or by default. Although many environmentalists pay lip service to reason, the leitmotif of environmentalism was made crystal-clear by a famous actress playing the role of Gaia, Mother Earth, on a major television network, "I applaud the irrational!"[51]

The widespread rejection of reason today did not and could not have happened spontaneously or overnight. Reason has been attacked or ignored throughout recorded history and probably before, but the systematic, wholesale, intellectual assault on reason began in modern times with the philosophy of Immanual Kant. In essence, Kant divorced true reality, which, he said, is unknowable, from the reality that we observe, which, he said, is an illusion. He divorced facts from judgements. He maintained that reason is a mechanism of distortion. He held that ideas could only be checked for internal consistency with other ideas but not validated by reality. Consequently, hypotheses could be proposed, validated, and accepted with never a reference to the facts; reason became impotent; intuition became supreme. In short, Kant's grand philosophical edifice suffered from all the invalid methods described above and was nothing more than an elaborate justification for preconceived, invalid ideas and irrational feelings.

Kantianism spawned the 20th century philosophies of logical positivism, pragmatism, and existentialism, all of which contribute to the philosophy of environmentalism. Along with eastern mysticism, these systems attempt to rationalize the rejection of reason. The environmental movement is the culmination of these corrupt philosophies, and, by rejecting reason, the ultimate effect of environmentalism will be the elimination of science and of civilized society along with it. The papers that follow in this volume give ample testimony that this process of decay is far advanced.

References

1. Edwards, J.G. 1991. DDT effects on bird abundance and reproduction. *This volume.* Hazeltine, W.E. 1973. Emotion vs. objectivity in pesticide usage. Talk presented at Entomological Society of America, Dallas, Texas, November 26, 1973. Sobelman, M. 1975. DDT Case Study for The National Academy of Sciences. Prepared for National Academy of Sciences, Environmental Studies Board for its study "Principles of decision making for chemicals in the environment" January 10, 1975, 56 pages.
2. Brookes, W. 1990. Acid rain's cost cloudburst. *Washington Times,* March 3, 1990.
3. Wissel, D. 1990. *Wall Street Journal,* May 25, 1990 (Estimates of the cost of the clean air amendment range from $967 million to $64 billion per year.)
4. Krug, E. C. 1990. Fish story: The great acid rain flimflam. *Policy Review* Spring 1990: 44–48. Kulp, J.L. 1990. Acid rain: Causes, effects, and control. *Regulation: The Cato Review of Business and Government* Winter, 1990, 13: 41–50.
5. Solomon, S. 1990. Progress towards a quantitative understanding of Antarctic ozone depletion. *Science* 347: 347–354.
6. Dobson, G.M.B. 1968. Forty years' research on atmospheric ozone at Oxford: A history. *Applied Optics* 7: 387–405. (The existence of the ozone "hole" is also a matter of definition. Defined in comparison to the Arctic, the Antarctic "hole" has always existed. However, defined in comparison to 1956–1973 data, it appeared relatively recently.)
7. Mitsch, W.J., & Gosselink, J.G. 1986. WETLANDS. Van Nostrand Reinhold, New York, 539 pages.
8. World Health Organization (WHO), *Tropical Diseases* 1990. p. 5 (Malaria accounts for 267 million cases and 1–2 million deaths per year). Gratz, N. 1978. Talk given to the American Mosquito Control Association, Chicago, Illinois, April 8, 1978, (cited 250 million cases per year as a "guestimate"). (Given uncertainties in estimates and a probable rise in incidence since 1978, the current number of cases is probably close to 500 million according to William Hazeltine, Director, Butte County Mosquito Abatement District, Oroville, California, personal communication, 1990.)
9. Rupp, R. 1990. RED OAKS AND BLACK BIRCHES: THE SCIENCE AND LORE OF TREES. Garden Way Publishing, Pownall, Vermont, p. 114ff, 176 pages.

10. Rind, D. 1988. The doubled CO_2 climate and the sensitivity of the modeled hydrologic cycle. *J. Geophysical Research* 93: 5385–5412.
11. The Nightly Business Report, February 5, 1990. (On that day President Bush proposed a 60% increase in spending for research on global climate change and indicated his desire that United States economy not be harmed by hasty actions. Dr. Rind then came on with the one-sentence quote, "When Pearl Harbor occurred, we didn't say, 'Well, we'll only take the sorts of action that won't affect our economy.'" In that one sentence broadcast to millions of viewers he managed to convey the idea that global warming is as certain, as sudden, and as devastating as Pearl Harbor. Plus he displayed a remarkable disregard for a free and healthy market economy.)
12. Kerr, R.A. 1989. Hansen vs. the world on the greenhouse threat. *Science* 244: 1041–1043.
13. Kukla, G., Gavin, J., & Karl, T.R. 1986. Urban Warming. *Journal of Climate and Applied Meteorology* 25: 1265–1270. Diaz, H.F. 1986. An analysis of twentieth century climate fluctuations in northern North America. *Journal of Climate and Applied Meteorology* 25: 1625–1657. Balling, R.C. & Idso, S.B. 1989. Historical temperature trends in the United States and the effect of urban population growth. *Journal of Geophysical Research* 94: 3359–3365.
14. National Cancer Institute (NCI), National Institute of Environmental Health Sciences (NIEHS) press release, *Draft Summary*, September 11, 1978.
15. Doll, R. & Peto, R. 1981. *Journal of the National Cancer Institute* 66: 1193–1308.
16. Ross, M. 1984. A survey of asbestos-related disease in trades and mining occupations and in factory and mining communities as a means of predicting health risks of nonoccupational exposure to fibrous minerals. *American Society for Testing and Materials, Special Technical Publication* 834: 51–104. Mossman, B.T., Bignon, J., Corn, M., Seaton, A., & Gee, J.B.L. 1990. Asbestos: scientific developments and implications for public policy: *Science* 247: 294–301.
17. Abelson, P.H. 1990. The asbestos removal fiasco. *Science* 247: 107.
18. U.S. Fish and Wildlife Service 1979. Classification of wetlands and deepwater habitats of the United States. Quoted in Mitsch, W.J. & Gosselink, J.G. 1986. WETLANDS. Van Nostrand Reinhold, New York, p. 18.
19. See footnote 16.
20. Bluss, L.J., Gish, C.D., Belisle, A.A., & Prouty, R.M 1972. Logarithmic relationship of DDE residues to eggshell thinning: *Nature* 235: 376–377. Bluss, L.J., Gish, C.D., Belisle, A.A., & Prouty, R.M 1972. Further analysis of the logarithmic relationship of DDE residues to eggshell thinning. *Nature* 240: 164–166. And see discussions by Hazeltine, W.E. *Nature* 239: 410–411 and Switzer, et al. *Nature* 240: 162–163.
21. O'Brien, S.J. & Mayr, E. 1991. Bureaucratic mischief: Recognizing endangered species and subspecies. *Science* 251: 1187–1188.
22. Lamb, H.H. 1977. CLIMATE, PRESENT, PAST, AND FUTURE. Methuen & Co., Barnes and Noble Books, New York, p. 423ff, 613 p.
23. See footnote 1.
24. Wicker, T. 1991. *The New York Times*. December 27, 1988. Kamm, T. 1991. *Wall Street Journal*. March 29, 1991.

25. Ames, B.N., Magaw, R., & Gold, L.S. 1987. Ranking possible carcinogenic hazards. *Science* 236: 271–280. Ames, B.N. 1989. Be wary of nature's own pesticides. *Los Angeles Times,* February 17, 1989. Ames, B.N. *this volume.* Fingerhut, M.A. et al. 1991. Cancer mortality in workers exposed to 2,3,7,8-tetrachlorodibenzo-p-dioxin. *New England Journal of Medicine* 324: 212–218. Ross, M. 1984 and Mossman, J. et al. 1990 (see footnote 16).

26. Mielke, J.E. 1990. Oil in the oceans: The short- and long-term impacts of a spill. *CRS Report for Congress, Congressional Research Service, Library of Congress,* July 24, 1990, 90-356 SPR, 34 p.

27. Travis, C.C. & Doty, C.B. 1990. Can contaminated aquifers at Superfund sites be remediated? *Environmental Science and Technology* 24: 1464–1466. Abelson, P.E. 1990. Inefficient remediation of ground-water pollution. *Science* 250: 733. Stipp, D. *Wall Street Journal,* May 15, 1991.

28. Yannacone, V.J. Jr. (a lawyer and one of the founders of the Environmental Defense Fund). Speech delivered in New York, May 20, 1970.

29. See footnote 8.

30. Graber, D.M. 1989. Mother Nature as a hothouse flower: *Los Angeles Times Book Review* October 22, 1989, p. 10.

31. Leopold, A, 1970. A SAND COUNTY ALMANAC. Sierra Club/Ballantine Books, New York, p. 246ff, 295 pages.

32. Schell, J. 1989. *Discover Magazine.* October, 1989, p. 47. (Quote from Stephen Schneider.)

33. Brooks, D. 1989. Journalists and others for saving the planet. *Wall Street Journal* Oct. 5, 1989.

34. Lee, M.A. & Solomon, N. 1991. . . . And that's the way it is. *E Magazine* January/February, 1991, p. 41.

35. Kant, I. 1929. CRITIQUE OF PURE REASON. Translated by N. Kemp Smith, Macmillan, London, 1929, second ed., p. 3–4. Quoted in Jones, W.T. 1969. A HISTORY OF MODERN PHILOSOPHY: KANT TO WITTGENSTEIN AND SARTRE. Second ed., Harcourt, Brace and World, New York, p. 66, 481 pages.

36. Durbin, P.R. 1968. PHILOSOPHY OF SCIENCE: AN INTRODUCTION. McGraw-Hill, New York, p. 3, 271 pages.

37. Jones, W.T. 1969. A HISTORY OF MODERN PHILOSOPHY: KANT TO WITTGENSTEIN AND SARTRE. Second ed., Harcourt, Brace and World, New York, p. 354, 481 pages.

38. Durbin. Preface, p. v.

39. George, M. 1991. *Denver Post,* April 28, 1991. (Mary George, environmental writer for the Denver Post, writes "Clouding the risk issue is recent research claiming that not all asbestos is equally deadly.")

40. Reilly, W.K. 1991. UPI press release, April 5, 1991.

41. Anon. 1989. Voters Polled on Greenhouse. *EOS* December 5 , 1989, p. 1531. (The news article describes results of a poll of 1200 registered voters by the Union of Concerned Scientists.)

42. Naess, A. 1990. ECOLOGY, COMMUNITY AND LIFESTYLE Translated and revised by David Rothenburg, Cambridge University Press, Cambridge, 223 pages.

43. Taylor, P.W., 1989. RESPECT FOR NATURE: A THEORY OF ENVIRON-MENTAL ETHICS. Princeton University Press, Princeton, p. 99, 329 pages.
44. Manes, C. 1990. GREEN RAGE. Little, Brown and Company, Boston, p. 24, 291 pages.
45. GREEN RAGE, p. 21.
46. GREEN RAGE, p. 22.
47. RESPECT FOR NATURE, p. 114–115.
48. GREEN RAGE, p. 18.
49. Schell, J. 1989. *Discover Magazine.* October, 1989, p. 47. (Quote from Stephen Schneider.)
50. Rand, A. 1961. The Objectivist Ethics. An essay in Rand, A. 1961. THE VIRTUE OF SELFISHNESS. New American Library, New York, p. 17. See also Peikoff, L. 1991. OBJECTIVISM: THE PHILOSOPHY OF AYN RAND. New American Library, New York, in press.
51. Dunaway, F. 1991. Voice of the Planet. Television series broadcast January, 1991, on Turner Broadcasting System.

Acid Rain

The Great Acid Rain Flimflam
Edward C. Krug

Acid Rain—The Whole Story to Date
John J. McKetta, Jr.

The Great Acid Rain Flimflam

Edward C. Krug, Ph.D.

Dr. Edward C. Krug is Director of Environmental Projects, Committee for a Constructive Tomorrow, P.O. Box 65722, Washington, D.C. 20035. He has been an associate scientist with the Illinois State Water Survey, University of Illinois in Champaign, Illinois; and an assistant soil scientist, Connecticut Agricultural Experiment Station in New Haven.

Dr. Krug received a B.Sc. in environmental science from Rutgers University in 1975 (valedictorian); an M.Sc. in soil chemistry from Rutgers in 1978; and a Ph.D. in soil science from Rutgers in 1981.

He has published more than 30 journal articles appearing in *Science, Nature, EOS* and others.

For more than a dozen years, the conventional wisdom among scientists, environmentalists, and politicians has blamed acid rain for the depletion of game fisheries in the Adirondacks and Nova Scotia and for substantial damage to forests from Vermont to North Carolina. The government of Canada certainly takes this view, and regard acid rainclouds from the Midwest as one of its principal sources of tension with the United States. President Carter endorsed a report by his Council on Environmental Quality calling acid rain one of the two most serious environmental problems of the century. It is largely to reduce the acid in rain that President Bush's Clean Air legislation calls for a 50 percent reduction in sulfur dioxide emissions (10 million tons) by the year 2000, at an estimated cost of $4 billion to $7 billion per year.

The concerns that acid rain legislation is intended to address are legitimate and understandable. A number of rivers in Nova Scotia have lost most of their salmon over the past 40 years as they have become more acidic. Red spruce are dying on top of Camel's Hump in Vermont and Mount Mitchell in North Carolina. And there is no doubt that some fisheries in the Adirondacks are in a bad way. From the turn of the century until the 1950s, for instance, Lake Colden was one of the best trout fly fishing lakes in the eastern United States. Teddy Roosevelt fished there often and was vacationing at Lake Colden when President McKinley was shot. Today, Lake Colden is highly acidic (pH 5.0) and nearly fishless. It is held up as the classic case of acid rain's destructiveness.

Recent research, however, suggests that acid rain has little or nothing to do with these problems. Surveys of lakes in New England and New York show much less acidity than anticipated, while other studies show that acid rain has very little effect on surface water acidity. Perhaps most intriguing, studies of the fossil records in lake sediments reveal that many lakes that are acidic today have been highly acidic for centuries, except for several decades in the late 19th century and early 20th century when they were unnaturally alkaline.

Memories of trout and salmon in now fishless lakes and streams apparently date from a period of ecological aberration. A number of lakes in the Adirondacks and Nova Scotia that are naturally acidic became more alkaline for several decades in the late 19th century and early 20th century—when massive cutting of trees and burning of stumps by lumberers reduced the acidity of the forest floor, and soil runoff made it possible for species such as trout and salmon to survive. After lumbering and burning came to an end, forests grew back, and the soil runoff, and hence the waters, returned to their natural acidity. These changes in land use often dwarf in importance the impact of acid rain.

There is similarly no evidence of widespread forest decline in North America related to acid rain. Indeed, U.S. Forest Service statistics indicate that northeastern forests appear to be the most robust in the country. The principal concern is that acid rain may have damaged red spruce in high-altitude spruce/fir forests. But these forests make up only a fraction of 1 percent of eastern forests, and even here, the influence of acid rain is uncertain. The Environmental Protection Agency's National Acid Precipitation Assessment Project (NAPAP) has determined that other stress factors (killing winters and severe droughts over the past 40 years, such as occurred at Camel's Hump and other mountains) have been more important. Also, many species of trees and bushes—among them red and black spruce, oaks, balsam fir, eastern hemlock, rhododendrons, and blueberries—depend on acidic soil for their survival.

Control Soot, Emit Acid

There is no question that the rain and snow over the Appalachians, New England, and Nova Scotia are more acidic than normal. Rain over the Ohio Valley and Adirondacks has a pH of 4.2, while that over Nova Scotia has a pH of 4.6 to 4.8—compared with a pH of 5.0 for normal rainfall over forested areas. (Chemists use the measure pH to describe the concentration of hydrogen ion released into water by an acid. It is a negative log scale, so the smaller the number the greater the acidity; thus pH 4.0 water is 10 times more acidic than pH 5.0 water. Lemon juice has a pH of 2.0 and our stomach juices are pH 1.0. Apples have a pH of 3.0—meaning they are 16 times as acidic as Adirondack rain.

Nor is there any question about the source of acid rain. It results from the combustion of fossil fuels, especially in the Midwest, and ironically it has been aggravated by environmental policies designed to reduce air pollution problems, especially soot, in areas such as Pittsburgh and Cleveland. Acid rain is created by burning fossil biomass. Carbon, nitrogen, and sulfur (present in all biological material) are converted to gases upon oxidation (burning). These gases combine with atmospheric water to form carbonic, nitric, and sulfuric acids.

When tall smokestacks were built to spread out emissions from the vicinity of Midwestern factories, acidic cloud formations were carried northeast by winds. The emissions also became more acidic as a result of measures such as particle precipitators and cleaner burning fuels to cut down on soot. (Soot contains alkaline substances that neutralize acid.) Efforts to alleviate local soot problems thus led to regional acid rain.

Sulfur Trouble

Widespread acid rain probably began by the 1940s in the Northeast. It peaked around 1978, when national sulfur dioxide emissions were measured at 31 million tons. Since then, sulfur dioxide emissions have fallen to 23 million tons in 1985, partly as a result of pollution controls and conservation, and replacement of older, more heavily polluting factories and power plants by state-of-the-art facilities as mandated under the Clean Air Act of 1970.

The major concern over acid rain has to do with sulfuric rather than nitric acid. This is because forest growth in eastern North America is limited by the availability of nitrogen as a nutrient. Essentially all of the nitric acid deposited in acid rain is absorbed as a nutrient by trees; little or none of it makes it to lakes and streams. However, deposition of sulfuric acid exceeds the nutritional requirements of the eastern forests and much, but not all, of the sulfate finds its way into lakes and streams.

Even normal rainfall in forested areas has a pH of 5.0, too acidic for most species of sport fish to survive in. Fish and many other species can survive in rainwater only because acids are naturally buffered by lime-like substances in rocks and mineral soils of lake and river drainage systems. Acid in watershed runoff is consumed by reaction with the alkalinity of lime-like substances that are found in many types of rocks and mineral soils that underlie watersheds and the bottoms of lakes and streams. The acidity of lakes in the Adirondacks and Nova Scotia results not from acid rain but from the absence of this natural buffering.

For Peat's Sake

Most lakes in the Northeast are not highly acidic, even though acid precipitation falls on the entire area. In 1980, before it had studied the situation, the EPA asserted that the acidity of northeastern lakes had increased 100-fold (a decrease of two pH units) as a result of acid rain. But a 1984 lake survey by NAPAP found that only 240 of New England's and New York's more than 7,000 lakes are "acid-dead"—that is, have a pH of 5.0 or lower. The survey found that in the whole eastern United States there are only 630 acid lakes, representing 35,000 of the approximately 200 million acres of water in the East, or less than one-fiftieth of 1 percent of the water. Over half the acid lake capacity—20,000 acres—is in Florida, which does not receive high rates of acid rain.

Further studies by NAPAP in 1988 and 1989 suggest that since 1850 there has been no increase in acidity in lakes with a pH above 5.5; these lakes have granite and gravel bottoms with sufficient lime to neutralize the acid in rain. For Adirondack lakes with a pH under 5.5, NAPAP concluded that there has been some acidification, but on average by less than half a pH unit.

Attention on acid rain now focuses primarily on the Adirondacks rather than on the entire Northeast, because studies show that 10 percent of Adirondack lakes (or 24 percent, depending on what is included as a lake) have a pH of 5.0 or lower.

There are much more important reasons than acid rain, however, to explain their acidity. The Adirondack lakes in question are in poorly buffered and, therefore, naturally acidic watersheds. Their rocks are poor in lime-like substances; their watersheds are mantled by highly acidic, very thick peaty forest floors, and leaching of water through the soil produces a low-nutrient environment where acid-producing trees and plants, such as sphagnum mosses, are common. In fact, a recently completed multimillion dollar biological and chemical survey of over 1,400 Adirondack lakes and ponds observed a striking correlation between biology and acidity—sphagnum mosses are associated with acid lakes. Oak Ridge National Laboratory determined that acid Adirondack lakes tend to be concentrated in watersheds having coniferous forests and the most acidic soils.

Bark Eaters

Perhaps the most important studies of the Adirondacks by NAPAP have involved the examination of fossil organisms and chemicals buried in lake sediments—what scientists call paleolimnology. The pH balances of earlier periods can be determined with considerable precision by examining algae fossils in conjunction with radio-isotope dating, because different species of algae are present with different pH levels. These studies reveal that high-altitude Adirondack lakes, including Lake Colden and Woods Lake, have been fishless for most of their history. They also reveal that these lakes temporarily lost some of their natural acidity during the mid-to-late 19th and early 20th centuries—and during this period were filled with fish. Woods Lake, for instance, was acid-dead (pH 5.0 or lower) in 1850, long before there was any acid rain. Beginning in the 1860s, the lake became gradually more alkaline, peaking at a pH of 5.7 in 1910. It has since reacidified, and now has a pH of 4.8.

These findings are consistent with what we know from history. The Indians never lived in the Adirondacks. The Iroquois word "Adirondack" means "bark eater," telling us that the food supply, including fish, was never plentiful. Most Adirondack waters have always been cold and unproductive. The few excellent fisheries were generally confined to the few large lowland lakes and rivers that the Indians traveled to pass through the Adirondacks.

Initial European settlement and activity damaged what fisheries there were. For example, Atlantic salmon were eliminated from Lake Champlain by the 1830s and lake trout have disappeared from the lake since 1880.

Since the 1800s, the Adirondacks have been well known to fishery experts for their massive stocking failures. Tens of millions of fish have been put into waters where they promptly died. Great Lakes whitefish were introduced as food for trout. But these whitefish grew too large for trout to eat. Competition from whitefish, along with the fisherman's preference to catch trout, reduced or eliminated trout populations. Introduction of bass and yellow perch outcompeted and eliminated trout fisheries from many ponds and streams.

Devastation on Land, Fish in the Lakes

However, there were successful stockings of brown and brook trout in the mid-to-late 19th century, when the Adirondacks became a major center for lumbering and paper pulp—and also for the destructive slash-and-burn methods that until recently were typical of logging. Cutting was invariably followed by fires. Fuel was plentiful—left-over cuttings, the luxuriant mosses and thick peaty forest floor exposed to the sun and air, became dry and highly flammable. In 1903, in a two-month conflagration rivaling the Yellowstone fires of 1988, Adirondack fires literally burned in their own updraft, like chimney fires sweeping up the mountainsides. Witnesses compared the burning mountains to "smoldering volcanoes." The sun became big and orange; stars were blocked out at night.

The cumulative devastation was horrible. According to an official history published by the Adirondack Museum, "Logging and fires changed the Adirondacks more drastically and rapidly than any factors since these mountains first rose above the sea." The landscape was denuded, and severe erosion washed away the underlying thin and rocky mineral soils. Runoff from the land became much more alkaline. Forests gone, spongy and water-absorbent mosses and the acidic peaty forest floor were burned off and replaced by alkaline ash. The ironic result, though, was that sport fish could now survive in lakes that had previously been uninhabitable.

Unprecedented disturbance in the Adirondacks was then followed by unprecedented protection. Plans for the "Forever Wild" Adirondack Park and Forest Preserve were established in 1892, and the state of New York gradually acquired parkland, much of it purchased literally at fire-sale prices from lumber and paper companies that were abandoning their devastated forests. Since about 1915, laws against cutting and burning have been seriously enforced. Forest fires are now put out quickly. As a result, the forests, acid peaty soils, and acid-requiring and acid-producing trees and mosses are coming back. And lakes that historically have been highly acidic are nearing their natural pH balances.

To add to the pressure on fisheries from natural acidification, the beauty of the mountains and improved public transportation resulted in a massive increase in the number of fishermen. Registration in Adirondack public campgrounds rose from 37,000 in 1927 to 950,000 in 1970. In addition, these campers were very uncomfortable with insects, particularly blackflies. DDT was mixed into cement bricks and deposited into the waters to slowly release DDT over the years to kill the flies. The elimination of flies contributed to the disappearance of the trout that relied on them for food.

The Key to Lox

The experience in Nova Scotia is similar to that of the Adirondacks. Nova Scotia has more acid lakes and streams than any other area of

North America; indeed nearly half of its lakes have a pH lower than 4.7. Salmon have been disappearing as lakes and streams have become more acidic, and fingers are pointing to acid rain as the culprit. The concern is perfectly understandable, but the rain over Nova Scotia is only one-third as acidic as that over the Adirondacks.

As in the Adirondacks, Nova Scotia has a history of slash-and-burn lumbering followed by reforestation, and a long history of fishery problems. A 1986 study of Nova Scotian fisheries by Environment Canada (the Canadian version of our EPA) noted the severe burning of forests in the province at the turn of the century, with 90 percent of southwestern Nova Scotia (where salmon depletion is most pronounced) then being mapped as treeless barrens. The soils—little more than peaty forest floor and moss—were literally burned down to the underlying granite bedrock leaving behind alkaline ash. By 1954, however, about half of the land was forested, and in 1970, 70 percent was forested. The effects on lakes were the same as in the Adirondacks: forest runoff became more alkaline, and then returned to its natural acidic state.

The process is confirmed in sediment studies, which again, as in the case of the Adirondacks, reveal that lakes are acidifying after a period when they were unnaturally alkaline.

Kejimkujik Lake, 10 square miles of water in the heart of the acid lake and river district of southwestern Nova Scotia, had a pH of 4.0 in 1850. In the decades thereafter, lake pH rose to about 5.0 with cutting and burning of the watershed. With forest recovery, the lake has reacidified slightly to pH 4.8. While the sediment record shows that Kejimkujik Lake has become somewhat more acidic over the past 30 years, it is still not as acidic as it was before cutting and burning.

Experience in Norway is also comparable. Acid rainclouds coming from Britain and Germany have been blamed for the recent depletion of trout in lakes in southern Norway. But this is not the first time these lakes had lost their fish. Viking legend is filled with stories of fishing lakes and streams that went barren as a result of "sinful" behavior. The medieval Norwegian language even had a word for lakes without fish, *fiskelostjern*. Sediment analyses of Lake Langtjern, the most studied acid lake in the country, show that it was more acidic 800 years ago (pH 4.3) than it is today (pH 4.7). Studies of two other Norwegian lakes show a sharp increase in acidity (from pH 5.5 to 4.5) between 1350 and 1500, when forests grew back because the plague of 1349 killed two-thirds of Norway's population. Of the 12 Norwegian lakes with published sediment records, 11 were acidic prior to the industrial era.

Between 1850 and 1900—long before tall smokestacks and particle precipitators—there was a sharp increase in the acidity of many Norwegian lakes and streams, and a corresponding decline in fisheries. The geologist Ivan Rosenqvist developed the theory of changing land use to explain the phenomenon. After 1850 an exodus of population to the New World, combined with modern agricultural and forestry practices, put an end to traditional slash-and-burn methods that had been used to burn off acid soils and acid-producing vegetation. Marginal rocky and mountainous lands were abandoned and afterward covered by lichens, moss, heath, and forest. Acidic peaty soils and acid-producing vegetation increasingly covered the landscape. According to Rosenqvist, it was these watershed changes that most affected the pH balance of surface waters.

The notion that acid rain is responsible for acidity in lakes and streams is also contradicted by the existence of highly acidic surface waters in regions without acid precipitation. Fraser Island, Cooloola National Park, and Tasmania in Australia, and the Westland area of New Zealand have no acid rain, yet are filled with highly acidic lakes and streams. Indeed the magnitude of acidic surface waters in areas without acid rain dwarfs that of areas supposedly "devastated" by acid rain. In the Amazon basin, a river system the size of the Mississippi, the Rio Negro, is naturally acidic and fishless. The naturalist and explorer Alexander von Humboldt wrote about these "rivers of hunger" nearly 200 years ago, definitely predating industrial activity in this part of the world.

Clear as Acid

The return to natural acidity in the lakes and streams of Nova Scotia and the Adirondacks is obviously a disappointment to sport fishermen. It is not clear, though, why this acidification should otherwise be considered an environmental problem. On the contrary, the acidity of lakes frequently contributes to their beauty, particularly the crystal clarity of acidic waters with little living in them. And swimmers do not have to worry about nuisances such as slimy green algae or leeches. Lake Colden and the other fishless High Peak lakes are the Adirondack Park's most visited. Similarly, the clear acidic lakes of the Cape Cod National Seashore (an area with almost as many acid lakes as the Adirondacks) are so popular that the National Park Service has deliberately underdeveloped swimming and hiking access to them to prevent overuse.

If, however, the loss of fish is considered a problem, then multibillion dollar controls on sulfur dioxide emissions are not the solution. Since lake acidification results much more from the absence of natural buffering than from acid rain, reductions in the sources of acid rain simply are not going to be effective in restoring fisheries. According to NAPAP's acidification model projections, a 30 percent reduction in rain acidity over 20 years would deacidify only 26 northeastern lakes.

A much less expensive and more effective solutions is to do what farmers and gardeners do with acid soil: add lime. Lime dropped from a helicopter buffers acids in watersheds in exactly the same way that cutting and burning, or limestone in rocks and gravel, does. A NAPAP study estimates that all Adirondack lakes and ponds more acidic than pH 5.7 can be limed for $170,000 per year. Extrapolating this study to the entire Northeast, all acid lakes in New England and New York could be limed for under $500,000 per year.

Over the past several years, scientists have added lime to Woods Lake. But, like most acidic headwater lakes, lime placed directly into the lake gets flushed out too quickly to be very useful. Therefore, in the fall of 1989, scientists limed the acidic soils surrounding Woods Lake so that the lime will slowly dissolve over time to wash alkalinity into the lake. It's a simple answer that is much less expensive than emission controls, and it has the advantage of working.

Similarly, the overall health of high-altitude red spruce forests can be improved by low-cost, low-rate application of a balanced neutral nutrient salt mix, as is being done in Europe. Liming is not appropriate in these forests because it can reduce acidity, thereby adversely affecting red spruce and similar trees that actually require acidity. Improved nutrient status enables forests to better resist all types of stress ranging from climate and disease to air pollution.

Of course, environmentalists concerned with preserving natural acidic aquatic ecosystems will be at odds with those who want to lime lakes and streams in order to go fishing. This conflict has come to a head in Cape Cod, where the National Park Service cites its mandate to "protect the natural ecosystems of its parks" in resisting pressure to "improve" acid lakes by liming. Liming would kill the sphagnum mosses that grow deep in the bottoms of these lakes. The question is whether we want sphagnum mosses or fish; we usually can't have both.

The response among most American and Canadian voters is almost certainly to be fish. If so, the central question for acid rain policy is this: Do we want to spend billions of dollars a year on emissions controls that won't put back fish in lakes and streams that used to have them? Or do we want to spend hundreds of thousands of dollars a year on a simple policy that will?

Reprinted, with permission, from
The Heritage Foundation, Policy Review,
Spring 1990, pp. 44–48

Acid Rain—The Whole Story To Date

John J. McKetta, Ph.D.

John J. McKetta holds the Joe C. Walter Chair of Chemical Engineering, The University of Texas, Austin, Texas.

He received a B.S. in chemical engineering from Tri-State University, Angola, Indiana, 1937; an M.S. in chemical engineering from the University of Michigan, Ann Arbor, 1943; and a Ph.D. in chemical engineering, University of Michigan, 1946.

Dr. McKetta has published 10 volumes, *Advances in Petrochemicals and Refining*, 36 (of 65) volumes of *Encyclopedia of Chemical Processing and Design* and 11 other books. He is the author of more than 400 journal articles and has contributed to *Industrial and Engineering Chemistry, Transactions of American Institute of Chemical Engineers,* and *Hydrocarbon Processing Magazine.*

He received the Lamme Award from American Institute of Engineering Education for "The Outstanding Professor of Engineering" 1984, and the Herbert Hoover Award (1989) for "unselfish service to society by a technical professional."

Introduction

"Acid rain" is one of the most abused, overused, misunderstood and dramatized terms since "Three Mile Island."

Many, many people are concerned about acid rain because they have been told that it is destroying the environment, killing fish, ruining lakes, deteriorating forests and crops, and is harmful to mankind. With increased coverage by members of the news media who lack the knowledge to deal with this complex topic objectively, the public concern is nationwide and not restricted to the northeast.

As a matter of fact, very little is known about acid rain. Actually, we've learned quite a lot during the past 4–5 years, but we still have much to learn. Many gaps still exist in our knowledge of this subject. We do not know how it is formed, where it comes from, where it goes, what it does, or what harm it can do, if any. This is precisely why we must expand and intensify research into these and many other questions. The acidic contents of rain and lakes must be studied carefully under scientifically controlled conditions. This is being done by a U.S.

Government group called the National Acid Precipitation Assessment Program.

We're always going to have acid rain, because 70 percent of the acid rain comes from nature. But the big problem is mostly local. For example, in the Northeast United States, they are having a considerable acid rain problem. Most of this is because of their own production of acid rain. The Northeast United States uses over 40 percent of the high sulfur fuel oil that is used in the United States and still the Northeast occupies only 4.5 percent of the land area. For example, there are over 35,000 apartment houses in New York City alone burning high sulfur oil. Man-made sulfur dioxide has been decreased about 40 percent since 1970, even though the use of coal has increased about 85 percent during that time.

The acid rain committee continues to monitor hurricanes. Most hurricanes come to the United States from the Atlantic Ocean area. The acid rain that falls from these hurricanes is about the same acidity as pears and this is more acid than the acidity that the Northeast United States is complaining about. This is mostly natural acid rain, because there are no industries out in the Atlantic Ocean.

What is acid rain?

The term "acid rain" itself is misleading. Most rainfall is acidic. Some alkaline rains have been reported from the midwest. This is probably a result of airborne soil being incorporated into the drops. The air contains 0.03% carbon dioxide (which forms carbonic acid). Over 99% of the carbon dioxide comes from nature. Therefore, natural acidity is present in all kinds of precipitation, whether it be in the dry form such as dust, or in the wet form, such as snow, rain, fog, and dew.

You can see from Figure 1 that acidity is measured on a pH scale from 1 to 14 (pH is a measure of the concentration of the hydrogen ions). This pH scale is used by chemists to measure the acidity of solutions. Any substance with a pH value below 7 is acidic. The lower the pH the more acidic is the substance. A substance with a pH value above 7 is called alkaline (or basic). Applying this scale to substances with which you are familiar, you can see that peas have a pH of about 6. (Incidentally, drinking water has approximately the same pH). Carrots have a pH of 5. The numbers from 1 to 14 are all logarithmic. This means that 5 is ten times more acid than 6, and that 4 is ten times more acid than 5, or 100 times more acid than 6. For example, pears, with a pH of 4, are ten times more acid than carrots with a pH of 5, but pears are 100 times more acid than peas with a pH of 6. It's interesting to note that the stomach juices' acidity is almost the same as the acid in your automobile battery. At the other extreme, note that soap is alkaline with a pH of 8 or more, while lye (sodium hydroxide) is extremely alkaline with a pH of 13. It's important to know that seawater has an alkaline pH of over 8. This makes it possible for the sea water to absorb carbon dioxide, sulfur dioxide, and other such gases from the air. In fact, the oceans are just like huge pumps, absorbing and expelling gases depending upon equilibrium conditions.

Because most of the carbon dioxide and oxides of nitrogen (which form nitric acid) come from nature, the natural rainfall that would result if we had no man-made pollution would be lower than 5.6. In fact, it seems reasonable that the natural pH of rain should be set at a value between 4.7 to 5.0.

When was acid rain first recognized?

We don't know when acid rain was first recognized but the first mention in recorded history was in 1848 by a Swedish scientist. Then in 1852, a French scientist measured the quantities of "nitric acid and nitrogen compounds" in the rain in Paris. In 1872, the acidic nature of rain was documented in a book covering the chemistry of English rain.

A great concern was raised in the late 1960's when Swedish scientists claimed that the cause of the acid lakes and acid rain was the sulfur dioxide emissions from industrial sources of Great Britain.

In the U.S.A., the controversy arose back in 1974 when a Cornell University researcher released reports (using data collected in the 60's and before for other purposes) concluding that rain acidity was increasing in the northeast and spreading in all directions. Since then others have speculated that acid rain is a post-World War II phenomenon caused by the increased use of fossil fuels for generating electricity, as well as the increased use of automobiles.

Is acid rain harmful?

The effect of acid rain on the environment is not clear. There are many claims that acid rain may harm the lakes, decrease fish population, reduce forest growth, decrease crop productiveness, decrease soil fertility, corrode buildings, and cause other detrimental effects. Yet scientific evidence does not substantiate these claims except in laboratory-size experiments. In fact, in some instances, crop yields have increased by using acid rain in laboratory experiments. Corn and tomatoes, for example, benefit greatly because of the fertilizing value of the extra nitrogen and sulfur in the rain.

Is acid rain increasing?

The acidity of rain must be measured for a long time, in fact, years, at the same location before reporting some meteorological average. That's because the average acidity of the rain during one single rainfall measured at the same time, at points only several hundred feet apart, may vary, plus or minus 200 to 400 percent. Therefore, to establish an accurate trend one must collect data over a 5–10 year period. When this has been done there has been no evidence that the rain is becoming more acid.

Some claim that coal-burning utilities in the midwest are the primary causes of acid rain in the northeast. However, they overlook the fact that a large percentage of domestic heating in the northeast comes from the use of fuel oil rather than coal. Many fuel oils contain as high as 2 to 3 percent sulfur.

Where does acid rain fall?

Acid rain falls everywhere. Natural rainfall is acidic. Natural rainfall has a pH averaging about 5. In the northeast, readings in the 4.0 to 4.5 range are not uncommon. Rainfall over most of the western states is closer to 5. But in some areas, such as San Francisco, Seattle, Denver, and Los Angeles, the rainfall has been measured at 4 pH. Acid rain pH of 4 has been measured at such remote spots as Samoa in the South Pacific, the tropical jungles of South America, the arctic coast of Alaska, as well as Hawaii and the islands in the mid Indian Ocean.

The three areas of the world where acid rain appears to be of the greatest current concern are: Southeastern Canada, the Northeastern United States, and Scandinavia.

Acidity from Natural Sources

It is now clear that the natural factors, such as organic acids, naturally emitted sulfur and nitrogen compounds also affect rain's normal acidity. More recent studies in remote parts of the world on natural sources of atmospheric acidity, suggest the unpolluted pH of rain is closer to 5.0 rather than 5.6 (5.6 pH is the acidity of rain saturated with natural carbonic acid. If the natural nutric acids are included, the natural pH drops to 4.8. The addition of natural SO_2 brings the rain natural acidity down to 4.4–4.6.)

Approximately 65% of the sulphur dioxide, 99% of the carbon dioxide and more than 99% of the total oxides of nitrogen come from nature on a world wide basis. All of these components make acid rain (sulfuric acid, carbonic acid, and nitric acid). It's possible that nature's contribution of these components in specific localities may not be as high as indicated but may be lower than 25 or 30% (see Appendix III).

Since the ratio of sulfates to nitrates is 2 to 1 in precipitation over Eastern North America, sulfur gases have been labeled as the major contributor to rain acidity. The ratio is reversed in the west. Moreover, in many instances the acidity of rain samples does not differ greatly between Eastern and Western United States.

In my own back yard in Austin, Texas, I measured the pH of rain throughout 1981 at an average of 4.3. The normal direction of the wind during all sampling was from the northwest. There is no coal burning plant within 1,000 miles northwest of my house. Also, the California Air Resources Board announced on March 4, 1981, that "rain more acid than vinegar is falling on California and may poison the lakes." CARB Chairwoman, Mary Nichols added "We've learned that the Sierra Lakes are especially vulnerable to acid because of the chemistry surrounding them." There are no coal burning electric generating plants upwind of this area.

Lightning's contribution to the acidity of rain is significant. Two strokes of lightning over 4/10th of a square mile (one square kilometer) will produce enough nitric acid to make 8/10th of an inch of rain with a pH of 3.5. One scientist calculated that lightning creates enough nitric acid so that annual rainfall over the world's land surfaces would average pH 5.0 without even accounting for contributions from other natural sources of acidity.

In the forest area of Brazil at the headlands of the Amazon River, an area remote from civilization, the monthly average pH of 100 rain events in the 1960's ranged from 4.3 to 5.0. One set of pH reading was as low as 3.6.

The rainfall from two hurricanes in September, 1979 sampled at six stations from Virginia to up-state New York averaged 4.5 pH, with one set of readings as low as 3.6 pH. This weather came directly from the Atlantic Ocean and was quite unlikely to have been affected by emissions from industrial activity. On the South Seas island of Pago Pago, some readings of pH as low as 4.3 were observed. In the heavy thunder storm activity at the start of the monsoon season in the remote northern territory of Australia, the rain averaged between 3.4 and 4.0 pH.

Recent ice pack analyses in the Antarctic and the Himalayas indicate that precipitation deposited hundreds and thousands of years ago in those pristine environments has not varied much from a value of 4.4 to 4.8. In fact, measurements have been made as low as 4.2 in these areas. This compares with the "average" pH of rain in Eastern United States, as well as in Scandinavia, of between 4.0 and 4.5.

Greenland ice pack analyses showed that many times in the last 7,000 years the acidity of the rain was as low as 4.4 pH. In some cases the periods of extremely high acidity lasted for a year or more.

Is rainfall increasing in acidity?

The United States Geological Survey collected rainfall samples in various locations in and near New York State during the period, March, 1965 to September, 1979. These data were collected at 22 locations; however, only 9 stations operated more or less continuously through that period. These data indicated that the long term level of the acidity was essentially constant.

The existing data and studies show that there has been no significant changes in acidity in Northeast U.S. precipitations since 1960. In fact, the new data show that the sulfate concentrations have decreased and the nitrate concentrations have increased. The U.S.G.S. has concluded that acidification of surface waters in the Northeast has "probably occurred long before the 1960's." They have also stated that the acidity of precipitation has been stable since the mid-fifties.

Man-Made Sources

It is well known that man puts many substances into the atmosphere. Many of these are acid forming. However, since 1960, emissions have declined. For example, there has been a decline of over 40% of sulfur dioxide emissions since 1960. At the same time, the use of coal has increased by 85%. There has been a similar decrease in man-made oxides of nitrogen.

There is a belief in some quarters that man-made emissions in other parts of the United States have increased the acidity of precipitation over Northeastern United States. The national program (see Appendix I) calls for conducting large-scale field studies to provide data needed to confirm model assumptions and model characteristics on flow behavior of pollutants. The Northeastern United States uses large amounts of residual fuel oil for domestic, commercial, and industrial purposes. In fact, the Northeastern United States uses 40% of the residual fuel oil, 35% of the distillate oil and 17% of the gasoline consumed in the entire country. Yet, the northeast comprises only 4.5% of the country's land area. Much of the fuel oil is high in sulfur. Dr. Kenneth Rahn, University of Rhode Island, is using trace elements to find the source of pollutants. His research data indicate that local pollution sources in New England are the main cause of acid rain and snow in that area. His research has not revealed sulfur compounds emitted from mid-western coal fired plants in the rain collected in the northeast. At St. Margaret's Bay, Nova Scotia, a study showed that 50% to 60% of the acid deposition came from the direction of Halifax, 15 miles to the east. A meteorological team at the University of Stockholm cautioned the Swedish people who blamed acid rain on the power plants in England, not to be so sure. This team's conclusion, after studying sulfur, nitrogen and water cycles, via long-term monitoring, was that much of the acid rain was local.

EPA scientists studying emissions from four large oil burning units in New York City found flue gasses from the boilers did indeed contain large amounts of both SO_2 and sulfuric acid. These flue gasses also contained traces of vanadium. Further analyses showed that vanadium was found in the oil, in the emissions from the boilers and incrusted in the lining of the boilers where combustion takes place. Vanadium is present in significant amounts in oil but is almost nil in coal. These EPA investigators concluded that more than half the winter time, sulfate emissions in New York City are attributable to local oil burning boilers. Some studies have indicated that the suspended particulates rarely travel over 300 miles (most often up to 100 miles). This means that the more than 35,000 oil fired boilers in apartment house in New York City play a dominant role in the elevated sulfate levels in that area.

The best available estimates of current interregional sulfate deposition are from the Advanced Statistical Trajectory Air Pollution (ASTRAP) model. This model shows that each region is its own largest source of deposited sulfate.

It's very discouraging when, in the face of these foregoing data, a politician such as Sen. Alfonse M. D'Amato (R-NY) says, "Regardless of the documented scientific data, I have found the effects of acid rain obviously detrimental . . . I am cosponsoring a bill, S. 2001, to tax large plants and factories according to how much pollution they emit . . ."

Does Acid Rain Affect Human Health?

Much research is going on concerning the relationship between acid rain and health problems. To date, no ill effects have been found. Naturally research will continue in this important area. We all know that many of the chemicals found in living plants and animals are acid. Muscles in the human body are mostly amino acids. Ascorbic acid or vitamin C is a dietary essential; malic acid gives apples their tangy taste. All of these have pH values within the range of rain term "acid" (3.0 to 5.0 pH). The pH of acidic deposition is well within the range normally tolerated by human skin and eyes. The statements in the press that some individuals die because of acid precipitation are unfounded and do not reflect the current state of knowledge. Almost all of us drink quarts of fluids daily that are 5,000 to 10,000 times more acid than milk or peas. Yes, soft drinks are of this acidity.

The world's outstanding epidemiologists who specialize in sulfur dioxide health effects in mankind deny that there are any adverse health effects. These include experts such as Dr. Arend Bouhyus (Chairman of the Cambridge Medical College), Dr. Robert Buechley, Dr. Merrill Eisenbud, Dr. Herbert Shimmel, Dr. Lawrence Hinkle, Dr. Battigelli, Dr. Thaddeus J. Murawski, and many others. Remember that the present law allows a maximum of 0.02 parts SO_2 per million parts air (0.00002%) in the ambient air.

What is the Effect of Acid Rain on Lakes and Fish?

For some lakes in sensitive regions, evidence indicates the lakes have been highly acidified and will not propagate fish life. The rate, character, and the full extent of these changes are scientific unknowns. These studies have not been made in sufficient detail to document the actual changes. We do not know that the vegetation and soil surrounding a lake and stream play a major role in determining the rate and nature of the water body's response to acid deposition. Many lakes in North America have complex watersheds where precipitation flows through forest canopies and soils, and is chemically modified before entering the lake. As the rain passes over the vegetation and through the soils, its acidity can be reduced or increased by many-fold.

The acidity of most of the waters involved are actually the greatest in the spring. The fish kills occur almost yearly in the Midwestern United States lakes such as in Wisconsin because of the interception of the light by ice and snow on the lakes so that green aquatic plants are not able to produce adequate oxygen. Then the fish simply suffocate.

Regardless of the scare stories of the media there are larger amounts, and record sizes, of fish caught in the New England lakes each year. In fact, just as an example, on January 1, 1984, the New York State Department of Environmental Conservation released the size and quantity records for freshwater fish, listing 34 specie(s). A review of this release reveals the following:

(1) In the period of 1979–83, 25 of the 34 records have been broken.
(2) In 1983 alone, 13 records were broken including:
  a. Brown Trout (23 lb. 12 oz.)
  b. Pink Salmon (1 lb. 9 oz.)
  c. Cisco Whitefish (2.97 lb.)
  d. Tiger Esocids (29 lb. 3 oz.)
  e. Bullhead Catfish (2 lb. 0 oz.)
  f. Channel Catfish (25 lb. 8 oz.)
  g. Bluegill (1.96 lb.)
  h. Rock Bass (1 lb. 4 oz.)
  i. American Shad (7 lb. 14 oz.)
  j. Angling Carp (40 lb. 4 oz.)
  k. Bow Carp (58 lb. 5 oz.)
  l. Freshwater Drum (18 lb. 4 oz.)
  m. White Sucker (1 lb. 6 oz.)

Except for four of these records from Lake Ontario, all others were from separate lakes or rivers. Of the 25 broken for the five year period, these occurred on 16 different lakes and rivers. These facts stand in stark contrast to media stories of impending doom for New England lakes.

Does Acid Rain Damage Trees and Crops?

Rainfall makes the grass in our yard grow faster and become greener than it would be if the grass were merely sprinkled using city water. The reason is that growing plants and trees require nitrates, ammonia, sulfates, magnesium, phosphorus, potassium, and other substances. The nitrates and the sulfates in rainfall are the ions which are the indicators of the major strong acid components in rain. Likens and Bormann pointed out way back in 1974 that the sulfur content of rain had decreased in New York State but that there was not a corresponding decrease in their rain acidity measurement. They concluded the observations might be due to the neutralization of sulfuric acid by particles in the air.

The best that scientists can say today is that they don't know what is causing the changes in growth of some tree species. Dr. Arthur Johnson of the University of Pennsylvania, one of the researchers looking into the problem, said this: "A lot of work needs to be done to understand whether it is a natural phenomenon or whether it is due to air pollution or a combination of circumstances."

Worldwatch, an environmental think-tank, says that "No single hypothesis can explain everything that's happening (to forests) everywhere."

So, those who raise the specter of dead forests and attribute their decline to SO_2 emissions from the Midwest coal-fired power plants and acid rain are misleading the public.

Even the forest products industry which has a larger stake than anyone else in the health of forests cannot document that there is any problem to forests. Incidentally, despite constant acid rain, the amount of standing timber in New York forests increased by 70% between 1952 and 1976.

Ex-EPA administrator William Ruckelshaus recently told a congressional committee that while the information about forest effects is troubling, the uncertainty about the cause "raises the possibility that if we act too quickly, we may control the wrong pollutant."

The May 1984 issue of the Smithsonian magazine devoted many pages on the terrible effect of the tussock and gypsy moths on the forests in various parts of the United States. It's interesting that essentially the same states cited in their article as complaining about the moth deforestation also claim that acid rain is killing their forests. Various pests and diseases are specifically responsible for much forest damage. The most notable of these includes Chestnut blight permanently altering eastern forests. Spruce budworm in Northern New England; Gypsy moths in eastern hardwoods; Pine bark beetles in the south and southwest; and Douglas fir tussock moths in the west.

To date, the evidence that acid deposition and associated man-made pollutants have contributed to observed forest declines is circumstantial and inferential rather than conclusive. A variety of complex causes are possible, and plans for accelerated work (see Appendix I) are underway to determine which factors—such as acid deposition, gaseous pollutants, insects, disease, and drought—contribute to the damage.

Even though it is well established that soil conditions affect the growth of vegetation and trees, the effects of acid precipitation on this growth process remain uncertain. Experiments conducted to determine the impact of acid precipitation on crops have produced mixed results. The EPA tested 38 varieties of plants under greenhouse conditions with artificial rain adjusted to pH levels of 3.0 to 4.0. Approximately 40% of the varieties of plants showed increased yield, about 20% showed no effect while the last 40% showed decreased yield.

A recent statement in the press indicated that acid rain made holes in leaves of tobacco plants. In investigating this statement, we found that in a laboratory test, two soil scientists used water with a pH level of 2.0 (0.01 normal hydrochloric acid) and observed that holes were made in the leaves of tobacco plants. They also indicated that a pH level of 2.5 had no observable effect. There is no record of any rainfall anywhere at concentrations of 2–2.5 pH. This was merely an extreme laboratory test.

The Interagency Task Force on Acid Precipitation stated in their annual report of 1982 that "While there is general agreement that unmanaged soils in forested and grassland areas in humid regions may be sensitive to acidification from acid precipitation, there is no indication to date that the soils have become acid because of it or that forest production is being affected."

Does Acid Rain Damage Stone and Metal Surfaces?

Damage to architectural stone surfaces has been connected with atmospheric acidity. But the major cause appears to be local sources of air pollutants such as heavy vehicular traffic and industrial activity. Moreover, impacts from air pollutions in general and impacts from acidic precipitation cannot be distinguished.

A flurry of news articles alleging an impact of acid rain on automobile finishes appeared in late 1980. But these stories were unfounded except as they related to acid smut fallout on vehicles parked near the offending chemical plants. Acid smut involves acidic particles quite unrelated to acid rain.

Another alleged hazard is the effect of acid waters on metal pipes. Both copper and lead pipes are relatively unaltered by moderately acid solutions, and none of the feared health impacts have been documented. In most cases, any precipitation would be partly or completely neutralized as it seeps from the surface through the soil to underground pools or wells.

All in all, the effects of acidity on the environment have not been found to be as severe as some have suggested. Moreover, the feared impacts have not been demonstrated outside the laboratory.

Are We the Cause of the High Acidity in Southeast Canada?

Canada's Foreign Minister Joseph Clark, told Secretary of State George P. Schultz the new Canadian administration "intends to keep raising the acid rain issue until the U.S. takes steps to reduce SO_2 emissions."

Mr. Allan Gotlieb, Ambassador of Canada to the United States, stated in a letter to the editor of the *New York Times* that the acid rain problem is recognized by all political parties in Canada reflecting deep and wide-spread public concern.

The *Louisville Courier-Journal* in late 1984 headlined an editorial. "The U.S. Policy On Acid Rain Just Insults Canada," and goes on with remarks like "The U.S. has no right to use Canadian skies as a sewer for air borne waste." A U.S. Senator, Daniel Moynihan has stated publicly that the U.S. is a very poor friend of the Canadians because we pour acid rain into Southeast Canada. The facts do not support these statements. Let us look at them.

The Canadian government passed out a 4-page "fact sheet" at the World's Fair in New Orleans indicating all the things the Canadian government has done and promised to do to decrease acid rain from 1978 through March of 1984. The indication in this presentation is that the United States government has not carried out its part of the bargain. However, nowhere in his report, did the Canadian Government point out that it has not put in a single scrubber to remove sulfur dioxide or oxides of nitrogen from any of Canada's coal burning plants or from any of its huge smelter operations. The huge INCO Smelter at Sudbury, Ontario, alone emits 1,950 tons of sulfur dioxide each day. All indications from the Canadian Environmental Minister are that Canada will install scrubbers by 1987 or they will remove "x" tons of sulfur dioxides and other pollutants by 1990. Many of the Canadian so-called "dead lakes" (lakes which no longer support fish stocks and other wildlife or vegetation) are located near these INCO smelters which annually send nearly 1 million tons of sulfur dioxide into the atmosphere. Sometimes in the heat of debate the Canadians seem to forget that the U.S. Clean Air act requires that our clean air laws are federally mandated and are enforced vigorously with civil penalties and the cutoff of federal funds. The U.S. already has 111 scrubbers in operation (51 more are under construction) at a cost of $5.5 billion, plus an annual operating cost of $2.1 billion. I do not know of a single scrubber in operation in a coal fired electric generating plant in Canada.

It's interesting that Canadians keep pointing out our lack of pollution control when they do not require pollution control devices even on their motor vehicles. There is no requirement for catalytic converters on cars in Canada. The total Canadian emissions of SO_2 are twice those of the United States on a per capita basis. The decline in the SO_2 emissions in Canada is mostly due to declining production by the copper smelting industry and to increasing use of nuclear and hydrogeneration of electricity. The decline is not due to governmental policies nor laws intended to improve the environment.

Earlier the press reported that Canada and nine European countries have signed an agreement to reduce sulfur dioxide emissions by 30% during the next ten years. The U.S. and Canadian press would have you believe that we are remiss in not being a member of the club. The fact is, they have just agreed to do in the future less than what the U.S. has already accomplished.

Canada—Acid Rain and Electricity Exports

Now, the Canadian say that if we will shut down the electric power plants in the Mid West in the United States, that they would sell us electricity to make up for that which we lost. In 1970 Canada exported to the U.S. 2.5 billion kilowatts of electricity. Fifteen years later, by 1985, that figure was upped to 40 billion kilowatts of electricity—the equivalent of fourteen 500-megawatt coal fired plants operating at 65% plant factor.

There is considerable concern regarding growing Canadian electricity exports to the United States. The Canadian Energy Research Institute estimates that by 1990 the total potential market for Canadian power could be 60 to 75 billion KW. Obviously, if the U.S. produced less power because of repressive constraints on generation, one beneficiary would be the power producers in Canada. They are building more nuclear and hydro plants. Adding insult to injury, the Canadians are today seeking U.S. financing for their nuclear plants. These plants will export nuclear power to the United States, where the anti-nuclear groups prevent the building of our own such plants.

It is interesting that 45 percent of the electricity that we purchase comes from nuclear power plants, which the U.S. resists in producing and 55 percent comes from fossil plants. Incidentally, again, none of these fossil plants have any scrubbers.

What Can We Do to Control Acid Rain?

Because there are still no accurate scientific data on conversion, transportation, accumulation, or transformation in cloud water of pollutants, we must continue research in these areas as outlined in Appendix I.

In the meantime, areas having local trouble should look at solutions on a local basis. The burning of high sulfur oil in the northeast should be curtailed. The oil should be desulfurized. This is a local problem. When one has a flat tire on an automobile he does not remove the entire body or engine. He only repairs the flat tire.

Some of the suggestions made by Washington politicians include placing scrubbers on all coal burning equipment in a 31-state area. The cost is estimated as high as $30 billion. The politicians say that this should be paid by the electric utilities. This ultimately means you and me! A high percent of the population does not understand this.

A test has been made in the Adirondack park region of New York State using powdered limestone in 51 small lakes that were acidic. The lime, being alkaline, reduces the acidity, raising the pH. Result: good fishing has been restored. The cost has been only between $15 to $30 each year for each acre of surface water. All known acid lakes in Adirondacks could be limed adequately for about $300,000 each year. Liming is also being used in Canada and Scandinavia. Liming is much cheaper than some politicians' suggestions to use expensive scrubbers.

There is no evidence to suggest that a reduction of sulfur dioxide emissions from one portion of the country (midwest, for example) will result in a proportional decrease in the acidity of rainfall in another portion of the country (for example, northeast). In fact, it's been predicted that a 90% decrease in the sulfur dioxide emission in the midwest would not increase the pH of the lakes in the northeast by much more than 4% (a pH increase of 0.2, i.e., 4.3 to 4.5). BUT—the electric light bills would double!

For those who feel that we should try scrubbers anyway to see whether they could help, I wish to point out that a SO_2 stack gas scrubber for a coal burning power plant has an initial cost of $100 to $300 million for the equipment alone. In addition, the scrubbers require energy to operate. Consequently 4 to 6 percent of a plant's power has to be put back directly into operation of the equipment. Other costs, including the purchase of chemicals and the disposal of sludge collected from the scrubbing, add further to the plant's operating cost. Disposing of that sludge actually imposes another problem for those concerned with the clean environment. These are all part of the $30 billion cost mentioned above.

Summary

In summary, the scientific and engineering community agrees that many gaps exist in our understanding of the acid lake—acid rain issue. We know what is known and we know the gaps of the unknowns. Fortunately, a large research effort is under way by the Acid Rain Task Force. Additional research is going on in private industrial and state laboratories. We do not yet have the basic information to determine the efficacy of any control strategy that so far has been recommended. As stated in the National Academy of Science Report of 1983, we do not believe it is practical at this time to rely upon currently available data and models to distinguish among alternative control strategies. The question, then, is "Should the government spend money on projects whose effectiveness cannot be judged in solving a problem which is still not well defined? "With out present knowledge this question cannot be answered by scientists and engineers.

If politicians insist on making a decision now, based on the meager information they presently have, the payment for this huge expenditure will be paid by the individual energy users—you and me. You know that decisions made in the past by politicians, trying to solve real problems, without the basic knowledge, have been expensive and in most cases wrong.

Reprinted, with permission, from
*National Council for Environmental Balance Inc., (NCEB)*1988

This article was prepared in 1990 before the final NAPAP report was completed.

Agricultural Chemicals

Pesticides: Helpful or Harmful?

Leonard T. Flynn, Ph.D.

Dr. Leonard T. Flynn has been a regulatory and scientific consultant since 1985 for the pharmaceutical, personal products and other manufacturing industries. His consulting practice involves preparation and review of regulatory documentation necessary for product or activity approval by government agencies such as the Food and Drug Administration, the Environmental Protection Agency, the Occupational Safety and Health Administration, the Consumer Product Safety Commission, and corresponding state authorities. Dr. Flynn also reviews materials, operations, advertising, and labeling for his clients including auditing of manufacturing facilities, laboratories, and paperwork to assure compliance with regulatory requirements.

Prior to being a consultant Dr. Flynn was Manager of Regulatory Affairs for Block Drug Company in Jersey City, New Jersey, and Director, Regulatory Affairs and Compliance for Organon, Inc. in West Orange, New Jersey. He was also Research Director for the Soap and Detergent Association in New York City.

His education includes a Ph.D. from Purdue University (1973) in organic chemistry and a Masters in business administration (1974) from the University of South Carolina. He graduated from Rutgers University with a Bachelor of Arts degree in chemistry in 1967.

Dr. Flynn is a charter member of the Regulatory Affairs Professionals Society, a Scientific and Policy Advisor to the American Council on Science and Health, and a member of The Nature Conservancy. He has given presentations on regulatory matters to the Proprietary Association's Manufacturing and Controls Seminar, the Center for Professional Advancement, and the Society of Cosmetic Chemists. Dr. Flynn has written several articles about regulatory and scientific topics and the impact of government regulations on the pharmaceutical, medical device, cosmetic, and pesticide industries.

Pesticide Development and Regulation

The Environmental Protection Agency (EPA) regulates pesticide products according to the Federal Insecticide Fungicide and Rodenticide Act (FIFRA). The regulations (40 CFR parts 152-180) set forth extensive controls that affect the research, development, distribution, promotion, handling, storage, disposal and use of pesticide products.

Early research involves screening new chemical compounds for biological activity and toxicological effect. Outdoor testing for effectiveness is performed initially on small plots. Long term toxicology studies are conducted to test for cellular or tissue abnormality, illness, cancer, birth defects (teratogenesis) or mutations in laboratory animals.

Field studies under actual farming conditions are conducted according to an EPA-granted Experimental Use Permit (EUP). To receive an EUP, the applicant must submit results from toxicology tests and studies done in small test plots, along with plans for the proposed test an a label for use of the product. If crops treated under the EUP are to be marketed, then adequate health and safety data must be submitted to EPA with a petition requesting a temporary tolerance. (If the temporary tolerance is not granted, the crops must be destroyed after the field tests are complete.)

Tolerances are levels of agrichemicals allowed to remain in food for human consumption or feed for livestock at the time the crop is harvested. EPA sets tolerances after public review and comment, then the FDA and other agencies enforce these tolerances on food and feed in commerce.

Residue levels in crops are determined through field tests. Safety of the levels is demonstrated by the toxicity data from animal feeding studies. A No Observable Adverse Effect Level (NOAEL) is determined to be the level at which the chemical has no harmful effect on the most sensitive test animal. By dividing the NOAEL by a safety factor of up to 100 or more, an Acceptable Daily Intake (ADI) is established. EPA plans to replace the ADI and implement a "reference dose" (RfD) approach to reach regulatory decisions about the significance of chemical exposures.[1]

After the above studies and evaluations of the pesticide are completed (about five to seven years), the registration package is submitted to EPA. Scientists within EPA review the submitted documents and evaluate the pesticide's ecological effects, residue data and toxicology studies.

Agency officials must consider the benefits of the pesticide in comparison to its potential acute, chronic and environmental risks. Feedback from the public, Congress and industry assists the agency in its evaluation—if the feedback is informed and reasonable. Otherwise, "when events move out of the realm of science and into the realm of public emotion, science may become foreclosed."[2] Although EPA's review process and criteria under FIFRA may be far from perfect[3] the system works best when science, not hysteria, is the basis for regulatory decisions.

Most farm chemicals and home use pesticides are "general use" products; that is, they can be used safely by anyone who follows label directions. However, some chemicals are classed as "restricted use" pesticides. They are only sold to and used by "certified applicators," persons who satisfy EPA and state training requirements. "Restricted use" pesticides are not considered suitable for use by the general public; that is, the label instructions alone are not considered adequate to assure safe and proper use for restricted use pesticides.

If EPA accepts the registration for the product, then the final rules on the registration and any tolerance petition are published in the Federal Register. At this point, the label of the pesticide becomes a legal document, and any deviations from the instructions on the label, unless provided for by the law (e.g., special local need and emergency exemptions, see below), subject the pesticide user to civil and criminal penalties. Note that the legal requirement to follow the label governs the use of all pesticides, whether general or restricted use.

Once the new pesticide is approved by EPA and begins to be used according to the label, further research will continue to uncover other major uses of the product. Each new use of or additional pest controlled by the pesticide requires further review and acceptance by EPA before it can be added to the pesticide label.

Often, the potential market for a product is too small to justify the manufacturer's time and expense to gain EPA approval for use on small acreage crops or localized pests. FIFRA provides for Special Local Need (SLN) Registration to cover such limited situations and permits an individual state to allow the use of EPA registered pesticides for specific purposes within that state under a permit which becomes part of the labeling.

In addition, FIFRA Section 18 allows emergency exemptions for new agricultural pests or diseases for which no EPA approved pesticide is available for control. An emergency exemption is granted only under highly restricted conditions including limited time, precisely defined area and rigid use restrictions.

Clearly, pesticides are pervasively regulated, and approval of a new pesticide is not simple, inexpensive or fast, Data submitted to EPA in support of a typical new agricultural chemical can represent as much as 25 million dollars of cost and more than seven years of laboratory, field, and environmental testing—an investment which cannot be recovered unless and until EPA approves the pesticide. The review of a registration document by EPA typically requires about three years for new pesticides; thus, the total time involved in development of the new product is about 10 years form initial testing to final marketing approval, a considerable period of time.

The development of pesticides operates on a progressive testing system which eliminates large numbers of candidate compounds during the initial years of toxicological testing. The study costs for these rejected substances must be included in the expenses pesticide manufacturers hope eventually to recover through approval of new products. Due to increased risk arising out of increasingly stringent requirements, new pesticide development is now primarily feasible only for the largest companies and for pesticides having wide agricultural markets.

Do we really need new pesticides? Are present pesticides "poisoning the environment" so that more of them would make things worse? The environmental issues related to pesticides will now be considered.

Pesticide Issues

Persistence

Most pesticides are non-persistent and are degraded fairly rapidly (a few weeks or less) by sunlight, soil microorganisms and moisture, so they do not remain in the environment for extended periods. A few pesticides—for example, some organochlorine insecticides—are considered persistent because they maintain their pesticidal potency for some time after application. This often represents an advantage in that fewer applications are needed. Costs for labor and materials can be lower compared to more frequent applications of rapidly degradable pesticides.

Fundamental to the issue of persistence are the toxicity and utility of the pesticide. For example, the organochlorine insecticides have been roundly criticized for their persistence; however, they are generally less toxic than many other insecticides, particularly in their toxicity toward mammals, including humans.

For example, two insecticides are used commonly in tropical areas against mosquitoes to combat malaria, the persistent organochlorine DDT and the non-persistent organophosphate malathion.[4] DDT is very stable and may remain active against insects for up to a year, although normally DDT is applied as a residual spray twice a year to the indoor surfaces of houses for mosquito control. Its toxicity to man is very low. Like organophosphates in general, malathion is more volatile than DDT and its active life is shorter (3–4 months average residual action) for indoor application. Malation, unlike most other organophosphates, is relatively low in human toxicity but some formulations undergo chemical degradation during storage under tropical conditions resulting in a much higher toxicity than expected. Thus, even with malathion, very strict safety precautions should be observed.

The application of malathion is about five times as expensive as DDT and the human safety edge, if any, for mosquito control is on the side of the persistent DDT, not the degradable malathion. The low cost and stability of the organochlorine insecticides such as DDT, lindane and dieldrin are "precisely the reason why they have been such useful pesticides in the last 35 years and the mainstay of vector-control programs."[5]

Persistence is essential for certain applications where frequent, repeated applications are not desirable. Since termites are relentless destroyers of wood, persistent pesticides are the preferred treatment to protect homes and farm buildings from destruction.

Biomagnification

A commonly held belief is that some pesticides biomagnify as one proceeds up the food chain as, for example, from algae to planktonic crustaceans to small fish to larger fish to predatory birds or mammals. The consumption of low levels of pesticides within each prey animal was presumed responsible for increased amounts in higher predators.

Careful research has revealed that this phenomenon is uncommon in nature and that pesticide concentrations in predators have little to do with biomagnification (increased pesticide levels) up the food chain.[6] As one review article[7] stated, "The popular conception of food chain biomagnification of chemical contaminants is not well substantiated," and "the role of biomagnification has recently been minimized as a significant contribution to the accumulations of residues." Another reviewer[8] concluded similarly: "biomagnification of contaminants is not a dramatic phenomenon in marine and freshwater foods webs." He continued, "Most of the evidence for the existence of nonexistence of biomagnification within aquatic food webs has come from highly circumstantial and/or marginally relevant data."

For most pesticides, their presence in the environment is short compared to the growing time of most organisms. In fact, they are designed not to biomagnify; hence, biomagnification is not even a theoretical problem.

Wildlife Effects

One important result of modern agriculture, including the use of pesticides and chemical fertilizers, is greater productivity per acre. Modern farmers can produce much more than their predecessors and require less acreage to do it. This situation releases land for other uses including conservation for wildlife habitat and human recreation such as camping, hiking, bird-watching, etc. Under less productive agricultural methods, more land must be utilized for farming—to the detriment of other land uses.

The presumed detrimental effect of pesticides on wildlife (e.g., as expressed in Silent Spring by Rachel Carson) has been widely accepted by the public but lacks scientific support. (Reduction in wildlife population is generally due to habitat destruction, overfishing and similar factors.) In fact, one author stated, "Wildlife populations all over the nation are bigger and healthier than ever, not in spite of pesticides, but in many cases because of them."[9] He cited surveys by fish and game authorities who reported widespread and increased abundance of waterfowl, deer, small and big game compared to decades ago before widespread use of pesticides.

The possible environmental effects of the pest to be controlled should not be ignored when weighing the overall environmental impact of a pesticide treatment. For example, imported fire ants "are highly competitive pests" which destroy "both harmful and beneficial species"[10] in addition to the effect of their painful stings on animals and man.

Resistance

Resistance of pests to pesticides is often used as an argument against pesticides and for the "natural is better" philosophy of pest control Actually, proper control of pests requires a more comprehensive view. Some pesticides evoke resistance problems but can still be suitable for use in specific situations; conversely, other less resistance prone pesticides can still lead to serious resistance in certain uses. "Consequently, resistance risk can only be assessed realistically in actual, complex situation,"[11] not on an oversimplified, chemical by chemical basis.

A recent review stated that, although the basic principles of resistance management apply to all major classes of pests (insects, fungi, weeds and rodents), "sweeping generalizations about the applicability or feasibility of specific tactics are not justified."[12] The review lists 15 tactics for resistance management and rates their suitability. Four methods are top rated as "very useful, generally supported by laboratory data and/or field experience" for resistance control for single pest classes. These include using local rather than areawide applications (insecticides), fewer or less frequent applications (fungicides), using less persistent pesticides (herbicides), and improved pesticide formulation technology (rodenticide baits).

However, two resistance management methods have top ratings for all four classes: (1) alternation, rotation or sequences of pesticide application and (2) discovery and development of new pesticides. The development of new pesticides is vital to continued pest control. Unfortunately, pesticide development is costly and slow (see Part 3).

One concern is the possible creation of "super pests" over time through development of resistant strains. Actually, pesticide resistance often carries some deficiencies in fitness, vigor, behavior or reproductive potential which make the resistant pest more susceptible to other control measure.[13] For example, resistant insects and mites must be at a reproductive disadvantage in the absence of pesticides or else resistance alleles (variants of a gene) would be more common prior to selection. Similarly, resistance alleles are usually deleterious for rodents in the absence of artificial selection with pesticides. In other words, the trait permitting resistance may be at the detriment of another trait more typical for the pest population as a whole.[14] Thus, resistant populations may be controlled with a less potent but more specific pesticide than the one to which resistance has developed.

Beneficial Species

Pesticides are criticized because they can kill beneficial species along with the undesirable ones. Predators and parasites are often portrayed as an efficient and ideal means to keep pest prey in check. Unfortunately, such "biological controls" are neither consistent nor predictable enough to adequately control pests except in a few isolated cases. T.H.C. Taylor, a scientists known for successfully introducing biocontrol of pest insects to the island of Fiji in the 1930s, concluded that biological control was the best of all control methods when it worked but that it seldom worked![15]

Insecticides create little or no hazard to beneficial insects in many situations because few beneficial species, if any, are present. When insecticides are used to control household, structural and industrial pests and when they are applied directly to control insects on host animals, harm to beneficial insects is rarely an issue.[16]

Predators are commonly given general credit for controlling rodent populations, but "the reverse is more accurate',[17] that is, the prey species must increase before the predator population can catch up and control the prey. Poison baits usually reduce rodent populations much more effectively than predators. For example, the average cat kills only about 25 to 30 rats a year—far too few to affect a colony's numbers.[18]

Farmers must deal with the whole pest picture when selecting control methods to protect their crops. If a farmer is confronted with three pests, each of which can devastate his crop, he may have to spray against all three, even if one was being controlled by predatory insects which the spray will harm.

Pesticide residues

As analytical methods are developed and improved, identification and measurement of smaller and smaller amounts of pesticides, pesticide metabolites and other trace chemicals become possible. Since the 1950s, analytical detectability has advanced from microgram (10^{-6} g.) to nanogram 10^{-9} g.) to even picogram (10^{-12} g.) amounts. As a result, residues previously reported in the parts per million (10^{-6}) range are now measurable in parts per billion (10^{-9}) or even parts per trillion (10^{-12}) concentrations.

With such incredibly minute quantities now being detectable, pesticides and other chemicals can be found almost anywhere in the environment, food, water or human or animal tissues. This too often results in fears of the "pesticide contamination" of the earth and in calls for more restrictions or bans on chemicals . . . a kind of "toxic terror."[19]

The ability to detect, however, has no relation to the biological effects of substances; that is, "residues only matter if they affect organisms."[20] As mentioned previously in Part 2, presence of minute pesticide quantities or other substances rarely presents even the slightest risk to human health. Since any biological effect is related to the size of the residues, the environment is similarly unaffected by minute residues. The extensive testing of an occupational exposure to much higher pesticide levels clearly demonstrate the lack of risk from miniscule amounts of these materials.

Indeed, an argument can be made that small amounts of toxic substances are often beneficial, according to the concept of hormesis[21] or "sufficient challenge."[22] It has been observed repeatedly in animal studies that the low dose animals often appear to be in better condition than the control (no dose) animals, e.g., by living longer, being larger, having fewer tumors, etc. The phenomenon of sufficient challenge was suggested in the historic "megamouse" study conducted by the National Center for Toxicological Research (NCTR), which was reviewed by a Special Committee of the Society of Toxicology. The study used 24,000 mice exposed to various amounts of the carcinogen 2-acetylaminofluorene (AAF). The Society's review noted that the results suggested "statistically significant evidence that low doses of a carcinogen are beneficial" and that if the extrapolation models are correct, "we must conclude that low doses of AAF protected the animals from bladder tumors"[23] (emphasis added).

Groundwater Pollution

Groundwater pollution rarely occurs when pesticides are properly applied. Groundwater is particularly critical for agricultural applications; nearly 70% of it is used annually for agricultural irrigation.[24] Obviously, farmers have a strong incentive to avoid poisoning their own water sources. Contamination of neighboring groundwater subjects a careless pesticide applicator to civil and criminal penalties plus lawsuits for damages.

Trace amounts of pesticides have been detected in groundwater, but this fundamental question has to be addressed: Are the trace amounts detected toxic to humans or animals or otherwise detrimental to the use of the groundwater? Unfortunately, the "sophistication of present-day analytical methods may have outstripped our ability to interpret what they reveal, our ability to determine the significance of low-concentrations (sic) of contaminants on the environment and on public health."[25]

One recent study of pesticides and groundwater[26] focused on the need for better understanding of toxicology associated with the discovery of trace amounts of pesticides in water. Public pressure to ban pesticides known to be contaminants can easily arise and this sentiment against pesticide usage can severely affect the agricultural sector for minor or specialty crops because the range of alternative for these crops is narrow.

To assure that regulatory actions reflect actual threats to public health or the environment, not thoughtless public panic, the EPA should establish realistic maximum contaminant levels (MCLs) or else provide health advisories to guide state and local officials who must respond to public concerns about groundwater pollution. Since EPA already requires submission of data regarding health effects in its pesticide review program, federal leadership in setting MCLs seems appropriate. The lack of federal MCLs is widely perceived as a critical impediment to state and local health protection programs. The widespread concern is that "public apprehension about groundwater contamination will grow to the point where statewide or national bans will become politically expedient, even in cases where pesticide contamination is a controllable, localized phenomenon."[27]

Pollution of surface or groundwater due to improper disposal of pesticide waste or from leaks or spills from pesticide containers can be a serious local problem. Since "in almost all situations, prevention of groundwater contamination is clearly much cheaper than restoration,"[28] every effort must be made to avoid such accidents or errors in handling concentrated pesticides. Fortunately, several approaches are available or under development for field scale disposal of pesticide wastes and spill residues, including evaporation beds, activated carbon absorption, incineration, water-soil degradation, UV-ozone degradation, abiotic hydrolysis and enzyme degradation.[29] Unfortunately, promising and viable treatment options can be paralyzed by ill-advised state and federal regulations. In one case, university pesticide treatment facilities in California, which had operated successfully and without incident for 10 years, were made illegal by the state's 1984 Toxic Pit Cleanup Act.[30]

"Balance of Nature"

One often hears that pesticides upset the "delicate balance of nature," but this charge is without substance. Actually, to restore the so-called "balance of nature" means returning to prehistoric times, the "caveman" era. One scientist stated: "The dominance of man, and his ability to survive in his present numbers has been the result of his success in bringing about an imbalance of nature—in his favor."[31]

Man "departed from natural processes when he domesticated his first animal and later when he first planted a seed."[32] In many places (e.g., Great Britain), it can be said, "there is no truly natural vegetation."[33] Modern agriculture involves vast fields of single, densely growing crops which can lead to explosive vast fields of single, densely growing crops which can lead to explosive spread of an insect pest or plant disease. But only through such "artificial" and intensive agriculture can enough food be produced to feed America and much of the rest of the world's population.

Although natural regulating mechanisms will usually maintain a reasonable balance between many organisms coexisting in the environment, devastating outbreaks of insects and other pests often do occur. More generally, the number of insects in certain ecosystems may well be within nature's normal balance, yet still far exceed acceptable numbers for efficient agriculture. For public health, nature's "normal" levels of mosquitoes, flies or cockroaches are unlikely to be acceptable today. Notions that society "will accept a natural balance alone" to control insects "must be disregarded as unrealistic."[34] Pest insects often must be suppressed. The only alternatives are "lower standards of living, comfort, and health."

Pesticides affect land management and the conservation of land resources in several different ways:

- rendering the land fit for human habitation, through the control of pest-borne human diseases;
- controlling the pests of domestic cattle on, for example, land suitable for grazing but not for cultivation;
- increasing agricultural production per unit area, thus releasing land for other purposes, or compensating for land losses already incurred;
- protecting that production in the period between harvest and consumption; and
- conserving the productive soil in the face of wind or water erosion[35]

If these obvious contributions to human welfare alter the "balance of nature," then let us welcome the improvements!

As for the alleged "delicate" state of nature and is great vulnerability to man's intrusions, the earth "is still a very large planet and its web of life is not nearly so vulnerable as it has been made out to be."[36] Those referring to a "fragile environmental balance" fundamentally misunderstand the resiliency of natural systems. Man's introduction of pollutants if "puny compared with that of nature herself" for the major global reservoirs—the atmosphere, oceans, terrain and the biota. Of course, man can and does pollute local environments if he uses pesticides incorrectly or if an accident occurs.

References

1. Barnes, D., et al.; "Reference Dose (RfD): Description and Use in Health Risk Assessments," Appendix A to the EPA Integrated Risk Information System (IRIS) available from Dr. Barnes at EPA's Office of Pesticides and Toxic Substances (TS-788), Washington, D.C. 20460.
2. Todhunter, p. 164
3. Retnakaran, A.: "Do Regulatory Agencies Unwittingly Favor Toxic Pesticides?" *Bulletin of the Entomological Society of America* 28:146 (1982).
4. Bruce-Chwatt, L.J.; *Essential Malariology;* Wiley Medical (1985) pp. 301 & 345.
5. Buchel, K.H.; *Chemistry of Pesticides;* John Wiley & Sons Inc. (1983) pp. 7–8.
6. Edwards, J.G. "The Myth of Food-Chain Biomagnification," *Agrichemical Age* (April 1980) pp. 10, 32–33.
7. Biddinger, G.R. and Gloss, S.P.; "The importance of trophic transfer in the bioaccumulation of chemical contaminants in aquatic ecosystems" in *Residue Reviews* 91:133 Springer-Verlag (1984) Francis A. Gunther, Editor.
8. Kay, S.H.; *Potential for Biomagnification of Contaminants within Marine and Freshwater Food Webs;* U.S. Army Corps of Engineers Technical Report D-84-7 (November 1984) p. 73.
9. Whitten, J.L.; *That We May Live;* Van Nostrand Co., Inc. (1966) p. 111.
10. Knipling, pp. 426–427.
11. Delp, C.J.; "Pesticide resistance management is a key to effective pest control, *BioScience* 36:101–102 (February 1986).
12. National Research Council (NRC) Committee on Strategies for the Management of Pesticide Resistant Pest Populations; *Pesticide Resistance;* National Academy Press (1986) pp. 315–325.
13. NRC Committee on Strategies for the Management of Pesticide Resistant Pest Populations, pp. 322, 260, & 237.
14. Claus, G. and Bolander, K.; *Ecological Sanity;* David McKay Co. (1977) p. 314.
15. Gunn & Stevens, p. 242.
16. Knipling, pp. 557–558.

17. Bohmont, B.L.; *The New Pesticide User's Guide*; Reston Publishing Co., Inc. Prentice Hall (1983) p. 65.
18. Canby & Stanfield, p. 78.
19. Whelan, E.M.; *Toxic Terror*; Jameson Books (1985).
20. Moriarty, F.; "Prediction of Ecological Effects of Pesticides" in F.H. Perring and K. Mellanby, Editors; *Ecological Effects of Pesticides*; Academic Press (1977) p. 171.
21. Stebbing, A.R.D.; "Hormesis—The Stimulation of Growth by Low Levels of Inhibitors," *The Science of the Total Environment* 22:213–234 (1982).
22. Smyth, H.F.; "Sufficient Challenge," *Food and Cosmetic Toxicology* 5:51–58 (1967).
23. Society of Toxicology ED01 Task Force; "Re-examination of the ED01 Study—Adjusting for Time on Study," *Fundamental and Applied Toxicology* 1:77 & 80 (January/February 1981).
24. Pye, V.I.; Patrick, R., and Quarles, J.; *Groundwater Contamination in the United States*; University of Pennsylvania Press, Philadelphia (1983) p. 21.
25. Pye, p. 10.
26. Holden, P.W.; *Pesticides and Groundwater Quality*, Issues and Problems in Four States; National Academy Press, Washington, D.C. (1986) pp. 51 & 53.
27. Holden, pp. 9–12,
28. NRC Geophysics Study Committee; *Groundwater Contamination*; National Academy Press, Washington, D.C. (1984) p. 17.
29. Kreuger, R.F. and Seiber, J.N., Editors; *Treatment and Disposal of Pesticide Wastes*; American Chemical Society (ACS) Symposium Series (1984).
30. Brosten, D.; "University and Applicators Grappling with Rinsate Morass," *Agrichemical Age* 31 (7):8, 15, & 18 (July 1987).
31. Beatty, R.G.; *The DDT Myth*; John Day Co. (1973) p. 118.
32. Whitten, p. 16.
33. Moriarty, F.; *Ecotoxicology: The Study of Pollutants in Ecosystems*; Academic Press Inc. (1983) p. 12.
34. Knipling, p. 1.
35. Gunn & Stevens, p. 207.
36. Claus & Bolander, p. 13.

Excerpt, with permission, from
Pesticides: Helpful or Harmful, a September 1988 Report
by the American Council on Science and Health.

Pesticides in Ground Water— Solving the Right Problem

Richard S. Fawcett, Ph.D.

Dr. Richard S. Fawcett is an Agricultural Consultant and Farm Journal Staff Environment Specialist. He has been a professor of Agronomy at Iowa State University and a professor of Agronomy at the University of Wisconsin.

Dr. Fawcett received his B.S. in agronomy from Iowa State University, 1970, and his Ph.D. in agronomy from the University of Illinois in 1974. He is the author of *Farmers Weed Control Handbook,* Doane Publishing, 1985; and has more than 40 journal articles published. He has been a contributor to *Weed Science, Agronomy Journal, and Journal of Environmental Quality.* Dr. Fawcett received the Iowa State University Excellence in Research and Extension Award in 1982.

Contamination of ground water by pesticides has been a controversial issue in Iowa. Why should ground water protection be controversial? Everyone is for it. The public wants safe drinking water and a clean environment. Farmers certainly want to protect ground water because their families are the first to be affected if ground water is contaminated. Farmers have an interest in maintaining the productivity of their farms. They are the closest to the land. Ag chemical companies and ag chemical dealers want to protect ground water. They are real people, too, with a sense of responsibility. Even if the public believes that national companies and the hometown dealer are out to make a profit at any cost they can't afford to allow contamination to occur (if only for reasons of legal liability). So what are all the disagreements about?

It's my belief that many of the controversies that have erupted over ground water protection are due to misunderstandings and disagreements about how ground water contamination occurs and sometimes are caused by differences in goals totally apart from ground water protection.

Nitrates get confused with pesticides. Point sources get confused with non-point sources. Surface water gets confused with ground water. And sometimes current knowledge is not complete enough. But, if we are to work together to protect ground water, surface water, and the rest of our environment, we must understand where contaminants come from and how they reach water. Otherwise we cannot design the best protection practices. We may spin our wheels, spending a lot of money, and not correct the real causes.

Even though contamination of ground water by nitrates is the most prevalent ground water problem in Iowa and will, in my opinion, be the hardest problem to solve, the public is clearly the most concerned about pesticides. That's understandable. These are "killers"—herbicides kill weeds, insecticides kill insects, fungicides kill fungi. Pesticide is the broad term that includes all pest killers. But some tremendous over-generalizations have taken place that only cloud the issue and make it harder to address the real problem. Pesticides vary tremendously from one product to another. Some are highly toxic and even trace levels of contamination are a concern. Others are very low in toxicity. Their physical properties vary. Some are not held tightly by soil and may leach, especially in sandy soils. Others are held so strongly by soil particles that very little movement occurs. We can often take advantage of these differences and protect ground water by changing what products we use and how we use them.

There are two general ways that pesticides or other contaminants can get to ground water. Much attention in Iowa has been placed on what are called non-point sources. This is where one can identify no localized problem such as a chemical plant or smokestack that is the source of the contaminant. Instead, for example, spraying a herbicide on a field might result in small amounts of the product moving through the soil over the whole field. This can happen for certain pesticides that are very persistent and/or weakly held by soil. Examples are aldicarb used on potatoes on sandy soils in Wisconsin and New York, and DBCP, a nematicide that formerly was used in California. These products are not used in Iowa, but I believe that non-point sources are in part the cause of the atrazine herbicide we can sometimes detect in shallow ground water.

But our data and the data of many other scientists across the country is pointing more and more to another very important cause of contamination, point sources: areas where herbicides have been stored and mixed and sprayers have been rinsed or dumped. The soil will normally hold and degrade herbicides and other pesticides, preventing significant leaching. But if extremely high rates are applied, that ability is overloaded and leaching can occur. Or if that activity takes place near a well and the well is not properly constructed or maintained, contaminated surface water enters the well. We have to realize that this kind of activity has gone on for 20 years, both at the ag chemical dealer and on the farm.

I very seldom see weeds growing in the driveway next to the farm well. Rinsing out the sprayer there for 20 years has done a job on weeds but may have also "done a job" on the well. We have to alert farmers to this risk and encourage them to haul water to the field and not to do the mixing and rinsing at the well where either high soil concentrations or even back-siphoning can contaminate the well. Most farmers and dealers I talk to just haven't really thought about this problem and are quick to change their practices when it's pointed out.

Several kinds of studies have led scientists to conclude that point sources are to blame for much ground water contamination. First, scientists have had a surprisingly hard time detecting most pesticides a few feet underneath treated fields. In Iowa studies, we can often detect atrazine in drainage tile effluent but only have detected certain other compounds for very brief periods of time under special conditions. When large rains closely follow herbicide application, we can sometimes measure detectable concentrations after the rain, which then disappear. We believe that macropore flow probably explains this transient appearance and have some exciting new studies that should better elucidate this mechanism.

Spring monitoring studies were also puzzling. While several different herbicides are occasionally detected in Big Spring in northeast Iowa where sinkholes can allow surface water and eroded soil carrying herbicides directly into the aquifer, another study has provided different results. We have monitored the Springdale Spring in eastern Iowa for the past two years. This is a small spring that drains a basin primarily planted to corn and soybeans. Depth to bedrock is about 20 to 40 feet, but there are no open sinkholes. Although the spring is high in nitrate (10 to 15 ppm nitrate-nitrogen), and we can usually detect subpart per billion levels of atrazine, no other pesticide has ever been detected in the spring, despite widespread use of many herbicides and insecticides in the basin.

Public water supply monitoring studies in Iowa and other states have very strongly implicated point sources as the cause of detections of most pesticides. In Iowa, 8 percent of all public water supplies utilizing wells as the water source had detected pesticide residues (all herbicides except for one case of an insecticide). Many of those wells contained subpart per billion concentrations of atrazine with no other pesticide detected. It's my belief that non-point source contamination caused at least some of these cases of atrazine contamination, as there clearly are wells with no known point source nearby. Often these wells are in shallow, sand and gravel aquifers along rivers. But it is a totally different story for all the other detected products. Over 80 percent of well systems that had detected pesticides other than atrazine had a known point source—usually the local ag chemical dealer—near one of the town wells, often a few hundred feet away.

A recent Illinois monitoring study is even more startling. At first the Illinois Environmental Protection agency randomly monitored 343 public wells for pesticides and found none. Monitoring efforts were than targeted at more vulnerable wells (based on depth, soil type, chemical use, and reported pesticide spill complaints). More than 450 wells have been sampled and analyzed for the presence of 34 pesticides with a 20 ppt detection limit (10 times more sensitive than the Iowa study). Only three wells were positive for pesticides and all three were near known point sources. Because detected contamination correlated so strongly with point sources, the Illinois Department of Public Health then zeroed in on ag chemical mixing-loading sites. Fifty-six wells definitely near these point sources were analyzed. Forty-three of 77 percent were positive.

So how does this understanding of how ground water contamination occurs help us in protecting ground water? We now know that point sources are a major cause of contamination. We must protect pesticide mixing-loading and storage areas with impervious dikes and pads. Iowa is one of the first states in the nation to require such containment due to the efforts of the agrichemical industry and the Iowa Department of Agriculture who proposed regulations in 1983. These rules were enacted in 1986, and pesticide dealers had until November 1988 to install these systems. Farmers need to be convinced to stop mixing herbicides and rinsing sprayers near wells. Mixing and rinsing in the field and using up excess spray mix rather than dumping it can eliminate much of our pesticide contamination problem. Point source contamination is solvable problem, one often more easily solved than non-point source contamination. We need to acknowledge this problem, get at it, and solve it.

There are especially vulnerable areas where farmers will have to do things differently or where certain products should not be used. Sinkholes and surface inlets of agricultural drainage wells allow surface water contaminated with pesticides to bypass the filtering ability of the soil. Clearly, alternative management practices such as filter strips, alternative land uses, or reductions or changes in pesticide use are needed around sinkholes. Alternative strategies to reduce contamination from agricultural drainage wells are also being studied. Very sandy soils can allow leaching of certain pesticides. Applicators need to observe the ground water advisory statements on pesticide labels and either use alternative products or other control methods in those areas.

The public perception and picture painted in the press seems to be: "Farmers use too many chemicals. Get them to use less and that will solve the problem." But the real world is not that simple. Our efforts to reduce pesticide use through scouting and use of Integrated Pest Management principles, band application of herbicides, proper rate selection and sprayer calibration, and alternative pest control strategies have a good economic and environmental basis, but should not be expected to reduce ground water contamination if leaching with normal use doesn't occur significantly for most of these products.

The concept that one could still use a pesticide and not contaminate ground water, that chemical use and clean water can coexist, has not always been well received. For some people who would prefer to see the use of all pesticides eliminated, this concept takes away a potent and popular argument for banning the use of pesticides. I believe that the goal of banning the use of pesticides has sometimes become entangled with the goal of protecting ground water. These are very different goals.

If we choose not to use pesticides it should not be because of someone's moral judgment that all pesticides are by definition bad, but because of real risks. In the real world, all pest control techniques have positive and negative impacts on the environment, the farmer, and the consumer. Tillage, the major alternative to herbicides, has the obvious environmental impact of causing soil erosion. Much of the progress made in soil conservation has been due to adoption of conservation tillage, made possible by herbicides. Tillage can be used in a properly constructed and balanced system, but there is a direct relationship between tillage and erosion.

Breeding crops for resistance to diseases and insects, whether by conventional or genetic engineering techniques is desirable for the farmer, because he/she does not have to buy a pesticide or be exposed to a sometimes significant health risk if handling precautions aren't adequately followed. But this technique also creates an unknown, though potentially large risk for the consumer. The reason the crop plant is resistant is often due to the presence of secondary metabolites, natural insecticides or fungicides. These chemicals are produced by the plant, are most likely found in the edible portions in significant concentrations, and may be highly toxic. Breeding for pest resistance selects for higher and higher concentrations of these chemicals, but until recently little thought has been given to potential effects on human health. A Canadian potato variety developed for excellent disease and insect resistance was ready for release in the United States when it was serendipitously discovered that eating one meal of the potatoes caused acute illness, due to high levels of chaconine and solanine.

In the end, ground water contamination will be solved by applying the scientific method, where hypotheses are rigorously tested and retested to get at the truth. We must correctly identify the on-site and off-site and the short and long-term impacts of agricultural practices. Science will identify the risks and benefits of these practices and provide us with alternatives. Then value judgments will legitimately come into play in the decisions that farmers, legislators, regulators and the public make. With better knowledge and intelligent use of that knowledge we can solve the right problems.

Reprinted, with permission, from
Ground Water Monitoring Review
Vol. IX, No. 4, Fall 1989

Regulatory Harassment of U.S. Agriculture

Robert M. Devlin, Ph.D.

Robert M. Devlin is professor of Plant Physiology and Weed Science at the University of Massachusetts, Laboratory of Experimental Biology, Cranberry Experiment Station, E. Wareham, Massachusetts. He formerly was an assistant professor, North Dakota State University, Fargo, North Dakota.

He received a B.S. in biology from State University of New York at Albany, 1959; an M.A. in plant physiology from Dartmouth College in 1961; and a Ph.D. in plant physiology, University of Maryland, 1963.

Dr. Devlin is the author of *Plant Physiology* (4 editions), published by the Wadsworth Publishing Company in 1966, 1969, 1975, and 1983; *Photosynthesis,* published by the Van Nostrand Company; and *A Biology of Human Concern,* published by J.B. Lippincott Company. He has more than 150 journal articles published and has contributed to *Physiologia Plantarum, Weed Science, Weed Research,* and the *Proceedings North Eastern Weed Sci. Soc.*

He is the recipient of an honorary degree of Doctor Honoris Causa awarded by the University of Agriculture at Szczecin, Poland, 1990.

The very wise American philosopher, Will Durant, once said: "Man's capacity for fretting is endless, and no matter how many difficulties we surmount, how many ideals we realize, there is a stealthy pleasure in rejecting mankind or the universe as unworthy of our approval." How well these words fit our present situation in agriculture today.

The greatest accomplishment of the twentieth century is not the invention of the airplane or the car, not the uncovering of nuclear power, not the discovery of wonder drugs—it is the modern miracle of American agriculture, which is the envy of the world. Consider, for example, that the same amount of acres harvested in 1910 in order to feed about 80 million Americans produced more than enough food and fiber to meet the needs of some 250 million Americans today. In fact, the American farmer produces a surplus of food and fiber which we give or sell to other nations unable to feed or clothe their growing populations. The amazing thing is that we produce all of this with less than 5 percent of our population. To quote the late Dr. Robert White-Stevens: "The towering structure of America rests primarily on the production of food and fiber. It is the basis of our standard of living and our national prosperity."[3]

With this miracle before our eyes, why are we trying our utmost to destroy the tremendously efficient productivity of our agriculture by literally strangling our farmers and their supporting industries with the red tape of bureaucratic regulations? Nothing could bring this nation to its knees more quickly and dramatically than the senseless interruption of our agricultural productivity by an ever-increasing number of government regulations—regulations which are, in many cases, born of news media-manufactured fear, emotional hysteria, and political expediency rather than from the expert assessment of problems and the analysis of thoughtful professional opinion.

People in America are constantly reminded by our daily newspapers and TV programs that the farmer and his suppliers—the pesticide and fertilizer industries—pollute the earth, water, and surrounding atmosphere. However, I think you will agree with me that the farmer is a better environmentalist than some Washington bureaucrat or New England bird watcher. The farmer is a professional environmentalist whose care for the land is a livelihood rather than an object for philosophical discussions and sensational revelations.

Environmentalists constantly remind the layman that before the days of pesticides and man-made fertilizers farming was an easy and wholesome occupation and that there were few insect and weed problems. This, of course, is a ridiculous myth. In nineteenth-century America and during the early years of the twentieth century a farmer commonly worked an 18-hour day and insect problems were so bad that whole communities were asked to pray for deliverance.

Another myth is the distinction made between natural and synthetic chemicals. Newspaper scientists and crusading environmentalists would have us believe that chemicals like pesticides do not occur naturally. If they really cared to find out, the entomologist could tell them about pyrethrins, the natural insecticides, and the plant physiologist could tell them of the allelopathic effect of certain plants on surrounding competing vegetation.

Finally, there is the cancer myth that environmentalists spread and the news media dramatize—that it is the synthetic chemicals that cause this dread disease. The truth is that numerous compounds, both natural and synthetic, can cause cancer. In fact, aflatoxin B, a natural compound, is one of the most potent carcinogens known to man. There are numerous other natural products that bedevil man. To pretend otherwise or make a clear distinction between natural and synthetic compounds based on their toxic properties is not valid.

A great number of regulatory agencies that have popped up in the last two decades are aimed at restricting the farmer's use of agricultural chemicals. The campaign against the use of these chemicals, which at times takes on the appearance of a religious crusade, had its beginning in the bestselling novel *Silent Spring*, by Rachel Carson.

In *Silent Spring* there are so many unsupported statements made, so many half truths told, and so much important material completely ignored that the book could almost be classified as "science fiction." Certainly Ms. Carson's prediction of the coming of a "Silent Spring," or anything remotely similar, was meant more to scare the uninformed than to draw the serious attention of thinking individuals.

This book, despite its numerous shortcomings, has become the "Bible" of the environmentalists who apparently don't really care if the material they read is factual as long as it supports their point of view. What is even worse, *Silent Spring* is required reading in many of our high schools and the children are told by well-meaning, but woefully uninformed teachers, that what they read in it is the "gospel" truth.

There is no doubt that *Silent Spring* led eventually to the banning of DDT which, in my opinion, is the safest and most efficient chemical for its purpose ever produced by man. I won't go into the reasons why DDT should **not** be banned, for they are well known to most of us. It is an easy chemical to defend. However, you should know that DDT alone has been responsible for saving more human lives than all the wonder drugs combined. It has been estimated that over 1 billion people are alive today because of DDT and that is something to think about!

The banning of other organochlorines such as aldrin, dieldrin, and chlordane soon followed the demise of DDT. In a short period of time the farmer found that many of his most useful agricultural chemicals were either banned or so tied up in regulations that profitable use of them was impossible. In addition, numerous agencies using almost every letter in the alphabet were spawned. And as you know, regulatory agencies have a vested interest in producing more and more regulations which, in turn, necessitates the hiring of more and more people to formulate and enforce these regulations—and we as taxpayers are forced to support their ever growing payrolls.

Is it possible that the average citizen is unaware of the unabated harassment of American agriculture by special interest groups that feed the insatiable hunger of the news media for sensationalism which, in turn, gives rise to a political climate that inevitably leads to more and more regulatory agencies? Regulations, most of which are frivolous or totally unnecessary, increase the price of agricultural chemicals, which increases the farmer's costs and again we, the consumers, pay.

Diethylstilbestrol (DES) has been banned, 2,4,5-T and 2,4,5-TP have been banned, the use of antibiotics in animal feeds has been severely restricted—the list goes on and on of useful agricultural chemicals that have been taken away from the farmer.

And as if the banning of chemicals is not enough, we now have Washington bureaucrats frowning on the farmer's use of labor-saving devices on his farm—presumably because farm workers will be displaced. At least that was the message I got from a statement made by Robert Bergland when he was our Secretary of Agriculture. The statement was: "I do not think the Federal funding for labor-saving devices is a proper use of Federal money. This is something that should be left to private enterprise and to the state universities, if they choose, in my view. But I will not put Federal money into any project that results in saving farm labor."

That statement is astounding when you consider the Secretary made known his interest in keeping the family farmer in business. Dr. Kendrick, former director of the California Agricultural Experiment Station, left no doubt how he felt Secretary Bergland's policy would affect the family farmer when he stated: "Most small family farmers eagerly adopt new technologies, many of them labor-saving, not only because they alleviate the physical drudgery of farming, but also because they increase productivity per worker. New technology is often the only thing that keeps the family farmer in the business of farming. Abruptly discontinuing public funding for technological advances, which might displace farm workers, could preserve a farm job but lose the farmer. In that case everybody loses, the family farmer, the farm worker who loses a job, the industry worker who handles the farm product, and the consumer who must now buy perhaps from another more profit-conscious source."[2]

Secretary Bergland grew up on a farm. What he doesn't seem to realize is that if it were not for labor-saving devices he most likely would still be on the farm. It was the labor-saving devices that allowed farmers' sons to choose careers in medicine, law, politics, etc.

When the average citizen stops taking food for granted, when he learns that food is not produced in the supermarket, but is the result of hard work in the field, then we might be able to attract the attention of our politicians who are ever alert to situations that might eventually lead to re-election problems.

I would like to end my discussion with a quote from a talk given by Dr. Wayland Hayes at a symposium on pesticides and their relation to man and the environment. Dr. Hayes stated: "We live in a chemical age. In the long run we have no choice but to adapt to it. There just are not enough people who will part willingly with the comforts such as increased food, shelter, and health, which modern chemistry has brought. By proper toxicological study it is possible to determine that some compounds are too dangerous for almost any use, while others are suitable for one use, but not for another. Unless these distinctions can be made at a professional level and enforced by the appropriate federal, state, and local agencies, without emotional appeals in the mass media, the people of this country may suffer a loss of confidence not only in their government, but in the organization of society. Through a nameless fear we may not only lose our position of leadership, but suffer needless privation. The problem is much broader than the fate of any particular pesticides, drug, or food additive that has been in the news recently."[1]

References

1. Hayes, W.J. 1970. How toxic is toxic? In: Background for Decision, J.V. Osmun (ed.) Symposium at Purdue University, 25–46.
2. Kendrick, J.B. 1980. Does the USDA have a new agricultural research policy? California Agriculture 34:2.
3. White-Steven, R.H. 1978. Agriculture, America's Brightest Star, Outshines its Quibbling Critics. Article distributed by the National Council for Environmental Balance.

Reprinted, with permission, from
the National Council for Environmental Balance, Inc.
Louisville, Kentucky

Alternative Agriculture—Proceed with Caution

Robert M. Devlin, Ph.D.

Please see biographical sketch for Dr. Devlin on page 78.

We all should take note that in the first quarter of the twentieth century, crop yields in the United States were comparable to that of other nations throughout the world. In the years that followed, especially in the 1940s, 1950s, and 1960s, university and U.S. Department of Agriculture scientists, government farm policies, and the pesticide industry helped the American farmer to far exceed the crop yields of his counterpart worldwide. In fact, the strong foundation upon which America stands has essentially been built by the hard labor and expertise of its farmers and agricultural scientists. We feel secure in the knowledge that we can feed ourselves—that we do not rely on other nations to keep the threat of malnutrition and starvation from our door. The question is, "Do we want to tamper with a system that has given us so much and has worked so well?"

The National Research Council (NRC) has published a report entitled Alternative Agriculture that stresses the importance of establishing alternative agricultural systems to replace our present system of farming. The NRC points out that conventional farming techniques are allowing alarming amounts of top soil erosion, polluting our water supplies, and contaminating our food supplies. According to the report, federal commodity programs encourage agricultural practices that aggravate these problems.

What the NRC says is mostly true. Soil erosion and water pollution are big problems that need to be attended to. Contamination of our food supply with pesticides and animal drugs is rare and not really a problem. What the NRC recommends to alleviate these problems are a variety of alternative farming systems variously referred to as organic, regenerative, low input, sustainable, and biological. They offer case studies that suggest that alternative agriculture can be competitive with conventional agriculture.

The NRC's observations in this respect are suspect. It has not been really proven that alternative agricultural methods are more profitable than conventional methods. Indeed, it seems the opposite is true. For example, if we were to prohibit cosmetic applications of pesticides, grading standards for our fruit and vegetables would have to be lowered. This, in turn, would not allow our produce to compete in the world market. Obviously, this low input method would be unprofitable for the American farmer.

We should be very careful of making any sweeping changes in our agricultural systems. Rather, let us proceed with caution and determination to protect our farmers as well as our environment. Both are important. Remember, not only our farmers' livelihood is at stake, but also the quality of life of the many people that rely on American agriculture to feed and clothe them.

Herbicide Concentrations in Ohio's Drinking Water Supplies: A Quantitative Exposure Assessment

David B. Baker, Ph.D. and
R. Peter Richards, Ph.D.

David B. Baker is Director, Water Quality Laboratory, Heidelberg College. His past affiliations include Professor of Biology, Heidelberg College; and Assistant Professor of Botany, Rutgers University.

Dr. Baker received a B.S. in Biology, Heidelberg College, 1958; an M.S. in Botany from University of Michigan in 1960; and a Ph.D. in Botany, University of Michigan, 1963.

He is the author of the book *Sediment, Nutrient and Pesticide Transport in Selected Lower Great Lakes Tributaries* (EPA-905/4-88-00, 1988); and has published more than 25 journal articles. He has been a contributor to the Journal of Soil and Water Conservation and received the Honor Award, Soil and Water Conservation Society, 1984.

R. Peter Richards is a water quality hydrologist, Water Quality Laboratory, Heidelberg College; and a former assistant professor of geology, Oberlin College.

Dr. Richards received a B.A. in geology, Oberlin College, 1965; an M.S. in geology, University of Chicago, 1968; and a Ph.D. in geology, University of Chicago, 1970.

He is the author of more than 25 journal articles and has contributed to *Journal of Great Lakes Research, Water Resources Bulletin,* and Water Resources Research. He is the recipient of the National Science Foundation Post-Doctorate Fellowship.

Abstract

The human health risks posed by pesticides in Ohio's drinking water supplies are contingent on two factors—pesticide concentrations and pesticide toxicities. Based on their quantities of use, their toxicity, and their concentrations in surface water, atrazine and alachlor appear to comprise more than three quarters of the total pesticide toxicity threat in Ohio's drinking water supplies. Data on concentrations of these two herbicides in various public and private water supplies in Ohio have been collected and organized to facilitate comparisons with both acute and lifetime drinking water health advisories. Concentrations in excess of acute standards were not observed in any supplies. Lifetime standards were exceeded in a small number of highly vulnerable private supplies. Overall, only 0.05% and 0.06% of Ohio's population are consuming atrazine and alachlor in excess of their lifetime standards, which are 3 ppb and 2 ppb respectively. Public water supplies derived either from rivers draining agricultural watersheds or from reservoirs associated with those rivers deliver the largest quantities of these herbicides into residents via drinking water. Pesticide concentrations in these surface water supplies are highly variable, and, during storm runoff periods following pesticide application, often temporarily exceed lifetime standards. However, on a multi-year basis, average concentrations in these supplies meet the lifetime health advisories established by the Environmental Protection Agency. The statewide population-weighted average concentrations of atrazine and alachlor are 0.28 ppb and 0.14 ppb. With rare exceptions, pesticide exposures from Ohio's public and private drinking water supplies do not appear to pose significant health problems for Ohio's residents.

Introduction

Application of risk assessment to the occurrence of pesticides in drinking water has been greatly facilitated by the U. S. EPA's recent development of drinking water standards for many pesticides. These standards include both proposed maximum contaminant levels (MCLs) for several major pesticides and drinking water health advisories (HAs) for an expanded list of currently used pesticides (U. S.

EPA 1989). Either MCLs or HAs are now available for compounds which constitute 92% by weight of the pesticides applied to Ohio's cropland. The MCLs are intended to support regulatory actions while the HAs are intended more for informational purposes. Both types of standards are largely based on an evaluation of both acute and chronic animal toxicity studies. For non-carcinogenic effects, safety factors on the order of 100- to 1,000-fold are incorporated into the standards to account for uncertainties in extrapolating from animal studies to humans. For pesticides considered to be non-carcinogens, MCLs and HAs indicate drinking water concentrations at or below which no adverse human health effects are expected. For pesticides which have caused cancers in animal testing, standards are set such that the additional cancer risks from exposures at or below the standard are deemed to be acceptably small (e.g. on the order of 1 in 100,000 to 1 in 1,000,000 additional cancer risk). HA levels have been established for short term (acute) exposures of one-day, ten-day, or seven year durations, and for lifetime (chronic) exposures. HAs and total amounts of application for several of the major pesticides used in Ohio are shown in Table 1.

Preliminary comparisons of pesticide concentrations in Ohio's surface water and groundwater with the acute and chronic standards indicate that concentrations in excess of the acute standards (one-day and ten-day HAs) have not been observed and are very unlikely to occur. Consequently, this exposure assessment will focus on evaluating herbicide concentrations in relation to chronic standards (lifetime HAs or MCLs). Further, we will assume that the appropriate characteristic of pesticide concentrations to compare with lifetime HAs is the long term (multi-year) average concentration. Thus, while concentrations during runoff events following pesticide application may exceed lifetime HAs in some supplies, so long as the annual average concentration falls below the HA, those supplies will be deemed safe for human consumption. This approach is consistent with procedures outlined in the 1986 amendments to the Safe Drinking Water Act where compliance is based on comparison between the lifetime HA or MCL and the running annual average of measurements taken quarterly or more often (Pontius 1990).

Two herbicides, atrazine and alachlor, make up about 44% by weight of the total amount of pesticides used on Ohio cropland (Waldron 1989). Based on their MCLs and HAs, these two herbicides are, on average, more toxic than other pesticides. When their quantity of use is normalized to their toxicity, these two compounds represent about 77% of the total pesticide "toxic load" applied on Ohio's cropland. That these two compounds represent "worst case" risks in Ohio's drinking water is further substantiated when average pesticide concentrations in rivers draining cropland are normalized by their drinking water standards (Table 2).

Table 1. Quantities Applied and Health Advisories for Major Pesticides Used in Ohio.

Pesticide	1986 Use on Ohio Cropland (metric tons)	One-and Ten Day Health Advisory (ppb)	Lifetime Health Advisory) (ppb)
Alachlor	2,635	100	2*
Atrazine	2,058	100	3
Metolachlor	1,761	2000	100
Cyanazine	657	100	10
Metribuzin	425	5000	200
Butylate	488	2000	360
Simazine	128	500	4
Terbufos	240	5	0.9
Carbofuran	100	50	40
Fonofos	140	20	10
Chlorpyrifos	214	30	20
Others	1,701		
Total 1986 use	10,547		

*For alachlor, a lifetime health advisory is not available and the proposed maximum contaminant level is listed in its place.

Unfortunately, most of the governmentally mandated pesticide monitoring at public water supplies has focused on the six pesticides for which National Interim Primary Drinking Water Regulations were adopted under the Safe Drinking Water Act of 1974. These six include 2,4-D, endrin, lindane, methoxychlor, toxaphene, and 2,4,5-TP (Silvex). The use of endrin, lindane and 2,4,5-TP is now banned in Ohio and the remainder comprise only 0.88% of the total weight of pesticides applied to Ohio's cropland (Waldron 1989). Since these compounds have higher HAs, on average, than many of the compounds used more frequently in Ohio, they comprise only about 0.2% of the toxic load of applied pesticides. The monitoring programs have rarely indicated even trace levels of these compounds and they have never approached concentrations close to the lifetime HA levels.

Methodology

For this exposure assessment, we have divided the state into eight major types of supplies, based on whether the supply is public or pri-

Table 2. Relative human risks from current generation pesticides as determined by observed average concentrations in the Sandusky River and the U.S. EPA's lifetime health advisories (LHA).

Pesticide	Long Term Average Concentration μg/L	Human LHA μg/L	Concentration/ Toxicity	Relative Human Health Risks*
Atrazine	1.73	3	0.576	1.000
Alachlor	0.63	2	0.315	0.547
Metolachlor	1.43	100	0.014	0.024
Cyanazine	0.36	10	0.036	0.062
Metribuzin	0.30	200	0.001	0.0002
Terbufos	0.009	0.9	0.010	0.017
Simazine	0.098	1	0.098	0.17

*Atrazine is arbitratily set as having a relative risk of 1.00.

vate and on the source of raw water. Public surface water supplies, which provide water for 55% of Ohio's 10,800,000 residents, are grouped into six types of supplies: Lake Erie, the Ohio River, rivers draining agricultural watersheds, inland lakes and reservoirs, upground pumped-storage reservoirs, and other surface waters. About 28% of the state's population derives water from public water supplies drawing upon groundwater; these supplies form our seventh group. The remaining 18% of the population are served by private water supplies which almost exclusively utilize groundwater; these supplies form our eighth group.

For each of the above types of water supplies available data on pesticide concentrations have been collected both from state mandated monitoring programs and from special studies. These data, together with information on the populations served by each type of supply, have been used for this exposure assessment.

Results and Discussion

Surface Water Supplies

More than half the population of Ohio uses drinking water drawn from lakes, rivers, and reservoirs, with Lake Erie and the Ohio River being the largest single sources. The populations served by the various types of surface water supplies and their associated average atrazine and alachlor concentrations are shown in Table 3.

Rivers draining watersheds with intensive row crop agriculture contain the highest average pesticide concentrations that have been observed in Ohio's water supplies. During runoff events following pesticide applications, peak pesticide concentrations often exceed 20 ppb for short durations (Baker 1988). Upground pumped storage reservoirs filled from these rivers can also have relatively high herbicide concentrations, particularly for atrazine which breaks down more slowly than alachlor. The Ohio River, which is also a major source of drinking water, has very low pesticide concentrations because its drainage area upstream from Ohio's water intakes is dominated by forested land use rather than cropland.

Lake Erie supplies drinking water for about 23% of Ohio's population. Extensive dilution of the herbicide concentrations from Ohio's

Table 3. Concentrations of atrazine and alachlor, and associated populations, in various public water supplies utilizing surface water.

Category	Population	Atrazine Concentration (μg/L)	Alachlor Concentration (μg/L)
Lake Erie	2,493,000	0.07	0.03
Ohio River	927,000	0.24	0.09
Agricultural Rivers			
Scioto	250,000	2.02	1.06
Sandusky	43,500	1.73	0.63
Maumee	62,000	1.50	0.46
Huron	2,500	0.89	0.35
Upground Reservoirs	221,000	2.19	0.27
Other Lakes			
and Reservoirs			
Lake Rockwell	397,100	0.56	0.06
Big Walnut Cr.	528,000	0.91	0.20
Alum Creek	33,000	0.62	0.21
Others	427,000	0.75	0.14
Other Surface Water	553,000	0.48	0.14
Total and Average	5,937,100	0.48	0.14

agricultural tributaries to Lake Erie by water from Lake Huron results in only trace levels of herbicides in these supplies. The herbicide concentrations in other inland lakes and reservoirs depends on the land use in their watersheds and their hydraulic flushing times. In Ohio, most of these types of supplies have relatively low herbicide concentrations.

Groundwater Based Supplies

About 45% of Ohio's population obtain their water supply from groundwater resources. Groundwater supplies were divided into public and private supplies for this assessment. The concentrations of atrazine and alachlor in these supplies are summarized in Table 4.

Private wells provide the water supply for most of the rural population of Ohio, and for some residents of small towns. These wells tend to be shallower and sometimes less well maintained than public wells. About 700,000 private wells provide the water supply for an estimated 18.5% of the total population of Ohio (97% of the rural population). As part of a statewide rural private well study, 610 private wells were sampled for pesticides in March 1988 or March 1989 (Baker et al. 1989). Of

Table 4. Concentrations of atrazine and alachlor, and associated populations, in various public and private water supplies utilizing groundwater.

	Atrazine		Alachor	
	Estimated Concentration (μg/L)	Estimated Population	Estimated Concentration (μg/L)	Estimated Population
Groundwater				
Private				
>HAL	5	4,918	5.16	6,557
1 - HAL	2	3,279	1.5	3,279
.2 - 1	0.6	34,426	0.6	4,918
.05 - .2	0.1	49,180	0.1	16,393
<.05	0.025	1,908,197	0.025	1,968,853
Public				
Circleville	—	—	0.4	17,050
Goble Village	—	—	0.9	155
Other	0.025	2,862,000	0.025	2,844,795

these supplies, three contained atrazine in excess of its lifetime HA and four contained alachlor in excess of its lifetime HA. Those supplies containing herbicides in excess of lifetime standards were from particularly vulnerable sources, such as springs or dug wells. Pesticide residues for more than 90% of the private wells were less than the detection limits of 0.05 ppb. Other sampling programs of private wells in Ohio, such as those conducted by the U. S. EPA as part of their National Pesticide Survey (Barrett 1990) or by Monsanto Chemical Co. as part of their National Alachlor Survey (Klein 1989), have failed to detect any pesticide residues.

About 1,200 community water supplies, serving 26.5% of Ohio's population, utilize wells as their source for drinking water. In Ohio, all groundwater based public supplies are required to analyze one sample per year for alachlor, but atrazine analyses are not yet required. Alachlor residues have been detected in two supplies, but in neither case did the concentration exceed its lifetime HA level.

Population Exposure Curves

To plot population exposure curves for the entire state, the various supplies in Tables 3 and 4 were ranked from the highest to lowest concentrations for both atrazine and alachlor. The supplies were then plotted as a variable width bar graph where the width of the bar reflects the percentage of the state's population served by that supply. The resulting graphs are shown in Figures 1 and 2. The lifetime health advisory

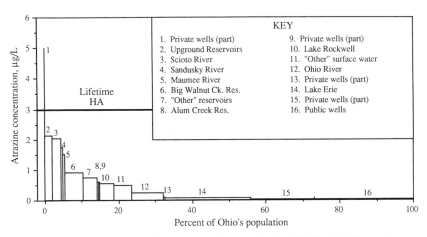

Figure 1. Population exceedency curve for atrazine in Ohio's drinking water, based on ranked, variable width histograms.

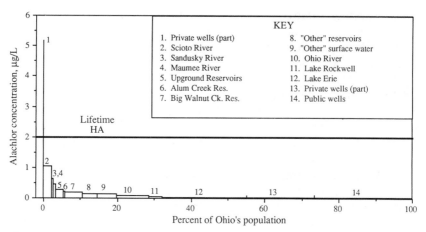

Figure 2. Population exceedency curve for alachlor in Ohio's drinking water, based on ranked, variable histograms.

for each herbicide is also shown on the graphs so that the concentrations in various supplies can be compared with the standards. For atrazine, about 0.05% of Ohio's population is consuming drinking water in excess of its lifetime health advisory, while the corresponding figure for alachlor is about 0.06%.

The exposure curves of Figures 1 and 2 are likely to overestimate true exposures for several reasons. When uncertainties existed in assigning concentrations to population components, we systematically chose concentrations toward the high end of the range of uncertainty. Furthermore, many surface water supplies, including Columbus, use activated carbon to reduce herbicide concentrations, as well as general taste and odor problems, on either a seasonal or a permanent basis. Reductions in exposures resulting from activated carbon treatment (Miltner et al. 1989, Richards and Baker 1989) were not considered in this analysis. Overestimation also probably results from assumptions which had to be made about actual concentrations in supplies for which analyses were below detection limits. This particularly affects groundwater based supplies. For private water supplies, we assumed a concentration of .025 µg/L, which is 50% of the detection limit of 0.05 µg/L. Public water supplies with concentrations below the detection limit were also assumed to contain 0.025 µg/L alachlor and atrazine. With 95 to 98% of public and private supplies having concentrations below the detection limit, it is very unlikely that the average concentrations in these samples would be as high as those used for these estimates.

Risk Distribution

While the human health risks posed by the occurrence of atrazine and alachlor in drinking water appear to be small, they are not shared equally by all Ohioans. The individuals exposed to the highest concentrations are a small proportion of private water supply users who happen to have contaminated supplies. Examination of data from these contaminated supplies indicates that they often represent particularly vulnerable situations, such as springs or "dug" wells.

Although the individual doses are lower, the largest quantities of pesticides entering Ohioans through drinking water occur in municipalities using rivers and reservoirs as water sources. Surface water based supplies account for 16 times more exposure to atrazine than do groundwater based supplies.

Monitoring Issues

The monitoring programs which have supported the above exposure assessments illustrate some of the problems that accompany efforts to derive accurate exposure assessments for pesticides in drinking water for large areas. It is particularly difficult to accurately characterize pesticide concentrations in supplies withdrawn from rivers. In such supplies, pesticide concentrations not only vary annually and from season to season, but also during runoff events, especially those following periods of pesticide application. Quarterly sampling, as required under the 1986 amendments of the Safe Drinking Water Act, is inadequate for assessing annual average pesticide concentrations for such supplies. Compositing of daily samples, with analyses at weekly or monthly intervals, may be necessary for accurate exposure assessment for these supplies.

As the detection limits for pesticides are lowered, the percentage of supplies having pesticide detections will increase. This is particularly the case for supplies based on groundwater or surface waters with very low concentrations. At the same time, it is likely that estimates of the amount of pesticide exposure will actually decrease, since assumptions regarding concentrations that are below detection limits will move to lower values. Thus, at the same time monitoring programs will indicate that contamination is more widespread, they will also indicate that total public health risks are smaller. The impacts of this evolving perception of pesticide exposures on pesticide policy remains to be seen.

Private well monitoring programs in Ohio have yet to identify what could be described as aquifer-level pesticide contamination, even in the most vulnerable aquifers where nitrate contamination is prevalent (Baker et al. 1989). Instead, pesticide contamination appears to be associated with individual wells and/or supplies which happen to be particularly vulnerable. Concentrations in excess of HA levels have largely been limited to supplies which owners describe as "dug" wells or springs, or which are located in the vicinity of pesticide handling or mixing sites. Efforts to identify Ohioans consuming pesticides in excess of HA levels could be focussed on these particular types of systems.

Pesticide Exposure Patterns in the Midwest

The exposure patterns observed for Ohio are by no means unique. It is becoming more and more evident throughout the midwest that pesticides are detected more frequently and at higher concentrations in surface waters than in groundwater. The U. S. Geological Survey has recently completed a study at 150 stream sites in 10 midwestern states that involved winter, spring, and fall collections at each site (U. S. Geological Survey 1989). During the winter collections, pesticides were detected in 55% of the samples, while during the spring and fall collections pesticides were detected in 90% and 79% respectively. During the spring sampling 35% of the samples exceeded the HA for alachlor and 56% exceeded the HA for atrazine. By contrast, in a statistically designed survey in which herbicides were monitored in 1,430 water supplies utilizing groundwater across U.S. corn and soybean production areas, herbicides were detected in 13% of the supplies and exceeded HAs in 0.11% of the supplies (Monsanto 1990). In Illinois, pesticides were found at all 30 surface water monitoring stations but in only 13 of 446 wells used for public water supplies (Taylor 1989). Results from the U. S. EPA's National Pesticide Survey indicate that less than 1% of either rural domestic wells or community water system wells contain any pesticides in excess of lifetime HAs (U. S. EPA 1990).

Approaches to Exposure Reduction

Reductions in drinking water exposure to pesticides can be achieved in several ways. Best management practices are available that can result in reduced concentrations of pesticides in surface waters. These practices result in either reduced use of pesticides, shifts in the types and amounts of pesticides used, or reduced transport of pesticides from fields into streams.

Since the bulk of the pesticide exposure occurs through a relatively small number of public water supplies using vulnerable surface waters, and since the exposure is largely seasonal, treatment to remove pesticides at the treatment plants is feasible. Costs of removal per population served are much lower at public water supplies than for individual private supplies. Either granular or powdered activated carbon can be used to lower pesticide concentration and thereby reduce exposures (Miltner et al. 1989, Richards and Baker 1989, Frank et al. 1990). Use of bottled water or point-of-use treatment systems are also options for reducing exposure.

The highest individual exposures occur from particularly vulnerable private water supplies. The development of alternate water supplies at such sites could reduce exposures at these worst-case situations. The major problem for this approach is to identify those particular supplies, since they comprise a very small proportion of the total number of private water supplies.

Conclusions

This assessment of herbicide exposures in Ohio's drinking water supplies suggests that:

1. Two pesticides, alachlor and atrazine, pose the bulk of the total pesticide threat to drinking water supplies in Ohio due to their heavy use, their relatively low HA levels, and their relatively high concentrations in surface waters.

2. Exposures in excess of one-day and ten-day health advisory levels have not been observed for any herbicides in public or private water systems of Ohio. Thus, concerns regarding pesticide exposures in Ohio can focus on assessing chronic rather than acute exposure patterns. If acute exposures were to occur, they would likely be found in private wells affected by point sources of herbicide contamination.

3. Greater amounts of drinking water exposure to agricultural pesticides are found in certain urban populations than among rural residents in farming areas.

4. Average herbicide exposures in excess of lifetime HA levels are absent in public water supplies but do occur in a small percentage of private water supplies. For the state as a whole, about 0.05% of the population consume atrazine in excess of lifetime HA levels and 0.06% consume alachlor in excess of lifetime HA levels.

5. Herbicide concentrations in excess of lifetime HA levels do occur seasonally in public water supplies withdrawn from rivers draining agricultural watersheds or in reservoirs associated with those rivers. While the multi-year average concentrations fall below the HA levels, variability in average annual concentrations can result in occa-

sional annual average concentrations in excess of the HA level for atrazine.

6. Public water supplies utilizing surface waters serve 55% of the state's population but account for 16 times more atrazine exposure than all groundwater based supplies. Surface water supplies also account for the bulk of the alachlor exposures.

7. Since the bulk of the pesticide exposures occur from public rather than private supplies, both accurate monitoring and treatment are more feasible and cost effective than if groundwater and private water supplies were the major pathway for exposure.

8. While atrazine and alachlor represent "worst case" situations in terms of human health risks posed by pesticides in drinking water, the actual risks posed by these two herbicides in Ohio's water supplies appear to be very small.

References

Baker, D.B. 1988. Sediment, nutrient and pesticides transport in selected lower Great Lakes tributaries. EPA-905/4-88-001. U.S. EPA, Great Lakes National Program Office, Chicago, Illinois.

Baker, D.B., L.K. Wallrabenstein, R.P. Richards, and N.L. Creamer. 1989. Nitrate and pesticides in private wells of Ohio: a state atlas. Water Quality Laboratory, Heidelberg College, Tiffin, Ohio.

Barrett, M. 1990. U.S. Environmental Protection Agency-GW-OPP, Washington DC. Personal communication.

Frank, R., B.S. Clegg, C. Sherman, and N.D. Chapman. 1990. Triazine and chloroacetamide herbicides in Sydenham River water and municipal drinking water, Dresden, Ontario, Canada, 1981–1987. Archives of Environmental Contamination and Toxicology 19:319–324.

Klein, A.J. 1989. Monsanto Agricultural Company, St. Louis, Missouri. Personal communication.

Miltner, R.J., D.B. Baker, T.F. Speth, and C.A. Fronk. 1989. Treatment of seasonal pesticides in surface waters. Journal American Water Works Association 81:43–52.

Monsanto Company. 1990. The National Alachlor Well Water Survey (NAWWS): Data Summary. Monsanto Technical Bulletin, July 1990. St. Louis, Missouri.

Pontius, F.W. 1990. Complying with the new drinking water quality regulations. Journal American Water Works Association 82:32–52.

Richards, R.P., and D.B. Baker. 1989. Potential for reducing human exposures to herbicides by selective treatment of storm runoff water at municipal water supplies. Pages 127–138 in D. Weigmann, editor. Proceedings of a national research conference. May 11–12, 1989, Blacksburg, Virginia.

Taylor, A.G. 1989. Illinois water quality sampling update—pesticides—1989. Illinois Environmental Protection Agency.

U.S. Environmental Protection Agency. 1989. Drinking water health advisory: Pesticides. Lewis Publishers, Chelsea, Michigan.

U. S. Environmental Protection Agency. 1990. National Pesticide Survey, Phase 1 Report: Executive Summary. Ofice of Pesticides and Toxic Substances. Washington, D. C.

U.S. Geological Survey. 1989. Reconnaissance sampling finds herbicides in streams in 10 midwestern states. News Release, Public Affairs Office, Reston Virginia.

Waldron, A.C. 1989. Pesticide use on major crops in the Ohio River Basin of Ohio and summary of state usage 1986. OARDC Special Circular 132. Ohio State University, Columbus.

Asbestos

Minerals and Health: The Asbestos Problem
Malcom Ross

The Asbestos Distortion
John E. Kinney

Asbestos
Natalie P. Robinson Sirkin and Gerald Sirkin

Minerals and Health: The Asbestos Problem

Malcolm Ross, Ph.D.

Dr. Malcolm Ross is a research mineralogist, U.S. Geological Survey.

He received a B.S. in zoology from Utah State University; an M.S. in physical chemistry from the University of Maryland; and a Ph.D. in geology from Harvard University.

Dr. Ross is the co-author of *Asbestos and Other Fibrous Materials,* published by the Oxford University Press. He has published more than 65 professional publications contributing to *Science, American Mineralogist,* and *Zeitschrift für Kristallographie.*

He is the recipient of the Distinguished Service Award, Department of Interior, 1986.

Abstract

Of the six forms of asbestos, only three have been used to any significant extent in commerce. These are chrysotile, crocidolite, and amosite. Between 1870 and 1980 approximately 100 million tonnes of asbestos was mined worldwide, of which more than 90 million tonnes was the chrysotile variety, about 2.7 million tonnes the crocidolite variety, and about 2.2 million tonnes the amosite variety.

The three principal diseases related to asbestos exposure are (1) lung cancer, (2) cancer of the pleural and peritoneal membranes (mesothelioma), and (3) asbestosis, a condition in which the lung tissues becomes fibrous and thus loses its ability to function. These diseases, however, are not equally prevalent in the various groups of asbestos workers that have been studied: the amount and type of disease depends on the duration of exposure, on the intensity of exposure, and, particularly, on the type or types of asbestos to which the individual has been exposed.

Lung cancer is caused by exposure to chrysotile, amosite, and crocidolite asbestos; however, increased risk of this disease is usually found only in those who also smoke. Asbestosis is caused by prolonged exposure to all forms of asbestos, whereas, mesothelioma is principally caused by exposure to crocidolite asbestos. There is good evidence that chrysotile asbestos does not cause any significant amount of mesothelioma mortality, even after very heavy exposure. The common non-asbestiform amphibole minerals cummingtonite, grunerite, tremolite, anthophyllite, and actinolite, which are often defined as "asbestos" for regulatory purposes, have not been shown to cause disease in miners.

Chrysotile asbestos, which accounts for about 95% of the asbestos used in the U.S., has been shown by extensive Canadian studies to be safe when exposures do not exceed 1 fiber per cm^3 for the working lifetime. Chrysotile dust levels found in buildings rarely exceed 0.001 fibers per cm^3. Such dust concentrations will have no measurable health effect on the building occupants.

Present and future controls of minerals defined as "asbestos" and removal of asbestos from buildings and homes may cost many billions of dollars. Much of this expense is unnecessary and even counterproductive. Asbestos abatement could in fact produce a health risk to those exposed to dusts during and after removal operations. Overly restrictive controls could force the use of poor or dangerous substitutes and could greatly affect the hard-rock mining industry.

Introduction

In our continuing search for new resources of industrial minerals and efforts for expanding their use, the possible health hazards related to exposure to mineral dusts has become an important factor in deciding which minerals can be safely mined, processed, and incorporated into commercial products.

A large number of experiments have been performed in which a variety of minerals and inorganic substances (such as fiberglass) have been implanted into the tissues of live animals. Often these animals developed cancer. These experiments strongly suggest that if ground up rock were similarly administered a significant number of animals would develop tumors. The ability is at hand to "prove" that the Earth is a carcinogen!

The prevailing cancer dogma in the United States espouses the "no threshold" theory for cancer genesis. It is said repeatedly by influential health specialists that since no one knows the minimum amount of a carcinogen required to initiate the growth of a cancer tumor, it must be assumed that **any amount of any carcinogen is unsafe.** The U.S. public has been led to believe that exposure to just one molecule of a chemical carcinogen can cause cancer. In regard to exposure to asbestos, this paradigm becomes "one asbestos fiber can kill you."

Unless we decide that living on a carcinogenic earth is unacceptable and move elsewhere, a perception and an observation must be reconciled: that (1) carcinogens produced by man, even in trace amounts, are responsible for most human cancer and (2) carcinogens produced naturally are ubiquitous in our environment. The public has not been made aware that a large percentage of the carcinogens taken into the human body are naturally occurring chemicals found in foods, beverages, and tobacco. Examples of some of these carcinogens are: *safrole, estragole,* and *methyleugenol* in edible plants; *piperine* in black pepper; *hydrazines* in edible mushrooms; *furocoumarins* in celery and parsnips; *aflatoxins* in corn, grain, nuts, cheese, peanut butter, bread and fruit; *solanine* and *chaconine* in potatoes; *caffeic acid* in coffee; *theobromine* in chocolate and tea; *acetaldehyde,* an oxidation product of ethyl alcohol, in alcoholic beverages; *allyl isothiocyanate* in mustard and horseradish; and *nitrosamines* formed from nitrates and nitrites found in beets, celery, lettuce, spinach, radishes, and rhubarb (Ames, 1983, 1984). The amount of these chemicals ingested by man is not trivial, for example, the common commercial mushroom *Agaricus bisporus* contains agaritine, a hydroxymethyl phenylhydrazine derivative, in a concentration of 3,000 parts per million or 45 mg per mushroom (Ames, 1984, p. 757)!

In contrast to naturally occurring chemicals, man-made chemicals which sometimes contaminate foods, air, and water make up a relatively small portion of the total human intake of carcinogens. For example, Ames (1983, p. 1258) estimates that the human intake of naturally occurring pesticides (chemicals formed by plants themselves to fight insects, fungi, etc.) is at least 10,000 times greater than the human intake of man-made pesticides. Dole (1983) estimates that avoidance of tobacco smoke and alcoholic drinks would reduce cancer mortality by 38% whereas avoidance of man-made carcinogens contaminating foods, air, and water would reduce cancer mortality by less than 1%.

The perception that small amounts of man-made chemicals are the major cause of human cancer mortality is blatantly false as proven by a massive amount of epidemiological data and summarized by Doll and Peto (1981). Yet this perception, exemplified by statements such as "one fiber or molecule can kill you," is so strong in the minds of the U.S. public that it will take a massive effort by responsible scientists and journalists to change it. Bruce Ames, Sir Richard Doll, Richard Peto, John Higginson, John Cairns, Gio Gori, John Weisburger, Ernst Wynder, Edith Efron, and Michael Bennett are some who are leading the way in bringing good science and good journalism to the public's attention.

In the following I will summarize the specialized topic "asbestos and health" but in context with the broader problem of human carcinogenesis and the public's perception of the causes of cancer. This summary is based on a long review paper published previously (Ross, 1984); for details and reference citations the reader is referred to this publication.

As I will try to bring out, the asbestos problem is not restricted to the mining, milling, and utilization of commercial asbestos; it includes much of the mining industry as well as the greatly expanding fiber and composite industry.

The Asbestos Problem

The widespread use of amphibole and serpentine asbestos by industrial society for such uses as service in brake and clutch facings, electrical and heat insulation, fireproofing materials, cement water pipe, tiles, filters, packings, and construction materials has contributed greatly to human safety and convenience. Yet, while our society was accruing these very tangible benefits, many asbestos workers were dying of asbestosis, lung cancer, and mesothelioma.

The hazards of certain forms of asbestos under certain conditions have been so great that several countries have taken extraordinary actions to greatly reduce or even ban their use. Recent experiments with animals demonstrate that the commercial asbestos minerals, as well as other fibrous materials, can cause tumors to form when the fibrous particles are implanted within the pleural tissue. These experiments have convinced some health specialists that asbestos-related diseases can be caused by many types of elongate particles: the mineral type, according to these health specialists, is not the important factor in the etiology of disease, but rather the size and shape of the particles that enter the human body.

At present, the most widely used definition of asbestos in the United States is taken from the notice of proposed rule-making for "Occupational Exposure to Asbestos" published in the *Federal Register* on 9 Oct. 1985 (pp. 47652–46760) by the U.S. Occupational Safety and Health Administration (OSHA). In this notice, the naturally occurring amphibole minerals amosite, crocidolite, anthophyllite, tremolite, and actinolite and the serpentine mineral chrysotile are classified as asbestos if the individual crystallites or crystal fragments have the following dimensions: a length greater than 5 μm, a maximum diameter less than 5 um, and a length-to-diameter ratio of three or greater. Any product containing any of these minerals in this size range is also defined as asbestos.

The crushing and milling of any rock usually produces some mineral particles that are within the size range specified in the OSHA rules. Thus, these regulations present a formidable problem to those analyzing for asbestos minerals in the multitude of materials and products in which they may be found in some amount, for not only must the size and shape of the asbestos particles be determined, but also an exact mineral identification must be made. A wide variety of amphiboles is found in many types of common rocks; many of these amphiboles might be considered asbestos depending upon the professional training of the person involved in their study and the methods used in mineral characterization.

If the definition of asbestos from the point of view of a health hazard does include the common nonfibrous forms of amphibole, particularly the hornblende and cummingtonite varieties, then we must recognize that asbestos is present in significant amounts in many types of igneous and metamorphic rocks covering perhaps 30 to 40 percent of the United States. Rocks within the serpentinite belts; rocks with the metamorphic belts higher in grade than the greenschist facies, including amphibolites and many gneissic rocks; and amphibole-bearing igneous rocks such as diabase, basalt, trap rock, and granite would be considered asbestos bearing. Many iron formations and copper deposits would be asbestos bearing, including deposits in the largest open-pit mine in the world at Bingham, Utah. Asbestos regulations would thus pertain to many of our country's mining operations, including much of the construction industry and its quarrying operations for concrete aggregate, dimension stone, road metal, railroad balast, riprap, and the like. The asbestos regulations would also pertain to the ceramic, paint, and cement industries, and to many other areas of endeavor where silicate minerals are used.

Geological Occurrence of Commercial Asbestos

Many minerals, including the amphiboles and some serpentines, are described variously as fibrous, asbestiform, acicular, filiform, or prismatic; these terms suggest an elongate habit. Although such minerals are extremely common, only in a relatively few places do they have physical and chemical properties that make them valuable as commercial asbestos. Locally, amphibole minerals may show an asbestiform habit, for example, in vein fillings and in areas of secondary alteration, but usually they do not appear in sufficient quantity to be profitably exploited.

Deposits of commercial asbestos are found in four types of rocks:

(a) Type I—alpine-type ultramafic rocks, including ophiolites (chrysotile, anthophyllite, and tremolite);

(b) Type II—stratiform ultramafic intrusions (chrysotile and tremolite);

(c) Type III—serpentinized limestone (chrysotile); and

(d) Type IV—banded ironstones (amosite and crocidolite).

Type I deposits are by far the most important and probably account for more than 85% of the asbestos ever mined. The most important Type I deposits are those in Quebec Province, Canada, and in the Ural Mountains of the Soviet Union.

Type II deposits are found mostly in South Africa, Swaziland, and Zimbabwe. These furnish mostly chrysotile asbestos. Type III deposits are small; the most notable of these are located in Globe, Arizona, and in the Carolina area of the Transvaal Province of South Africa. Type IV deposits are found only in Precambrian banded ironstones located in the Transvaal and Cape Provinces of South Africa and in Western Australia. Only the South African deposits are still in production. A complete review of the geological occurrences of commercial asbestos is given by Ross (1981).

Since the first recorded use of asbestos by Stone Age man more than 100 million tonnes of asbestos had been mined throughout the world; of this, more than 90% was chrysotile and more than 5% crocidolite and amosite. Nearly 40 million tonnes of this total world production was chrysotile mined in Quebec Province near the towns of Thetford Mines and Asbestos. The total production of anthophyllite asbestos to date is probably no more than 400,000 tonnes, 350,000 tonnes being produced by Finland alone. The production of tremolite asbestos has been sporadic, and it has been mined in various parts of the world for short periods of time. The total production to date of this form of asbestos is probably no more than a few thousand tonnes. Commercial exploitation of actinolite asbestos is practically unknown.

The world asbestos production for 1981 was 4.34 million metric tons—the U.S.S.R. led with 48.5%, Canada second with 25.9% of the world's output (Clifton, 1983). Both countries mine mostly chrysolite asbestos, and most of this comes from the Ural Mountains and Quebec Province.

The Nature of Asbestos Related Disease

The three principal diseases related to asbestos are (1) lung cancer, (2) cancer of the pleural and peritoneal membranes (mesothelioma), and (3) asbestosis, a condition in which the lung tissue becomes fibrous and thus loses its ability to function. These diseases, however, are not equally prevalent in the various groups of asbestos workers that have been studied: the amount and type of disease depends on the duration of exposure, on the intensity of exposure, and, particularly, on the type or types of asbestos to which the individual has been exposed.

Chrysotile or "White" Asbestos

Chrysotile asbestos, sometimes referred to in the trade as "white" asbestos, is the form that is usually used in the United States—as wall coatings, in brake linings, as pipe insulation, and in other uses. About 95% of the asbestos in place in the United States is the chrysotile variety, and a large percentage of this was mined and milled in Quebec Province, Canada. Epidemiological studies of the chrysotile asbestos miners and millers of Quebec undertaken by medical researchers in Canada show that, for men exposed for more than 20 years to chrysotile dust averaging 20 fibers/cm^3, the total mortality was less than expected (620 observed deaths, 659 expected deaths). The risk of lung cancer was slightly increased: 48 deaths observed, 42 expected. Exposures to 20 fibers/cm^3 are an order of magnitude greater than those experienced now (which are generally less than 2 fibers/cm^3); thus, chrysotile miners working a lifetime under these present dust levels should not be expected to suffer any measurable excess cancer. A similar mortality picture is reported for Italian chrysotile miners and millers.

Mesothelioma incidence among those working only with chrysotile asbestos is very low. Thus far, about 16 deaths due to this disease have been reported among chrysotile asbestos miners and millers and one among chrysotile trades workers. In addition, 6 deaths among sons and daughters of chrysotile miners and millers and 2 among others living in chrysotile asbestos mining localities have been reported as being due to mesothelioma.

Four epidemiological studies of the female residents of the Quebec chrysotile mining localities show no statistically significant evidence that their lifelong exposure to asbestos dust from the nearby mines and mills has caused excess disease.

Crocidolite or "Blue" Asbestos

Crocidolite, usually referred to in the trade as "blue" asbestos, was first imported into the United States in 1911 or 1912. By 1930, 35,000 short tons of crude blue fiber had entered the country, and by 1946, an additional 21,000 tons had been imported. In addition to these imports, much crocidolite has come in the United States as manufactured products, such as yarns, tapes, and pipe coverings. Almost all of the imported crocidolite has come from South Africa.

Epidemiological studies of groups that worked only with crocidolite asbestos show that rather short periods of exposure, or even relatively light exposure, causes excessive mortality due to lung cancer, mesothelioma, and asbestosis. This is evident not only in those exposed to crocidolite during gas mask fabrication and building construction but in those employed in the crocidolite mines.

There are only two mining regions in the world where mesothelioma is a statistically significant cause of death. These are the crocidolite mining districts of Cape Province, South Africa, and of Wittenoom, Western Australia. Prevalence studies in Cape Province report that at least 278 people have died of mesothelioma as a result of exposure to crocidolite; 161 of these people worked in the mines and mills and 117 others lived in the vicinity of the mines but were not employed in the asbestos industry.

Thirty-one men who had worked in the small crocidolite industry (total production—155,000 tonnes) at Wittenoom, Western Australia, have died of mesothelioma. Of these, 13 had worked for less than twelve months and 9 had had light to medium exposure to blue asbestos. Sixty miners and millers at Wittenoom have died of lung cancer; 34 of these men had worked in the industry for less than twelve months and 19 had had light to medium exposure to the crocidolite dust. In addition to this occupationally related mortality, 6 others who lived near but did not work in the mines or mills have died of mesothelioma.

Amosite or "Brown" Asbestos

All amosite asbestos comes from the Transvaall Province of South Africa, where between 1917 and 1979 approximately 2.2 million tonnes have been mined. Importation of amosite into the United States started in the 1930s.

Two complete epidemiological studies of asbestos trades workers exposed mainly to amosite asbestos have been published. The incidence of asbestos-associated disease in these two groups of men formerly employed at factories in London, England, and Patterson, New Jersey, was excessive, there being a 18.4% lung cancer mortality (117 cases), and 2.8% mesothelioma mortality (8 cases), and 4.2% asbestosis mortality (27 cases). Only prevalence studies have been made of amosite miners and millers; 2 individuals have died of mesothelioma. One resident of an amosite mining district has been reported as having died of this disease.

The rock-forming amphibole minerals grunerite and cummingtonite, which are isostructural and chemically similar to amosite, are considered (incorrectly) by some to be forms of asbestos. Health studies of miners working ores that contain these minerals as gangue do not show any indication of asbestos-related mortality, for example, the Homestake gold mines and the Reserve iron ore miners.

Anthophyllite Asbestos

This form of asbestos has been mined sporadically in many localities, but the only major production has been at Paakkila, Finland, where approximately 350,000 tonnes was mined between 1918 and 1975. The only health study of individuals exposed predominantly to anthophyllite asbestos is that of the Paakkila miners. This group showed a 67% excess of lung cancer and a large mortality due to tuberculosis and asbestosis. There were no deaths from mesothelioma. Because anthophyllite was and is used so little in commerce, no additional health studies appear to be possible except for follow-up studies of the Paakkila miners.

Discussion

The Relative Hazards of the Asbestos Minerals

Must the use of all commercial asbestos be stopped? The answer is an emphatic no—but with qualifications that are presented here.

Nonoccupational exposure to chrysotile asbestos, despite its wide dissemination in urban environments throughout the world, has not been shown by epidemiological studies to be a significant health hazard. If it were, the women of Thetford Mines and Asbestos, Quebec, where over 40 million tonnes of chrysotile asbestos has been mined, would be dying of asbestos-related diseases. They are not. Health studies accomplished in Canada show that populations can safely breathe air and drink water that contains significant amounts of chrysotile fiber. These studies also show that there is a "threshold" value for chrysotile asbestos exposure below which no measurable health effects will occur.

The same fiber dose-disease response relationships observed for chrysotile asbestos do not hold for crocidolite asbestos. Health studies of those exposed to crocidolite only show it to be much more hazardous than chrysotile. No study has been reported, comparable to that made for chrysotile, which would indicate what a safe level of exposure to crocidolite would be. The danger of crocidolite dust is particularly emphasized by the many mesothelioma deaths occurring among the residents of the crocidolite mining districts of Cape Province, South Africa, whose only exposure was in a nonoccupational setting. Such mortality is practically unknown among residents of the chrysotile mining localities of Quebec Province.

The hazards of amosite asbestos are more difficult to assess. The amosite factory employees of London, England, and Patterson, New Jersey, who generally worked under very dusty conditions, have experienced a great deal of excess mortality due to lung cancer, asbestosis, and mesothelioma. In contrast to these factory works, amosite miners and millers, at least with regard to mesothelioma, do not appear to be at much risk.

The fear caused by alarmist statements such as "one fiber can kill you" and by the much exaggerated predictions of the amount of asbestos-related mortality expected in the next 20 or 30 years has generated great political pressure to remove asbestos from our environment and to reduce greatly or even stop its use. An example of this is the concerted effort in several industrial nations, including the United States, to remove asbestos from schools, public buildings, homes, ships, appliances, and so forth. This is being done, even though most asbestos in the United States is of the chrysotile variety and even though asbestos dust levels in schools, public buildings, and city streets are much lower than those found in chrysotile asbestos mining communities, where little asbestos-related disease appears in the nonoccupationally exposed residents. The impetus for these costly removals and appliance recalls (hair dryers, for example) apparently comes from propagandizing of the "one fiber can kill you" concept. Not only is this program costly—it could be dangerous if the removal of blue asbestos is not accomplished with great care. In most cases, asbestos coatings and insulation, where necessary, can be repaired at no risk and at a fraction of the cost of complete removal.

Substitutes for Asbestos

If all use of asbestos were to be discontinued, substitutes would have to be developed to meet many diverse requirements for materials, such as nonflammability, high strength, flexibility, reasonable cost, and safety. With respect to safety, the substitutes must not induce disease in those exposed to them and also must not endanger lives in other ways by having inferior strength or durability, increased flammability, or other undesirable characteristics. A good substitute must not have so high a cost that it forces the use of an inadequate replacement. Possible problems can occur with substitutes, for example, with the replacement of chrysotile asbestos in drum brake linings. The chance of increased automobile accidents due to a possible inferior substitute material must be weighed against the probability of anyone being harmed by the small amounts of chrysotile asbestos emitted from drum brakes. Also, the health effects of emissions from substitute brake linings must be considered.

The requirements of strength and flexibility make it necessary that asbestos substitutes be fibrous. Generally, the thinner and longer the fibers, the stronger, more flexible, and useful they are. However, fibers longer than 4 μm, and less than 1.5 μm in diameter are capable of producing malignant neoplasms when implanted into the pleura of rats. Test fibers used in these studies have included aluminum oxide, fiberglass, wollastonite ($CaSiO_3$), silicon carbide, dawsonite ($NaAl\,CO_3OH$), and potassium octatitanate.

Man-Made Mineral Fibers

In addition to the very common exposure of man to naturally occurring fibrous minerals, he is also exposed to many different types of synthetically made fibers. More and more of these substances are being developed each year in the greatly expanding fiber and composite industries. Our knowledge of the health effects of most of these fibers is minuscule or non-existent. The health effects of those exposed to man made vitreous fibers, however, have been studied intensively and statistically significant studies are just now being reported. These glassy fibers, usually referred to as man-made mineral fibers (MMMF), include slag wool, rock wool, glass wool, fiberglass, and continuous filament products. Two separate reports, one of 25,146 workers at 13 plants in Europe (Saracci et al., 1984), and one of 16,730 workers at 17 plants in the United States (Enterline and Marsh, 1985) have been published. Table 1 gives a summary of the lung cancer mortality for male workers in the European and United States MMMF industries who lived at least 30 years since first exposure. A statistically significant excess of lung cancer (52%) is seen in these MMMF workers. Exposure was generally less than 1 fiber per cm^3 and most commonly in the range of 0.01 to 0.1 fibers per cm^3. In contrast, the Quebec chryostile asbestos miners and millers with at least 20 years service in the industry and exposed to between 10 and 21 fibers per cm^3 showed an excess of lung cancer of 12% (Ross, 1984, p. 93).

Non-Asbestiform Rock-Forming Amphiboles

Some health and regulatory specialists classify the common rock-forming amphiboles grunerite, cummingtonite, actinolite, tremolite, and

Table 1. Lung cancer mortality, males, at least 30 years observation since first employment in MMMF industry

Europe (13 factories)	Lung Cancer		
	Observed	Expected	Excess
Rock wool	11	5.7	93%
Glass wool	4	2.6	54%
Continuous filament	2	0.6	233%
Subtotal	17	8.9	91%
United States (17 factories)	**Observed**	**Expected**	**Excess**
Glass wool	47	36.0	31%
Slag wool	45	28.1	60%
Rock wool	14	8.1	73%
Subtotal	106	72.2	47%
Totals	**123**	**81.1**	**52%**

anthophyllite as asbestos even though they do not possess the physical properties requisite to be valuable commercially. Such a classification has been made in the case of taconite mining by the courts (United States District Court for Minnesota, 380 F. Supp. 11) and by the U.S. Environmental Protection Agency (Reserve Mining vs. EPA, U.S. Court of Appeals Eighth Circuit, 14 March 1975); in the latter case, the U.S. Environmental Protection Agency (EPA) sued to prevent the Reserve Mining Co. from dumping taconite tailings into Lake Superior because of the perception that these tailings contain "amosite asbestos" and thus constitute a threat to public health. For a complete review of the case see 514 *Federal Reporter*, 2d Series, 492–542, 1975; and 256 *Northwestern Reporter*, 2d Series, 808–852, 1977.

The possible health effects of exposure to rock dust containing one or more of the many amphibole minerals is an important issue. Amphiboles are contained within the gangue (waste rock) of many hard-rock mines; gold, vermiculite, talc, iron ore, crushed stone and aggregate, copper, etc. Certainly control of most dust, regardless of the mineral content, is necessary for past heavy exposure to dusts containing crystalline silica (SiO_2), slate, coal, talc, and radioactive minerals have caused significant disease. Unusually tight control of amphibole mineral particles such as proposed by the National Institute for Occupational Safety and Health (NIOSH) in 1976 for asbestos fibers (0.1 fibers per cm^3), however, could stop major mining in the United States.

Several epidemiological studies have been performed on hard-rock miners who were exposed to non-commercial amphibole dusts. One example is a cohort of gold miners at the Homestake Gold Mine, Lead, South Dakota, who were exposed to rock dust containing very significant amounts of amphibole belonging to the cummingtonite-grunerite series. The 861 observed deaths included 43 lung cancer deaths (43 expected) and no mesothelioma deaths. Non-malignant respiratory disease due to quartz dust, however, was excessive. This cohort thus showed no evidence of mortality that could be related to amphibole dust (David P. Brown et al., International Symposium, Chapel Hill, North Carolina, April 5, 1984, unpublished preprint). Studies of Reserve taconite (iron ore) miners who were exposed to cummingtonite amphibole contained in the mine dusts also show no amphibole-related disease. Of the 298 deaths (344 expected) there were 15 lung cancer deaths, 18 were expected. There were three excess deaths due to pneumoconiosis, perhaps due to quartz dust (Ross, 1984, p. 75). Despite no indication of asbestos-related disease in these miners the Reserve Mining Company was required by the U.S. Courts (see above) to spend 300 million dollars to build an inland dump site for waste rock.

Commentary

The "no threshold" cancer dogma that has been repeatedly foisted upon an unaware American public is generating a national crisis of such proportions that our economy could be very adversely affected. The economic consequences are particularly apparent in regard to the requirement by the Environmental Protection Agency that administrators report if asbestos is present in their schools. What is the school administrator to do when he is required by law to post signs in his schools stating that asbestos is present? He has no option—he has to have it removed for our cancer dogma translates the words on his building sign to "a little child breathing just one asbestos fiber may get mesothelioma 30 years later." The school administrator, the teachers, the parents, are in a box. The only way out of the box is to call for full ripout of the asbestos even though well informed experts know that many asbestos removal programs, perhaps most, will put more fiber into the air than if it were left in place. Building owners, school systems, and local and state governments are suing many businesses for costs and damages related to asbestos removal, insurance companies are dropping liability coverage on workers removing asbestos, and occupants of buildings are suing over health effects or potential effects alleged to be caused by exposure to asbestos during renovation. The asbestos removal workers themselves, I predict, will soon be suing their contractors because they too will believe that their health is threatened by exposure to asbestos.

And what are we to do about asbestos substitutes that also may be carcinogenic in man—for example the man-made mineral fiber products such as rock wool and fiberglass? Since these products apparently cause excess lung cancer in long-term production workers a logical and honest cancer policy would require that they also be removed from schools. Will the whole "asbestos in schools" scenario then be replayed, the main actor being fiberglass products rather than asbestos?

Fibers are not restricted to the school environment, many public and commercial buildings, churches, and private homes contain asbestos and also the more modern fiberglass products. Are we to go through the asbestos removal program for these structures also? Where is the money to come from to cover the multi-billion dollar costs? How do we insure the workers? Is there any assurance at all that the occupants of the buildings undergoing asbestos ripout will not receive more exposure to fibers than if the material were left in place?

And lastly, what is to be done to control mineral dusts perceived to be asbestos or asbestos-like, for example the rock-forming amphiboles, that occur in most hard rock mines? Will all mine and mill waste piles be considered toxic dumps? The challenge to the mining industry is here and it is serious.

References

Ames, Bruce N. (1983) Dietary carcinogens and anticarcinogens. *Science,* vol. 221, p. 1256–1264.

Ames, Bruce N. (1984) *Letters,* cancer and diet. *Science,* vol. 224, p. 656–760.

Clifton, Robert A. (1983) Asbestos. U.S. Bureau of Mines Yearbook, vol. 1, Metals and Minerals, p. 113–118.

Doll, Richard (1983) Prospects for the prevention of cancer. Clinical Radiology, vol. 34, p. 609–623.

Doll, Richard and Peto, Richard (1981) The cause of cancer: Quantitative estimates of avoidable risks in cancer in the United States today. *Journal National Cancer Institute,* vol. 66, p. 1193–1308.

Enterline, P.E. and Marsh, G.M. The health of workers in the MMMF industry. In: Guthe T., ed. Biological effects of man-made mineral fibres: Proceedings of a WHO/IARC conference in association with JEMRB and TIMA Copenhagen, 20–22 April 1982. Volume 1. World Health Organization,

Regional Office for Europe, Copenhagen 1984, pp. 311–339.

Ross, Malcolm (1981) The geologic occurrences and health hazards of amphibole and serpentine asbestos. *In:* Reviews in Mineralogy, vol. 9A, Amphiboles and Other Hydrous Pryiboles—Mineralogy, D.R. Veblen, Ed., Mineralogical Society of America, Washington, D.C., p. 279–323.

Ross, Malcolm (1984) A survey of asbestos-related disease in trades and mining occupations and in factory and mining communities as a means of predicting health risks of nonoccupational exposure to fibrous minerals. *In:* Definitions for Asbestos and Other Health-Related Silicates, ASTM STP 834, B. Levadie, Ed., American Society for Testing and Materials, Philadelphia, PA, p. 51–104.

Saracci, R. et al. (1984) Mortality and incidence of cancer of workers in the man made vitreous fibres producing industry: an international investigation of 13 European plants. *British Journal Industrial Medicine*, vol. 41, p. 425–436.

Reprinted, with permission, from
Proceedings of the 21st Forum on the Geology of Industrial Minerals,
1987, Published by the Arizona Bureau of Geology and Mineral Technology

The Asbestos Distortion

John E. Kinney, P.E., D.E.E.

John (Jack) Kinney received a BCE degree from Manhattan College in 1941 and an MCE from Cornell in 1942 majoring in sanitary engineering. His efforts have been testing in the field the theories produced by the researcher and interpreting what the researcher and the professor write about for public and for legislative hearings, to present an analysis of the present and future impacts. His presentations number about 200 and include testimony before House and Senate Committees, and state legislative committees. He has been a member for more than 15 years on the USGS Advisory Committee on Water Data for Public Use and the Chamber of Commerce of the U.S. committees on natural resources and environment. He is a registered professional engineer, a Diplomate in the American Academy of Environmental Engineers, and has been involved in expert testimony.

According to EPA all asbestos is equal in potential to cause cancer, that exposure to a single fiber is all that is required, that children are especially vulnerable so all asbestos should be removed from schools. Secretary of Health Califano, in a major speech in 1978, forecast 67,000 excess cancer deaths a year from exposure to asbestos. Panic ensued.

There are many forms of asbestos but three are of prime importance. Crocidolite, or blue asbestos, found in South Africa and Australia, is deadly. As little as three months exposure in the mines has caused mesothelioma, a fatal cancer. Amosite is also of African origin and is similar to crocidolite in physical and chemical characteristics and also requires stringent workplace standards. The third is chrysotile which is mined in Canada and was in the southwest USA until court cases encouraged by EPA and Justice Department made it impossible for mines to operate.

Crocidolite and amosite are amphiboles that once in the lungs, remain there. Chrysotile is a hollow tube-like fiber that dissolves in the lungs and is expelled. There was a small excess of lung cancer among Canadian miners with heavy exposure for many years. Since 1926 the only cancer victim in the Globe, Arizona area where asbestos was mined, was an itinerant smoker. About 95% of the asbestos used in the United States has been chrysotile. What EPA did with the assistance of the National Academy of Science was to assert that all forms of asbestos are the same in causing cancer. That was all some US Senators needed to adopt legislation requiring removal.

Four federal agencies commissioned the Academy to establish a panel to evaluate asbestos. The panel was at the point of typing its report when it was disbanded. It appears that a request from EPA caused its dissolution, and the appointment of a new panel. This new group rushed a report that was so sloppy it had to be recalled. The report finally released stated that the asbestos concern was real, and that all forms have to be considered as having the same potential to cause cancer, but made no mention of some 20 very important papers that scientifically contradict the NAS conclusions. EPA then announced its decision was based on the Academy recommendation. The EPA had come to its conclusion long before the Academy report and apparently conned the Academy into endorsing it since the second panel was entirely funded by EPA. EPA then changed the individuals in charge of the section on toxics and the new people claimed their predecessors had the information and had made the decisions, that they were not knowledgeable.

The estimated cost of the program of asbestos removal is upward of $100 to $134 billion. The benefits are nil or negative for both the economy and safety. The effect, however, could be disastrous if schools contain crocidolite. The concentration in a school before ripout is about 0.001 fibers per cc (f/cc). After ripout the fibers get entrained in the air system and months later have been measured at 20 to 40 f/cc. The workplace limit was 2 fibers and has been reduced to 0.5. EPA will not set a permissible limit. EPA says that the level after ripout should not be higher than before. If it is ok after ripout what was wrong with it before ripout?

One effect of this continues to be closure of many schools which can not afford asbestos removal. I have encountered individuals who are delighted with the possibility that this will cause the closure of private and parochial schools. Doesn't this make you wonder about such demands for a better environment?

Emphasis is now on requiring removal of asbestos from office and public buildings. EPA published a notice (March 18, 1991 in the *Federal Register*) of proposed rule making to require removal from public and commercial buildings.

New Jersey is seeking legislation that would require every home-owner to certify there is no asbestos in the house before it can be sold. It is going to be a real problem when these panic-mongers learn that data on work-place cancer rates in the manufacture of the man-made fibers being used as replacement are higher than in the asbestos work-place. Outside of mining and textile workers those working with chrysotile are showing no effect. This includes roofing felts and tars, brake and clutch lining and cement.

Cat lovers beware. Kitty litter you buy consists of attapulgite which qualifies as an EPA asbestiform fiber. Is your cat endangered?

The latest on this fiasco is the report that Catholic bishops are now questioning the Administrator on the information EPA used to determine that the asbestos should be removed from the schools. Administrator William Reilly is denying that EPA ever called for the removal of the asbestos and is blaming the hysteria on the "cantankerous" press. This denial is equivalent to yelling fire in a crowded theater and then denying responsibility for the panic and loss of life by saying that he never told the people that they should leave the theater.

EPA published a "blue book"—*Guidance for Controlling Friable Asbestos-Containing Materials in Buildings* (1983) and with it a film showing EPA inspectors dressed in space suits inspecting a grammar school. The children sat amazed at the visitors. When the EPA and its consultant met with the Michigan school administrators in Detroit to show the film and distribute the book, they told these people that exposure to a single fiber of asbestos was sufficient to cause cancer; that, as the book said, children and young adults exposed to asbestos have a greater chance of getting cancer than older people; and that the responsibility was up to the administrators to determine whether asbestos is present, notify all teachers and parents if it is and proceed with a program to protect the children. The administrator would be responsible if children entered the room and got cancer. EPA provided no information on the relative risks of different kinds of asbestos, nor did it promote a reasoned course of analysis and control. EPA's estimate, based on the Califano speech of 67,000 excess deaths, was the data base. EPA did not tell the administrators the concentration of asbestos in the air outside the hotel in which they were meeting was as high or higher than in school buildings. Wearing space suits in the school and removing them in the street proved EPA was interested in creating a sense of fear and theater, not a true concern about asbestos in the air.

At the insistence of the House of Representatives Appropriation Committee, EPA published a revised report—the "purple book" — because it could not substantiate the hysteria included in the blue book. However, although the blue book was widely circulated, one had to know about the second edition and request it. EPA did not try to calm the hysteria; it met with school boards and repeated its concern about asbestos and left it up to Congress, state legislatures and school boards to order the removal. The EPA action is most charitably described as contemptible. If the school contained crocidolite, the residual long term exposure of children can result in the mesothelioma they would not have gotten if the asbestos had not been ripped out. EPA may then be classified as murderer of school children and the immigrants hired to rip out the asbestos for the mestothelioma is fatal. The diversion of funds from necessary education for the ripout is another onus for EPA and the legislators. Some bureaucrats would seem to have no conscience. But they enjoy their power to decree.

To add insult to injury, EPA inserted a disclaimer in the blue book: "Neither the United States Government nor any of its employees . . . assumes any legal liability or responsibility for any third party's use of or the results of the use of any information . . . in this report."

England and Canada did in-depth appraisals of the asbestos question. Both agreed that crocidolite (blue asbestos) should be banned, amosite (brown) used only under very stringent control standards, and chrysotile (white) continued in use and controlled to 1 f/cc. The last was adopted because not only is there a very reduced chance of harm to health, but also the alternatives do not provide equivalent safety (e.g. insulation in buildings to protect against fire, brake linings in cars and trucks, insulation in clothes, etc.), utility (e.g. substitutes are not as satisfactory in terms of life span and effectiveness), and cost. Our government—EPA and the NAS panel—refused to consider these studies or these aspects.

Reprinted, with permission, from
National Council for Environmental Balance, (NCEB) 1990.

Asbestos

Natalie P. Robinson Sirkin
and Gerald Sirkin, Ph.D.

Natalie P. Robinson Sirkin has been a columnist with the *Citizen News* in New Fairfield, Connecticut, since December 1977.

She received a B.S. from Simmons College in 1942 and an M.A. from Columbia University in 1950.

She is the author/co-author of several published articles in British Indian History of the early 19th century.

Gerald Sirkin, now retired, has been Professor of Economics, the City College of the City University of New York; and an Assistant Professor, Yale University.

Dr. Sirkin received a B.A. from Harvard in 1942, an M.A. from Columbia in 1948, and a Ph.D. from Columbia in 1956.

He is the author of *Introduction to Macroeconomic Theory* (three editions) published in 1961, 1965, 1970; and *The Visible Hand: the Fundamentals of Economic Planning* published in 1968.

Dr. Sirkin has more than 12 published articles in economics, and is co-author with his wife, Natalie, of several articles in Indian History. He has contributed to *Quarterly Journal of Economics, Review of Economics and Statistics,* and *Business History Review.* He has been awarded Fulbright and Rockefeller Foundation grants.

"Asbestos is like a big sleeping dog. If not stirred up, it does no harm." . . . Leave the sleeping dog alone. It will save lives and perhaps hundreds of billions of dollars.

—Michael Fumento

. . . I think the public doesn't realize that it was probably . . . the ban on asbestos that caused the Challenger disaster. It wasn't the O rings themselves that failed. It was the putty that held the O rings in place. And up until . . . the time of the Challenger, that putty had had asbestos in it to strengthen it and make it fire-retardant. When the asbestos was removed, it was the putty that gave way.

—Dixy Lee Ray, Interview, C-SPAN, June 16, 1991

The "environmentalists" who restored malaria to its preeminent position as a killer by their hysterical campaign against DDT are back again, spreading sickness and death with their campaign against asbestos.

The Asbestos Hazard Emergency Response Act (AHERA), which breezed through Congress by voice vote after little discussion and

119

became effective in late 1987, mandates that school building—public and private—must be inspected for asbestos. For those containing asbestos, an abatement-plan was required by October, 1988. Protective action had to be begun by July, 1989. Similar abatement legislation for all public buildings is pending in Congress.

If It Ain't Broke, Break It

A sensible person might think the intelligent approach would be to monitor the air to see if asbestos particles present a hazard. But that isn't how the bureaucratic mind works. The Environmental Protection Agency maintains that visual inspection shall be the determinant. If it "looks" dangerous, it must be treated, whether it is in the air or not. But visual inspection is actually a highly inaccurate predictor of air-borne asbestos levels.

The sensible person might also think that the intelligent method of treatment would be to seal off the asbestos, called "encapsulization." But no. Hysteria will not permit encapsulization: The asbestos must be ripped out of a building.

The cure creates the disease. Asbestos, which was bothering no one, becomes an airborne hazard when it is ripped, pulled, and scraped out.

The principal risk is to the workers who remove the asbestos. Although safety equipment is prescribed by Federal regulations it does not work perfectly even with the most expert and careful use. Worse, a great many workers and removal contractors are neither expert nor careful. Hundreds of lives will be needlessly sacrificed to the ignorance, nuttiness, or scheming of the environmental panic-mongers.

What Dangers?

The only cases of lung cancer known to be attributable to asbestos are in workers who were subjected to high concentrations of asbestos while working with it. For the general population, exposed to very small amounts of asbestos, the fear of lung cancer—groundless—is born of the kind of illogic that says: If smoking three packs a day creates a risk of cancer, then smoking one pack a year must create some risk.

The difference between high and low dosage has still not penetrated the American skull. A substance that is hazardous in high doses can be harmless in low doses because the human body has natural defenses against harmful substances. So long as the amount is not large enough to overwhelm these defenses, no harm is done.

In the case of asbestos, the continuously moving mucuous lining of the lungs removes fibres that are inhaled. If any get beyond that defense into the lower respiratory tract, other natural defenses can remove them.

That is how humans survive; otherwise, asbestos would have killed us all because it is a naturally occurring pollutant. Asbestos is in rock formations and from there it gets into the air and water. "Regardless of where we work or attend school, all of us are exposed to some level of the mineral every day," notes Michael Fumento in his excellent, lengthy article in the October, 1989, issue of *The American Spectator.* In a sample of 43 Federal buildings, the EPA found that the highest levels of airborne asbestos were no higher than those found outside the buildings.

The natural defenses of the body explain why innumerable recent scientific studies find no detectable hazard from low levels of asbestos. For example, a Canadian Royal Commission study of the use of asbestos in Ontario concludes, "There are no documented cases of lung cancer associated with low-level asbestos exposure over a lifetime."

The idea that asbestos is highly dangerous probably derives from the record of the short-fiber amphibole asbestos, a type that does not occur in and is not used in the United States. The asbestos used in the United States is the long-fiber chrysotile, which, being easily expelled from the lungs, is dangerous only in occupations entailing prolonged exposure to extremely heavy concentrations.

Insulation, once provided by asbestos, is now being provided chiefly by vitreous or "glassy" fibres. Studies indicate that, to workers producing or applying insulation, these man-made fibres are more hazardous than asbestos.

The Staggering Costs

As yet we have only rough estimates of the cost of asbestos abatement. These estimates generally turn out to be understated, as in the case of a San Francisco high school where the estimate of $10 million turned into $18 million.

Recent estimates of the direct costs run from $100 billion to $200 billion. If we add the indirect costs—like shutting down buildings, laying off workers, and disrupting production during the ripping-out of the asbestos—the total costs could be several times the $200 billion. These costs are enough to eat up the potential increase in U.S. disposable income for the next decade. If anyone wonders why the increase in disposable income has been slow in recent years, a good place to look for an answer is extreme environmental regulations.

These immense sums will be spent to raise levels of airborne asbestos and create risks of illness and death where none now exist. The organized opportunists and ignoramuses—the National Education Association, the Sierra Club, Nader's public-interest group, and the rest—have worked through the ignoramuses and cowards of Congress to bestow this deadly blessing upon us.

First published in *Citizen News*
November 8, 1989
New Fairfield, Connecticut

Biomagnification

The Myth of Food-Chain Biomagnification
J. Gordon Edwards

The Myth of Food-Chain Biomagnification

J. Gordon Edwards, Ph.D.

Dr. J. Gordon Edwards is professor of biology at San Jose State University. He served as a ranger naturalist for nine summers in Glacier National Park, and has been a graduate assistant at The Ohio State University in Entomology and Ornithology.

He received a B.S. in botany/zoology from Butler University in 1942; an M.S. in entomology (1947) and a Ph.D. in entomology from The Ohio State University (1949).

Dr. Edwards is the author of *Coleoptera East of the great Plains* published in 1949; *Climbers Guide to Glacier National Park, Montana* published by the Sierra Club Press in 1961 (last revision May 1991); and has contributed chapters in many books. He has written and published numerous journal articles contributing to *Wasmann Journal of Biology, Coleopterists Bulletin, Systematic Zoology, and Pan Pacific Entomologist*.

He is a Fellow, California Academy of Science (300 life fellows only).

One of the most remarkable anti-pesticide allegations used by environmental extremists has been the claim that pesticide levels are "magnified" at each step of the food chain. Scientist who are aware of the physiological processes at work in living creatures, however, realize that the concept of "biomagnification," as touted by the anti-pesticide industry, is untenable; it is refuted by experimental data as well as by field analyses.

In 1968, Virgil H. Freed described an experiment in which fish living in two tanks of water with identical DDT concentrations were fed different diets.[1] One group of fish got pesticide-free food, while the other was fed high levels of DDT in its food. The amount of DDT residue accumulated in the fish tissues was almost the same in the two tanks of fish, demonstrating that more than 80% of the DDT in the fish had entered the body via the gills, rather than being extracted from the food.

J.L. Hamelink wrote in 1971 that "a hypothesis that biological magnification of pesticides was dependent upon passage of the residues through a food chain was rejected and a hypothesis that accumulation depends upon adsorption and solubility differences was proposed . . ."[2] Dr. Frank Moriarty observed in 1972, "It is commonly believed that such persistent substances accumulate or concentrate as they pass along or up the food chain . . . There is, however, little good evidence

125

that pollutants in general, or organochlorine insecticides in particular, do concentrate along the food-chain."[3] He also pointed out errors he found in the 1960 article by Hunt and Bischoff[4] on the subject of residues in grebes, fish and plankton at Clear Lake, California.

D.L. Gunn, in his 1977 presidential address to the Royal Entomological Society of London, observed, "Here are some samples of scares that deceived laymen . . . the oft-quoted food-chain scare in general, often untrue or at least dubious, and the Clear Lake food-chain story, which was false."

In 1973 a committee of the National Academy of Sciences, chaired by John Kanwisher of the Woods Hole Oceanographic Laboratories, concluded that "the absence of concentration increase going up the food chain can be due to either varying solubilities or to biodegradation in a pool of pesticides relatively sequested from their surroundings." In August of that same year another NAS committee, headed by George Harvey (also of Woods Hole) pointed out that "The measured concentrations of DDT in various organisms of the open Atlantic and the Gulf of Mexico give no support for food web concentration as a general phenomenon." The committee questioned that more data would resolve the problem of food web magnification of DDT.

A Misleading Article

An article which has been cited by environmentalists as supporting the concept of biological magnification is one written by J.J. Hickey.[5]

Referring to the data in that article, the following statement was published in the mid-1970s as part of the training material for individuals involved in research, development and application of pesticides. That training was offered at a university cooperative extension service, which makes it a cause for much more concern than similar statements that have been published in pseudo-environmental publications. That statement was:

> "DDT in lake Michigan is probably one of the best examples of biological magnification. The bottom sediments were found to contain an average of 0.0085 ppm. Small invertebrates averaged 0.41 ppm while predaceous fish such as coho salmon and lake trout contained 3 to 8 ppm. Herring gulls, at the top of the food chain, contained as much as 3,177 ppm of DDT. In this case, the overall magnifications from the bottom sediment is about 370,000 fold."

The author of that propaganda used many of the exact figures given by Hickey, but other facts and figures were altered and significant data were omitted. Since the training material used the Lake Michigan study as "one of the best examples of biological magnification," I shall also use it to expose some of the deceptions which are to so typical of anti-pesticide literature.

Food, Chemical Accumulation

If there were no fecal elimination of ingested materials, no metabolism of chemicals in the body and no excretion of those chemicals and their metabolites, then animals would accumulate all of the chemicals they ever swallowed (including their food), plus all chemicals entering through their skin, gills or lungs. Obviously those processes do occur, therefore the long-term retention of those chemicals does not occur.

Unfortunately the popular press and some semi-scientific journals have been crammed with "biomagnification" allegations for many years, and anti-pesticide activists have made profitable use of that myth. Environmental organizations welcomed the radical concept and used it to frighten the public into donating more money to "help fight pesticides."

In the following discussion, DDT plus its breakdown products will be referred to as DDTR. Occasionally all such compounds are lumped together as DDT, and sometimes it is specified that DDT does not include DDD, DDE and other residues. The difference often leads to confusion.

In his Lake Michigan article, Hickey pointed out that the study area was along the coast of Door County, Wisconsin, where thousands of acres of cherries were being grown. He stated that the annual insecticide use there included 50 tons of lead arsenate, 30 tons of DDT, 15 tons of TDE (DDD), 15 tons of methoxychlor and 7.5 to 22.5 tons of malathion. He also wrote that the bay "receives pollutants from the heavily used Fox and Wolf River drainage systems," and that the water quality there is "quite atypical of that found elsewhere in Lake Michigan." The extent of pollution in that part of the lake was not mentioned. The emphasis was always on DDT and the alleged "biomagnification" of DDTR in the Lake Michigan food chain.

Water and Mud Concentrations

Advocates of the "biomagnification" theory often compare the concentrations of pesticides in bodies of water with those in bottom sediments, and consider the differences to be "biomagnification." They also compare the concentrations in those sediments with those in animals living there and call those differences "biomagnification" as well. Obviously those differences cannot be biological magnification because the chemicals are not yet in any biological system. Hickey stated that the concentration in his Lake Michigan sediments "was running 0.014 ppm of DDT," but the extension service writer claimed that the sediments "contained an average of 0.0085 ppm."

This excessively low figure was actually the single lowest reading of all the analyses made during Hickey's lake study and was omitted by Hickey from his own conclusions. By comparing that unique low figure (0.0085) with the 0.41 ppm in the amphipods living in the sediment, the writer implied that was a 500-fold "biological magnification" between the sediment and the amphipods. Hickey had specifically stated that "if the material is attached to soil particles in lake sediments or in muddy streams and lakes, it is not yet in the biological system."

Fish Concentrations

The extension service author stated that the Hickey study showed that "coho salmon and lake trout contained 3 to 8 ppm" of DDT. He thus implied there might be a threat to people eating those prized game fish from the lake. Actually the study made no mention of coho or lake trout. The only fish mentioned there were club, alewife and whitefish; the highest level of DDTR found in any fish was 7.47 ppm, which was in only one chub (the chub average was only 4.52 ppm). Hickey wrote that 10 of the 11 chub were collected far from the study area and at depths greater than 120 feet. They had certainly not been eating any of the amphipods with 0.41 ppm DDTR in the study area; therefore any apparent correlation in DDT levels must be purely coincidental. Only the alewives had been eating significant numbers of the amphipods, and Hickey wrote that "overall mean and standard error of 3.35 ± 0.65 ppm . . . represents our best estimate of pesticide residues and variance in the alewives."

J.C. Davis found that trout pass all the blood in their body through the gills every 64 seconds, causing a rapid transfer of waterborn chemicals into the body tissues.[6] It is therefore not surprising that Chadwick and Brockson discovered only 16% of the pesticide residue in fish is acquired from their food, while the rest is taken in over their gills.[7] The most DDT attributable to the food of the alewives would be about 0.5 ppm (16% of their total residue of 3.35 ppm DDTR). The difference between the 0.40 ppm in the amphipods and the 0.5 ppm attributable to food intake of DDTR by alewives is certainly much too small to be heralded as "biological magnification up food chains."

Bird Concentrations

Many controlled studies have revealed that most ingested chlorinated hydrocarbon insecticides pass quickly through bird digestive tracts and are voided with the feces. For example, U.S. Fish and Wildlife biologists fed heptachlor-saturated earthworms to captive birds (woodcock) for 60 days. During this time the birds only absorbed 20% of the insecticide they ingested, the rest was passed with the feces. Metabolic loss and excretion of pesticide residues was then 2.8% daily, thereafter.[8]

Elsewhere, D.J. Jefferies and B.N. Davis recorded a *decrease* of dieldrin residues at each step up the food chain.[9] They stated: "A content of 25 ppm [of dieldrin] in the soil resulted in a range of 18.4 to 24.9 ppm in the earthworms after 20 days of exposure. After six weeks of diets ranging from 0.32 to 5.69 ppm dieldrin thrushes showed total body residues of 0.09 to 4.03 ppm." The authors concluded, "It was clear that there was no concentration effect." Incidentally, Davis had earlier reported that "the highest dieldrin residues found in earthworms were under 1 ppm" near cultivated sites that had been heavily sprayed.

Tissue Variability

In Hickey's Lake Michigan study, the fat, muscle and brain tissues of the birds were analyzed separately. The extension service author mentioned only the fat tissue levels of DDTR, which the study showed to be 40 times higher than levels in muscles and 150 times higher than in brain tissue of the same birds. That proportional variance between residues in different tissues was also demonstrated by U.S. Fish & Wildlife studies in 1965.[10] Bald eagles were fed 5 ppm of DDT in their daily diet for 210 days, after which their tissues were analyzed for DDTR. Fat contained 35.68 ppm while muscle contained only 2.58 ppm. Lesser amounts were found in liver and brain tissue.

Referring to the Lake Michigan ducks, the extension service material compared *only* the fat tissue of those birds with the whole body concentrations of DDTR in fish and amphoids. To be meaningful, comparisons between different animals in the food chain must be based on analyses of similar tissues. In most fish, fat tissues make up less than 5%

of the total body weight, so it is not valid to compare whole-body residues of fish with pure fat tissue residues in birds unless an adjustment is made to compensate for the 30-fold difference in residue levels expected between the two tissue types. Furthermore, Stickel, *et al*[11] documented as much as eight-fold increase in the pesticide concentration in bird eggs as they lost water while developing. Adjustments must be made to compensate for that dehydration factor in other tissue as well. The pesticide level in Hickey's dehydrated bird tissues should not be compared with concentrations in the non-dehydrated, whole-body, wet-weight fish samples which Hickey also analyzed.

Amphipods Compared With Ducks

Of all the birds analyzed in the Lake Michigan study, only the old squaw ducks actually belong in the food chain studied. Ellarson found that the amphipods formed 50 to 96% of the stomach contents of those ducks.[12] The old squaws contained only 2 to 6 ppm of DDTR in their muscle tissue, and after the necessary adjustments are made to compensate for dehydration and for variability between tissues, the pesticide concentration indicated for fresh, whole, wet-weight duck would be less than 1 ppm of DDT residues. From 0.41 ppm in the amphipods to the concentrations (adjusted) of about 0.5 ppm in the old squaw ducks at the top of the food chain, there was no significant increase in pesticide residue.

Herring Gulls Not In Food Chain

The extension service writer also attempted to include herring gulls in his "biomagnification" scenario, even though those four birds were not in the food chain which includes the amphipods, alewives and old squaw ducks. The levels of DDTR in the fat tissues of those gulls had little relationship to the aquatic food chain studied. The gulls were notorious garbage feeders in dumps around Green Bay cities. There they certainly ate large amounts of DDT residues. The gulls did not eat the amphipods or the ducks, but the extension service writer placed them at the top of that food chain anyway. He then implied that the difference between the DDTR level in the single lowest mud sample and the highest level imaginable in any gull fat tissue should be considered as "biological magnification up the food chain." In that manner he arrived at his fantastic estimate of a 370,000-fold increase up the food chain with a 500-fold difference between the mud and the lowest living creatures and a final irrelevant comparison between the duck *muscle* and the *fat* tissue of those four wandering herring gulls. His inflated estimate was obviously nonsensical.

Epilogue

Scientists and other professional people who are aware of the truth should forcefully refute the "biomagnification" propaganda at every opportunity. As long as news writers, teachers, legislators and laymen continue to be uninformed or misinformed about the facts, we cannot blame them much for being unreasonably concerned about traces of pesticides in the environment and for fearing that they might "build up" in the food chain until ultimately they could even harm people.

Whenever allegations of biological magnification are encountered the propagandists should be challenged to produce the actual data from analyses of each step of the actual food chain. They should also be required to specify which methods of analysis were used, which tissues were analyzed, how old the samples were and whether they were wet weight or dry weight. Of course, there must also be valid evidence that the animals really were involved primarily in the food chain under study.

When all of these factors are considered, it becomes evident that there is really little or no increase in pesticide concentration attributable to "biological magnification up the food chain."

References

1. Freed, V.H. 1968. Toxicology of pesticides in the environment. Annual progress report of Environmental Health Science Center, Oregon State University.
2. Hamelink, J.L. 1971. Trans. Amer. Fish. Soc. 109:207–214.
3. Moriarty, F. 1972. New Scientist. 53:708, pp. 594–596.
4. Hunt, E.G. and A.I. Bischoff. Calif. Fish & Game. 46:91–106.
5. Hickey, J.J. 1966. J. Applied Ecology. 3:supplement, pp. 141–154.
6. Davis, J.C. 1970. J. Fisheries Research Board, Canada. 27:204–214.
7. Chadwick, G.G. and R.W. Brockson, 1970. J. Fisheries Research Board, Canada, 1970. 27:1850.
8. Stickel, W.H. et al. 1965. U.S.F.W.S. Circular 226, p. 165.
9. Jefferies, D.J. and B.N. Davis. 1968. J. Wildlife Management. 32:441–456.
10. Anon. 1966. Trans. 31st N.A. Wildlife Conf. pp. 191–199.
11. Stickel, Lucille, et al. 1973. Bull. Environ. Contam. & Toxicology. 9:193–196.
12. Ellarson, 1956, PhD thesis, University of Wisconsin.

Reprinted, with permission, from
Agrichemical Age, April 1980.

Biotechnology

The Hazards of Biotechnology: Facts and Fancy
Thomas H. Jukes

Those Terrifying Cows
Reprinted with permission of The Wall Street Journal

The Hazards of Biotechnology: Facts and Fancy

Thomas H. Jukes, Ph.D.

Thomas H. Jukes is professor of Biophysics at the University of California in Berkeley. He received a Ph.D. in biochemistry from the University of Toronto in 1933.

Dr. Jukes is the author of *Molecules and Evolution* published in 1966; *Antibiotics in Nutrition* published in 1955; and *B Vitamins for Blood Formation* published in 1952. He has more than 400 journal articles published and has contributed to *Journal of Molecular Evolution, Journal of Nutrition,* and *Nature* and other leading scientific publications.

He received the D.Sc. honoris causa from the University of Guelph in 1972.

Introduction

Biotechnology is a broad term for putting biological processes to practical use. Its roots are in the distant past, such as the repeated discovery of fermentation to produce alcoholic beverages. Other familiar examples of its use are hybrid corn, artificial insemination and vaccines. Today, it is genetic engineering and especially recombinant DNA that are in the spotlight. An historian, Professor John Franklin at Duke University, said, speaking of the new DNA-based biology, 'I don't suppose there are any comparable events. You'd have to go back to the discovery of fire.'

How has the present discovery been greeted? What are its dangers, real or imaginary? I believe that imaginary rather than actual dangers are what has primarily influenced public perception and regulation of recombinant DNA technology.

As with all technologies, genetic engineering can unquestionably be misused: So can chemistry and microbiology. One of the greatest crimes in history, Hitler's holocaust, was perpetrated through inorganic chemistry—cyanide. Microbiology could be used for an evil purpose in germ warfare, but objections to microbiology as a science are not heard. Perhaps we owe some of this acceptance of microbiology to the genius of Louis Pasteur, not only as a scientist, but as a public figure. His words spoken 85 years ago are still timely:

'Young men, have confidence in those powerful and safe methods, of which we do not yet know all the secrets. And, whatever your career may be, do not let yourselves become tainted by a deprecating and barren

135

skepticism . . . Live until the time comes when you have the immense happiness of thinking that you have contributed in some way to the progress and to the good of humanity.'

Public apprehensive about genetic engineering includes questions about changes in human genes, and this is understandable. I read a newspaper report of 19 September 1987, in which seven people, ages 20 to 28, were asked in the street, 'Would you want a gene-altered baby?' Six gave a variety of thoughtful reasons for saying no; including one who said that 'A lot of individuality would become extinct.'

The main potential of recombinant DNA is not altering the genetic make-up of people, but in combating disease and hunger. The opportunities are very large. What should we do to assure that genetic engineering is undertaken only to benefit society?

First, let us consider hazard. For a hazard to be real, it must actually occur or exist. By this definition, the actual hazards of biotechnology are the regular risks in biological laboratory, including infections by known pathogens that are being used for genetic engineering. The US Commissioner of Food and Drugs said in 1987, 'There has not been a single significant safety problem during the more than ten years that new biotechnological techniques have been used in laboratories or applied by industry.'

In contrast, the imaginary hazards are limitless. Manufacture of them has become a minor industry that has led to the proliferation of regulatory agencies.

Protecting the Environment

My first-hand experience with environmentalist organizations goes back to when I became a life member of the Sierra Club in 1939. In the intervening time, environmental societies have metamorphosed from groups of amateurs into organizations that mimic big businesses. Typically they have staffs of full-time employees, fund-raising campaigns which are analogous to sales campaigns, and for which they need causes, corresponding to products, and, to stay in business, they need large annual budgets. As I pointed out some years ago, it would be fiscal lunacy for the Audubon Society not to campaign against pesticides. Fund-raising is carried out by asking members and sympathizers to contribute to campaigns against real or perceived dangers to the environment. Unfortunately, perceived dangers are often more easy to describe in glowing terms than are the real ones.

On previous occasions, environmentalists had started campaigns against opponents who defended themselves, such as oil companies and builders of nuclear energy plants. But with recombinant DNA, for the first time it was the practitioners who said what they were doing was dangerous. This self-condemnation took place in an organized way, at a large meeting at Asilomar, California, 1975. The environmentalists did not even have to concoct stories about dangers, the molecular biologists supplied them themselves. It did not matter that the stories were imaginary. So in one sense, the title of this symposium: *Hazards of Biotechnology—Real or Imaginary* should be qualified by pointing out that the effect of the imaginary hazards is what we must consider, because real hazards have not yet appeared.

The events following Asilomar have been exhaustively chronicled by Jim Watson and John Tooze in their volume *The DNA Story* 1981).[1] In this book, they have reproduced hundreds of documents, some describing efforts to hamper research on recombinant DNA, or even to stop it altogether, together with other documents chronicling the struggle of scientists against being kept from freedom to do experiments. Three of the most prominent environmentalist opponents were Friends of the Earth, Environmental Defense Fund and Natural Resources Defense Council. The Sierra Club board of directors passed a resolution calling for a halt to recombinant DNA work, but did not conduct a long campaign of harassment such as that mounted by Friends of the Earth and Environmental Defense Fund. Probably the most eloquent defender of science was James Watson; for example, he said, 'Recombinant DNA-induced diseases to me fall in the category of UFOs or witches ... We must make clear to the public that there is no more reason to fear recombinant DNA than there is to panic about the Loch Ness Monster.' He also voiced his discomfiture at the behavior of environmentalists:

'Until the last year, I never thought much about my allegiances ... People who went on bird trips or camped in the national forests and wanted to save Mineral King were the right sort, while those who owned big yachts or stripped the rolling fields of Ohio for coal were the bad guys whom we must get laws to stop ... Now, however, I [don't] respond to Robert Redford's latest appeal ... He must be unaware he and I are, for practical purposes, real enemies. For some of the money he raises for the Environmental Defense Fund is being used to try to stop the experiments we do with "recombinant DNA"...

... I most certainly am a Friend of DNA and want work with recombinant DNA to go as fast as possible. In the old days, the impulse would generally be viewed as good for the earth. Now, however, there exist highly vocal groups who think I'm a danger to the world. The Friends of the earth, the Sierra Club and the Natural Resources Defense Council, as well as the Environmental Defense Fund, all say that our experiments ... must be constrained by their new breed of environmental lawyers.'[1]

Eventually, the successes of biotechnology, especially the commercial companies, together with the complete absence of measurable harmfulness turned the tide and many of the public became more interested in the stock market prices of biotechnology companies than in the environmentalist propaganda against genetic engineering. This does not mean that the opponents have given up, or have become ineffective. My own participation in the controversy came rather late. At first, I regarded the attacks on recombinant DNA, and especially the intricate, complicated and unnecessary regulations as a lot of nonsense, and I steered clear of the issues, except for having to make an official declaration that no recombinant DNA work was going on in my laboratory. I was precipitated into the argument because I work in a University of California building, of which one floor is leased to a biotechnology company, Advanced Genetic Sciences, who are developing an iceminus strain of *Pseudomonas* under the name Frostban. On 15 January 1986 I found a large crowd of demonstrators at the front door. They were making various nonsensical statements, and one of them carried a placard saying, 'We don't want no designer genes'. Most importantly, a large group of reporters and television cameras was recording what they said. I tried to voice some rebuttal, but I was ignored except for one gesture that I made. A similar performance by protesters took place a year later, and was picked up by four television networks. Needless to say, in the intervening 12 months, no injurious effects of a tangible nature had been produced by Frostban bacteria. The issue raised by the opponents was entirely mythical, they claimed that external release of a strain of *Pseudomonas* with one gene removed would present serious dangers to the environment and to human beings. All their claims were completely unsubstantiated, and did not permit of substantiation, but this has not stopped them.

On 1 December 1987, the environmental terrorist organization 'Earth First' announced:

> 'Eco-saboteurs broke into the AGS test site in the early hours of Monday, 30 November, and used salt, a herbicide and ammonia to damage strawberry plants scheduled for treatment with the genetically engineered "Frostban" bacteria. This is the third such test of genetically engineered bacteria in California and the third of the tests to be sabotaged.'

The Media

Transmission of scientific information to the public by the media is governed primarily by consideration of whether it will sell newspapers or attract television viewers. We have one small edge: many of the public are interested in science. The same is not true of reporters and commentators. Usually their eyes become glazed when one attempts to explain something scientific to them. When the Sierra Club was trying successfully to drum up votes for Proposition 65 in the 1986 California elections, their advisors pointed out that scientists were boring, and advised that movie stars would be the best vote-getters. As a result, aided by Jane Fonda and her fellow thespians, Proposition 65 passed by a large majority and the state is faced with the burden of enforcing it. It may well be that all female vertebrates in California are living an illegal existence because of their content of estradiol, which is officially classified as a carcinogen.

Yet, in spite of all this, a survey conducted by the U.S. Congress Office of Technology Assessment showed that a majority of individuals questioned believed that the risks associated with genetic engineering have been greatly exaggerated, and 12% believe that small-scale field tests of genetically engineered organisms should be permitted. Less than 43% find the news media credible. However, 65% said they would believe an environmental group rather than the government in a situation in which the Federal Government claims there is no risk from a biotechnology release. The survey revealed a great need for reliable public information to be disseminated. To achieve this, we must overcome the misinformation that is being spread. We can expect very little help from regulatory agencies, because they usually are anxious to protect themselves from criticism by articulate activists.

The Debate, 1974 and Subsequently

Many of the participants, especially J.D. Watson, immediately realized the predicament into which their research had been precipitated by Asilomar, and a period of recrimination and retrenchment followed. As I noted above, most molecular biologists were sympathetic with the environmentalist movement, not realizing that environmentalists regard technology based on science as one of their principal enemies. Utilization of scientific discoveries tends to increase the production of energy and of food, and the prevention of disease, all of which tend to accelerate the population explosion against which environmentalists struggle. For example, Lynton Caldwell, the author of the National Environmental Policy Act has been quoted as saying that the population of the United States should preferably be reduced to 100 million. Leading environmentalist organizations such as Friends of the Earth are too sophisticated to oppose recombinant DNA head-on; instead they set up organizations to find fault with the details of guidelines and laboratory procedures, thus mounting a very effective delaying action.

Although environmentalists have campaigned vociferously for 25 years against chemical pesticides, they nevertheless also oppose a new technology, recombinant DNA, a means of replacing chemical pesticides. Because it is scientific; whatever it is, they're against it.

Biotechnology and Doomsday

One of the leaders of the doomsday group is George Wald, whose visibility is greatly enhanced by his Novel Prize. According to George, the end of the world is close at hand, principally because of American capitalism. In a 1975 article, 'There Isn't Much Time,' Wald found it difficult to 'see how the human race will get itself much past the year 2000.' Towards the end of this article, he updated doomsday to 1985. He blamed the forthcoming disaster on Western Christian society. Next year, 1976, he attacked recombinant DNA technology (*The Sciences* 16(6) 1976), saying that it would place in human hands the capacity to redesign living organisms, the product of three billion years of evolution, and the results would be 'essentially new organisms, self-perpetuating and hence permanent . . . Once created, they cannot be recalled.' Since Wald estimates that life on earth may end before the year 2000, the new organisms will have only a short time for enjoyment of their newly found permanence.

Another prominent critic who opposes recombinant DNA is Barry Commoner, who warned a biotechnology meeting in California on 3 November 1987 of its dangers, primarily because it is a product of the American economic system.

Outstanding among the apocalyptics in his flair for picturesque vituperation is Erwin Chargaff, famed both for his early work on nucleotide ratios in DNA and for his distaste for all subsequent work by others with it. In 1987, Chargaff foresaw the emergence of a new branch of biotechnology: 'a gigantic slaughterhouse, a molecular Auschwitz, in which valuable enzymes, hormones and so on will be extracted instead of gold teeth.' It appears, however, that the biotechnological victims of Chargaff's molecular Auschwitz are, and will be, bacteria that have been trained to produce enzymes and hormones, and do not possess gold teeth. Chargaff shows originality in comparing biotechnology with The Holocaust rather than with nuclear war.

Environmental Release: *Pseudomonas*

By 1981, opposition to recombinant DNA had died down to the point where Watson and Tooze were able to say in *The DNA Story,*[1] 'Politics and politicking preoccupied the first years of the recombinant DNA story, but that phase, fortunately, is fast becoming history. This book is our epitaph to that extraordinary episode in the story of modern biology.' The episode to which they refer is the eight years during which recombinant DNA research was placed under constraints that were crippling. Watson and Tooze concluded that the proliferation and success of new biotechnology companies showed that the battle for recombinant DNA had been won. Four years later, in *The Gene Splicing Wars*,[2] Norton Zinder said that protests had dwindled to the point where they were inconsequential. These authors had reckoned without realizing the resourcefulness of Mr. Jeremy Rifkin, who selected as his next target the release of genetically engineered bacteria into the open environment.

The public has for many years been aware of the science fiction concept of a 'Green Plague' or 'Andromeda Strain' of deadly microorganisms raining down from outer space, usually enveloping and destroying the biosphere, an idea that has been successfully exploited by writers who know that people will buy a book that makes their flesh creep. This vision set the stage for Rifkin and his followers to affix the pestilence label to one of the more harmless and widely distributed of bacterial genera, *Pseudomonas,* species of which commonly occur in soil, drinking water and food. It is true that some strains of *Pseudomonas* are pathogenic and can produce, for example, urinary tract infections, but this is also true of *Esherichia coli*. It is also true that the ice-minus strains of *Pseudomonas syringae* and *Pseudomonas fluorescens* that were proposed for use in preventing frost injury and had been thoroughly tested with animals and found to be safe and that similar strains, also ice-minus, were found freely living in the natural environment. Moreover, the genetic complement of the ice-minus bacteria had been diminished, rather than increased, by removing the bacterial ice nucleation gene. Ice-plus strains, called 'Sno-max' are approved for use in snow-making machines on ski slopes. The activists who opposed use of ice-minus bacteria resorted to pure fantasy, saying that the modified bacteria may 'decrease rainfall', and that an experiment with them 'is very similar to the first atmospheric tests of the atom bomb'. The magic words were 'genetically engineered', which transmogrified ordinary microorganisms into dangerous ones. In 1987, Rifkin boasted, 'Our opposition to the first few trials with frost resistant genetically altered bacteria halted a university and a company for four years'. When a field test of frost-minus *Pseudomonas* was finally approved in 1987, a pregnant woman told a reporter that she and her family had abandoned their home four miles from the test site and planned to live in a hotel until their money ran out. The Environmental Protection Agency (EPA) did its best to see that these fears were reinforced. The EPA decided that the test of ice-minus bacteria should be governed by their code book for testing new pesticides, using the tortured reasoning that frost was a pest. (One could ask, 'Is drought a pest?') Accordingly, the bacterial culture had to be applied to plants by someone wearing what has been termed a 'moon suit'. But, on 25 April 1987, Dr. Julie Lindemann put on her moon suit as required by law and duly sprayed *Pseudomonas* on some strawberry plants. No protective clothing was worn by the dozens of reporters and cameramen who were standing nearby. The picture, which went around the world, was sufficient to convince practically everybody who say it that ice-minus bacteria must be quite deadly, or why should the person dispensing them be enveloped in a moon suit and wear goggles and a mask?

Six weeks later, it became evident that the EPA did not believe in the precautions on which they had insisted. Miraculously, ice-minus bacteria had become harmless. Clad in blue jeans and a short-sleeved sports shirt and unprotected. Dr. Lindemann was photographed on 4 June when she dug up the sprayed plants to take them back to the laboratory.

The next episode involved the same bacterial species, *Pseudomonas syringae*, in an attempt in Montana to protect trees from Dutch Elm Disease. The wild type bacterium produces an antifungal agent that inhibits the fungus that causes Dutch Elm Disease. Professor Barry Strobel used plasmid conjugation, rather than recombinant DNA, in an attempt to increase the effectiveness of a strain of *Pseudomonas*. Strobel injected his organism into elm trees without asking for approval by the EPA. Apparently, he wanted to save a year of postponing his tests. The combined wrath of regulatory bureaucracy fell upon him, and investigations were started by four agencies. The fact that he had injected trees with pseudomonads ensured a reaction, because the same crime against trees had been committed in California in 1986, resulting in a national uproar. His colleagues at Montana State University joined sides with the regulators. *The Wall Street Journal sarcastically* commented, 'We should have known that any regulatory power called the Environmental Protection Agency would some day feel compelled by the logic of its laws to burn elm trees ... It's time to choose sides: one may either join the lawyers and the biosafety committee as they lay torch to these trees, or one may join what few researchers are willing to save the Strobel elms and the self-respect of American science'. Strobel cut down his trees himself without waiting for the torch of the regulators.

Evolution and Recombinant DNA

Robert Sinsheimer, erudite, eminent and eloquent, is a conspicuous critic of genetic engineering, which he feels may recapitulate the horrors of the atomic bomb. He says, 'Ideally, I would like to see such research and development done on the Moon'. One of his main concerns in 1977 was the possible emergence of new, self-propagating organisms that once released 'will be with us, potentially, forever', and he advances 'the concept that science itself—the simple pursuit of truth, the exploration of Nature—could in itself be dangerous . . . to the entire planet.' In a US Senate hearing in 1976, he was asked, 'Are you saying that . . . we now have the power to change the evolutionary process? He replied 'Yes'.

I maintain that his fears of new organisms are rooted in the Andromeda Strain fiction. His imaginary danger is contrary to what we know about evolution. There are many examples of rapid spread of species introduced to new environments, such as rabbits in Australia, and starlings and gypsy moths in the USA. These organisms are products of long evolution; rabbits are 80 million years away from the common mammalian ancestor. During the ensuing period they have made countless genetic adjustments to living in a complex environment, so that they are able to take advantage of the new, favorable ecological niche in Australia, and also to survive elsewhere as a species, when times become adverse. We cannot expect such toughness from new genetically engineered organisms. We do not even expect it from our domesticated animals and plants, for they persist only as long as we protect them.

The three major sources of the world's food are rice, maize and wheat. All of them, just like domestic animals, are products of genetic selection by human beings. None of these grains has escaped from cultivation to spread in the wild, even though they have had every opportunity to do so. Maize, especially, is maintained only by careful nurture; it cannot even spread its own seeds. Thousands of other genetically manipulated organisms have been introduced without causing problems.

Selected strains of nitrogen-fixing bacteria have been used in agriculture for many decades without spreading. Fears that ice-minus bacteria would spread when released were shown to be groundless. One of the most axiomatic principles of evolution is that most mutations are deleterious changes and are eliminated by nature. Adaptive mutations are very rare and, to be adaptive, they must fit into the general makeup of the organism. I do not see genetically engineered species 'taking over'. The record with classical methods of genetics leads me to expect that new strains that escape from the laboratory will find themselves the victims of merciless and experienced competitors which have been successful in the struggle for existence for million of years.

Bernard Davis points out that the fuss about releasing *Pseudomonas* had several causes: first, activism and obstruction by Jerry Rifkin and the German Greens; next, regulation by uninformed bureaucrats, and third, opposition by ecologists in spite of recommendation by microbiologists. The regulations are based on several unwarranted assumptions: that changing genes in common saprophytic bacteria makes them dangerous, that planned release may be catastrophic, that recombination between distantly related organisms increases the dangers, and that because certain introduced species, such as chestnut blight and gypsy moth have become pests, engineered bacteria are likely to do the same.

About a million bacteria are normally present per gram of soil, and are constantly mutating. Pathogenicity is the only real cause for concern, and this depends on evolution and co-adaptation. A committee of the US National Academy of Sciences pointed out in August 1987 that 'assessment of the risk of introducing DNA-engineered organisms into the environment should be based on the nature of the organisms and the environment into which it will be introduced'.

Dr. Frank Young, US Commissioner of Food and Drugs, formerly worked with *Bacillus subtilis*. He said in 1987:

> 'One thing that brought me to FDA was the opportunity to leave the research lab and struggle with how to bring the fruits of new biotechnology into the marketplace. I had been involved my whole career in microbiology and genetics and was involved in the exciting research that led to a revolution in cloning techniques... There should not be any scientific controversy over field trials of the recombinant DNA-manipulated ice-minus *Pseudomonas*.'

References

1. Watson, J.D. & Tooze, J. *The DNA Story*, W.H. Freeman & Co., San Francisco (1981).
2. Zilinskas, R.A. & Zimmerman, B.K. (eds) *The Gene Splicing Wars*, Macmillan, New York (1986).

Abridged and reprinted, with permission, from
J. Chem. Tech. Biotechnol.,
43, 245–255, 1988.

Those Terrifying Cows

Editorial, Wall Street Journal,
January 7, 1991

No modern advance is more vulnerable to damaging public assault today than agricultural biotechnology. It promises to produce a more bountiful, cheaper food supply. But for years the promise has had to confront demagogic scaremongering about the science itself, which in turn frightens consumers, which in turn causes not-very-courageous supermarket executives to repudiate the new technology. Consider the story of BST.

Since 1985, many dairy farmers have been hoping that the Food and Drug Administration would approve the use of a genetically engineered version of the growth hormone that cows secrete naturally. This hormone is bovine somatotropin, or BST, and four U.S. companies have been trying to bring it to market. When given to cows, they produce up to 25% more milk on the same amount of feed.

The FDA determined as far back as 1985 that milk from cows treated with BST is safe for human consumption. And the productivity burst from wide use of BST holds out the hope for lower prices for one of man's most basic foods. This sort of extraordinary innovation is of course one way to make life easier for people living on tight budgets, but the opposition won't stand for it.

In April, Wisconsin's Governor Tommy Thompson signed legislation to bar the use of BST there until June 1, 1991. Minnesota has enacted a similar ban. A moratorium, Governor Thompson explained, would provide "sufficient time to allow for additional farmer and consumer education on the use of" BST. Unfortunately, it is also providing time for some miseducation.

At the forefront of the anti-BST movement is free-media specialist Jeremy Rifkin, filing petitions with the FDA and raising the possibility of time-consuming litigation. Ben & Jerry's ice cream, serving a market that can afford to pay extra to feel better about living well, has already announced that it won't attach its name to dairy products made of milk from BST-treated cows. Last month, Consumers Union, which feels obliged to launch a left-wing policy offensive periodically, objected to BST. Five major grocery chains have announced their unwillingness to handle BST milk so long as the controversy lasts.

Opponents of BST make two arguments: that cheaper milk will force smaller dairy farms to close, and that BST renders cows more susceptible to infection and thus endangers the safety of the country's milk supply. Both claims are false.

Competition dooms not the smallest producers but the least efficient producers. In the dairy business, the most efficient producers are also often among the smallest. As an official from one of the companies that makes BST told a congressional subcommittee in 1986: "The average size of the highest yielding herds (top 20%) in the state of New York is approximately 60. The average size of the lowest yielding herds (bottom 20%) is also 60—exactly the same." The effect of BST is to cut all dairy farmers' fodder costs—the saving is proportionally the same for a big farm as for a small one.

Worries over the safety of the milk supply are a more serious issue. Happily, a panel of doctors and scientists convened by the National Institutes of Health unanimously concluded on December 7 that milk from BST cows is safe for human consumption. The panel confirmed the observation made by the deputy director of the FDA's Center for Veterinary Medicine in December 1989 that "BST is one of the safest products we've ever looked at."

So if BST is so great, why would anyone want to keep it off the market? BST is the first big agricultural biotechnology breakthrough. If the fear-of-science movement can kill so safe and beneficial a product, they—and the biotechnology companies—will know that they can stop other products.

For biotech's opponents, this is D-Day: If BST gets past them, the way is clear for innovation after innovation. There can be little doubt that this would be tough on the least efficient farmers, who would prefer the status quo. For the more politicized opponents, the game is undoubtedly about power. Once the large corporations bringing BST to market have proved that genetic engineering is a safe technology, government regulators will spend less time with the subject. That in turn diminishes whatever leverage the "public-interest" groups have been able to accumulate over this area of the country's economic life.

In 1950, before artificial insemination and new feeding methods became common on American dairy farms, the average cow yielded some 5,314 pounds of milk a year. Today, the average cow produces more than 14,300. That is the achievement of scientific farming. Better achievements are still to come—if the public and policymakers are willing to stand up to the scaremongers.

HARVARD UNIVERSITY
SCHOOL OF PUBLIC HEALTH

DEPARTMENT OF NUTRITION
665 HUNTINGTON AVENUE
TEL. (617) 432-1154
BOSTON, MASSACHUSETTS 02115-9915
CABLE ADDRESS: NUTHARV, BOSTON
U.S.A.

January 9, 1991

Letter to the Editor
THE WALL STREET JOURNAL
200 Liberty Street
New York, New York 10281

Congratulations for standing up to the scaremongers in your excellent editorial titled, "Those Terrifying Cows," issue of January 7. I wish more of the media would have the courage to do so, not only for milk from cows treated with BST (bovine somatotropin) but also to the presumed risks to human health from minute amounts of pesticide, insecticide, and other "chemical" residues on many of our foods.

Fredrick J. Stare, Ph.D., M.D.
Professor of Nutrition Emeritus
Founder, Harvard's Department
of Nutrition

FJS:jca

Cancer/ Carcinogenesis

Environmental Pollution and Cancer: Some Misconceptions

Bruce N. Ames, Ph.D. and Lois Swirsky Gold, Ph.D.

Dr. Bruce N. Ames is professor, Biochemistry and Molecular Biology at the University of California, Berkeley; and director, NIEHS Environmental Health Sciences Center.

Dr. Ames received a B.A. in Chemistry from Cornell University; and a Ph.D. in Biochemistry from California Institute of Technology. His 290 scientific publications resulted in his being the 23rd most-cited scientist (in all fields) 1973–1984. He has been elected to the National Academy of Sciences and the Royal Swedish Academy of Sciences. He has received numerous awards including the G.M. Cancer Research Foundation Prize, the Tyler Foundation Prize in Environmental Achievement, and the Gold Medal of the American Institute of Chemists.

Dr. Lois Swirsky Gold is director, Carcinogenic Potency Project, Cell and Molecular Biology Division, Lawrence Berkeley Laboratory; and specialist, Molecular and Cell Biology Division, University of California, Berkeley. She is a past lecturer, School of Public Policy, University of California, Berkeley; and senior fellow, Carnegie Commission on the Future of Higher Education.

Dr. Gold received a B.A. from Goucher College in 1963 and a Ph.D. from Stanford University in 1968. She has published more than 35 journal articles. She is the recipient of the Phi Beta Kappa Award.

The public has numerous misconceptions about the relationship between environmental pollution and human cancer. Underlying these misconceptions is an erroneous belief that nature is benign. Below we highlight 8 of these misconceptions and describe the scientific information that undermines each one.

Misconception No. 1: Cancer Rates Are Soaring

Cancer death rates in the United States (after adjusting the rates for age and smoking) are staying steady or decreasing. According to the latest update from the National Cancer Institute (Feb. 1988), "the age adjusted mortality rate for all cancers combined except lung cancer has been declining since 1950 for all individual age groups except 85 and above." (That represents a 13-percent decrease overall, 44,000 deaths below expected, and a 0.1-percent increase in the over-85 group.)

The types of cancer deaths that have been decreasing during this period are primarily stomach (by 75 percent, 37,000 deaths below expected), cervical (by 73 percent, 11,000 deaths below expected), uterine (by 60 percent, 9,000 deaths below expected), and rectal (by 65 percent, 13,000 deaths below expected). The types of cancer deaths that are increasing are primarily lung cancer (by 247 percent, 91,000 deaths above expected), which is due to smoking (as is 30 percent of all U.S. cancer deaths), and non-Hodgkin's lymphoma (by 100 percent, 8,000 deaths above expected).

Changes in incidence rates and effects of treatment are also relevant in interpreting the changes in mortality rates. Incidence rates have been increasing for some types of cancer. Sir Richard Doll and Richard Peto of Oxford University, two of the world's leading epidemiologists, in their definitive study on cancer trends point out that although incidence rates are of interest, they should not be taken in isolation, because trends in the recorded incidence rates are biased by improvements in the level of registration and diagnosis. Even if particular types of cancer can be shown to be increasing or decreasing, establishing a causal relation among the many changing aspects of our lives is difficult. There is no persuasive evidence that life in the modern industrial world has in general contributed to cancer deaths.

Cancer is fundamentally a degenerative disease of old age, although exogenous factors can increase cancer rates (e.g. cigarette smoking in humans) or decrease them (e.g. caloric restriction in rodents). For mammalian species, cumulative cancer risk increases with approximately the fourth power of age, both in short-lived species such as rats and mice (about 30% have cancer by the end of their 2-year life span) and in long-lived species such as humans (about 30% have cancer by the end of their 85 year life span).

Life expectancy is steadily increasing in the United States and other industrial countries. Infant mortality is decreasing. Although the statistics are less adequate on birth defects, there is no evidence that they are increasing. Conclusion: Americans are healthier now than they have been in their history.

Misconception No. 2:
Cancer Risks to Humans Can Be Assessed by Testing Chemicals at High Doses in Rodents.

Results from animal cancer tests, which are conducted at near toxic doses of the test chemical, cannot predict the cancer risk to humans at the usually low levels of human exposures. Knowledge of the mechanisms of carcinogenesis is necessary for prediction, and is now progressing rapidly. Evidence is accumulating that it may be the high dose itself, rather than the chemical *per se*, that is the risk factor for cancer. High doses of chemicals—whether natural or synthetic—cause chronic wounding of tissues, cell death, and consequent chronic cell division of neighboring cells, which is a risk factor for cancer. At the low exposures to which humans are normally exposed, such increased cell division does not occur. So the very low levels of chemicals to which humans are exposed through water pollution or synthetic pesticide residues are likely to pose little or no cancer risks.

Misconception No. 3:
Most Carcinogens and Other Toxins
are Synthetic

About 99.99% of all pesticides in the human diet are natural pesticides from plants. All plants produce toxins to protect themselves against fungi, insects, and animal predators such as man. Tens of thousands of these natural pesticides have been discovered, and every species of plant contains its own set of different toxins, usually a few dozen. When plants are stressed or damaged, e.g., during a pest attack, they increase their levels of natural pesticides manyfold, occasionally to levels that are acutely toxic to humans. We estimate that Americans eat about 1,500 mg per person per day of natural pesticides, which is 10,000 times more than they eat of synthetic pesticide residues. The concentration of natural pesticides is usually measured in parts per million (ppm), rather than parts per billion (ppb), which is the usual concentration of synthetic pesticide residues or of water pollutants. We also estimate that a person ingests annually about 5,000 to 10,000 different natural pesticides and their breakdown products. Lima beans contain a different array of 23 natural toxins that, in stressed plants, range in concentration from 0.2 to 33 parts per thousand fresh weight: none appears to have been tested for carcinogenicity or teratogenicity (the ability to cause birth defects). A large literature has examined the toxicity of many of these compounds to herbivorous animals, such as humans and domestic animals.

Surprisingly few plant toxins have been tested in animal cancer tests, but among those tested in at least one species, about half (27/52) are carcinogenic. A search in plant foods for the presence of just these 27 natural-pesticide rodent carcinogens indicates that they occur naturally in the following foods (those at concentrations greater than 10,000 ppb of a single carcinogen are listed in italics): *anise, apple,* banana, *basil,* broccoli, *Brussels sprouts, cabbage,* cantaloupe, *caraway, carrot, cauliflower, celery, cherry,* cinnamon, cloves, cocoa, *coffee (brewed), comfrey tea, dill, eggplant, endive, fennel, grapefruit juice, grape, honey,* honeydew melon, *horseradish,* kale, *lettuce, mace, mango, mushroom, mustard (brown), nutmeg, orange juice, parsley, parsnip,* peach, *pear, pepper (black),* pineapple, *plum, potato,* radish, raspberry, *rosemary, sage, sesame seeds (heated),* strawberry, *tarragon, thyme,* and turnip.

Thus, it is probable that almost every plant product in the supermarket contains natural carcinogens. The ppm levels of the known natural carcinogens in the above plants are commonly thousands of times higher than the ppb levels of manmade pesticides. The occurrence in the diet of natural pesticides that are rodent carcinogens should be interpreted cautiously. We need not be alarmed by the presence of low doses of synthetic toxins and a plethora of natural toxins in our food. As will be discussed below, humans are well protected against low doses of toxins by many layers of inducible, general defenses that do not distinguish between synthetic and natural toxins.

Dietary exposures to natural toxins are not necessarily of much relevance to human cancer. Indeed, a diet rich in fruit and vegetables is associated with lower cancer rates. This may be because anticarcinogenic vitamins and antioxidants come from plants. What is important in our analysis is that chronic exposures to natural rodent carcinogens may cast doubt on the relevance of far lower levels of exposures to synthetic rodent carcinogens.

Teratogens and clastogens are common. It is also reasonable to assume that a sizable percentage of both natural and synthetic chemicals will be reproductive toxins at high doses because a high proportion of positives is reported for rodent teratogenicity tests (tests to determine the potential to cause reproductive damage). One-third of the 2,800 chemicals tested in laboratory animals have been shown to cause reproductive damage in the standard, high dose protocol.

Results from other types of tests also indicate that the natural world should not be ignored, and that positive results are commonly observed in high-dose tests. Ishidate reviewed experiments on the clastogenicity (chromosome breakage) of 951 chemicals in mammalian cell cultures. (A clastogen is one type of mutagen.) Of these 951 chemicals, we identified 72 as natural plant pesticides. Among these 72, 35 (48%) were positive for clastogenicity in some or all tests. This is similar to the results of the remaining 879 chemicals; 467 (53%) were positive in some or all tests. Thus, about half of the chemicals tested—whether synthetic or natural—have been shown to break chromosomes at some dose. These *in vitro* experiments do not necessarily simulate *in vivo* conditions, and chromosome breakage is probably much less extensive in tissues of the body than in laboratory tissue cultures.

Cooking food. The cooking of food is also a major dietary source of potential rodent carcinogens. Cooking produces about 2000 mg per person per day of mostly untested burnt material that contains many rodent carcinogens. Roasted coffee, for example, is known to contain over 800 volatile chemicals. Only 21 have been tested, and 16 are rodent carcinogens that total at least 9 mg/cup (40,000 ppb). (There is some, but not sufficient evidence to conclude that coffee causes cancer in humans.) When proteins or amino acids are heated, certain mutagens known as heterocyclic amines are sometimes produced.

The total amount of browned and burnt material consumed per person in a typical day is at least several hundred times more than that inhaled in a day from severe outdoor air pollution. Three mutagenic nitropyrenes present in diesel exhaust have now been shown to be rodent carcinogens, but the intake of these carcinogenic nitropyrenes has been estimated to be much higher from grilled chicken than from air pollution. Gas flames generate NO_2, which can form both carcinogenic nitropyrenes and nitrosamines in foods that are cooked in gas ovens. Food cooked in gas ovens may be a major source of dietary nitropyrenes and nitrosamines.

Residues of manmade pesticides. By contrast, human exposures to manmade pesticide residues are minuscule. The Food and Drug Administration assayed food residues of the 200 synthetic compound, thought to be of greatest importance, including most synthetic pesticides and a few industrial chemicals. The FDA estimates that the intake of these residues averages about 0.09 mg per person per day. For comparison, we estimate that the intake of natural pesticides averages about 1500 mg per person per day. About half of the intake of synthetic residues is composed of four chemicals (ethylhexyl diphenyl phosphate, dicloran, malathion, and chlorpropham) that were not carcinogenic in rodent tests. Thus, the intake of carcinogens from synthetic residues (0.05 mg a day, if one assumes that all the other residues are carcinogenic, which is unlikely) is extremely tiny relative to the background of natural substances; this 0.05 mg intake is equivalent to about 60 ppb of synthetic residues in plant food consumed daily.

Misconception No. 4:
Synthetic Toxins Pose Greater
Carcinogenic Risks than Natural Toxins.

The possible carcinogenic hazards from synthetic pesticides (at normal exposures) are minimal compared with the background hazards of nature's pesticides. Even though the overwhelming number of the chemicals that humans eat are natural, the natural world of chemicals has never been tested systematically. Synthetic chemicals account for 350 (82%) of the 427 chemicals tested chronically at high doses in both rats and mice. Of the 77 natural chemicals tested, the proportion carcinogenic is about half (37/77), i.e., similar to that of synthetic chemicals (212/350). It is unlikely that the high proportion of carcinogens in rodent studies is due simply to selection of suspicious chemical structures: while some synthetic or natural chemicals were selected precisely because of suspect structures, most chemicals were selected because they were widely used industrially, e.g., they were high-volume chemicals, pesticides, drugs, dyes, or food additives. The natural world of chemicals has never been looked at systematically.

In recent years, we have tried to formulate a method of setting priorities among possible carcinogenic hazards. The potencies of different carcinogens vary more than ten-million fold in rodent tests, and the comparison of possible hazards from various carcinogens ingested by humans must take this into account. We have analyzed animal cancer tests from our Carcinogenic Potency Database, and, for each chemical, have calculated the TD_{50} (Tumorigenic Dose 50), which is essentially the daily dose of the chemical estimated to give half of the animals tumors. We have constructed an index to rank possible carcinogenic hazards: first, we estimate a reasonable daily lifetime human exposure to each chemical, and express that as milligrams (of the chemical) per kilogram of body weight. Then that mg/kg human exposure is expressed as a percentage of the rodent TD_{50} dose (mg/kg) for each carcinogen. We call this percentage HERP—Human Exposure dose/Rodent Potency dose. Because rodent data are all calculated on the basis of lifetime exposure at the indicated daily dose rates, the human exposure data are similarly expressed as lifelong daily exposure rates, even though the human exposure is likely to be less than daily for a lifetime.

The HERP values do not estimate human risk directly, because extrapolation to low doses are inherently inaccurate, but they do offer a way of comparing possible hazards, and thus of putting exposures into a relative context so that priorities can be more reasonably set. (Carcinogens clearly do not all work in the same way, and as we learn more about mechanisms, HERP comparisons can be refined, as can risk assessments.) Our results suggest that alcohol at moderate doses should be high on our priority list for epidemiological studies on cancer. The HERP analysis further suggests that the possible carcinogenic hazard of synthetic chemicals that humans ingest from pesticide residues or water pollution appears to be trivial relative to the background of carcinogenic hazards from natural chemicals and chemicals formed by cooking food.

Alar. To put the carcinogenic hazard of Alar in perspective, we estimate that the possible hazard from UDMH (due to a carcinogenic hydrazine, the breakdown product of Alar) in a daily glass (6 oz.) of apple juice for life is more than ten times lower than the possible hazard from the naturally-occurring carcinogenic hydrazines consumed in a daily mushroom or from the aflatoxin in a daily peanut butter sandwich. It is also lower than the possible hazards from the carcinogenic nitrosamines in a daily portion (100 g) of bacon. The possible hazard of UDMH in a daily apple is 1/10 that of a daily glass of apple juice. Apple juice has been reported to contain 137 natural volatile chemicals, of which only 5 have been tested for carcinogenicity; 3 of these—benzyl acetate, ethanol, and acetaldehyde—have been found to be carcinogenic in animals.

Water pollution. The possible hazards form carcinogens in contaminated well water in places like California's Santa Clara ("Silicon") Valley or Woburn, Massachusetts should be compared with the possible hazards of ordinary tap water. Of the 35 wells that were shut down in Santa Clara Valley because of a supposed carcinogenic hazard to humans (low traces of trichloroethylene), only two were of a possible hazard greater than ordinary tap water. Well water is not usually chlorinated and therefore lacks the 83 ppb chloroform present in average chlorinated tap water in the U.S. Water from the most polluted well in the Santa Clara Valley had a relative hazard that was orders of magnitude less than that for an equal volume of coffee, beer or wine. The consumption of tap water is only about one or two liters per day, and animal evidence provides no good reason to expect that either the chloroform produced in water by chlorination or the current levels of synthetic pollutants in water would pose a significant carcinogenic hazard. Natural arsenic appears to be the most significant carcinogen in both well water and tap water, and is often present at quite high levels. Arsenic is a known human carcinogen.

The trace amounts of chemicals found in polluted wells are likely to be a negligible cause of birth defects, in comparison to the background level of known teratogens such as alcohol. The important risk factors for birth defects and reproductive damage in humans are the age of the mother, her consumption of alcohol, her smoking habits, and her exposure to the rubella virus.

TCDD (dioxin) compared with broccoli and alcohol. Cabbage and broccoli contain a chemical whose breakdown products appear to act on the body like dioxin (TCDD)—one of the most feared industrial contaminants. If we compare the potential carcinogenic and teratogenic hazards of broccoli and dioxin, we estimate that eating a portion of broccoli daily poses a possible hazard one thousand times that of being exposed to the EPA's allowable dose of dioxin. It seems likely that many more of these natural "dioxin simulators" will be discovered in the future.

If TCDD is compared with alcohol it seems of minor interest as a teratogen or carcinogen. Alcohol is the most important known human chemical teratogen. In contrast, there is no persuasive evidence that TCDD is either carcinogenic or teratogenic in humans, although it is both at near-toxic doses in rodents. If one compares the teratogenic potential of TCDD to that of alcohol for causing birth defects (after adjusting for their respective potency as determined in rodent tests), then a daily consumption of the EPA reference dose of TCDD (6 fg) would be equivalent in teratogenic potential to a daily consumption of alcohol from 1/3,000,000 of a beer. That is equivalent to drinking a single beer (15 g ethyl alcohol) over a period of 8,000 years.

In humans alcoholic beverages are carcinogenic as well as teratogenic. A comparison of the rodent carcinogenic potential of TCDD with that of alcohol (adjusting for the potency in rodents) shows that ingesting the TCDD reference dose of 6 fg per kilogram per day is equivalent to a man ingesting one beer every 345 years. Since the average consumption of alcohol in the United States is equivalent to more than one beer per person/day, and since 5 drinks a day are a carcinogenic risk in humans, the experimental evidence does not of itself seem to justify the great concern over TCDD at levels in the range of the reference dose.

Misconception No. 5:
The Toxicology of Manmade Chemicals
is Different from That of Natural Chemicals.

It is often assumed that, because plants are part of human evolutionary history, whereas industrial chemicals are not, the mechanisms that animals have evolved to cope with the toxicity of natural chemicals will succeed in protecting them against natural chemicals, yet will fail to protect against synthetic chemicals: "For the first time in the history of the world, every human being is now subjected to contact with dangerous chemicals, from the moment of conception until death." Rachel Carson, *Silent Spring*, 1962. We find this assumption flawed for several reasons.

Defenses that animals have evolved are mostly of a general type, as might be expected, since the number of natural chemicals that might have toxic effects is so large. General defenses offer protection not only against natural but also against synthetic chemicals, making humans well buffered against toxins. These defenses include the following: (a)

The continuous shedding of cells exposed to toxins: the surface layers of the mouth, esophagus, stomach, intestine, colon, skin, and lungs are discarded every few days. (b) The induction of a wide variety of general detoxifying enzymes, such as antioxidant enzymes of the glutathione transferases for detoxifying alkylating agents: human cells that are exposed to small doses of an oxidant, such as radiation or hydrogen peroxide, induce antioxidant defenses and become more resistant to higher doses. These defenses can be induced by both synthetic oxidants (e.g., the herbicide paraquat) and by natural oxidants, and are effective against both. (c) The active excretion of planar hydrophobic molecules (natural or synthetic) out of liver and intestinal cells. (d) DNA repair: this is effective against DNA adducts formed from both synthetic and natural chemicals, and is inducible in response to DNA damage. (e) Animals' olfactory and gustatory perception of bitter, acrid, astringent, and pungent chemicals: these defenses warn against a wide range of toxins, and could possible be more effective in warning against some natural toxins that have been important in food toxicity during evolution, than against some synthetic toxins. However, it seems likely that these stimuli are also general defenses and are monitoring particular structures correlated with toxicity; some synthetic toxic compounds are also pungent, acrid, or astringent. Even though mustard, pepper, garlic, onions, etc., have some of these attributes, humans often ignore the warnings.

The fact that defenses are usually general, rather than specific for each chemical, makes good evolutionary sense. The reason that predators of plants evolved general defenses against toxins is presumably to be prepared to counter a diverse and ever-changing array of plant toxins in an evolving world; if a herbivore had defenses only against a set of specific toxins it would be at a great disadvantage in obtaining new plant foods when favored plant foods became scarce or evolved new toxins.

Various natural toxins, some of which have been present throughout vertebrate evolutionary history, nevertheless cause cancer in vertebrates. Mold aflatoxins, for example, have been shown to cause cancer in trout, rats, mice, monkeys and, possibly, humans. Eleven mold toxins have been reported to be carcinogenic and nineteen mold toxins have been shown to be clastogenic. Many of the common elements are carcinogenic (e.g., salts of lead, cadmium, beryllium, nickel, chromium, selenium and arsenic) or clastogenic at high doses, despite their presence throughout evolution.

Furthermore, epidemiological studies from various parts of the world show that certain natural chemicals in food may be carcinogenic risks to humans: the chewing of betel nuts with tobacco around the world has been correlated with oral cancer. The phorbol esters present in the Euphorbiacea, some of which are used as folk remedies or herb teas, are potent mitogens (inducers of cell proliferation) that are thought to be a cause of nasopharyngeal cancer in China and esophageal cancer in Curacao. Pyrrolidizine toxins are mutagens that are found in comfrey tea, various herbal medicines, and some foods; they are hepatocarcinogens in rats, and may cause liver cirrhosis and other pathologies in humans.

Plants have been evolving and refining their chemical weapons for at least 500 million years and incur large fitness costs in producing these chemicals. If these chemicals were not effective in deterring predators, plants would not have been naturally selected to produce them.

Humans have not had time to evolve into a "toxic harmony" with all of the plants in their diet. Indeed, very few of the plants that humans eat would have been present in an African hunter-gatherer's diet. The human diet has changed drastically in the last few thousand years, and people are eating many recently introduced plants that their ancestors did not, e.g. coffee, cocoa, tea, potatoes, tomatoes, corn, avocados, mangoes, olives, and kiwi fruit. In addition, cruciferous vegetables such as cabbage, broccoli, kale, cauliflower, and mustard were used in ancient times "primarily for medicinal purposes" and were spread as foods across Europe only in the middle ages. Natural selection works far too slowly for humans to have evolved specific resistance to the food toxins in these newly introduced plants.

Poisoning from plant toxins in the milk of foraging animals was quite common in previous centuries. In non-industrial societies, cow or goat milk and other ingested dairy products were contaminated by the natural toxins from plants that were eaten by foraging animals, because toxins that are absorbed through the animal's gut are often secreted in the milk. Since the plants foraged by cows vary from place to place and are usually inedible for human consumption, the plant toxins that are secreted in the milk are, in general, not toxins to which humans could have easily adapted. Abraham Lincoln's mother, for example, died from drinking cow's milk that had been contaminated with toxins from the snakeroot plant. When cows and goats forage on lupine, their offspring may have teratogenic abnormalities, such as "crooked calf" syndrome caused by the anagyrine in lupine. Such significant amounts of these teratogens can be transferred to the animals' milk that drinking the milk during pregnancy is teratogenic for humans: in one rural California family, a baby boy, a litter of puppies, and goat kids all had a "crooked" bone birth-defect. The pregnant woman and the pregnant dog had both been drinking milk obtained from the family goats that had been foraging on lupine, the main forage in winter.

Anticarcinogenic chemicals in the diet may help to protect humans equally well against synthetic and natural carcinogens. Although plants contain anticarcinogenic chemicals (e.g. antioxidants) that may protect against carcinogens, these anticarcinogens do not distinguish whether carcinogens are synthetic or natural in origin.

It has been argued that synergism between synthetic carcinogens could multiply hazards; however, this is also true of natural chemicals, which are by far the major source of chemicals in the diet.

Although the synthetic pesticide DDT bioconcentrates in the food chain due to its unusual fat solubility, natural toxins can also bioconcentrate. DDT is often viewed as the typically dangerous synthetic pesticide because it persists for years; it was representative of a class of chlorinated pesticides. Natural pesticides, however, also bioconcentrate if fat soluble: the teratogens solanine (and its aglycone solanidine) and chaconine, for example, are found in the tissues of potato eaters. Although DDT was unusual with respect to bioconcentration, it was remarkably non-toxic to mammals, saved millions of lives, and has not been shown to cause harm to humans. To a large extent DDT, the first major synthetic insecticide, replaced lead arsenate, a major carcinogenic pesticide used before the modern era; lead arsenate is even more persistent than DDT. When the undesirable bioconcentration and persistence of DDT and its lethal effects on some birds were recognized it was prudently phased out, and less persistent chemicals were developed to replace it. Examples of these newer chemicals are the synthetic pyrethroids that disrupt the same sodium-channel in insects as DDT, are degraded rapidly in the environment, and can often be used at a concentration as low as a few grams per acre.

Misconception No. 6:
Storks Bring Babies and Pollution Causes Cancer and Birth Defects.

The number of storks in Europe has been decreasing for decades. At the same time, the European birth rate also has been decreasing. We would be foolish to accept this high correlation as evidence that storks bring babies. The science of epidemiology tries to sort out the meaningful correlations from the numerous chance correlations. That is, epidemiology attempts to determine correlations that may indicate cause and effect. However, it is not easy to obtain persuasive cause-

and-effect evidence by epidemiological methods, because of inherent methodological difficulties. There are many sources of bias in observational data, and chance variation is also important. For example, because there are so many different types of cancer and birth defects, by chance alone might expect some of them to occur at a high frequency in a small community here and there. Toxicology provides evidence that can help us decide whether an observed correlation might be causal or accidental.

There is no persuasive evidence from epidemiology or toxicology that pollution is a significant cause of birth defects or cancer. For example, the epidemiological studies of the Love Canal toxic waste dump in Niagara Falls, New York, or of dioxin in Agent Orange, or of pollutants produced by the refineries in Contra Costa County, California, or of the contaminants in the wells of Silicon Valley or Woburn, Massachusetts, or of the now-banned pesticide DDT, provide no persuasive evidence that pollution was the cause of human cancer in any of these well-publicized exposures. At Love Canal, where people were living next to a toxic waste dump, the epidemiological evidence for an effect on public health is equivocal. Analyses of the toxicology data on many of these cases suggest that the amounts of the chemicals involved were much too low relative to the background of naturally occurring carcinogens and carcinogens from cooking food to be credible sources of increased cancer in humans. With respect to birth defects, a comparative analysis of teratogens using a HERP-type index, which would express the human exposure level as a percentage of the dose level known to cause reproductive damage in rodents, would be of interest. Such an analysis has not been done in a systematic way.

Environmental exposures to industrial chemical pollutants are thousands of times lower than some occupational exposures to these same agents. Thus, if ppb levels of these pollutants were causing cancer or birth defects, one might expect to see an effect in the workplace. So far, however, epidemiological studies on these chemicals do not suggest an association with cancer.

Historically, for chemicals that have been shown to increase cancer in the workplace, exposures were at high levels. For example, in California the levels of the fumigant ethylene dibromide (EDB) that workers were allowed to breathe in were once shockingly high. We testified in 1981 that our calculations showed that the workers were allowed to breathe in a dose higher than the dose that gave half of the test rats cancer. California lowered the permissible worker exposures more than a hundred-fold. Despite the fact that the epidemiology on EDB in highly exposed workers does not show any significant effect, the uncertainties of our knowledge make it important to have strict rules about workers, because they can be exposed chronically to extremely high doses.

Misconception No. 7:
Tradeoffs Are Not Necessary in
Eliminating Pesticides.

Since no plot of land is immune to attack by insects, plants need chemical defenses—either natural or synthetic—in order to survive pest attack. "It has been suggested that one consequence of crop plant domestication is the deliberate or inadvertent selection for reduced levels of secondary compounds that are distasteful or toxic. Insofar as many of these chemicals are involved in the defense of plants against their enemies, the reduction due to artificial selection in these defenses may account at least in part for the increased susceptibility of crop plants to herbivores and pathogens..." Thus, there is a tradeoff between nature's pesticides and manmade pesticides.

Cultivated plant foods commonly contain fewer natural toxins than do their wild counterparts. For example, the wild potato, the progenitor of cultivated strains of potato, has a glycoalkaloid content about 3 times that of cultivated strains and is more toxic. The leaves of the wild cabbage (the progenitor of cabbage, broccoli, and cauliflower) contain about twice as many glucosinolates as cultivated cabbage. The wild bean contains about 3 times as many cyanogenic glucosides as does the cultivated bean. Similar reductions in toxicity through agriculture have been reported in lettuce, lima beans, mango, and cassava.

One consequence of the disproportionate concern about synthetic pesticide residues is that some plant breeders are currently developing plants that are more insect-resistant, and, thus, higher in natural toxins. Two recent cases illustrate the potential hazards of this approach to pest control. 1) When a major grower introduced a new variety of highly insect-resistant celery into commerce, a flurry of complaints were made to the Centers for Disease Control from all over the country because people who handled the celery developed rashes when they were subsequently exposed to sunlight. Some detective work found that the pest-resistant celery contained 6,200 ppb of carcinogenic (and mutagenic) psoralens instead of the 800 ppb present in normal celery. It is not known whether other natural pesticides in the celery were increased as well. The celery is still on the market. 2) A new potato, developed at a cost of millions of dollars, had to be withdrawn from the market because of its acute toxicity to humans when grown under particular soil conditions—a consequence of higher levels of the natural toxins solanine and chaconine. Solanine and chaconine inhibit cholinesterase, thereby blocking nerve transmission, and are known rodent teratogens. They were widely introduced into the world diet about 400 years ago with the dissemination of the potato from the Andes. Total toxins are present in normal potatoes at a level of 15 mg per 200-g potato (75 ppm), which is less than a ten-fold safety margin from the measurably-toxic daily dose level for humans. Neither solanine nor chaconine has been tested for carcinogenicity. In contrast, the cholinesterase inhibitor malathion, the main synthetic organophosphate pesticide residue in our diet (0.006 mg per day), has been tested and is not a carcinogen in rats or mice.

Certain cultivated crops have become popular in developing countries because they thrive without costly synthetic pesticides. However, the tradeoffs of cultivating some of these naturally pest-resistant crops are that they are highly toxic and require extensive processing to detoxify them. For example, cassava root, a major food crop in Africa and South America, is quite resistant to pests and disease; however, it contains cyanide at such high levels that only a laborious process of washing, grinding, fermenting, and heating can make it edible; ataxia due to chronic cyanide poisoning is endemic in many of the cassava-eating areas of Africa. In one part of India, the pest-resistant grain *Lathyrus sativus* is cultivated to make some types of dahl. Its seeds contain the neurotoxin beta-N-oxalyl aminoalanine, which causes a crippling nervous system disorder, neurolathyrism.

As an alternative to synthetic pesticides, it is legal for "organic farmers" to use the natural pesticides from one plant species against pests that attack a different plant species, e.g. rotenone (which Indians used to poison fish) or the pyrethrins from chrysanthemum plants. These naturally-derived pesticides have not been tested as extensively for carcinogenicity (rotenone is negative, however), mutagenicity or teratogenicity as have synthetic pesticides; therefore, their safety compared to synthetically-derived pesticides should not be prematurely assumed.

Efforts to prevent hypothetical carcinogenic risks of 1 in a million could be counterproductive if the risks of the alternatives are worse. For example, Alar was withdrawn from the market after the EPA proposed cancellation hearings on it and after the Natural Resources Defense Council (NRDC) went to the media to get the process accelerated. However, we incur various risks by withdrawing Alar, and these risks should be addressed. Alar a is growth regulator that delays ripening of apples so that they do not drop prematurely, and it also delays over-ripening in storage. Alar plays a role in reducing pesticide use on some types of apples, particularly in the Northeast. Without Alar, the danger of fruit fall from the pests known as leafminers is greater, and more pesticides are required to control pests. When fruit falls prematurely, pests on the apples remain in the orchard to attack the crop the next summer, and more pesticides must be used. Since Alar produces healthier apples that stay on the trees, Alar-treated fruit is less susceptible to molds. Therefore, it is likely that the amounts and variety of mold toxins present in apple juice, e.g., patulin, will be higher in juice made from untreated apples. The carcinogenicity of patulin has not been adequately examined. Another trade-off of eliminating Alar is decreasing the availability of domestically grown, fresh apples throughout the year, and increasing the price of apples, which might lead consumers to substitute less healthy foods.

Synthetic pesticides have markedly lowered the cost of plant food, thus promoting increased consumption. Eating more fruits and vegetables, and less fat, may be the best way to lower risk of cancer and heart disease (other than giving up smoking).

Misconception No. 8:
Technology Is Harmful to Public Health

Modern technologies are almost always replacing older, more hazardous technologies. Billions of pounds of TCE (one of the most important non-flammable, industrial solvents) and PERC (the main drycleaning solvent in the United States) are used because they are low in toxicity and are not flammable. Chlorinated solvents replaced flammable solvents in industry and in dry cleaning; this was a major advance in fire safety, with a minor tradeoff of an occasional ppb contamination in water.

Eliminating a carcinogen may have unwanted effects. For example, EDB, the main fumigant in the United States before it was banned in 1984, was present in trivial amounts (about 0.4 ppb) in our food: the average daily intake was a possible carcinogenic hazard about one tenth that of the aflatoxin in the average peanut butter sandwich, itself a minimal possible hazard. It is possible that the elimination of EDB fumigation will result in greater insect infestation and contamination of grain by carcinogen-producing molds. This would result in a reduction in public health, not an advance, and would greatly increase costs. Furthermore, alternative fumigants to replace EDB do not appear satisfactory, and may be more hazardous and expensive.

Similarly, modern synthetic pesticides replaced more hazardous substances, such as lead arsenate, one of the major pesticides before the modern era. Lead and arsenic are both natural, highly toxic, and carcinogenic. Pesticides have increased crop yields and have brought down the price of foods, a major public health advance. Each new generation of synthetic pesticides is more environmentally and toxicologically benign.

Every living thing and every industry "pollutes" to some extent. The fact that scientists have developed methods to measure part per billion levels of chemicals, and are developing methods to measure part per trillion, makes us more aware of toxicity, but does not mean that exposure to toxins is necessarily increasing, or that detected chemicals are causing human disease. Minimizing pollution is clearly desirable for other reasons, but is a separate issue from cancer prevention; getting the most pollution reduction for the lowest economic cost is, of course, important.

Focusing on minor rather than major health risks is counterproductive. If we divert too much of our attention to traces of pollution as a public health concern we do not improve public health—and, in the confusion, the important hazards may be neglected: e.g., smoking (400,000 deaths per year), alcohol (100,000 deaths per year), eating unbalanced diets (such as too much saturated fat and cholesterol, and too few fruits and vegetables), AIDS, radioactive radon released from the soil into homes, and high-dose occupational exposures to chemicals.

It is the inexorable progress of modern technology and scientific research that is likely to lead to a decrease in cancer death rates, a decrease in the incidence of birth defects, and an increase in the average human lifespan.

Acknowledgments

This work was supported by National Cancer Institute Outstanding Investigator Grant CA39910, by National Institute of Environmental Health Sciences Center Grant ES01896, and by U.S. Environmental Protection agency Grant No. DE-AC03-76SF00098. This paper has been partially adapted from References 1–4; from B.N. Ames, "What Are the Major Carcinogens in the Etiology of Human Cancer? Environmental Pollution, Natural Carcinogens, and the Causes of Human Cancer: Six Errors," in *Important Advances in Oncology*, 1989, V.T. De Vita, Jr., S. Hellman, and S.A. Rosenberg, eds. (J.B. Lippincott, Philadelphia, 1989), pp. 237–247; from B.N. Ames and L.S. Gold, "Misconceptions Regarding Environmental Pollution and Cancer Causation," in *Health Risks and the Press: Perspectives on Media Coverage of Risk Assessment and Health*, M. Moore, ed. (The Media Institute, Washington, D.C., 1989), pp. 19–34; B.N. Ames and L.S. Gold, "Dietary Carcinogens, Environmental Pollution, and Cancer: Some Misconceptions," *Med. Oncol. and Tumor Pharmacother.* 7, pp. 69–85; and B.N. Ames and L.S. Gold, "Environmental Pollution and Cancer: Some Misconceptions," in *Science and the Law*, P. Huber, ed., In press.

References

1. Ames, B.N. and Gold, L.S. (1990). I. Chemical Carcinogenesis: Too Many Rodent Carcinogens. *Proc. Natl. Acad. Sci. USA*, 87, 7772–7776.
2. Ames, B.N., Profet, M., and Gold, L.S. (1990). II. Dietary Pesticides (99.99% All Natural). *Proc. Natl. Acad. Sci. USA*, 87, 7777–7781.
3. Ames, B.N., Profet, M., and Gold, L.S. (1990). III. Nature's Chemicals and Synthetic Chemicals: Comparative Toxicology. *Proc. Natl. Acad. Sci. USA*, 87, 7782–7786.
4. Ames, B.N., Magaw, R. and Gold, L.S. (1987). Ranking Possible Carcinogenic Hazards. *Science 236*, 271–280.

Abridged, with permission, from
original version appearing in *Science and the Law*,
Peter Huber, editor.

Cell Proliferation in Carcinogenesis

*Samuel M. Cohen, M.D., Ph.D.
and Leon B. Ellwein, Ph.D.*

Dr. Samuel M. Cohen is professor and vice chairman, University of Nebraska Medical Center, Department of Pathology/Microbiology. Previously he was an associate professor, University of Massachusetts Medical School, and staff pathologist at St. Vincent Hospital.

He received an M.D. and Ph.D. in Oncology from the University of Wisconsin in 1972.

Dr. Cohen is the author of *The Pathology of Bladder Cancer*, Volume I and II, published by the CRC Press in 1983. He has published more than 150 journal articles and contributed to *Cancer Research, Carcinogenesis*, and *Food and Chemical Toxicology*.

Dr. Cohen received the Outstanding Research and Creativity Award from the University of Nebraska Medical Center in 1990.

Dr. Leon B. Ellwein is professor and associate dean, University of Nebraska Medical Center. He was previously affiliated with the Science Applications International Corporation, and the National Institutes of Health.

He received a B.S.M.E. (1964) and an M.S. (1966) from the South Dakota State University; and a Ph.D. from Stanford University in 1970.

Dr. Ellwein has published more than 40 articles in peer-reviewed journals contributing to *Science, Toxicology and Applied Pharmacology,* and *Risk Analysis.*

Chemicals that induce cancer at high doses in animal bioassays often fail to fit the traditional characterization of genotoxins. Many of these nongenotoxic compounds (such as sodium saccharin) have in common the property that they increase cell proliferation in the target organ. A biologically based, computerized description of carcinogenesis was used to show that the increase in cell proliferation can account for the carcinogenicity of nongenotoxic compounds. The carcinogenic dose-response relationship for genotoxic chemicals (such as 2-acetylaminofluorene) was also due in part to increased cell proliferation. Mechanistic information is required for determination of the existence of a threshold for the proliferative (and carcinogenic) response of nongenotoxic chemicals and the estimation of risk for human exposure.

Certain chemicals have long been associated with cancer in humans, and animal models have been developed to study processes involved in the transition from a normal to a cancer cell[1]. During the past two decades, emphasis has been shifting from the use of animal models primarily for the study of carcinogenic mechanisms to the use of animals to assay for carcinogenic potential of chemicals[2]. Research has been directed more at quantitatively estimating the risk to humans. Traditionally, risk assessments have entailed the use of various mathematical and statistical formulations to extrapolate from, results of high-dose animal bioassays to estimates of risk at low doses[3]. However, high-dose tumor response data are inadequate for this purpose, as is most evident when efforts are made to predict a threshold below which there is no effect. These limitations indicate the need to base risk assessments on knowledge of the biology, of tumor formation.

We have developed a model of carcinogenesis, based on biological data and principles, that we originally used as an analytical tool to interpret results of experiments with the bladder carcinogen N-[4-(5-nitro-2-furyl)-2-thiazolyl]formamide (FANFT) in rats[4]. We demonstrated quantitatively that the tumorigenic effects of FANFT administration result from its dose-dependent genotoxic and proliferative effects, and that the proliferative effects operated only, at the highest doses employed[4,5].

The model can be viewed as an assembly of dynamic relationships between variables that contribute to tumor production (Figure 1), and incorporates several biological suppositions. A fundamental assump-

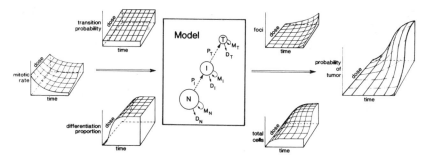

Figure 1. A mathematical model of carcinogenesis that entails two irreversible transitions, from normal (N), to initiated (I), to transformed (T) cell populations. Population mitotic rates, M_N, M_I, and M_T, respectively, and cellular differentiation (and death) rates D_N, D_I, and D_T are primary model inputs. The interaction of these rates determines the size of cell populations. Initiation and transformation transitions occur randomly during cell replication, represented by the probabilities p_I and p_T. Model inputs are dependent on dose and animal age. Model outputs that can be validated with experimental data include target organ size (total number of cells), number of initiated cell foci (hyperplastic foci in the liver), and the probability of a visible tumor. The model is implemented computationally using stochastic simulation.

tion is that cells exist within one of three states, normal, initiated (intermediate), or transformed, and that transitions between states occur or are irreversibly fixed only in replicating cells. These transitions are assumed to take place in a stochastic fashion and represent genetic changes introduced during cell replication, possibly with the involvement of oncogenes or tumor suppressor genes[6]. Transformed cells are those that are malignant, not cells in benign lesions. In the absence of a genotoxic exposure, the probability of a transition occurring is small but not zero (thus accounting for spontaneous tumors). The likelihood of producing a cancerous cell is increased if either the probability of a genetic transition or the rate of cell replication is increased.

Another model that also incorporates the effect of cell proliferation and was validated using human epidemiology data lends further support for a two-event hypothesis for carcinogenesis[7]. Although based on similar biological parameters, our model uses a different mathematical construct. To represent the biological dynamics within the target organ, we resorted to a recursive simulation. Beginning with its early development period, the status of the cell population in the target organ was computed in simulated time using the probabilities for each possible event (mitosis, genetic transition, or death) facing each cell within each of a series of specific time intervals. Calculations for each subsequent time interval incorporate the results of the preceding interval. The probabilities of mitosis or death are estimated by observation of cell proliferation and cell number at various times, and the probabilities of genetic transition were inferred by a comparison of model outcomes with the observed time course of tumor development at the particular dose being simulated. Although this simulation approach precludes the possibility of directly estimating genetic transition probabilities and other experimentally unobservable model parameters using statistical inference, it does not risk the mathematical oversimplification required for the derivation of a computationally tractable expression that would relate tumor incidence to cellular proliferation and genetic transition variables. The quest for closed-form expressions is problematical because of the multiplicity of cellular states and the time- and dose-varying nature of the numerous cell behavior variables.

To illustrate the critical role of cell proliferation in carcinogenesis, we discuss here two prototypical compounds: a genotoxic carcinogen, 2-acetylaminofluorene (2-AAF), and a nongenotoxic agent, sodium saccharin.

2-Acetylaminofluorene (2-AAF)

To determine the tumorigenicity of 2-AAF at low doses, more than 24,000 female BALB/C mice were fed different doses (30 to 150 ppm) of 2-AAF for different periods of time (9 to 33 months) and killed at various intervals between 9 and 33 months of study[8]. This "megamouse" experiment was designed to detect a 1% increase in the prevalence of tumors (thus is referred to as the ED_{01} study) in two target organs, liver and urinary bladder. Rather than demonstrating how to extrapolate to low doses, this study raised additional questions[8-10]. The dose-response curve for the liver was nearly linear down to the lowest amount administered, 30 ppm. In contrast, the dose-response curve for the bladder was nonlinear. At doses below 60 ppm, there was no detectable increase in bladder tumor prevalence compared to controls, whereas prevalence increased sharply at doses above 60 ppm. Examination of tumor response as a function of time complicated the issue further[9].

Initially, investigators postulated that the differences in dose-response curves between liver and bladder could be explained by differences in 2-AAF toxicokinetics, and that binding of 2-AAF to DNA would not occur in the bladder below some threshold, whereas in liver even the lowest doses would have an effect. However, the administration of 2-AAF to BALB/C mice at similar and lower doses (5 to 150 ppm) produces a linear dose-response relationship for DNA adduct formation in both the liver and bladder[11].

The Armitage-Doll multi-stage model was also applied to explain the differences in 2-AAF response between liver and bladder tissues, leading to the postulation of a one-hit carcinogenic phenomenon for the liver and a three-hit process for the bladder[11]. By accounting for the proliferative effects of 2-AAF in addition to its effects on DNA, which the Armitage-Doll model is unable to do, we are able to explain both dose-response curves using a two-event model of carcinogenesis[10].

Liver Response To 2-AAF

In normal hepatocytes, 2-AAF is metabolized to its active, N-sulfated metabolite, which forms DNA adducts[11-13]. This is reflected in our model by raising the probability of the first genetic event (p_I) above background. In contrast, cells in hyperplastic foci do not metabolize 2-AAF as readily, and considerably fewer DNA adducts are formed[12]. Apparently, 2-AAF has a negligible effect on the probability, of the second generic event (p_T). At doses utilized in the ED_{01} study, enlargement of the liver is not observed[8], providing evidence of no increased hepatocyte proliferation. Thus, the only apparent impact of 2-AAF on the liver was an increase in (p_I) over background levels; p_T and hepatocyte mitotic rates remained at background levels and were not affected by 2-AAF administration.

Mitotic rates in the normal adult liver are relatively low (labeling index ≤ 0.1 %). During the high proliferative phase of organ development, occasional cells are likely to become initiated, even with a low, background value for p_I. The remainder of the animal's life can then provide sufficient opportunity for at least one of these initiated cells to progress to a transformed cell, and then proliferate to a tumor of detectable size. In the ED_{01} study, spontaneous liver neoplasms were observed in 2.3% (n = 383) of control mice sacrificed at 24 months and 34.8% (n = 23) of mice sacrificed at 33 months[8], illustrating the influence of elapsed time on tumor development.

With a potent genotoxic compound such as 2-AAF, the relatively small number of cells initiated spontaneously during organ development is insignificant compared to the number initiated by reaction with 2-AAF metabolites (because of the increased p_I). The large number of initiated cells after exposure to 2-AAF, in combination with subsequent proliferation and transformation at background rates, results in an increased prevalence of liver tumors, particularly as the animal ages beyond 2 years (Figure 2). At doses higher than those used in the ED_{01}

study 2-AAF also increases compensatory proliferation of surviving hepatocytes and sharply increases tumor prevalence as early as 6 months[13].

Bladder Response To 2-AAF

Metabolism of 2-AAF in the liver also involves production of the N-glucuronide, which accumulates in the urine and is hydrolyzed to an electrophile that can react with both normal and initiated urothelial cells[11,14]. Thus, 2-AAF affects both p_I and p_T in the bladder. The relationship between 2-AAF dose and DNA adduct formation is apparently linear within the 5 to 150 ppm range[11]. In contrast to the situation in liver, 2-AAF induces urothelial hyperplasia at doses $\geq$ 60 ppm (Figure 3)[8]. Modeling the interaction of these responses to 2-AAF effectively duplicates the in vivo results[8,10] (Figure 2). Below 60 ppm, the apparent lack of increase in tumor prevalence reflects the minimum experimental detection limit (1%) rather than the absence of tumors. At the higher doses, increased cell proliferation has an impact, and an increase in tumor formation occurs. From our modeling analyses of hypothetical situations, we calculated that if 2-AAF influenced only p_I and p_T in the bladder, tumor prevalence at 24 months would be 4% at a dose of 150 ppm, whereas, if only cell proliferative effects were present,

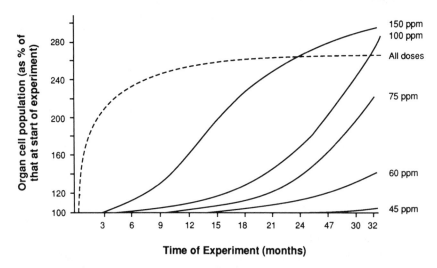

Figure 2. Model results for effects of duration of exposure (18, 24, or 33 months) and 2-AAF dose on live tumor (----) and bladder tumor (——) prevalence in mice. These analytical results have been demonstrated as being consistent with actual data from the ED_{01} study[8,10].

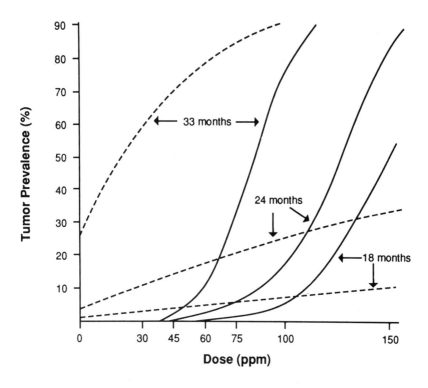

Figure 3. Effect of normal growth, duration of exposure, and 2-AAF dose on total number of liver hepatocytes (----) and bladder urothelial cells (——) in mice. The increase in number of hepatocytes parallels the normal growth of the liver [10]. The increase in bladder cell number caused by 2-AAF is quantified from histopathology information from the ED_{01} study [8,10]. 2-AAF administration began at approximately 1 month of age.

the corresponding tumor prevalence would be 6%. The prevalence with both operating simultaneously is 88%, suggesting a synergistic effect between genotoxicity and proliferation.

Sodium Saccharin

Dietary administration of high doses of sodium saccharin (NaS) to rats over two generations results in a significant increase in the frequency of bladder cancer, particularly in males [15,16]. In these two-generation studies, NaS feeding begins in the dam, is continued through gestation and lactation periods, then through the lifetime of the offspring. Subsequent experiments have shown that NaS administration beginning at birth results in essentially the same tumor prevalence as with NaS administration from conception[16], but NaS administration started after weaning usually produces an insignificant response[15,16]. However, if the post-weaning rat is first treated with a short regimen of a bladder carcinogen, such as FANFT, N-butyl-N-(4-hydroxybutyl)nitrosamine (BBN), or N-methyl-N-nitrosourea (MNU), followed by NaS, tumors result[5,17]. Unlike 2-AAF, saccharin is nucleophilic, is not metabolized to a reactive electrophile, does not react with DNA, and is not mutagenic in most short-term assays[17]. However when NaS is administered to the rat at high dietary doses, proliferation in the urothelium increases, resulting in mild focal hyperplasia[17].

Role of cell proliferation. Modeling analyses demonstrate that NaS-induced cell proliferation is sufficient to account for the increase in bladder tumor prevalence after exposure to NaS[18]. In the FANFT-NaS experiments, tumors are produced by the stimulating effect of NaS on the dynamics of a pool of FANFT-initiated cells. Because a nonzero probability of spontaneous genetic transformation (p_T) is associated with each mitosis of an initiated cell, an increase in the mitotic rate after exposure to NaS increases the number of opportunities for transformation.

In studies where NaS administration is not preceded by initiation with a genotoxic compound, it is possible to produce an increased number of initiated cells strictly by the increase in proliferation that occurs when NaS administration is begun early in the developmental period. Because the bladder already has a maximally proliferating epithelium during gestation (labeling index approximately 10%), NaS administration during the in utero period does not further increase the proliferation rate[17]. However, during the 3 weeks after birth, the labeling index normally declines to <0.1%. Although relatively brief, this 3-week period is of disproportionate biological importance because approximately one-third of the total number of cell divisions in a rat's 2-year life-span occur during this period[18]. A significant increase in cell proliferation rates during the 3 weeks after birth, coupled with the background probability of spontaneous genomic errors, can substantially increase the number of initiated cells. In assessing the carcinogenicity of nongenotoxic chemicals such as NaS, it is critical to consider the increased number of initiated cells generated during fetal and neonatal development and the resulting increase in tumor prevalence to experimentally detectable levels[17].

An increase in the number of initiated cells caused only by excess proliferation has also been demonstrated in male rat bladders after weaning. The epithelium was ulcerated by freezing, and the resultant burst of mitotic activity was comparable to that seen during fetal development[19]. Within 3 to 4 weeks the epithelium healed and returned to mitotic quiescence and normal morphology. Nevertheless, if high doses of NaS are subsequently administered, bladder tumors result. In terms of our model, a sufficient number of initiated cells are generated spontaneously during the regenerative hyperplasia such that the increased and sustained proliferative activity induced by NaS generates tumors[18,19].

Proliferative Mechanism and Threshold

Utilizing traditional risk assessment methods, the results described above in male rats with extremely high doses of NaS can be extrapolated to arrive at an approximate calculated risk for humans exposed to low doses of NaS[20]. However, there is clearly a need to understand the underlying mechanisms of carcinogenesis by nongenotoxic compounds before any rational estimate of human risk can be made. The complexity of the task in risk assessment is indicated by the finding that female rats are much less susceptible to bladder tumorigenesis in response to NaS than males, and mice, hamsters, and monkeys are resistant even at high doses[15,17].

The different salt forms of saccharin produce markedly different urothelial proliferative responses[21]. Potassium saccharin somewhat increases urothelial proliferation relative to controls, but less than does NaS. Urothelial proliferation after treatment with calcium saccharin and acid saccharin is statistically indistinguishable from controls; thus it might be assumed that neither calcium saccharin nor acid saccharin would be carcinogenic in the rat model. Absorption and urinary excretion of the saccharin anion is similar regardless of which form of saccharin is administered, but the physiological changes in the urine associated with the high loads of the different salts produce marked differences in urinary pH, ion concentrations, volume, and osmolality. The changes in pH and salt concentrations do not alter the ionic structure of saccharin, and there is no evidence that saccharin interacts directly with a urothelial cell receptor[17]. A similar increased proliferative and tumorigenic activity in the male rat urothelium following chemical initiation is seen with high doses of several other sodium salts of weak to moderate organic acids, many of which are naturally occurring and essential for the well-being of living organisms, including vitamin C, glutamate, and bicarbonate[5,17]. No tumorigenicity is observed when the acid form of these chemicals was tested[5,17].

We have recently observed that, in addition to the normally present $MgNH_4PO_4$ crystals, many crystals in the urine of rats fed high doses of NaS contain. silicate, and a large amount of a flocculent precipitate that contains silicate is also present[22]. The silicate crystals and precipitate appear to act as microabrasives or cytotoxic material for urothelial cells, resulting in focal necrosis and consequent regenerative hyperplasia. Silicate precipitate and crystals require protein for their formation[23]. Saccharin binds to urinary protein, particularly (α_{2u}-globulin[24], thus enhancing the precipitation and crystallization that only occasionally occurs in control male rats[25]. Urinary acidification inhibits silicate precipitation and inhibits the proliferative effects of NaS. High levels of urinary sodium and protein enhance silicate precipitation[23]. The principal factor that appears to predispose the male rat to silicate crystal formation following NaS feeding is the presence of large quantities of normally occurring urinary protein, especially the protein specific to the male rat, α_{2u}-globulin[24]. The female rat has much less urinary protein than the male and is less responsive to the proliferative and tumorigenic effects of NaS on the urothelium. The mouse, a species that is not responsive to saccharin even at NaS levels of 7.5% of the diet (at least three times the apparent threshold level in male rats), has low levels of urinary protein and did not form silicate crystals when fed NaS[25].

The multiple physical-chemical parameters in the male rat suggest that a fairly high threshold exists for NaS dose in producing silicate crystals. It is extremely unlikely that the silicate precipitates and crystals would form in humans under normal conditions of NaS ingestion, since human urine has very little protein and has less sodium than rat urine. This is consistent with the general lack of an association in humans between NaS ingestion and bladder cancer or hyperplasia[20,26,27].

Classification of Chemicals for Human Risk Assessment

The current practice is to classify chemicals as initiators, promoters, complete carcinogens, or progressing agents. In light of the demonstrated ability of compounds to increase the risk of cancer by either directly altering DNA, increasing cell proliferation, or both, distinctions blur and traditional terminology is inadequate. We feel it is useful to classify chemical carcinogens into those that interact with DNA (genotoxic) and those that do not (nongenotoxic) (Figure 4)[28]. Many of the latter chemicals act by increasing cell proliferation, either by direct mitogenesis of the target cell population or by cytotoxicity and consequent regenerative proliferation. Genotoxic chemicals (2-AAF and numerous others, such as diethylnitrosamine, dimethylnitrosamine, and FANFT) usually do not exhibit a threshold for the interaction with DNA, and, at higher doses, may cause cell death resulting in cell proliferation[5,29]. This dual effect of genotoxic chemicals frequently leads to a

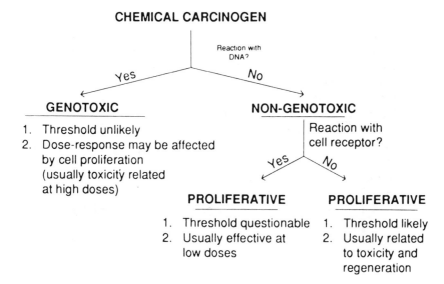

Figure 4. **Proposed classification scheme for carcinogens. The effect of genotoxic chemicals can be accentuated if cell proliferative effects are
also present. Nongenotoxic chemicals act by increasing cell proliferation directly or indirectly, either through interaction with a
specific cell receptor or nonspecifically by (i) a direct mitogenic
stimulus; (ii) causing toxicity with consequent regeneration; or (iii)
interrupting physiological process. Examples of the latter mechanism include TSH stimulation of thyroid cell proliferation after
toxic damage to the thyroid, and viral stimulation of proliferation
after immunosuppression.**

dose-response curve similar to that of 2-AAF in the bladder described
above. A modest rate of increase in tumor prevalence at low doses is
due only to a genotoxic effect, and a much greater rate of increase at
higher doses is due to the synergistic influence of increased cell proliferation. The actual dose- and time-response for a chemical is dependent
on whether the compound has a genotoxic effect, a proliferative effect,
or both, and whether it affects normal or initiated cells, or both.

The nongenotoxic chemicals can be further categorized by their
mechanisms of action, if known. For example, phorbol esters, dioxin,
and hormones each interact with a cellular receptor[30], whereas NaS[17],
antioxidants[31], thin films, hepatotoxins, and nephrotoxins[28] act through
a non-receptor mechanism. Cytotoxicity, direct mitogenesis, or both
can also occur with chemicals acting through cell receptors (such as the
phorbol esters)[28,30]. Compounds acting through specific receptors tend

to be active at low doses, and it is unclear whether a no-effect threshold could be ascertained for these compounds. Similarly, chemicals that are directly mitogenic to target cells may or may not have a threshold. In contrast, most, if not all, compounds that act solely through a cytotoxic mechanism would be expected to have a no-effect threshold above which cytotoxicity becomes apparent. Below the threshold, cytotoxicity and increased cell proliferation would not occur, and there would be no increased risk of tumors. Interpretation of long-term bioassays for nongenotoxic chemicals must take into account aspects of nonreceptor mechanisms.

Examples of a dose-response threshold occur with uracil and melamine[32]. If sufficiently high doses of either of these nongenotoxic chemicals are fed to rats or mice, urinary calculi, urothelial proliferation, and tumors occur. If the dose is below the minimum at which calculi occur, there is no increased cell proliferation or tumor formation.

Cell Proliferation as a Predictor of Carcinogenesis

Despite the importance of cell proliferation in carcinogenesis, short-term assays of increased cell proliferation in response to nongenotoxic chemicals are likely to prove as inadequate as short-term genotoxicity assays for predicting carcinogenicity. Some chemicals induce only a temporary or mild increase in proliferation that may not be adequate to produce a detectable increase in tumor prevalence within the lifetime of the experimental animal. Also, increased proliferation must occur in cells susceptible to cancer development, rather than in nonsusceptible cells, such as terminally differentiated cells, that may also be present in the target organ. For example, turpentine can cause proliferation of the skin, but is a very weak skin tumor promoter[33]. Turpentine primarily increases proliferation of the keratinocytes rather than the dark basal cells that are the apparent precursors of skin tumors.

Confusion can also arise with chemicals such as cyclophosphamide[34]. Although it is extremely cytotoxic to the bladder epithelium, leading to a marked regenerative hyperplasia, it is also cytotoxic to any bladder tumor cells that might form. If cyclophosphamide is administered at doses high enough to be genotoxic but below those that are cytotoxic, the prevalence of bladder tumors is increased in animals and humans. At higher cytotoxic doses, regenerative hyperplasia occurs but no tumors are produced.

There are numerous indications in humans that prolonged, increased cell proliferation is necessary for the development of tumors, particularly for hormonally related tumors such as estrogen-related endo metrial carcinomas[35]. It appears that most virally related human tumors are also a result of sustained increased proliferation. For example, Epstein-Barr virus (EBV) stimulates B lymphocyte proliferation. When a patient is immunosuppressed, whether due to heredity, immunosuppressive drugs associated with transplantation, or AIDS, the B-cell proliferation cannot be controlled, and there is an appreciable increase in the risk of B-cell lymphomas[36]. Hepatitis B virus (HBV) can produce chronic hepatitis and cirrhosis, characterized by persistent necrosis and regenerative hyperplasia, and is also associated with an increased incidence of hepatoma[37].

It would appear that increased cell proliferation also contributes to the development of tumors secondary to various chemical exposures in humans. For example, cigarette smoking is known to cause bladder cancer in humans, perhaps due to a hyperplastic effect on the urothelium of many cigarette smokers, in addition to the probable genotoxic damage that occurs[27].

As the mechanisms of carcinogenesis become more thoroughly understood, a more rational approach can be taken for extrapolation from high dose experimental data in animals to low dose natural exposure and assessment of the risk faced by human populations exposed to chemical agents. The effects of toxicity and consequent cell proliferation are particularly critical for nongenotoxic agents, because a threshold effect is likely.

References

1. E.C. Miller and J.A. Miller, *Cancer* 47, 2327 (1981).
2. L.S. Gold *et al., Environ. Health Perspec.* 58, 9 (1984); L.S. Gold *et al.,ibid.* 67, 161 (1986); L.S. Gold *et al., ibid.* 74. 237 (1987); L.S. Gold *et al., ibid.* 84, 215 (1990).
3. J. Van Ryzin, *Biometrics (Suppl.)* 38, 130 (1982).
4. R.E. Greenfield, L.B. Ellwein, S.M. Cohen, *Carcinogensis* 5, 437 (1984); L.B. Ellwein and S.M. Cohen, in *Biologically-Based Methods for Cancer Risk Assessment*, C. Travis, Ed. (Plenum, New York, 1989), pp. 181–192.
5. S.M. Cohen and L.B. Ellwein, *Toxicol. Lett.* 43, 151 (1988).
6. R.A. Weinberg, *Cancer Res.* 49, 3713 (1989).
7. A.G. Knudson, *Proc. Natl. Acad. Sci. U.S.A.* 68, 820 (1971); S.H. Moolgavkar and D.J. Venzon, *Math. Biosi.* 47, 55 (1979); S.H. Moolgavkar and A.G. Knudson, Jr., *J. Natl. Cancer Inst.* 66, 1037 (1981).
8. J.A. Straffa and M.A. Mehlman, *J. Environ. Path Toxicol.* 3, 1 (1980).
9. F.W. Carlborg, *Food Cosmet. Toxicol.* 19, 367 (1981); Society of Toxicology, *Fundam. Appl. Toxicol.* 3, 26 (1983); D.H. Hughes *et al., ibid.*, p. 129; P. Shubik, *ibid.*, p. 137; D. Krewski *et al, ibid.*, p. 140; L.N. Park and R.D. Snee, *ibid.*, p. 320; K.G. Brown and D.G. Hoel, *ibid.*, p. 458; *ibid.*, p. 470; R.L. Koddell *et al., ibid.*, p. 9a; R.D. Bruce *et al., ibid.*, p. 9a.

10. S.M. Cohen and L.B. Ellwein, *Toxicol. Appl. Pharm.* 104, 79 (1990).

11. F.A. Beland, N.F. Fullerton, T. Kinouchi, M.C. Poirier, *IARC (Int. Agency Res. Cancer) Sci. Publ. No. 89* (1988), p. 175.

12. E. Farber, S. Parker, M. Gruenstein, *Cancer Res.* 36, 3878 (1976); R.C. Gupta, K. Earley, F.F. Becker, *ibid*, 48, 5270 (1988); C.C. Lai, J.A. Miller, E.C. Miller, A. Liem, *Carcinogenesis* 6, 1037 (1985).

13. N.A. Littlefield, C. Cipiano, Jr., A.K. Davis, K. Medlock, J. *Toxicol. Env. Health* 1, 25 (1975).

14. F.F. Kadlubar, J.A. Miller, E.C. Miller, *Cancer Res.* 37, 805 (1977).

15. D.L. Arnold *et al.*, *Toxicol. Appl. Pharmacol.* 52, 113 (1980); *IARC (Int. Agency Res. Cancer) Monogr. Eval. Carcinog. Risk Chem. Hum.* 22, 111 (1980).

16. G.P. Schoenig *et al.*, *Food Chem. Toxicol.* 23, 475 (1985).

17. L.B. Ellwein and S.M. Cohen, *Crit. Rev. Toxicol.* 20, 311 (1990).

18. ———, *Risk Analysis* 8, 215 (1988).

19. G. Murasaki and S.M. Cohen, *Cancer Res.* 43, 182 (1983); R. Hasegawa, R.E. Greenfield, G. Murasaki. T. Suzuki, S.M. Cohen, *ibid* 45, 1469 (1985).

20. B.K. Armstrong, *IARC (Int. Agency Res. Cancer) Sci. Publ. No. 65* (1985), p. 129; D. Krewski, *ibid.*, p. 145; R.W. Morgan and O. Wong, *Food Chem. Toxicol.* 23, 529 (1985); F.W. Carlborg, *ibid.*, p. 499.

21. R. Hasegawa and S.M. Cohen, *Cancer Lett.* 30, 161 (1986).

22. S.M. Cohen, M. Cano, E.M. Garland, R.A. Earl, *Proc. Am. Assoc. Cancer Res.* 30, 204 (1989).

23. C.B. Bailey, *Can . J. Biochem.* 50, 305 (1972); C.J. Schreier and R.J. Emerick, *J. Nutr.* 116, 823 (1986); R.J. Emerick, *Nutr. Rep. Inst.* 34, 907 (1986); *J. Nutr.* 117, 1924 (1987).

24. J.A. Swenberg, B. Short, B. Borghoff, J. Strasser, M. Charbonneau, *Toxicol. Appl. Pharm.* 97, 35 (1989).

25. S.M. Cohen *et al.*, unpublished observations.

26. R.N. Hoover and P.H. Strasser, *Lancet* i, 837 (1980).

27. O. Auerbach and L. Garfinkel, *Cancer* 64, 983 (1989).

28. B.E. Butterworth and T.J. Slaga, "Nongenotoxic Mechanisms in Carcinogenesis," *Banbury Report No. 25* (1987).

29. R. Peto, R. Gray, P. Benton, P. Grasso, *IARC (Int. Agency Res. Cancer) Sci. Publ. No. 57* (1985), p. 627.

30. A. Poland and E. Glover, *Mol. Pharmacol.* 17, 86 (1980); V. Solanki and T.J. Slaga, in *Mechanisms of Tumor Promotion*, T.J. Slaga, Ed. (CRC Press, Boca Raton, FL, 1984), vol. 2, 97; A.L. Brooks, S.W. Jordan, K.K. Bose, J. Smith, D.C. Allison, *Cell Biol. Toxicol.* 4, 31 (1988); R.N. Hill, *Fundam. Appl. Toxicol.* 12, 629 (1989).

31. N. Ito and M. Hirose, *Adv. Cancer Res.* 53, 247 (1989).

32. J.W. Jull, *Cancer Lett.* 6, 21 (1979); T. Shirai, E. Ikawa, S. Fukushima, T. Masui, N. Ito, *Cancer Res.* 46, 2062 (1986); H. D'A. Heck and R.W. Tyl, *Regulat. Toxicol. Pharmacol.* 5, 294 (1985).

33. R.K. Boutwell, *Prog. Exp. Tumor Res.* 4, 207 (1964).

34. MS. Soloway, *Cancer* 36, 333 (1975); L.A. Levine and J.P. Richie, *J. Urol.* 141, 1063 (1989).

35. A. Paganini-Hill, R.K. Ross, B.E. Henderson, *Br. J. Cancer* 59, 445 (1989).

36. D.T. Purtilo and T. Osato, *AIDS Res.* 2, 1 (1986).
37. R.P. Beasley, *Cancer* 61, 1942 (1988).
38. We thank the late R. Greenfield, our colleagues, and technologists for their contributions, and G. Philbrick for assistance with this manuscript. Supported by grants CA32513, CA28015, and CA36727 from the National Cancer Institute, by the Department of Health, State of Nebraska, and by the International Life Sciences Institute-Nutrition Foundation.

Too Many Rodent Carcinogens: Mitogenesis Increases Mutagenesis

Bruce N. Ames, Ph.D. and
Lois Swirsky Gold, Ph.D.

Please see biographical sketches for Dr. Ames and Dr. Gold on page 151.

A clarification of the mechanism of carcinogenesis is developing at a rapid rate. This new understanding undermines many assumptions of current regulatory policy toward rodent carcinogens and necessitates rethinking the utility and meaning of routine animal cancer tests. At a recent watershed meeting on carcinogenesis, much evidence was presented suggesting that mitogenesis (induced cell division) plays a dominant role in carcinogenesis[1]. The work of Cohen and Ellwein in this issue[2] is illustrative. Our own rethinking of mechanism was prompted by our findings that: (i) spontaneous DNA damage caused by endogenous oxidants is remarkably frequent[3] and (ii) in chronic testing at the maximum tolerated dose (MTD), more than half of all chemicals tested (both natural and synthetic) are carcinogens in rodents, and a high percentage of these carcinogens are not mutagens[4].

Mitogenesis increases mutagenesis. Many "promoters" of carcinogenesis have been described and have been thought to increase mitogenesis or selective growth of preneoplastic cells, or both. The concept of promotion, however, has been fuzzy compared to the clearer understanding of the role of mutagenesis in carcinogenesis. The idea that mitogenesis increases mutagenesis helps to explain promotion and other aspects of carcinogenesis[2,5].

A dividing cell is much more at risk of mutating than a quiescent cell[4]. Mutagens are often thought to be only exogenous agents[4], but endogenous mutagens cause massive DNA damage (by formation of oxidative and other adducts) that can be converted to stable mutations during cell division. We estimate that the DNA hits per cell per day from endogenous oxidants are normally ~ 10^5 in the rat and ~ 10^4 in the human[3]. This promutagenic damage is effectively but not perfectly repaired; for example, the normal steady-state level of 8-hydroxy-deoxyguanosine (1 of about 20 known oxidative DNA adducts) in rat DNA has been measured as 1 per 130,000 bases, or about 90,000 per cell[3]. We have argued that this oxidative DNA damage is a major contributor to aging and to the degenerative diseases associated with aging, such as cancer. Thus, any agent causing chronic mitogenesis can be indirectly mutagenic (and consequently carcinogenic) because it increases the probability of converting endogenous DNA damage into mutations. Nongenotoxic agents [for example, saccharin[2]] can be carcinogens at high doses just by causing chronic mitogenesis and inflammation, and the dose response would be expected to show a threshold. Genotoxic chemicals [for example, N-2-fluorenylacetamide (2-AAF)[2]] are even more effective than nongenotoxic chemicals at causing mitogenesis at high doses (as a result of cell killing and cell replacement). Since genotoxic chemicals also act as mutagens, they can produce a multiplicative interaction not found at low doses, leading to an upward curving dose response for carcinogenicity. Furthermore, endogenous rates of DNA damage are so high that it may be difficult for exogenous mutagens to increase them significantly at low doses that do not increase mitogenesis. Therefore, mitogenesis, which can be increased by high doses of chemicals, is indirectly mutagenic, and seems to explain much of carcinogenesis[1,4,5]. Nevertheless, the potent mutagen 2-AAF induces liver tumors at moderate doses in the presence of only background rates of mitogenesis. Detailed studies of mechanism, particularly in the case of apparent exceptions, are critically important.

Causes of human cancer. Henderson and co-workers[6], and others[4], have discussed the importance of chronic mitogenesis for many, if not most, of the known causes of human cancer, for example, the importance of hormones in breast cancer, hepatitis B[7] or C viruses or alcohol in liver cancer, high salt or *Helicobacter (Campylobacter)* infection in stomach cancer, papilloma virus in cervical cancer, asbestos or tobacco smoke in lung cancer, and excess animal fat and low calcium in colon cancer. For chemical carcinogens associated with occupational cancer, worker exposure has been primarily at high, near-toxic doses that might be expected to induce mitogenesis.

Epidemiologists are frequently discovering clues about the causes of human cancer, and their hypotheses are then refined by animal and metabolic studies. During the next decade, it appears likely that this approach will lead to an understanding of the causes of the major human cancers[8]. Cancer clusters in small areas are expected to be common by chance alone, and epidemiology lacks the power to establish causality in these cases[9]. It is important to show that pollution exposure that purportedly causes a cancer cluster is significantly higher than the background of exposures to naturally occurring rodent carcinogens[4].

Causes of cancer in animal tests. Animal cancer tests are conducted at near toxic doses (the maximum tolerated dose, MTD) of the test chemical, for long periods of time, which can cause chronic mitogenesis[1]. Chronic dosing at the MTD can be thought of as a chronic wounding, which is known to be both a promoter of carcinogenesis in animals and a risk factor for cancer in humans. Thus, a high percentage of all chemicals might be expected to be carcinogenic at chronic, near toxic doses and this is exactly what is found. About half of all chemicals tested chronically at the MTD are carcinogens[4].

Synthetic chemicals account for 82% (350/427) of the chemicals adequately tested in both rats and mice[4]. Despite the fact that humans eat vastly more natural than synthetic chemicals, the world of natural chemicals has never been tested systematically. Of the natural chemicals tested, approximately half (37/77) are carcinogens, which is approximately the same as has been found for synthetic chemicals (212/350). It is unlikely that the high proportion of carcinogens in rodent studies is due simply to selection of suspicious chemical structures; most chemicals were selected because of their use as industrial compounds, pesticides, drugs, or food additives.

One major group of natural chemicals in the human diet are the chemicals that plants produce to defend themselves, the natural pesticides[4]. We calculate that 99.99% (by weight) of the pesticides in our diet are natural. Few natural pesticides have been tested in at least one rodent species, and again about *half* (27/52) are rodent carcinogens. These 27 occur commonly in plant foods (10). The human diet contains thousands of natural pesticides and we estimate that the average intake is about 1500 mg per person per day[4]. This compares to a total of 0.09 mg per person per day of residues of about 100 synthetic pesticides[4]. In addition, of the mold toxins tested at the MTD (including aflatoxin) 11 out of 16 are rodent carcinogens.

The cooking of food produces thousands of pyrolysis products, and we estimate that dietary intake of these products is roughly 2000 mg per person per day. Few of these have been tested; for example, of 826 volatile chemicals that have been identified in roasted coffee, only 21 have been tested chronically, and 16 are rodent carcinogens; caffeic acid, a non-volatile carcinogen, is also present. A cup of coffee contains at least 10 mg (40 ppm) of rodent carcinogens (mostly caffeic acid, catechol, furfural, hydrogen peroxide, and hydroquinone)[4]. Thus, very low exposures to pesticide residues or other synthetic chemicals should be compared to the enormous background of natural substances.

In the evolutionary war between plants and animals, animals have developed layers of general defenses, almost all inducible, against toxic chemicals[4]. This means that humans are well buffered against toxicity at low doses from both man-made and natural chemicals. Given the high proportion of carcinogens among those natural chemicals tested, human exposure to rodent carcinogens is far more common than generally thought; however, at the low doses of most human exposures (where cell-killing and mitogenesis do not occur), the hazards may be much lower than is commonly assumed and often will be zero[4]. Thus, without studies of the mechanism of carcinogenesis, the fact that a chemical is a carcinogen at the MTD in rodents provides no information about low-dose risk to humans.

Trade-offs. Pesticide residues (or water pollution) must be put in the context of the enormous background of natural substances, and there is no convincing evidence from either epidemiology or toxicology that they are of interest at causes of human cancer[4,9]. Minimizing pollution is a separate issue, and is clearly desirable for reasons other than effects on public health. Efforts to regulate synthetic pesticides or other synthetic chemicals at the parts per billion level because these chemicals are rodent carcinogens must include an understanding of the economic and health-related trade-offs. For example, synthetic pesticides have markedly lowered the cost of food from plant sources, thus encouraging increased consumption. Increased consumption of fruits and vegetables, along with decreased consumption of fat, may be the best way to lower risks of cancer and heart disease, other than giving up smoking. Also, some of the vitamins, antioxidants, and fiber found in many plant foods are anticarcinogenic.

The control of the major cancer risks that have been reliably identified should be a major focus, and attention should not be diverted from these major causes by a succession of highly publicized scares about low levels of synthetic chemicals that may be of little or no importance as causes of human disease. Moreover, we must increase research to identify more major cancer risks, and to better understand the hormonal determinants of breast cancer, the viral determinants of cervical cancer, and the dietary determinants of stomach and colon cancer. In this context, the most important contribution that animal studies can offer is insight into carcinogenesis mechanisms and into the complex natural world in which we live.

References

1. B.E. Butterworth and T. Slaga, Eds. *Chemically Induced Cell Proliferation: Implications for Risk Assessment* (Wiley-Liss, New York, in press).
2. S.M. Cohen and L.B. Ellwein, *Science* 249, 1007 (1990).
3. B.N. Ames, *Free Rad. Res. Commun.* 7, 121 (1989); C.G. Fraga, M.K. Shigenaga, J.-W. Park, P. Degan, B.N. Ames, *Proc. Natl. Acad. Sci. U.S.A.* 87, 4533 (1990).
4. B.N. Ames, M. Profet, L.S. Gold, *Proc. Natl. Acad. Sci. U.S.A.*, 87, 7777–7781; B.N. Ames, M. Profet, L.S. Gold, *ibid.*, 7782–7786; B.N. Ames and L.S. Gold, *ibid.*, 7772–7776; *Med. Oncol. Tumor Pharmacother* 7, 69 (1990); B.N. Ames, *Environ. Mol. Mutagen.* 14,66 (1989); ———, R. Magaw, L.S. Gold, *Science* 236, 271 (1987); L.S. Gold et al., *Environ. Health Perspect.* 81, 211 (1989).
5. J.E. Trosko, J. *Am. Coll. Toxicol.* 8, 1121 (1989); ———, C.C. Chang, B.V. Madhukar, S.Y. Oh, *In Vitro Toxicol.* 3, 9, 1990; Trosko has proposed that suppression of gap junctional intercellular communication in contact-inhibited cells could lead to cell proliferation by cell death, cell removal, promoting chemicals, specific oncogenic products, growth factors, and hormones.
6. B.E. Henderson, R. Ross, L. Bernstein, *Cancer Res.* 48, 246 (1988); S. Preston-Martin et al., in *Chemically Induced Cell Proliferation: Implications for Risk Assessment*, B. E. Butterworth and T. Slaga, Eds. (Liss, New York, in press).
7. H.A. Dunsford, S. Sell, F.V. Chisari, *Cancer Res.* 50, 3400 (1990).
8. Current epidemiologic data point to these risk factors for human cancer: cigarette smoking (which is responsible for 30% of cancer deaths), dietary imbalances, hormones, viruses, and occupation. "The age-adjusted mortality rate for all cancers combined except lung cancer has been declining since 1950 for all individual age groups except 85 and above" [National Cancer Institute, 1987 *Annual Cancer Statistics Review Including Cancer Trends: 1950–1985*," NIH Publication 88-2789 (National Institutes of Health, Bethesda, MD, 1988), p. II.3] Although incidence rates for some cancers have been rising, trends in recorded incidence rates may be biased by improved registration and diagnosis. Even if particular cancers can be shown to be increasing (for example, non-Hodgkins lymphoma and melanoma) or decreasing (for example, stomach, cervical, and rectal cancer), establishing causes remains difficult because of the many changing aspects of our life-style. Life expectancy continues to increase every year.
9. J. Higginson, *Cancer Res.* 48, 1381 (1988).
10. A search in foods for the presence of just these 27 natural pesticide rodent carcinogens indicates that they occur naturally in the following (those at levels over 10 ppm of a single carcinogen are listed in italics): *anise, apple,*

banana, *basil,* broccoli, *brussels sprouts, cabbage,* cantaloupe, *caraway, carrot, cauliflower, celery, cherry,* cinnamon, cloves, cocoa, *coffee (brewed), comfrey tea, dill, eggplant, endive, fennel, grapefruit juice, grape, honey,* honeydew melon, *horseradish,* kale, *lettuce, mace, mango,* mushroom, *mustard (brown), nutmeg, orange juice, parsley, parsnip,* peach, *pear, pepper (black),* pineapple, *plum, potato,* radish, raspberry, *rosemary, sage, sesame seeds (heated)* strawberry, *tarragon, thyme,* and turnip (4). Particular natural pesticides that are carcinogenic in rodents can be bred out of crops if studies of mechanism indicate that they may be significant hazards to humans.

11. This work was supported by National Cancer Institute Outstanding Investigator grant CA39910, by National Institute of Environmental Health Sciences Center grant ES01896 and by DOE Contract DE-AC03-76SF00098. We thank M. Profet, S. Linn, B. Butterworth, and R. Peto for criticisms.

This paper has been reprinted, with permission, from
Science, Vol. 249, pp. 970–971.
Copyright © 1990 by the American Association
for the Advancement of Science (AAAS).

Worried About TCE? Have Another Cup of Coffee

Jane M. Orient, M.D.

Dr. Jane M. Orient is executive director, Association of American Physicians and Surgeons, and an assistant clinical lecturer in, internal medicine, University of Arizona. She received a B.S. in mathematics and a B.A. in chemistry from the University of Arizona in 1967, and an M.D. from Columbia University in 1974.

Dr. Orient is the contributing editor of *The Art and Science of Bedside Diagnosis* by Joseph D. Sapira, published by Urban and Schwarzenberg, Baltimore, 1990. She has published more than 50 professional, contributing to the *Annals of Internal Medicine, Southern Medical Journal,* and the Journal of the *American Medical Association.*

Before you drink that cup of coffee, there's something you should know.

The FDA tolerance levels for trichloroethylene (TCE) in coffee are 25 parts per million (ppm) for ground coffee and 10 ppm for instant coffee. That's 25,000 parts per billion (ppb) and 10,000 ppb, respectively.

The worst Tucson water well was contaminated with 123 ppb TCE.

Now that the Environmental Protection Agency has declared TCE a "probable human carcinogen," there have been calls for the resignation of the director of public health here, and news that the Air Force plans to spend $25 million (for starters) to clean up the plume of TCE.

Just how serious is the situation, and how much money should we spend on it?

What does it mean to say that something is a "probable human carcinogen"? It means that maybe the stuff causes cancer in humans, and maybe it doesn't.

If TCE causes cancer, the number of cases is so low that the effect is difficult to prove. That's because cancer is a very common disease. It causes about 180,000 out of every 1 million deaths. An exposure that caused an additional incidence of 10 in a million would raise the figure to 180,010 out of 1 million. There are so many factors involved in cancer that it would be hard to prove that substance X was responsible for that small difference.

So how much do we know about TCE? How do we know that it isn't in fact a very powerful carcinogen?

TCE is still very commonly used for degreasing and dry cleaning, and was formerly used for decaffeinating coffee. One of the reasons is that it is much safer than the alternatives: carbon tetrachloride, for example. The limit for occupational exposure to TCE is 100 ppm (or 100,000 ppb). About 290,000 U.S. workers are potentially exposed to this solvent.

There's no doubt that sniffing high concentrations of TCE (or other solvents such as model airplane glue) is bad for you. It can give you cirrhosis of the liver in the long run. If the concentration is high enough, it can cause seizures or unconsciousness. In fact, TCE can even be used as a general anesthetic.

But TCE has not been shown to be a public health hazard, although it has been found in water supplies quite frequently. In Bedford, Mass., 80 percent of the drinking water was contaminated with up to 500 ppb TCE. That's before **coffee** was brewed with it. With instant coffee added, the level might have been 25,500 ppb.

Of course, you'd rather drink the cleanest water possible. But there's a price tag. How much should we be willing to spend? It's important to remember that money spent for one purpose cannot be used for something else. If our main goal is to save lives, then we need to put the risks in perspective and concentrate our effort where it will do the most good.

The following table shows some of the ways that we could spend money, and the number of lives that $1 million might save:

Item	Lives saved per $1 million
Sulfur scrubbers in power plants	2
Screening for breast cancer	12.5
Rescue helicopters	15.4
Smoke alarms in homes	16.7
Highway maintenance	50.0
Screening for colon cancer	100.0
Expanded immunization in Indonesia	10,000.0
Reducing TCE from 100 ppb to 5 ppb	0.0

It is likely that nobody has died yet from drinking water contaminated with the levels of TCE that have been found. On the other hand, between 5,000 and 30,000 people may die each year from breathing radon in the air in their houses. The radon gas comes from natural sources, and is concentrated in the house by energy conservation measures. Better ventilation of houses would prevent many cases of lung cancer in nonsmokers. Yet this problem has not been in the headlines, or in health department budgets.

The $25 million planned for cleanup of the drinking water plume could also be applied to other projects related to public health and safety: assuring a future water supply for Tucson; research into better ways of detoxifying wastes (using bioengineered bacteria, for example); defending Davis-Monthan against nuclear warheads (especially if the money is from the Air Force budget); or educating children not to sniff solvents or abuse other substances.

Any of the above uses has more life-savings potential than do crusades against TCE. That's why I believe that drinking a little TCE—and directing $25 million to other programs—can save lives.

I wonder if they serve coffee at the meetings about the hazards of this useful cleaning agent?

Reprinted, with permission, from
the *Tucson Citizen*, Tucson, Arizona,
December 6, 1985

Some Convinced TCE 'Guilty' Despite Lack of Evidence

Jane M. Orient, M.D.

Please see biographical sketch for Dr. Orient on page 186.

Several relatively rare cancers appeared in a small community within a short period of time.

The drinking water in that community was contaminated with small amounts of a suspected animal carcinogen.

Did the second event cause the first?

Maybe, but there's no evidence for it.

Clusters of rare diseases are only clues. An epidemiologist can study the cases and try to find out what the victims had in common. Then he can formulate a hypothesis, which must be tested further. Say that all the victims were fond of red chili. Should they immediately sue the farmer who grew the chili?

Wait a minute. Further testing would probably show that most red chili lovers *don't* have Disease X, and that they are no more likely to get that disease than folks who never touch chili peppers. The researchers would have to go back to the lab.

Outbreaks of diseases have often been traced to bacteria or viruses. For example, cholera occurred in Londoners who drank from a certain well. Other possibilities include hereditary or environmental factors. Tobacco, asbestos, vinyl chloride, arsenic, soot, tar, and diethystilbestrol are established carcinogens, discovered through epidemiological evidence.

But some clusters of diseases *don't* have a common cause—they just happen. It is not necessarily abnormal for a neighborhood to have a "higher than average" incidence of a particular disease—just as it's normal for some students to get a "higher than average" score on a test. The normal curve covers a wide range. A lot of cases fall near the middle or "average," and some fall near the ends.

Furthermore, not every class of students will have exactly the same distribution of test scores. Some have a higher concentration of brighter students, by chance alone. Some classes have more tall students, some have more sickly children, and some may have four or more children whose birthdays fall on the very same date.

If we do have a community with what appears to be an unusual incidence of a particular disease, we should first ask just *how* unusual it is. The answer to that question is often expressed in terms of "statistical significance." If the observed incidence of a disease is "significantly" higher than "expected," say "at the .05 level," that means that only 5 out of 100 similar populations would have an incidence that high by *chance alone.*

However, since there are lots of rare diseases, the chance that a small community will have "more than its fair share" of at least a few of them is actually quite high. The more diseases one looks at, and the larger the number of subgroups considered (age groups, for example), the more likely one is to "discover" a higher than "normal" incidence of something or other.

It's also likely that a community will have "less than its fair share" of some other diseases—but that tends to cause less excitement. And it certainly wouldn't prove that TCE, or chili peppers, or whatever is under suspicion, prevents those diseases!

To put the probabilities in perspective, it turns out that a man's chance of getting testicular cancer in a certain year is about the same as the chance that he will draw a full house in a single hand of Five Card Stud Poker. In a population of 10,000 men, the number of cases expected in 20 years is about the same as the number of full houses in 200,000 hands of Five Card Stud. What if a few extra are observed?

A run of good luck in poker might be just that. Or it might have another explanation—such as cheating.

It's possible that a health problem on Tucson's South Side (if one is shown to exist), like a good poker hand, is *not* just due to chance. If that is so, determining the actual cause of the problem is not easy. To assume that TCE must be the culprit is like convicting somebody of murder because he happened to be acquainted with all three recent victims. The evidence is circumstantial.

TCE was in the water, and there is some evidence that this substance *might* cause cancer in animals. Even if it does, animal carcinogens do not necessarily cause human cancers too. And there are a lot of other suspects. The list of potential carcinogens includes chili peppers as well as many—most, really—other foods. To name just a few: mushrooms, some types of honey, corn oil, moldy sweet potatoes, soybeans, many green vegetables, cheese, eggs and meat.

Notice that these are natural substances, like the one responsible for the greatest number of cancers (tobacco). There is, of course, less indignation about natural carcinogens. Could that be because the "almighty dollar" is not involved? After all, you can't sue the Almighty God for putting a carcinogen in your chili, whereas you could sue a corporation for allowing a carcinogen into your water. The "vested interests" that complicate the identification of the culprit (if any) are on *both* sides of this question.

Would you like to bet that TCE causes human cancer? It's a hunch—but you could be right.

You might also win the lottery, and your horse might come in first, or you might get more than your fair share of winning poker hands.

Could you *prove* your good fortune to be due to chance alone?

TCE—or red chili—will never be proved innocent. But before calling the lawyers, we ought to try to find some evidence for its guilt.

Reprinted, with permission, from
the *Tuscon Citizen*, June 18, 1985

DDT

DDT Effects on Bird Abundance and Reproduction
J. Gordon Edwards

The Tragedy of DDT
Thomas H. Jukes

DDT Effects on Bird Abundance and Reproduction

J. Gordon Edwards, Ph.D.

Please see biographical sketch for Dr. Edwards on page 125.

Early in the DDT controversy so many allegations were made that it was difficult to respond to them all in the available time. Hundreds of scientists patiently refuted allegations, however, and their contributions are summarized in many more recent books and articles.[1-10]

The news media were informed that they had been misinformed by some environmentalists, and a few carried stories that were more in agreement with the facts. They also became aware that most scientists did not support the claims that DDT caused cancer, mutations, hepatitis, poliomyelitis or other human illnesses.[8,11] It was made clear that DDT residues do NOT "build up" in animal food-chains, because they are metabolized or excreted by fish, birds and mammals.[12-15]

Bird Populations Threatened?

Attempts were made by some persons to link bird declines with the heavy usage of DDT between 1946 and 1960 (and to imply that birds later increased after DDT was banned). When those claims were investigated with objectivity it became evident that most declines had either occurred before DDT was present, OR after levels of DDT had waned.[3-6,15] The Environmental Defense Fund commented on the abundance of birds during the DDT years, with "increasing numbers of pheasants, quail, doves, turkeys and other game species."[17] To the list of increasing bird populations, the Audubon Society added eagles, gulls, ravens, herons, egrets, swallows, robins, grackles, red-winged blackbirds, cowbirds, and starlings.[19] The annual Audubon Christmas Bird Counts between 1941 (pre-DDT) and 1960 (after DDT use had waned) showed that at least 26 different kinds of birds became more numerous during those decades of greatest DDT usage. They documented an overall increase from 90 birds seen per observer in 1941 to 971 birds seen per observer in 1960.[18,19] Statistical analyses of the Audubon data confirmed the perceived increases.

At Hawk Mountain, Pennsylvania, counts of migrating raptors were made daily by teams of ornithologists for over 40 years. The results were published annually by the Hawk Mountain Sanctuary Association and revealed great increases (not decreases) of most kinds of hawks during the DDT years.[20] Total numbers of hawks counted there increased from 9,291 in 1946 to 13,616 in 1956 and 20,196 by 1967. Sparrow hawks seen were as follows: 1946—98; 1956—192; 1965—408; and 1967—666. Ospreys (which environmentalists kept saying were in danger of extinction by DDT) increased as follows: 1946—191; 1956—288; 1961—352; 1967—457; 1969—529 ; and 1972—630.

J. J. Hickey (1942) had reported that 70% of the eastern ospreys had recently been killed by pole traps around fish hatcheries[21] and there were many reports of high levels of mercury in the fish upon which ospreys depended for food (mercury causes drastic thinning of shells in eggs produced).[34–37,121–125] It was alleged, however, that DDT and DDE had contaminated the fish world-wide, and it was because of that that fish-eating birds could not reproduce successfully! Chesapeake Bay ospreys had no apparent problem with DDT or DDE, but "lipids (in their eggs) contained PCB levels from 545 to 2270 parts per million."[130] PCBs were later shown to cause drastic thinning of eggshells, as well as many other adverse effects on birds, yet environmentalists continued to place the blame on DDT despite the fact that feeding birds high levels of that pesticide did NOT cause birds to produce such thin eggshells.

Further evidence that the allegations were unfounded abound. The fish-eating gannets of Funk Island (in the North Atlantic) increased from 200 pairs in 1945 (when DDT use began) to 2,00 pairs in 1958 and 3,000 pairs by 1970 (before DDT was banned). Murres increased there also, from 15,000 pairs in 1945 to 150,000 pairs in 1956 and a million and a half pairs by 1971.[23]

Many other kinds of birds multiplied so well in the presence of DDT and DDE that they became destructive pests. In North Carolina, six million blackbirds ruined the town of Scotland Neck in 1970, polluting streams, depositing nine inches of droppings on the ground, and fouling the forests where they roosted at night.[26] Near Fort Campbell, Kentucky, 77 million blackbirds were roosting, and the resulting threat of human disease (histoplasmosis) was of great concern.[25] Audubon Magazine reported: "Today in a small area of northern Ohio ten million redwings mill about in the cornfields after the nesting season."[26] They destroyed the crops, while farmers begged for avicidal relief. In Virginia the Department of Agriculture stated: "We can no longer tolerate the damage caused by redwings . . . 15 million tons of grain are destroyed annually—enough to feed 90 million people."[27] Near Nantucket Sound gulls had increased from 2,000 pairs in 1941 to over 35,000 pairs in 1971. Even though the gulls were on the state's "protected list," Audubon Society bird-lovers poisoned 30,000 of them on Tern Island, because they were threatening the terns which the "birders" preferred. William Drury, the Audubon leader, stated that "it's kind of like weeding a garden."[28]

Obviously, heavy DDT usage was not eradicating birds of any kind . . . but what caused the great increase of birdlife in areas that had been sprayed by DDT? The reduction in numbers of blood-sucking insects provided birds with protection from avian malaria, rickettsialpox, bronchitis, Newcastle disease and encephalitis. Also, more seeds and fruits were available for birds after phytophagous insects had been killed by the insecticides. Perhaps even more important, DDT is known to trigger the induction of hepatic enzymes that detoxify potent carcinogens (such as aflatoxins) which abound in the natural foods eaten by birds. Aflatoxins are carcinogenic at levels of 0.03 to 0.08 ppm in the diets of birds, fish and mammals.[29] Many humans in third world countries also die because of aflatoxins in their diet, and some health authorities in United States observed that when they ingest DDT with their food they are less likely to suffer from cancer.[29,30]

Brown Pelicans

California brown pelicans experienced NO difficulty during all the years of heavy DDT usage[31–33] but suffered a sudden catastrophic reproductive failure ten years later . . . only two months after the nearby Santa Barbara Oil Spill surrounded their breeding island of Anacapa.[32,33] That oil, through which the pelicans were diving for food, was not even mentioned in pelican reports issued by federal or state fish and wildlife agencies even though oil, when ingested by birds, causes them to produce thin-shelled eggs, and oil on normal eggs is known to kill the embryos within.[38–44] Dieter found that just 5 microliters of oil on eggs caused 76% to 98% mortality of embryos, and "More importantly, flight feathers failed to develop normally" and "liver atrophy and splenic atrophy were evidence of the pathological effects of the oil."[46] Szaro (USFWS) reported that single low doses of oil in food caused kidney nephrosis, pancreatic degeneration, pneumonia, and extreme dehydration.[43]

Articles by government biologists revealed that they collected 72% of the intact pelican eggs on Anacapa (for analysis) and that they shotgunned incubating pelicans on their nests there.[46] The thickness of their eggshells bore no good correlation with the levels of DDT residues in the eggs.[47,48] Inconsistencies involving the data were exposed by Switzer during the EPA Hearings in Washington.[47] A similar lack of correlation has been reported by scientists working with ospreys, bald eagles, gulls, hawks and gallinaceous birds. Hazeltine discusses the inverse correlation between residues and shell thickness.[48]

In late 1968 (just before the first California "pelican problems") Jehl and Keith wrote that "Sick pelicans were observed and hundreds of dead pelicans were found along the northwestern shoreline of the Gulf of California."[51] At that time colonies along the Mexican coast were producing misshapen, soft-shelled eggs and southern California soon thereafter experienced a major epidemic of Newcastle Disease. Millions of domestic birds were killed in southern California in the process of eradicating that disease.[52] A major symptom of Newcastle is the production of misshapen, thin-shelled eggs identical to those along the Mexican coast and on Anacapa Island in 1969.[53,54] The seriously ill pelicans were never mentioned again, and Anacapa pelicans with Newcastle Disease were not mentioned publicly by any fish and wildlife biologists. Neither were the hundreds of large rats in the Anacapa pelican colony or the feral cats that had been left on that island by the former resident who was employed by the National Park Service.[32,33] Instead, the biologists devoted all of their efforts to convincing the public, via the news media, that DDT and only DDT was the cause of the thin shells and low reproductive rate of the Anacapa pelicans.

Dawson, in his BIRDS OF CALIFORNIA (1923) described how just the brief presence of a human on Anacapa Island caused the adult pelicans to spring from their nests, breaking their eggs underfoot as they did so, and how the exposed eggs and nestlings were then quickly devoured by marauding gulls.[55] Despite such warnings, fish and game biologists landed helicopters repeatedly in the breeding areas in 1969 and 1970 and one of them (Franklin Gress) spent days actually sitting in old nests in the colonies.[3,4,32] He observed the pelicans deserting the colonies, while he sat there, but blamed it all on DDT![56] National Park rangers were appalled by such activities and by allegations that the Anacapa pelicans were threatened with extinction because of DDT. Superintendent Donald Robinson told reporters that "This business with the pelicans is all the doing of only one man's research, and he's a rabid opponent of DDT."[57]

Even though the destruction and desertion of the Anacapa colonies could therefore not have been unexpected by the paid government biologists, they sought to blame it entirely on traces of DDT in the fish eaten by the birds. Significantly, no feeding experiments dosing captive pelicans with DDT or DDE were ever shown to cause any eggshell distortions. The EPA ban on DDT was based in large measure on the alleged "environmental harm to birds caused by DDT."

At the request of the National Park Service, Richard Main and J. G. Edwards drew up protective regulations for the Anacapa pelican colonies in 1970. The rangers then enforced the regulations, keeping state and federal biologists out of the colonies during the breeding season. Despite a few illegal invasions, the birds quickly increased their productivity that year, even though there had certainly been no decrease in the amounts of DDT and DDE present in the water, the fish, or the pelicans.[33]

In Texas the brown pelican population had declined from 5,000 birds in 1918 to a low of 200 in 1941 (three years before any DDT was used in U. S.).[58-60] An 82% decrease in pelicans there between 1918 and 1935 was documented by Audubon ornithologists.[61]

In 1939 Gustafson attributed the disappearance of Texas pelicans to fishermen and hunters, saying the population had been reduced from 5,000 birds to 500 and that "the future of the brown pelican in those areas is uncertain."[62] Acting as if they were unaware of the data, and ignoring dozens of ornithological articles and books, federal wildlife biologists repeatedly stated in speeches and reports that along the Texas coast "a population of over 50,000 brown pelicans has all but disappeared since 1961 without any ornithologist even having investigated the cause."(emphasis added)[46,63] The cause. they indicated, was DDT. That erroneous allegation (and many others) was publicized during the U. S. Congressional Committee hearings in 1971 and 1973.[3,4] U. S. Fish and Wildlife spokesman R. B. Finley later privately retracted the damaging allegations in a letter to committee chairman W. R. Poage, in which he wrote that "The year 1961 was merely a hasty approximation of an unknown time" and that "After reviewing the evidence, I think now that I should have said they disappeared *by* 1961" (rather than *since* 1961).[64] The significance of Finley's written retraction of that damaging anti-DDT allegation was great, but the Service has never publicly corrected their ridiculous figure of 50,000 Texas pelicans or the implication that they were suddenly eradicated by DDT, *after* 1961! That false allegation has been repeated dozens of times by the news media and by the pseudoenvironmental organizations that elicited millions of dollars in donations "to save the environment from DDT," and it still appears in books, magazines and newspapers.

Bald Eagles

In 1921 Ecology magazine stated that bald eagles were "threatened with extinction."[65] From 1917 to 1952 Alaska paid over $100,000 in bounties on about 115,000 of those noble birds, and they were still abundant there.[66] Bent wrote in 1937 that "The bald eagle probably nested at one time over much of New England, but there are no recent authentic records of its nesting in the three southern states of New England."[67] (That was eight years before DDT was ever used, and the eagles surely could not have disappeared in anticipation of DDT!) After 15 years of heavy and widespread usage of DDT Audubon ornithologists still counted 25% more eagles per observer in 1960 than during the pre-DDT 1942 Christmas bird census.[18,19] Spencer (1976) provided updated details of migratory bald eagles in his well-documented book.[68]

Postupalsky found no significant correlation between DDT or DDE residues and shell thicknesses in a large series of bald eagle eggs.[50] Krantz et al. analyzed bald eagle eggs from three states,[49] and found that Florida eggshells were thinnest (0.50), followed by those from Maine (0.53) and Wisconsin (0.55). Analyses of their DDT residues revealed an inverse correlation, with the Maine eggs containing 21.76 ppm of DDE, twice as much as those from Florida and five times as much as those from Wisconsin. Obviously DDT and DDE did not cause shell-thinning! In Wisconsin and Michigan, an increase in bald eagle productivity caused increases from 51 young produced in 1964 to 107 in 1970.[69]

In 1966 Fish and Wildlife biologists fed large doses of DDT to captive bald eagles for 112 days, then concluded that "DDT residues encountered by eagles in the environment would not adversely affect eagles or their eggs."[70] Other wildlife authorities attributed bald eagle population reductions to "a widespread loss of suitable habitat" but said "illegal shooting continues to be the leading cause of direct mortality in both adult and immature eagles."[71] USFWS biologists analyzed every bald eagle found dead in the United States from 1961 to 1977 (226 birds) and reported no adverse effects caused by DDT or its residues.[71-74] They warned that "The analyses revealed many unidentified compounds that can interfere with the gas chromatograph analyses of endrin, DDT and DDD" and observed that "PCBs occurred in the same order of magnitude as DDE."

In 1972 the eagles contained twice as much PCB as DDT, DDD and DDE combined, and their brains had four times as much PCB. Their conclusion was that "the role of pesticides has been greatly exaggerated."[73] Bagley found 19 different PCB compounds in extracts from two bald eagles.[76]

Because of the high levels of mercury in the fish that eagles were eating, eggshell thinning should have been anticipated, but it was not until much later that FWS biologists admitted that mercury inhibited eagle reproduction.[77] In 1984 the National Wildlife Federation still ranked shooting as the leading cause of eagle deaths, followed by powerline electrocution, collisions in flight, and poisoning from eating ducks that contained lead shot.[78]

The great increase in bald eagle numbers in the contiguous 48 states has ben reported with great enthusiasm by the anti-DDT collaborators, but rather than cite authentic explanatory details, they almost always attribute the increase to the 1972 ban on DDT. Any myth, if repeated often enough, can obviously attain the stature of a "fact," no matter how little evidence actually supports it!

Peregrine Falcons

In Britain the war-time shooting of about 600 peregrines by the Air Ministry campaign of 1940–1945 (to protect carrier pigeons) reduced the raptors to half their pre-1939 level.[79] Ratcliffe had earlier estimated that British peregrine shells had become "almost 20% thinner" by 1947, but he never actually measured the thickness of any eggshells.[80] Instead he used an unreliable "shell thickness index" based on the egg weight, which was soon discredited. In 1970, Ratcliffe stated that "The halt to the crash from 1963 onwards resulted from the implementation of partial bans on the use of seed dressings with dieldrin, aldrin and heptachlor in 1961."[81] In 1980 he absolved DDT of guilt and provided much valuable information.[79] Newton (1974) reported that "Dieldrin has been identified as the chief cause of the peregrine's decline in Britain, not DDT, which was always used less there."[82] A three-year study by the British government (The Wilson Report) revealed that the British decline ended in 1966 even though DDT was as abundant as ever, and concluded "There is no close correlation between the decline in populations of predatory birds, particularly the peregrine falcon and the sparrow hawk, and the use of DDT."[83]

DDT was widely blamed for decimating peregrine falcon populations in many parts of the world. Such charges were, however, directly refuted by all available data from the field and from laboratory studies.

The great peregrine decline in eastern United States occurred long before any DDT was present.[84–91] The disappearance of the passenger pigeon certainly played a major role before the turn of the century, but in 1913 Dr. William Hornaday of the New York Zoological Society referred to the "duck hawks" (peregrines) as undesirable birds which "deserve death but are so rare that we need not take them into account."[87] Persons who found a nest of them were urged to "shoot the parents and destroy the eggs or young in the nest." Berger later wrote that although falconers had caused the desertion of a great many eyries in eastern U. S. "a second, and quite possibly even more important factor, was the prevalence, for 70 years, of fanatic egg collection."[88] Oologists often specialized in peregrine eggs, for which they found a ready market. Rice noted that 53 sets of eggs were taken in Vermont before 1934 and most eyries were deserted by 1940.89 Roger Tory Peterson wrote: "One collector in Philadelphia showed me a cabinet full of them, drawer upon drawer. This awesome display represented years when the eyries in eastern Pennsylvania fledged scarcely a single young bird. A Boston oologist was reputed to have 180 sets—which means more than 700 peregrine eggs! Sixty, eighty and a hundred sets were frequent, and one collector told me that he secured the eggs at one eyrie each year for 29 years but now, 'due to civilization, that site is no longer occupied.' He claims he caused less damage than the falconers, who would often climb the cliffs first, rough up the newly-laid eggs, and daub India ink on them to make them worthless to collectors."[90]

Peregrines disappeared gradually from western United States, and Hickey commented that "We are now convinced that this is due to a climate change, with increasing annual temperatures, decreasing precipitation, and a shrinkage of the aquatic habitats on lake edges which peregrines probably use in that region."[91]

In Canada, peregrines were "reproducing normally" in the 1960s even though their tissues contained thirty times more DDT. DDD and DDE than the midwestern peregrines that were allegedly being extirpated by those chemicals.[94] In fact, the peregrine bearing the very highest DDT residues ever found (2,435 ppm) was efficiently feeding her three healthy young![92] Fyfe admitted that there was no decline in the far north between 1950 and 1967, despite DDT and DDE being present in birds and eggs.[93]

As soon as U. S. peregrines were officially declared to be "endangered," environmental activists were paid by the government to "study" them. They trapped brooding birds on their nests, removed fat samples from live birds for analysis[95] and operated time-lapse cameras beside the nests night and day.[96] They also collected a large portion of the eggs annually and killed dozens of nestlings by suffocation (for later analysis).[97] They then pointed with alarm at "declining numbers" of the persecuted peregrines.) Cade (1959) warned that when disturbed, the parent will leave the eggs unprotected and that may result in embryo deaths during cold weather. Of 21 disturbed nests in Alaska, he found that six batches of eggs were chilled so badly that they did not hatch.[99] In Alaska, that harrassment, predation and destruction by government-financed biologists continued during the 1950s and 1960s.[86,94–98,100] Beebe (1971) devoted many pages of his books to an expose of the shameful manipulations by members of the environmental clique who had cornered the lucrative market on peregrine studies and propaganda in Canada and Alaska. He quoted from letters and documents written by those men (including Cade) and exposed details of their controversial studies along the Yukon and Colville Rivers.[86,100] Shooting, egg collecting, killing of nestlings, and constant harrassment of nesting peregrines there were the real causes of the eventual decline of those populations, and the primary guilt must be placed on the paid "environmentalists."[94–100]

Despite the rarity of peregrine falcons, and their official status as an "endangered species," dozens of U. S. falconers continued to harrass and trap them. Falconer Steve Herman bemoaned the fact that peregrines suffer a high mortality rate while in the traps.[101] The Bureau of Sports Fisheries and Wildlife indicted 37 falconers in 1984 and fined 14 of them a total of $89,000 (many of them were still awaiting sentencing). This culminated a three-year undercover operation and it was estimated that falconers had taken 2,380 raptors illegally during those three years. Early in 1985 nineteen more were arrested in Canada for illegal trafficking in birds of prey.[102] Officials assert that more than 100 wild raptors, primarily peregrines and gyrfalcons, were removed from nests in northern Canada and shipped to buyers in the Middle East, West Germany, Japan, and the United States. Hundreds of thousands of dollars worth of birds were sold by those dealers.[103] The bulk of the peregrine trade remains under control of a few members of the old clique, which is now headed by Cade and his colleagues. They have received hundreds of thousands of dollars in government financing for their elaborate "Peregrine Falcon Recovery" organization.

Causes of Eggshell Thinning

It has been demonstrated repeatedly in caged experiments that DDT, DDD and DDE do not cause significant shell-thinning, even at levels many hundreds of times greater than wild birds would ever accumulate.[34,36,37,104–109,112] During many years of carefully controlled feeding experiments Scott et al. "found no tremors, no mortality, no thinning of eggshells and no interference with reproduction caused by levels of DDT which were as high as those reported to be present in most of the wild birds where 'catastrophic' decreases in shell quality and reproduction have been claimed." Also, "DDT did not have any deleterious effect upon the sex hormones involved in egg production and may indeed have had a beneficial effect upon egg shell quality."[108]

Older birds produce thinner shells than normal,[111,135] as do very young birds. Eggshells become 5% to 10% thinner as the developing embryo withdraws calcium for bone development.[136–138] Stress from noise, fear, or excitement,[136–141] and numerous diseases of birds also cause them to produce thin-shelled eggs.[53,54] Dehydration,[142] temperature extremes,[98] decreased illumination,[115–119] and intrusion of humans or predators into the breeding areas always cause birds to form thinner eggshells. Simple restraint interferes with the transport of calcium throughout the body, preventing it from reaching the shell gland and forming good shells.[139–141]

The most notorious cause of thin eggshells is the deficiency of calcium in the diet.[110–113,131] Calcium-starved birds often eat their own eggs (a syndrome that is then of course also blamed on DDT). It was outrageous for researchers at Patuxent to deliberately feed their birds only calcium-deficient food (0.5% rather than the necessary 2.5% calcium) and then attribute all shell problems to the DDT and DDE they had added to that calcium-deficient diet.[113,114] After much criticism for their use of calcium deficient diets that they knew would give the false impression that DDT caused shell-thinning, Bitman et al. repeated their tests with DDT and DDE, but put adequate calcium in the birds' diet. They thus proved that with sufficient calcium in their food the quail produce eggs without thinned shells.[104] Science would not publish such results, so they were published in Poultry Digest instead. There were no reviews or comments in other journals or by news media who were sent dozens of copies of the article.

Peakall abruptly reduced the illumination in his dove cages from 16 hours daily to just 8 hours daily . . . at the same time he began feeding the birds high levels of DDT in all of their food. He then alleged that DDT had inhibited shell formation, rather than the physiological stress and decreased food consumption by birds in the dark.[115,116] Even chickens are so severely affected by decreased illumination that they either stop laying or produce thinner eggshells.[117–119] Peakall's doves, however, continued to produce good shells. He finally was able to induce his desired eggshell thinning by injecting large amounts of DDE directly into each bird![116] He then blamed the thinned shells on DDT!

Details of all those tendentious "experiments" were exposed and discussed in detail,[1–6,106] however the media, the pseudoenvironmentalists, and the well-paid anti-DDT "researchers" failed to publicize the frauds. As a result, the general public remains misinformed about the causes of eggshell thinning, and the environmental industry continues to accumulate money and power as a result.

There are many environmental contaminants that do cause shell-thinning.[108,111,112,127,128] Oil, lead, mercury, cadmium, lithium, manganese, selenium and certain sulphur compounds have been shown to have adverse effects upon birds, including severe shell-thinning. Fimreite reported "elevated mercury levels (0.315 to 0.568 ppm) were in four peregrine eggs from the Northwest Territories"[123] and Stoewsand demonstrated that just a few ppm of mercury in the diet results in eggs with scarcely any shell at all.[35]

Polychlorinated biphenyls (PCBs) came into usage in the 1930s as plasticizers, paint extenders and heat exchange fluids in transformers. Every fluorescent light ballast contained liquid PCB, so gas chromatograph analyses nearby distorted all testing for DDT. So did the plastic tubing in the apparatus, and any plastic that had been in contact with the samples being analyzed. PCBs occurred in most fish and wildlife samples at higher concentrations than DDT residues.[129,130,154,156] When ingested by birds, they are known to cause infertility, eggshell thinning, non-hatchability, and severe physiological damage.[131-133] Curley (1971) stated that "PCBs throw off hormonal levels in birds, affecting calcium reserves, which in turn result in thin eggshells."[133] Herring gulls from the Great Lakes contained an average of 2,224 ppm of PCB, and "signs of eggshell thinning were found in nine of thirteen species in the region."[134]

Scott (1975) observed that "Many reports relating reproductive declines of wild birds to DDT and DDE were based on analytical procedures that did not distinguish between DDT and PCBs."[108] Because PCBs cause severe effects on fish, birds, and wildlife, it was noteworthy that so great a proportion of the "DDT and residues" alleged by the anti-DDT literature was later admitted to have really been PCBs.

If the misidentifications that were so sensationalized by anti-DDT activists were actually unintentional, the persons responsible should have sought to correct the record, rather than continuing to conceal their errors, but very few did. In 1969 Anderson, Hickey and Risebrough reanalyzed the five samples of egg-pools which in 1965 they claimed had high levels of DDT. They belatedly reported that three of the five samples actually had no DDT at all, and the other two had only a fraction as much as touted in their original anti-DDT articles.[156] Scott noted that those authors led the campaign to ban DDT because of its supposedly "proven" effects on eggshell quality, but now they seek to prove that PCBs were really the cause.[131]

Even if all analyses of chlorinated hydrocarbon insecticides had been accurate,there should not have been such hasty and unreasonable incrimination of DDT. Many more obvious environmental hazards should have been considered. W. H. Stickel (a senior Patuxent biologist) said in 1969: "Let us look seriously at other factors, too! It is not likely that pesticides will explain everything . . . One could explain the observed phenomena at least as well by a theory of chronic poisoning by metals such as lead or mercury."[158] It is unfortunate for DDT (and millions of humans) that Stickel's advice was ignored.

Eggshell Measurements

Early egg-collectors routinely saved only the thicker-shelled eggs. Comparisons between the shell thickness of museum eggs and the more recently collected ones are obviously misleading, and have led to the propaganda that bird eggs during the "DDT years" had much thinner shells than those produced before DDT appeared.

Hickey and Anderson measured the shells of hundreds of museum eggshells. They reported that red-tailed hawk eggs just before DDT was used had much thinner shells than did eggs produced 10 years earlier. Then, during the years of heavy DDT usage, hawks produced shells that were 6% thicker.[157] Golden eagle eggshells during the DDT years were 5% thicker than those produced before DDT was present in the environment. The authors concluded that for five of the six species of raptors in their study "the data do not permit a precise delineation of the time of onset" of the change in shell thickness.

Faulty Analyses by Gas Chromatography

Although the gas chromatograph, which was universally used for pesticide analyses since the mid-1960s, was very sensitive, it did not differentiate between DDT residues and many non-pesticide chemicals.[144–152,154]

Frazier (1970) analyzed soil samples that had been sealed in glass jars since they were collected in 1911. The gas chromatograph indicated that five kinds of chlorinated hydrocarbon insecticides were in that soil, even though none of them were in existence until 30 years after those samples were sealed. They concluded that "the apparent insecticides were actually misidentifications caused by the presence of co-extracted naturally-occurring soil components."[149] Bowman (1965) analyzed soil that was sealed in 1940 (years before DDT) and reported that "a naturally-occurring extraneous substance appearing in pre-DDT soil gave the same retention times as DDT."[150]

Glotfelty and Caro (1970) commented that misidentifications of chlorinated hydrocarbon insecticides resulted from interference by "pigment-related natural products found in photosynthetic tissues."[145] Coon confirmed that material he and his colleagues at the WARF Institute earlier believed to contain DDT residues actually had none at all.[146] He also reported on a gibbon that was collected in Burma in 1935 and sealed in a tight container for 30 years before being analyzed. The analysis indicated DDE in the kidneys, testes, liver and fat tissues, even though DDE did not exist anywhere on earth until six years after the gibbon was preserved. J. J. Sims (a plant pathologist and biochemist) discovered that marine algae produce halogen compounds that were misidentified as DDT by the gas chromatograph.[147] He explained that "DDT's chain of atoms contain sequences in which chlorine and carbon are linked, and it is this carbon-chlorine bond that is very common in natural marine environments. Halogen compounds containing bromine or iodine, rather than chlorine, may also falsely register on the gas chromatograph as DDT."

J.J. Hickey sent to the WARF Institute, for analysis, a robin that had been collected and stored in formalin in 1938. Although it was collected seven years before DDT was ever used, the analysis indicated that the robin contained DDT and DDD![144] Dozens of other robins were similarly analyzed in Michigan and Wisconsin before it became evident that mercury in the worms eaten by those birds actually caused the symptoms attributed to DDT poisoning, including tremors, ataxia, and death. Massive amounts of DDT fed to caged robins failed to make them ill (except for diarrhea) and even nestling robins were not sickened after they had been fed only food that was spiked with DDT.[153] Even in samples from Antarctica scientists overestimated the DDT concentration in fish by more than 100-fold because the plastic bags in which they were stored for six months before analysis produced false "DDT" readings as a result of the PCBs in the plastic.[155]

Sherman called attention to these difficulties, and observed that "This negated the putative data that were the basis for previous alleged sensational charges against DDT . . . For thirty years DDT has been a scapegoat for artifacts and mimics of that pesticide."[144]

Despite the absence of any confirmation that wild bird illness or mortality was caused by DDT, a few pseudoenvironmentalists succeeded in halting the use of DDT on American elm trees, following which thousands of the stately trees died in five midwestern states because of the Dutch Elm Disease that was transmitted by the uncontrolled European Bark Beetles. Years later, millions of eastern oak trees were killed by gypsy moth larvae. which had previously been safely controlled by a single spray of DDT. The environmentalists appeared unconcerned by the needless destruction of forests, and fought desperately to prevent the use of insecticides to save trees. NOW they extoll the virtues of trees, and urge everyone to plant more of them! Even if every American planted dozens of trees annually, they could not recoup the losses caused by those quixotic eastern pseudoenvironmentalists!

Although these errors were pointed out repeatedly by other scientists, neither the news media, the EPA, the legislators, nor the ecoterrorist oganizations ever acknowledged them. Instead, they continued to blame benign DDT and DDE as the causes of numerous difficulties which they should have attributed to the other chemicals and environmental factors that were discussed above.

The Final Hypocrisy

Hazeltine observed that "To support a hypothesis, all of the available data must be considered, and they must be affirmative: to refute a hypothesis, however, only one piece of solid negative evidence is required." Where DDT was involved, hypotheses were never considered in this manner and the much-heralded "scientific method" was generally disregarded.

The practice of not correcting misstatements appealed to the scientists, journalists, politicians and professional environmental groups involved in the anti-pesticide campaigns. Consequently, DDT was eventually placed on trial by the Environmental Defense Fund and their cohorts in the EPA. After seven months of EPA hearings, Judge Edmund Sweeney specified (26 April 1972) that "DDT is not a carcinogenic hazard to man . . . DDT is not a mutagenic or teratogenic hazard to man . . . The uses of DDT under the regulations involved here do not have a deleterious effect on freshwater fish, estuarine organisms, wild birds or other wildlife . . . The evidence in this proceeding supports the conclusion that there is a present need for the essential uses of DDT."[159] Incredibly, EPA Administrator William Ruckelshaus (an attorney, with no knowledge of the scientific facts) who never attended a single day of the seven months of EPA hearings, and who admittedly had not even read the transcript of those vital hearings, overruled his hearing examiner (Judge Sweeney) and personally banned DDT.[160] Several of the serious errors and discrepancies in the Ruckelshaus "Opinion and Decision" concerning DDT were inserted into the Congressional Record by Senator Barry Goldwater in 1972.[161] They were never reported by the news media (to whom dozens of copies were mailed) and the environmental industry never revealed the details to their supporters and contributors.

In a letter to Allan Grant (the president of the American Farm Bureau Federation) Ruckelshaus later wrote: "Decisions by the government involving the use of toxic substances are political with a small 'p'. Science has a role to play, but the ultimate judgement remains political, (and) the power to make this judgement has been delegated to the Administrator of EPA."[162] At that time, Ruckelshaus was senior vice president of the Weyerhauser Lumber Company, but he is now running a garbage disposal company and is seeking permission to dump trash and garbage into a pleasant little stream in Apanolia Canyon, near Half Moon Bay California. Environmentalists object, charging that it would destroy irreplaceable wetlands, riparian habitat and an ecological complex of great diversity."[163]

John Quarles' 3 June 1982 affidavit to the U. S. District Court for Northern Alabama is of interest, since it bears on the Ruckelshaus ban of DDT.[164] Quarles served as General Council for Mr. Ruckelshaus in 1971 and 1972. He testified: "After seven months of hearings, the EPA Hearing Examiner made findings generally supportive of the position that DDT did not cause undue harm and that an adequate basis did not exist for cancelling the uses of DDT." The Hearing Examiner's decision came to Mr. Ruckelshaus on appeal by the Environmental Defense Fund, of which Mr. Ruckelshaus was a member and for which he had solicited donations under his own letterhead. "Because of the importance of the question, rather than refer it to the judicial officer, Mr. Ruckelshaus decided to rule on the appeal himself." Quarles continued: "I am quite certain that Mr. Ruckelshaus had no idea at the time that the decision would be regarded as having any controlling influence on questions that might occur in specific cases in the federal courts." Quarles added that "There were no findings that DDT had caused harm or would cause harm under a specific set of circumstances or at any particular time or place."

References

1. Claus, G. & K. Bolander 1977. ECOLOGICAL SANITY. David McKay Co., N.Y., 592 pages (277–592 deal with DDT)
2. Claus, G. 1984. A critical review of the evidence against DDT Columbia University Seminar Series, Vol, 16, pages 69–104.
3. Edwards, J. G. 1971. Testimony and affidavit, U. S. Congressional Committee on Agriculture, Washington D. C. 18 March 1971, Published in S.N. 92-A, pages 575–594.
4. Edwards, J. G. 1973. Testimony and affidavit, U. S. Congressional Committee on Agriculture, Washington D. C. 24 August 1973, 24 pages.
5. Hazeltine, W. E. 1971. Statement to Secretary of State's Advisory Committee: U. N. Conference on Human Environment, publ.16 March 72, 36 pp.
6. Hazeltine, W. E. 1974. Statement at EPA Hearing on Tussock Moth Control, Portland Oregon, 14 January 1974, 54 pages.
7. Jukes, T. H. 1963. Pests and pesticides. *Amer. Scientist* 51: 355–361.
8. Jukes, T. H. 1974. Insecticides in health, agriculture and the environment. *Naturwissenschaften* 61: 6–16 (Springer Verlag)
9. EPA Consolidated DDT Hearings, 1971–1972, Washington D. C. (More than 9,200 pages of official transcript.)
10. Simmons, S. W. (Editor) 1959. DDT AND ITS SIGNIFICANCE: VOL. II, HUMAN AND VETERINARY MEDICINE. Birkhauser Verlag, Basel, 570 pages.

11. Edwards, J. G. 1978. Testimony and affidavit, U. S. D A. Hearings on Japanese Beetle Quarantine, San Francisco, 22 Pages. (Cites more than 74 articles on the non-carcinogenicity of DDT.)
12. Edwards, J. G. 1980. The myth of food chain biomagnification. *Agrichemical Age* April 1980: 32–33.
13. Gunn, D. L. 1972. *Annals of Applied Biology* 72: 105–127.
14. Harvey, G. R. 1973. DDT in the marine environment. National Academy of Sciences Committee Report, 9 August 1973, 8 pages.
15. Kanwisher, J. W. 1973. DDT in the marine environment. National Academy of Sciences Committee Report, 30 January 1973, 9 pages.
16. Marvin, P. R. 1964. Birds on the rise. *Bull. Entomol. Soc. Amer.* 10(3): 184–186.
17. Wurster, C. F. 1969. *Congressional Record* S4599, 5 May 1969.
18. Anon. 1942. The 42nd Annual Christmas Bird Census. *Audubon* Magazine Jan/Feb 44: 1–75.
19. Cruickshank, A. D. (Editor) 1961. The 61st Annual Christmas Bird Census. *Audubon Field Notes* 15(2): 84–300.
20. Taylor, J . W. Summaries of Hawk Mountain migrations of raptors, 1934 to 1970. In *Hawk Mtn. Sanctuary Assn Newsletters* (Pennsylvania)
21. Hickey, J. J. 1943. In GUIDE TO BIRD WATCHING.
22. Taylor, J. W. 1971. *Hawk Mtn Sanctuary Newsletter* 43: 2.
23. Bruemmer, F. 1971. *Animals Magazine* April 1971, page 555.
24. Associated Press Release, 18 March 1970.
25. Anon, 1975 . *Louisville Courier-Journal*, December 1975.
26. Graham, F. 1971. Bye-bye Blackbirds? *Audubon Magazine* Sept., pp 29–35.
27. Anon. 1967. *Bull. Virginia Department of Agriculture*, May 1967.
28. Graham, F. 1985 . *Audubon Magazine*, January 1985, page 17.
29. McLean, A . E. & E. K. McLean 1969. (Aflatoxin toxicity prevented by DDT in diet) *British Med. Bull.* 25 (3): 278–281.
30. Silinskas, K. C. & A. B. Okey 1975. (Protection by DDT against tumors and leukemia) *J. Natl. Cancer Inst.* 55(3):653–657.
31. Banks, R. C. 1966. *Trans. San Diego Soc. Nat. Hist.* 14: 173–188
32. Edwards, J. G. 1971. Statement and affidavit, California Water Quality Control Board, Los Angeles, 20 February 1971.
33. Main, R. E. (U.S.N.P.S.) 1970–1973. Annual nesting surveys for Channel Island National Monument. (Also documents the history, ecology and reproductive success of brown pelicans in U. S., in 300 page MS)
34. Scott, J. L. et al. 1975. Effects of PCBs, DDT and mercury upon egg production, hatchability & shell quality. *Poultry Sci.* 54: 350–368.
35. Stoewsand, G. S. et al. 1971. Shell-thinning in quail fed mercuric chloride. *Science* 173: 1030–1031.
36. Tucker, R. K. 1971. (Effects of many chemicals on shell thickness) *Utah Science* June 1971: 47–49.
37. Tucker, R. K. & H. A. Haegele 1970. *Bull. Environ. Contam.* Toxicol. 5: 191–194.
38. Anon.1979. Embryonic mortality from oil on feathers of adult birds) National Wildlife Federation, *Conservation News* 15 October, pp 6–10

39. Hartung, R. 1965. (Oil on eggs reduces hatchability by 68%) J. Wildlife Management 29: 872–874.

40. Libby, E. E. 1978. Fish, wildlife, and oil. *Ecolibrium* 2 (4): 7–10

41. King, K. A. et al. 1979. (Oil a probable cause of pelican mortality for six weeks after spill) *Bull. Environ. Contam.Toxicol.* 23: 800–802.

42. Dieter, M. P. 1977. (5 microliters of oil on fertile eggs kill 76 to 98% of embryos within; birds ingesting oil produce 70% tc 100% less eggs than normal; offspring failed to develop normal flight feathers) Interagency Energy-Environment Research and Development Program Report, pages 35–42.

43. Szaro, R. C. 1977. *Proc. of 42nd N. Amer. Wildlife & Nat. Resources Conference,* pp. 375–376.

44. Albers, P. H. 1977. FATE AND EFFECTS OF PETROLIUM HYDROCARBONS IN MARINE ECOSYSTEMS, Pergamon Press, N.Y. (Chapters 15 & 16)

45. Risebrough, R. W. et al. 1970. (Residues and egg-shell thicknesses of 210 pelican eggs) (No strong correlation) Wilson Bull. 82: 15–28.

46. Keith, J. O. et al. 1970. *Proc. 35th N. Amer. Wildlife Conference* (Repeated verbatim in Finley's statements of l971)

47. Switzer, B. 1972. Consolidated EPA Hearings Transcript, pp 8212–8336.

48. Hazeltine, W. E. 1972. Why pelican eggshells are thin. *Nature* 239: 410–412.

49. Krantz, W. C. 1970. (Analysis of bald eagle eggs shows no correlation between shell-thinning and pesticide residues in eggs) *Pesticides Monitoring J.* 4 (3): 136–141.

50. Postupalsky, S. 1971. (No correlation between shell-thinning and DDT in eggs of bald eagles and cormorants) *Canadian Wildlife Service* manuscript, dated 8 April 1971.

51. Jehl, J. R. 1970. (In Keith, J. O. et al.) (Sick and dead pelicans along Mexican coast in late 1968) *Trans. 35th N. Amer. Wildlife Conference,* pages 56–64.

52. UPI Press release, April 1972, (Newcastle disease epidemic in Calif.)

53. Pomeroy, B. S. & C. A. Brandly 1953. (Newcastle disease symptoms and egg deformities) *Univ. Minn. Agric. Expt. Sta. Bull.* 419, 22 pages.

54. Hofstad, M. C. et al. 1972. DISEASES OF POULTRY, Iowa State Univ. Press

55. Dawson, W. L. 1923. BIRDS OF CALIFORNIA, South Moulton Co., Los Angeles, 2,121 pages in three volumes.

56. Gress, F. 1970. Status of the California Brown Pelican in 1970. Calif. Dept. Fish & Game Admin. Rept. (unpublished)

57. Anon. 1971. *Washington Post* story, 22 July 1971 (Also in L.A. Times)

58. Pearson, T. G. 1919. *Review of Reviews,* May 1919, pages 509–511.

59. Pearson, T. G. 1934. ADVENTURES IN BIRD PROTECTION, Appleton-Century Co., page 332 (1918 pelican survey of Gulf of Mexico, from Key West to Rio Grande River, revealed 65,000 pelicans, including 5,000 in Texas and Louisiana)

60. Pearson, T. G. 1934. (Discussion of the 1918 survey) *National Geographic Magazine,* March 1934, pages 299–302.

61. Allen, R. P. 1935. (Update of Pearson's pelican survey) Auk 52: p. 199

62. Gustafson, A. F. et al. 1939. CONSERVATION IN THE UNITED STATES,

Comstock Publ. Co., Ithaca. (Repeated in U. S. Fish & Wildlife Research Report No. 1, 1970)

63. Finley, R. B. 1971. Statement before California Water Quality Control Board, Los Angeles, 20 February 1971. (Exact duplicate of portion of article by J. O. Keith in *Trans. 35th N. Amer. Wildlife Conf.*)

64. Finley, R. B. 1971. Letter to chairman W. R. Poage of the U. S. Congressional Committee on Agriculture, dated 2 August 1971.

65. Van Name, W. G. 1921. *Ecology* 2: 76.

66. Anon. 1943. *Science News Letter*, 3 July 1943.

67. Bent, A. C. 1937. Raptorial Birds of America. *U.S.National Museum Bull.* 167: 321–349. '''

68. Spencer, D. A. 1976. WINTERING OF THE MIGRANT BALD EAGLE, 170 pages.

69. U. S. Forest Service, Milwaukee Wisconsin 1970. *Annual Report on Bald Eagle Status.*

70. Stickel, L. et al. 1966. Bald eagle pesticide relationships. *Trans. 31st N. Amer. Wildlife Conference,* pages 190–200.

71. Anon. 1978. U. S. Fish & Wildlife Service, *Endangered Species Tech. Bull.* 3: 8–9.

72. Reichel, W. L. et al. 1969. (Pesticide residues in 45 bald eagles found dead in U. S., 1964–1965) Pesticides Monitoring Journal 3 (3): 142–144.

73. Belisle, A. A. et al. 1972. (Pesticide residues and PCBs and mercury, in bald eagles found dead in U. S.,1969–1970) *Pesticides Monitoring Journal* 6 (3): 133–138.

74. Cromartie, E. et al. 1974. (Organochlorine pesticides and PCBs in 37 bald eagles found dead in U. S., 1971–1972) (4 dead of dieldrin, 9 of thallium, none of DDT residues) *Pesticides Monitoring Journal* 9: 11–14.

75. Coon, N. C. 1970. (Causes of bald eagle mortality in U. S., 1960–1965) *Journal of Wildlife Diseases* 6: 72–76.

76. Bagley, G. E. et al. 1970. J. Assn. Official Analytical Chem. 53: 251–261.

77. Spann, J. W., Heath, R. G., Kreitzer, J. F. & L. N. Locke 1972. (Lethal and reproductive effects of mercury on birds) *Science* 175: 328–331.

78. Anon. 1984. National Wlldlife Federation publication. (Eagle deaths)

79. Ratcliffe, D. H. 1980. THE PEREGRINE FALCON, Buteo Books, Vermillion S. Dakota, 416 pages.

80. Ratcliffe, D. A. 1967. Decrease in eggshell weights in certain birds of prey. *Nature* 215: 208–210. (Formula for "shell thickness index" is given here, but no shell thicknesses were actually measured)

81. Ratcliffe, D. H. 1970. *J. Applied Ecology* 7: 67.

82. Newton, I. 1974. In PROCEEDINGS OF FALCON RECOVERY CONFERENCE, Audubon Conservation Report Series No. 4, 43 pages.

83. Wilson Report, 1969. Review of organochlorine pesticides in Britain. Report by Advisory Committee on toxic chemicals. Department of Education and Science.

84. Hickey, J. J. 1942. (Only 170 pairs of peregrines in eastern U. S. in 1940, before DDT) Auk 59: 176.

85. Hickey, J. J. 1971. Testimony at DDT Hearings before EPA Hearing Exam-

iner. (350 pre-DDT peregrines claimed in eastern U. S., with 28 of the females sterile)

86. Beebe, F. L. 1971. THE MYTH OF THE VANISHING PEREGRINE FALCON: A study in manipulation of public and official attitudes. *Canadian Raptor Society Publication*, 31 pages.

87. Hornaday, W. T. 1913. OUR VANISHING WILD LIFE, New York Zoological Society, page 226.

88. Berger, D. D., Sindelar, C. R. & K. E. Gamble 1969. In PEREGRINE FALCON POPULATIONS, pages 165–173.

89. Rice, J. N. 1969. In PEREGRINE FALCON POPULATIONS, Univ. of Wisconsin Press, pages 155–164.

90. Peterson, R. T. 1948. BIRDS OVER AMERICA, Dodd Mead & Co., N. Y., pages 135–151.

91. Hickey, J. J. (editor) 1969. PEREGRINE FALCON POPULATIONS. Univ. of Wisconsin Press, 596 pages.

92. Enderson, J. H. et al. 1968. (Pesticide residues in Alaska and Yukon Territory) Auk 85: 383–384.

93. Fyfe, R. W. 1968. Auk 85: 683.

94. Enderson, J. H. & D. D. Berger 1968. (Chlorinated hydrocarbons in peregrines from Northern Canada) *Condor* 70: 149–153.

95. Seidensticker, J. C. 1970. (Biopsy to remove tissue from falcons for analysis) *Bull. Environ. Contam.. Toxicol.* 5 (5): 443–451.

96. Enderson, J. H. et al. 1972. (Time lapse photography in peregrine nests.) *Living Bird* 11: 113–128.

97. Cade, T. J. 1968. *Condor* 70: 170–178.

98. Cade, T. J. 1960. Ecology of the peregrine and gyrfalcon populations in Alaska. *Univ. Calif. Publ. Zool.* 63 (3). 151–290.

99. Anon. 1944. THE DUCKHAWK AND THE FALCONERS. Pamphlet by Emergency Conservation Committee, New York.

100. Beebe, F. L. 1975. *Brit. Columbia Provincial Museum Occas. Paper*, No. 17, pages 126–144.

101. Herman, S. et al.1968. (Deaths high among peregrines trapped for falconry) *Nature* 220: 1098.

102. Anon. 1984. (Citing article from National Wildlife Federation) Audubon Action December 1984.

103. Anon. 1985. *New York Times* story, 19 February 1985.

104. Cecil, H. C., Bitman, J. & S. J. Harris 1971. (No effects on eggshells, if adequate calcium is in DDT diet) *Poultry Science* 50: 656–659.

105. Chang, E. S. & E. L. R. Stokstad 1975. (No effects of DDT on shells or on carbonic anhydrase) *Poultry Science* 54: 3–10. (Also in Univ. of California Dateline No. 21, 1971)

106. Edwards, J. G. 1971. (Summary of shell-thinning allegations, and refutations presented by revealing all data) *Chem. & Engineering News*, 16 August 1971, pages 6 & 59

107. Jefferies, D. J. 1969. (Shells 7% thicker, after two years on DDT diet) *J. Wildlife Management* 32: 441–456.

108. Scott, M. L., Zimmermann, J. R., Marinsky, P. A., Mullenhoff, P. A., Rum-

sey, G. L. & R. W. Rice 1975. (Egg production, hatchability and shell quality depend on calcium, and are not affected by DDT and its metabolites) *Poultry Science* 54: 350–368.

109. Speers, G. & P. Waibel 1972. (Neither egg weight nor shell thickness affected, by 300 ppm DDT in daily diet) *Minn. Science* 28 (3): 4–5.

110. Greely, F. 1960. (Effects of calcium deficiency) *J. Wildlife Management* 70: 149–153.

111. Romanoff, A. L. & A. J. Romanoff 1949. THE AVIAN EGG, Wiley & Sons.

112. Tucker, R. K. & H. A. Haegele 1970. *Bull. Environ. Contam.. Toxicol.* 5 (3): 191–194. (Also Tucker, R. K. 1971. in Utah Science June: 47–49.)

113. Bitman, J., Cecil, H. C. et al. 1969. (Low calcium diet caused shell-thinning in Japanese quail) *Nature* 224: 44–46.

114. Bitman, J., Cecil, H. C. & G. F. Fries 1970. (Quail were fed either low calcium OR adequate calcium diets: some produced thinned shells, but there was no indication by the authors as to which group of birds produced which types of shells) *Science* 168: 594–595.

115. Peakall, D. B. 1970. (Shells not thinned even after illumination was abruptly reduced from 16 hours daily to just 8 hours daily, AND high DDT dosage began at that same time!) *Science* 168: 592–594.

116. Peakall, D. B. 1970. (Pesticides and the reproduction of birds) *Sci. Amer.* 222: 72–78.

117. Day, E. J. 1971. (Importance of even illumination on laying birds) *Farm Technology*, Fall 1971.

118. Morris, T. R. et al. 1964. (The most critical area of illumination for birds is that between 6 hours and 8 hours daily) *British Poultry Science* 5: 133–147.

119. Ward, P. 1972. (Physiological importance of photoperiod in bird experiments) *Ibis* 114: 275.

120. Bellrose, R. C. 1959. (Lead poisoning in wildlife) *Illinois Nat. Hist. Survey Bull.* 27: 235–288.

121. Spann, J. W. et al. 1972. (Lethal and reproductive effects of mercury on birds) *Science* 175: 328–331.

122. Scott, M. L. et al. 1971. *Poultry Science* 50: 1055–1063.

123. Fimreite, N. Fyfe, R. W. & J. A. Keith 1970. (Mercury contamination of Canadian seedeaters and their avian predators.) *Canadian Field-Naturalist* 84: 269–276.

124. Main, R. E. 1972. Mercury in the Ecosystem. MS Thesis, San Jose State University, 107 pages.

125. Tejning, S. 1967. (Effects of methyl mercury) *Oikos* (suppl.) 8: 7–34.

126. D'Itri, F. M. & P. B. Trost 1970. International Conference on Mercury Contamination, Ann Arbor, 30 September 1970.

127. Mueller, W. J. & R. M. Leach, Jr. 1974. (Effects of chemicals on eggshell formation) *Ann. Rev. Pharmacology* 14: 289–303. (142 references cited)

128. Genest, P. & R. Bernard 1945. (Effects of sulfonamides, causing 40% to 60% decrease in shell thickness) *Rev. Can. Biology* 4: 172–192. (also *Science* 101: 617–618.)

129. Risebrough, R. W. et al. 1968. Polychlorinated biphenyls in the global ecosystem. *Nature* 220: 1098–1102.

130. Risebrough, R. W. & R. R. Spitzer 1971. *Chem. & Eng. News* 12 Dec 71, pp 32–33.
131. Scott, M. L. et al. 1975. (Effects of PCBs, etc. on egg production, hatchability and shell quality) *Poultry Science* 54: 350–368.
132. Anon. 1971. *Hawk Chalk* 10 (3): 47–57.
133. Curley, A. 1971. (PCBs affect hormonal and calcium reserves in birds) Presented at American Chemical Society Conference in Los Angeles, 24 March 1971. (See also *Environ. Res.* 4: 481–495.)
134. Boyle, R. H. 1978. *Audubon Magazine,* November 1978, pages 150–151.
135. Sunde, M. L,. 1971. (Older birds produce thinner shells) *Farm Technology,* Fall 1971.
136. Romanoff, A. L. & A. J. Romanoff 1967. BIOCHEMISTRY OF THE AVIAN EMBRYO, Wiley & Sons, N. Y.
137. Simkiss, K. 1967. (Shells thinned by embryo development within) In CALCIUM IN REPRODUCTIVE PHYSIOLOGY, Reinhold, N.Y., pages 198–213.
138. Kreitzer, J. F. 1973. (Shell thinning during embryonation) *Poultry Science* (in press) ALSO *Bull. Envir. Contam. & Toxicol.* 9: 261–286.
139. Scott, H. M. et al. 1944. (Physiological stress thins shells) *Poultry Science* 23: 446–453.
140. Draper, M. H. & P. E. Lake 1967. Effects of stress and defensive responses. In ENVIRONMENTAL CONTROL IN POULTRY PRODUCTION. Oliver & Boyd, London.
141. Reid, B. L. 1971. (Effects of stress on laying birds) *Farm Technology* Fall 1971.
142. Tucker, R. K. & H. A. Haegele 1970. (30% thinner shells formed after quail were kept from water for 36 hours) *Bull. Environ. Contam. Toxicol.* 5 (3): 191–194.
143. Enderson, J. H. & D. D. Berger 1968. Chlorinated hydrocarbons in peregrines from Northern Canada. *Condor* 70: 170–178.
144. Sherman, R. W. 1973. Artifacts and mimics of DDT and other insecticides. *J. New York Entomol. Soc.* 81: 152–163.
145. Glotfelty, D. E. & J. H. Caro 1970. *Anal. Chem.* 42: 82–84.
146. Coon, F. B. 1966. Electron capture gas chromatographic analyses of selected samples of authentic pre-DDT origin. Presented at the Conference of American Chemical Society in New York.
147. Sims, J. J. 1977. Press release, 15 June 1977 (and letter to J. G. Edwards, 29 June 1977).
148. Hylin, J. W., Spenger, R. E. & F. A. Gunther 1969. Potential interference in certain pesticide residue analyses from organochlorine compounds occurring in plants. *Residue Reviews* 26: 127.
149. Frazier, B. E., Chesters, G, & G. B Lee 1970. "Apparent" organochlorine insecticide content of soils sampled and sealed in 1910. *Pesticides Monitoring Journal* 4 (2): 67–70.
150. Bowman, M. C., Young, H. C. & W. E. Barthel 1965. (Soil sealed in jars in 1940 contained modern pesticide mimics when analyzed in 1965) *J. Econ. Entomology* 58: 896–902.

151. Gustafson, C. G. 1970. (Chromatogram peaks of PCBs compared with identical peaks of chlorinated hydrocarbon insecticides, including DDT, DDD and DDE) *Environ. Sci. Technology* 4 (10): 814–819.

152. Lisk, D. J. 1970. Analysis of pesticide residues: methods and problems. *Science* 170: 589–593.

153. Mitchell, R. T. et al. 1953. Effects of DDT upon the survival and growth of nestling songbirds. *J. Wildlife Management* 17: 45–54.

154. Wolff, E. W. & D. A. Peel 1985. Global pollution in polar snow and ice. (More misidentifications of DDT cited) *Nature* 313: 535–540

155. George, J. L. & D. E. H. Frear 1966. Pesticides in the Antarctic. *J. Appld. Ecology* 3 (suppl): 155–167. (samples placed in plastic bags, stored at – 60°C for six months prior to analysis)

156. Anderson, D. W., Hickey, J. J., Risebrough, R. W., Hughes, D. F. & R. E. Christensen 1969. *Canadian Field-Naturalist* 83: 91–112.

157. Hickey, J. J. & D. W. Anderson 1968. (Analysis of hundreds of shells.) *Science* 162: 271–273.

158. Stickel, W. H. 1969. In PEREGRINE FALCON POPULATIONS, page 538.

159. Sweeney, E. M. 1972. EPA Hearing Examiner's recommendations and findings concerning DDT hearings, 25 April 1972 (40 CFR 164.32, 113 pages). Summarized in *Barrons* 1 May 1972 & *Oregonian* 26 Apr 72

160. Ruckelshaus, W. D. 1972. "Opinion" of EPA Administrator concerning the seven months of hearings, and banning DDT) 2 June 1972, 40 pages.

161. Edwards, J. G. 1972. The infamous Ruckelshaus DDT decision. *Congressional Record,* 24 July 1972, pages S11545–11547. (via Senator Barry Goldwater)

162. Letter from William Ruckelshaus to Donald Grant, President of the American Farm Bureau Federation, dated 26 Apr 1979.

163. Wildermuth, J. 1990. Peninsula Dumpsite Opposed. San Francisco Chronicle, 14 April 1990

164. Quarles, J. 1982. Affidavit to U. S. District Court for Northern Alabama, 3 June 1982.

The Tragedy of DDT

Thomas H. Jukes, Ph.D.

Please see biographical sketch for Dr. Jukes on page 135.

In 1962, the position of DDT as an insecticide seemed impregnable. It had eradicated malaria from many areas of the world, thus saving millions of lives and bringing health to hundreds of millions of people who would otherwise have been victims of "the monarch of diseases."

DDT had stopped massive epidemics of typhus fever and had helped to control many other arthropod-borne diseases such as plague and "river blindness" (onchocerciasis).

Whole populations had 10 percent DDT dust blown into their clothing as they wore it. The World Health Organization commented that "the only confirmed cases of injury have been the result of massive accidental or suicidal ingestion." Its effects on "non-target species" occurred only in misuse, such as killing certain fish when sprayed on water.

The National Audubon Society, in 1961, had declared DDT harmless to birds when used at 1 pound per acre for control of the gypsy moth in Pennsylvania. Its usefulness in the global malaria eradication program continued despite the appearance of resistance. In 1962, Dr. R. Pal stated that the average life span in India was now 47 years compared with 32 years before the program, and that malaria in India had been reduced from 75 million cases to less than 1 million. Populations freed from malaria needed more food, and DDT played a second and major role in protecting crops to supply this.

Silent Spring and the Big Lie

The opening gun of the successful campaign against DDT was Rachel Carson's *Silent Spring,* in 1962. As a single example of its inaccuracy, the book stated that the American robin was on the verge of extinction. In 1963, however, Roger Tory Peterson said the robin was most likely the most numerous North American bird. Another outstanding piece of nonsense was the story that DDT in the oceans would kill all the algae and would bring an end to the world's supply of oxygen. This fear was echoed by the head of the United Nations, even though the original research showed no effect on algae by DDT at saturation levels in sea water.

Many of the alleged environmental ill effects of DDT probably resulted from PCBs or mercury, but this was ignored, and the campaign rolled on, aided by the mass media.

In June 1972, DDT became a victim of the big lie. It was banned in the United States for practically all uses by Environmental Protection Agency (EPA) administrator William Ruckelshaus. He declared in 1970 that DDT was safe and that "the carcinogenic claims concerning DDT are unproved speculation," but in 1972, he changed his tune. The Ruckelshaus ban overturned the recommendation of his examiner, Administrative Law Judge Edmund Sweeney, who had conducted nine months of public hearings on DDT for the EPA.

Judge Sweeney brought a fresh viewpoint without previous involvement with environmental or medical disputes. The record shows that he was fair. His impartiality infuriated the opponents of DDT, who apparently felt that he should favor them. When he admonished some of them for their poor evidentiary procedures, they became so enraged that they refused to appear the next day.

The opponents of DDT seemed to have a pipeline to *Science* magazine and to the *New York Times,* both of which attempted to prejudice the hearings by criticizing the judge, alleging that he lacked environmental expertise. So far as I know, neither publication sent a reporter to the hearings.

Appearing for the Montrose Chemical Company in defense of DDT were two vigorous and forceful attorneys, Robert Ackerly and Charles O'Connor. As they learned the facts about DDT, they became and remained strong supporters of its use. Indeed, Robert Ackerly has since written scholarly articles on DDT for *Chemical Times & Trends,* 1982. Their commitment gave unexpected pique to the hearings. During the cross-examination of DDT opponent Dr. George Woodwell, they elicited his admission that, in a study to show the DDT content of an area, he had taken soil samples where a DDT spray truck had been standing and he had reported these high values without correction.

Another tidbit came when Professor Joseph Hickey was reminded that he had stated in a 1942 publication that the number of peregrine falcons in the eastern United States had been declining since 1890, and in 1940 had dwindled to the precariously low level of not more than 140 mating pairs, long before the discovery of chlorinated hydrocarbon insecticides. Even today, mishaps to the eggs of peregrine falcons are blamed on DDT without an analytical test, and any mention of this picturesque bird is usually accompanied by remarks about the devastating effect (unsubstantiated, of course) of DDT on it.

More Lies

The question of DDT and cancer was also aired during the hearings. Cancer occurs in certain susceptible strains of mice fed DDT. There is no evidence linking cancer in human beings with prolonged and high levels of exposure to DDT, and former Surgeon General Jesse Steinfeld so testified in the hearings. Similar conclusions have been since reached by the National Institutes of Health.

One of the pro-DDT witnesses was Nobel laureate Dr. Norman Borlaug, who came to the defense because of his deep interest in prevention of hunger and disease in the developing countries. After the hearings, he was called a paid liar by an official of the National Audubon Society as reported in the *New York Times* Aug. 14, 1972. Scientists Gordon Edwards, Bob White-Stevens, and myself were simultaneously so named.

We sued for libel, and won a jury verdict in U.S. District Court in 1976, only to have the decision reversed on appeal by Judge Irving Kaufman, who, according to the *Village Voice*, March 5, 1984, is a close friend of the *New York Times.*

In 1982, the Olin Corporation settled out of court for $24 million, against claims by residents of Triana, Ala., largely based on DDT residues in local fish. The main measurable effect was an increase in a blood enzyme, gamma glutamyl transpeptidase, and the U.S. Public Health Service said, "the effect is small and probably does not affect well being." As for the fish, they were being caught alive and well with DDT contents ranging up to 627,000 parts per billion, despite the stories told about DDT killing fish in *Silent Spring*'s Chapter 9, "Rivers of Death."

The Olin Corporation stated, correctly, "we live in a time when the popular perceptions regarding a chemical are inconsistent with the scientific facts," and that to have pursued the matter in courts would have involved years of protracted trials and appeals. Such is the tragedy of DDT, which has been transformed from a savior of lives to a dangerous poison by means of the big lie.

Various reasons are possible for the impassioned virulence, the misrepresentations, and the distortions used against DDT. Entomologist Gordon Edwards suggests that DDT is resented for its major role in accelerating population growth. I have proposed that attacking DDT led to the greatest bonanza ever for environmentalist organizations. With financial success, they have used, and still use, the DDT issue to represent themselves as saviors of birds, people, and the entire biosphere from extinction at the hands of what they call the irresponsible and greedy "agribusiness complex." The most eloquent reason is given by Dr. Richard Rappolt, who said: "only DDT has the press and romance: only DDT was associated with a Nobel prize and success against typhus and malaria . . . Reputations are made by killing heroes and, to some of the Lee Harvey Oswalds of the scientific world, DDT fulfills this imperial aura and 'when you strike at a king, you must kill him.' "

The best epitaph for DDT was written by Dr. Samuel Simmons in 1959 before the environmentalist attack started. He pointed out: "the total value of DDT to mankind is inestimable and is composed of health, economic, and social benefits . . . Most of the peoples of the globe have received benefit from it, either directly by protection from infectious diseases and pestiferous insects, or indirectly by better nutrition, cleaner food, and increased disease resistance . . . The discovery of DDT will always remain an historic event in the fields of public health and agriculture."

The defense of DDT was, from the beginning, a lost cause. A few of us vainly hoped that science would prevail. We soon found that Gresham's Law, which states that bad currency drives out good currency, applies to science as well as to economics.

I dedicate this brief essay to the memories of Bob White-Stevens, our tireless and silver-tongued humanitarian; Max Sobelman, manufacturer of DDT, who was dedicated to the fight against malaria and was an encyclopedic source of information; and George Claus, whose erudite and brilliant mind contributed so greatly to exposing fallacies by opponents of DDT.[1]

References

1. G. Caus and G. Bolander, *Ecological Sanity* (New York; David MacKay, 1977).

Reprinted, with permission, from
Farm Chemicals, 1988

Dioxin

Deadly Dioxin?
Elizabeth Whelan

Taking the Die Out of Dioxin
Natalie P. Robinson Sirkin and Gerald Sirkin

Deadly Dioxin?

Elizabeth M. Whelan, M.P.H., D.Sc.

Dr. Elizabeth M. Whelan is president of the American Council on Science and Health, New York, New York.

She received a B.A. from Connecticut College in 1965; an M.P.H. from Yale University 1967; an M.S. from Harvard University in 1968; and a Sc.D. from Harvard in 1971.

She is the author of over two dozen books including *Panic in Pantry* published in 1975; *Toxic Terror* published in 1985; and *Balanced Nutrition* published in 1989; and has published more than 200 journal articles. She received the Walter Alverez Award for Distinguished Medical Writing.

The Charges

An ultratoxic herbicide was rained down leaving an endless harvest of genetic defects and cancer; tens of millions of Vietnamese and 2.4 million American soldiers are estimated as having been contaminated with Agent Orange in Vietnam. . . .

Finally, in 1971 when mounting evidence of the harm Agent Orange was doing became too obvious to ignore, the government halted Agent Orange warfare—too late for the army of its civilian and military victims.

The Poison Conspiracy[1]

The enemy is 2,4,5-T, a powerful phenoxy herbicide contaminated with dioxin, generally considered the deadliest substance ever created by chemists. The story of 2,4,5-T symbolizes many environmental issues, pitting a powerful corporation against a grassroots movement unaided by the government agencies designed to protect the environment and the public health. . . .

. . . the evidence on 2,4,5-T raises deep doubts about its effects on the health of the people living in or near the forest land. These are killer chemicals. . . .

The dead and deformed children of Orleans and Denny are the innocent victims of the herbicide wars, casualties of the struggle to determine our values in this chemical age.

Who's Poisoning America[2]

. . . the herbicide 2,4,5-T, whose chief danger is in its frequent contamination with dioxin, [is] one of the deadliest substances known. . . . 2,4,5-T is a component of Agent Orange, the defoliant used in Vietnam which was later implicated as the cause of birth defects and infant deaths.

Pills, Pesticides & Profits[3]

Dioxin . . . [is] the most toxic known chemical . . .

223

You can't go around spraying deadly chemicals on people's property, where they keep animals and kids. I thought if you could go to the moon you ought to be able to control something like that.

Hazardous Waste in America[4]

Dioxin can cause severely adverse health effects, and death, at the lowest doses imaginable. . . . millions of pounds of the two ingredients of Agent Orange are still being sprayed, despite the enormous harm they are known to have caused to human health, wildlife and the environment. . . .

The disposal of dioxin wastes is another time bomb ticking across the country . . . .

No one knows the damage that dioxin is causing to public health, but the pervasive presence in the environment and food chain of a variety of toxic chemicals known to cause cancer may play a significant role in the current cancer epidemic . . . .

Nor can anyone say what will be the long-term effects on the American people—the chemical industry's ultimate guinea pigs. By the time the final results are in, it may be too late to take corrective action.

"Across America, Dioxin"
New York Times op-ed piece[5]

The so-called phenoxy herbicides 2,4,5-T, silvex and 2,4-D, used here to combat aquatic weeds that can clog rice fields, are contaminated by dioxins, one of which is the most toxic compound synthesized by man.

New York Times 6

Humans who have been exposed to it [dioxin] near Love Canal, in Vietnam or in situations such as train wrecks or factory explosions have suffered a variety of severe health problems, including kidney and liver ailments, birth defects and cancer.

Ibid. [7]

The Facts

- Dioxin is highly toxic to some species of animals but less so to others.

- Dioxin is a potent animal carcinogen.

- As far as we know, no human has ever died or become chronically ill from environmental exposure to dioxin in the United States. The only human illnesses so far proved to occur from exposure to dioxin are chloracne, a severe acnelike skin disorder and short-term reversible nerve dysfunction. Acute dioxin poisoning can also be toxic to the liver and kidneys.

- There is no conclusive evidence that 2,4,5-T or dioxin causes human cancer, spontaneous abortion or birth defects.

- Followup studies on accidental exposure to high levels of dioxin (for example, Monsanto workers exposed in a 1949 accident) indicate no long-term adverse health effects.

- Some studies, yet to be confirmed. have suggested that exposure to 2,4,5-T and dioxin increases the risk of soft-tissue sarcomas in heavily exposed industrial workers. Even if this link with occupational exposure were confirmed, it would have no bearing on the allegations of the myriad of other complaints, including miscarriage. being made by those objecting to traces of dioxin. Further, if the causal link were made for those occupationally exposed to high doses of dioxin containing herbicides over many years, it would be of little relevance to those of us who might be only occasionally exposed to trace amounts in the general environment.

Hardly a day goes by without an unsettling media report citing yet another contaminated dump with measurable amounts of what is routinely called "the most toxic chemical known to man." Dioxin, many Americans suppose, is synonymous with skull and crossbones. signifying illness, cancer, and death. Remarkable it is. and indeed a tribute to the power of the media, that a substance of such infamous reputation has never been shown to cause any deaths or serious harm to humans.

The dioxin saga consists of three separate but related chapters. The first focuses on a fear of dioxin itself; that is, its presence in places like Times Beach, Missouri, the Love Canal neighborhood. and Newark, New Jersey. The question here is: Is the mere presence of dioxin an imminent health hazard and a reason for grave concern?

The second chapter relates to anxiety over a previously widely used agricultural chemical that has traces of dioxin as an unavoidable contaminant. The herbicide 2,4,5-T has in the past few years been charged with causing a variety of human ills, and, in particular, been cited in some preliminary (but since discredited) studies as causing miscarriage and other reproductive difficulties. As a result of these concerns, 2,4,5-T is now partially banned as a commercial herbicide. But the question remains: Is there evidence that this herbicide itself, or in conjunction with its contaminant dioxin, causes human reproductive failure? And the third chapter of the dioxin saga involves Agent Orange, that now infamous defoliant used in Vietnam during the period 1962–70. Veterans claim that a wide spectrum of diseases they now experience and birth defects in their children are the result of their exposure to this chemical (Agent Orange was actually a combination of the herbicides 2,4,5-T and 2,4-D). And here too the question begs an answer: Is there scientific evidence that the wartime use of Agent Orange was responsible for the health problems now alleged by veterans?

As will be evident from the material in this chapter, the answer to all three questions is no. Obviously there is nothing beneficial about dioxin. And given the chemical's varied toxicity to humans and animals, there is just reason for concern and calm, reasoned and efficient remedial action to clean up the contaminated areas. Its presence in the environment serves absolutely no useful purpose. [1] But its mere presence in the environment does not translate into a massive public health disaster. And the fact remains that, although public and political pressure was such that the federal government bought out the town of Times Beach, Missouri, when roadside dioxin contamination was discovered in spring 1983 (costing some $33 million), there was no documented case of death or serious illness caused by the dioxin among the residents. The charges of health effects related to 2,4,5-T and Agent Orange are both confusing for the laymen and very volatile. It takes a callous individual not to have sympathy for the veteran who has developed cancer or has fathered children with one or more birth defects. Similarly, the passionate appeal of a woman from Oregon who tells of the trauma of experiencing a miscarriage clearly can draw an emotional response from almost anyone. But emotional outbursts are not adequate substitutes for scientific evidence and serve only to cloud the real issues. In determining causality between dioxin, 2,4,5-T, Agent Orange, or any other environmental factor on the one hand and human illness on the other, we must have data that indicate that the adverse health effects are appearing at a rate more frequent than that in the "unexposed" population—that is, women who were not living in an area sprayed with 2,4,5-T or American men who did not serve in Vietnam. The unfortunate reality of life is that individuals from all walks of life. from all parts of the country, have children with birth defects, suffer premature deaths from cancer, and report the full gamut of human ills. And as we will see, no convincing evidence exists that Vietnam War veterans are experiencing more than their share of disease and other health misfortunes, nor is there evidence to indicate correlative miscarriage rates in women living in areas where 2,4,5-T has been sprayed. What has happened instead is that the public has become increasingly anxious—sometimes hysterical—as a result of a media blitz of half truths, distortions, and nonscientific fantasies about dioxin, 2,4,5-T and Agent Orange.

Many representatives of the media would have us believe that the dioxin in the environment, both at present and during the exposures addressed in this chapter, has been proved guilty without a shadow of a doubt. The crime: death, cancers, birth defects, and numberless other ailments to exposed segments of the population. The jury in the case is the scientific community. But, as we shall see, there is nothing resembling a significant number of scientists and public health officials who feel that the dioxin and dioxin-containing herbicides have been responsible for the crimes with which they have been charged. To the television news programs, however, the scientific community's consensus is offered by a select group of "experts," over and over again.

Rarely do we see coverage about the expert who finds, for example, that dioxin in soil samples does not pose much of a health threat to humans in the area. What we do see on television news programs are interviews with the few same scientists preaching horror stories and doomsday prophecies. During early 1983 almost every television report. documentary, and print story on the subject cited University of Illinois' Dr. Samuel Epstein and Harvard's Dr. Matthew Meselson. Epstein, in particular, has frequently been quoted in absolute terms when the data on which he bases his statements are not accepted by the scientific community, such as this excerpt from the Washington Post: "The evidence is overwhelming, that dioxin is carcinogenic in humans." [8] The media want sensationalism, and they know where they can get it.

As Joan Beck of the *Chicago Tribune* astutely wrote in her column of June 30, 1983: "Newspaper editors and broadcasters, who help set the national agenda by what they choose to report. have hyped the dioxin danger, made it more dramatic by their choice of human interest stories, ignored much of the scientific evidence and used quotes from some scientists whose conclusions weren't justified by their research." [9]

What Is Dioxin?

Seventy-five phenoxy acid compounds constitute the group labeled dioxins. The most acutely toxic of these is 2,3,7,8-tetrachlorodibenzo-p-dioxin, or TCDD. A stable compound, dioxin breaks down slowly in the environment when protected from the ultraviolet rays of the sun. Ultraviolet radiation does cause the chemical to degrade quickly, but once it finds its way into the soil, it persists for a long time, remaining, tightly bound to soil particles. Since dioxin is highly insoluble in water, it tends not to migrate very fast in soil.

Dioxins are not manufactured directly for use in any product. Rather, they are formed as unwanted byproducts in the production of herbicides such as silvex and 2,4,5-T and in wood preservatives made from chlorophenols. Recent study has revealed that dioxins are also formed naturally by the incomplete combustion of wood products and industrial and municipal wastes (although the 2,3,7,8 form is less predominant among, the dioxins formed in this way). Dioxin has also been found in the exhaust from diesel-powered vehicles, suggesting that it is formed during the combustion of diesel fuel.

Herbicide 2,4,5-T

In 1945, 2,4,5-T was developed as an herbicide and three years later the product was registered for use with the U.S. Department of Agriculture as a pesticide under the provisions of the federal Insecticide, Fungicide, and Rodenticide Act. It was not until 1957 that dioxin was identified as an unavoidable contaminant in 2,4,5-T.

Unlike other members of the dioxin group, the contaminant in 2,4,5-T, 2,3,7,8-TCDD, is a potent toxin; small amounts of the chemical can cause extensive damage to some species of animals. Although in the 1960s some samples of commercially supplied 2,4,5-T were reported to have contained 70 ppm of this dioxin, 2,4,5-T currently sold in the U.S. contains an average of only 20 ppb or three thousand times less. As used in agricultural and forestry applications. 2,4,5-T is usually mixed with oil, which dilutes its dioxin concentration even further. Thus, health risk estimates based on the earlier dioxin concentrations are overstated.

Since it was registered in 1948, 2,4,5-T has proven successful in controlling the growth of undesirable plants, especially broadleaf weeds and brush. The herbicide has been used to control broadleaf weeds and brush along highways and railways, in rangeland and forests, and in wheat, rice, corn, and sugarcane fields.

The extraordinary feature about the phenoxy herbicides is their selectivity. They can kill undesirable plants while leaving desirable ones unharmed. They allow for efficient production of foods as well as paper and other forest products. The most popular phenoxy herbicide used in the U.S. is 2,4-dichlorophenoxvacetic acid, or 2,4-D, because of its low cost and effectiveness. It is manufactured by a different process from that of 2,4.5-T and contains no dioxin as a contaminant. For those shrubs and plants that are resistant to 2,4-D, other herbicides such as 2,4,5-T are substituted.

Current use of 2,4,5-T is restricted to rangeland and rice fields. Annual use of 2,4,5-T on open land will kill sagebrush, chaparral oak, and other hardwoods. In time these plants will be replaced by grasses. thus making previously unusable land suitable for grazing livestock. In rice fields 2,4,5-T is used to control the growth of weeds. In southern states where weeds such as the curly indigo are particularly resistant to other phenoxy herbicides, 2,4,5-T is the only effective method for controlling these weeds.

In some cases, alternatives to 2,4,5-T can be adequately substituted. Mechanical tillage, manual labor, or the use of other phenoxy compounds and other herbicides can be used to control weeds. However, in many applications 2,4,5-T is the most effective weed-control method. It is a highly selective herbicide when used properly in agricultural and forestry management and is both inexpensive and efficient.

Half-Baked Science:
The Alsea Studies

The current partial ban on 2,4,5-T was based largely on an epidemiological study in the Alsea, Oregon, area: the study linked spraying, of the herbicide with human miscarriages. The alleged problem which prompted the first epidemiological study was brought to the attention of the EPA when Alsea resident Bonnie Hill and eight other Alsea women suffered miscarriages which they thought might have been caused by exposure to 2,4,5-T. At their request the EPA began a preliminary study.

Initially ten miscarriages were reported by eight women. The final total, however, was thirteen miscarriages experienced by nine women. Twelve of the miscarriages occurred within the first twenty weeks of pregnancy.

The preliminary investigation of the EPA concluded that there was a seasonal variation in the miscarriage rate among the Alsea women compared with a control group from the nearby city of Corvallis, but that there was insufficient evidence to prove a relationship between the miscarriages and 2,4,5-T spraying. This study came to be known as Alsea I.

But the matter did not end here. Indeed, anxieties soon intensified. The so-called Alsea II study was a retrospective evaluation of the earlier epidemiological investigation. Using hospital records, EPA scientists collected information on miscarriages from three study areas: the Alsea region along the Oregon coast; the city of Corvallis, immediately inland from Alsea; and a sparsely populated rural area near the Idaho border. The Corvallis and rural areas were to serve as controls for the Alsea study period from 1972 through 1977.

Unlike Alsea 1, Alsea 11, released in February 1979, suggested that there was a relationship between the use of 2,4,5-T and an increase in miscarriages, it concluded, "The agency's systematic survey of the occurrence of spontaneous abortions in an area of 2,4,5-T use indicates that there was an unusually high number of spontaneous abortions in the area, and that the incidence of spontaneous abortions may be related to 2,4,5-T in that area."

Based on the findings of this report, the EPA immediately issued an emergency suspension of all remaining uses of 2,4,5-T except for rangeland clearing and weed control in rice fields.

Understandably, when the EPA partial ban on 2,4,5-T was announced, much scientific attention was thrust on the Alsea II study, which had not, at that point, undergone any peer review. [2] In the months that followed, at least eighteen reviews concluded that the EPA data did not support the EPA conclusions. [10]

Criticism of the study was extensive, consistent, and virtually unanimous. Those reviewing it represented a number of universities in the U.S., as well as agencies from Australia. Canada, and New Zealand. James A. Witt, a member of the team from Oregon State University which published the most extensive critique of the study, told the Northeastern Weed Science Society seminar in January 1980 that what the EPA alleged to be a yearly "June peak" of miscarriages was in fact a single large excursion from the norm, which occurred only in one year. Furthermore, there was no "significant" correlation between the amount of 2,4,5-T used and the number of pregnancy losses, and, except for June 1976, there was not even an "insignificant" correlation. [11]

A scientific advisory panel set up under the auspices of the federal Insecticide, Fungicide, and Rodenticide Act found neither immediate nor substantial threat to human health or the environment when Silvex or 2,4,5-T was applied to rice, rangeland, orchards, or sugarcane, and when used for certain noncrop purposes. [12] And a study by S. H. Lamm of Tabershaw Associates concluded that both statistically and epidemiologicallv, the EPA report did not support any relationship between herbicide spraying, and miscarriages. [13]

While the scientific community was systematically discrediting the results of Alsea 11, citing it as just another case of a half-baked, non-peer-reviewed government agency attempt at science, the Alsea women were being fashioned into folk heroes by the American press and media. Portrayed by many as helpless victims of corporate neglect and greed, the women became symbols of the natural, environmental struggle to free our country from the onslaught of the giant poisoners.

On October 2, 1979, for instance, WGBH in Boston aired "A Plague on Our Children," a documentary about the "victims" of 2,4,5-T spraying in Oregon. Consider these excerpts:

About one hundred miles south of Debby Marano's house is Eve

DeRock's valley and farm. . . . She is still sick two years after 2,4,5-T was sprayed there . . .

While the animal studies continue, the spraying continues. And Oregon residents are convinced that their health problems are linked to it. The Hoedads plant trees after a site is sprayed. They may come in contact with a poison by breathing it in, drinking contaminated water or by absorption through the skin.[14]

The program goes on to quote Edwin Johnson, director of pesticide programs of the EPA (who did not work on the Alsea II study), as holding to the "official" EPA position (generated largely by the then deputy administrator Barbara Blum).

There's never been much question that it causes problems in animals. . . . The Alsea II study . . . demonstrated that women were, in fact, having the same sort of effects in the study area where 2,4,5-T is used as one would expect them to have from test animals.[15]

How quickly the scaremongers accept matters such as the Alsea II conclusions as fact. No mention was made of the peer review findings published during the months before the broadcast, findings which showed that the study was statistically flawed and meaningless.

KRON-TV of San Francisco also managed to distort the public's view of the facts in its irresponsible documentary "The Politics of Poison." This 1979 attack on the herbicide industry was filled with scare tactics and half truths. One need go no further than the opening statement by the narrator:

What if you and your children were receiving tiny doses of a terrible poison—a synthetic chemical so powerful that an ounce could wipe out a million people?, Is that too incredible, too bizarre to believe? And if that were happening. wouldn't the government do something about it? These unbelievable questions face millions of Americans who are exposed to phenoxy herbicides every day. They are in the front lines of a symbolic battle—a struggle over what values will prevail in this chemical age. This film is their story.[16]

The message here is, you and your children are receiving tiny doses of a terrible poison, and, no, it is not too incredible and bizarre to believe. These accusations are so easy to make. The credibility of the accusers is rarely if ever questioned—they apparently come from concerned citizens whose aim is the protection of public health and the environment. But we must look beyond their concern, even if the concern is genuine. For it is accusations like those cited above that are incredible and bizarre.

Many reasons move people to make such outrageous, nonscientific statements about things they don't understand. First, the herbicides and other agricultural chemicals are manufactured by giant corporations whose sales total billions of dollars annually. Perhaps envy causes some to resent another's business successes. Second, it is always easier to blame life's misfortunes on someone else, especially when that someone is as enormous and anonymous as the federal government or the chemical companies. And third, there is a tendency for the general public to accept anecdotal information (especially as it pertains to something as emotional as cancer, birth defects. or other incidents of bad health) as undeniable proof. It is one thing for a Bonnie Hill of Alsea, Oregon, to be curious as to why she and some of her neighbors suffered miscarriages; it is something quite different for her observation to be equated with proof of a causal relationship between 2,4,5-T and miscarriages. The facts are, women within and outside areas sprayed with herbicides suffer miscarriages, and several factors, including age, parity (how many children one has had), and cigarette smoking, increase the risk of having one. Spontaneous abortions are common, not rare, among women: estimates are that 20 percent to 60 percent of all pregnancies are spontaneously aborted. Upon review of the incidence of miscarriage in the Alsea area, the scientific community overwhelmingly ruled out the spraying of herbicide 2,4,5-T as a contributor to miscarriage incidence.

Since EPA's 1979 regulatory action on 2,4,5-T use, Dow Chemical has spent millions of dollars in an attempt to reverse the administrative cancellation proceedings. But the dioxin debate has grown so large and controversial that in October 1983 Dow executives decided that continuation of the legal battle would no longer be productive. Dow spokesmen have expressed the hope that in the coming months, EPA resources will be devoted to a more objective and efficient sorting of fears from facts.

Health Effects of Dioxin and the Herbicide 2,4,5-T

Animal Studies

Several animal species involving pure forms of 2,4,5-T, dioxin, and a combination of the herbicide combined with trace amounts of the contaminant dioxin, have been studied to determine any health effects of 2,4,5-T and dioxin. Monkeys, sheep, dogs, rabbits. guinea pigs, hamsters, rats. and mice have all been tested to study the effects of these chemicals on cells, tissues, organs, and whole animals.

The results of the studies have been as varied as the species of animals tested. There is a great deal of species specificity in response to dioxin exposure. Physiological and biochemical differences among animal species also produce differing test results. Dogs, for example, are far more sensitive to the effects of 2,4,5-T and dioxin than mice. However, even within a single species, different strains or breeds vary in their response to 2,4,5-T and dioxin.

The many animal studies of pure 2,4,5-T without dioxin have shown that excessively high doses will cause muscle weakness, weight loss, and tissue changes. Equivalent doses to humans are not even closely approached where 2,4,5-T is used commercially. In contrast, animals exposed to pure dioxin will develop damage to the thymus gland, lymphoid tissue, and certain blood-forming tissues such as the bone marrow, as well as death in high enough doses. These findings back up the statement that dioxin is extremely toxic. The hazard from 2,4,5-T in the ambient environment lies in the dioxin component. However, even with dioxin's high toxicity, the truly tiny amount contained in commercial 2,4,5-T at the current time is so very low that the public hazard is negligible.

Perhaps the biggest health question is, Do 2,4,5-T and dioxin cause cancer? A number of animal studies using mice and rats have addressed this question, producing toxicological evidence that 2,4,5-T causes cancer.

Some argue that dioxin may not be a direct cancer-causing chemical, or cancer initiator. Instead, they interpret the test evidence as suggesting that dioxin accelerates. or promotes, the carcinogenic action of other chemicals. Other scientists argue that dioxin is a cancer initiator rather than a promoter.

The potential hazard posed by dioxin exposure must not be overlooked. The chemical is highly toxic and a proven animal carcinogen. The animal data assembled on this chemical are extensive and overwhelmingly support the claim that dioxin causes a number of health effects in many species of animals. However, a high degree of species variability is observed, with guinea pigs having the greatest amount of sensitivity.

Human Exposure

There is little evidence on the lone-term adverse effects of 2,4,5-T and dioxin in man because most human exposure to these chemicals has been accidental rather than experimental, and the dioxin-containing products have been manufactured only since the end of World War II. However, from the evidence that has been gathered, no conclusive relationship has been established between 2,4,5-T or dioxin on the one hand and human cancer, spontaneous abortion or birth defects on the other. The only human illnesses so far proven to occur from exposure to these chemicals are chloracne (a severe acnelike skin disorder) and reversible nerve disorders. Skin changes caused by a liver metabolism disorder, called porphyria cutanea tarda, have also been observed. These conditions are caused solely by dioxin and not by 2,4,5-T.

One of the most publicized chemical accidents in recent years, one which gives us considerable assurance that dioxin is not the health villain the press makes it out to be, occurred on July 10, 1976, in Seveso, Italy. Following the explosion of a chemical reaction chamber at the ICMESA chemical company, a section of the densely populated Seveso community was contaminated by an estimated one to four pounds of dioxin. The dioxin was produced during the manufacture of hexachlorophene, an antibacterial agent.

Plants, birds, rabbits, and chickens died soon after the accident. In addition, children and adults exposed to the chemical dust complained of nausea, nervous symptoms, and chloracne like skin disorders accompanied by redness and swelling. Following these reports, the local authorities evacuated some five thousand persons from the contaminated area.

An extensive health surveillance system was put into effect to record the short- and long-term effects of dioxin exposure on the population. Medical examinations and laboratory tests were performed, pregnant women were closely monitored to record miscarriages and birth defects, and a cancer registry was created to track any new cases of cancer among exposed individuals.

A thorough analysis of the health data gathered from the Seveso population has been conducted by the Instituto Scientifico per lo Studio e la Cura dei Tumori in Genova, Italy. The results of the study showed that Seveso residents developed chloracne and minor, temporary nerve damage. Some studies uncovered early effects on the liver and blood lipids. The net effect of these conditions does not appear to be significantly harmful. No other organs or body functions were found to be affected despite the fact that the Seveso population was enveloped by a cloud containing a few pounds of "the most toxic chemical known to man." Having claimed in "The Politics of Poison" that one ounce of dioxin could kill one million people. the staff at KRON-TV might be somewhat interested in this fact. Importantly, there was no increase in the number of miscarriages, birth defects, or infant deaths that could be linked to dioxin, although researchers noted a slight decrease in the overall fertility in the years following the accident. The study period was not long enough to uncover any cases of cancer related to the accident.

The Seveso experience does suggest that the oral ingestion of dioxin is the most toxic route of exposure. Many animals and birds died from the accident—most of them were herbivorous. Contrast the facts about Seveso with the account rendered by John G. Fuller in his ludicrous book The *Poison That Fell from the Sky*. On the inside cover we read of the residents of the town whose "children have sickened with a disfiguring and life-threatening disease; old people have died and autopsies have revealed that the chemical fatally attacked them; pregnant women have borne birth-defective babies. . . . At first these symptoms were limited to 'Zone A,' the area closest to the contamination. Soon there were signs that the residents of 'Zone B' were affected; the contamination is apparently still spreading."[17] On what does Mr. Fuller base these claims? He is not a doctor, public health specialist, epidemiologist, chemist, or any other type of scientist. For if he was, he would certainly be wise enough to consider the work and opinions of his peers. It is clear that this man does not believe in the work of the medical and scientific communities. And if he did take the time to read the literature, he simply ignored it.

Rather than study the incident, Mr. Fuller chose to rush to the scene and churn out a bestseller while the story was still hot. One can only wonder about the hypocrisy involved when someone promotes fear and doom by inaccurate, profit-generating reporting while criticizing the "giant corporate profiteers" who produce the hazardous chemicals as a byproduct of their operations.

At least Thomas Whiteside waited a few years before publishing *The Pendulum and the Toxic Cloud*. The doomsday theme runs strong in this book as well, although the author does acknowledge that the episode at Seveso did not result in the kind of human suffering predicted by Fuller. Rather, this fortunate fact is treated nearly as an omen that the worst is yet to come. "In view of the mounting evidence of the potential dangers of dioxin to people, the relatively light scattering of ill-effects so far manifested among the Seveso population tends to illustrate not the lack of dioxin's toxicity to humans, but the mysterious— one might say devilishly capricious—manner in which it can strike and how little is as Yet understood about this substance."[18] In other words, the disaster has not occurred—yet. They want a disaster, and they will wait for it if necessary rather than admit the errors of their prophecies.

In addition to the reassuring results of the followup study of the exposed population at Seveso, a "natural experiment" has yielded useful information on the health effects of dioxin. In 1949, 288 employees of a Monsanto chemical firm were exposed to dioxin in a Nitro, West Virginia plant accident and 122 of them developed the classic symptom of dioxin exposure, chloracne. Studies over the next thirty-four years on the exposed group of workers did not, however, detect any increased rates of cancer, birth defects, or mortality in general. Nervous disorders were reported in some of the men just after exposure, but, as in Seveso, these subsided without any lasting effect.

The Nitro accident provided medical researchers with perhaps the only sample on which an assessment can be made of the long-term effects of dioxin among people exposed to a level of the chemical sufficient to cause acute symptoms such as chloracne. The absence of increased rates of chronic diseases such as cancer in the group affected ought to be quite reassuring to residents of areas like Times Beach, Missouri, and the Love Canal in New York, as environmental trace levels of dioxin did not even approach the level of exposure in the Monsanto plant.

Because so little is known about the safe versus unsafe levels of dioxin and because it is potentially a hazard that will affect all Americans, if not their health then at least financially, it is clear how important the above-mentioned episodes are. But, as usual, the extremists neglect to consider the comforting information. It's almost as if they would rather resist the good news, at considerable financial cost to society and psychological stress to many of us, than accept the reality that our well-being is not threatened. It seems that the alarmists have developed an emotional issue on the subject of dioxin and have committed a lot of time and energy to keeping that issue alive. The desire to keep tensions high seems to make them unwilling to acknowledge any data not consistent with their doomsday hypothesis.

Does Dioxin Cause Human Cancer?

A number of studies of individuals heavily exposed to dioxin-contaminated herbicides have yielded conflicting results on the question whether these compounds can increase the risk of certain types of human cancer.

Studies conducted in Sweden in 1977 on heavily exposed lumberjacks suggested that there was a relationship between phenoxy herbicides and chlorophenols on the one hand and a very rare form of cancer on the other, a cancer called soft-tissue sarcoma (STS). Similarly, the United States National Institute of Occupational Safety and Health (NIOSH) published in 1983 the results of its study of mortality patterns of over four thousand workers who manufactured these products at a number of occupational sites. The individuals involved in the sample population all had been exposed to the dioxin-containing products for ten to twenty years, and some showed evidence of chloracne. NIOSH reported five cases of STS, which statistically amounts to twenty-one times the expected number of cases.

But these studies need some perspective. First, a 1982 New Zealand investigation which focused on 102 STS patients recorded in the New Zealand Cancer Registry between 1976 and 1980 found "no excess for the occupational group involving agriculture and forestry in spite of the fact that phenoxy herbicides have been used extensively for many years in New Zealand . . ."[19] Second, although the NIOSH data are fascinating, they provide us with good examples of why statistics need to be interpreted cautiously. A twenty-one-fold increase sounds substantial enough to merit the description epidemic. but though the significance of five cases of STS is not to be minimized or disregarded, and indeed the data suggest the need for intensive followup studies. one must remember that in dealing with such a rare disease as soft-tissue sarcoma. it is statistically easy to achieve a twenty-one-fold increase with just a few cases—in this instance, five.

Some evidence has been presented suggesting a higher than expected mortality rate among white females in Midland County, Michigan, because of STS. Dow Chemical, manufacturer of 2,4,5-T and other herbicides, is based in Midland and has been operating a large plant there for decades. In response to this concern, Dow has offered to assist financially in a study to be conducted by the Michigan Department of Public Health (MDPH). The project will look at the elevated rates and attempt to determine their possible causes.

Does exposure to dioxin-containing herbicides increase the risk of cancers in agricultural workers who are exposed to large doses of it. Unfortunately, we cannot now answer yes or no. Commenting in the dioxin STS data, Dr. Edward Brandt, assistant secretary for health of the Department of Health and Human Services, summed up the scientific consensus on this subject: "There is an increasing body of evidence that there may be an association in workers between exposure to products containing dioxin and soft-tissue sarcomas; however, further studies are necessary before we can determine whether or not this association is causal."[20]

Whether or not the dioxin-herbicide-STS cancer link is established, two points are relevant here in evaluating the scaremonger approach to dioxin and the herbicide 2,4,5-T. First, even if the link between STS and occupational exposure were confirmed in later years, the finding would have no bearing on the allegations of myriad other complaints, including miscarriage, being presented by those who maintain that America is being poisoned by dioxin and agricultural chemicals. Second, if the causal link were made for those occupationally exposed to high doses of dioxin-containing herbicides over many years, this circumstance would bear little relevance to those of us who might be occasionally exposed to trace amounts in the general environment.

Times Beach—Health Hazard
or Political Panic

> There are two dangers in toxic wastes. One is the very real threat to health posed by the chemicals themselves. The second is that a hysterical exaggeration of that threat will needlessly frighten people and drive them from their homes.[21]

The *Wall Street Journal* editorial from which that quote was taken was written in the days following the EPA decision to spend $30 million to buy all of the homes of Times Beach. Missouri. What led to this decision?

About ten years ago, some two thousand gallons of a combination of oil and industrial wastes were sprayed on a horse arena near Times Beach. The wastes were contaminated with dioxin. Soon after the spraying, numerous birds, rodents, and horses in the area died. Two children became ill and required hospitalization after playing in the area. By 1975 it was apparent that the animals had died from dioxin poisoning traced to the spraying, which was present at approximately 30 ppm. In addition, unpaved roads in and around Times Beach were later sprayed with the same material to decrease dust.

Times Beach is a little town of two thousand or so residents located about twenty-five miles southwest of St. Louis. Flood waters swept through the community after severe rains in December 1982. Before the flooding, EPA officials found dioxin in residues at a chemical plant and alongside roads. They feared that the flood waters spread the dioxin throughout the town. What followed was a sequence of events. characterized by agency panic, resulting in a $33 million bill in the absence of any observable or measurable health effects.

At the time of the incident. the EPA and its administrator, Anne Burford, were under severe and frequent attack by the press and environmentalists throughout the country. Much of the criticism had to do with the possible mismanagement of Superfund money the fund created to clean up the nation's hazardous waste dumps. Leading members of the EPA administration had recently been fired, and Burford herself had been cited for contempt of Congress when she refused to provide certain requested documents.

The residents of Times Beach were scared; they did not know if their health was being jeopardized by the presence of dioxin in the soil. The Centers for Disease Control's recommended lifetime human exposure limit of 1 ppb was exceeded by some of the soil samples, which measured more than 100 ppb in some areas. CDC officials therefore recommended evacuation. But this course of action was not widely suggested. Dioxin in soil is not equivalent to the hazard posed by dioxin in food or water—that is, dioxin which is ingested. The compound adheres tightly to soil particles, and therefore human exposure is not likely. In a June 24, 1983, editorial, the editors of Science wrote that "while ironclad proof of a null effect is missing, so too is a basis for believing that TCDD is a dangerous carcinogen in humans. It is clear that. when administered orally, TCDD is highly toxic. but when bound to soil it does not pose much of a hazard."[22]

So, why the buyout? Only the decisionmakers at the EPA can give us the real answer. But one can't help noticing that soil analysis was the sole criterion used in determining the course of action. Again it is critical to emphasize that the presence of dioxin in soil does not provide acceptable evidence of the existence of significant human exposure. The decision was forced too quickly to allow the development of adequate scientific evidence [23]

Were the citizens of Times Beach at risk? Although it will take decades before any increase in cancers can develop, we can at this point say that on the acute level, they were not. After screening 100 residents of Times Beach and two other contaminated sites, the Missouri director of the Division of Health said, "We've seen nothing to alarm us or to make us believe Missourians are feeling acute health effects."[24] And unlike the episode in Seveso, there was no chloracne among the residents, indicating that the dioxin was remaining in the soil, making exposure negligible.

In June 1983, shortly after the Times Beach buyout. the American Medical Association made its position known concerning the attention given dioxin by the media. The backbone of the resolution passed at a convention held in Chicago was that there is no scientific evidence that dioxin poses a direct threat to human health. The AMA was reacting to the recent coverage of the Times Beach incident. It voted to "adopt an active public information campaign . . . to prevent irrational reaction and unjustified public fright and to prevent the dissemination of possibly erroneous information about the health hazards of dioxin."[25]

Times Beach, like Love Canal, is an environmental problem turned into an environmental fiasco. Decisions and subsequent actions were based as much on political considerations as on public health realities. The reality was that (a) dioxin was present and (b) the public was not being exposed to health-threatening levels. We are left with a feeling of frustration and bewilderment—frustration that such carelessness can occur and lead to the dispersal of chemical toxins in our environment; bewilderment over how an agency presumably equipped to deal responsibly with these kinds of problems can be backed into a hasty, panicstricken approach to solving them. The Wall Street Journal editorial summed up the feelings of many Americans: "We'd like to be able to read in a newspaper some day that the EPA has gone out and cleaned up a toxic hazard problem without its having been turned into a political circus. When that happens we will mark another milestone on the road to political maturity. "[26]

Agent Orange Is A Red Herring

The ingredients were unique: equal measures of chemical phobia; anecdotal accounts of human tragedy; guilt about the Vietnam War; and the inevitable media hype. Also unprecedented was the yield: $180 million in compensation for thousands of veterans who asserted that the herbicide Agent Orange harmed them and their families. But the background leading to that out-of-court settlement had an underlying taste that was distinctly familiar: the distortion of science to indict and convict an environmental factor as a cause of disease and death, couched in that emerging doctrine "unless you prove otherwise, my bad luck is your fault."

As part of the Vietnam War effort, a number of herbicides were used by Americans to clear the vegetation that provided cover for the enemy. The use of herbicides is widely credited with saving the lives of thousands of American soldiers. The herbicides were also designed to destroy enemy food crops. The herbicides were stored in color-coded drums from which their names were derived, and included Agents Orange, Orange 2, Purple, Pink, Green. White, and Blue.

Agent Orange was the military code name for a fifty-to-fifty mixture of the herbicides 2,4,5-T, and 2,4,D used during the Vietnam War, from 1965 to 1970. These herbicides had been used singly and in combination, throughout the world since 1949 with no apparent deleterious human effects. Thus, these were not new chemicals developed specifically for use in Vietnam, although the manner in which they were employed in the war differed from the domestic agricultural use which had primarily been for highway weed control, forest management, and agricultural and residential landscaping.

In the early 1960s it was discovered that the contaminant dioxin was unavoidably formed during the manufacture of 2,4,5-T, but when spraying began in 1966, all parties concerned accepted the prevailing scientific opinion that trace amounts of dioxin in the herbicide did not represent a health hazard.

In 1969, scientists reported a high rate of birth defects in experimental animals exposed to 2,4,5-T and 2, 4-D, and in 1970, citing these laboratory findings, the U.S. Department of Defense canceled all uses of herbicides in South Vietnam.

The current Agent Orange saga began on the notepad of a claims counselor, Maude de Victor, in the Veteran Administration's Chicago office. As she puts it, "It all started when I picked up the phone and a sobbing woman told me her husband was dying. He had served in Nam and was convinced that his cancer had been caused by chemicals sprayed there." De Victor began keeping track of other vets who called to tell her their woes, ranging from liver disorders to miscarriages suffered by their wives. When she thought her "data collection" was complete, she approached a CBS TV station in Chicago, and in March 1978 the documentary "Agent Orange: Vietnam's Deadly Fog," a collection of horror stories about "those chemicals," was aired, not only winning an Emmy, but generating an enormous amount of fear and anger among veterans. In July of that year, ABC's "20/20" picked up where the Chicago show left off, with another set of segments on devastating terminal illnesses among vets. The crucifixion of Agent Orange had begun. The media, quite simply, had a new toy, and wanted everyone to play with it.

Not surprisingly, there was an immediate increase in complaints at VA offices around the country, veterans now convinced that the defoliants used during the war were responsible for everything that ailed them. In September 1978 Paul Reutershan, a Vietnam vet dying from liver cancer, filed a $10 million federal suit, claiming that his disease was the result of exposure to Agent Orange. Some months later lawyer Victor Yannacone, Jr. filed a new suit as a class action, saying forty thousand veterans and their families may have been involved. Five chemical companies—Dow, Monsanto, Hercules, Diamond Shamrock, and Thompson-Hayward—were the targets of the suit on behalf of "all American servicemen whose health has been damaged because of contact with Agent Orange."[27]

Just after the filing of the class-action suit, new headlines about the domestic use of the herbicide 2,4,5-T gave momentum to the veterans' legal push against Agent Orange. A study of women in Alsea, Oregon claimed a causal link between spraying, and an elevated rate of spontaneous abortion, and as a result of that study, citing an "alarming" rate of miscarriage, the EPA banned most uses of 2,4,5-T. Unfortunately, the media gave little coverage to the scientific community's peer review of that study and its later dismissal of it as yet another case of half-baked government science, as a result, the Alsea women joined ranks with the veterans, fashioned into folk heroes by the media.

Where's the Evidence?

From a scientific point of view the question of whether Vietnam vets were victims of Agent Orange relies on two points: First, is the incidence of cancer, birth defects, liver disorders, and the like higher than expected in exposed veterans? And the answer to that is, we do not know because we do not have that type of information.

Dr. Felix Moore, professor emeritus of biostatistics at the University of Michigan, has, however, made some calculations that remind us of some basic realities. He notes that the expected number of deaths from all causes for a population similar in age distribution to the population of Vietnam War veterans would be 521 deaths from cancer and nearly 800 from cardiovascular disease. Similarly, the March of Dimes estimates that birth defects strike more than 250,000 babies each year. Thus these figures demonstrate a point frequently overlooked in press reports, usually laden with emotional stories of human tragedy: while we all feel an innate compassion for those veterans or anyone else we see in agony and despair, any population having the size and age distribution of Vietnam Veterans has normal and expected mortality. To prove an association between exposure to Agent Orange and any health parameter requires an observed incidence that exceeds that expected. This proof has never been presented.

Second, if such an elevation of mortality and morbidity were ever documented among vets, the question would be what caused it. The open-minded epidemiologist certainly would not focus on only one variable like herbicides. Marijuana use, for example, would present an interesting research hypothesis.

But if Agent Orange exposure were on the suspect list, the most logical query would be, What evidence do we have of the effects of high-level human exposure to the herbicide and dioxin? And here we do have data, none of which would advance the case of the ailing vet. As was mentioned earlier, studies of individuals exposed to high levels of the herbicide and dioxin in occupational settings, as a result of accidents resulting in discharges of dioxin, as happened in Seveso, Italy, in 1976 and at Monsanto's Nitro, West Virginia plant in 1949, could identify no ill effects other than chloracne (a severe skin disorder) and some temporary nerve problems.

Perhaps the most significant studies completed to date as a means of testing the legitimacy of claims made by some veterans, were the two so-Called Ranch Hand investigations. named after the Air Force spraying team that applied Agent Orange during the war. The twelve hundred or so men who flew the plane and sprayed the herbicide were those most heavily exposed, handling large amounts of it, frequently reporting accumulation of Agent Orange on their clothing. If Agent Orange is indeed the silent killer, then these highly exposed teams would certainly be affected first and most severely.

Neither study could find statistically significant differences in health status between those who worked daily with Agent Orange in Vietnam and those veterans who did not, a finding that does little to substantiate the claims of those who blame a host of cancers. birth defects. and other chronic illnesses on exposure to the defoliant.

Bruce E. Herbert, deputy director of the Center for International Security, summed up the state of our knowledge about Agent Orange and health:

> After nearly five years of almost constant publicity, as of March 1, 1983, only 16,821 veterans have even filed claims with the VA for suspected Agent Orange damage. Of this number, less than 8,400 present any certifiable medical condition, whether or not these disabilities can ever scientifically be linked to exposure. Three thousandths of one percent of the 2.4 million men who could have been exposed to Agent Orange in Vietnam is hardly a compelling statistic on which to make assumptions about "unusually large numbers" of veterans suffering latent Agent Orange-induced health impairments.[28]

Those are the scientific facts. But then these are days when emotion, not reality, prevails in litigation, and anything can happen. On May 7, 1984, the much publicized Agent Orange trial was scheduled to begin. Judge Jack B. Weinstein bluntly warned both sides about the problems they faced in the impending trial. He told the chemical company and insurance lawyers that they were taking, a major risk by entering the courtroom. even though they were scientifically correct in their argument that Agent Orange has never been shown to be a health hazard. Faced with a Brooklyn jury that was certain to be sympathetic to the veterans, the odds against the chemical companies were stacked. On the other hand, the judge warned the veterans' lawyers that their case was a hodgepodge of flaws and inconsistencies, and that ultimately the claims of the veterans would never hold up in an appeals court. Thus the $180 million out-of-court settlement—all of which will ultimately be paid for by the American consumer, in higher costs of goods and services logically passed on by the corporations that settled. As S. Maynard Turk, vice president and general counsel of Hercules, put it,

> Who ultimately pays for this litigation, settlement cost and all? Sad but true—you do. It's the public who pays. . . . All business costs must be passed on in the price of the company's products. A company has two choices: (1) pass on the costs or (2) if it isn't able to pass them on, go out of business. In either case, the public pays the higher costs or lost jobs.

The Impact of the Misuse of Science

Consider the Vietnam veteran who has been reading about the toxic time bomb that supposedly ticks within him. Many of these young men have had a hard time adjusting to American society in the aftermath of our country's most unpopular war. Many others have suffered from the infirmities described, cancer, birth defects, and other health defects, which, as we have stressed, occur in all segments of the human population. A disturbed or sick veteran is going to find it very easy to believe the words of a Barry Commoner or other "expert" who writes of the toxicological time bomb wreaking havoc on the veterans and their families. And they are likely to step forward to detail their misery. These are the kinds of conditions from which big headline controversies are made.

Agent Orange has left its stamp on the lives of many Vietnam veterans, and on those of their families and friends. But the evidence increasingly points to the impact as being created more by an army of emotional extremists, the professional scaremongers and their media friends. than by the armed forces of the U.S. And our society as a whole has been swept up in the wind of the movement, as described by Dr. Archie B. Blackburn: "It may be more acceptable for us, our society, and our leadership to believe that Agent Orange has damaged our youth, rather than to recognize and deal with the logical consequences of warfare."[29]

Where Do We Go From Here?

The dioxin alarm has been sounded. And it has been heard. The scientific community, government, and private industry are committing new resources (finances and manpower) to the effort to understand the relationship between dioxin and human health. But it is essential that while the work is being, done, our citizens should understand that their lives and the lives of future generations do not appear threatened by the most infamous of chemicals, dioxin.

Yes, we can blame "deadly dioxin" for the chloracne it has caused. And we may one day finger the chemical for actually causing a small number of deaths among STS victims who were exposed over a long period of time while on the job. But to imply that the chemical is "the most hazardous compound ever made by man" is simply a gross distortion. Toxicity is not the same thing as hazard. If a toxin, in this case dioxin, does not have a means of entering our bodies, then the public health hazard exciting the extremists is overstated. A *Lancet* editorial accurately summarized:

> It is no use blaming a chemical for the fact that children have been burnt and starved in Vietnam, or that an American community is, rightly, dissatisfied about the way in which commercial and agricultural and civic interests lead to spraying of insecticides and pesticides on citizens. These are more political than toxicological issues. For the amount of dioxin con-

taminating the 2,4,5-T used in our environment there is no evidence of harm—only grounds for care in manufacturing and use of a socially valuable substance.[30]

A small portion of the American public has been exposed to trace amounts of dioxin. Future studies will tell us if that exposure has, in fact, caused any disease or death. But the majority of. if not all. Americans have been exposed not to dioxin, but to dioxin coverage, courtesy of the media. Many distortions and misconceptions have been conveyed to the public. The *Wall Street Journal* did much to bring the issue into proper perspective in a column titled "Dioxin Hysteria":

> Clearly, there are reasons to be worried about the stuff. High concentrations in the soil—30,000 times greater than the CDC evacuation standard—killed horses at Times Beach.... What is known, however, suggests that most of the scare stories are exaggerated. Human exposure to dioxin has so far been scientifically linked to only one health problem—a skin disease called chloracne, which tends to go away fairly rapidly. This is true even at fairly high exposure levels.... None of this means that dioxin is harmless.... But there are a lot of other threats to worry more about, such as tobacco, marijuana, drunk driving or street crime, where the evidence of threats to health is clear. The notion that dioxin is a doomsday menace is based less on medical evidence than on some kind of psychological phenomenon.[31]

Footnotes

[1] Oddly enough, however, despite being a powerful animal carcinogen, dioxin has also been found to be a potent inhibitor of tumor formation caused by two known carcinogens. BAP and DMPA, when low levels are administered to rodents, according to a report appearing in the October 1979 *Cancer Research.*
[2] The process by which the scientific community judges the merit and validity of the results and conclusions of a study.

References

1. Karl Grossman, *The Poison Conspiracy* (Sag Harbor, N.Y.: The Permanent Press, 1983), pp. 55, 61.
2. Ralph Nader et al., *Who's Poisoning America* (San Francisco: Sierra Club Books, 1981), pp. 240–41, 250, 258.
3. Ruth Norris, ed., *Pills, Pesticides & Profits: The International Trade in Toxic Substances* (Croton-on-Hudson, N.Y.: North River Press, 1982), pp. 32, 33.
4. Samuel Epstein et al., *Hazardous Waste in America* (San Francisco: Sierra Club Books, 1982), pp. 26, 140.
5. Lewis Regenstein, "Across America, Dioxin," *New York Times*, March 8, 1983.

6. *New York Times,* April 9, 1983.

7. *New York Times,* January 23, 1983.

8. Pete Earley, "Dioxin Is Still a Mystery," *Washington Post,* February 27, 1983.

9. Joan Beck, "Have the Media Hyped the Danger of Dioxin?" *Detroit Free Press,* June 30, 1983.

10. James M. Witt, "A Discussion of the Suspension of 2,4,5-T and the EPA Alsea II Study," Northeastern Weed Science Society, January 8, 1980.

11. *Ibid.*

12. American Medical Association (AMA), *Agent Orange Dioxin* (1981), p. 24.

13. *Ibid.*

14. "A Plague on Our Children," WGBH Educational Foundation, 1979.

15. *Ibid.*

16. KRON-TV, "The Politics of Poison," 1979.

17. John G. Fuller, *The Poison That Fell from the Sky* (New York: Random House, 1977), front and back flap.

18. Thomas Whiteside, *The Pendulum and the Toxic Cloud* (New York: Yale University Press, 1979), p. 133.

19. Smith et al., "Do Agricultural Chemicals Cause Soft Tissue Sarcoma?" Initial Findings of a Case-Control Study in New Zealand," *Community Health Studies,* vol. 6, no. 2, p. 114.

20. *Pesticide & Toxic Chemical News,* August 3, 1983, p. 21.

21. "The Dioxin Scare," *Wall Street Journal,* February 25, 1983.

22. "Chlorinated Dioxins," *Science,* June 24, 1983.

23. Council for Agricultural Sciences and Technology, *The Missouri Dioxin Controversy: Scientific Overview,* April 1983, p. 3.

24. Mark Edgar, "Times Beach Residents Pass 1st Dioxin Tests," *St. Louis Globe-Democrat,* July 1, 1983.

25. Philip J. Hilts, "AMA Votes to Fight Dioxin 'Witch Hunt'," *Washington Post,* June 23, 1983.

26. *Wall Street Journal,* February 25, 1983.

27. AMA, *Agent Orange,* p. 3.

28. Bruce Herbert, "Agent Orange, a Media Myth," *Chicago Tribune,* March 31, 1983.

29. Archie B. Blackburn, "Review of the Effects of Agent Orange: A Psychiatric Perspective on the Controversy," *Military Medicine,* April 1983, p. 339.

30. "Editorial: 2,4,5-T—Where Next?" *Lancet,* 1979, pp. 114ff.

31. "Dioxin Hysteria," *Wall Street Journal,* May 31, 1983.

Reprinted, with permission, from
Toxic Terror, published by Jameson Books, 1985

Taking the Die Out of Dioxin

Natalie P. Robinson Sirkin
and Gerald Sirkin, Ph.D.

Please see biographical sketch for Natalie P. Robinson Sirkin on page 119.
Please see biographical sketch for Dr. Gerald Sirkin on page 119.

A long-term follow-up study of an industrial incident (involving dioxin) that occurred in 1953 encompassed a population that was large enough for a twofold excess cancer mortality to have been detected. There was no increase in death from cancer during the entire 34-year follow-up . . .

Vernon N. Houk, M.D., Assistant Surgeon General,
speech, May 21, 1991

Dioxin still lurks as a terror in the minds of sincere chemophobes and in the tool-kits of insincere environmentalists-for-a-weaker-America. Both are fighting to stop waste incinerators on the ground that they emit dioxin.

We are being told that fish containing PCBs also contain dioxin. Scientific studies having produced no evidence that PCBs in the amounts found in fish are harmful to humans, dioxin is being called in to keep fear alive.

Dioxin is never intentionally produced. It is an unavoidable contaminant in the manufacture of trichlorophenol, 2,3,7,8-tetra-chloro-dibenzo-p-dioxin, usually abbreviated to TCDD. It also results from some forms of combustion. (In this paper, instead of "TCDD" we shall use the word "dioxin," though it covers 75 different compounds.)

Evidence from Industrial Accidents

We now have a considerable body of evidence on dioxin's health effects drawn from studies of accidents during the manufacturing prcoess. These cases of heavy exposure should tell us the worst we can expect from dioxin.

The first explosion occurred in 1949 at the Monsanto chemical plant in Nitro, West Virginia. Within minutes, workers entered the building to begin the clean-up. Within a week or two, they developed chloracne, a skin rash that is transitory, though severe cases leave scars and discoloration. (Chloracne indicated that the dosage was heavy because a light application even directly on the skin does not produce chloracne.) Other effects followed a few weeks later, including headaches, nausea, muscle pains, and fatigue, almost all of which disappeared with time. Years of study of the Nitro workers followed. The conclusion is that the workers suffered no higher rate of deaths, cancer, or other diseases than the population at large.

In Seveso, Italy, population 37,000, an accident in a chemical plant in 1976 released a dioxin-bearing cloud. The quantity of dioxin—one to four pounds—was substantial. Plants and some species of animals died. But the medical studies of the exposed population have so far found no adverse health effect except chloracne and nausea among 184 people.

Twenty-six pregnant Seveso women, fearing birth defects, went to Switzerland for abortions. "No abnormalities even in the chromosomes were found in the aborted fetuses (Rehder et al., *Schweiz. Med. Wocheuschr.* 1978, 108, 1617)."

Eight other accidents over the next thirty years in Germany, France, the Netherlands, Czechoslovakia, England, and Michigan, yielded the same conclusion: There is no evidence of adverse health-effects from the dioxin-exposures. As biologist Michael Gough, who worked on dioxin in the Office of Technology Assessment, says, "Although many were clearly exposed to dioxin because they came down with chloracne, life-threatening and life-shortening diseases are no more frequent among them than expected in the unexposed population."

Be Glad You're Not a Rat

Regulators depend on animal tests to evaluate the toxicity of a substance. They proceed on the assumption that if it is toxic to a laboratory animal, it is toxic to a human. Lacking human-dosage data that are precise, they extrapolate animal data to humans though in so doing they overestimate the hazard to humans by a factor of thousands.

The scientists in the regulatory agencies understand from the evidence from human exposure that their standards for humans are wildly overestimated. Even among laboratory animals, sensitivity varies widely. Guinea pigs are highly sensitive to dioxin, hamsters highly insensitive. The dose (relative to body weight) that kills half a test-population of hamsters is 5,000 times the guinea-pig dose. Despite all the evidence that the human body is not very sensitive to dioxin, nevertheless the EPA sets a dioxin standard for humans as if they were rats or guinea pigs.

Connecticut's Department of Environmental Protection goes further. It boasts that its standard of emissions of dioxin from waste-to-energy plants treats humans as if they were 2,000 times more sensitive to dioxin than the most sensitive laboratory animal, the guinea pig. A man is not a grown-up guinea pig, but a regulator-bureaucrat sees no reason to risk a charge of laxity just to be scientifically rational.

Agent Orange

During the Vietnam War, the Air Force, to remove the enemy's cover, defoliated trees, using the herbicide Agent Orange, which contained dioxin. Study after study of the effects of Agent Orange have found no evidence that it caused cancer or any other illness among our troops who may have been exposed to it. The Air Force, in its "Ranch Hand" study of 1200 military personnel involved in the spraying, found no evidence of adverse health effects.

Investigations of cancer among Vietnam veterans culminated in the release on March 29, 1990, of a five-year $10-million study by the Centers for Disease Control. A most-carefully designed and thorough research project, it found no significant incidence of cancer among the six types studied except for non-Hodgkin's lymphoma, which affected, not Vietnam veterans based on land, but Navy vets on sea duty off the coast of Vietnam, far from exposure to Agent Orange.

A study of dioxin reported in the *New England Journal of Medicine* of Jan. 24, 1991, covers 5,172 workers occupationally exposed to dioxin in 12 chemical plants in the U.S. Deaths from cancer of workers exposed for less than one year were not significantly different from deaths expected on the basis of rates in the U.S. population. For workers exposed more than one year, the deaths (many years later) from soft-tissue sarcoma, and cancers of the respiratory tract were significantly higher than expected, but no increased mortality was found from most types of cancer previously suspected of being linked to dioxin: Hodgkins' disease, non-Hodgkins Lymphoma, liver, stomach, and nasal cancers.

The results are inconclusive. But should prolonged exposure of occupational workers to high-levels of dioxin be ultimately found to be linked to cancer, that finding could not be linearly projected to low-levels of dioxin. Nor would that finding be relevant to the effect of trace amounts of dioxin to which one may be occasionally exposed in the general environment.

Can We End Toxic Terror?

Scientific evidence has piled up, building an overwhelming case against fear of dioxin. Yet the panic continues. EPA standards remain preposterously strict, and the environmental activists use the dioxin argument to block waste-to-energy plants.

Dioxin authority Fred H. Tschirley comments that the initial reports of its acute toxicity based on animal tests "have instilled a public fear that probably cannot be dispelled even by adequate information about the countervailing experience with human beings." Yet till we try to educate the public, we cannot know that it cannot be done. We have certainly not yet tried. EPA, which should be the chief educator, has been a major inciter of groundless fear of dioxin as it has been of PCBs, pesticides, Alar, asbestos, and radon, about which toxic terror has spread. Other organizations and individuals with an interest in sustaining toxic terror have been the most active—indeed nearly the only—participants in public discussions and actions.

Over the last forty years, U.S. cancer death rates have been decreasing or staying the same except chiefly smoking-related cancers. Nevertheless, the government alone has spent over a billion dollars on dioxin research. Toxic fear should not be allowed to continue diverting our scarce resources from vital research to demonstrating repeatedly what we already know: the absence of serious long-term effects of dioxin on human beings. The dioxin lunacy may eventually peter out, as public lunacies have in the past, but some educational aid from society's saner elements could save us waste and worry.

Reprinted with permission from *Citizen News*
July 18, 1990
New Fairfield, Connecticut

Electromagnetic Fields

Electromagnetic Fields and VDT-itis
Petr Beckmann

Electromagnetic Fields and VDT-itis

Petr Beckmann, Ph.D.

Dr. Petr Beckmann is professor emeritus at the University of Colorado in the Department of Electrical Engineering and publisher/editor of *Access to Energy*. He was previously affiliated with the Institute of Radio Engineering and Electronics, Czechoslovak Academy of Sciences.

He received a M.Sc. from Prague Technical University, 1949; a Dr.Sc. from Czechoslovak Academy of Science, 1961; and a Ph.D. from Prague Technical University in 1955.

Dr. Beckmann is the author of *The Scattering of Electromagnetic Waves from Rough Surfaces,* published by Pergamon in 1963; *The Health Hazards of Not Going Nuclear,* published by Golem in 1976; and Einstein Plus Two, published by Golem in 1987. He has published more than 60 journal articles contributing to *Radio Science, IEEE Transactions Antennas and Propagation, and Galilean Electrodynamics.*

Millions of Americans now spend considerable time near computer video display terminals (VDTs), and the resulting risk of cancer, nervous disorders, miscarriages, and minor afflictions such as persistent headaches has never been refuted.

But neither has this risk ever been demonstrated.

Similarly, there are few Americans left who do not spend much of the day near the electric currents flowing through transmission lines, residential wiring, industrial equipment and home appliances. A recent study commissioned by the State of New York has raised suspicions that the fields accompanying such currents are correlated with child leukemia and possibly other diseases.

Yet the authors of this study themselves warn that these results are only suggestive; no risk has been demonstrated.

These same fields are also known to affect the human body, particularly its biological clocks and rhythms.

Yet the very scientists who work in this field and have proved this link, also warn that it does not imply any health effects.

These electromagnetic fields differ only in frequency (if it were sound, we would call it "pitch") from another type of electromagnetic fields, those of microwaves, to which Americans are exposed near radar beams, telephone relay lines, radio and TV transmission antennas, and microwave ovens. Now here a risk has certainly been identified: prolonged exposure to microwaves at high levels can lead to damage of the retina (the sensitive "projection screen" in the eye) and to male sterility.

But at lower levels no such effects have been demonstrated.

So what is a layman to make of these bewildering reports?

First of all, let me tell you that in a certain sense I am a layman myself. I am not a physician, and I will never tell you what a certain type of electromagnetic field does or does not do to the human body. But I have spent a lifetime in electromagnetics as my major subject and I can tell you what the comparisons are: why sleeping next to an electric clock is negligible compared with living above a grounding loop, why VDTs have negligible electromagnetic fields, and what they do have instead, though nobody seems to worry about it. Perhaps equally important to my qualifications, for the last 20 years I have intensively studied the way in which health hazards, genuine or alleged, are being abused by politicians, ideologues and social engineers as a way in which to frighten laymen into an agenda from which these scaremongers profit in political power while wearing the false aura of crusaders for justice, morality and a healthy environment.

Why should you believe me?

You shouldn't. In our time, degrees and academic careers are no longer guarantees of honesty, truthfulness, or even competence. You should use your own head in reading what I have to say; you should judge whether what I say makes sense IF my data are correct, and you should check out those data and everything else by what is probably the best method: go to those who claim the opposite—in the media, among the "greens" and among the social engineers. In many cases they do not lie outright; they just don't tell you the full truth. Am I playing the same game? Ask them what I have left out; ask them whether my data are false; ask them whether my conclusions are flawed. Confront the two sets of information; get more on your own if necessary; and then make up your own mind—not whom to parrot, but whom to give greater credence with the greatest gift ever given to you: your own mind.

Electromagnetic Fields

Electromagnetic fields are parts of space where an electric charge (such as an electron) is acted on by a force. They generally oscillate, meaning that this force alternates in direction, and they are characterized by their frequency of oscillations in hertz (Hz), or cycles per second. (A static field is constant in time and does not oscillate, but it can be included as one with zero frequency.) Depending on their frequency range, electromagnetic fields behave very differently, and that includes their effects on the human body. For example, it is only in the range of wavelengths from about 0.35 to about 0.7 microns (millionths of one meter) that they stimulate the retinas of our eyes, and we know that range as (visible) "light;" the different wavelengths correspond to different colors.

An important division of electromagnetic fields can be made by their ability to ionize, that is, to tear off an electron from an electrically neutral atom. The frequency has to be sufficiently high for this to happen, and only two types of radiation can ionize cells in body tissue: X-rays and gamma rays, the penetrating type of radioactive radiation. In small doses, they are harmless, and by the latest evidence probably beneficial, but this is irrelevant, for there is virtually no ionizing radiation emanating from either VDTs or power lines.

There is a good reason why ionizing radiation can practically be ruled out for VDTs and TV sets manufactured since the early 70s.

X-rays are emitted when high-velocity electrons are suddenly stopped. But even in older picture tubes the velocity of the electrons impinging on the fluorescent screen was not very high. The intensity of the resulting X-rays was therefore quite low—highly addicted TV fans got a dose of less than 1 millirem per year, or 1/200 of the natural radioactive background. But in the early 70s more sensitive screens and the replacement of vacuum tubes by semiconductors led to lower electron velocities, and the X-ray level was therefore reduced from even that insignificant level to virtual absence.

The picture tubes and associated circuits in VDTs are quite similar to those in TV sets, but the viewer usually sits closer to a VDT than to a TV set, and in many cases spends more time in its vicinity.

As for power lines, at the extreme low frequency end of the electromagnetic spectrum, they never did emit any ionizing radiation.[1]

Effects of Non-Ionizing EM Fields

Like the neighborhood of other electric appliances, that of VDTs is permeated by alternating electric and magnetic fields. An electric field makes electric charges move, and a magnetic field curves their paths when they are in motion.

Such electric charges and fields occur naturally in the body. An electrocardiograph, for example, measures the voltage generated at certain points of the human body in the rhythm of its heart beat. More to the point here, there are natural electric fields across the membranes of cells. Since they are some 100 times higher than any that can be induced by common power fields, it was thought unlikely that external electric fields could affect them. Yet it has been shown that small changes in the external field can trigger detectable biological changes, such as changes in the body's biological clocks ("circadian" rhythms), though all investigators are careful to point out that at present there is no evidence that this is harmful.[2]

But if the electromagnetic fields of VDTs do have an effect, then other appliances and electric wiring must have the same effect, and in at least some of them the effect would have to be more pronounced, for their fields are more intense, and the daily duration of exposure to them is comparable or larger.

Indeed, the VDT anxiety arose only some time after a scare campaign against power lines had been waged, which in turn led serious investigators to the conclusion that appliances (including VDTs) would have to have a comparable or larger effect.

Even the San Francisco ordinance of December 1990, pushed through by political activists to restrict the use of VDTs, makes no claim of electromagnetic radiation or fields (though the press missed no opportunity in reporting it together with the alleged "link" between such fields and cancer); it is entirely limited to issues of eye strain and posture stress—neither of which, incidentally, I am qualified to judge and will not comment on.

The possible health effects of the fields surrounding high-voltage power transmission lines have been investigated for decades in both Europe and North America. They never turned up any conclusive evidence for any health hazards. In the early 60s a Soviet source alleged that linemen suffered from reduced sexual libido, and this juicy bit of news was not only eagerly picked up by the press, but triggered further studies in the West, none of which succeeded in confirming the Soviet allegation.

But with the advent of environmental paranoia, such shoddy papers were given a new lease on life, and in 1979 a paper alleging a link between child leukemia and power lines made it into the American Journal of Epidemiology. It was authored by Nancy Wertheimer, a psychologist working at the University of Colorado. The paper is often quoted to this day, though the methods of measurement were flawed, the researchers had no access to the houses investigated, and the statistics were incorrectly claimed to be significant. Child leukemia is a rare disease, afflicting an average of 3.1 of 100,000 US children under 10, and for such tiny fractions the sample investigated was nowhere near the size that would have made the results conclusive.

Nevertheless, a shoddy paper does not disprove its conclusions, and in 1988 a team of scientists commissioned by New York State found no evidence of effects from electric fields (as Wertheimer had claimed), but did find suggestions of a possible link between magnetic fields and child leukemia.

In its report, this team led by David Savitz, an epidemiologist at the University of North Carolina, was careful to warn that the findings were "suggestive, but inconclusive (individually and in the aggregate)."

The other members of this team were equally reluctant to draw conclusions. One of them, Prof. Frank S. Barnes of the University of Colorado, who has spent decades studying the effect of electromagnetic fields on living cells, writes,[3]

> "One of the problems with these and other epidemiologic studies is that it is virtually impossible to obtain an unexposed control group, and other environmental confounders are hard to eliminate. For example, in the Savitz study, corrections for variations in the measured fields with time of day and the season of the year were not made, nor were possible correlations with population density and traffic patterns. Childhood cancer is a relatively rate disease, so that the number of cases which could be studied from a population of about 1.5 million people in the metropolitan area around Denver was small and the spread in the data is very wide. Thus it is very hard to draw firm conclusions from this study with high confidence."

When a scientist makes such statements, they are often thought to be overcautious, especially in view of the "conclusive" doubling of the expected value near a power line or a nuclear plant or whatever else is accused of causing the "cancer cluster." With a child leukemia mortality of 3.1 per 100,000 children under 10, the expected value in a group of 100,000 such children is 3.1, and to have 6 children in such a "large" group die of leukemia looks persuasive to many who have not studied statistical inference. It is easy enough to show mathematically that for such small probabilities (0.000031), double the expected value in a group of 100,000 is nothing extraordinary; but this does not impress laymen, who often would rather agree with Mark Twain's comparatives "liar, damned liar, statistics."

What all this boils down to is that for very small probabilities you need enormous samples to draw any valid conclusions. Using a program I designed for my personal computer, I easily got 35% discrepancy for even one million children. But Savitz (let alone Wertheimer) did not test 1,000,000 children, or even 100,000. Savitz was therefore well advised to warn that his results were inconclusive.

But such warnings are brushed aside by the press and the politicians. In May 1989 the Congressional Office of Technology Assessment issued a sober and credible report concluding that no risk has been shown, which does not mean there is none; and it also made the important point that if there is a risk, other sources than power lines could play a far greater role in any health problem. Yet a mere week later (June 1989) the report was used by a Florida judge to justify his prohibition to let school children play in a school yard near overhead power lines ("We are talking catastrophic health risk," said one of the plaintiffs). To dramatize the dangers, people had themselves photographed holding up unconnected, shining fluorescent lights under power lines—as if a high electric field were needed to light them.

The superstitious soon found a prestigious ally. In December 1990 the Environmental Protection Agency put an aura of respectability on this type of pseudoscience by identifying "60 Hz magnetic fields as a possible but not proven cause of cancer in humans."

VDTs

Yet all this information had an obvious flaw: people receive a bigger dose (level of exposure times duration) from electric wiring and appliances in the home than they do from power lines, even close ones, (see Table I). It is one thing to incite people against the utilities and stop them building power liens; it is quite another to ask them to stop using television sets, refrigerators, electric clocks, blankets and other electrically powered conveniences.

The VDTs is therefore a well chosen target, since in the cases of longest exposure it is not used in the home, but most often belongs to the employer company—those profiteers who cynically exploit their workers without a shred of social responsibility.

In reality, the fields of VDTs are so small that no even long exposure to them can make up for the dose from some short-term appliances: data published by the Electric Power Research Institute show that for exposure to the magnetic field, 1 minute of electric shaving can equal 8 hours at a VDT. Besides, exposure to some electric appliances is not short, and exposure to grounding loops is continual.

None of this is to say that the electromagnetic field of VDTs are necessarily harmless; but it is to say that their electromagnetic field is weaker than that of many other sources that have been around for decades. (I am not discussing eye strain, posture stress or other non-electromagnetic factors). And it seems strange that electric clocks and alarm clocks, with magnetic fields of comparable intensity and used up

Table 1. Magnetic fields from home appliances

| Appliance | Field in milligauss | |
	Typical	Maxim.
Blow dryer	1–752	125
Ceiling fan	1–11	125
Clothes dryer	1–2	36
Clothes washer	1–24	93
Computer	1–25	1,875
Dishwasher	1–15	712
Electric alarm clock	1–12	450
Freezer	1–3	6
Kitchen range	1–80	625
Make-up mirror	1–29	125
Microwave oven	3–40	812
Oven	1–8	67
Refrigerator	1–8	187
Shaver	50–300	6,875
Stereo	4–100	500
Toaster	2–6	9
TV	1–3	100
Vacuum cleaner	1–11	60

Compiled from "Power Frequency Magnetic Fields in the Home," *IEEE Transactions on Power Delivery*, Vol. 4, p. 465–478, Jan. 1989.

to 8 hours a day in the close vicinity of the sleeper, should produce effects that have gone unnoticed for the last 50 years and are not demonstrable even now.

Government officials, claimed the New Yorker's Paul Brodeur, failed to point out that the magnetic field from household appliances

"... falls off rapidly with distance, and that most appliances, including hair dryers and food mixers, are not sources of chronic exposure, because they are used sporadically. Similar claims by the utilities industry, which was anxious to exonerate power lines as sources of chronic exposure to magnetic fields, found ready acceptance by gullible government health officials and undiscerning scientist."

Both of his claims, distance dependence and exposure, are untrue. There are sound theoretical reasons, amply confirmed by measurement, why magnetic fields from power lines fall off more rapidly than from wiring in the home. An electric current is continuous: as much of it must flow from the source as returns to it. The magnetic field of the out and back currents have opposite directions and they partially cancel if the conductors are parallel, as they are in power lines. Moreover, the exposure from outdoor power lines takes place at a distance much larger than the distance between the conductors. Both circumstances make the magnetic field small.

To the contrary, magnetic fields are strong when the current forms a loop: inside the loop the fields from current elements reinforce each other instead of canceling. And there very often is such a loop in the typical American house, which uses three-phase current (even though each phase singly for different parts of the house) and grounds the system by connecting it to the water plumbing. The loads on the phases are rarely exactly equal, and the resistance from the plumbing to the ground wire of the power line is rarely the same in neighboring houses. The result is a loop with a magnetic field in which people live constantly and which is about three times larger than that from either appliances or power lines.[4] I am not saying that this field is dangerous, and I am not saying it is harmless; I am saying that it makes the field of an electric clock (Brodeur's favorite bugaboo) laughable in comparison.

The remedy advocated by the corporation and utility baiters is to put the distribution lines underground and anything else that will "punish" them. This will not reduce the magnetic field linked to the ground current by a single milligauss. The remedy is plastic inserts in the plumbing, and direct grounding for a direct return current. Neither is of the slightest use to the utility baiters, because it would have to be done at the owner's expense.

As for one suggested remedy in the case of VDTs, "sit no closer than 28 inches from a VDT," it is characteristic of this type of pseudoscience. If the writers were genuinely afraid of radiation, they would ask for a standard limiting its power flow, say, in watts per square centimeter, or if worried only about the magnetic field, they would promote a standard expressed in microtesla or milligauss. Many experts must surely have advised them on that, but the alarmists cannot accept that: the radiation is zero watt/cm^2, and the magnetic field smaller than from sources that have never caused a complaint.

But the remedy by distance is not just fuzzy; it is also self-defeating; if you move away from the screen of a VDT, you have to make the screen bigger, so as to maintain the same angle of view and the same readability. That means scaling up the electron acceleration, the deflection currents and most other causes of alleged dangers to health, so that one would have to spend a lot of money without counteracting the complaint.

The Greater Plausibility Goes Unnoticed

But ideological motivations aside, where do the miscarriages by computer operators, their cancers and headaches come from?

First, I am not convinced that the statistics are meaningful. I have seen no studies backed by control experiments (with subjects under similar circumstances, but without exposure to VDTs) with statistically significant results.

Second, there are well known psychosomatic effects: when people think they ought to get a headache, they get one.

Finally, there is something in VDTs that is absent from electric clocks and other electric appliances: ultrasound.

Like ordinary sound, ultrasound propagates in air and other matter as waves of oscillating pressure caused by a mechanical vibration in the source. Is only difference from ordinary sound is the frequency of these vibrations, too high to be heard by the human ear.

Most people can hear sounds up to about 16,000 hertz or 16 kilohertz (kHz), though this varies with ages, sex, and particular individuals; some children can hear as high as 20 kHz. Beyond this upper frequency limit this "inaudible sound" is called ultrasound. It can be produced up to several millions of cycles per second (megahertz) for industrial purposes that do not concern us here; but it can also be produced inadvertently by unintended electro-acoustic coupling.

Perhaps the simplest case of electro-acoustic coupling, or electricity/sound conversion, is a headphone. Its essential components are a coil wound on an iron core, and an elastic, metallic membrane. The electric current flows in the rhythm of the original sound waves as converted by a microphone. When the current is stronger, it attracts the elastic membrane more; when it is weaker, it attracts it less. The membrane thus vibrates in the rhythm of the current, compresses or rarefies the air in its vicinity with each vibration, and thus becomes a source of sound reproducing the oscillations of the current. Most loudspeakers work on a similar principle, except that the membrane is replaced by a paper cone attached to a coil carrying the current and vibrating in the field of a permanent magnet.

Both examples use the magnetic field of a current to turn its signals into sound (there are also other ways), and both do so intentionally. Could the same thing happen inadvertently?

Easily. A transformer has coils wound on an iron core consisting of thin laminations that are bolted together to form a solid frame. As in the headphone, the magnetic field acts on the laminations, compressing and decompressing them in the 60 Hz rhythm of the mains voltage. Ideally, the bolts should prevent the laminations form vibrating, but they cannot do so completely and everywhere. They will vibrate a little and become an unwanted source of sound, producing the well known 60 Hz "mains hum." The mechanism is not limited to transformers; other metal parts can also vibrate in the magnetic field of a 60 Hz alternating current.

No one has suggested that this 60 Hz hum might have any health effects, and it was explained here only as an example of inadvertent electro-acoustic coupling.

But VDTs and TV receivers have comparatively strong currents that will, by the same mechanism, cause vibrations at a frequency of about 31.5 kilohertz. They are associated with the 525 lines on the screen traced by an electron beam in a TV picture tube every two seconds. The most common type of VDT uses the same frequency, others use frequencies that result in mechanical vibrations of 44 kHz and 63 kHz—all well above the upper limit of audible sound.

Moreover, unlike the 60 Hz currents in a transformer, the currents controlling the horizontal scan of the electron beam are not sinusoidal, but saw-tooth shaped: they increase slowly as the beam moves across the screen, but drop back quickly during the "fly-back" on the return journey to start the next line. Such a saw-tooth waveform is rich in higher frequencies. The resulting ultrasound is then generated by the common type of VDT not only at 31.5 kHz, but also at its integral multiples 63 kHz, etc. Though TV sets and VDTs have no visibly moving parts, they may have components prone to mechanical vibrations under magnetic forces at these frequencies; the deflection coils and high-voltage transformers, for example, would be well worth investigating.

At 31.5 kHz and higher, the emitted sound is completely inaudible, but it is sound nevertheless; the receiver's eardrum, skull and other parts of the body will vibrate when the ultrasound radiation hits it (yes, this time it really is radiation, i.e. the propagation of waves—though they are not electromagnetic waves, but waves of acoustic pressure). Some terminals may be more prone to these vibrations than others, and of course, they may affect some operators more than others.

What does long exposure to ultrasound at 31 to 100 kHz do to the human body?

I don't know, for I am not a physician. I do know, however, that some people cannot stand art exhibitions where the security system uses ultrasound, because they detect its presence by getting strong headaches, just as some professors get a strong headache when they demonstrate ultrasound sources in class. So the idea is not completely absurd and is worth testing, though I fully realize that such tests may well be negative.

As of now, such a refutation is absent. While the EPA, local governments, the judiciary, and the press have been beating the alarm, what investigations were carried out to say the idea is absurd?

VDTs are not nuclear power or offshore drilling, where the majority, however misinformed, overrules the minority. This is a case of individual choice. If you buy a microwave oven, you take your own risk and endanger nobody else. Foods preserved by irradiation are clearly marked and as yet you may buy them at your risk. And if you are afraid of electromagnetic fields, you need only ask your utility to disconnect you without affecting other people's power, for you risk more (if anything) from the fields inside your home than from those of power transmission lines.

The remedy is to seek the truth eagerly, vigorously, and independently; for only the truth will keep us free.

References

1. High electric fields are able to shock-ionize gases by causing collisions of particles during discharges (arcs), which are either purposely arranged (fluorescent lights) or due to inadvertent faults (coronas). Neither has any relevance to ionizing radiation.
2. B.W. Wilson, R.G. Stevens, L.E. Anderson, *Extremely Low Frequency Fields: The Question of Cancer,* Battelle Press, Columbus, Ohio, 1990.
3. F.S. Barnes, "The effects of time varying fields on biological materials," *IEEE Transactions on Magnetics,* Sept. 1990.
4. See D.L. Mader, and others, "A simple model for calculating residential electromagnetic fields," *Bioelectromagnetics,* vol. 11, pp. 283–296, Nov., 1990. It contains some very revealing direct measurements of the magnetic field correlated with the ground current.

Abridged from *Different Drummer,*
Booklet #10, The Golem Press, Box 1342,
Boulder, CO 80306

Environmental Economics

The Free Market and the Environment
Richard L. Stroup and Jane S. Shaw

In Whose Interest?
Jo Ann Kwong

Chemophobia:
Is the Cure More Deadly Than the Disease?
Richard L. Stroup

Foolish Environmental Rules Concerning
Gasoline Substitutions
John E. Kinney

Environmental Tyranny—A Threat to Democracy
Raphael G. Kazmann

Chasing A Receding Zero: Impact of the Zero Threshold
Concept on Actions of Regulatory Officials
Thomas H. Jukes

The Free Market and the Environment

Richard L. Stroup, Ph.D. and Jane S. Shaw

Richard L. Stroup is professor of economics, Montana State University, and senior associate, Political Economy Research Center. From 1982 to 1984, he was the director, Office of Policy Analysis, U.S. Department of Interior.

Dr. Stroup received his Ph.D. from the University of Washington. He is a co-author of *Economics: Private and Public Choice* (with James Gwartney) and *Natural Resources: Myths and Management* (with John Baden). He has published more than 20 journal articles and has contributed to *American Economic Review, Journal of Law and Economics, and Economic Inquiry*.

Jane S. Shaw is a senior associate at the Political Economy Research Center in Bozeman, Montana. From 1981 to 1984 she was an associate economics editor, *Business Week.*

She received her B.A. from Wellesley College in 1965. Shaw is the author of several hundred articles published in magazines and newspapers including *Business Week, The Wall Street Journal,* and *Policy Review.* She is a senior editor with Liberty, and a member of the Editorial Advisory Board of Regulation.

Conventional economic wisdom, in a theory first propounded by Nobel laureate Paul Samuelson, holds that the unregulated market cannot be expected to protect the environment. In this theory, clean air and water are "public goods" whose value is not well reflected by market processes. Potential polluters do not consider the social costs of their action, but only the costs to themselves; in addition, since efforts to maintain a clean environment benefit even those who do not help fund them, each individual faces a strong temptation to avoid footing the bill. Similarly, when markets cannot immediately reflect the benefits of preserving biotic diversity or of saving individual species, private landowners may not be willing to pay for socially beneficial preservation. In the absence of government intervention, the argument goes, the environment will therefore be insufficiently protected.

This analysis has become so accepted that many people now see no alternative to the system of government environmental regulation and control that has been pieced together over the past two decades. This system, however, is beset with difficulties; for when environmental goals and controls are politically determined, they are subject to a process that is often driven by groundless accusations, supported by public fear, and legislated with special interests in mind. Populist

267

sentiment and pork-barrel politics, rather than actual environmental dangers, currently determine priorities.

We should therefore be prepared to reconsider the free-market solution to environmental pollution, which has worked in the past and could be made to work better now. Over the long run, private ownership is the most effective protector of the environment—provided ownership is transferable and backed by courts that make people liable when their pollutants invade the person or property of others. This system of private ownership would protect the environment for the same reason that it protects other kinds of property: because it encourages good stewardship.

Love Canal

The contrast between the effectiveness of public and private environmental protection is well illustrated by the case of the most notorious toxic-waste dump in American history—Love Canal. The story, which was not fully known until Eric Zuesse uncovered it in Reason in 1981, begins in 1942. The Hooker Electrochemical Company started dumping chemical waste into the abandoned Love Canal, after lining the canal with impermeable clay to prevent the chemicals' escape. After the canal was filled, a clay cap was installed over the top, sealing the chemicals' "tomb" so that rainwater could not wash the chemicals out. Hooker's precautions were extremely thorough. In fact, the Chief of Hazardous Waste Implementation for the Environmental Protection Agency said in June 1980 that they would have met 1980 Resource Conservation and Recovery Act regulations.

Around 1950, the Niagara Falls school board, searching for a site for a new school, inquired about Love Canal. Hooker warned it about the chemicals, and gave school-board representatives a tour of the site, even making test borings into the ground to demonstrate that chemicals were indeed present. Despite warnings of liability from its own attorney, however, the school board, eager to get the site, prepared for eminent-domain proceedings. Under these conditions, in 1952 Hooker donated the site to the school board in return for $1. Hooker insisted on noting in the transfer papers the presence of and potential danger from the chemicals sealed below. Private ownership, and the liability of the owners, had ended.

The school board subsequently built the school and scraped away part of the clay cap to provide fill dirt for other school sites. The scraping was sufficiently extensive that construction plans had to be changed in order to avoid the partially exposed chemicals. Rain could now get into the dump.

In 1957, over Hooker's strong public objections, the board tried to sell the remaining land. Hooker won that fight, and the land was retained temporarily. At about the same time, however, apparently without Hooker's knowledge, the city constructed a sewer line, surrounded by permeable gravel, that punctured both the clay walls and the cover of the canal. A storm sewer was placed through one wall of the canal in 1960, again in a bed of gravel. These and later punctures that occurred when the state built an expressway through the site meant that incoming rainwater could carry the stored chemicals freely through nearby neighborhoods along the gravel beds of the sewer pipes.

The area around Love Canal, sparsely populated when Hooker used the dumpsite, gradually became residential. The school board sold the south end of the site, where the chemical wastes were concentrated, for residential development. The escaped wastes began to invade the neighborhoods, and the disaster hit the national press. [1]

But the press paid little attention to the history of the property rights. Hooker was judged guilty by the press and the public, and Superfund was born. This law, enacted in 1980, created a $1.6-billion fund—financed by a tax on chemical and oil producers—for cleaning up abandoned waste sites. Expanded by $8.5 billion in 1986, Superfund has become an expensive, wasteful pork barrel that has accomplished little cleanup. "All sides agree that the Superfund program for cleaning up hazardous wastes sites is not working as intended and that progress on permanent cleanups has been painfully slow," former Environmental Protection Agency administrator Russell Train wrote in 1989. A Rand Corporation study came to the same conclusion.

Had Love Canal remained in Hooker's ownership, the chemicals would probably never have escaped—or if they had, they would have been rapidly contained— and a national trauma and a multi-billion-dollar tax would have been avoided. When Hooker owned the site, company managers had a powerful economic incentive to act responsibly. Liability laws made clear that Hooker would be liable for harm caused by the waste; and since stock prices reflect expected profits and losses, both stockholders and the responsible managers (through the impact on their professional reputations) would pay financially and personally as soon as it was evident that their actions caused harm. But the members of the school board, like public officials generally, did not have the same personal or financial accountability for their actions. Perhaps that is why they allowed their desire for a cheaper school and their desire to sell surplus land for development to override whatever personal concern or caution they may have had.

Property Rights and Accountability

When backed by effective liability laws, private property rights tend to work well. Because well-tended property increases its value, private owners generally take care not to despoil their land. This safeguard works even when owners care only for themselves, not for their heirs. For at the very first signs of poor stewardship—the first indications of land erosion, for instance—appraisers and potential buyers can project the results into the future, and the value of the property declines immediately.

With an effective liability system, these pressures can also keep corporations from despoiling land or property that they do not own. Although disputes occur, the obligations of those who harm others' property are so widely accepted that people may not even have to go to court when their cars are damaged; insurance companies generally handle such cases routinely.

Unfortunately, environmental damage is often not as recognizable as a dented fender. Common law requires plaintiffs to prove damages and identify the responsible parties, and though the standard of proof is not as high as in criminal cases, it remains substantial; in order to sue you successfully for polluting my lungs, I must show that I suffered the damage for which I am demanding compensation, that the cause of the damage was air pollution, and that your factory was the source of the pollution. Without reliable information on which to base such damage suits, owners cannot adequately defend their property rights in court. If the air could have been contaminated by many different sources that may be hard to identify, or if the health effects are hard to measure, I may have trouble convincing a court that you should compensate me.

Thus the nature of emissions can make liability laws unenforceable, particularly in the case of air pollution. The difficulty of obtaining satisfaction in court was, in fact, an important factor creating pressure for government intervention to control pollution. But government intervention does not eliminate the need for accurate information. Like private individuals, the government has trouble knowing the source and effect of pollutants. Unfortunately, it has therefore tended to adopt standards that do not demand solid evidence connecting emissions with harm. Under today's regime, the mere suspicion of harm, combined with educated guesses as to the source of pollution, are driving policies that have enormous costs. Los Angeles, for example, is about to impose measures to require reformulation of products such as deodorants and paints and conversion of cars so that they run on methanol rather than gasoline.

Not only does the government lack the necessary information for controlling pollution, but politicians often have little incentive to obtain it. Politicians find it easier and more popular with most constituencies simply to adopt a stance of outrage against polluters. In fact, generating outrage is an effective way to generate votes. The passage of Superfund boosted the careers of a number of congressmen, even though it resulted from misinformation about Love Canal and the incorrect implication that every town had a potential disaster in its backyard.

Problems with Government Control

The political pressures that dominate government work against taking the long view. Government officials are legally barred from personally capturing any value that they help create; correspondingly, they pay no financial penalty for property that deteriorates. By contrast, a private owner of land will see its value change immediately after a major investment, because the value reflects future benefits and costs stemming from his action. Since no such "capitalized value" exists in the government setting, government officials are more interested in maximizing political power than economic value. Their interests often figure in controversial decisions. Henry Clay's remark that he would "rather be right than be President" is famous because it represents an unusual elevation of principle above personal success in politics; and, of course, Clay never did become President.

It is true that government officials are usually well-intentioned. But pursuing their professional mission almost inevitably means disregarding some goals important to the public interest and catering instead to specific individuals and groups. The dedicated (but narrow) professionalism of public officials—together with their need for political allies at budget time—often leads to their agencies being "captured" by the special interests that they ostensibly regulate. Which special interest prevails usually dpends on how the political winds blow.

The results can be seen in the way that agencies manage the properties entrusted to them. For example, Forest Service foresters tend to be highly committed to harvesting and replanting trees, often neglecting the potential value of the national forests for recreation. This commitment has led the Forest Service to log extensively areas such as the Rocky Mountains, where the timber value of the trees is low and where the environmental harm from extensive cutting can be severe. A perverse result is that the harvested trees command prices lower than the cost to the taxpayer of cutting them down. According to environmental consultant Randal O'Toole, for example, the Gallatin National Forest in Montana loses between $2 million and $4 million per year on timber sales. Environmental organizations such as the

Wilderness Society have lobbied against such pointless destruction for some years, but the political power of sawmills and their workers, backed by Forest Service officials' commitment to timber cutting, has prevented change.

Politics also affects our national parks. The National Park Service generally follows the views of the leaders of prestigious environmental groups, even though the policies that this small minority espouses are not necessarily those that most Americans want. In his book Playing God in Yellowstone, Alston Chase has documented the deterioration of rangeland and the disappearance of wildlife in Yellowstone National Park that has resulted from the "natural regulation" or "hands-off" management of most of the park's land. "Hands-off" management has resulted in the largest elk population in history, to the point that the land cannot sustain animals such as the deer, beaver, and grizzly bear. Similarly, the decision to allow fires to burn in spite of decades' worth of fuel buildup led much of Yellowstone to be devastated in the summer of 1988. While environmental leaders endorse these policies because they minimize human intervention, the disappearance of wildlife such as the beaver and the grizzly bear disturbs many people. One reason the harm is so severe is that it follows decades of the opposite extreme—extreme intervention, during which park rangers killed off Yellowstone's wolves and suppressed all fires.

The point of these examples is that government ownership and control leads to decisions determined by political muscle, not necessarily by wise judgment. Other examples of destructive or at least questionable government actions abound. For many years, the Bureau of Reclamation built costly dams that flooded thousands of acres of habitat. Today, feral horses and burros are harming federally owned rangelands, but they cannot be controlled because of opposition from animal-rights groups. And until recently, the Bureau of Land Management, which oversees rangeland, was routinely using crawler tractors to pull up bushes and small trees in large stretches of grazing land, despite a low cost-benefit ratio on much of the land.

Improving the Common Law

In short, the government record in controlling pollution and preserving habitats is not impressive. To control pollution, our chief priority should be to improve the common law. The common law, of course, has its flaws; nevertheless, its rules of evidence and its history of even-handed protection of individual rights make it in most cases the best vehicle for holding accountable those who damage the environment.

We should begin by recognizing that many of the common law's failings were introduced by the legal activists who have been working to change the system since the 1950s. According to a number of analysts, courts today tend to compensate victims from whatever "deep pocket" might be found, even if the deep pocket (such as Hooker Chemical) acted responsibly. This approach destroys the link between liability and responsibility, and thereby reduces the incentive to take costly steps to avoid damaging others.

In recent years the search for deep pockets has led some courts to void insurance-contract clauses that explicitly specified and thus limited coverage. Partly for this reason, a number of insurance companies have stopped insuring hazardous-waste handlers, depriving industry of an important source of expertise. Similarly, when property has been sold, and the liability for hazardous materials simultaneously transferred by contract, the courts have sometimes voided that transfer, making all parties liable. This disrespect for contract reduces incentives to put hazardous properties into the hands of competent, solvent specialists.

One remedial step would be to restore the sanctity of contract, and to let insurers help control the risks from unintended pollution. Insurance companies have enforced safety in many industries while at the same time making safety cost-effective. In addition, governments could rely less heavily on direct regulation and instead require environmentally risky ventures, such as hazardous-waste dumps, to be bonded or insured; both bonds and insurance can provide the accountability that is otherwise absent when insolvency or bankruptcy prevents companies from compensating victims. A firm that has posted a large bond to guarantee solvency in case of liability claims will have a much stronger incentive to handle its hazardous materials safely and efficiently.

Increased emphasis on accountability through the common law could lead to other salutary developments. For example, chemicals that might escape into the water or air might be "branded" by dyes or radioactive isotopes to help identify their source. Responsible companies could protect themselves with branding, because they would be in the clear if contaminants that caused damage did not carry their brands. In addition, faced with laws that assure the solvency of potential polluters and that make liability more certain for anyone whose contaminants invade the property of others, insurers and others responsible for potential damages would provide a bull market for the development of better forensic technology, as well as better containment and decontamination procedures. When general accountability— rather than specific behavior—is stressed, the incentive to avoid damage is greater.

Private Protection of the Environment

When it comes to maintaining environmental quality, protecting natural beauty, and preserving wildlife habitat, private organizations have often done a better job than government. One reason for their effectiveness is that their actions do not have to reflect majoritarian views, which often change.

Private conservation, undertaken by voluntary associations, began long before the American public developed today's environmental consciousness or enlisted the government to protect endangered species and enforce cleanups. Hawk Mountain Sanctuary Association in eastern Pennsylvania, for example, was formed privately in 1934, at a time when hawks were considered vermin because they ate chickens. Sea Lion Caves, a tourist attraction on the coast of Oregon, began protecting sea lions in the 1920s, when the state of Oregon had a $5 bounty on each sea lion. At that time, the animals were viewed as pests because they ate fish and harmed the salmon industry; Sea Lion Caves provided a haven until public opinion changed and laws were passed to protect sea lions.

Even today, when the government is supposed to control the environment, private groups are responsible for much of the effective protection of wildlife. Hundreds of organizations—some operating nationwide, like the Nature Conservancy, and others more local, such as the Montana Land Reliance—act in a myriad of ways. The Nature Conservancy has more than a thousand nature sanctuaries, and since its founding in 1951 it has preserved some 2.4 million acres. The Conservancy carefully inventories land in order to target the most ecologically critical areas, which it then acquires and either manages directly or gives to government preservation agencies. The newer Montana Land Reliance, formed in 1978, keeps agricultural land from development, primarily through donated conservation easements.

Many private organizations manage sanctuaries and preserves. The National Audubon Society, for example, has more than sixty preserves, covering more than 250,000 acres. Ducks Unlimited protects more than a million acres of wildlands each year through easements that preserve waterfowl habitats. Operation Stronghold is a national association of private landowners committed to managing their land in a way that protects or enhances wildlife habitat. There are hundreds of other such sites in the U.S., providing refuge and habitat for all sorts of flora and fauna. There is even a preserve in Maryland that studies and protects butterflies, moths, and wasps.

The beauty of such private efforts is that people who do not care for wasps or egrets need not pay for their upkeep, as taxpayers do when the government is in control. Also, since private organizations do not use funds coerced from other people, but rather rely primarily on donations, they tend to target their efforts efficiently. And they can act quickly, without having to obtain government approval. Indeed, the owners of Hawk Mountain and the Sea Lion Caves were acting against both the conventional wisdom and the public policies in place at the time of their most important deeds.

So in spite of the "free rider" problem decried by Paul Samuelson, plenty of environmental groups—motivated by altruism, scientific curiosity, or love of natural beauty—are preserving scenic areas and wildlife habitats. While it is difficult to say whether "enough" environmental protection is provided, it is certain that a lot of protection is going on. Indeed, the free-rider problem is probably greater with each public policy aimed at correcting an environmental problem than it is with private environmental activity. The lobbying needed to create and maintain policies that reflect the public interest, rather than pork-barrel special interests, is a "public good" of the sort identified by Samuelson.

While private, nonprofit groups carry out much environmental preservation, environmental protection can also be a good commercial investment. International Paper Company, one of the largest private landowners in America, does much more than raise trees in its forests. It has an active wildlife and recreation program that offers camping and hunting facilities to individuals and clubs. The company employs wildlife biologists who work with foresters to improve animal habitats through such techniques as controlled burning and the creation of stream "buffer zones." The revenue from these efforts makes the company wary of disturbing the ecology, and willing to invest large sums to protect and enhance its wilderness and wildlife resources.

As more people seek out scenic surroundings, protecting the environment will be an increasingly good investment. Big Sky of Montana, for instance, is a private resort that was created in the 1970s to provide a beautiful setting for homeowners and visitors in the northern Rockies. To maintain the natural beauty, the developers bought up more land than they intended to develop. They then resold the land with legally binding covenants that will keep the valley from being tarnished by unsightly buildings or incomplete construction. By protecting most of the valley this way, they increased the value of both the land that they developed and the land that they sold. Because they could buy and sell the land, a public good—environmental amenities in a large mountain valley—coincided with a private good—property values.

Public Pressure in the Past

Many people, of course, see private industry as the source of current pollution problems, not of their possible solutions. They point, for example, to the capitalists who cut down forests near the Great Lakes and failed to replant them. But what they fail to recognize is the high value of cleared land at that time, and the role of changing popular preferences. The same emerging desires that led to the Wilderness Act and the creation of the Environmental Protection Agency also energized the private sector—and brought benefits changes more rapidly than the government has.

Strong public pressure to control pollution is a relatively recent phenomenon; we often forget that smokestacks used to symbolize economic vitality, not irresponsible business. In the past, when people generally were poorer, they were less interested in preserving wildlands than in having food to eat and places to live. T.J. Iijima, in a paper prepared for the Political Economy Research Center, explains the destruction of the forests around the Great Lakes in terms of the popular demand for farmland and for lumber; understandably, there was little desire to keep virgin forestland. Even if the government had controlled the land, therefore, there is every reason to think that the forests would still have been cleared. (In fact, virgin forests in Michigan's Porcupine Mountains were nearly logged under Department of Defense contracts during World War II.)

As our standard of living has improved, our tastes for environmental amenities have increased as well. We can expect this demand for natural beauty to continue to grow as our national income increases, for attention to the environment is correlated with higher income. (Thus the advertisements in magazines published by environmental groups are decidedly aimed at the affluent—who are their readers, as the groups' own demographic surveys show.) We can expect the private sector—both profit and nonprofit—to continue to take the lead in meeting the increasing environmental demands whenever it is allowed to do so.

That does not mean that private organizations will solve all environmental problems. Where property rights are nonexistent, ill-defined, or unenforceable, there will be no owner to insist on protection. Rather than abandoning private management in favor of direct governmental control, however, we should try to find ways to establish accountability (along with the freedom and incentive to innovate) by establishing or strengthening property rights. We need to compare the problems stemming from imperfect property rights with the "solutions" put into effect by imperfect government. The evidence suggests that the political process has all too frequently caused waste and destruction.

Footnotes

[1] To this day, despite many detailed health studies, there is no generally accepted evidence of a threat to long-term human health from living in the neighborhood, despite the obvious presence of noxious pollutants. It is still possible that serious long-term risks will eventually turn up, though the evidence to date is against that possibility.

An abridged version
Reprinted from *The Public Interest* no. 97, Fall 1989
© 1989 National Affairs Inc., Washington, D.C.

In Whose Interest?

Jo Ann Kwong, Ph.D.

Jo Ann Kwong is an environmental research associate with the Atlas Economic Research Foundation and a research assistant professor of George Mason University in Fairfax, Virginia. She has been affilated with the Capital Research Center, Washington, D.C.; the Institute for Humane Studies, Fairfax, Virginia; and the Political Economy Research Center, Bozeman, Montana.

Dr. Kwong received her B.A. in biology from Brown University in 1979; an M.U.P. masters in urban planning from the University of Michigan in 1981; and a Ph.D. in natural resource economics and management from the University of Michigan in 1986.

Dr. Kwong is the author of *Protecting the Environment: Old Rhetoric, New Imperatives, Capital Research Center, 1990; and Market Environmentalism, Lessons for Hong Kong,* Hong Kong Centre for Economic Research, 1990. She is a well known lecturer and has contributed numerous articles to major publications.

Given the influence enjoyed by environmental groups, both individually and as a movement, it is important to consider the degree to which these groups fulfill their claimed "public interest" role. Despite the "public defender" rhetoric, it would be naive to suggest that the environmental movement—or any other, for that matter—is capable of representing anything as diverse as the American people.

Virtually everyone is in favor of a clean, healthy environment; but there is no reason to assume that most people favor extremist, uncompromising environmentalism at the expense of all other considerations.

A May 1989 survey by the New Jersey-based Opinion Research Corporation, for example, noted that even though people expressed heightened concern for the environment following the *Exxon Valdez* oil spill, support for stricter environmental regulations was tempered by the probable impact of such regulations on several factors: higher product costs, higher taxes, difficulty for American companies in competing, job losses, and more U.S. dependency on foreign oil.[1]

For the most part, however, environmental activist groups devote a tremendous amount of their budgets to promoting such essentially extreme causes as utopian legislation, more and more absolute wilderness, and costly development delays under the perennial guise of endangered species; and activities mounted by these groups have been protested vigorously by local people who are part of the "public," whether the environmental establishment cares to acknowledge them as such or not.

The Sierra Club persists in suggesting that its leaders are promoting the "public interest" amidst intense popular hostility. In the rugged Burr Trail region which borders five major roadless areas in southwestern Utah, Jim Catlin, conservation chairman of the Utah chapter, complains that "wilderness advocates in Garfield County cannot speak out without inviting social ostracism and economic retribution"; and in the forests of the northwest, where opening federal lands to industry is often seen as an issue of economic survival, "Dr. John Osborn is pretty unpopular in Idaho these days" because "when you're obsessed with preserving the wilderness of the American West . . . you're likely to offend people."

When environmental concerns clash with job security, people invariably value jobs more than environmental preservation. The controversy over the northern spotted owl is inseparable from job survival. The 1989 cascade of revelations about surreptitious burning and dumping of radioactive and toxic wastes at the Energy Department's Rocky Flats, Colorado, plutonium plant also has been met with tremendous local objection. In an economically troubled state, environmental hazards inevitably will be weighted against potential job loss. Colorado College economics professor William J. Weida cautions that in addition to the positions held by the plant's 6,000 employees, another 6,000 jobs are threatened by an economic slowdown; but to Melissa Kassen, a lawyer for the Environmental Defense Fund, the state's governor still "is a lot more concerned about the economy than the environment."[2]

Elitism

Environmental groups typically equate vehement opposition with public ignorance. By portraying their unpopular activities as victorious achievements, the Sierra Club, Wilderness Society, and other "public defenders" seem to be saying that because people do not know any better, they need self-appointed spokesmen to tell them what is good for them.

Environmental organizations are fundamentally elitist, with memberships that represent an affluent, highly educated class. A 1989 survey ("accurate within 1% to 4% and enjoy[ing] a 95% confidence level") revealed that

- 17 percent of the Sierra Club's members are in the $100,000-plus income bracket (seven times the national figure).
- 49 percent claim over $50,000 in annual income.
- The average value of homes owned by members ($187,300) is more than twice the U.S. average.
- 35 percent own homes worth more than $200,000.
- More than 70 percent hold jobs, of which 70 percent are professional, managerial, or technical, compared to 30 percent nationally.

Of course, most groups vehemently deny any suggestion that they tend to represent middle-class or upper-class interests. Such an admission would undermine the desired grassroots image. But the reality, as admitted by a Sierra Club writer, is that "it's getting tougher all the time to spot the tree-hugger lurking within the well-educated, well-compensated, middle-aged, professional whose image the survey conjures up."[3]

In addition, a glance at any of the leading environmental magazines shows advertisement after advertisement geared to the nation's "leisure class." "Colorado Ranch Vacations," "Arctic Treks," "East Africa voyages," "Pentecost Island Land Diving," and "White Water World wide explorations" all can be arranged through the pages of *Wilderness, Sierra, Audubon,* and similar magazines, although some advertisements indicate that not all members of these groups are seeking pure, rugged, roadless wilderness: "4 Wild Days, 3 Cozy Nights. Group walk exploring Oregon's Wild Rogue Canyon. Stay in comfortable lodges. Great food and no hassles with gear."[4]

William Tucker, author of *Progress and Privilege,* views environmentalism as a protection of entrenched privilege: "People who have reached a certain level of affluence and privilege in society inevitably turn their efforts away from the accumulation of more wealth and privilege, and toward denying the same benefits to others."[5] He notes in particular the undeniable lack of minority participation in the movement.

At least since 1970, when white college students buried a new car to celebrate Earth Day, blacks and hispanics have argued that environmentalists ignore the need to improve the lot of the poor. Perhaps in response to such allegations, Environmental Policy Institute/Friends of the Earth executive director Michael Clark has said that, in reaching out to the minority community, "we will be drawn more into questions dealing with poverty, housing, income, food and shelter." Although Clark acknowledges that the environmental movement has been built largely on a white, upper-class, well educated, affluent base, he hopes EPI/FOE will change that through this new "wholistic approach."[6]

One effort to smooth over these differences instead ended up reinforcing minority hostility toward environmentalism. In 1979, Vernon Jordan, director of the National Urban League, was asked to attend a joint conference on urban and environmental affairs. He responded with these remarks:

> Walk down Twelfth Street [in Washington, D.C.] and ask the proverbial man on the street what he thinks about the snail darter and you are likely to get the blankest look you ever experienced. Ask him what he thinks about the basic urban environmental problem is and he'll tell you jobs. I don't intend to raise the simple-minded equation of snail darters and jobs, but that does symbolize an implicit divergence of interests between some segments of the environmental movement and the bulk of black and urban people. . . .[7]

Image and Reality

From comfortable surroundings in the United States, environmentalists have campaigned against destruction of tropical rain forests. One of the newest strategies for rain forest protection is "debt for nature swaps" by which environmental groups purchase Third World debt in exchange for local land conservation projects.

After Ecuadoran officials signed the largest debt-for-nature swap arranged by the Nature Conservancy, World Wildlife Fund, and Fundacion Natura, however, Brazilian president Jose Sarney rejected similar proposals as part of an "alarmist international campaign" to halt development of Brazil's Amazon rain forests.[8] "The Amazon is ours . . . After all, it is situated in our territory," Sarney said, excoriating "great powers or international organizations . . . that would come to dictate to us how to defend what is ours to defend"[9] and further pointing out that it is industrialized countries like the United States that most harm the global environment.[10]

Another issue that has been framed as pitting "haves" against "have nots" stems from the growing debate over international ozone accords. In March 1989, at a 123-nation conference on the ozone problem held in London, environmental activists sought to limit production of chlorofluorocarbons to reduce destruction of the atmospheric ozone layer; but some in the Third World contend that United States and Western European environmentalists are trying to impose international restrictions that would impede development. Such restrictions, they argue, make it even harder for their countries to become first-class economic achievers.[11]

Despite the altruistic image that environmentalists like to project, there are serious questions about the sincerity of their professed concern for the American people. The drive to lock land in wilderness, for example,[12] cannot possibly be backed by people unable to afford the costs of travel to the back country or the vacation time to venture into wilderness without vehicular assistance; by people with handicaps or health conditions that limit their access by foot; or by the multitudes of urban people who prefer vacation opportunities closer to home. (Even among environmentalists, back packing, one of the few permissible activities in wilderness areas, is on the decline. According to National Park Service statistics, backcountry use peaked in the 1970s.)

Other evidence suggesting a substantial difference between the wants of environmentalists and those of the people can be observed in consumer preferences. While environmental activists fight for pristine air, water, and land, how clean is clean for the average person? What trade-offs are people willing to make in terms of consumer convenience, jobs, and other tangible costs?

Product manufacturers, for example, must struggle to meet the demands of these special interests while at the same time producing goods that consumers will buy:

- PepsiCo sells some of its sodas in the three-liter plastic bottles, partly to cut down on packaging and trash, but consumers will not buy because the bottles are too large for refrigerators.

- Wendy's International is testing a paper plate to replace its foam plates; but, according to company officials, consumers do not want it, and previous experience with paper cups has made the company mindful of what consumers will and will not buy.

- Proctor & Gamble offers its Downy fabric softener in concentrated form, which requires less packaging than ready-to-use products; but the concentrate must be mixed with water, and sales have been poor.[13]

These sales problems suggest that, as they do with jobs, the American people value convenience and practicality more than environmental quality. Human existence since the beginning has created pollution and changed the environment. The challenge is to determine the acceptable levels of trade-offs involved in different ways of living. Problems arise when one special interest presumes to speak in behalf of all in determining the level and direction of these trade-offs.

Leaders vs. Members

These considerations suggest that environmental activists may be motivated only partially by a concern for the individual citizen. In addition, there is the equally serious question of the extent to which environmental leaders, no matter how successful they may be at pursuing their own narrow interests, actually speak even for the members of their organizations.

The Downing and Brady model of citizen interest groups[14] predicts that as a CIG succeeds in moving pollution policy toward the preferred position of its median member, it will lose members and become more extreme in its preferred position. The model further predicts that the CIG will be more extreme in its preferred environmental policy than is the population generally. On several fronts, this model seems to be an accurate predictor of environmental interest group behavior.

Has the political sophistication of the movement driven a wedge between the national headquarters of the leading groups and their local affiliates? The ceaseless promotion of increased regulatory control and government spending has led some affiliates to express fears that national representatives are no longer concerned about the environmental issues that hit home. To Lauri Maddy, a grass-roots activist who handcuffed herself to a chair in Kansas governor Mike Hayden's office in November 1988 to protest the state's lack of concern about dangerous emissions from a local plant, "The bigger movement may be caught up in making enough money to cover their budgets, but nobody pays us for forty hours a week, or even eighty. Because most of us living in high-chemical-impact areas came to fighting this out of self-preservation. We lived it, we feel it, and we still hurt."[15]

Similarly, Environmental Defense Fund staff scientist Michael Oppenheimer worries that the national groups will become mesmerized by the possibilities of expanding their influence in Washington "and forget where their strength comes from." Even Barry Commoner cautions that national leaders are developing "a congenital arrogance that they are in charge."[16]

Gaps between leaders and members afflict both older and newer groups. We have seen how supporters of the Natural Resources

Defense Council and Environmental Defense Fund felt compelled to disassociate themselves from these groups once the political nature of their agendas with regard to DNA research became clear[17] and how environmental leaders chose to pursue their battle against DNA research, despite the loss of scientific support, on the basically elitist ground that they represent an unbiased voice uniquely able to speak for the "public."[18]

Ruptures also have developed within some of the older, more established environmental groups as their missions likewise have become increasingly politicized. It appears that not all modern environmentalists support the notion of bigger government and stronger regulation. Some see value in conservation as a wise-use movement and shun the preservationist, or no-use, attitude; but advocates of more balanced approaches all too frequently find themselves drowned out by "mainstream" leaders.

While many "mainstream" groups have been created since Earth Day, several were firmly established in prior decades. In general, the "old-line" groups differ from the newer ones in having more tightly focused missions. The Sierra Club, founded in 1892, was established as a mountaineering group; the National Audubon Society was founded in 1905 for the protection of wild birds and animals; the Wilderness Society was established in 1935 to focus on government land management; and the National Wildlife Federation was originally incorporated in 1936 to "make effective progress in restoring and conserving the vanishing wildlife resources of a continent." Yet each, with the exception of the Wilderness Society, has been touched fundamentally by the changes that have come with transitions in the movement.

National Wildlife Federation

The National Wildlife Federation, like old-line conservation groups, originated with a strong dedication to promoting wildlife conservation through public education. Today, it has amassed a widespread education network and teaches millions about the wonders of nature. To a great extent, it has been effective in accomplishing this central mission.

On the other hand, the Federation has branched out to join the host of other environmental groups working to bring about more government, stronger regulation, and greater limitation of individual stewardship.[19] Its current activities extend beyond education and into the political arena,[20] and this change has not come without its share of trade-offs. In 1988, the organization engaged in a bitter battle that pitted some of the old line conservationists against the newer breed of environmentalists.

In 1975, Dr. Claude Moore, honorary president of the National Wildlife Federation, donated a 357-acre tract of land near Washington, D.C., to the group for use as a conservation center. A guidebook published by the Smithsonian Institution calls the property, once home to 185 species of birds, "a gem of a wildlife area."[21]

In 1986, contrary to the wishes of its benefactor, the Federation sold the Moore sanctuary for $8,500,000 to a developer who planned to put up 1,350 housing units. Dr. Moore alleged that the Federation broke a contract explicitly restricting the preserve's use in perpetuity. He filed a lawsuit demanding that the Federation reverse the sale or build a new nature preserve elsewhere. In 1988, a Virginia court issued an opinion that found no evidence of fraud on the part of the Federation; but the animosity arising from the Federation's actions may yet prove to be even more costly than its unwillingness to meet the wishes of a donor.

Some observers have called the case a classic example of the metamorphosis of conservationism into preservationism. The National Wildlife Federation, like many other environmental groups, "began to change in the 1980s as highly paid careerists replaced the old leadership and the focus shifted from conservation to lobbying." By 1988, it had an annual budget of $63,000,000 and 17 staff lawyers, and a seven-story, $40,000,000 office building was being built in downtown Washington. Ray Arnett, a former Federation president, said he was "shocked" at its actions: "NWF was known as the largest conservation education association in the world. Now they have moved more to advocacy, lobbying."[22]

Sierra Club

The Sierra Club has a sophisticated network of activists, a growing range of political campaigns, a political action committee (the Sierra Club Committee on Political Education), and its own "public interest" law firm (the Sierra Club Legal Defense Fund) to ensure that it remains at the forefront of environmental litigation and lobbying. As with other groups, however, the Club's growing political sophistication has come with its share of trade-offs. As it devotes more and more resources to staffing itself with layers, lobbyists, and professional fundraisers, it has had to make significant cutbacks in other areas.

In late 1988, the Club's board of directors imposed a ban on mountain climbing, reportedly to avoid a skyrocketing insurance bill. The ban outlaws all Club-sponsored climbing trips risky enough to require the use of ropes or ice axes. This decision severs the Club from its original purpose. In 1892, when John Muir established the Sierra Club, he attracted potential members by sponsoring outings to the magnificent high mountain wilderness of the Sierra Nevadas; "new members had the best of all motives for joining the Club: their mountaineering spirit and appreciation of the wilderness they already had experienced at first hand."[23]

The decision to break away from mountaineering has added another chapter to the tumultuous history of the Sierra Club. The Club reportedly has received hundreds of angry letters from members who have threatened to resign, particularly in Los Angeles and Orange counties where the Angeles Chapter has dismantled part of its popular mountaineering training course and dropped about 300 mountaineering trips from its extensive outings program.[24]

Just as the Claude Moore controversy caused dissension between old-line conservationists and the new breed of environmentalists at the National Wildlife Federation, the mountaineering ban has pitted some of the Sierra Club's rugged outdoorsmen against ruling political leaders. Outing devotees contend that the ban illustrates how the Club has changed in recent decades as it has broadened its membership and focus, becoming one of the nation's most politically influential environmental organizations; some board members dismiss such allegations as based on misguided ideas of the Club's purpose.[25]

The mountaineers, however, appear to be correct, judging from the Club's original articles of incorporation, which stated in part that its purpose was to "explore, enjoy and render accessible the mountain regions of the Pacific Coast" and "to enlist the support and co-operation of the people and government in preserving the forests and other natural features of the Sierra Nevada . . ."[26] Approximately, Sierra later replaced accessibility with preservation to reflect its changing philosophy.

According to a Club history,[27] during its first five decades (1892–1940), the leadership tended to focus on one or two major conservation campaigns per year. Eventually, members grew increasingly concerned about the role of the organization. By 1949, William Colby, who had served as a director for nearly half a century, resigned from the board because he feared the Club no longer represented Muir's founding desire to open the beauty of the mountains to everyone.

A major parting came when several Club directors blocked a proposal for a road into the upper Kings River country in the Sierra. Previously, the organization had proposed a number of roads across the mountain passes; but now the growing ranks of the environmental elite objected to increased access. It was during this battle that the directors voted to amend the by-laws to delete any reference to rendering the mountains "accessible."

In 1952, several years after Colby's departure, David Brower was appointed executive director of the Sierra Club. Under his tenure, the Club continued to develop, pursue, and refine its political network. By the end of the 1950s, it had established a conservation committee with sections in northern and southern California to serve as a watchdog on issues and to encourage cooperation among Sierra Club chapters and offices. It also had established the East Coast and Pacific Northwest chapters.

This growing national network participated in blocking construction of an Echo Park dam in Dinosaur National Monument. By the 1960s, the Sierra Club was fighting against sonic booms from airplanes, urban sprawl, excessive use of pesticides, coastal development, wilderness loss, and alleged misuse of recreational facilities. Today, it openly acknowledges the radically changed nature of this "unique role":

> Let's start with the five fields of action in which the Sierra Club is actively involved. Other environmental organizations might do one or two of these things, maybe three. But we do all five: we educate the public; we influence public policy through legislative action; we are active in electoral campaigns; we have an aggressive litigation program; and we take direct action leading outings, maintaining trails, walking precincts, phone-banking.

National Audubon Society

The National Audubon Society experienced its transition during the 1970s and 1980s as "the old conservation community confronted the growing challenge of pollution," a challenge that caused the Society to learn "that to be effective we had to help develop and shape public policy—a process that transformed us from conservationists into environmentalists." Since joining the ranks of the environmentalists, Audubon has tackled a growing range of issues:

> While a passion for birds is still the force that most often brings Audubon members together, the Society has long since become involved in the full spectrum of environmental issues, including such global concerns as air and water quality, population, energy policy, climate disruption, and management of wildlife refuges.

With this broader focus, the Audubon Society has become a forceful political voice in the environmental movement, complete with a sophisticated activist network. Political activity is encouraged and aided by a system of "Action Alerts" sent to activists to inform them of issues that need urgent attention. Elizabeth Raisbeck, Vice President for Regional and Government Affairs, has urged members to "get political."

Audubon president Peter Berle acknowledges the strength of the Society's newer focus, describing five "high-priority campaigns that were chosen because they required an organization with Audubon's clout": protection of the Platte River, preservation of the remaining ancient forests of the Pacific Northwest, safeguards against acid rain and other toxic pollution, protection of the wilderness values of the Arctic National Wildlife Refuge, and conservation of wetlands. All obviously presuppose political, not private, control.

The leaders of the National Audubon Society apparently have managed to retain enough of its founding mission to survive the transition to a political organization, possibly because the sanctuary system that was an integral part of its original purpose remains central to its operations today.[28]

The dimensions of the gaps between leaders and members are difficult to gauge; but rifts within the movement itself provide ample evidence that it is difficult, if not impossible, to say that any one group, or coalition of groups, fully represents the American public.

Environmentalists have a political agenda to implement. They cultivate public opinion to boost their base of support. They are not unbiased disseminators of information, but carefully calculating activists that compete for Congressional funding, private foundation grants, individual contributions, or other scarce goods. In short, no matter how exalted the rhetoric, they operate in exactly the same manner as does any other special interest concerned with enhancing its political effectiveness and power.

References

1. *Business and Public Affairs Fortnightly,* June 15, 1989, p. 7.
2. Matthew Wald, "Colorado, Needing Jobs, Tiptoes on Nuclear-Plant,"*New York Times,* June 21, 1989, p. B-5.
3. See *Sierra,* January/February 1989, pp. 22–23, for additional highlights of this survey. The Sierra Subscriber Profile, 1988, conducted by Mediamark Research, Inc., revealed that Sierra Club members own Volvo, Saab, Acura, and Porsche automobiles at a rate ten times greater than the national average. *Sierra,* May/June 1988.
4. *Wilderness,* Winter 1988, p. 56.
5. William Tucker, "Environmentalism: The Newest Toryism," *Policy Review,* Fall 1980, p. 46.
6. When the interviewer commented on this broad redefinition of environmental issues, Clark responded, "FOE stands out in that regard. It has

always taken a wholistic approach to dealing with environmental problems." *Not Man Apart*, Nov. 88–Jan. 89, p. 5.

7. Tucker, *Progress and Privilege*, p. 37.

8. Jennifer Spevacek, "Nations resist pressure for debt swaps," *Washington Times*, April 11, 1989.

9. Eugene Robinson, "Brazil Angrily Unveils Plan for the Amazon," *Washington Post*, April 7, 1989, p. Al.

10. Philip Shabecoff, "What standing does the US have to seek a global effort against pollution?," *New York Times*, May 2, 1989, p. A-22.

11. Frederic A. Moritz, "Third World and Ozone 'Blackmail'," *Christian Science Monitor*, March 23, 1989, p. 18.

12. "More than 90 million acres of federal public lands in national parks, forests, and a small amount of other federal land have been included in the Wilderness System. Yet another 90 million acres of wildlands, prime wildlife habitat, and fragile ecosystems . . . are still in dispute, still unprotected, still at risk. This is the challenge that lies before us."

13. Alecia Swasy, "For Consumers, Ecology Comes Second," *Wall Street Journal*, August 23, 1989.

14. Downing and Brady, "The Role of Citizen Interest Groups," in White, ed., *Nonprofit Firms in a Three Sector Economy*, pp. 61–93.

15. Dick Russell, "We Are All Losing the War," *The Nation*, March 27, 1989, p. 403.

16. *Ibid.*

17. Marshall, "Environmental Groups Lose Friends."

18. See Section Three, *supra*.

19. The National Wildlife Federation, for example, is one of several environmental groups that have criticized a presidential order requiring federal agencies to analyze whether their regulatory actions might interfere with private property rights. Telephone inquiry to Don Barry, House Merchant Marine and Fisheries Committee, August 22, 1989, and H. Jane Lehman, "Environmentalists Attack 'Taking' Order," *Washington Post*, July 1, 1989, p. F1.

20. President Jay Hair responded indignantly to the possibility that Congress will limit judicial review of resource-management agency decisions: "The result would be to cut off citizens' fundamental democratic right to go to court to challenge agencies that have tremendous influence over the fate of our ancient forest." Hair, "The competition between endangered species."

21. "Exploiting the Environment," *Wall Street Journal*, April 25, 1988, p. 26.

22. *Ibid.*

23. Strong, "The Sierra Club—A History."

24. Bettina Boxall, "Ban on Mountaineering Causes Rift in Sierra Club," *Los Angeles Times*, February 17, 1989.

25. *Ibid.*

26. Strong, "The Sierra Club—A History."

27. *Ibid.*

28. The Massachusetts Audubon Society recently launched a multimillion dollar conservation effort in Belize to protect critical wintering grounds of

migrating songbirds. The heart of the $7,000,000 effort, called Program for Belize, will be the purchase of 110,000 acres of forest rich in wildlife along the Rio Bravo in the western part of the country. The Coca-Cola Company, which owns land in the area, will donate an additional 42,000 acres of forest, savanna, and wetlands. Another 100,000 acres will be retained by a private owner and managed as part of the program. Philip Shabecoff, "An Audubon Group Finds Its Interest Extend Far to South," *New York Times,* July 11, 1989, p. C4.

Reprinted, by permission, from
Jo Kwong Echard, *Protecting the Environment: Old Rhetonic,*
New Imperatives, Studies in Organization Trends #5
(Washington: Capebal Research Center, 1990), pp, 65–74.

Chemophobia:
Is the Cure More Deadly
Than the Disease?

Richard L. Stroup, Ph.D.

Please see biographical sketch for Dr. Stroup on page 267.

The Non-Crisis Dominating
Today's Environmental Policy

There is no question that as we become wealthier and more technologically advanced, the potential for man-caused danger increases. New chemicals are introduced constantly, and there is no question that chemophobia—the fear of chemicals—is increasing also. Environmental activists tell us that if the government does not take immediate control of the situation, results will be catastrophic. As in so much of environmental policy, a crisis atmosphere dominates.[1] The results are anything but constructive. In this chapter, we will look at the source of these fears, see why they are so greatly exaggerated, and examine the counterproductive policies that have resulted. We will see also that the procedures inherent in traditional property rights and the common law have the potential to protect us well, as indeed they did for most of the preceding century and more.

Governmental intervention in environmental affairs poses serious dangers to our safety and material well-being, as well as to our liberties. Take, for example, the case of Times Beach. In February 1983, this small Missouri town became nationally famous as newspaper headlines reported that its streets had been contaminated with the chemical dioxin more than 10 years earlier. Environmental Protection Agency administrator Anne Gorsuch (later Burford) announced that the EPA would pay $33 million to buy the town's homes and businesses. This decision was made even though there was no evidence that anyone in Times Beach had ever been harmed by the dioxin. Administrator Gorsuch traveled to Times Beach to announce the buyout personally and to demonstrate to the national media her concern over the risks of hazardous waste. But less than two weeks later, she was forced to resign—accused of, among other things, a callous disregard for such risks. There is still no evidence that human health is endangered, but Times Beach remains a ghost town.

Crisis Secures Power and Funding

While public opinion has fueled most of today's federal anti-pollution policies, organized special interests have been quick to take advantage of them. Members of Congress, after all, do not have national constituencies, and helping important constituent groups is the name of the legislative game. It is not surprising that many of the costly policies of the Clean Air Act have been used to fight regional economic battles and to protect the powerful eastern coal industry and its unions.[3]

Those whose careers are bound up in "going after" polluters are not necessarily acting selfishly or in bad faith. Many are undoubtedly drawn to their work precisely because they believe deeply in the "anti-pollution" mission of their agencies. But the fact is that additional public fear and outrage expands their agency budgets and thus their career possibilities. The bureaucratic leadership forms one leg of what political scientists call the "Iron Triangle." Special interests, each seeking a competitive edge—or trying to avoid being put at a severe disadvantage through the political process—are willing and able to use their lobbying power to help the bureau, which in turn sees to it that their clients' interests are not forgotten as bureau strategies and tactics are decided and carried out.

The politicians are the third leg of the iron triangle, operating as brokers between special interests, from whom they draw campaign support, and bureaus, which, with carefully manipulated budgets and regulatory authority, can work with politicians to do much, at public expense, for the politically organized special interests.

The power to tax and transfer society's wealth, which is given to government in order to protect the public, is given much more easily when the public is convinced that a crisis is at hand and that disaster may occur without government action. Therefore, members of the "Iron Triangle" have little incentive to stress or publicize facts that might reduce the fear or the uninformed outrage of the public. For this and other reasons, the general public remains badly uninformed about the risks actually posed by man-made chemicals.

In the case of environmental politics, non-profit environmental groups have become almost an adjunct of the bureaucracy. The major environmental groups receive large sums of money from government agencies, and contributions from their donors partly depend on donors' fear, outrage, or even panic about their favorite environmental issues.[4]

Risk, Fear, and the Role of Environmental Activists

Environmental activism took hold in the U.S. in the 1960s as politically active groups shifted their attention from atmospheric nuclear testing to chemical pollutants. The movement began with the publication of Rachel Carson's *Silent Spring* in 1962. As in much environmental literature since that time, the rhetoric in *Silent Spring* was eloquent and full of strong emotions, but the supporting evidence—against DDT, for example—was considered by many scientists to be one-sided and misleading. The book elicited fear and indignation and moved people to demand political action against those who would pollute and plunder the planet. In fact, the fight against DDT was the origin of the Environmental Defense Fund, which from 1967 to 1969 received a reported 19,000 column inches of press coverage on this, the organization's first major issue.[5] By playing on and feeding public fears of such environmental threats, environmental leaders attract thousands of members and hundreds of millions of dollars each year to fight what they view as "the good fight" against potential harms. These harms certainly include chemicals, but for some, the fight is an ideological battle against economic growth.[6] (As we will see below, slowing economic growth itself has some important negative effects on human health and safety.)

Silent Spring inaugurated the view that a whole new regime of governmental controls was required to protect the earth and its inhabitants from technological disasters. This view is still common, and when an environmental crisis stirs public concern, it is the driving force behind public policy. As a result, we are increasingly operating in a new regulatory climate. Among the important changes is the heavy burden of proof on anything new. New activities are often allowed only if shown to be almost "risk-free"—as exemplified by the Delaney clause of the Food, Drug, and Cosmetic Act, which forbids the use of any food additive that can be carcinogenic for humans or for animals, no matter how weak the carcinogenic effect. Faced with the potential risk of new chemicals, drugs, or genetically altered micro-organisms, politicians and bureaucrats are inclined to ban use until safety can be proven. But as AIDs sufferers and the families of those who have died waiting for federal approval of new drugs can attest, "zero risk" is actually a very dangerous strategy.

Although tragically unsafe, this narrow-minded approach is understandable in the government setting. Regulators may not be blamed for the problems faced by people denied the new substance—most of whom will never know about the benefits the government held back—while they most assuredly would be blamed for any untoward effect of a substance they approved. Thus, a drug that could save thousands of lives might be banned to prevent a single death from a carcinogenic side-effect.

Contrast this one-sided pressure on regulators with the balanced set of incentives facing a private firm, such as a drug maker. The firm wants to protect its reputation and avoid liability for selling dangerous drugs. But it also wants to enhance its reputation and its profits by getting out new "miracle drugs" that can help people. Without formal regulation, the company faces the trade-off squarely: Will the added profit from selling new and possibly dangerous drugs to willing buyers (properly warned by the firm, to avoid extra liability problems) be sufficient to offset the likely losses from lawsuits and reputational harm caused by selling the drug? It is highly ironic that in reality, the search for profit provides the private firm with a more balanced view of "the public interest" than that brought to bear on government regulators. Clearly, political safety regulation can be dangerous to our health.

The unintended side-effects of "uncompromisingly tough" environmental standards can be dangerous, too. When DDT was banned because its misuses harmed some wildlife, more dangerous pesticides replaced it. (Elsewhere, the situation was worse: Sri Lanka withdrew from the World Health Organization's large-scale DDT spraying programme, and the number of cases of malaria rebounded from a low of 110 cases in 1961 to 2.5 million cases in 1968 and 1969.) We cannot know just how many farm workers have died from DDT substitutes, but since the replacement pesticides are more harmful to people and must be more frequently applied, banning DDT to avoid risk actually introduced far more human risk.

In general, attempts to achieve "zero risk" in any single arena are dangerous to society because they stifle technical and economic progress. Most research and development efforts in biotechnology have shown no clear and present dangers and, indeed, offer the promise of more effective waste cleanups and less environmentally threatening means of controlling agricultural pests and weeds. But again, the atmosphere of fear and even of crisis has meant that research, testing, and use of these techniques has been hampered needlessly by a web of regulations created by the EPA and the U.S. Department of Agriculture.[7]

Wealthier Is Healthier

A common attitude among environmental activists is that we should ignore the economic costs of formulating health, safety, and environmental regulations because safety is an absolute. As Lori Mott of the Environmental Defense Fund puts it, there is "no room for consideration of the benefits of pesticides."[8] Yet from the point of view of health and safety, it's hard to imagine worse advice. Empirical evidence strongly suggests that higher incomes (both for populations of countries and for individuals within those populations) contribute more to good health and life expectancy than whatever risks are introduced in raising those incomes. In general, the higher our incomes, the more options we have—to change out lifestyles, regulate our diets, and choose more selectively among the risks we do take.[9] In addition, when earthquakes, floods, hurricanes, and other natural disasters occur, richer nations can mobilize the resources necessary to save large numbers of citizens in distress.

People in more developed countries have considerably higher life expectancies than people at lower levels of economic development, despite—and in part, because of—the greater use of chemicals. What is true of whole societies is also true of individuals within societies. In England, death from cancer among males in the highest socioeconomic class is 25 percent below the national average, and death from respiratory disease is 63 percent below the national average.[10] In contrast, deaths from cancer and respiratory disease are 31 percent and 87 percent above the national average respectively among males in the lowest socioeconomic class.

Similar evidence exists for the United States. One study of mortality and income for U.S. counties found that each one percent increase in income reduces mortality by 0.05 percent.[11] Based on this study, it is estimated that, for a 45-year-old man working in manufacturing, a 15-percent increase in income would have about the same risk-reducing value as eliminating every single hazard from his workplace.[12] With higher incomes, individuals drive larger, safer cars, fly instead of drive, live in safer houses, eat better, and in a myriad of other ways live safer lives.

In the light of these findings, it appears that government regulation in general, and health and safety regulation in particular, has done a great deal of harm to counter whatever good it has accomplished, even when measured solely in terms of its effects on health. Between 1959 and 1969, productivity in U.S. manufacturing increased by almost 1 percent annually.[13] Between 1973 and 1978, however, manufacturing productivity fell by more than one half of one percent annually.

There is evidence that a significant portion of this drop in productivity was caused by regulations imposed by the Occupational Safety and Health Administration (OSHA) and the Environmental Protection Agency (EPA). One estimate is that 31 percent of the overall drop in manufacturing productivity was due to regulatory burdens created during the 1970s by OSHA and EPA.[14] Nineteen percent of the drop in productivity growth was attributed to OSHA regulations and 12 percent to regulation by the EPA. Moreover, the productivity drop between 1973 and 1978 did not affect all industries equally. Productivity fell by more than 2 percent per year in highly regulated industries, yet it actually rose in the same period in less regulated ones.

Increases in worker incomes are roughly equal to increases in productivity, so whatever the positive impacts of OSHA and EPA may have been, it appears that the damage to health and safety that they have caused by reducing income growth is substantial.

The Risk of Cancer

Fears of chemicals have focused on one potential outcome—cancer. A popular belief is that an epidemic of cancer is sweeping the modern world as a result of increasing chemical inputs into the human body. This "common knowledge" is false. Although new chemicals can, of course, be deadly, chemical production and increasing human exposure to them has been going on for many decades. In their 1981 comprehensive survey "The Causes of Cancer," sponsored by the congressional Office of Technology Assessment and published in the *Journal of the National Institute,* Richard Doll and Richard Peto conclude that except for the increase in lung and skin cancer, "examination of the trends in American mortality from cancer over the last decade provides no reason to suppose that any major new hazards were introduced in the preceding decades."[15] Despite claims of a cancer epidemic from toxic chemicals, there is no such thing. Taking age into account, cancer is not increasing in the U.S. Also, our best estimates place the percentage of cancers caused by all man-made chemical and radiational sources at less than 5 percent.[16]

Still, nothing seems to frighten the voting public more, or bring more hurried action from elected officials, than the potential risk of cancer. In the political arena, expressions of outrage are easy since the people demanding action expect others to pay the costs. Often "big corporations" or other faceless entities are thought to be picking up the tab for politically established regulations.

These same citizens, however, act quite differently in accepting and avoiding risk when they control the degree of risk through their own behaviour, and when they bear the direct costs and reap the direct benefits of their actions. Epidemiologists estimate that at least 70 percent of human cancer is avoidable in principle through changes in human behaviour.

Individuals also remain calm and careful in the face of newly revealed dangers such as the natural cancer risk from radon gas in their homes, despite the fact that this danger can be considerable. As many as one million homes are believed to be generating radon decay exposure levels higher than those received by uranium miners, and as much as 10 percent of lung cancer in the United States has been tentatively attributed to radon pollution in houses.[17] Such revelations have not caused panic, but have instead created a market for detection devices, allowing people who are exposed to take cost-effective actions to reduce their exposure to such dangers.

While the risk of cancer may be greatly exaggerated, it is always present. Carcinogens are everywhere. Without any help from man, carcinogens are naturally present in almost every meal. They are present in mushrooms, parsley, basil, celery, cola, wine, beer, mustard, peanut butter, bread, lima beans, and hundreds of other everyday foods. Human beings also produce carcinogens through everyday activities. Baking bread, browning meat, cooking bacon and eggs—all of these activities cause chemical reactions that produce carcinogens. Allowing a sliced apple to become slightly brown involves an oxidation reaction that produces carcinogenic peroxides. Carcinogens also occur naturally inside the human body.

California's Proposition 65

Equally disturbing to those who believe that active environmental policy can be established rationally is the enactment of California's Proposition 65, formally called the Safe Drinking Water and Toxics Enforcement Act of 1986. Criticized by pro-business forces as a drag on the growth of the state's economy and praised by a high-profile anti-toxics movement as an effective way to curb consumer demand for chemicals that allegedly cause cancer, Proposition 65 was passed by a voter initiative, receiving 63 percent of the vote. Drafted by the Environmental Defense Fund and the Sierra Club, backed by Tom Hayden, his wife Jane Fonda, and other Hollywood celebrities, Proposition 65 is the most sweeping chemical regulatory law ever enacted by a state government.

Among other provisions, Proposition 65 bans the discharge into drinking water of chemicals "known" to cause cancer or reproductive harm and requires warnings to individuals exposed to these chemicals.[21] As noted earlier, the list of things that cause cancer (in at least some animals, at sufficiently high laboratory doses) is very large. Almost from day one, the list of chemicals covered by the initiative became a political battleground. Environmental and consumer advocates wanted more chemicals added to the official list and businesses struggled to stay abreast of the growing requirements for fulfilling the public's right to know.[22] Interestingly enough, only private water sources were covered.[23] Municipalities, supplying most of the water Californians drink, were exempted.

As of July 1, 1988, 216 substances were listed as carcinogens and 15 substances were listed as reproductive toxins. That list could grow considerably. California officials are considering chemical substances ranging from cocaine to aspirin, and ultimately Californians may discover that it is impossible to enter a retail store or place of work without seeing warning labels.[24]

Public employees are required to notify the news media when they discover violations, and are subject to criminal penalties if they do not disclose the violations that they discover. The law does not merely open the door to numerous lawsuits, it encourages them through bounty-hunter provisions that allow private citizens to collect 25 percent of the fines imposed if they initiate successful suits against violators of the act.

To be guilty of violating Proposition 65, you do not have to actually harm anyone, or even put anyone at risk. Instead, the standards for violations are entirely hypothetical. In the case of carcinogens, a violation has occurred if you expose someone to a chemical and that person would have been at a significant risk if the exposure level had been maintained over the whole of the person's lifetime. In the case of reproductive toxins, a violation has occurred if the chemical would have produced a detectable risk if the person were exposed over the course of a lifetime at a level 1,000 times the level at which the person was actually exposed. Even if we can determine what would have happened, what constitutes a significant risk? "Significant risk" is not a scientific term. There is nothing in science, for example, that says that a one-in-a-million risk of cancer is not significant, while a two-in-a-million risk of cancer is significant.

Under California law there is no penalty for an unnecessary warning. There is a penalty, however, for mistakenly failing to warn. Moreover, those accused of a failure to warn bear the burden to prove that a chemical exposure did not put anyone at a significant risk, a burden that is scientifically impossible to meet. What Proposition 65 does is produce a line-up of suspects that will forever be just that—suspects. Without a viable means of proving innocence, businesses will tend to compound the problem by posting more warnings than necessary. Indeed, they may label all their products just to be on the safe side.[25] Already, homebuilders are posting warnings on all new houses just to play it safe.[26]

Yet if everything is labelled, especially if all labels contain the same warning, then warning labels lose their value. A warning label will affect behaviour only if consumers can distinguish a few especially dangerous risks out of the thousands of minor risks they take in everyday activities. Putting a warning label on every product robs any meaning from the warning label on a truly dangerous product.[27] California law also has the potential to dangerously misdirect attention from noncarcinogenic risks and dangers that we should be concerned about and toward trivial cancer risks. For example, chlorine used in processing milk leads to the production of chloroform, one of the chemicals listed in California as a carcinogen. Yet chlorine helps prevent a much more serious danger—the risk to children of death by food poisoning from milk made without the use of chlorine. The spirit of the California law would appear to require a warning label on milk made with chlorine, but no warning label for the much more dangerous chlorine-free milk.

In Conclusion: A Historical Parallel

Ecologist William Clark has pointed out that in the 16th and 17th centuries half a million people—ostensibly witches—were burned at the stake.[28] These centuries, like most times, were riddled with grievous social ills, such as plagues, mysterious livestock deaths, and disastrous crop failures. No cause could be found for these ill omens, yet the established authorities of the time, church officials, felt obligated to take some action for the benefit of the public good. The blame was placed on "witches."

It was essentially impossible for accused witches to demonstrate that they did not pose a real threat to the community. They could deny that they were harmful, but they certainly could not prove it. With all the terrible societal risks, how could a political authority (or church official) justify not acting against an accused witch? And so it is today with persons using new chemicals, biotechnology, or other innovations. An elected official often is now, as were the ecclesiastical authorities, strongly pressured by an outraged but not very well-informed public for action against polluters, actual and potential.

There is another similarity between the witch-hunts then and governmental programmes today. When witches were burned, their property was confiscated by the authorities. Today, when businesses are "convicted" politically, without evidence or trials, of endangering the public via chemicals, billions of dollars are taken from them (as in Superfund taxes and cleanup provisions), partly to augment the budgets of the agencies in charge of "prosecuting" them.

In sum, a crisis mentality spurs today's environmental regulations, bolstered by well-meaning government officials whose careers benefit from crisis, just as church officials profited four centuries ago. The frequent result, as we have seen, is bad policy that reduces health and safety. Recognizing the value of a property rights regime backed by common law is a first step to correcting this situation.

The introduction of some detrimental chemicals or hazardous wastes will inevitably occur as man progresses, but their presence does not imply a crisis. With a properly functioning system, the creation of overall improvements for human life is also inevitable.

References

1. See Richard L. Stroup, "Environmental Policy," *Regulation*, 1988, No. 3, for a more complete exposition of this section's topic.
2. See "Unfinished Business," an internal EPA document examining EPA programme priorities. It found relatively little correlation between expenditures and their potential for improving the health of U.S. citizens.
3. See, for example, Robert W. Crandall, "Clean Air and Regional Protectionism," *Brookings Review* (Fall, 1981) pp. 17–20.
4. See James T. Bennett and Thomas J. DiLorenzo, *Destroying Democracy: How Government Funds Partisan Politics* (Washington, Cato Institute, 1985), espe-

cially Chapter VII, for some details about how dependent many large environmental groups are on government funding.

5. See Elizabeth Whelan, *Toxic Terror* (Ottawa, Ill. Jameson Books, 1985) pp. 16, 73.

6. See Edith Efron, *The Apocalyptics* (New York, Simon and Schuster, 1984) for an account of how ideology has driven many of the media-reported accounts from scientists warning of disaster from man-made chemicals in the environment and the workplace.

7. On the topic of biotechnology and problems in its regulation, see L.R. Batra and W. Klassen, eds., *Public Perceptions of Biotechnology* (Bethesda, Agricultural Research Institute, 1987).

8. Reported in Tom Hazlett, "Ingredients of a Food Phobia," *The Wall Street Journal*, August 5, 1988.

9. For a fuller treatment of this topic, see Aaron Wildavsky's chapter on "Richer is Safer vs. Richer is sicker," in his important book *Searching for Safety* (New Brunswick, Transaction Books, 1988).

10. *Ibid.*, Table 2, p. 63.

11. Jack Hadley and Anthony Osei, "Does Income Affect Mortality? An Analyses of the Effects of Different Types of Income on Age/Sex/Race-Specific Mortality Rates in the United States," *Medical Care*, September 1982, Vol. XX., No.9.

12. Peter Huber, "The Market for Risk," *Regulation*, March/April 1984, p. 37.

13. Wayne B. Gray, "The Cost of Regulation: OSHA, EPA and the Productivity Slowdown," *American Economic Review*, Vol. 77, No. 5, December 1987, pp. 998–1006.

14. *Ibid.*

15. Richard Doll and Richard Peto, "The Causes of Cancer," *Journal of the National Cancer Institute*, Vol. 66, June 1981, p. 1256.

16. *Ibid.*

17. Bruce M. Ames, Renae Magaw and Lois Swirsky Gold, "Ranking Possible Carcinogenic Hazards," *Science*, Vol. 236, April 17, 1987, pp. 274–275.

18. Superfund, or technically the Comprehensive Environmental Response, Compensation, and Liability Act of 1980 (CERCLA) provided for "liability, compensation, cleanup, and emergency response for hazardous substances released into the environment and the cleanup of inactive hazardous waste disposal sites." It provided $1.6 billion to clean up abandoned sites. The Superfund Amendments and Reauthorization Act (SARA), passed in 1986, authorized an additional $8.5 billion to finance the Superfund site cleanup effort. In addition, SARA enlarged the number of enforcement authorities for the purpose of compelling private cleanups. It intends also to shift waste management practices toward long-term prevention, rather than containment of wastes.

19. The origins of the Love Canal crisis are described in Eric Zuesse, "Love Canal: The Truth Seeps Out," *Reason*, Feb. 1981, pp. 16–33.

20. Dante Picciano, "A Pilot Cytogenetic Study of the Residents Living Near Love Canal, A Hazardous Waste Site," *Mammalian Chromosome Newsletter*, 1980, 21 (3).

21. For a summary of the key provisions of Proposition 65, see Richard J. Denny, Jr., "California's Proposition 65: Coming Soon to Your Neighborhood," *Toxics Law Reporter,* Dec. 17, 1986, pp. 789–794; and Jerome H. Heckman, "California's Proposition 65: A Federal Supremacy and States' Rights Conflict in the Health and Safety Arena," *Food Drug Cosmetic Law Journal,* Vol. 43, 1988, pp. 269–282.

22. Paul Jacobs, "27 Chemicals Added to the List," *Los Angeles Times,* June 19, 1987; and Richard C. Paddock, "56 Chemicals to be Added," *Los Angeles Times,* Feb. 2, 1988.

23. See Robert Gottlieb, *A Life of Its Own* (Orlando: Harcourt, Brace, Jovanovich, 1988), Chapter 7.

24. Jerome Heckman, "Proposition 65—A Legal Viewpoint: Reflections on the Political Science of How Not To Do It," paper presented to the American Industrial Hygiene Association's Toxicology Symposium on Aug. 8, 1988, in Williamsburg, VA., p. 4.

25. David Roe of the Environmental Defense Fund, a staunch proponent of Proposition 65 and defender of warning requirements, admits that business may trivialize the law with unnecessary warnings. Yet he believes that withdrawing a few products from the markets is more than enough compensation. see Michael deCourcy Hinds, "As Warning Labels Multiply, Messages are Often Ignored," *The New York Times,* March 5, 1988.

26. "Houses with Warning Labels," *The Sacramento Bee,* July 25, 1988.

27. The authors and major supporters of Proposition 65 have always viewed warnings as an intermediate step, or compromise, toward a more radical goal: the banning of all carcinogenic and reproductive toxins. See the statement attributed to Tom Hayden in Heckman, "Proposition 65," op. cit., p. 271, n. 24.

28. See William C. Clark, "Witches, Floods, and Wonder Drugs: Historical Perspectives on Risk Management," in *Societal Risk Assessment: How Safe is Safe Enough?,* Richard C. Schwing and Walter Albers, Jr., eds. (New York: Plenum Press, 1980), pp. 287–313.

Foolish Environmental Rules Concerning Gasoline Substitutions

John E. Kinney, P.E., D.E.E.

Please see biographical sketch for John Kinney on page 115.

Congress's Clean Air Act requires alternative fuels for automobiles. Attention now is turning to ethyl alcohol from corn, but principal attention has been given to use of methanol that might be manufactured from domestic coal reserves. Existing methanol production is from natural gas. Media promotion has been singularly quiet on some problems widespread use of methanol will create. For example, one ounce may be fatal; it is not only acutely toxic to humans—it affects the central nervous system, especially the optic nerve—but it is also a cumulative poison for it takes a long time for removal from the body. Methanol is highly soluble in water and has only a faint odor. So if there is a spill to stream, lake or underground aquifer, damage would be great and long term. Oil and gasoline spills, regardless of their magnitude and the attention they receive, have a limited time of damage. Neither has much solubility in water.

Also, methanol can not be blended alone directly with gasoline. Methanol dissolves in water and any water present would result in two fuels—a water-methanol-rich phase and a gasoline-rich phase. Although engines can not run on the water-methanol phase, this mixture can cause serious corrosion of the fuel system. So expensive cosolvents have to be added. Questions remain on combustion controls to prevent unacceptable discharges of harmful products such as aldehydes, which have very high smog forming potential.

Moreover, since the fuel value is less, it will take 1.8 gallons of methanol fuel to equal the effectiveness of 1.0 gasoline. Watch gas mileage drop.

There is a similar situation with ethanol, the additive now used in a product called "gasohol." It, too, has less than 75% of the energy of gasoline, requires much more energy to produce because it is essentially distilled alcohol. That distillation is a big energy user. Little is said about this energy requirement to convert corn to alcohol. It may well

301

require more heat to make the fuel than will be returned by the fuel to the car. A quite peculiar conservation approach. EPA estimates that the wholesale cost of ethanol would likely be between $1.00 and $1.50 per gallon and that the engine to burn it would cost several hundred dollars more. The corn crop in the United States, even if used totally for this purpose, is insufficient to answer the need. At best, it is not a replacement for gasoline. Flexible-fueled engines to run on either ethanol or gasoline would not be properly tuned to get the advantage of ethanol. Best described as a bureaucratic scam, this program ignores the effect on food supply, for diversion of corn would increase cost. Also, every effort would be made to increase production with its resultant wearing out of the land and use of marginal land.

Congress now wants alternative fuels required in cities not meeting air standards. Imagine the problems with operating a car with various fuels and fuel effects. The Smokey Mountains do not meet the air standards. Terpines given off by the trees are the cause; they cause the smokey effect. I have seen no data suggesting it is unhealthy to live there.

Reprinted, with permission, from
National Council for Environmental Balance, (NCEB) 1990.

Environmental Tyranny—
A Threat to Democracy

Raphael G. Kazmann

Raphael G. Kazmann is a professor emeritus of Civil Engineering, Louisiana State University. His past affiliations include professor of Civil Engineering, Louisiana State University; consulting engineer, Stuttgart, Arkansas; and in the Ground Water Division, USGS, Washington, D.C.

He received a B.S. in civil engineering from Carnegie-Mellon University.

Kazmann is the author of *Modern Hydrology*, published by Harper and Row, in 1965; with the Third Edition, published by NWWA in 1988. He is the author of more than 75 articles published in journals such as Journal of *Ground Water*, Envir. Geol. and Water Science; and in the proceedings of the American Society of Civil Engineers.

Introduction

The push for a cleaner environment has been accompanied by use of a significant form of governmental coercion. This has probably become necessary because the "environmentalists" have found it difficult to persuade people to voluntarily act in the way in an "evironmentally correct" way. After all, except for direct threats to human life or property, it is not easy to persuade someone that a recommendation based primarily on a computer simulation will really produce the desired results.

The principal method used is write laws "mandating" actions by individuals, corporations and local political bodies, without appropriating the money to get the job done. Thus the legislators' budgets do not show any expenditures! This is a form of involuntary servitude, in the sense that individual citizens are forced to spend money and resources to accomplish something that they otherwise wouldn't do and would not vote to tax themselves to do. At the same time, paradoxically, the lawmakers don't ever know exactly what they have done. There are no Hitlers, Stalins, or Saddam Husseins in this process, just congressmen and bureaucrats who create and enforce legislation that is almost invariably predicated on a complete misapprehension of how the world works, both in a physical sense and in an economic sense. The politicians claim to do everything for the best of reasons and the highest of motives—and who could possibly dispute this?

On the surface the goals are unexceptionable: who could be against them? All of us desire to live in clean surroundings, breathe uncontaminated air, drink potable water—it would be difficult to fault the objectives. And it is just these objectives that successfully conceal the underlying trend to an all-encompassing political dictatorship wherein the individual's desires are subordinated to the state to the point where his freedom of choice will be reduced almost to zero.

Let us look, for a moment, at the real world. Every living being, every industrial process, and many natural phenomena produce undesirable residues or materials. We have biodegradable human wastes, debris from construction and debris resulting from replacing the old with the new: roads, buildings, factories, dams, bridges—all facets of engineering construction. We have wastes from industrial plants, especially chemical and petro-chemical plants that produce the base materials for most of our conveniences, the fuel for our cars, the fertilizers, the herbicides and pesticides. We have the acid residues of forest: the dead branches and leaves; we have mineral springs; and "acid" mine wastes. The last results from the unavoidable slow collapse of abandoned mines that produced coal and other essential minerals, including metals—these voids first fill with water and later collapse forcing the mineralized water into the streams. Our surface waters contain unintentional contributions of fertilizers and pesticides from our food-producing areas. Our air receives residues from combustion: power plants, industrial plants, the heating of homes, and the operation of trucks and cars. We live and produce residues—there is no other way.

There are other sources of undesirable materials: volcanoes and other vents add sulfates, methane, carbon dioxide and dust to the atmosphere. Lightning storms add nitrogenous compounds. Methane is also generated by termites in megaton quantities and by cattle and other living beings, including humans. Deserts and semi-deserts put dust into the air during every storm. The natural erosion of the land, augmented in part by farming practices and construction, puts sand, silt, and clay into our streams. This erosion may be expected to continue as long as the rain falls and the winds blow. Add to this the occurrence of radioactive materials in the clays used for the production of bricks and you have a small sample of the unfriendly natural environment.

From outer space comes a perpetual planetfall of meteorites and icy materials, and a variety of deleterious hard radiation from x-rays to cosmic rays. From the sun the ultraviolet spectrum causes changes in human metabolism that may culminate as skin cancer.

Part of the environment is the habitat of a wide variety of microorganisms, friendly and noxious bacteria and viruses, all of which mutate rapidly with changes in the climate and the availability and resistance of host creatures, insect, animal, and human. We, as individuals, are in a constant battle for survival with our tiny enemies: the bacteriological warfare is the reason that the North American continent is primarily populated by Europeans. The natives had no resistance to the "childhood" diseases that the European invaders brought with them. This process will continue, for the environment is a constantly changing battleground for the human race. If we fail to meet the challenge we will disappear, at one with the dinosaurs.

When you were born you unavoidably signed a contract to die. Whether you die because your internal clock decrees it or because your immune system fails to prevail against bacteria or viruses, whether by accident or by exposure to radiation, there is no such thing as complete safety or immortality. The individual's survival is enhanced by the maintenance of his individual environment: diet and exercise and in the avoidance of accidents by recognizing natural hazards and taking the proper precautions.

The human being is not fragile: the race has survived bubonic plague, smallpox, tuberculosis, pneumonia, mumps, typhoid, typhus, influenza in all of its varieties, dysentery, sleeping sickness, schistomaisis, and a multitude of other diseases both bacterial and viral. We actually eat vegetables containing oxylates, drink fluids containing high concentrations of carcinogens, fill our lungs with smoke that contains everything from radioactive polonium to coal tar products (and it takes two or three, sometimes four, decades before this exposure causes sickness and death), we eat crops that contain natural pesticides in large concentrations. That isn't all: we find that bathing in water containing radioactive materials is sometimes beneficial; we enjoy garnishing our meat, fowl, or fish with funguses (mushrooms) some forms of which are among our most persistent enemies. We endure extremes of cold and heat, dryness and humidity better than any other species on the face of the earth because we tend to learn from useful deaths and avoid making the same mistakes. We also have learned to cooperate and thus increase our survivability beyond that of the individual or small group.

There is another facet to the "fragile" human species: it is not fragile because of the wide range in human physical and biochemical abilities. The biochemical variability is extreme as are the physical and mental capabilities. Thus there is a wide response by individuals to the conditions under which they live. Some individuals are disease resistant, some are susceptible; some are well coordinated others are not; some are fast, others are slow; some are geniuses, others morons. Let us not press the point: we can express the distribution of any characteristic in the human race as a bell-shaped curve (called a "normal" distribution). So under any condition of stress a certain percentage of the population should survive and from their offspring the race will continue. This observation does not subsume some of the major physical phenomena: a meteorite strike may well kill the entire population of a large area; earthquakes, volcanic eruptions, floods, droughts, and fires are capable of annihilating entire human populations. Something killed off the dinosaurs and we don't know what it was. The world, in a nutshell, is not a safe place and it never will be. Before man arrived on the scene the planet was hotter and drier; colder and wetter: sometimes with more carbon dioxide in the atmosphere than we have now, sometimes with less. Sometimes there was more volcanic dust, sometimes less. And I haven't even mentioned the influence of the sun spot cycle on weather: the absence of sunspots leads to wintry conditions. All in all, this is not a static world, nor is survival guaranteed despite our best efforts.

What Environment?

In order to make sense in any discussion we need to define terms. Most Congressional legislation, for example, defines the key terms used in the law. Nowhere, however, will you find a definition of "environment" although the focus of many laws is to "protect and restore" the environment. But no one ever tells us exactly what we are protecting and what we will restore and what are the actual goals to be achieved. Let me, therefore, propose a crude definition of "environment."

The environment of any living creature, human or otherwise, can be pictured as the total of the areas subsumed in a number of concentric circles of increasing diameter surrounding a center, where the living being exists. The first circle is the man (and this includes women) himself: he needs food, shelter, and clothing. A "good" environment is one that supplies him with these needs and provides him with time to satisfy his other need: the nurture of family. This companionship, essential to his future, is enclosed in the second circle. The third circle encompasses his society and the interrelationships between the people and organizations within it. It includes national defense, establishment of justice, domestic tranquillity and other objectives outlined in the Constitution of the United States. Outside the third circle is the real world, a physical world indifferent to the survival of any species. In the final analysis it requires that we adapt, as well as we can, to its ever-changing conditions. We do not really understand how the real world operates, although in certain limited areas we have learned through bitter experience, the range of response to our feeble activities.

Water

It is in the area outside the three circles that the "environmentalist" strives to operate. The environment, as they visualize it, is an idealized place where the water is virtually distilled, contains no harmful micro-organisms or other beasts, little if any silt, is neither acid nor alkaline and is neither too cold or too hot. Through efforts of the environmentalists the Congress has passed legislation to control pollution. Thus, the goals of the anti-pollution legislation can be taken as representative of what the "environmentalists" wish to do. Let us turn, then, to the latest re-writing of the anti-pollution legislation which is now embodied in the "Water Quality Act of 1987," the reincarnation of the old Water Pollution Control Act:

> The objective of this chapter is to **restore** and **maintain** the chemical, physical, and biological integrity of the Nation's waters. In order to achieve this objective it is hereby declared that . . . [my emphasis—RGK]
>
> (1) it is the national goal that the discharge of pollutants into the navigable waters be eliminated by 1985 [It wasn't, but except for the date, the goal stands—RGK]
>
> (2) it is the national goal that wherever possible, an interim goal of water quality which provides for the protection and propagation of fish, shellfish, and wildlife and provides for recreation in and on the water be achieved by July 1, 1983 [It wasn't, but except for the date, the goal stands—RGK]
>
> (3) it is the national policy that toxic pollutants in toxic amounts be prohibited.
>
> * * *
>
> (7) it is the national policy that programs for the control of nonpoint sources of pollution be devised and implemented in an expeditious manner so as to enable the goals of this chapter to be met through the control of both point and nonpoint sources of pollution.

In order to understand what this law is about, we turn to the official list of definitions:

> (6) The term "pollutant" means dredged spoil, solid waste, incinerator residue, sewage, garbage, sewage sludge, munitions, chemical wastes, biological materials, radioactive materials, heat, wrecked or discarded equipment, rock, sand, cellar dirt and industrial, municipal and agricultural waste discharge into water . . .
>
> (13) The term "toxic pollutant" means those pollutants, or combinations of pollutants, including disease-causing agents, which, after discharge and upon exposure, ingestion, inhalation or assimilation into **any organism** [my emphasis—RGK], either directly from the environment or

indirectly by ingestion through food chains, will, on the basis of information available to the Administrator, cause death, disease, behavioral
abnormalities, cancer, genetic mutations, physiological malfunctions
(including malfunctions in reproduction) or physical deformations, in
such organisms or their offspring.

(14) The term "point source" means any discernible, confined and
discrete conveyance, including but not limited to any pipe, ditch, channel,
tunnel, conduit, well, discrete fissure, container, rolling stock, concentrated animal feeding operation, or other floating craft, from which pollutants are or may be discharged. This term does not include agricultural
stormwater discharges and return flows from irrigated agriculture.

(16) The term "discharge" when used without qualification includes
discharge of a pollutant and a discharge of pollutants.

(19) The term "pollution" means the man-made or man-induced
alteration of the chemical, physical, biological, and radiological integrity
of water.

Let us try to state the goals in simple English: we're going to
restore and maintain the integrity of the Nation's waters. If I were to
ask you what that means, specifically at St. Louis and New Orleans on
the Mississippi, at Sandy Creek just south of Canton, Ohio, at Omaha
on the Missouri, and El Paso on the Rio Grande, neither you nor any-
body else in the world can tell me. There is no standard to repair to for
any of these places. And there is every reason to believe that in the past
both the quality and discharge of the rivers at these places changed
from time to time even before the advent of man. In short the stated
goals cannot be defined and to commit the Nation's resources to
achieve an undefined goal is irrational as our resources are limited, rich
as we may be in comparison to other nations.

The major goal of cleaning up our surface waters is to be achieved
by eliminating the discharge of pollutants. And if anyone can find anything that is not subsumed in the definition of pollutant, item (6), the
writer will gladly eat it all. In short, only distilled water may be discharged as a waste and, under the law possibly the water should be
triple-distilled, suitable for boilers producing high pressure steam.

But even the legislators realized that the goal should be achieved
in a step-by-step manner, so the law provides for **interim goals**. These,
for municipalities, are essentially to provide secondary treatment for all
sewage (in some instances, subject to the whim of the Administrator,
only tertiary treatment will do). And, as technology progresses, the
standards must be raised, little by little so that the goal may be
achieved. At whose expense? In the beginning the Federal government
aided with the capital cost. Now? Federal funds have run out and the
cities must do it at their own expense. Who pays the operating cost?

The taxpayers of the cities, of course. This sewage treatment is "mandated" under threat of serious fines. Other sanctions are available. If the municipality does not comply and pay the fine: the EPA can prohibit the connection of new sewer and water facilities, that is, stop all new construction. If that does not work they can stop the flow of federal funds for any and all purposes and in this way coerce the citizens to do their bidding—it's all in the law. Police protection may break down, fire departments may be eliminated, education neglected, roads turned into potholed trails—sewage must be treated even if the tax-base erodes and the citizens suffer. The new idol must be supported even if civilized life collapses: an abstraction called "environment" must be protected! As in the Middle Ages, we seem to have a Holy Grail to pursue.

There is political fallout from the worship of the idol: the government is forcing the expenditure of the individual's funds for its own purposes, purposes that are, presumably, beneficial to the entire population. The purpose is not to improve health—this could be measured. It is to protect an undefined abstraction, the environment. It is for this that capital expenditures in excess of $100 billion were made between 1974 and 1981 by municipalities and industries for the construction of plants for the secondary treatment of sewage.

Two papers dealing with the subject have appeared. One of them, by Henry M. Peskin, appeared in *Environment*, May, 1986, and it concluded that only 15 percent of all pollutants originate from sewer pipes, industrial and municipal. The other, by Richard A. Smith and his colleagues, appeared in *Science* in March, 1987 and it showed that, overall, less than 2 percent of the river miles have been measurably affected by the aforementioned construction and operation of secondary treatment plants. So even if all sewer pipes discharged distilled water, the pollution of the rivers, for all practical purposes, would be the same. Conclusion: using the forces at the command of the national government we have coerced a substantial fraction of the population to waste its labors, labors that could have been productive. Can this be considered "involuntary servitude," prohibited under the Constitution?

To be perfectly fair it may be difficult to take issue with item (3) of the goals: don't discharge toxic pollutants in toxic amounts. If something is toxic to mosquitoes, under the definition of toxic pollutants (13), it is prohibited. And if you'll pardon a technical question: what is a "toxic amount"? Enough to kill one mosquito or to affect the life cycle of coliform bacteria? Or enough, so that if you drink the water you'll get sick? By the last definition, the discharge of ten pounds a day of a toxic material into the Mississippi at Memphis would not be a toxic amount because of the large volume of flow whereas a similar discharge into Sandy Creek in Ohio would be, because the discharge of Sandy Creek is tiny. Manifestly, a law to control human action and applied throughout the nation must be uniform. But when it is applied to the actions of people on the physical world under a wide range of hydrologic and geologic conditions what makes sense in one place may create a terrible injustice in another. A bureaucracy can only be concerned with the letter of the law, not its spirit. There can be no exceptions and no discretion except as the law provides for them. Conflicts must be settled in court and exceptions, if granted, must originate in the legal process.

Our legislators may not have wanted to protect vermin and microbes, but they did. They may have wanted to improve the drinking water (on the assumption that local people are not smart enough to produce potable water), but while protecting some few individuals it has caused "externalities," needless expenditures with no commensurate benefit to the vast majority of the popula tion.

In short, in curing a difficulty the legislators have imposed what the economists term "externalities" on the entire population. The legislation to protect the environment, as embodied in the Water Quality Act of 1987, proved to be off-target from scientific, technical, and operational standpoints. It also shows no awareness of its probable side-effects on the environment and the human and marine populations, were it stringently enforced throughout the country.

As early as 1979, Phillip West, Boyd Professor of Chemistry at Louisiana State University, had this to say:

> Environmental contamination, as defined and restricted to the water environment, in no way implies impairment. Instead, water in the pristine state will seldom have adequate nutritional value and the introduction of essential nutrients merely fortifies or supplements. These waters obviously contain necessary dissolved oxygen and carbon dioxide but are most unlikely to contain a balance of iron, manganese, zinc and other essential trace elements. They are unlikely to contain enough organic matter to provide food for aquatic organisms. If water is to play a role in providing food, and millions of people depend on this, then supplemental contaminants are necessary and a case can even be made for a bit of transitory pollution.
>
> Before readers become upset, let it be said that some waters should be valued for their aesthetic and recreational properties only. Such waters should be clear and pure enough for fishing and swimming. In this connection, some fishermen will point out that fish are found in clear mountain streams in pristine areas. Admittedly, there are enough trout in Colorado to satisfy sportsmen but certainly there are not enough fish to feed the people in Denver, or even in Aspen. It is also interesting to note here that the trout get much of their food, directly or indirectly, from the

air—in the form of insects and their larvae. The fishing in such "pure" mountain streams is done by fly-casting.

Turning now to productive waters, it is quite apparent that absolute purity is out of the question. If the Mississippi River passing Baton Rouge and New Orleans consisted of distilled water there would be no seafood industry such as we now have in Louisiana. Without copper "contaminating" the water there would be no oysters. Traces of iron, manganese, cobalt, copper and zinc are essential for the crabs, snapper, flounder, shrimp and other creatures that abound in the Gulf waters. As unpleasant as it sounds, even the run-off from the fertilized fields of the heartlands and the sewage discharges into the Missouri, Ohio, and Mississippi River systems pollute and thus ultimately nourish the waters.

Another piece of legislation dealing with the liquid environment is the Safe Drinking Water Act, which also deals with ground water. Interestingly enough, there is no statement of purpose: the legislation starts off with definitions. I guess, since it deals with public water systems the Congress didn't think it necessary to state what it had in mind as the objective. What the legislation really says is that in the opinion of the lawmakers, the existing methods of producing potable, safe, water throughout the United States have been a failure and that only Federal intervention stands between the public and disasterous effects attributable to unsafe water. Although the actual record of waterborne disease does not bear out that opinion, the Congress has constituted itself as the supreme authority in these matters as well as many others.

The most interesting part of the legislation is the list of potentially hazardous contaminants to be prepared by the Administrator. Now in the days of "wet chemistry" contaminant concentrations in parts per million could be determined—and were. With the advent of the mass spectrometer it is now possible to measure the amount of dissolved materials in water in the parts per billion and, in some instances, parts per trillion range. (Detecting one part per trillion is like detecting the vodka content in New York water if the total city water supply for five [5] days had a jigger of vodka somehow mixed with it in a gigantic glass.) There are, nevertheless, some small technical problems: when dealing with chlorinated hydrocarbons there is a distressing inability to replicate the results: if a representative sample of contaminated water is divided into four parts and each part tested separately, four results will be obtained and only by accident will any two coincide to within 10 percent. It is not uncommon for the variance to be a factor of two or four. So if the Administrator sets a "maximum contaminant level" and one of the samples is above the level and three are below it by differing percentages, does that mean that people drinking the water are risking their health?

The standard for assessing the hazard is that if you ingest 2 liters a day for seventy years and your risk of cancer, for example, increases by more than one part in a million (from 0.2 to 0.200001), the hazardous concentration must be reduced. The costs of removal are never mentioned, nor are priorities set, for surely one substance is a greater risk than another. The standard is based on the interpretation of animal tests and tests of mutagens in the laboratory. By way of comparison, Richard Wilson in *Technology Review* of February, 1979, showed what this means by publishing the following table:

Risks which increase the chance of death by one in one million

Activity	Cause of Death
Smoking 1.4 cigarettes	Cancer, heart disease
Living 2 days in New York or Boston	Air pollution
Living 2 months in average stone or brick building	Cancer caused by natural radioactivity
Traveling 6 minutes by canoe	Accident
Traveling 300 miles by car	Accident
One chest x-ray	Cancer (radiation caused)

Executive Alert, May–June, 1989, pointed out that the Federal Government has a tendency to adopt policies that eliminate specific risks while ignoring other, related risks. For example the government banned EDB as a fumigant because it was found to be a mild carcinogen in rodents. Yet EDB is the safest known way of combating molds (which contain the natural carcinogen, aflatoxin) and now infect a significant percentage of the nation's grain crop.

It should be noted, again, that natural carcinogens are found everywhere. They are present in mushrooms, parsley, basil, celery, cola, wine, beer, mustard, peanut butter, bread, lima beans and hundreds of other foods. Compared to natural carcinogens, man-made chemicals pose a very small risk. Yet, by virtue of a Congressional mandate, without let or hindrance, tunnel-vision governs the actions of the Environmental Protection Agency. The Nation's treasure is poured forth to combat trivial risks while reducing the funds available for very real problems, such as road and bridge maintenance, sewer line repairs, water main repairs, and traffic control, not to mention education and a multitude of other uses for the scarce funds. Most of the funds expended are not found in the Agency budget since it can **mandate** the expenditure of private and public funds in accomplishing its mission, thus it affects local expenditures and local taxes directly. Moreover we find that such everyday processes as cooking add substantially to the carcinogen load that people are exposed to. Takashi Sigimura, director of the National Cancer Center in Tokyo used the Ames test to detect mutagens and carcinogens in raw foods. He also conducted experiments on the effects of elevated temperatures on proteins and amino acids and has observed the resulting formation of strong mutagens and carcinogens. He has pointed to differences in food preparation as the reason why incidence of stomach cancer in Japan is twice that in the United States. It is likely that research in food preparation and an educational campaign based on the findings would be far more effective in reducing cancer risk than the present expenditures on removing minute traces of substances from the water.

John Higgenson, founding director of the World Health organization's International Agency for Research on Cancer (IARC) is the man who gave Americans the idea that carcinogens lurk in everything we eat, drink, and breathe. In an interview with *Science* (9/28/79) he said that his statement had been misinterpreted,". . . primarily by the chemical carcinogen people and especially by the occupational people." He continued, ". . . many people have failed to distinguish between the environmental origin of cancers with clearly defined etiology—for example, smoking, alcohol, and occupation—and the large group of digestive and endocrine dependent tumors whose environmental cause can be inferred only circumstantially . . . I'm not saying that one shouldn't clean up the environment; of course you should, people shouldn't be exposed unnecessarily. [But the simplistic approach] has prevented possible acceptance of the idea that there may be doses of carcinogens which, for practical purposes, are unimportant. If you consider smoking plus asbestos, or smoking plus uranium , those combinations lead to more lung cancer as everybody knows. But the corollary, then, that small bits of 20 different carcinogens add up to produce a cancer—there are simply no experimental or human data that this is the case . . . The problem, in a nutshell, is that no thought has been given to the costs of protecting against poorly identified disease-causing substances as compared to using the same funds where an actual health-benefit can be expected to occur."

At this writing there are vigorous campaigns mandated by the EPA against trihalomethanes (the chloroform-like substances produced when the organic material in treated and filtered water is exposed to the disinfectant, chlorine) and lead in the taps of private homes. Whereas the first of these has produced no epidemiological data of any sort for no one has been demonstrably hurt by ingesting the trace amounts found in public supplies, and since public utilities have avoided the use of lead for decades, the only source of lead is in the homeowner's water supply system. Usually, if the water is slightly alkaline, little, if any, lead will be dissolved in it. In any event, running the tap for a short while before using the water in the morning will get rid of most of the lead. That utilities must sample and analyze the water from taps, over which they have no control, is an example of bureaucracy run wild with the spending of other people's money.

Land

If the quest for the Holy Grail of surface water is enough to stagger the imagination of a physical scientist or engineer, the effort to protect the land surface and the subsurface against residues is equally awe-inspiring. There is more than one piece of legislation involved in this endeavor. We might divide this problem into two facets: repairing the ravages of past practice and adopting policies to avoid repeating the mistakes.

The heritage of the past

The disposal of residues has not been the center of attention for the population or the engineering profession. The simplest way to get rid of wastes, household or industrial, has been to take them out of town and put them in a trench, alternating layers of waste with layers of dirt. Liquid wastes were usually placed into a hole, whether the underlying material was pervious or impervious.

As the population grew and the tonnage of household wastes increased, the area of "sanitary" landfills (sanitary because, if they were properly designed, rodents were kept out and insects were unable to multiply in them) has increased. Some of the soluble materials in the landfill were leached out by rainfall and, in certain areas, reached the ground water body. In addition, as decay progressed methane was generated and here and there a landfill would catch fire due to spontaneous combustion and the presence of methane. For liquid household sewage, after the sewage had been treated, the residual material was either burned, spread on the land, or dumped at sea.

Industrial wastes, in many instances, received equally cavalier treat ment. These wastes, not normally subject to bio-degradation because they were either inorganic, like the pickling liquors of the steel industry, or because they were chlorinated or fluoridated hydrocarbons, tended to stay in one place if they were solid and not soluble in water and to leach out if they were soluble. Liquid wastes would leak out of the pits if the underlying sediments were at all permeable—and almost all of them are permeable.

Interestingly enough, although the potential for some sort of health problems was there, actual health problems were remarkably scarce. The publicized Love Canal "disaster" was not the result of the improper disposal of industrial wastes—the industry was well aware of what had gone into the fill and had covered it with a clay blanket to prevent the infiltration of water. Then, when the land was sold to the School Board for $1, a playground was placed on the fill and there were no ill-effects on anyone. It was only after the School Board sold the property to a developer and sewers were built that exposed the buried wastes to the air, that the wastes began to come into contact with the home owners. The Love Canal waste pit was properly handled by industry in line with the best practice at the time and should never have been developed into home sites. It was fine as a playground. The School Board should have been held responsible, individually and collectively, for criminal negligence in selling the land to a developer.

The Superfund legislation is calculated to restore all of the old waste disposal sites that contain "hazardous" waste to their original condition—with no mention being made of costs. It turns out, that in a great percentage of the sites, it is not physically feasible to restore the land to its original condition and that the best solution would be to cover the site over with more or less impervious material (the more impervious, the better), and surround the area with a barrier, to prevent access. On the perimeter, if the site were underlain by a water-bearing formation, monitor wells would be drilled and maintained. If fluid leached out of the site and into the ground water, appropriate wells would be pumped and the fluid removed for treatment. Such an approach would do less damage than the "restore to original condition" requirement and would probably cost less than the interest on the bonds that would have to be sold for the restoration project. We might actually clean up the Superfund sites in from 10 to 20 years without bankrupting the fund and the original generators of the wastes—not to mention the great reduction in legal and court costs.

Old sites of sanitary landfills could be handled in a similar manner: expeditiously and economically. But this cannot happen under the present legislative rubric, which was designed by the legal minds of Congress without reference to the physical realities. It was designed to blame and punish, not to cure. It incorrectly assumed an immediate hazard to the health of the surrounding population and paid no attention to economics since the generators of waste, if these could be established, were to pay and it would serve them right.

As might have been expected the Superfund, which was to achieve the unachievable regardless of costs, promptly became a new form of pork-barrel. The beneficiaries were the various members of Congress who were fortunate enough to have some designated sites in their districts and the lawyers, consulting engineers, and contractors who had the proper "contacts."

As to the Future

There are a number of strategies for dealing with the ultimate disposal of solid wastes including burial, using the material to produce steam or power, and incineration. Recycling the wastes is only a partial answer to the problem and, while it should be used, this will depend on local conditions for economic viability. The destruction of the waste material by incineration is undoubtedly the ultimate answer to the problem of waste accumulation but this, again, is not always economically feasible. The situation is similar with respect to using the combustible material to produce power: it is a solution that is usable under certain economic conditions—and as these change (if the cost of gas, oil, and coal rise) then it will become feasible.

At this writing, however, burial is the most feasible disposal method in most locations. And the technology is such that the wastes can be substantially protected from rainfall and any leaching into the ground water that does occur can be detected, the leachate collected and treated, and the waste pile confined to a designated location. It is not clear that Federal legislation is needed to set standards and force everyone to meet them.

Air

The engineering and scientific complexities associated with the control of wastes in the water and land, including the subsurface, are child's play when compared to those encountered in the global atmosphere. Not only is the data difficult to obtain, but the atmosphere is a gigantic chemical, or even bio-chemical, reactor whose capabilities are virtually unknown. Even obtaining representative samples of air to determine the presence of materials like methane, sulfur in its various forms, carbon dioxide, carbon monoxide, and the various complex nitrogen compounds in the air, not to mention the obscure petrochemical compounds emitted by vehicles is difficult. And the same compounds may have both natural (biogenic) and artificial (abiogenic) sources. The relative influence on the global climate of artificially occurring and naturally occurring compounds has never been determined.

The writer has never seen a convincing world-wide mass balance of such materials as carbon dioxide or sulfur compounds, which also originate in nature. We can add to these the methane and carbon monoxide [CO] from natural sources (and CO is food for certain microorganisms and as it becomes available from any source, these microorganisms multiply and remove it from the atmosphere) so the uncertainty increases still more. [A mass balance simply sums up all of the sources of a compound and then sums up all of the material that is disposed of naturally: in the sea and land and by microorganisms, plants and other living beings. They should be equal unless the concentration in the atmosphere of one or another of these materials is rising.] If we take sulfur, for example, the tonnages produced by man plus the tonnages spewed into the atmosphere by volcanoes and other vents must equal the uptake by plants, the lands and the seas, unless the concentration of sulfur compounds in the atmosphere is rising. Some attempts have been made to produce a global mass-balance, but these have not been convincing, partly because the data base is inadequate and the measurements do not go very far back and partly because the atmospheric reactions on which the calculations are based are still obscure.

Nonetheless, the Congress has decided that man-made emissions are the root of all evil, that these must be stopped, and damn the costs. So we have the catalytic converter for autos, the lime-slurry uptake of sulfur compounds from the stacks of power plants, the prohibition of freon as the working material for air-conditioners, etc. etc. Some of the substances that they want to ban may be deleterious, others may cause little damage and their replacements may be both expensive and even more deleterious.

For example, in carrying out the will of these well-intentioned congressmen, the EPA has set forth its fight against ozone-producing emissions in great detail—and many of the major cities in the United States, especially on the West Coast, are out of what they have arbitrarily termed "compliance." This simply means that they do not meet the standards approximately .01 per cent of the time, and that 99.99 percent of the time these cities are in compliance. The fight against smog is extremely local and national standards may, or may not, be germane. Moreover the costs imposed for round-the-clock monitoring of air quality all over the country whether the air is "bad" or not can be very expensive, although much less than devising and enforcing some ill-founded measures in the cities that are found to be out of compliance. After all, the Blue Ridge Mountains of Virginia were named that way because of the hydrocarbon emissions from the pine forests and the abiogenic (man-made) contribution to smog in that area is low.

The latest idolatry is the idea that the recorded increase in carbon dioxide over the past hundred years, if continued for another 100 years, will raise world wide temperatures by, possibly, 5 degrees Fahrenheit. The glaciers of Antarctica will then melt, the oceans will get warmer and the net result will be a rise in sea level of from 1 to 4-feet which will flood many of our low-lying coastal areas. To avert this looming disaster we have to drastically reduce or eliminate gases that produce the "greenhouse effect" (the sun's heat gets in but the gases, in essence, prevent it from leaving): carbon dioxide and methane, to name two. Close examination of temperature records by a number of researchers has revealed a possible, but not certain, world-wide increase in temperature of possibly half a degree in the last 100 years despite a rise in the carbon dioxide concentration by 25%—and the temperature should have risen, according to the latest model, by from 2 to 3 degrees. So it looks as though the model is wrong and that the great hubbub concerning carbon dioxide and methane may be unwarranted. Some people have said that the increase in sulfates in the atmosphere has increased the cloud cover (which reflects the sun's heat) and neutralized the greenhouse effect. If this theory is correct, we should not be spending so much effort in curbing sulfate emissions and we don't have to be so concerned with the other gases.

Presumably the global warming that should accompany the greenhouse effect could be monitored by measuring the advance or retreat of glaciers. The position of alpine glaciers has been monitored for many years; some glaciers have been monitored for centuries. Fred B. Woods of the Office of Technology Assessment reported, in a study that appeared in "Arctic and Alpine Research in 1988," that advancing alpine glaciers increased from 6.3 percent in 1960 to 57.4 percent in 1980, based on a study of from 400 to 450 glaciers. This might be interpreted as saying that the world is getting colder, surely it is not getting hotter. A study of the Greenland glacier, published in *Science* of December, 1989, reported that the elevation of the surface of the glacier increased between 1978 and 1986 (ice accumulated) at the rate of almost 8 inches a year. What price the greenhouse effect? The Department of Agriculture has (1990) issued a new "plant hardiness map" which supercedes the one in use that was issued in 1965. Commercial nurseries and home gardeners use these maps to select varieties that will survive and thrive in their areas. The new maps show that the temperatures are from 5 to 10 degrees Fahrenheit cooler in the winter than in the old maps: the temperature in each zone is getting colder. There is evidence that the climate is getting colder and nothing more than scientific speculations based on computer programs that the climate will get warmer.

The only certainty is that the present models of atmospheric circulation and the interchange of heat and moisture between the oceans, the atmosphere, and the land do not produce results that can be corroborated in the field. Our understanding of weather and climate is tested every day by weather forecasts—and while one and two-day forecasts are useful, forecasts of the weather for longer periods have been far from reliable. The whole matter of air circulation and roles played by the oceans, the ocean currents, and continents, not to mention biogenic sources and sinks, is not well understood. The creation, movement, and presence of clouds in the atmosphere has not yet been taken properly into account. Nor are the sources of biogenic materials reaching the atmosphere, for example the methane produced by termites and cattle, taken into account in any useful way. Present models are extremely limited in value and cannot be used as a guide to major actions. We need more data and more study before the citizens who produced machine civilization are forced to bankrupt themselves to satisfy some unproven hypothesis that has been sold by the environmentalists to the group of professional lawyers who constitute the Congress of the United States.

Protection—Against What?

In areas of science and technology, efforts by the Congress to "protect the public" have been singularly futile. The Alar scare, which followed the cranberry scare, which followed the mercury scare, which followed the children's flammable nightgown scare, which was followed by the bis-scare (bis is the chemical used to reduce the flammability of nightgowns), etc. etc. These scares, resulting from bureaucratic actions called for by Federal mandates to protect the health of individuals, proved to be unjustified by the evidence and forced large expenditures on great numbers of people with no discernible benefits. They were the direct result of agencies created by the Congress which simply carried out the congressional mandates. The bureaucrats played their usual game of CYA. We are now in the midst of the asbestos scare: it has been shown that aside from asbestos miners and processors who smoke, the greatest threat to human life is experienced by the workers hired to remove asbestos from buildings. The threat to those who occupy the buildings can be removed by spraying the asbestos so that particles cannot be detached from the asbestos board or the asbestos that has been sprayed on beams and ceilings. The expenditures by School Boards and owners of offices throughout the country have been huge and have diverted moneys that could better have been spent elsewhere. The threat due to asbestos particles produced by automobile brakes is trivial, but manufacturers are now frantically searching for an economic substitute for the asbestos used in brake drums. The threat to human health due to the use of water that has flowed through asbestos-cement pipes is, based on the 17 year record at Duluth, Minnesota, zero. The whole asbestos scare is an environmental scam, to put it mildly.

The list of protections and mandated actions could go on almost indefinitely. The mandated actions have been expensive and ineffective. The bureaucracies have flourished and have increasingly constricted the activities of individuals in their every-day lives. A particularly good example is found in the Federal Food and Drug Administration, which passes on the efficacy of any new drugs produced by the manufacturers. The average time between the submission of a new drug to the agency and receipt of permission to distribute it commercially is now between 7 and 9 years. And there is no guarantee that the tested drug will not produce some unforeseen and undesirable side-effect despite all of the testing. This agency, too, is trying to accomplish the impossible: making sure that any new drug will be both effective and safe for everyone. In view of the range of biochemical differences in people, there are bound to be individuals for whom some substances, even milk or aspirin, will produce life-threatening effects.

In the final analysis, no amount of testing performed on microorganisms, rodents, monkeys, or small groups of people will produce medicines or drugs that are both safe and effective for everyone. When we ingest an unknown substance we put ourselves at risk, regardless of what anyone may assert. All that the agencies do is prevent individuals from doing what they think best while giving them a false sense of security. The thalidomide controversy is pertinent: thalidomide was widely used in Europe as a sleeping pill. Millions, if not billions, of doses were successfully used for insomnia over a 10-year period with no reported ill effects on the users. It was only when deformed babies began to appear that thalidomide came under suspicion. But as we know, pregnant women should not take any drugs during pregnancy, not even alcohol or aspirin. If all pregnant women had practiced this essential precaution, thalidomide would still be the sleeping pill of choice, probably throughout the world. The only conclusion is that the individual must be sole judge of what he may do with impunity. The role of the government, if any, is to make information readily available at minimum expense to the user. Then the individual can reach his own decision. The system of trying to protect everyone against every possible danger has not worked, and the costs to society are undoubtedly far higher than they would be under another, less coercive, more informational system (although rigorous studies to demonstrate this have yet to be performed).

Any drug, for example, that is submitted for testing by the FDA may be generally effective or may be ineffective. If, after 7 or 8 years of extensive testing it proves effective, many people will have died during the testing period who might have been saved if the manufacturer had made it immediately available to the medical profession. If there were a central registry where doctors could report the adverse reactions to drugs, it would be a simple matter for any physician to access the data bank and inform the patient of the dangers posed by taking it. Under such circumstances the number of people suffering ill effects could be compared to those who would have died if they had delayed taking the drug until it had been approved by the FDA. Many AIDS sufferers have realized this and since the alternative is a certain and lingering death, if any drug shows promise some of them are more than willing to take a chance on the untested drug. Under enormous public pressure the FDA has slowly moved to permit this or, at least, not to prosecute those participating in such trials.

We should remember that drug manufacturers have a keen interest in selling effective drugs, not drugs that don't work. They may be relied upon to react promptly to reports of drugs that do not work. If the government must intervene it should be on the basis of collecting information on the results of real-world tests of all drugs, old as well as new. A law might well be passed telling doctors to keep accurate and detailed records of the results that they have observed and to make those results available to the data banks of whatever organization has been designated to collect and evaluate the results, possibly state health departments, the U.S. Public Health Service, or a central drug registry. Each month a publication could be issued on the results of the real world testing. Actually, private data banks might be made repositories of the information and they might charge a modest subscription price for their findings.

Members of Congress do not seem to have realized that each reduction in the individual's ability to make meaningful decisions is a restriction on his liberty. Such decisions are costly to society in many ways: new approaches are not tried, individuals incorrectly believe that they are protected by some government agency from danger and do not examine their individual circumstances with care. It has become well-nigh impossible to avoid making technical violations of some rule promulgated in the Federal Register by an agency, a rule that has the force of law. The restrictive milieu has also given rise to an unprecedented temptation to bureaucrats for corruption: by inaction on a legitimate request the bureaucrat, even at a low level, can cause financial losses of great magnitude to citizens and businesses. There is no way to force a bureaucrat to act promptly through the legal system: either you get a Congressman to intervene (a campaign contribution should have already been made) or you slip the bureaucrat a few bucks on the side. He's not doing anything illegal by granting you the permit (assuming that you have complied with all of the regulations), but he's doing it faster than he would otherwise have done.

We have already seen such a procedure in action. In Florida when the Interstate was being constructed it was revealed that one of the biggest road contractors was paying $15 to $25 a month to the inspectors and a little more than that to their supervisor. A great hue and cry went up that the concrete was not up to specifications or that the surface roughness was not as specified or some other physical problem, due to cost cutting procedures made possible by bribing the inspectors. After a lengthy investigation, the construction was found to meet or exceed the construction specifications: the concrete was stronger, there was an adequate amount of steel, the surface was smoother than it had to be, etc. etc. It was also revealed that the inspectors willingly worked overtime when requested by the contractor: they helped the contractor operate efficiently and by so doing saved him money. It was for the timely inspection that the inspectors were being paid. They were doing their job at the convenience of the contractor to enable him to use his equipment, and move his equipment, in the most efficient manner. They, at times, worked several hours a week more than the 40 hours that they were supposed to—it was for this that they were being compensated: to do their work properly, at the right time.

The writer encountered the same circumstances in trying to get an exit permit from Saigon in 1960. Had I waited for the routine procedures, I would not have been able to use my plane reservation even though I allowed plenty of time for getting the exit visa. Some discreet inquiries revealed that for the equivalent of $10 I could get the visa immediately. I paid it and I got the visa and was able to use my plane reservation. The clerk had simply done his job promptly at my convenience, not his. He did not break any rules. You might call it a "gray bribe" as contrasted with a "white bribe" (when you enter a bid at an auction), or a "black bribe" (when someone pays a bribe for an illegal action—fixing a traffic ticket, or not ticketing a DWI). Anyone who has worked in a Third World country, will understand from personal experience: "gray" bribes are an elaborate form of "tipping." Amusingly enough, the Foreign Corrupt Practices Act, enacted in 1977, specifically okays payments to officials to "facilitate" routine government action—everything from processing visas and licences (what I have termed "gray bribes") to providing water, electricity, phone service, and police protection.

It is certain that corruption will show up in one "protective" agency after another as the bureaucrats learn their way around the regulations and realize the impact of enforcement on the members of the population that they "protect." There will be many "gray" bribes and a significant percentage of "black" bribes consisting of payments commensurate with the risk.

The number of agencies striving manfully to protect people is large and their initials are becoming ever more familiar: OSHA, EPA, FDA, and SEC come readily to mind. In terms of impact, the EPA is the most menacing, with the FDA, OSHA, and the SEC following in that order. Let it be understood that the writer has no quarrel with these agencies or any other governmental agency—the quarrel is with the Congress and the Courts. After all, the bureaucrats have been hired by the Congress to carry out its desires. Their independent judgment of the validity of the mission has not been requested—just carrying out orders, sorry about that. There is a vaguely ominous ring to these words.

What Now?

There is an old axiom that "You can't beat something with nothing." This is undoubtedly true, so we are obligated to answer the question, "If what you say is true, how do we go about curing things?" Of course, considering the great number of regulatory agencies and the volumes that are added each year to the Federal Register, this is a very broad question. It is not a cop-out to say that, in a short essay it is only possible to set out some general principles, with some illustrative examples and to leave it to others to do the additional research that will be needed before reaching conclusions on an issue by issue basis.

Let us start with the legislative process: in law after law the Congress does not spell out the regulations—and this is particularly true in the scientific and technical areas. The laws leave it to unelected bureaucrats to publish regulations, that have the force of law. Only the Courts may invalidate them and this in only some instances. So the first reform is in the legislative area:

1. If the legislature cannot pass the law, including its regulations, it may not bestow the law-making authority (via the Federal Register) on bureaucrats who have not been elected, are protected by Civil Service regulations, and are not answerable to the general public in any way except under the most egregious of circumstances.

Essentially, if the Congress can't write the detailed rules and regulations, after all the Hearings and the advice that the Congressmen receive, the legislation should not be introduced, much less passed. Better no legislation than legislation by unelected persons: the American Revolution was not fought to give random, unelected people the power of economic life and death over us. Confine law-making to elected officials.

The argument may be made that there are so many things needing adjustment that the Congress must make laws on almost everything and it lacks the time to complete the tasks and so must delegate the law-making power to the unelected bureaucrats. The answer must be that unless an issue is serious enough for the Congress to attend to it in a complete and serious manner, the Congress should not involve itself. The Congress is the analog to a Board of Directors, not an executive committee group of bureau-chiefs.

The Congress must make the policy decisions on major issues and embody these decisions in the details of law. The proper model is found in the current legislative struggles on the abortion issue where a major point of argument is whether or not to include victims of incest or rape and whether or not abortion should be prohibited or shall be permitted and funded during the first trimester of pregnancy. By way of contrast, in similar circumstances, details of the law have been left for bureaucratic determination as, in secondary treatment of sewage, the EPA has set the 5 day biological oxygen demand (BOD) of the effluent at a maximum of 30 ppm. This is an arbitrary figure. Why not 50 or 200? Considering the costs involved in meeting and maintaining this standard, surely the Congress could have made the decision instead of the EPA.

Observance of this stricture would keep the focus of the Congress on the major issues and minimize political micro-management of the total economy of the United States, something that it is attempting to do on a larger and larger scale. This, combined with fields of endeavor far removed from the legal profession, has resulted in legislation with totally unrealistic goals, goals which, had they been set by an individual, would have been immediately and devestatingly rejected. If the Congress had to write the detailed rules and regulations, the goals might well have been modified and expensive and ineffective programs might have been avoided.

2. Dismantle and abolish those portions of agencies that promulgate and enforce laws ("rules"). Charge the remainder with the collection, interpretation, and publicizing of information connected with their original mission. Presumably the Congress can write laws pertaining to the collection, analysis and publication of information including the obligation of various sectors of the population to furnish it to the Agency in a timely manner. If the present members of Congress cannot write the laws to accomplish this, let them resign and we can elect someone who will.

For example, if the EPA is concerned with air quality, enable it to set up stations to obtain data, let its staff analyze and publish the data and make such recommendations as seem fitting in the areas of public health and convenience. If the FDA is interested in drugs, authorize it to collect data from any source, make experiments, and publish the findings, including in the publication any recommendations that seem supported by the analysis in the areas of efficacy, safety, and public health.

3. Legislation should embody the details of laws and when the law is inadequate, the bureaucrats should have to have corrections made by legislation, not by publication of the proposed changes in the Federal Register.

If there are problems that cut across state lines, there is nothing to prohibit the signing of interstate compacts between states that have mutual extrinsic problems whether it be sewage treatment or groundwater withdrawals and these compacts would, of course, be ratified by the Congress. There are many precedents for this procedure and they do not involve legislation by federal civil servants. Problems that are even more local can be handled by individual states or subdivisions thereof.

Conclusions and Tentative Recommendations
(What do we do now?)

In the Middle Ages it was believed that a man or woman, and sometimes even little children, could be "possessed" by the devil or an evil spirit. The subject would act peculiarly, irrationally, not in accord with his best interests, and might even fail to recognize old friends and forget his family in pursuit of strange goals. We even have a phrase, "He acted like a man possessed!"

You might say that when a person acts against what are obviously his best interests and there is no rational explanation, he is acting under duress. The source of the duress may or may not be apparent: if he is responding to a death threat by some mobster or drug lord, with courage he can usually find help from the authorities. But what if the authorities, themselves, are the source of the threat: fines, imprisonment, or possible bankruptcy due to the legal costs of defending himself in court?

As long as laws were passed that governed the relationships between human beings, whether they might apply to inheritence, divorce, adoption, stealing, murder, drunken driving, abortion, property ownership, trade, investment, or the multitude of activities that people engage in between themselves, the legislature, courts, and the legal system could set the rules of the game and provide for hardship cases.

But when a law is passed that forces people to take actions to obtain the same physical results regardless of geography, geology, hydrology, and biology, without reference to costs and benefits, public or private, the citizen and all of his financial and personal relationships are "possessed" by, or the person is under complete control of, the agency that is to enforce the law. In the final analysis each individual has effectively been "possessed" by the Congress that promulgated the laws. He has been forced to work for the benefit of others, sometimes for the benefit of no one, animal, vegetable, or mineral.

The preceding paragraph is an objective statement of the outcome of "environmental" legislation, an outcome that was inherent in the terms and goals of each law passed by the Congress and enforced by the Environmental Protection Agency (or other Agency). We've seen the same results from legislation mandating safety in the workplace by OSHA or similar legislation mandating safe and effective drugs by the FDA. When legislation is passed with the mere statement of objectives and how the first monies are to be distributed (the sugar coating on the pill) leaving the actual laws to be written and enforced by some bureaucratic agency, the citizens become "possessed" and lose their freedom of action and, sometimes, even their freedom of thought. Researchers have documented the expenditures of hundreds of billions of dollars on sewage treatment by municipalities and industries. These have produced no discernible improvement in public health or other commensurate benefit. Manufacturers have spent millions to produce fire-resistant clothing only to find that the fire-proofing substance did not meet the approval of another agency, school boards are now being forced to spend hundreds of millions of dollars removing asbestos only to find that the threat is negligible and that they have put the workers to great risk from respiratory disease—the list can be lengthened if need be.

The country is being led down a blind alley by Congressmen who do not understand science and technology. They respond to those who can create the greatest fuss and so exert pressure on them, in this instance the "environmentalists." The latter are composed of a vast majority of ignorant and well-meaning people led by political figures with a patina of scientific and technical expertise (not necessarily in the field that they are discussing) who have their own agenda: unspoken for the most part but real, nevertheless. What the agenda is, the writer will leave to others to ferret out. The question is, "How do we get out of this mess?"

Of course the Congress should stop bestowing law-making authority on the agencies that it creates. But adoption of this procedure will only help in the future and the legacy of past legislation is still with us and will be with us until it is repealed or replaced by something better.

As a general principle, the tangible benefits from any procedure over its lifetime should be greater than its cost over the same lifetime. Legislation that does not acknowledge that there are costs connected with mandated expenditures by citizens and corporations should be prohibited. One cannot defend the present Congressional stance with regard to environmentalism: that the benefits of the mandated expenditures will always result and will always exceed the costs.

The second principle is that the goal of any legislation must be one that does not put the resources of the entire society at risk by setting goals that are impossible to achieve because they ignore natural laws which state how the real world behaves. They ignore simple rules such as "water runs downhill and will erode land," "coastal sediments, especially at the mouths of major rivers, tend to compact and reduce the land surface elevation," "different concentrations of minerals, natural and man-made, when ingested, will produce different results in the same person and a wide range of results in a population," "our knowledge of the local and global climates is very superficial and any of our computer models is bound to be inaccurate—so don't force people into a major restructuring of their lives in order to influence a desired outcome, let them bet on race horses instead—they have a better chance of winning," and "the time-frame of the costs and benefits may not be long enough to properly evaluate the project, for example, the navigation and flood control program on the Mississippi-Missouri did enhance navigation and reduce flooding for the first half century after the 1927 flood—but the loss of wetlands and habitat in coastal Louisiana, some 40 square miles a year at present, is the direct result of flood prevention and the accumulation of sediment in the reservoirs on the Missouri and Arkansas Rivers. The effect of flood control projects in reducing the sediment load in the Mississippi River was ignored."

The third principle is really an instruction that will require years to fully effectuate: Congress shall set as its goal the abolition of agencies whose authorizing legislation mandates that they do something that is physically impossible. The goal should be to abolish at least three major agencies at each session of Congress. Each abolished agency should be replaced with a fact-finding agency with instructions to publish and publicize the facts and any analyses that seem pertinent. Over the long run, States, or other subdivisions, must be assumed to act rationally or their constituents will suffer and the political leadership that erred will be replaced. The Federal government's role must be confined to information collection and dissemination and persuasion—not the use of financial blackmail, like the withholding of taxes collected by the government for use on the road system, to force compliance with an unrelated regulation.

In general, where the states now have organizations that more or less track the present Federal organizations, the states should receive "primacy" as the federal agency or division or branch is abolished. That is, the states would have the power to do what they thought was pressing. During a two or three year transition period, the states might receive some financial help while they increased their staffs to do whatever job, in their opinion, needed to be done.

We might expect careful scrutiny of the existing standards and regulations and revisions of these standards in line with the physical circumstances. If a state decided that "hazardous" meant a greater chance of disease than one part in a million over a 70 year lifetime, it could decide the level of risk. If experience with the subsurface injection of liquid wastes was good, then only needed standards of construction and operation would be enforced and the injection could proceed. There would be no nation-wide prohibition of the practice. All of the legislation pertaining to matters other than the interactions between people would be tailored to the physical conditions in relatively small areas and there might be a chance that these could be both workable and economic. At least the local citizens would contribute to the decision making and could effectively bring pressure to bear on the lawmakers if experience showed that a grievous error had been made.

Despite my fear of duplicating a previous principle, the fourth principle is not to embark on major social and economic programs without overwhelming evidence, from the real world (not the laboratory or the computer model), that there is a real disaster in the making. For example, if a large meteorite or small planetoid were computed to make a landfall, it would be reasonable for a world-wide effort to be made for the purpose of diverting or destroying the large rock. Creeping disasters like the alleged heating from the greenhouse effect can be studied until enough information from field observations is available to reach a conclusion. At present, if we base ourselves, for example, on the retreat or advance of alpine glaciers to determine whether the world is heating up or cooling down, there is the following information: in 1960 only 5.6 percent of glaciers were advancing (the rest were either not changing in size or were retreating); in 1980 57.4 percent of the glaciers were advancing. And between 1978 and 1986 snow and ice were accumulating on the ice mass of Greenland, raising the elevation of the "land surface." Unless you believe that glaciers advance and accumulate snow during warm periods, the field data indicates that the world is getting slightly cooler at the very moment when the carbon dioxide content of the atmosphere is rising. So some thing may be counteracting the greenhouse effect: maybe the increased concentrations of sulfates in the atmosphere (which tends to increase cloud cover) has overwhelmed the influence of carbon dioxide, CFC's, and methane. Until we understand what we are doing, basing ourselves on field observations, there is no need to burden the population with direct taxes and the indirect taxes of "mandated" changes.

Chasing a Receding Zero: Impact of the Zero Threshold Concept on Actions of Regulatory Officials

Thomas H. Jukes, Ph.D.

Please see biographical sketch for Dr. Jukes on page 135.

The principle of toxicological threshold was stated in 1564 by Paracelsus as "All things are poisonous, yet nothing is poisonous. Dosage alone determines poisoning." This principle remains valid, even with consideration of today's public concern with low-dosage effects of compounds that are categorized as carcinogens, mutagens and teratogens. Modern laboratory methods enable detection of increasingly small traces of such compounds in foods. The search for these has been termed "Chasing a receding zero," and this quest has been stimulated by statements that no one knows how small an amount of a carcinogen, taken for how short a time, can induce cancer, and that even one molecule of a carcinogen, acting on a single cell, can transform a normal cell into a cancer cell.

This latter proposal is a stochastic impossibility. Every cell contains millions of carcinogenic molecules. Examples are arsenic, cadmium and chromium. In addition, there are estrogens and other steroid hormones present in each cell and needed for normal bodily functions. These substances are regarded as carcinogens because of their effect at high levels on increasing the cancer rate in experimental animals. Obviously, there must be a threshold for this effect. An estimate has been made by Dinman that "a threshold for biological activity exists within a cell at 10,000 atoms."

It is also obvious that there is a threshold for deficiencies that induce cancer, such as that of iodine. Iodine deficiency is mimicked by administration of goitrogens, and these are present naturally at low to moderate levels in many common foods.

The threshold principle is stated by Claus and Bolander to be a law of nature, valid in many circumstances, and governing the fact that a causative agent must be present in a quantity exceeding a definite minimum in order to produce the effect.

I have been involved in the question of a threshold for carcinogens starting in November 1959. This was the date when the Secretary of Health, Education and Welfare went on the radio to warn the public not to eat cranberries because they might contain traces of aminotriazole, which could cause cancer. The FDA had been informed by American Cyanamid Company that large doses of aminotriazole caused enlargement and adenoma of the thyroid gland in rats. The findings were made by Hazleton Laboratories under contract with American Cyanamid. The finding took place shortly after the passage of the Delaney Clause in the Food Additives Amendment which does not recognize the possibility of a threshold for carcinogens.

As we pointed out at the time, aminotriazole was excellent as an example inferring the existence of a threshold[1]. The reason for this is as follows: its effect was that of a goitrogen. Goitrogens inhibit the uptake of iodine by the thyroid gland. As a result, the pituitary gland works overtime to stimulate the thyroid to resume hormone production at normal levels. The stimulation of the thyroid by the pituitary hormone, thyrotropin, results in enlargement of the thyroid, and, in today's climate, any enlargement or hyperplasia is regarded as a potential cancer. However, the same effect is produced by withdrawing iodine from the diet of rats. Therefore, if a dietary deficiency can cause cancer, there must be some threshold of intake of iodine above which there is no carcinogenic effect. The same phenomenon may be true for vitamin A and for selenium. In the case of vitamin A, Claus and Bolander[2] pointed out that Johannes Fibiger, a Danish pathologist, was awarded the Nobel Prize in medicine for his discovery that a parasitic nematode caused cancer in rats, by the effect of cockroaches eaten by the rats. Soon afterwards, it was shown that the actual cause of cancer in his rats was vitamin A deficiency in the diet used by Fibiger. This is a second case of the indisputable existence of a threshold.

But let us return to aminotriazole. Goitrogenic substances are present naturally in foods such as soybeans, turnips and cabbage in physiologically active quantities. Synthetic goitrogens are used in medicine to control over-activity of the thyroid gland, and aminotriazole itself was in use as an experimental antithyroid drug in the late Dr. Edwin Astwood's clinic at Tufts University Medical School in 1959, when the cranberry incident took place. Adenoma of the thyroid gland in rats can eventually result in carcinomatous changes when the animals have received a goitrogen for many months. Aminotriazole had been introduced in the spring of 1959 as a herbicide for weed control in various crops, including cranberry bogs. The permitted use called for stopping the spraying several months before harvest time. A few growers did not take this precaution, and, as a consequence, small traces of aminotriazole were detected in less than 0.34% of the cranberry crop. Mr. Arthur Flemming, Secretary of Health, Education and Welfare, on this basis threatened the public with cancer unless they avoided cranberries at Thanksgiving. The results was a panic from which the United States consumers have never fully recovered. The interpretation was that a chemical company was planning to make money by giving people cancer.

We pointed out in vain in 1959 that any effect of eating cranberries containing small traces of aminotriazole was probably no greater than what would result from eating raw cabbage. Nevertheless, hundred of tons of cranberries that contained no aminotriazole were bulldozed underground. A writer denounced the "poisoned cranberries of Madison Avenue." Ever since the cranberry incident, pesticide residues in foods have been widely regarded as carcinogenic, and the possibility of a threshold has been represented as nonexistent.

During the years following the cranberry incident, there have been some monumental disputes about the question of a threshold for carcinogenic effects of residues in foods originating from human-made chemical substances. Carcinogens naturally present, or arising from pyrolysis have been largely ignored. For example, Lijinsky and Shubik[3] showed in 1964 that several identifiable carcinogens of the benzpyrene type, derived from burning fat, entered meat during barbecuing over a charcoal fire. This discovery was not followed up even by Lijinsky, in spite of his preoccupation with the possibility of carcinogens in foods. More recently, an entirely different class of carcinogens derived from pyrolysis has been discovered by Sugimura and his colleagues in charred meat and other protein foods[4]. Such discoveries, including the recent finding that parsnips contain a carcinogen, seem to be treated as an unavoidable fact of life, and therefore increased efforts should be made to remove "artificial" carcinogens. Nevertheless, carcinogens naturally present in foods far exceed introduced contaminants in quantity.

First of all, let us examine the "one molecule" hypothesis of carcinogenic effect.

The concept that a single molecule could start a train of events leading to cancer received attention following studies with ionizing radiation. These studies led to the idea that a single impact causing a break in one strand of a DNA molecule might lead to a mutation in an important cell. It was subsequently argued that a molecule of a mutagenic chemical, changing one nucleotide at an important locus in a chromosome of one cell, could trigger a chain of events leading to cancer. It was not known just how many key nucleotides were present in the three billion base pairs in a single mammalian cell.

This imaginative hypothesis has been used to impress Congressional hearing committees with the wonders of science, and the dangers of carcinogens.

This was reaffirmed by Dr. Arthur Upton, former director of the National Cancer Institute as follows[5].

> "No scientific method currently available can accurately determine how much of a carcinogen, if any, may be safely added to our food supply without increasing the risk of cancer in the human population. Many scientists believe there is no "safe" level of a carcinogen. Transformation of a normal cell into a cancer cell could conceivably occur with one molecule of a carcinogen acting on a single cell."

Does Dr. Upton mean one molecule per person, or one molecule per cell? Each person has billions of molecules of carcinogens present naturally, including radioactive carbon and potassium, heavy metals, such as uranium (10,000 atoms per cell), steroid hormones and numerous other carcinogens naturally present in foods. Perhaps he means one molecule per cell. By "could conceivably occur," perhaps he means there is a one-chance-in-a-million possibility of one molecule causing cancer. Surely the chance could not be much more remote than this, if it is worth scaring people about.

Estimates of the number of atoms of different elements present in each mammalian cell were made by G.E. Hutchinson in an address at the 100th anniversary meeting of the National Academy of Sciences.[6] The number for the heavy metals ranged between 10^4 and 10^8 atoms per cell. There are about 10^{14} cells in the human body, according to calculations by Dr. George Claus and others. I shall use this figure, which is a conservative one.

Arsenic, cadmium and chromium are considered to be carcinogenic. The approximate amounts of these present in the body are respectively 4.4 mg, 30 mg and 6 mg per person. These amounts would supply respectively 10^5, 2×10^6, and 0.7×10^6 molecules per cell, and these calculations are within the range calculated by Hutchinson for the concentration of heavy metals per cell.

The arsenic content of a normal, healthy human being is 4.4 mg, which is 9×10^{18} molecules of arsenic as As_4. This fact makes it obvious, without further elaboration, that the "one-molecule" hypothesis is preposterous. But let us try some calculations, anyway.

If each molecule has a one-in-a-million chance of transforming a normal cell into a virulent cancer cell, then the odds are 9×10^{12} to 1 that any and every human being will get cancer from arsenic normally present in the body. In other words, **everyone in the world is going to die of cancer caused by arsenic**. The chances are about 1,000 to 1 against the possibility that even one person in the 4×10^9 people in the world will be able to escape this fate.

Maybe Dr. Upton mans that each cell in the body must receive one molecule of arsenic before anyone is endangered. There are 10^{14} cells in the body, and each cell contains 100,000 molecules of arsenic. How many of these cells are capable of being transformed into a dangerous cancer cell? Should we say one-billionth of the total? This would be 10^5 cells. each of these has a 10 to 1 chance of being transformed, even if the odds are 10,000 to 1 against a possibility of any single arsenic molecule in the cell carrying out the transformation. So there are still odds of a million to one that everyone will get cancer from the arsenic normally in their bodies, if there is anything to Dr. Upton's theory.[7]

About 8 mg of nitrite are excreted in the saliva per day. No figures are available on the total nitrite content of the body, but it would seem reasonable to estimate that this is not less than the amount excreted per day in the saliva. Eight mg in the body corresponds to 100 million molecules per cell.

The basic principle in biology was stated by Paracelsus in 1574 as "All things are poisonous and yet there is nothing that is poisonous; it is only the dose that makes a thing poisonous." The threshold principle is well known in physiology, for example, subliminal stimuli applied to a nerve are without effect upon the muscle that is innervated.

In the case of "single molecule theory," I have noted that we should all be dead of cancer from the millions of molecules of arsenic, cadmium and chromium in each one of our cells if the theory were valid.

Nitrites

The public, after recovering somewhat from the cranberry incident, was startled in 1978 to learn from the FDA that nitrites used in treatment of bacon, ham, salami and other common foods were carcinogenic. The

Congressional Record, June 25, 1981, has reprinted an article with the headline, "The Day Bacon was Declared Poison," by Philip Hilts[8]. This describes a day in August, 1978 when FDA and USDA announced that nitrites were carcinogenic and in consequence, nine billion pounds of meat became suspected poison. The article states that FDA Commissioner Kennedy called Carol Tucker Foreman, the Assistant Secretary of Agriculture in charge of food quality and consumer services.

> "Her activist approach to food and health issues had already earned her the title of "dragon lady" among pork producers of the Midwest. They believed that 'whenever she opened her mouth, pork prices dropped,' one meat industry spokesman said.
>
> Foreman had been wrestling for a few years with the problem of nitrosamines found in bacon. At just about the time Don Kennedy called, she thought she had finally wrestled it to the mat, arriving at a technical compromise with the meat industry. She answered the phone in her large second-floor office. `I have some very bad news for you,' (Don) Kennedy told her. He said he had a study showing nitrite could cause cancer directly, never mind the nitrosamines.
>
> 'Oh (expletive deleted),' Foreman said, and sank through the same rapid, dismal thoughts that other regulators say they also had when they heard the news. 'All this struggle over the nitrosamines, and now this guy comes in and says that nitrites themselves are the bad actors. This was obviously something much bigger, much worse.'"

Evidently, Mrs. Foreman accepted without question the information that nitrites in meat cause cancer.

Nitrates and nitrites have been present in the environment for at least 2 billion years, when photosynthesis first became widespread. Following this, oxygen became a major component of the atmosphere and nitrates (and simultaneously nitrites) were formed by thunderstorms. Ever since then, all living organisms have been exposed to nitrates, and also to nitrites formed biologically by reduction of nitrates. All of us eat nitrates every day and these are converted in the body to nitrites. Nitrates are synthesized in our bodies as shown recently by Steven Tannebaum and his collaborators.

Given an elementary knowledge of biochemistry, including the fact that nitrites are always present in saliva, no one would jump to the conclusion that nitrites used in meat-curing had to be banned without taking a second good, hard look at the question. Mrs. Foreman's expletive should have been directed at the flimsiness of the statement, rather than being an expression of dismay.

In June, 1979, after the "bacon incident," another food was declared poison and once again FDA and USDA participated. This food was beef from cattle that had been treated with DES according to procedures previously approved for more than 15 years. As a consequence, billions of McDonald's hamburgers and other sources of beef muscle became retrospectively carcinogenic, bearing in mind a latent period of 20 years for cancer and the widespread use of DES in beef production. The annual consumption of beef in the USA is about 10 million tons. The ban was announced in June, 1979 and imposed in October, 1979. Some beef producers who violated the ban, according to Dr. Lester Crawford, who was then Director of the FDA Bureau of Veterinary Medicine, did so because Americans are "enshrouded in a number of myths that will presage more lawlessness."[9] He included in these "the myth of the dose-response curve, which lives in part because of a pervasive national mistrust." Some of us thought that dose-response curves were derived from experimental results. The relation of dosage of carcinogens to response has been extensively documented by Druckrey. Indeed, a linear dose-response to carcinogens by DES is postulated by FDA as discussed below.

Alexander Schmidt, as Commission of FDA stated in 1976.[10]

"The Catch-22 in the Delaney Clause is the DES clause. Delaney says you can't use a chemical if it causes cancer when fed to animals or man. The DES exemption says: unless it can be used in such a way that there is no residue in the edible tissue of the animal, when looked for by methods approved by the Secretary. The Catch-22 is 'the methods approved by the Secretary.'

This means that we will be chasing a 'receding zero,' and some idiot in some lab will come up with something sensitive to parts per quintillion, and our policy says we will adopt it."

Ever since Dr. Schmidt said this, FDA has diligently pursued the receding zero. They ran it to ground in a trial held in Wichita, Kansas in November, 1980 in which the combined forces of FDA, USDA and the Department of Justice were brought to bear on 77 tons of boned beef, obtained from cattle that had been implanted with DES according to a procedure that was legal for 15 years, but had been used after the October 1979 deadline. The FDA's main thesis was that one part per trillion of DES in beef constituted a definite carcinogenic hazard to consumers.

The FDA's calculation is based on the dose-response curve relating to DES dietary level to mammary cancer in C34 mice in the publication by Gass and co-workers, as follows[11].

Additions to Diet	Percent with Cancer
None (controls)	33
100 ppb DES	65.6

Therefore, 100 ppb produced 32.6 additional cases of cancer percent, and one part per trillion would produce 32.5×10^{-5} cases in 100 individuals, or one case per ~ 300,000. The FDA then assumed that the intake of beef muscle was 500 g per day, representing about one-third of the average per capita daily food intake. Actually, the consumption of beef averages 35 g daily. Their assumption leads to one case of cancer per million consumers caused by eating beef from implanted cattle.

FDA assumes that human beings and C3H mice are equally susceptible to cancer caused by an estrogen. This ignores the fact that C3H mice carry the mammary tumor virus and have a spontaneous cancer rate of 30 to 40%. Their extreme susceptibility to estrogens is well-documented.

In my opinion, the FDA estimate is worthless because results from an FDA laboratory support the conclusion that a threshold exists in the DES-C3H mouse test. The results were not made available during the court trial in Wichita and were in press at that time. The conclusion by the authors is that 10 ppb DES was without detectable effect in MTV + mice (Highman *et al.*, published November 1980, accepted November 1979[12]). The conclusion supports my statement that a threshold level for DES in cancer production in C3H mice extends from 0 to 25 ppb of diet (Jukes, Feedstuffs 49: 22–23, 1977). I therefore conclude that 1 part of DES per trillion of diet is far below the threshold.

Similar conclusions were drawn by Elwood Jensen in testimony at Wichita. The U.S. District Court decided against FDA and USDA, who subsequently withdrew their appeal from this decision. One of the remarkable things about the allegation was that DES pills are still permitted by FDA to be sold for administration to human beings. A pill containing 1 mg of DES would be equivalent to 1,000 tons of the indicted beef containing one part per trillion. Furthermore, any carcinogenic effects of DES are inseparable from its estrogenicity, according to leading authorities in the field.

The emphasis by FDA on DES in beef production obviously reflects public concern with the announcements in 1971 by Dr. Arthur Herbst. He reported that several young women developed cancer almost 15 years after their mothers had been given massive doses of DES during pregnancy, as a treatment to prevent threatened miscarriage. Consumerists have ever since clamored for a ban on DES in beef production. However, one series of women each received a total of about 12 g of DES during pregnancy.[13] No cancer cases were reported in this series, but precancerous vaginal changes were found in some of the daughters. Twelve grams of DES corresponds to 12 million tons of beef containing 1 part per trillion. No wonder Dr. Lester Crawford prefers to regard the dose-response curve as a myth.

Selenium

One of the most interesting illustrations of thresholds is found in the effects of selenium, which, prior to 1957, was regarded exclusively as a toxic element. Excessive quantities of selenium in the soil in certain areas, such as the Dakotas and Manitoba, result in toxic levels of selenium in field crops. Farm animals in these areas often suffer from selenium toxicity, including birth defects. Selenium became regarded as a carcinogen from studies with rats receiving levels as low as 4.3 ppm of selenium.

In 1957, it was discovered by two groups of investigators that selenium was an essential nutritional trace element at levels in the range of 0.1 to 0.2 ppm of diet. The finding was made by Schwartz and Foltz at the National Institutes of Health[14], and by my colleagues Patterson, Milstrey and Stokstad at Lederle Laboratories[15]. Eggert, also in our group, noted that selenium deficiency in pigs produced a marked necrosis of the liver, with fibrotic changes, perhaps suggesting a precancerous state[16]. Other findings have also suggested a connection between selenium deficiency and cancer (Underwood[17]).

The categorization of selenium as a carcinogen presented a harrowing dilemma to the FDA. How could a carcinogen be essential in nutrition? How could it use be permitted under the grim shadow of the Delaney Clause? The FDA agonized for 15 years over this question, because selenium is of great importance in the nutrition of farm animals, and selenium deficiency is common, and severe in its consequences. One way to evade the issue was by using fish meal as a feed supplement, because selenium in fish meal is natural, and therefore legal.

Nutritional requirements for selenium are satisfied by about 0.1 ppm in the diet. The threshold for toxicity appears to be in the neighborhood of 2.0 ppm. Selenium therefore resembles estrogens in being essential to life, yet possibly carcinogenic at high levels.

Summary

The problem of achieving consensus in the matter of thresholds for carcinogens has been eloquently addressed by Claus and Boland[2]. I quote from their unpublished manuscript:

"With the discovery that more and more elements such as selenium . . . chromium and arsenic—all of which are proven carcinogens—are essential elements for animal life, a curious twist in the thinking of some scientists has arisen. It is also known that certain hormones—especially the sex hormones—are carcinogenic and simultaneously essential. Most oncologists have accepted these findings as evidence for threshold levels for essential substances that produce cancer only in excessive quantities. However, the defenders of the dogma of 'no threshold for any carcinogens' have interpreted these findings as a reinforcement of their position.

They say that the essential daily intake of such trace elements does not represent safe levels of carcinogens, but is dangerous, and 'is the price we have to pay in order to stay alive.' (Lijinsky[18])

This argument, carried to its ultimate conclusion, leads to an old saying . . . 'The cause of cancer is life.'

On April 24, 1610, Galileo invited a number of his colleagues to his home to demonstrate to them through his telescope his recent discovery of the moons of Jupiter. Some of the scholars looked into the instrument and either denied seeing anything or described what they saw as optical illusion. A second group refused to look into the telescope, declaring both the contraption and Galileo's purported findings to be the work of the devil. The behavior of many experimenters involved in feeding chemical carcinogens to animals seems reminiscent to that of Galileo's learned guests. The first group, when reminded of the phenomenon of threshold, respond by stating that even if the principle does operate for carcinogens, it is illusory to define its values (Rall[19]).

The second group behaves similarly to Galileo's guests who said the telescope was the work of the devil. These scientists state that the necessary intake of compounds that are both essential for life and capable of acting as carcinogens, involves the risk of developing cancer. In short, they imply that the devil is exacting his fee."

References

1. Jukes, T.H. and Shaffer, C.B.: Antithyroid effects of aminotriazole. Science 132: 296–297 (1960).
2. Claus, G. and Bolander, K.: The threshold principle: A law of nature. Environmental Carcinogenesis, R.E. Olson (Ed.) Marcel Dekker, NY (in press).
3. Lijinsky, W. and Shubik, P.: Benzo(a)pyrene and other polynuclear hydrocarbons in charcoal-broiled meat. Science 145: 53–55 (1964).
4. Matsukura, N., Kawachi, T., Morino, K., Ohgaki, H., Sugimura, T. and Takayama S.: Carcinogenicity in mice of mutagenic compounds from a tryptophan pyrolyzate. Science 213: 346–347 (1981).
5. Upton, A. quoted in Reflections on the Delaney Clause. ACSH News & News 1 (2): 4 (1980).
6. Hutchinson, G.E.: The influence of the environment. Proc. Natl. Acad. Sci. (USA) 51: 930–934 (1964).
7. Dinman, B.D.: "Non-concept" of "No-threshold:" Chemicals in the environment. Stochastic determinants impose a lower limit on the dose-response relationship between cells and chemicals. Science 175: 495–497 (1972).

8. Hilts, P.: The day bacon was declared poison. Washington Post, 26 April 1981. Congressional Record-Senate 25, June 1981, S7153–S7156.
9. Crawford, L.M., quoted in an editorial "The continuing DES Saga." FDA Consumer 14 (6) :2, July/August (1980).
10. Schmidt, A.M.: Speech to the National Advisory Food and Drug Committee Meeting, June 29–30, 1976, Washington, D.C. (1976).
11. Gass, G.H., Coats, D. and Graham, N.J. Natl. Cancer Inst. 33: 971–977 (1964).
12. Highman, B., Greenman, D.L. Norvell, M.J., Farmer, J., Shellenberger, T.E.: Neoplastic and preneoplastic lesions induced in female C3H mice by diets containing diethylstibestrol or 17 ß-estradiol. J. Environ. Pathol. Toxicol. 4: 81–95 (1980).
13. Herbst, A.L. Poskanzer, D.C. Robboy, S.J., Friendlander, L. and Scully, R.E.: Prenatal exposure to stibestrol: A prospective comparison of exposed female offspring with unexposed controls. N. Engl. J. Med. 292: 334 (1975).
14. Schwartz K. and Foltz, C.M.: J. Am. Chem. Soc. 79: 3292 (1957).
15. Patterson, E.L., Milstrey, R. and Stokstad, E.L.R.: Effect of selenium in preventing exudative diathesis in chicks. Proc. Soc. Exp. Biol. Med. 95: 617–620 (1957).
16. Eggert, R.G., Patterson, E.L. Akers, W.T. and Stokstad, E.L.R.: The role of vitamin E and selenium in the nutrition of the pig. J. Anim. Sci. 16: 1037 (1957).
17. Underwood, E.J.: Trace Elements in Human and Animal Nutrition, 3rd Ed., Academic Press, NY. 543 pp. (1971).
18. Lijinsky, W.: Direct Testimony before the U.S. Dept. of Labor, Assoc. Secretary of Labor, OSHA, OSHA Docket No. H-090, 33 pp. (1978).
19. Rall, D.P.: Thresholds? Environ. Hlth. Perspect. 22: 163–165 (1978).

Reprinted, with permission, from
J. Am. Coll. Toxicol. 2(3) 147–160, 1983.

Environmentalism: What's Real, What's Not

Don't Believe Everything You Read
John J. McKetta

Just Maybe . . . The Sky Isn't Falling
Edward C. Krug

Root Causes of Extremist Activity
Max Singer

A Different Look at Environmentalism
Jay H. Lehr

Don't Believe Everything You Read

John J. McKetta, Ph.D.

Please see biographical sketch for Dr. McKetta on page 44.

I'd like to tell you today about some of the exaggerated claims concerning our environment, and I'd like to explode some of these myths because most of them are myths. The reason that I'm so interested is not because my main interest is environment. (Although I'm extremely interested in the environment,) my concern is the energy problem of the U.S.A. This country has been put on her knees, in fact, on her tail-end, because of the demands of the extreme environmentalists, the pressures that they've made in Congress and the laws and regulations that have been passed that have put us where we are. You're going to find out soon and first-hand that we're in a position where we're not going to get out of this serious energy problem.

I'd like to get started by asking you not to believe anything you read or hear, and that may even include what you're going to hear today. I mean this, regardless of the background of the speaker or the writer. I'd like to start this afternoon by telling you a little bit about something you well know, and that is the fastest living creature on earth—the deerfly.

How Fast Can a Deerfly Fly?

Way back in the early 20's, a great educated professor, an entomologist, with a very, very high educational background (his name was Dr. Charles Townsend and he graduated from a huge school in the east—at least huge in academic standing)—and he wrote an article in the very respected *Scientific Monthly* about speed. He was talking about the possibility of flying around the earth in a single day. You remember way back in the 20's our airplanes were flying about 100 miles an hour. He brought up the fact that the fastest creature on earth was the deerfly. He said the deerfly could fly somewhere in the neighborhood of 400 yards a second. He published this information, and this, of course, caught the eyes of the scientists and mathematicians all over the world. Everyone recognized right away that the deerfly was indeed the fastest living creature on earth. The male deerfly could fly 818 miles an hour! This is faster than the speed of sound. The female could only fly 600. You know why, of course.

Dr. Irving Langmuir, the great Nobel award winner in chemistry in 1932—saw this story and thought it sounded like it was phony. He made some calculations and he found that at 818 miles an hour, this little deerfly would have to generate over ½ horsepower of energy, and it would have to consumed 1.5 times its weight in fuel every second in order to go 818 miles an hour. At 818 miles an hour, the force into the wind would crush the poor, little deerfly to death. Whereupon he measured the speed of the deerfly and found out that sure enough it was fast, but it only flew about 25 miles an hour. That's the actual speed of the deerfly. But don't let that worry you. The highly respected entomology book published in 1959—*General Entomology*—still carries the speed of the deerfly at 818 miles an hour. Even *Webster's Third International Dictionary*, published in 1961, clocks this deerfly at over 800 miles an hour. So, when some of these things get into literature, you'll never get them out.

Scientists Wrong 90% of The Time

I don't blame Dr. Charles Townsend, who started this thing. Unbeknownst to himself, he got himself in the literature. The reason I don't blame him, and I don't blame me or my fellow engineers and scientists—we are wrong 90% of the time. Maybe not intentionally, but we are very often factually wrong. What we do is this—this is how we live, and it's our job to do this . . . We do these kinds of things because we don't believe we read. We make up new hypotheses, then we try to prove we are wrong. Let me give you an example. I ask my graduate students to work on my hypothesis to prove that I'm wrong, and they'll do everything they can to prove that I'm wrong. If they find out that McKetta isn't wrong, then we have a theory. We'll go to a meeting in New Orleans or in Louisville, or somewhere, and we meet with experts in our field, and we say, "We've got a new hypothesis. Here's what it is." I tell them our hypothesis.

Let's take an example. Here I am a professor—McKetta—and I've noticed that everywhere I bite myself, there's pain. Did you ever try this? Under the arm, under the leg, behind the rear end, everywhere, every time I bite, there's pain. So I write this up as a little hypothesis. I have my students all biting each other, and they find that's really true. That's the McKetta hypothesis. And we go to this big meeting, and we tell our colleagues what it's all about. They say, "I've never thought of that; this is a new hypothesis. They go to their laboratories and sure enough, I get a letter from some guy, and he says, "Johnny, you're wrong. We found that if you bite your fingernail or your toenail, there isn't any pain." My hypothesis is shot. It isn't true that everything you bite on your body will produce pain.

Headlines are Misleading

You see, this is how we live. We're wrong almost all the time. Now let me tell you what has changed. We used to sit in a meeting room alone—just 50 or 60 of us colleagues—but today there's someone else in the room. There's a journalist in the crowd! In the front seat, some youngster who just got out of college and got a little degree in journalism with very little knowledge of science, and he's going to write headlines. He's from *The New York Times* or "The Podunk News." He writes all these things down. All right, you ready for my new hypothesis? This is what he hears:

> "In our laboratory, we found out that if we took an ivory elephant tusk and stuck it under the skin of a mouse, the mouse skin got red and bled and a little tumor was formed. We took another mouse and we stuck that elephant tusk in the thing, and sure enough, the second mouse . . . and in fact, colleagues, of the 25 mice that we tried, 19 of them bled and formed a tumor."

Now what does this journalist write in the magazine or in a newspaper?

"Elephant Tusk Causes Cancer!"

This sort of reporting brings about a lot of the problems that you and I are facing. I want to show you a few of these headlines and discuss a couple of them.

The first is from the *National Observer*. It says, "Scientists fear spray cans will be the death of mankind." Now, who reads any further than the headline. What does it mean? Does it mean if you get hit in the head with a spray can, you're going to die? No! It tells how you will die when the ozone is removed from the Stratosphere. This is a hypothesis—so far unproven.

Here's one that says, "Fertilizers threaten ozone destruction." What you're going to do is use fertilizer on your grounds, and then it will go right up into the air somehow (I don't know how) and it removes the ozone and then the sun shines down and it melts the ice caps. I'm getting to like this one because when the ice caps melt, the first two states to be inundated will be Maine and Massachusetts and we get rid of Muskie and Kennedy at one fell swoop.

This headline says, "Morning coffee could be fatal." Another says, "Scientists fear man may be heating up this world with carbon dioxide." Now I know of 3 scientific committees that are studying this problem—you know when you make carbon dioxide (by burning coal or anything that contains carbon) you are going to form what they call a greenhouse effect and one committee says you'll heat up the world. But, the second committee thinks you're going to cool the world. Now these are both scientific committees, one thinks it's going to heat and the other thinks it's going to cool. There is a third committee. What do you think they think? They think there really won't be much effect. Now these are not a bunch of kids you get off the street—these are great scientists and engineers. Some even know how to write—books. Don't believe everything you hear or read!

Here's another headline. "Cleaning Fluid Causes Cancer." That's why I'm wearing this dirty white suit. Another—this is the one I like. Did you know, they finally found out—after 3 million dollars of research—that "Research Backs Breast Feeding." They claim that the breast-fed babies tend to be more healthy. Isn't that strange? My mother knew this 63 years ago. But we finally found out it's so. Goody! Maybe now I'll live.

Let me read this one to you. "Coal pollution is blamed for 21,000 deaths every year." When you burn coal, some pollutants go into the atmosphere. Now, the article headline scares the reader, but it's not until you get way down deep into the reading (only foolish guys like me will read that far), in the 14th paragraph it says that following—"The study emphasizes that the figure of 21,000 deaths from sulphur pollutants is only an estimate." The study says that "it can only be inferred that the deaths were caused from pollution since they had no other explained cause." How do you like that? These are scientists??

There is a great (?) scientist (?) whose name is Jacques Costeau? Many of us spend Sunday afternoon watching his beautiful pictures on TV—He does a tremendous job of taking colored pictures under the water. He wrote me a letter which I thought was a personal letter. His letter said, "The Oceans are dying." It went on to say if I would send $14 that in 25 years he could stop the oceans from dying. You know, as I was writing out my check for $14, I thought the guy has nothing to lose. In 25 years, he's going to say, "You save the ocean with $14." Then I found out that 12 million people received the same letter that I did. And I don't know how many of you sent him the money. I did not.

Let me tell you about Jacques Costeau. Of course, he may be a physicist. I say may be! I'm not sure. I don't even care. But he doesn't understand pollution. Jacques Costeau said, "When effluents from a paper mill can be drunk and exhaust from factory smokestacks can be breathed, only then will man have done a good job in saving the environment. What we want is zero toxicity—zero effluents."

The Perfect Plant—the Human Body

Ladies and gentlemen, if Jacques Costeau breathed perfectly clean exhaust from a smokestack, he could die after a few whiffs because this exhaust would contain only nitrogen and carbon dioxide. Jacques Costeau is talking about pollution from man-made chemical plants. Jacques Costeau lives in the perfect—the most perfect—plant, the human body. This is a chemical plant that God made. And it's the most efficient plant that was ever made. Jacques Costeau, who lives in this perfect body, breathes air that contains 21% oxygen, and he expels air that contains only 18% oxygen, and his exhaust air contains CO_2 which "pollutes the air." He doesn't mind his pollution of the environment. let me go further. He drinks water that contains 200 parts per million solids and he expels water that contains 30,000 parts per million solids. This guy is getting millions of people to believe that you can have zero pollution but he pollutes air and streams each day. Your Senate unanimously passed a zero pollution bill. And the bill is impossible! I don't mean it's hard to do—it's impossible to have zero pollution. Nature produces 60% of the particulates, 65% of the sulfur dioxide, 75% of the hydrocarbons, 93% of the carbon monoxide and over 95% of the oxides of nitrogen. But you're stuck with this zero pollution bill. So far you've lost DDT; you lost cyclamates; you're losing fertilizers. People are complaining about the fact that they found PCB in the mothers' milk somewhere, and you are reading now that all chemicals are carcinogens.

What They Don't Tell You

There are a whole bunch of myths. Today I want to go through one myth. Attached is my talk on the "Eight Surprises" (Editors note: See *Chino World*, Vol. 4, Nos. 18, 19, 20, 1975.) which gives you some of the extreme exaggerations that are made and the fact that they are really myths.

First, I would like to read to you a statement that was made in 1975 by the Executive Director of the Committee of Natural Resources of The National Academy of Science—Dr. Richard Carpenter. "On environmental issues, the National Academy of Engineering and the National

Academy of Science Advisory Committee finds that the factual data base is incomplete, inaccurate and difficult to interpret and suspect of bias. The environment consists of complex interactions of organisms, matter and energy with few well understood cause-and-effect relationships." So when you read about the fact that "coffee is going to kill you" or that "bacon is the most dangerous food in the supermarket." you won't be so worried. Let's take this latter topic about bacon.

"Bacon is the most dangerous food in the supermarket." This was taken from the Food Day Conference, and it was sponsored by Ralph Nader's group—the Center for Science—and immediately after that was published, there was so much pressure put on the Agriculture Department that plans were made to propose regulations to limit the use of nitrites in curing meat because of its possible cancer-causing effects. Maybe I ought to mention to you that everything that these "scientists" put under the skin of a rat so far causes, by the Delaney "definition," cancer in rats. So I would suggest you quit eating bread, drinking milk, etc., or at least you shouldn't inject it under your skin. Incidentally, some scientists put sterile dimes under the skin of mice, and what do you think happened? Money causes cancer! Oh Hum!

Let's go back to these nitrites. It isn't nitrite they're worried about, but nitrites are further oxidized to form nitrosoamines and nitrosoamines have been found to cause cancer in rats. In bacon, the nitrosoamines are present in concentrations as high as 5 to 25 parts per billion. What they didn't say in their story was that you would have to eat 4,200 pounds of bacon every day for 70 years of your life to get as much nitrite as they had in that one crystal they stuck under the skin of a rat. What they didn't mention is that nitrosoamines are necessary for digestion of your food. In his great wisdom, God put a little gland up near your tonsil area, so when you chew you secrete saliva that contains nitrosoamines that flow down into your stomach and this helps you to digest your food. How do you like that? It's all right for God's nitrosoamines to be taken in, but it's no good if you get it out of bacon. So don't worry so much about your bacon problems. Incidentally, cabbage, carrots, and most other vegetables contain much more nitrites than your bacon.

You know, of course, that penicillin may be fatal to a few people. (I call penicillin the second greatest chemical made by man, the first is DDT. DDT has save more lives than all the other chemicals put together—over 100 million people are alive today who wouldn't have been alive if there hadn't been DDT that had been made by man. But today you aren't allowed to use it in the U.S.A. In India, and many other countries in the world, they're using DDT. But your Congress took it off the market.) Well, penicillin has saved millions of lives also;

but do you know that penicillin upsets one out of every 10 people—one out of every 2,500 people become seriously ill and about 1 out of 250,000 people die from penicillin. They're allergic to it and they die. You know some people wear a little medallion around their neck that says, "I am allergic to penicillin."

Let me tell you about one that hurts me most, and that's lima beans. I like lima beans! But I want to tell you that lima beans contain hydrogen cyanide. Hydrogen cyanide is the poison used when a person is executed by gas. 80 milligrams of hydrogen cyanide will kill you; well, 3.7 pounds of lima beans contain 80 milligrams of hydrogen cyanide. Now you know why no restaurant will feed you 3.7 pounds of lima beans at one sitting. Not only will this kill you, but look what it will do to all of the people sitting around you.

The Myth of DDT

Let me go back to DDT. A headline in one national magazine said, "DDT is the worst thing that ever happened to mankind." Do you know that we don't have a single case on record where a person has died by ingesting DDT? Do you know that we have had some scientists in several places throughout the country—Arizona, California, Illinois and Pennsylvania—who have been feeding volunteers one gram of DDT internally each day for over a two-year period of time with no bad health effect? Do you know that in 8,800 pages of testimony (taken at hearings covering the removal of DDT from the market) there was no proof brought out that DDT was harmful to mankind. They did pretty well prove to the satisfaction of Ruckelshaus (who was administrator of the EPA at that time) that it's possible that DDT may have thinned the shells of bird eggs. For this reason they took DDT off the market! Now I want to tell you a little about DDT. Suppose it's true—right now we think it's really not true that bird shell thinning was caused by DDT—but suppose it were true, suppose the brown pelican shells were thinned because of DDT, it's quite possible the desirable properties of DDT so greatly outnumber the undesirable ones that maybe we made a grave mistake by taking this remarkable chemical off the market. Some insect-borne diseases, such as encephalitis, malaria, yellow fever and typhus fever were almost extinct because of DDT. I mentioned earlier the World Health Organization estimated that over 100 million people are alive today who wouldn't have been alive if it hadn't been for DDT. But we took DDT off the market in the 1970's. In 1972, the deaths (from malaria alone) increased from 200,000 to over 3 million people in one year in India alone. You see, you have a trade-off when you make these big decisions. You have a choice of variables, a decision to be made—birds or people—your government chose birds! I hope you and your legislators are proud!

Now you read in the paper regularly that mothers' milk has been tested and they found it to contain DDT, PCB, 2-4-D, and other chemicals.

But if you want to, you can find anything in any thing. For example, at the University of Wisconsin, where they have the world's finest analytical laboratory, the chemists in area were given 34 soil samples and they were asked to find out how much DDT was in the samples. Thirty-two of the 34 samples were found to contain DDT. What the chemists didn't know was that those soil samples were selected and hermetically sealed in 1902, and DDT was not made until 1940. Of course, you can find DDT in mothers' milk and you can find anything in any thing with the untra sensitive analytical equipment we have nowadays. But be careful not believe the results!

In conclusion, I want to tell you we're going to live—not only are we going to live but the children that are born in the year 2000 and the year 2010 may have a life expectancy of over 100 years of age. So we're going to live and by year 2000 we're going to find out that all these fallacies that these people have been scaring the general public with over the years are only fallacies and myths. They've been crying wolf, and so far, you've been believing it. But you're going to continue to hear this, year after year, after year.

But ladies and gentlemen, we're going to make it! We're going to live! We're not on the brink of ecological disaster—our oxygen is not disappearing—there will be no dangerous buildup of carbon monoxide or carbon dioxide, or oxides of nitrogen—DDT will not kill us—the waters can be pure again—food additives won't poison you—we'll still be using them—disappearance of the species is natural—radiation from nuclear power plants will not annihilate earth—a large percentage of the pollution is natural—most of the pollution in the air is natural—and it will be here whether you make it or not—but you are going to make it! We cannot solve our real problems unless we attack them on the basis of what we know, rather than what we don't know, and I ask you, and plead with the environmentalists, to let us use our knowledge and not our fears, nor our emotions, to solve the real problems of our environment.

Just Maybe . . .
The Sky Isn't Falling

Edward C. Krug, Ph.D.

Please see biographical sketch for Dr. Krug on page 35.

Industry's Credibility Crisis

Surveys show Americans think environmental conditions are worsening and would sacrifice economic growth to protect it.

A majority—67%—agree with the statement that "threats to the environment are as serious as the environmental groups say they are." Only 6% said they trust scientists seen as representing industry (*Washington Times*, August 15, 1990).

Scientific truth is not established by vote but political reality is. Government has been responding to this public mandate through legislation that is strangling material extraction and production. Similarly, environmentalists are winning almost all litigation largely through precedent-setting cases such as Sierra Club v. Morton (1972), where environmental groups are legally recognized as representing the public interest. This seriously weakens recognition of the public interests served by industry.

In the ongoing struggle over public opinion, environmental legislation and litigation, environmentalists are portrayed as the idealistic defenders of the public interest and Mother Earth. Their opponents are viewed as Darth Vader in a three-piece suit—special interests opposed to the common good.

A key mistake made by industry is examining issues one by one rather than analyzing the underlying ideology of environmentalism. By responding to accusations and always being on the defensive, industry gives the appearance of always hiding something. This supports the environmentalist image of good against bad.

There are those who have decided that if you can't break 'em join 'em. Over half of the financial support ever received by environmental organizations has come from the great private-sector foundations. As Ron Arnold, the environmental pioneer, observed this attempt "to co-opt the environmental movement with massive foundation grants . . . only sold rope to the hangman" (Arnold, 1987, p. 10, 53).

By not grasping the true nature of the conflict, industry's effort has been turned against itself.

The Environmental Party

Environmentalism has emerged to become the preeminent American political party. Twelve organizations comprise the base of support for the Environmental party. "All told, the Environmental party has an operating budget of $336.3 million (1988) and a donor base of 12,959,000. That's nearly $250 million more than the Republican and Democratic parties combined and a donor base of some 10 million persons more!" (Rep. Dannemeyer of CA, September 1, 1990). About 90% of funds go to support political activities.

The Environmental party attracts well-meaning citizens who are honestly concerned about pollution and conservation—wise use of the Earth's resources.

Having established the moral high ground, a large and highly-motivated corps of volunteers and professionals, and overwhelming financial resources, the Environmental party has become the proverbial 800-lb gorilla.

As if this is not enough, government and mass media are natural allies of the Environmental party.

To help understand government's role as an environmental ally, think of government as the **business of regulation.** Like all businesses, the business of regulation seeks growth. However, even governmental growth is not risk-free. It has the inherent liability of incurring the displeasure of the consumer (taxpayer) by increasing the cost (taxes) of the product. Environmental legislation, however, is like manna from heaven: a free ride in that it is an **off-budget tax.** The government is thus perceived as working literally for free for the common good.

Industry has few friends in the media. A report of the Media Institute shows that "Television almost never portrays business as a socially useful . . . activity." Americans love the underdog. Stories of the "common citizens" battling against the industrial giants are popular and, like government's environmental involvement, pays big dividends with little risk. The myth of the great lobbying power of the "Timber Barons," "Oil Monopolies," and "Mining Kings" support the image of a David v. Goliath conflict.

Objectives

Very few of our fellow travelers on Spaceship Earth realize the fundamental goals of the Environmental party. This is partly because well known environmental organizations, such as the Sierra Club and National Audubon Society, started out as bona fide organizations promoting wise use (conservation) of the land. However, as Ray Arnett (former National Wildlife Federation President—the largest of the 12 organizations) noted, the "National Wildlife Federation was known as the largest conservation education association in the world. Now they have moved into advocacy, lobbying" (Warren Brookes, *Detroit News*, March 11, 1990).

This cover is used to distract from the fact that environmental advocacy is merely a means to an end which lies in the domain of social engineering. As the best-known environmentalist of our time, Dennis Hayes, observed for the first Earthday, "We feel that Earthday has failed if it stops at pollution, if it doesn't serve as a catalyst in the values of society" (*Newsweek*, April 13, 1970).

Environmental advocacy has long relied on emotive arguments. To this already impressive arsenal is being added an ever-increasing array of the super-firepower weapons of science advocacy. To quote Jonathan Shell, who theorized nuclear winter:

> "We have to offer up scary scenarios, make simplified, dramatic statements, and make little mention of any doubts that we might have" (*Discover*, October, 1989).

Environmental Science: Upon These Conclusions Our Facts Are Based

Advocacy is the anathema of science. As students of law learn in forensic debate, advocacy is not objective.

Skepticism, not advocacy, is the heart of the scientific method (Chamberlin, 1965). Accordingly, SOCIETY BESTOWS SCIENTISTS WITH THE PRIVILEGE OF BEING PRINCIPALLY SELF-POLICING SINCE IT IS OUR RESPONSIBILITY TO BE SKEPTICS WHO ESCHEW ATTACHMENT TO THEORIES, CAUSES, AND OTHER FORMS OF SELF-INTEREST IN THE OUTCOME OF SCIENCE. Because it is obvious that self-interest creates conflict-of-interest, rarely are human endeavors so exclusively self-policing as science is.

With the no-holds-barred advocacy of environmentalism, the end justifies the means. Advocacy disguised as science is the most powerful modern weapon of persuasion. People of all political beliefs regard science as being objective "Truth."

Clearly, the growth of scientific advocacy—be it nuclear winter, acid rain, or global warming—is a cancer on science. Such activity is science in name only. When science stops, so stops the progress of modern civilization. And so goes the respect and privileges that society bestows on us scientists.

Global Warming

A call to 1-800-TOO-WARM connects us with the likes of William Shatner (Star Trek), Lloyd Bridges (Sea Hunt) and John Ritter (Three's Company) who tell us about the horrors of global warming. This is just parts of "the Sierra Club informational campaign . . . designed to move that process along with the urgency normally felt as one's feet approach the flame" (*Sierra*, May/June, 1990, p. 19). Unfortunately, there are plenty of "scientists" who are making glorious careers out of also giving us the Sierra Club hot foot.

In the summer of 1988 a NASA scientist, James Hansen, turned the world on its head by stating the climate record of the 1980's shows that it is 99% certain that greenhouse warming is here. Dire consequences of temperature changes were predicted for forests and agricultural crops.

That NASA's own data show no global warming was somehow overlooked by this NASA expert. Strangely, when NASA's data are brought up "greenhouse experts" point out that the NASA data cover too short a time period to establish trends in temperature even though these NASA data go back to the 1970's and are more comprehensive and are of higher quality than the data used by Hansen to start the greenhouse scare in the first place.

Changing temperatures are having a negative impact on forests and crops. But the bad effects are those of increasing cold, not warmth.

USDA climate growing zone maps have required changes to show cooling not warming. For example, citrus crops were successfully grown up to Charleston, South Carolina around the turn-of-the-century. The citrus belt has subsequently shrunk out of South Carolina, then Georgia, then northern Florida, and is now moving south out of central Florida.

Mountain forests are also becoming increasingly more damaged by winter weather. However, increasing cold damage is being blamed on acid rain, not increasing cold.

"Greenhouse experts" do not advertise the fact that the World Meteorological Organization was established in the late 1970's because of concern that we are heading into a new ice age. In fact, given the periodicity of the last 7 ice ages, we are due, or even overdue for the next one.

Assessment of the Problem

We continually hear global warming merits draconian sacrifices to avert global disaster.

Nevertheless, world-class agronomists, geologists and environmental scientists of the United Nation's Intergovernmental Panel on

Climate Change (IPCC) find that if global warming occurs to the degree and extent that the doomsday global warming models predict, it will be of great benefit to the world (Table 1). What these IPCC data mean for the U.S. alone is an increase of $12 billion per year in food production, $30–50 billion per year in water resources, an increase in wood of 80 billion cubic feet, more than $500 billion in wood by 2050.

Archeology and the biological fossil record support the IPCC conclusions by showing that the world was a much better place to live in when it was warmer. That is why archaeologists and natural scientists call the several thousand year warm era prior to the present "Little Ice Age" the **"Climate Optimum."**

However, the IPCC estimates are underestimates of the benefits of global warming. During the Climate Optimum, continental interiors were wetter not drier, as predicted by the models. For example, many of the American Plains Indians were not Plains Indians prior to about 1300 A.D. They were forest Indians who planted corn and hunted elk and deer, not buffalo. When it got colder, it got drier, the forests shrank, the plains and desert expanded. Forest Indians became Plains Indians. And Plains Indians of the Far West left when plains became desert, driving out and killing their neighbors, the cliff city dwelling Indians.

Also, CO_2 is the ultimate plant fertilizer. Over 1,000 laboratory and field experiments show that, on average, a doubling of CO_2 increases plant productivity by about one-third. Commercial greenhouse growers, whether growers of tomatoes or tropical orchids, elevate concentrations of CO_2 in greenhouses to increase yield.

Table 1. The Effects of Doubled CO_2 by 2050.*

Country	Temp. Rise	Food Prod.	Water Res.	Forestry Volume	Shoreline Costs
USA	2.5	+15%	+ 9%	+10%	$25.0
USSR	3.0	+40%	− 1%	+20%	1.0
EEC	3.0	+20%	+ 6%	+10%	25.0
China	2.5	+20%	− 1%	0%	5.0
Australia	2.5	+25%	+15%	0%	2.0
Brazil	2.0	+15%	+17%	− 3%	3.0

*Temperature rise in °C. Shoreline costs in billions assuming a 50 cm. rise in sealevel. EEC is the European Economic Community.

Source — *Climate Impact Response Functions*, September 11-14, 1989, Coolfont, W.V. United Nations Intergovernmental Panel on Climate Change (IPCC) and National Climate Program Office, NCAA.

Science indicates that increases in temperature, moisture, and CO_2 inherent to the global warming scenario will transform the Earth into a Garden of Eden and not a den of death as we are led to believe.

It is frightening to read that even in light of the IPCC's very conservative assessment to the benefits of global warming (if it will even occur), the IPCC is recommending draconian sacrifices to stop this make-believe catastrophe. It is even more frightening to read that the stated fundamental goal of the formal negotiations toward an international convention on global warming is no less than the development of a statist world government which will control every activity of life in excruciating detail (Nitze, 1990).

Those who manipulate science as a tool of persuasion do not respect the sanctity of science. Nor do they respect the sanctity of an individual's right to self-determination. Their belief is that given the truth, we the people do not have the ability to make the "correct" decisions. Only they have the ability to come to "correct" decisions.

They view the present world order—which places power in the hands of the people—as an act in bad faith. They must feel ironic validation of their low opinion of us in using our own institutions to force the world through a form of boot camp in which they break us down and remake us in a new green image.

Environmentalism and Social Responsibility

On the cover of the February 10, 1986 issue of *Newsweek* is a dramatic picture of the space shuttle Challenger exploding in midair with the caption, "WHAT WENT WRONG?"

The Answer—O-ring failure.

The Problem—O-rings worked fine for 15 years until asbestos was taken out of them. Then with the new "safe" asbestos-free O-rings, 4 of 12 shuttle flights showed O-ring erosion. In the fateful thirteenth flight, O-ring failure was catastrophic killing all 7 astronauts aboard the Challenger.

Challenger blew up because of the environmental crusade to take all asbestos products off the market. The EPA considered all forms of asbestos to have the toxicity of the most toxic asbestos mineral, crocidolite, which was used in only 2% of commercial asbestos.

But Challenger is just the tip of the iceberg. Consider the over $100 billion to unnecessarily remove asbestos from buildings and the resulting unnecessary exposures to it. Consider what must be over a trillion dollars in property devaluation because of such perceived hazard and financial liability and what this is doing to our economic vitality, including the S & L's and other financial institutions who hold the mortgages and the EPA-manufactured risk. Now consider the death and disease in Third World countries who cannot use asbestos piping for sanitation because they accept our aid.

Asbestos is one of the many materials that the Environmental party is attacking.

Risk

Starting with the first federal water pollution law in 1948, our government has given itself increasing authority to regulate everything from noise to wildlife. With the passage of RCRA (the Resources Conservation and Recovery Act of 1976) and "Superfund" the Comprehensive Environmental Response, Compensation, and Liability Act of 1980), government now has the authority to regulate literally every substance in existence (Arnold, 1987, p. 15).

Government also has the means to label anything a carcinogen. By definition, a substance shown in any study to cause cancer at any concentration is defined as causing cancer at all concentrations. The definition of carcinogen does not recognize thresholds.

Take a common substance, like quartz, and add massive enough amounts of it to damage lung tissue. Such damage causes the body to respond by inducing rapid cell division to make up for the dead cells. Rapid cell division is a condition which, in and of itself, increases the likelihood of cancer. Just one study out of hundreds showing cancer— or even benign tumors—gets that substance labeled a carcinogen. Negative results do not count.

It was such testing that now allows the EPA to label quartz a carcinogen and rocks containing more than 0.1% quartz as toxic material. Such idiocy also means that metals such as iron, chromium, and selenium are toxic at any concentration—remember threshold values are defined as not to exist. However, these "carcinogens" if deficient result in anemia (iron), diabetes (chromium), and heart disease (selenium).

Thus, the Environmental party has the means to subdue any resource extraction and production activity.

The risk that the Environmental party has is that the public is educated enough to know that common beach sand and iron are not normally cancer-causing agents. However, this risk of offending public sensibilities does not exist for most things because, in our post-industrial society, few of us are directly involved in resource extraction and production. Technical people are also pretty ignorant of the overall picture because increasing specialization is making our expertise thumb-wide and mile-deep. Add to these demographic factors the shaping of public perception by environmental "education" and the risk of offending public sensibilities becomes quite small.

The media/government/green triad of self-interest can prevail over the rational objections of the impacted minority, even one as large as the agricultural community.

The Alar scare is a case in point. The media Alar scare was based on a Natural Resources Defense Council report which asserted that alar sprayed on apples causes cancer in children.

Alar is a plant growth hormone that when sprayed early in the growth of apples promotes the growth of thick stems. Thick stems means that the farmer does little or no late-season spraying of pesticides to fight infections of mites and leaf miners which weaken stems and make apples fall off prematurely. Alar also suppresses the production of a natural pesticide produced by apples—a natural pesticide 40 times more toxic than alar.

In fact, 99.99% of pesticides ingested are natural pesticides, the remaining 0.01% are man-made (Ames and Gold, 1990). Overall:

"The risk of pesticide residues to consumers is effectively zero. This is what some 14 scientific societies representing over 100,000 microbiologists, toxicologists and food scientists said at the time of the ridiculous Alar scare. But we were ignored" (Sanford Miller, Univ. of Texas Health Science Center, *The Detroit News*, February 26, 1990).

Social Responsibility: To Shut Us Down

When the Environmental party began in the late 60's and early 70's corporate social responsibility meant such things as: do you hire the poor and disadvantaged; provide adequate health insurance and retirement benefits, and; provide opportunities for employee advancement through education and on-the-job training. Since then, the Environmental party has learned to use financial clout to directly motivate industry to be "socially responsible" through a series of investment funds, $450 billion (and growing) in government pension funds, and a "green" version of United Way which is supposed to hit at the end of the year.

However, the adds for these funds indicate a new definition of "social responsibility," one which has little to do with people and much to do with sea turtles, dolphins, acid rain, global warming, the ozone hole, and "toxic" chemicals.

This brand of "social responsibility" focuses on minute or "manufactured" risks that are useful in shutting down some resource extraction or production activity while overlooking very real and serious problems.

Such "solutions" often create even bigger risks. For example, fungicides are being banned from use on food. However, the molds and fungi that fungicides control produce some of the most potent poisons known.

There is great concern about minute amounts of man-made pesticide but none about natural pesticides, which can comprise up to 5% by weight (50 million parts per billion) of the food. Calls for breeding of more pest-resistant strains of fruit and vegetables invariably means breeding for greater pesticide content of food. Because natural pesticide is part of the food it cannot be avoided. Most farmer-applied pesticides are avoided by washing and timing of application.

There are calls to spend $ billions on acid rain to prevent a hypothetical risk to about 300,000 acres of high altitude spruce-fir forest. But little is being done for the remaining 18 million acres of spruce-fir forest which are getting chewed up by imported insect pests—90% of the mature fir trees in the Southern Appalachians are dead or dying because of the woolly adelgid and the spruce budworm has eaten millions of acres of red spruce in Maine and many more millions of acres in Canada.

Then there are the 150 million acres of eastern hardwoods that have been assaulted by a variety of imported diseases and pests. Before the chestnut was eliminated from our forests by blight, an enterprising squirrel could have gone from Maine to Georgia on the branches of chestnut trees coming down only to cross rivers.

President Bush planted an elm in Indianapolis this year as a symbol of Earthday. This elm, like the others, will die of Dutch elm disease.

Oak, which has replaced much of the elm and chestnut, is being devastated by the gypsy moth. And now beech bark disease—a "gift" from Canada—is making its way across the country after being introduced at Halifax, Nova Scotia.

If the Environmental party was truly interested in the forest it would not be preoccupied with acid rain.

Environmentalists in the Midwest bemoan the 4,000 acres of acid-dead water which they assume is the result of acid rain. But we hear little from them about the zebra mussel—a "gift" from Norway—which has no natural enemies and threatens to eat the planktonic base of the Great Lakes food chain to the severe detriment of the productivity of over 180 million acres of water.

I submit to you that there is a pattern here.

The real agenda is energy, not forests and fish. Increasingly strict environmental regulation is choking our use of nuclear energy and coal even though we are the world's "Saudi Arabia of coal." Thus, between 1987 and 1989 alone, American utilities increased their oil consumption 34%. But this additional oil must be imported.

Increasingly strict environmental regulation is also choking our production of oil. We were once the world's leading oil producing nation. But no more. Current oil production has dropped to 1942 levels.

The federal government owns over one-third of the country—762 million acres. From the earliest days, U.S. policy has been to dispose of public domain lands and put them into private hands. But these lands, in which lie 50% of our proven energy reserves, are being sealed off from use. Indeed, the federal government has now reversed its philosophy and history by acquiring about 160 million acres of private land. The government is steadily reducing our resource base.

We are being prohibited from exploring offshore for fear of killing seagulls. We cannot drill in the Alaska Wildlife preserve for fear of disturbing birthing caribou.

The Environmental party tells us to conserve energy by doing such things as going to smaller cars. But analysis by the Insurance Institute for Highway Safety shows that for every 1 mpg improvement there is a 4% increase in the accident mortality rate. Air bags help out with direct front end collisions, the least likely type of car accident. Big cars with air bags are safer than small cars with air bags. Put this together with the political stance of "we'd rather fight than drill" and you get the idea that wildlife is more precious than human life.

The Environmental party tells us to go to alternative, clean, non-polluting forms of power generation.

But try putting up a dam. And geothermal plants have stopped in the West and in Hawaii on the basis aesthetics and interfering with volcano worship. The ground under California windmills are littered with the dead birds that fly into the spinning blades.

Is the Environmental party really for solutions? If they were sincere, you would expect cheering from the rooftops at the announcement of an inexpensive, cheap, and virtually unlimited source of unpolluting power. But when such a possibility was announced with low-temperature fusion, the reaction was quite the opposite. Paul Ehrlich (*The Population Bomb*) stated, "the prospect of cheap, inexhaustible power from fusion is 'like giving a machine gun to an idiot child'". Jeremy Rifkin stated that " 'It's the worst thing that could happen to our planet'" (*Los Angeles Times*, April 19, 1989).

The answer to the question is—even if we can achieve the impossible goal of perfectly risk-free, unpolluting activity, it is not good enough. Apparently they just want us to drop dead.

Ron Arnold's assessment of the Environmental party appears to be correct:

> "America's new-found sensitivity to nature came packaged in a strongly anti-industry, anti-people wrapper. It came with a gut feeling that people are no damn good, that everything we do damages nature and that we must be stopped before we totally destroy the earth" (Arnold, 1987, p. 10).

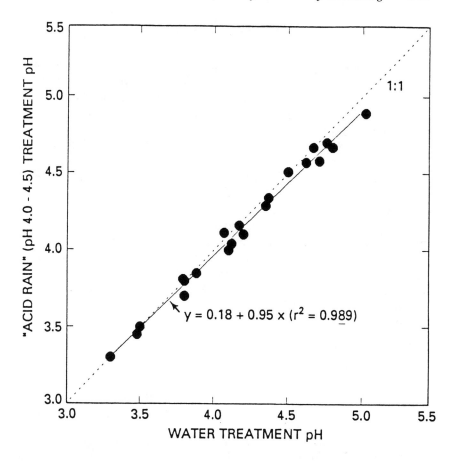

Figure 1. Relationship between the pH of water from "acid rain" and "clean rain" drained or separated from soil. Data from Cronan, 1978; 1980; Abrahamsen and Stuanes, 1980; Mulder, 1980; Wiklander, 1980; Jones et al., 1983; Chang and Alexander, 1984; Wieder and Lang, 1986; David et al., 1989.

What To Do

Businesses must not attempt to take on the Environmental party directly. Regardless of the merit of your arguments, your motives are immediately suspect because you exist for a profit. Secondly, the American public does not yet have the ability to distinguish the environment from the environmentalist.

The private sector has created a "Frankenstein monster." Its support has enabled the Environmental party to become a powerful self-supporting grass-roots citizen advocacy movement. The private sector must stop supporting its mortal enemy.

If we are to maintain a vital economy and a functioning government your support must be switched to the free-enterprise advocates to create a grass-roots citizen support movement.

Your fate is in your own hands. You have friends but we need your help. The free-enterprise organization I support is: CFACT (Committee for a Constructive Tomorrow), P.O. Box 65722, Washington, D.C. 20035. Phone: (202) 319-0104.

Additional Reading

Ames, B.N. and L.S. Gold. 1990. Too many rodent carcinogens: mitogenesis increases mutagenes. *Science* 249:970–971.

Arnold, R. 1987. *Ecology Wars*. Free Enterprise Press, Bellevue, WA. 182 p.

Brady, N.C. 1974. *The Nature and Properties of Soils, 8th Edition*. MacMillan Publishing Co., Inc., New York. 639 p.

Chamberlin, T.C. 1965 (republished from *Science*, 1890). The method of multiple working hypotheses. *Science* 148:754–759.

Detroit News. 1990. The environment: risk and reality. Editorial Page Reprint, Special Issue, February 25, 1990—March 11, 1990. The *Detroit News*, c/o Editorial Page, 615 Lafayette Blvd, Detroit, MI 48226.

Krug, E.C. 1989. Assessment of the theory and hypotheses of the acidification of watersheds. Illinois State Water Survey Contract Report 457. 252 p.

Krug. E.C. 1990. Fish Story. *Policy Review* (50):44–48.

Nitze, A.W. 1990. A proposed structure for an international convention on climate change. *Science* 249:607–609.

Rosenqvist, I. Th. 1990. From rain to lake: pathways and chemical changes. *J. Hydrol.* 116:3–10.

Root Causes of Extremist Activity

Max Singer, J.D.

Max Singer is president of The Potomac Organization, Inc., Chevy Chase, Maryland. He is a founder and former president of the Hudson Institute.

He received a B.A. in social sciences from Columbia College in 1953, and a J.D. degree from Harvard Law School in 1956.

Singer is the author of *Passage to a Human World*, published by Hudson Institute, and in paper by *Transection* in 1989. He has published more than 10 journal articles and has contributed to *Public Interest, National Interest,* and *Commentary.*

Issues That Go Beyond Environmental Risks

There is another set of perspectives that is crucial to the way we deal with subtle risks—and with other issues as well. Although these perspectives are rarely discussed explicitly, they underlie much of the debate as well as the reactions of many people to subtle risks and a wide range of issues. Therefore it is important that we formulate and consider this set of influential perspectives.

Safety vs. Profit

If the choice is between safety and profit, the public interest is clear. For many people this formulation leaps to mind because of the ideas they have about profit, business, and the American experience.

The following are some of widely accepted partial truths that influence public and political attitudes about risk policy decisions: The purpose of production, construction, and other business activities is to make a profit. When these activities create risks, society must balance these risks against the corporate interest in its right to make profits. In a dispute that is too complicated and technical to understand, the only way one can decide which side to be on is to evaluate the people or organizations on the two sides. If the people on one side are paid by corporations and are motivated directly or indirectly by a concern for corporate profits, and the people on the other side represent the public interest in safety either through the government or a public interest group of some kind, then it is better to side with safety against profit.

One of the major reasons why many people in the United States and around the world find it difficult to feel good about the United States is that much of American life is supposed to be based upon the profit motive. Much American activity at home and abroad is carried on by large corporations, which are also supposed to be organized and operated according to the profit motive. [1] The profit motive is selfish. In view of our wealth and all the problems and suffering in the world, how can one think well of people who dedicate their lives to getting richer? How can one expect decent behavior from an organization devoted to accumulating more wealth? How can one feel committed to defending, supporting, and getting moral and emotional sustenance from allegiance to a country dominated by a selfish pursuit of profit? [2]

"Good Guys" vs. "Bad Guys"

Many people in their personal life (including business) make it a practice to rely on people they can trust. They want to hire, work for, or do business with people they feel are on their side (or at least share their basic values): members of the family, friends, alumni of the same school, people from the same neighborhood, members of the same national or religious, age, or sex group; or whatever. It is an everyday experience, producing much folk wisdom, that one cannot understand all the issues and know all the necessary facts, so one decides by lining up with or following the advice of those whom one trusts. And trust of this kind is based primarily on judgment about motivation. For many people it is very hard to understand how people and organizations who admit they are motivated by profit can be doing things that benefit somebody else and are in the public interest. Abroad, and even among some Americans, that wisdom leads to great suspicion of the United States.

These attitudes are important for risk policy because a large share of the actions that raise risk issues are actions that are taken by big corporations or by government. American government, particularly the branches of it that do things and build things, doesn't get much more trust than big business in many circles in America. So, often, risk issues come up in the context of a government agency or corporation wanting to do something and being challenged by non-profit organizations resisting action on the grounds that it is too risky. In these situations many people decide their vote by choosing between the "good guys" and the "bad guys."

A few words are necessary about this good guys-bad guys perspective. It is possible to talk in a high-minded way about balancing conflicting interests and increasing risk efficiency, etc.; but if we are going to be realistic it is important to recognize that the public is right: there are bad guys. They can be defined as people who (1) selfishly pursue their own interest without due regard to the interests of anyone else, and/or (2) try to get what they want by using improper means, such as lying, bribing, forging documents, or otherwise deliberately using phony evidence and arguments they know to be specious.

There are two problems about voting for the good guys. First, the good guys can be on either side of the risk issue [3] even if it is between a corporation and a public interest group. Civil servants, politicians, and public interest group leaders also have personal interests that can be pursued as selfishly and cynically and with as much disregard for the interests of others as those of corporate profit-seekers. They are also equally capable of using illegitimate tactics.

More fundamentally, it is not necessarily true, even in a case where the good guys are one side and the bad guys on the other, that it is in the public interest for the good guys to win. There can be a case, for example where a corporation wants to sell a product that is in fact safe, where they try to get the sale approved by using improper methods without due regard for assurance of its safety, and are opposed by genuinely public-spirited groups genuinely concerned about the safety issue. It is quite possible for there to be a genuine, even careful, concern about the danger from a product which is in fact safe. In such a case the public interest is for the product to be allowed even if the good guys are against and the bad guys are for it. (The opposite situation is of course also possible.)

Punishing the Guilty vs. Helping the Needy

Another common attitude that hurts policy making about risks is the preference for condemning and trying to punish the guilty rather than helping the needy. This attitude is more an undertone, less central to most of the risk policy debates than the ones discussed above. Essentially the same spirit (greater enthusiasm for punishing than helping) influences a variety of policy issues, such as tax policy, policy about human rights violations in other countries, concern about welfare cheats, anti-discrimination programs, and many other policy areas. (On the other hand, despite the fact that concern for punishing the wicked can be carried to such excess that it hurts those it is designed to help, such punishment often serves important public functions, such as deterrence or expression of public values, even when the immediate victim is not helped.)

Fundamentally, risk policy decisions are a matter of balancing—conflicting evidence, advantages and disadvantages, and various costs and risks to different parties—so when the problem is approached as a search for villains its essential nature is distorted.

Of course there are villains, of various kinds and degrees; and a single-minded search for villains and for intentionally hidden dangers is a necessary part of the process of protecting the community. But carrying this spirit to excess produces three kinds of costs to the community. First, it can lead to wrong decisions, that is decisions that sacrifice too much benefit to the community because of excessive attention and concern about potential harms and suspicion that the advocates of action conceal or don't understand the harms. Second, it can interfere with relationships in the regulatory and decision making process. It gets harder for the producers to cooperate with the regulators if the producers feel that the regulators see them as villains and are pursuing them vengefully, and that therefore they need to protect themselves. The process becomes healthier and more effective the more each group sees the other as having an equally respectable role to play in a common community effort to reach the best balance of action and risk avoidance. This view is of course too idealistic to be fully achieved. But it is the right ideal to hold up, and tendencies that too strongly fly in the face of this ideal and weaken it more than necessary are harmful. Third, operating the policymaking process in a way that teaches the community that villainy is a large part of the risk problem, and that there are lots of villains, hurts community morale.

Wallowing in Villainy

The sense of villainy is a persistent undercurrent in modern social science and policy debate. [4] There is a strong tendency to write with anger and accusation, and to assume that most troubles—like poverty or pollution or war—are the result of misdeeds by greedy businessmen or by governments that are not "progressive."

The health and safety branch of the environmental movement in the United States has demonstrated the result of the predisposition to focus on villainy. For many years they have missed one of the most dangerous environmental hazards—radon seepage into homes—because it is nobody's fault. Radon comes from nature not from industrialists.

Over the last twenty years both the government agencies, and the non-governmental "watchdogs," responsible for protecting people from environmental hazards have created scores of minicrises about manmade "hazards" in our food supply, work place, and the air we breathe. Tens of billions of dollars have been spent evaluating and debating, and then removing "hazards" whose casualties have been either theoretical or probably less than a few thousand people a generation. (Of course the environmental-industrial complex has also done useful work against a few substantial killers.) It took too long for the government to recognize the threat of radon in homes, although, by the government's usual methods of estimating risk, radon is much more dangerous than most of the "hazards" pursued at great fuss and expense. Perhaps the reason the radon problem took so long to discover is that there was no villain and the solution was so simple.

The tone of the discussion of dangers to human life—as well as most of its content—gives the impression that modern science and industry, abetted by business villainy and government failure or corruption, is making our lives more dangerous. But if we ask what are the substantial killers in our lives today, the answers are: smoking, drinking, driving, drug use, and unhealthy lifestyles. Compared to these hazards, nuclear reactors, hazardous waste dumps, all artificial food contaminants, and all work place hazards, together produce tiny amounts of human harm. But these political targets produce immense amounts of accusations, bad feeling, legislation, interference with productive activity, and expenditures.

The conclusion seems to be that those who claim to be trying to protect us from environmental hazards are too bound up in the search for villains to do a good job. [5]

The field of economic development is as misdirected as the field of protection against environmental hazards. The recurring themes are: lack of progress and hopelessness, insufficient foreign aid appropriations, disastrous population growth, exploitation by the advanced countries and by multinational companies, remnants of colonialism, unfair terms of trade, and crushing foreign debt.

What are the basic truths? The background is tremendous progress and almost certain success—although perhaps fairly slowly. The real obstacles are the normal resistance of human beings and human societies to change, the difficulty of maintaining reasonably stable political environments, and—most significant—the strong tendency of governments to pursue harmful policies. (Not only policies that deliberately sacrifice the nation's welfare to the interests of those in power, but perhaps even more often, policies chosen as a result of misunderstanding.) Again the community of experts and idealists are led astray by their basic view of the world.

The problem is that the real solutions are dull. Poverty is abolished by people learning how to work better, producing per capita growth rates of 2% per year for a century. The first steps to industrialization are rural development, which results from farmers being allowed to sell their crops for their market value. The most important jobs of the government in speeding this process are: keeping peace, providing a legal system that permits contracts to be made and enforced, maintaining a stable currency and building transportation and communication infrastructure. Its toughest task is to limit the number of government employees, redtape, and corruption. But there does not seem to be any romance in these useful efforts. Idealists, especially young ones, are too often captivated by pursuit of dramatic successes and of villains who are thought to be blocking progress—even if progress is not blocked at all, but coming at quite a reasonable pace.

The biggest enemy of poor people may be the reluctance of idealists and intellectuals to work for dull ideas.

Footnotes

[1] We will not here get into the question of how much corporations are really dominated by the profit motive, nor will we address the deeper question of the extent to which, in fact, people who do actively seek profit act out of a selfish spirit. George Gilder, for example, argues that entrepreneurs are the true creators and that creation involves giving, or at least generosity of spirit. I do not wish to take a position here on any of these issues. Here we are talking about the feelings of those who find the profit motive distasteful, and it doesn't matter too much whether they are accurate in their understanding of those they dislike. In presenting the view of the distaste-for-profit people I do not endorse them.

[2] This is not the place to deal with this set of ideas. In brief they have two problems. First they are terribly crude about the relationship between selfishness, profit, and the system. Second, they ignore the immense gains for society and people from the use of profit.

[3] For a nice example of Ralph Nader's understanding of excessive or mistaken concern for safety, see his defense of sodium azide quoted in the Wall Street Journal, Sept. 20, 1977, p. 20 (cited in Julian Simon's The Ultimate Resource, Princeton, 1982).

[4] I should use the modifier "American" because that is the only literature with which I am at all acquainted, but I don't want to assert that American writers and thinkers are worse than their foreign counterparts.

[5] Ironically, in the one major area of policy where something very like villainy is at the heart of much of the problem—the role of Soviet Union in international politics—the same intellectual mainstream rejects villainy as a subject of serious attention. There is contempt for a President who talks about the "evil empire." And anti-communism is called "conservatism," or "hard-line, right-wing ideology," and regarded as outside the bounds of sophisticated discussion.

Adapted from a section of the author's book,
Passage to a Human World

A Different Look at Environmentalism

Jay H. Lehr, Ph.D.

Please see biographical sketch for Dr. Lehr on page v.

First Things First

I'd like to start this article by making a few things abundantly clear. I am an environmentalist. By virtue of your jobs, you are all environmentalists. None of us see redeeming qualities in contaminating air, water, or soil. We all want pollution reduced and eventually eliminated in a cost-efficient manner. No water professional could see any value in allowing chemical contaminants to be foolishly, unwittingly, or maliciously placed in our atmosphere or biosphere. Through education and regulation, we will all work to curtail these activities. We all want to methodically clean up what can be reasonably cleaned up; in fact, most of us will earn our livings in this endeavor for the better part of our professional lives. The smarter among us will guide our children into similar career pursuits.

Having said this, and with no desire to waffle on the issue or be disingenuous, it is time to stop scaring the !?#*!?! out of the American public with fraudulent, malicious, and totally inaccurate risk assessments of the latest environmental screwups. Neither apple polish, cow-antibiotics, gas emanating from rocks, mundane landfill leachate, leaking tanks, chemical spills, or the ubiquitous industrial degreasers pose an imminent or terminal threat to our lives or, in most cases, even to our daily health. But the public, inadequately schooled in science and totally untrained in risk assessment, is being held captive by an "environmental industry" reaping its economic and political rewards by malicious overstatement. The American media's sales and rating-point incentives encourage many reporters to walk in lock-step with those who would enslave our population in needless fear.

We have just witnessed a miracle in our lifetime as the people of eastern Europe have thrown off the yoke of tyranny and tasted the fresh air of freedom. Do not take lightly our own subtle enslavement by an overzealous environmental industry that would have us return to a life of cowering under one environmental threat after another.

369

Today, the environmental industry employs tens of thousands of talented people making excellent salaries, no longer in the mold of their once spartan, stoic leader Ralph Nader. They have the support of 20 million U.S. citizens (up from only four million a decade ago), most of us being counted among them, who contribute to their various causes. They have a business strategy as clear as any Fortune 500 company and major financial support from many of those companies.

Of course, access to big money doesn't discredit global environmentalism. In fact, it ought to make the movement more honest. Environmentalists pretend—and much of the media promote the pretense—that they are somehow above financial self-interest and possess an inherent moral superiority to those who fail to share their vision. They are, however, an interest group who, like any other interest group, work hard to promote their perquisites and politics. Motherhood and apple pie just happen to appear to color their every move.

On the surface, the goals of the environmental industry are very reasonable: who could be against them? All of us desire to live in clean surroundings, breathe uncontaminated air, drink potable water—it would be difficult to fault their objectives. But it is just these objectives that "successfully conceal the underlying trend toward an all-encompassing political dictatorship wherein the individual's desires are subordinated to the state to the point where his freedom of choice will be reduced almost to zero" (Kazmann 1992). It is time to throw off our naivete about those who profess to be acting in our best interests.

With all the good—no!—fantastic news coming from the world in matters of politics, the media, ever believing that ultimately only the worst news sells, are redoubling their efforts to show gloom, doom, and outrage at industry's manhandling of our water, air, and food. A day rarely passes that a dour newscaster does not attempt to scare the wits out of viewers with the latest "possible carcinogen" found in *trace* amounts somewhere in our environment.

After an intensive study of the first decade of cancer prevention in this country, Edith Efron, in her book, *The Apocalyptics* (1984), concluded that the antiscience and antitechnology trends in our liberal society have distorted environmental cancer research and saturated society with theories of cancer causation and prevention that are pure myths. Her book contains explosive information about scientific misrepresentation as well as manipulation of the press.

Efron points out that for more than 20 years, apocalyptic scientists have focused on predictions that the modern industrial system will ultimately bring about the destruction of life on earth. A major form of that destruction will be cancer, they say, and proceed to make their case by ignoring data (some from as far back as the early 1960s) which contradict their prophecies.

The American press has been nourished on bad science since the inception of our cancer prevention program. It was trained, as one trains a circus dog, to view doomsayers in and out of government as fountainheads of scientific truth. The media has shown a consummate credulity in the face of arbitrary edicts brandished by the policymakers as the voice of science; it has been taught by scientists seeking the spotlight to treat secretive or baseless assertions from scientific sources like scoops. The press doesn't seem to know that science does not operate by assertions, by leaks, by off-the-record briefings, by Xerox machines spitting out documents with release dates geared to the evening news, or by documents with no authors' names at all. No one has taught the press that the appearance of such phenomena means that what one is hearing is not science at all.

Undoubtedly, some journalists have been idealogically receptive to the "apocalyptic axioms." Some have enjoyed the excitement of a new kind of war between good and evil. Some have unquestionably seen themselves as righteous adjuncts of the regulatory process while most have never known the meaning of the scientific words they were so eagerly transmitting to the public. I cannot, however, indict the press for failing to understand what it takes years to learn. To write this series of editorials, I had to review over 200 papers and a dozen books. Reporters must write swiftly; they often cover the daily news and may have insufficient time to do their "homework." The inadequacy of the media coverage is more correctly the inadequacy of their informants. Above all, it is the inadequacy of thousands of scientists who have been fully aware that this country has been fed politically corrupt science but have remained silent in the face of the cultural malpractice of their colleagues—perhaps out of self-interest but, unquestionably, out of fear and helplessness.

The Media

Nationally renowned author Ben Wattenberg offers valuable insight about the popular environmental movement in his book, *The Good News Is The Bad News Is Wrong* (Wattenberg, 1984).

> "In terms of specifics, the environmental viewpoint is often valuable as a comprehensive vision of our time; however, it is in my judgment, typically both wrong and damaging. . . . No sane people are antienvironmental; we are all for improving the quality of life. No one proposes to swim in polluted waters; no one gobbles carcinogens just for the hell of it; and surely, some of the alarm bells that have sounded point to real problems.
>
> But that is not the point. The point is that the environmental movement has argued that the overall quality of our lives is poor and getting worse. Such a view is incorrect."

Why are the media so eager to give us bad news? Why do the same scientists confirm each apparent disaster? Why aren't mainstream American scientists so outraged that they contact the media? There are no simple answers to any of these questions. Sensationalized bad news is apparently what the media think the public wants. George Will, *Newsweek* columnist, once said, "Only man is perverse enough to feel most alive when the news is most lurid."

The news media thrive on conflict. Never mind if the fight is between a Nobel Laureate in geophysics and the president of the Flat Earth Society; credentials mean little to the media. A fight is a fight. Environmental reporters are often in too much of a hurry to listen to the pros and cons of an issue. The journalists, under pressure, want the scientists who can offer the best one-liners, regardless of the fact that ethical scientists describing serious work will rarely risk over simplification or quick summaries. Furthermore, environmental alarmists appear to be experts on everything, willing to pontificate on nearly any subject while most scientists are specialists, rarely willing to speak authoritatively outside their field.

Our best role models are the men and women of the U.S. Centers for Disease Control (CDC) in Atlanta, Georgia—arguably this country's most reliable and accredible scientific agency. As scientists, we should get in the habit of accepting cancer analyses and related statistics only when they are determined by the Center of Disease Control. The CDC staff, some 5200 strong, are remarkably dedicated and known widely for its lack of turf battles and budget fights that so often disrupt the work of many other federal agencies. A CDC official recently described his agency as a place that allows one to keep his or her idealism well into middle age. State and local health officials who assist the CDC through cooperative grants also revere the agency and their fair and comprehensive efforts. Unfortunately, their thorough studies, that so frequently diffuse environmental crises, take years to complete, and ultimately receive page 19 coverage from the press. Such has been the treatment of their information concerning compounds like EDB, PCB, dioxin, and many other targeted, terrifying substances. Their pronouncements of the scant or absent evidence supporting the public terror is invariably buried where few readers go.

Risk and Fear

When presented in dramatic form with skilled, passionate rhetoric, risks from hazardous wastes, chemical pesticides, and other environmental dangers can certainly generate outrage. Once the outrage is created, the objective facts—the actual dangers under alternative policies—become much less important. Extreme measures to protect the public from a single danger seem justified.

To be sure, we recognize that environmentalists do ask important questions about the ethical, economic, and social implications of new technologies. The problem is that because they commonly present their case in such a shrill and, at times, unscrupulous manner, the debates they encourage are filled with fear and anger rather than information and reasoned judgment.

Increasingly, regulation of chemicals is being governed by political responses to public fear and hysteria rather than by careful, objective evaluations of the actual risks and benefits posed by the chemicals.

"Risk" is often interpreted as bad—something to be avoided. Yet all economic and technological progress requires that human beings take risks. It is precisely because our ancestors took risks that we enjoy healthier, longer lives than they did. Aaron Wildavsky in his book, *Searching for Safety*, persuasively argued that "there can be no safety without risk" (1988).

Immunization against childhood diseases is a good example. Each year, three and one-half million children receive vaccines against whooping cough, diphtheria, and tetanus. Twenty-five thousand of these children come down with high fevers; 9000 of them become seriously ill; 50 are brain-damaged, and as many as 20 may die (Wildavsky, 1988). And yet, quite correctly, our public health services find these risks preferable to the frequent incidents of disease and death that would occur were these vaccines not administered.

Frightened people truly suffer due to distorted information. For the most part, they do not know what to do about their situation other than to modify their lifestyles to the extent possible. They can live without smoking, for example, but they cannot live without breathing—no matter how great they are told the imminent threat of air pollution may be.

Fear can be a healthy response when the danger is real and when fear prompts us to proper action. But at least in relation to chemical technology, fear is often of the irrational, phobic variety. It is the outcome of information that is inaccurate, exaggerated, or presented in a way that is calculated to alarm.

A magazine showing an artist's fanciful rendition of a human head sinking into a cesspool and losing his skin on the way down does not serve the public well. It is easy to understand, however, that reporters want their stories to be as close to the front page or the beginning of the newscast as possible. These reporters have a vested interest in catastrophes and seem to find them with ease when chemicals are involved. But the role of the reporter should be comparable to that of the person assigned to inform people of a fire in a theater. Jackson Browning described it well when he said, "That person becomes a risk communicator with all the responsibility that the phrase should imply" (1987). If the person fails to inform people of the danger, he fails in his role. If he goes into the theater and frightens people by screaming at the top of his lungs, he also fails. The objective should be to achieve an orderly withdrawal from the fire in a method that suits the situation. Excited, sensational reporting has no place when people's safety and welfare are concerned.

In general terms, chemical companies who may be guilty of contamination tend to err by saying too little, while the media err by saying too much. Most people are intelligent and perceptive individuals who, even without scientific education, are completely capable of understanding scientific facts relating to subjects that are vital to their health and well-being.

Bruce Ames (1987) has labored intelligently to translate the relative meaning of risk in a myriad of substances ingested on a regular basis by the human population. He developed a simple, yet undisputed analytical system, abbreviated HERP for daily HUMAN EXPOSURE dose/RODENT POTENCY dose. He takes the estimated daily dose of a chemical that will cause cancer in one-half of a group of test animals and compares it with the estimated daily dose that humans would receive of a given chemical. The result is a percentage that gives the carcinogenic danger of the chemical. Table 1 in the Appendix to this editorial offers a significant list of comparative HERP indices using one liter of chlorinated tap water as the base of one against which all other substances are compared. Table 2 compares the relatively trivial risk of a number of everyday activities, and Table 3 describes those activities in which a serious number of lives are lost each year.

The data and evidence cited in these tables are not secret. They are available to everyone in the scientific literature, but the legislative process at both federal and state levels seems to have given little weight to these facts. Instead, R.L. Stroup and J.C. Goodman of the National Center for Policy Analysis (1989) described it thus: "The outrage of citizens, uninformed about toxicology and swayed by articulate and passionate rhetoric condemning each potential danger—usually without regard to the problems of alternative courses of action —has led from the Love Canal tragedy to the Superfund fiasco, and from largely phantom carcinogenic chemical threats in California to Proposition 65."

Public Interest

The alliance of attention-hungry, public-interest groups and sensation-seeking media is a very powerful one. As consumer groups have become more adept at manipulation, the media have become more

Table 1. Risk of Getting Cancer (Relative to Drinking Tap Water)

Relative risk[1]	Source/daily human exposure	Carcinogen
	Water	
1.0	Tap water, 1 liter	Chloroform
4.0	Well water, 1 liter (worst well in Silicon Valley)	Trichloroethylene
	Risks Created by Mother Nature	
30.0	Peanut butter, 1 sandwich	Aflatoxin
100.0	Mushroom, 1 raw	Hydrazines, etc.
2,800.0	Beer, 12 oz.	Ethyl alcohol
4,700.0	Wine, 1 glass	Ethyl alcohol
0.3	Coffee, 1 cup	Hydrogen peroxide
30.0	Comfrey Herbal tea, 1 cup	Symphytine
400.0	Bread, 2 slices	Formaldehyde
2,700.0	Cola, 1	Formaldehyde
90.0	Shrimp, 100 g	Formaldehyde
9.0	Cooked bacon, 100 g	Dimethylnitrosamine, diethylnitrosamine
60.0	Cooked fish or squid, broiled in a gas oven, 54 g	Dimethylnitrosamine
70.0	Brown mustard, 5 g	Allyl isothiocyanate
100.0	Basil, 1 g of dried leaf	Estragole
20.0	All cooked food, average U.S. diet	Heterocyclic amines
200.0	Natural root beer, 12 oz. (now banned)	Safrole
	Food Additives and Pesticides	
60.0	Diet cola, 12 oz.	Saccharin
0.4	Bread and grain products, average U.S. diet	Ethylene dibromide
0.5	Other food with pesticides, average U.S. diet	PCBs, DDE/DDT
	Risks Around the Home	
604.0	Breathing air in a conventional home, 14 hrs.	Formaldehyde, Benzene
2,100.0	Breathing air in a mobile home, 14 hrs.	Formaldehyde
8.0	Swimming pool, 1 hr. (for a child)	Chloroform
	Risks at Work	
5,800.0	Breathing air at work, U.S. average	Formaldehyde
	Commonly Used Drugs	
16,000.0	Sleeping pill (Phenobarbital), 60 mg	Phenobarbital
300.0	Pain relief pill (Phenacetin), 300 mg	Phenacetin

Source: Bruce N. Ames, Renae Magaw, Lois Swirsky Gold, "Ranking Possible Carcinogenic Hazards," *Science*, Vol. 236, April 17, 1987, pp-271-236.

[1]The underlying measure of risk used here is a HERP value: Human Exposure dose divided by Roden Potency dose. The measure of rodent potency is the milligrams of substance per kilogram of rodent body weight necessary to produce cancer in one-half the rodents, given daily exposure over the rodents' lifetime. Human exposure is measured by the daily consumption indicated in the table per kilogram of human body weight. In the table above, the HERP values have been normalized with respect to the HERP value for water.

Table 2. Risk Which Increase the Chance of Death by One in One Million

Activity	Cause of Death
Smoking 1.4 cigarettes	Cancer, heart disease
Drinking ½ liter of wine	Cirrhosis of the liver
Living 2 days in New York or Boston	Air pollution
Living 2 months in Denver on vacation from NY	Cancer caused by cosmic radiation
Living 2 months in average stone or brick building	Cancer caused by natural radioactivity
Traveling 6 minutes by canoe	Accident
Traveling 10 miles by bicycle	Accident
Traveling 300 miles by car	Accident
Flying 1000 miles by jet	Accident
Flying 8000 miles by jet	Cancer caused by cosmic radiation
One chest x-ray	Cancer caused by radiation
Eating 40 tablespoons of peanut butter	Liver cancer caused by aflatoxin B
Drinking Miami drinking water for 1 year	Cancer caused by chloroform
Drinking 30 12-oz. cans of diet soda	Cancer caused by saccharin
Eating 100 charcoal-broiled steaks	Cancer from benzopyrene

Source: Richard Wilson, "Analyzing the Daily Risks of Life, *Technology Review*, February, 1979, p.45.

Table 3. Annual Fatality Rates per 100,000 Persons at Risk

Activity/event	Death Rate
Motorcycling	2,000
Aerial acrobatics (planes)	500
Smoking (all causes)	300
Sport parachuting	200
Smoking (cancer)	120
Fire fighting	80
Hang gliding	80
Coal mining	63
Farming	63
Motor vehicles	24
Police work (non-clerical)	22
Boating	5
Rodeo performer	3
Hunting	3
Fires	2.8
1 diet drink per day (saccharin)	1.0
4 tbs. peanut butter per day (aflatoxin)	0.8
Floods	0.06
Lightning	0.05
Meteorite	0.000006

Source: Adapted from E.L. Crouch and R. Wilson, *Risk/Benefit Analysis* (Cambridge: Balinger, 1982). Reported in Paul Slovic, "Informing and Educating the Public About Risk," *Risk Analysis*, Vol. 6, No. 4, 1986, Table 1, p. 407.

willing accomplices in publicizing the latest scare. The newsmen and women who consider themselves skeptics when dealing with almost all other sources of information, and particularly when dealing with big business, appear to accept the pronouncements of "public interest" groups and "experts" with the flimsiest of qualifications as though they were written on tablets of stone. No better example can be found than actress Meryl Streep. Sadly, if one person in one million dies from the ingestion of a product, or even increases their risk of contracting cancer because of something in the product, that one person becomes a news story. The other 999,999 who use the product safely are not bad news. This simple truth almost inevitably leads to an exaggeration of the problem, both by the media and thus in the minds of the public.

Fifty-one years ago, a radio play about giant, slimey, lizard-like creatures from Mars sent the U.S. into a panic. Churches all over the country filled with people praying. Cars jammed highways in a desperate race to get our of the towns. Looters appeared from nowhere. Men took out their shotguns and lined up to protect their homes. The radio play was Orson Welles' famous *War of the Worlds.* The show was completely outrageous, but because it was presented like a news broadcast, the public believed every word. Fortunately, Mr. Wells and CBS were back on the air offering apologies before too much damage was done. The great disaster became a great joke, something people would tell their grandchildren.

This story is relevant today because it so closely parallels the action on another Sunday night CBS broadcast which invaded our public consciousness. I refer to the "60 Minutes" broadcast that set off the Alar debacle. Unfortunately, by the time our leading scientific agencies got together to assure people that our world wasn't coming to an end, it was too late. The concerns about children, apple juice, cancer, and pesticides had gained a momentum that still has not entirely subsided. It is unfortunate how single-mindedly the American people tend to believe what they see on television and hear on the radio, no matter how much it goes against common sense or their personal experience.

Alar and the NRDC

The NRDC (Natural Resources Defense Council), a New York based group with a multimillion dollar budget and a long record of lobbying and litigation on subjects ranging from air pollution to nuclear energy wrote, produced, and directed our nation's epic Alar scare. With the help of a public relations agency the NRDC had, for months, been orchestrating a big publicity program for its report "Intolerable Risk: Pesticides in Our Children's Food," (Sewell and Whyatt, 1989) which was released early last year.

The NRDC gave CBS an exclusive on the report which was not released until the day after "60 Minutes" aired. That served two purposes: (1) it kept the report out of the hands of scientists and medical experts who later found stunning flaws and misrepresentations in it; and (2) it made reporters anxious to get the report and do a story on it as soon as it was released following the broadcast. The media totally lost sight of basic journalistic standards in an effort to produce their stories. While CBS gave great play to the NRDC report, it did not give independent scientists an opportunity to either examine or evaluate it.

The hysteria created by NRDC and "60 Minutes" had authorities pulling apple products from schools around the country and brought the sale of apples to a virtual standstill.

What makes all this more nauseating is the fact that to reach the exposure level that produced an ill-effect in laboratory animals, a person would have to consume thousands of pounds of Alar-treated apples every day for 70 years! Yet it took three federal agencies—EPA, FDA, and USDA—to calm the panic over apples and apple products, only to have it fueled again a month later by cover stories in both *Time* and *Newsweek*.

The report allegedly contains scientific proof that the use of eight common pesticides on fruit and vegetable crops was responsible for causing cancer in the nation's preschool population. Not surprisingly, the national news media reacted with an almost religious fervor to this dire prediction. Unfortunately, what was overlooked in the media hype was the lack of scientific evidence in the report itself. When toxicologists and pharmacologists began to take a close look at the NRDC report, their response was virtually unanimous. The report contained serious scientific flaws and factual errors that made it difficult to take its conclusions seriously.

One of the strongest allegations against the report is that it was not adequately peer-reviewed. The lack of proper peer-review led to basic inaccuracies making it impossible for scientists to reconstruct or fairly evaluate NRDC's risk assessment model. The numbers used in the NRDC's report were poorly referenced, says former EPA official, Dr. Christine Chason (personal communication, 1989): "If they had submitted this report to any peer-review group, they would have gotten slapped on the wrist for not being able to track back to the numbers used," she said. NRDC also assumed that the chemicals were genotoxic (capable of causing genetic damage). It turned out that most of the pesticides they studied were, in fact, not genotoxic. This was a serious flaw in their whole risk assessment procedure and led to a gross overestimation of risk.

A Look at Serious Science

About the time Ms. Streep and her friends had school superintendents throwing apples in dumpsters all across the country, other scientists were publishing reports and studies on the same subject. The National Research Council, an arm of the National Academy of Sciences, released an exhaustive study (1989) of the available data on disease risk and the American diet. It recommended that to cut the occurrence of cancer and heart disease Americans ought to eat more fruits and vegetables. As to the chemicals that obsess the NRDC, the report said, "Exposure to nonnutritive chemicals individually in the minute quantities normally present in the average diet is unlikely to make a major contribution to the overall cancer risk to humans in the United States."

Somewhat earlier, the science magazine *Nature* published an article by researchers from Carnegie Mellon, Case Western Reserve, and the University of Washington raising questions about the usefulness of rodent bioassay studies. "Extrapolating from one series to another," they wrote, "is fraught with uncertainty" (Lave et al., 1988). However, none of these serious scientific revelations slowed the NRDC in their race to the press. Evidently, today, the end justifies the means.

I have reprinted here some of the equations used in the NRDC report to attempt to extrapolate cancer risk in humans from cancer risk in rodents. "The risk in humans will be equal to that in rats if their exposure in $MG/KG^{2/3}$ equals the rat exposure in $mg/kg^{2/3}$. Thus, for these circumstances:

$$R \text{ (human)} = R \text{ (rodent)} = q_1 \text{ (rodent)} \times MG/KG^{2/3} \times 1/kg^{1/3}$$

or:

$$R \text{ (human)} = q_1^* \text{ (rodent)} \times MG/KG \times KG^{1/3}/kg^{1/3} = q1^* \text{ (human)} \times MG/KG$$

if

$$q_1^* \text{ (human)} = q_1^* \text{ (rodent)} \times K/G^{1/3}/kg^{1/3}.\text{"}$$

These equations appear in the NRDC's report as part of their Appendix 3, "Methodology for Estimating Pre-School or Cancer Risk from Carcinogenic Pesticides in Food." Appendix 3 didn't get much mention when the public, nodding off in front of televisions all across America, first learned from TV news about the problem with apples ("Cancer in Apples?—More at 11"). It's probably asking too much to expect news anchors to take an extra 20 seconds to also discuss the fact that the study is based on something known as a "lifetime rodent bioassay." It wouldn't be too much, however, to ask everyone who conveys the results of these food-cancer studies to put the rats equals humans arguments into some perspective. In other words, maybe public policy would be better served if the public were given the actual facts to consider, rather than conclusions that encourage panic.

What Can You Do?

The point of all this is that most of the information on risk assessment is funneled through the media, local news sources more than national ones. Most local reporters have little knowledge of, or background in, technical matters. Most of them tend to parrot things they are told. They primarily look for victims. They look for the smoking gun, the body count, the rocket's red glare. You have seen it all a hundred times, and likely you will see it a hundred more. Therefore, YOU have to educate the media. YOU have a responsibility to become a participant. The media must not view the environmental groups any differently than they would an industrial group. They don't accept press releases from industry as being the gospel, and they shouldn't accept reports from the NRDC, EDF, NWF, the Sierra Club, or the Audubon Society any differently.

When the chemical industry spokespeople talk openly about risk and elaborate their outstanding safety record, or the program they have in place to deal with emergency situations, everyone views such communications with skepticism. People wonder what ulterior motive or hidden agenda they have in communicating this type of information. And to be sure, they have reasons beyond serving the public interest for effectively communicating risk. But so does everyone else in the risk communication area—whether the objectives are to sell more products, secure additional research funding, draw public attention, attract more readers or viewers, acquire additional political clout, or merely beat the competition to the punch. Everyone who communicates risk is serving more than just the public interest by doing so.

Tom Vacor recommended "that the press should look more to the middle. What tends to be missing in media coverage," he said, "is enough attention to sources who are saying 'Well, the chemical is dangerous but not as dangerous as some other things,' and give a reasoned, view," (1987). This is the kind of viewpoint most of you, the readers of *Ground Water*, would give. But people with such views are least likely to want to appear in the media; they (YOU) are the most reclusive sources. Yet the recommendation I would make to the media is to seek you out and give the middle as much attention as the two extremes. And the recommendation I would give YOU is to seek out the media.

Uncertainty

It is important to understand that uncertainties are not unique to matters of risk. Uncertainty drives all of science! If there were certainty, there would be no science. Science is an endlessly changing series of approximations. Science is about testable conclusions based on often disputable facts. It is about a community of professionals coming to reasonable agreements on matters riddled with uncertainties and incomplete data. It is in that context, underlain by an absence of prejudice, that science has a critical role in both the perception and the communication of risk.

If our goal is to convey useful information about risk, a better warning would be one which relates the risk involved to risks associated with everyday activities such as those I have included in the Appendix. This type of warning would underscore, for example, the silliness of such things as the requirements of California's Proposition 65. In most cases, the risk of consuming a product is lower than the risk of driving to the store to buy it. If, as is required in Proposition 65, everything must contain a basic warning—especially the same warning—then warning labels lose their value. A warning will affect behavior only if consumers can distinguish the few especially dangerous risks from the thousands of minor risks they take every day. California's Proposition 65 has the potential to misdirect our attention away from noncarcinogenic risks and dangers that we should be concerned about, and toward trivial risks that are no greater than that which we experience from increased exposure to natural radiation when we travel on an airplane (Wilson, 1979), which itself has been unbelievably dramatized in the press despite the fact that the concern is for those of us who travel twice weekly across a continent for 20 years.

To the most radical supporters of California's Proposition 65, posting warnings is just an intermediate step. The long-range goal is clearly to ban carcinogenic substances altogether. At a workshop sponsored by the Environmental Defense Fund, Sierra Club, and the NRDC, Tom Hayden, a California legislator, said that he hoped the state would "lead other states down the path that will ultimately lead to legislation that will eliminate all carcinogens and toxic substances that the American people are subjected to" (Heckman, 1988).

We must honestly inform the public that we do not fully understand what risk many situations carry, and that we are still collecting exposure data. Unfortunately, when we admit this, we create problems of social distrust. People want to believe that experts have the information and understand the danger. Thus, the more we admit we don't know, the greater will be the perception that we are not in control. This is not an easily solved dilemma, especially in light of the public's propensity for irrationality. While people stopped buying apples because of Alar, they continue to oppose gun control. They worry about radon but clamor for the return of the 65-mile-an-hour speed limit. Let's face it, Americans are not rational about risks.

Truly Serious Problems

No one questions that environmental pollution in some places is egregious. Behind the old Iron Curtain, we are provided with ample examples. The USSR itself is full of them. In the Ukrainian town of Zaporozhye, plants emit 400,000 tons a year of toxic emissions such as benzol, phenol, formaldehyde, fluoride, nitrogen dioxide, and lead. Zaporozhye is an ecological disaster zone, according to Anatoly Nesterenko, the Deputy Mayor, who said the pollution there is about 10 times higher than Soviet norms (*Wall Street Journal* October 1989).

Official statistics show that cancer, bronchitis, and other diseases are much more common in Zaporozhyme than in other parts of the Soviet Union. The infant mortality rate and the number of miscarriages are also way above average. Ukrainian health experts recently warned that life expectancies will fall by several years in the near future unless urgent measures are taken to clean up the environment.

Mr. Gorbachev has set out to attack the pollution problem, but like so many of his initiatives, this one hasn't been forceful enough. He set up a State Committee for nature protection to monitor and enforce standards, but it has little political or economic clout and hasn't had any impact. Perestroika has thus far failed to clean up what comes out of Zaporozhye's factories.

The story is much the same in Rumania where frenzied industrialization ran roughshod over the landscape, transforming many rural towns into slag heaps with foul-smelling smokestacks. An ink factory near the city of Sideu has so polluted the atmosphere that surrounding trees and houses are black. In a town called Turda, a cement works expels a beige mist from a forest of smokestacks. Strip mining has destroyed much of the rural landscape and polluted waterways. There is scant hope for quick improvement. The environment there is a disgrace.

The total environmental cleanup of all of eastern Europe will be one of many economic burdens the West must share. While now out from the cloud of secrecy, eastern Europe is still deep under the cloud of air, water, and soil heavily laden with contaminants—an order of magnitude worse than anything Americans can imagine. Could we but harness the well-meaning energies of American environmentalist armies, we could help our newfound European friends clean up their surroundings in about a decade. It will take nothing less.

Environmental groups would make a far greater contribution to society if they would concentrate more on helping these parts of the world achieve the success we in America already enjoy instead of continuing to chase the smaller and smaller amounts of less and less important substances in our own environment. The very same groups that needlessly rattle our chains today are, indeed, largely responsible for our current dedication to the environment—the strides we have already made and the advances we are yet destined to make. Now they should address more attention to portions of society in much greater need.

Truly Unserious Problems EDB

The EDB crisis of the early 80s was not a public health crisis. It was, in fact, a crisis of communication. The message that the U.S. EPA regulators tried to convey was that the food contamination risk from EDB was a long-term rather than a short-term risk. They were frustrated because they could not induce the media to accept and disseminate that message. Instead, public perception was dominated by images of squad cars rushing with their sirens wailing to remove contaminated muffin mixes from supermarket shelves, and of chemical workers being carried off to hospitals. The media transformed the EDB incident from a general quality issue into a national crisis.

Asbestos

Yale medical professor, Dr. Bernard Gee, in a recent article in the *New England Journal of Medicine* (Gee and Mossman, 1989) concluded that there is no evidence that environmental exposure to asbestos is a public health hazard. "A friend," he says, "calls it paratoxicology," referring to the type of evidence used by the antiasbestos crowd.

Greenhouse

NASA scientist James Hansen is widely credited with launching the highly politicized crisis atmosphere around the greenhouse question. Hansen went before a Congressional committee and said he was 99 percent sure that the earth was getting warmer, and that he had a high degree of confidence that warming was caused by the greenhouse effect. This, of course, got the desired response—tremendous press play. But some of Hansen's scientific colleagues were dismayed. In an article entitled "Hansen vs. the World on the Greenhouse Threat" (Kerr, 1989) the journal *Science* reported that Hansen's colleagues found his greenhouse assertions unforgivable, largely because of their absolutist certitude. It seems unfortunate that absolutism is a commandment of modern environmentalism.

Global Warning

Three MIT scientists, Reginal Newell, Jane Shiung, and Woo Zhongxiang, recently processed ocean temperature data taken all over the world by merchant mariners since the mid-19th century. Their results were summarized in *Technology Review* (1989). One of the most striking results suggested by the data is that there appears to have been little or no global warming over the past century. The computer models that foretell, a greenhouse effect indicate that there already should have been about a 1.8 degree rise in global temperature, but that hasn't happened. Unfortunately, the no-nonsense MIT report has been virtually ignored. Science may still be about surveying all the available facts but, increasingly, public policy is not.

Today's Activists

The Animal Rights Movement is a textbook example of how activist groups press their agendas into today's political system. It hardly matters, for instance, that an American Medical Association poll (Harvey and Shubat, 1989) found that 77 percent of adults think that using animals in medical research is necessary, since few people want to volunteer to replace the critters. Those people answered the phone and went back to their daily lives, working at real jobs and raising families. Meanwhile the professional activists—animal rights, antinukers, fringe environmentalists, Hollywood actresses—descend on the people to create issues in America. They elicit sympathetic free publicity from newspapers and magazines. They do Donahue and Oprah. They beat on politicians and bureaucrats. They create a kind of nonstop, twilight zone of issues about which most American voters are barely aware. They do this because such tactics have succeeded so many times before.

If we in the U.S. are forced to work under the constant burden of all these varieties of public issue nonsense, we can never hope to realize continued gains in either human welfare or our international competitiveness.

Today, much public policy practiced by many environmental advocates is mainly about making doubters or opponents reluctant to challenge the consensus. Strobe Talbot of *Time* magazine, for example, recently announced that "no respectable scientist denies the greenhouse phenomenon." Do you have the nerve to deny it?

There is no doubt that participants of all stripes in the policy game these days have become frustrated at their inability to enact their agendas. What seems sometimes to work, though, is whipping up a kind of mass media fervor behind one's ideas. The danger in this is that it may cause the public to think that science is now primarily about politics, and in politics about half the people usually think that you are not telling the truth.

An example of the environmental industry's desire to shut off scientific debate and keep the world focused on fear appears in a statement delivered by Ellen Silbergeld (1987), a senior scientist with the Environmental Defense Fund, at the 1986 National Conference on Risk Communication in Washington, D.C. "I am not sure how I feel about this well-attended conference," she said, "because to a certain extent I look upon the subject that we are here to talk about as a result of the destruction of consensus on environmental and other risk areas, which has occurred over the last decade in this country. In fact, I would describe the topic as a shield for inaction." In other words, groups like EDF don't want educated people discussing proper communications or education about the meaning of risk; rather, they want to keep everyone in a state of fear, interested only in eliminating all risk.

Another example of contorted environmental concern was provided by Jane Hathaway, an attorney for the NRDC who said, "Allowing the EPA to condone continued use of a chemical whenever the benefits outweigh the risks is absolutely an anathema to the environmental community" (*Wall Street Journal*, November 7, 1989). Some environmentalists have violently opposed using cost-benefit analysis to weigh any regulatory program. For them, this sort of thing is the thin edge of the wedge. Cost-benefit analysis directly threatens some of their most treasured policies. The people might start asking: "Is this worth it?" when it comes to pesticide policy, for instance. Then, they soon might ask the same cost questions of urban smog policy or ozone. Instead of cost-benefit judgments, some environmentalists would rather rely on the pseudoreligious exhortations that have carried the day for them so far and kept the Congress virtually hostage to their threat of the month.

In areas of science and technology, efforts by the Congress to protect the public have been singularly futile. We have, in past decades, had a plethora of unfounded environmental scares including cranberries, mercury, flammable nightgowns, nonflammable nightgowns, BID, Alar, and asbestos to name but a few. They were a result of bureaucratic action, apparently mandated by federal health directives. They proved to be unjustified by the evidence but forced large expenditures on great numbers of people with no discernible benefits. In the words of Ray Kazmann (1992).

> "Police protection may break down, fire departments be eliminated, education neglected, roads turned into pothole trails, but the latest unproven environmental concern must be dealt with even if the tax base erodes and the citizens suffer. The new idol must be supported even if civilized life collapses. An abstraction called 'environment' must be protected. As in the middle ages, we have a holy grail to pursue."

The Harshest View of Environmental Extremism

Let me emphasize that many environmentalists are not alarmists. I'm referring only to people who make claims with no basis in science.

Alarmists fancy themselves defenders of the public interest while constantly predicting the end of the world through technological disaster. It appears they are not representing the interests of most Americans.

Listening to overzealous environmentalist rhetoric and observing their targets proves that these individuals are antagonistic toward American industry. If there were ever a way of solving a problem without economically penalizing an industry, they show no interest in that alternative. Ron Arnold, author of *At the Eye of the Storm: James Watt and the Environmentalists* (1982) minces no words when he charges extreme environmentalism with the "desire to drastically reduce or dismantle industrial civilization and to impose a fundamentally coercive form of government on America through the implementation of a wide range of environmental laws and controls."

We must, therefore, consider that to environmental agitators environmentalism is a means to an end. Their primary interest may not be preserving health and the quality of life, but rather changing our political and economic system by reducing corporate influence and substituting greater governmental control in order to redistribute the country's wealth. What is so amazing is how easy it is to achieve these ends. If you picket a nuclear power plant, condemn a pesticide, or charge the chemical industry with polluting air and water, your views are quickly accepted by the media and the public as honest expressions of a sensitive individual.

If, however, after examining the facts as I have, one concludes that nuclear power plants are safe, food additives and pesticides are necessary, and our drinking water is of good quality, you are in trouble.

One so outspoken will quickly be accused of being a paid mouthpiece with a vested interest. But I am not, and have none, and in truth much of the aforementioned malicious foolishness has feathered the bed of all of us working in ground water science.

The truly concerned public health specialist works to build a world without illnesses and unnecessary deaths. Drug abuse, suicide, heart disease, and tobacco-induced mortality are the primary agents of human misery affecting our lives. These, not parts per billion of pesticide residues, are the real causes of human pain and suffering.

We can all be proud of the strides we have made in improving our environment. The technological breakthroughs of the 20th century have produced the healthiest population ever to live on this planet.

Advances have not come without risks, but the benefits have vastly outweighed the costs.

It's time for knowledgeable scientists to speak out against the steady diet of misinformation espoused by the holier-than-thou activists and lobbyists who would have us believe that technology is destroying our health and well-being. Dr. Phillip Handler, former president of the National Academy of Science, once observed "if the scientific community will not unfrock the charlatans, the public will not discern the difference—science and the nation will suffer" (1980).

Environmental health problems do exist. Many have been addressed and others warrant strong commitments in the future. No one is in favor of pollution. The public health must be protected from harm, and it is reasonable that we should expect to pay for this benefit. However, we should know what we are paying for with our regulatory dollars and the true health benefits we are getting in return.

By tightening environmental standards, would we be paying for the reduction of real or hypothetical risks? Is it desirable to pay a great deal of money for very strict regulations based on the most cautious interpretations of ambiguous or flawed data?

There are no simple solutions to these problems. Some of us feel that it is worthwhile to protect against any possible health effects, regardless of the cost. Others will decide that the cost must be justified by some evidence of benefits. The economic consequences of these regulations are real. We must be sure that the benefits of regulations intended to protect human health are equally real.

Perhaps it is time for environmental regulations to reflect both the economic and the health needs of our population. I do not advocate the protection of the public pocketbook at the expense of the public's health. But the toxic terrorists have over-reacted to a frightening extent.

You, the scientists and engineers, educated in a way which the general public is not, must begin to help your friends, neighbors, and clients understand. In a world where toxic chemicals are essential, and extremists are a reality, the government, the media, and all of us will have to learn how to evaluate risks more accurately to avoid overreacting or underreacting, and to better determine which is which.

References

Ames, Bruce N., Renae Magaw, and Lois Swirsky Gold. 1987. Ranking possible carcinogenic hazards. *Science*, v. 236, April 17.

Arnold, Ron. 1982. At the Eye of the Storm: James Watt and the Environmentalists. Regency Gateway, Chicago. p. 210.

Browning, Jackson. 1987. Future Challenges for Risk Communicators. Risk Communication: The Conservation Foundation.

Chason, Christine. 1989. Technical assessment systems, Washington, D.C. Personal communication.

Efron, Edith. 1985. *The Apocalyptics*. Simon & Schuster, New York. p. 589.

Gee, J. Bernard and Brooke T. Mossman. 1989. Asbestos-related diseases. *New England Journal of Medicine*, v. 320, issue 26, June 29, p. 1721.

Handler, Philip. 1980. Science and the American future. Speech at Duke University.

Harvey, L.K. and S.C. Shubat. 1989. Physician and public attitudes on health care issues. American Medical Association.

Heckman, Jerome. 1988. Proposition 65—a legal viewpoint: reflections on the political science of how not to do it. American Industrial Hygienic Association. Toxicology Symposium, Williamsburg, Virginia. August.

Kazmann, Ray. *Rational Readings on Environmental Concern*. 1992. Van Nostrand Reinhold, 1992.

Kerr, Richard A. 1989. Hansen vs. the world on the greenhouse threat. *Science*, v. 244, June 2, pp. 1041–1043.

Lave, L.B., et al. 1988. Information value of the rodent bioassay. Commentary Section, *Nature* v. 336, p. 631.

National Research Council. 1989. Committee on Diet and Health Report. National Academy Press.

Sewell, B.F. and R.M. Whyatt. 1989. Intolerable risk: pesticides in our children's food. Natural Resources Defense Council.

Silbergeld, E. 1987. Responsibilities of risk communicators. Risk Communication, The Conservation Foundation.

Stroup, R.L. and J.L. Goodwin. 1989. Making the world less safe: the unhealthy trend in health, safety, and environmental regulation. National Center for Policy Analysis Report No. 137.

Technology Review. 1989. Has the globe really warmed. v. 92, no. 8, Nov./Dec., p. 80.

Vacor, Tom. 1987. Trust and credibility: the central issue? Risk Communication, The Conservation Foundation.

Wall Street Journal. 1989. Environment and common cents. November 7.

Wall Street Journal. 1989. Rumania's environmental misery. October 19.

Wattenberg, Ben. 1984. The good news is the bad news is wrong. *Reader's Digest*. April.

Wildavsky, Aaron. 1988. *Searching for Safety*. Transaction Publishers, New Brunswick.

Will, George. *Newsweek* columnist.

Wilson, Richard. 1979. Analyzing the daily risks of life. *Technology Review*. February.

Reprinted, with permission, from
Ground Water, Vol. 28, No. 3, May–June 1990.

Greenhouse/ Global Warming

Global Climate Change: Facts and Fiction
S. Fred Singer

The Great Greenhouse Debate
Hugh W. Ellsaesser

Carbon Dioxide and Global Change
Sherwood B. Idso

It's the Water
Peter E. Black

Global Climate Change: Facts and Fiction

S. Fred Singer, Ph.D.

S. Fred Singer is director, Science and Environmental Policy Project, Washington Institute for Values in Public Policy. His past affiliations include professor of Environmental Sciences, University of Virginia; chief scientist, U.S. Department of Transportation; and director, National Weather Satellite Center, U.S. Department of Commerce.

Dr. Singer received a B.E.E. (electrical engineering) from Ohio State University in 1943; an A.M. in physics from Princeton in 1944; and a Ph.D. in physics from Princeton in 1948.

He is the author of *Global Climate Change* published by Paragon House in 1989; *Free Market Energy,* Universe Books, 1984; and *Is There an Optimum Level of Population?* by McGraw-Hill, 1971. Dr. Singer has published more than 400 journal articles and has contributed to *Journal of Geophysical Research, Science,* and *Nature.*

He is the recipient of an honorary D.Sc. from the Ohio State University in 1970.

Abstract

Greenhouse gases have been increasing in the atmosphere, largely as a result of human activities. However, the climate record does not show the temperature increase and other telltale signs of the expected greenhouse effect. The mathematical models used for predicting such effects are evidently not complete enough to encompass all of the relevant physical processes in the atmosphere, thus throwing grave doubt on the drastic warming hypothesized for the next century. Furthermore, any modes warming is likely to have beneficial consequences to human existence on the planet. For all of these reasons, drastic actions to control energy use are likely to be harmful, whether applied unilaterally or by international agreement. On the other hand, energy conservation and substitution of nonfossil-fuel energy for coal, oil, and gas are very much in order, as long as they make economic sense.

Greenhouse warming (GHW) has emerged as the issue of the 90's. The easing of international tension with the Soviet Union could make GHW a leading foreign policy issue, along with other global environmental concerns. Wide acceptance of the Montreal Protocol, which limits and rolls back the manufacture of chlorofluorocarbons (CFCs) has encouraged environmental activists, at conferences in Toronto (1988) and The Hague (1989), to call for similar controls on carbon dioxide (CO_2) from fossil-fuel burning. They have expressed disappointment with the White House for not supporting immediate action. But should the United States assume "leadership" in a campaign that could cripple its economy, or would it be more prudent to assure first, through scientific research, that the problem is both real and urgent?

We can sum up our conclusions in a simple message: The scientific base for a greenhouse warming is too uncertain to justify drastic action at this time. There is little risk in delaying policy responses to this century-old problem, since there is every expectation that scientific understanding will be substantially improved within a few years. Instead of panicky, premature, and likely ineffective actions that only slow down, but not stop the further growth of CO_2, we may prefer to use the same resources—a few trillion dollars, by some estimates—to increase economic resilience so that we can then apply specific remedies as necessary. That is not to say that prudent steps cannot be taken now; indeed, many kinds of energy conservation and efficiency increases make economic sense even without the threat of greenhouse warming.

The scientific base for GHW includes some facts, lots of uncertainty, and just plain ignorance—requiring more observations, better theories, and more extensive calculations. Specifically, there is consensus about the increase in so-called greenhouse gases in the earth's atmosphere as a result of human activities. There is some uncertainty, however, about the strength of sources and sinks for these gases, i.e., their rate of generation and the rates of removal. There is major uncertainty and disagreement about whether this increase has caused a change in the climate during the last 100 years. There is also disagreement in the scientific community about predicted changes as a result of further increases in greenhouse gases; the models used to calculate future climate are not yet good enough. As a consequence, we cannot be sure whether the next century will bring a warming that is negligible or a warming that is significant. Finally, even if there is a global warming and associated climate changes, it is debatable whether the consequences will be good or bad; likely we would get some of each.

Greenhouse Gases (GHG)

It has been common knowledge for about a century that the burning of fossil fuels—coal, oil, and gas—would increase the normal atmospheric content of carbon dioxide (CO_2), causing an enhancement of the natural greenhouse effect and a warming of the global climate. Advances in spectroscopy in the last century produced evidence that CO_2—and other molecules made up of more than two atoms—absorb infrared radiation and thereby would impede the escape of such heat radiation from the earth's surface. In fact, it is the greenhouse effect from naturally occurring CO_2 and water vapor (H_2O) that has warmed the earth's surface for billions of years; without the natural greenhouse effect ours would be a frozen planet without life.

The issue now is whether the nearly 30% increase in CO_2, mainly since World War II, calls for immediate and drastic action. Taking account of increases in the other trace gases that produce greenhouse effects, we have already gone halfway to an effective doubling—something that cannot be reversed in our lifetime—and locked in a temperature increase of 1.5 degrees Celsius, according to the prevailing theory.

Precise measurements of the increase in the atmospheric CO_2 date to the International Geophysical Year of 1957–58. More recently it has been discovered that other greenhouse gases (GHG), i.e. gases that absorb strongly in the infrared, have also been increasing, at least partly as a result of human activities. They currently produce a greenhouse effect nearly equal to that of CO_2, but could soon outdistance it.

- Methane (CH_4) is produced in large part by sources that relate to population growth; among these are rice paddies, cattle, and oil field operations. Indeed, methane, now 20% of the greenhouse gas effect but growing twice as fast as CO_2, has more than doubled since pre-industrial times; it would become the most important greenhouse gas, if CO_2 emissions were to stop.

- Nitrous oxide (N_2O) has increased by only 10%, most likely because of soil bacterial action promoted by the increased use of nitrogen fertilizers.

- Ozone (O_3) from urban air pollution adds about 10% to the global greenhouse effect. It may decrease in the US as a result of Clean Air legislation but increase in other parts of the world.

- Chlorofluorocarbons (CFC's or "freons"), manufactured for use in refrigeration, air conditioning, and industrial processes, can make an important contribution but will soon be replaced by less-polluting substitutes.

- The most effective greenhouse gas by far turns out to be water vapor (H_2O)! It is not manmade at all, but is assumed to amplify the warming effects of the manmade gases. We don't really know whether H_2O has increased in the atmosphere or whether it will increase in the future—although that's what all the model calculations assume. Indeed, predictions of future warming depend not only on the amount but also on the horizontal and especially the vertical distribution of H_2O, and whether it will be in the atmosphere in the form of a gas or as liquid cloud droplets or as ice particles. The current computer models are not refined enough to test these crucial points.

The Climate Record

Has there been a climate effect caused by the increase of greenhouse gases in the last decades? The data are ambiguous to say the least. Advocates of immediate action profess to see a global warming of about 0.5 C since 1880, and point to record temperatures experienced in the 1980's. (Fig. 1). Most scientists tend to be more cautious; they call attention to the fact that the strongest increase occurred before the major rise in greenhouse gas concentration; it was followed by a quarter-century decrease, between 1940 and 1965, when concern arose about an approaching ice age! Following a sharp increase during 1975–80 there has been none during the eighties—in spite of record GHG increases. Similarly, global atmospheric (rather than surface) temperatures measured by Tiros satellites show no trend in the last

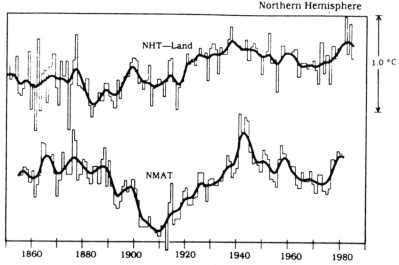

Figure 1. Northern hemisphere land air temperatures. Note the major increase before 1938 (when greenhouse gas concentrations were low), the sustained cooling to 1975, and the sharp rise to 1980 (more rapid than predicted by greenhouse models). P.D. Jones, T.M.L. Wigley, P.B. Wright, Nature 322, 0. 430, 1986.

decade.

NOAA scientists Kirby Hanson, Tom Karl, and George Maul find no overall warming in the US temperature record (Fig. 2)—contrary to the global record assembled by NASA's James Hansen. Patrick Michaels and colleagues at the University of Virginia, using a technique that eliminates urban "heat islands" and other local distorting effects, confirm a temperature rise before 1950, followed by a decline. Reginald Newell and colleagues at MIT report no substantial change in the global sear surface temperature in the past century; yet the ocean, because of its much greater heat inertia, should control any atmospheric climate change.

Perhaps most interesting are the studies of Tom Karl that document a relative rise in night temperatures in the US in the last 60 years, while daytime values stay the same or decline. This is just what one would expect from the increase in atmospheric greenhouse gas concentration. But its consequences, as Michaels and others have pointed out, are benign: a longer growing season, fewer frosts, no increase in soil evaporation.

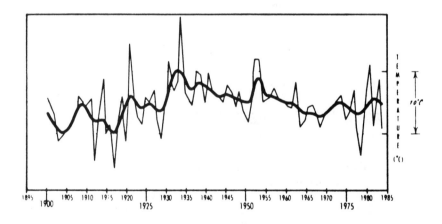

Figure 2. US temperature record since 1900. Annual average and smoothed temperatures, areally weighted and corrected for "urbanization" effects (that would show a warming trend simply because the fixed recording stations are being enveloped by growing urban areas).

Note that the highest annual averages in the US occurred in the 1930s rather than in the 1980s. Note also the great year-to-year variability compared to the global average temperature. Data from T.R. Karl, National Climate Data Center, NOAA.

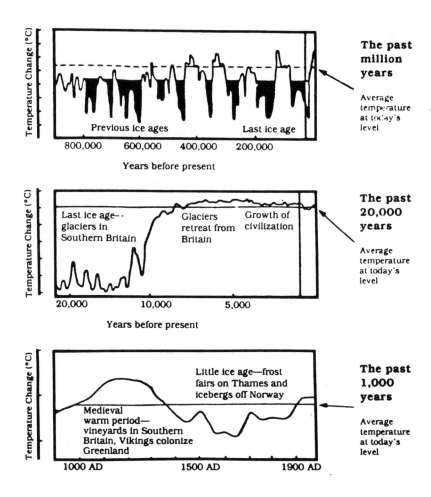

Figure 3. Global climate over different time scales. The British
Meteorological Office, which edited the IPCC report, issued a
glossy brochure last year showing these natural climate changes.
During the "little ice age" the Thames frequently froze over, as
recently as the mid-1800s, and European mountain glaciers
reached their greatest extent.

398

It is therefore fair to say that we haven't seen the huge greenhouse warming expected from some theories and played up by the media. But why not? This scientific puzzle has many suggested solutions:

- The warming has been "soaked up" by the ocean and will appear after a delay of some decades. Plausible—but there is no evidence to support this theory until deep-ocean temperatures are measured on a routine basis, as suggested by Scripps oceanographer Walter Munk.

- The warming exists as predicted, but has been hidden by offsetting climate changes caused by volcanoes, solar variations, or other causes as yet unspecified—such as the cooling from an approaching ice age. Others, like Robert Balling of Arizona State University, consider the warming before 1940 to be a recovery from the "Little Ice Age" that prevailed from 1600 to about 1850. Each hypothesis has vocal proponents—and opponents—in the scientific community; but the jury is out until better data become available.

- The warming has been overestimated by the existing models. Meteorologists Hugh Ellsaesser (Livermore National Laboratory) and Richard Lindzen (MIT) propose that the models do not take proper account of tropical convection and thereby overestimate the amplifying effects of water vapor over the important part of the globe. Other atmospheric scientists suggest that the extent of cloudiness may increase as ocean temperatures try to rise and as evaporation increases. Clouds reflect incoming solar radiation; the resultant cooling could offset much of the greenhouse warming. Most intriguing has been the suggestion by British researchers that sulfates from smoke stacks—the precursors of acid rain—may have played a role in producing an increase in bright stratocumulus clouds.

Mathematical Models

Indeed, there is much to complain about when it comes to predictions of future climate, but there is really no alternative to global climate models. Half a dozen of these General Circulation Models (GCM) are now running, mostly in the United States. Even though they use similar basic atmospheric physics they give different results. There is general agreement amongst them that there should be global warming; if the effective greenhouse gases double, the calculated average global increase ranges between 1.5 and 4.5 C. These values were unchanged for many years, then crept up, and have recently dropped back. Just during 1989 modelers cut their predictions in half as they tried to include clouds and ocean currents in a better way. But there is serious disagreement amongst the models on the regional distribution of this warming and on where the increased precipitation will go.

The models are "tuned" to give the right mean temperature and seasonal temperature variation, but they fall short of modeling other important atmospheric processes, such as the poleward transport of energy. Nor do they encompass longer-scale processes that involve the oceans or the ice and snow in the earth's cryosphere, nor fine-scale processes that involve convection, cloud formation, boundary layers, or that depend on the earth's detailed topography.

There are serious disagreements also between model results and the actual experience from the climate record of the past decade. Existing models predict a strong warming of the polar regions and of the tropical upper atmosphere, and less warming in the southern hemisphere than the northern—all contrary to observations.

Yet there is hope that research, including satellite observations and ocean data, will provide many of the answers within this decade. And faster computers will have higher resolution and incorporate the detailed and more complicated interactions that are now neglected.

Impacts of Climate Change

But assume the most likely outcome—a modest general warming of perhaps one degree Celsius in the next century, mostly at high latitudes and in the winter: Is this necessarily bad? One should perhaps recall that only a decade ago when climate cooling was a looming issue, National Academy of Sciences/National Research Council economists calculated a huge national cost associated with such cooling. More to the point perhaps, actual climate cooling, experienced during the Little Ice Age or in the famous 1816 New England "year without a summer," caused large agricultural losses and even famines.

If cooling is bad then warming should be good, it would seem—provided the warming is slow enough so that adjustment is easy and relatively cost-free. Even though crop varieties are available that can benefit from higher temperatures with either more or less moisture, the soils themselves may not be able to adjust that quickly. But agriculturalists expect that with increased atmospheric CO_2—which is, after all, plant food—plants will grow faster and need less water. The warmer night temperatures suggested by Tom Karl's findings translate to longer growing seasons and fewer frosts. And increased global precipitation should also be beneficial to plant growth.

Keep in mind also that year-to-year changes at any location are far greater and more rapid than what might be expected from greenhouse warming; and nature, crops, and people are already adapted to such changes. It is the extreme climate events that cause the great ecological and economic problems: crippling winters, persistent droughts, killer hurricanes, and the like. But there is no indication from modeling or from actual experience that such extreme events would become more frequent, as claimed, if greenhouse warming ever becomes appreciable. The exception might be tropical cyclones, which—Balling and Richard Cerveney argue—would be more frequent but weaker, cool vast areas of the ocean surface and increase annual rainfall. In sum, climate models predict that global precipitation should increase by 10—15%, and polar temperatures should warm the most, thus reducing the driving force for severe winter weather events.

There is finally the question of sea level rise as glaciers melt—and fear of the catastrophic flooding so often discussed in the tabloids. The cryosphere certainly contains enough ice to raise sea level by 100 meters; and, conversely, during recent ice ages enough ice accumulated to drop sea level 100 meters below the present value. But these are extreme possibilities; tidal gauge records of the past century suggest that sea level has risen modestly, about 0.3 meters. Further, the gauges measure only relative sea level, and many of their locations have dropped because of land subsidence. Besides, the locations are too highly concentrated geographically, mostly on the US east coast, to permit global conclusions. The situation will improve greatly, however, in the next few years as precise absolute global data become available from a variety of satellite systems.

In the meantime, satellite radar-altimeters have already given a surprising result. As reported by NASA scientists Jay Zwally in a recent issue of *Science,* Greenland ice-sheets are gaining in thickness —a net increase in the ice stored in the cryosphere and an inferred drop in sea level—leading to somewhat uncertain predictions about future sea level. It is clearly important to verify these results by other techniques and also get more direct data on current sea-level changes.

Summarizing the available evidence, we conclude that even if significant warming were to occur in the next century, the net impact may well be beneficial. This would be even more true if the long-anticipated ice age were on its way.

What To Do?

In view of the uncertainties about the degree of warming, and the even greater uncertainty about its possible impact—what should we do? During the time that an expanded research program reduces or eliminates these uncertainties, we can be putting into effect policies and pursue approaches that make sense even if the greenhouse effect did not exist. These include:

Conserve Energy By Discouraging Wasteful Use Globally.

Conservation can best be achieved by pricing rather than by command-and-control methods. If the price can include the external costs that are avoided by the user and loaded onto someone else, this strengthens the argument for proper pricing. The idea is to have the polluter or the beneficiary pay the cost. An example would be peak-pricing for electric power. Yet another example, appropriate to the greenhouse discussion, is to increase the tax on gasoline to make it a true highway user fee— instead of having most capital and maintenance costs paid by the various state taxes, as is done now. Congress has lacked the courage for such a direct approach, preferring instead regulation that is ineffective and produces large indirect costs for the consumer.

Improve efficiency in energy use
Energy efficiency should be attainable without much intervention, provided it pays for itself. A good rule of thumb: If it isn't economic, then it probably wastes energy in the process and we shouldn't be doing it; over-conservation can waste as much energy as under-conservation. But provided that energy is properly priced, the job for government is to remove the institutional and other road blocks:

- Provide information to consumers, especially on life-cycle costs for home heating, lighting, refrigerators and other appliances;

- Encourage the turnover and replacement of older, less efficient (and often more polluting) capital equipment: cars, machinery, power-plants. Some existing policies that make new equipment too costly go counter to this goal.

- Stimulate the development by the private sector of more efficient systems, such as combined-cycle powerplants or a more efficient internal combustion engine.

Use non-fossil fuel energy sources wherever this makes economic sense
Nuclear power is competitive now, and in many countries is cheaper than fossil-fuel power—yet is often opposed on environmental grounds. The problems cited against nuclear energy, such as disposal of spent nuclear fuel, are mostly political and ideological rather than technical. Nuclear energy from fusion rather than uranium fission may be a longer-term possibility, but the time horizon is uncertain.

Solar energy, and other forms of renewable energy, should also become more competitive as their costs drop and as fossil-fuel prices rise. Solar energy applications are restricted not only by cost; solar energy is both highly variable and very dilute; it takes a football field of solar cells to supply the total energy allocated to the average US household. Wind energy and biomass are other forms of solar energy, competitive in certain applications. Schemes to extract energy from temperature differences in the ocean have been suggested as inexhaustible sources of non-polluting hydrogen fuel, provided we can solve the daunting technical problems.

If greenhouse warming ever becomes a problem, there are proposals for removing CO_2 from the atmosphere. Afforestation is widely talked about, but probably not cost-effective; yet natural expansion of boreal forests in a warming climate would sequester atmospheric CO_2. A novel idea, proposed by California oceanographer John Martin, is to fertilize the Antarctic Ocean and let plankton growth do the job of converting CO_2 into bio-material. The limiting trace nutrient may be iron, which could be supplied and dispersed economically.

And if all else fails, there is always the possibility of putting satellites into earth orbit to modulate the amount of sunshine reaching the earth. By absorbing or reflecting solar radiation they could counteract climate warming or cooling. These satellites could also generate electric power and beam it to the earth, as originally suggested by Peter Glaser and others. Such schemes may sound farfetched, but so did many other futuristic projects in the past—and in the present, like covering the Sahara with solar cells or Australia with trees.

Conclusion

Drastic, precipitous—and, especially, unilateral—steps to delay the putative greenhouse impacts can cost jobs and prosperity, without being effective. Stringent controls enacted now would be economically devastating without being able to affect greatly the growth of greenhouse gases in the atmosphere. Yale economist William Nordhaus, one of the few who has been trying to deal quantitatively with the economics of the greenhouse effect, has pointed out that "... those who argue for strong measures to slow greenhouse warming have reached their conclusion without any discernible analysis of the costs and benefits ..." It would be prudent to complete the ongoing and recently expanded research so that we will know what we are doing before we act. "Look before you leap" may still be good advice.

Reprinted, with permission, from
World Climate Change Report, Volume 2, Number 4, December 1990

The Great Greenhouse Debate

Hugh W. Ellsaesser, Ph.D.

Dr. Hugh Ellsaesser is a participating guest scientist at the Lawrence Livermore National Laboratory (LLNL). He previously was a physicist at LLNL and has been a weather officer for the United States Air Force.

He received an S.B. in meteorology from the University of Chicago in 1943; an M.S. in meteorology from UCLA in 1947; and a Ph.D. in dyn. meteor. from the University of Chicago in 1964.

Dr. Ellsaesser is the author of *Global 2000 Revisited,* presently in press with Paragon House; *Proceedings Workshop on Stratospheric Analysis and Forecasting,* U.S. Weather Bureau, 1956. He has published more than 100 journal articles and has contributed to *Atmospheric Environment, Reviews of Geophysics,* and *Monthly Weather Review.*

He is a Phi Beta Kappa.

Greenhouse Warming

To be sure that you do not miss my message, let me state it quite clearly. After 48 years of studying the atmosphere and how it behaves, including 30 years of trying to model it on computers and 15 years pondering over the greenhouse effect itself, I strongly believe that greenhouse warming has been greatly exaggerated and its effects distorted, largely for the same purposes that motivated the tailors of "The Emperor's New Clothes."

I believe that the climate models are over estimating the amount of warming for a doubling of carbon dioxide by at least 2- to 3-fold and that there are good reasons to believe that the exaggeration is even greater than this. I believe that most of the warming which does occur will occur in mid and high latitudes in winter and will show up primarily as increases in night-time minimum temperatures, i.e., as effects that most people are likely to prefer to our present climate.

Secondly, I believe that, independent of its climatic effects, increased carbon dioxide will exert primarily beneficial effects on the biosphere, both acting as a fertilizer for plants and improving their efficiency in the use of water. Carbon dioxide, after all, is by far the main food of most, if not all, plants. And much of the water loss by plants occurs while they are trying to ingest their food—carbon dioxide. Since increased concentration of carbon dioxide will make it easier to get their food; they will lose less water in the process. In his recent book,

404

Idso (1989) pointed out that thousands of commercial nurseries were using carbon dioxide enrichment 50 years ago and are continuing to do so today—because it pays to do so.

The Temperature Record

I recently spent about five years doing almost nothing else except reviewing what we know about the past temperature record of our planet. This led to a lengthy review article published in the *Reviews of Geophysics* (Ellsaesser et al., 1986) in 1986.

Recovery from "the little ice age"

In my opinion there is no question but what the available meteorological and historical record shows a progressive, if not continuous, warming since the middle of the 18th century (roughly 1750). The best number for this warming is, in my opinion, about 0.5°C (0.9°F) over the past century—possibly a little less. The least controversial explanation of this warming to date is that it is a recovery from a recognized colder period, termed "The Little Ice Age" (Grove, 1988), which began circa 1300 AD.

The 2,500-year temperature cycle of the holocene

At present we do not know why "The Little Ice Age" began nor why it ended—but only that it was a continuation of an approximately 2,500-year-long cycle of warmings and coolings which have been recognized as extending throughout the Holocene, or our present interglacial (Denton and Karlen, 1973). This temperature cycle is illustrated in the lower left panel of Figure 1, which is a reconstruction of our current estimate of the terrestrial temperature record of the past million years that appeared in *The National Geographic* of November 1976.

What the record tells us to expect over the next century or so

Now look at the blow-up of the last 1,000 years of the record in the lower right panel. It is obvious that if this cycle were to continue in isolation, the global mean surface temperature would not merely recover to the average temperature for the whole cycle. On reaching the level of this average, the temperature would presumably be rising at its maximum rate into the next warm part of the cycle—comparable to the "Medieval Little Optimum," centered about 1100 AD. It was then that the ice melted back in the North Atlantic and the Norsemen were able to colonize Iceland and Greenland and perhaps explored North America. The ensuing "Little Ice Age" apparently wiped out the colony on Greenland and produced considerable historically documented glacier

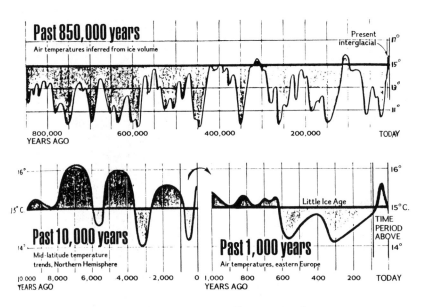

Figure 1. Reconstruction of the Earth's climate over the most recent 850,000 years. (Composited from Matthews, 1976).

advance, cooling and general hardship both in Iceland and in many parts of Europe.

If we are not, in the long term, now going into an extended warmer period, irregardless of any greenhouse warming, there will have to be a severe distortion of the cyclic nature of our present understanding of the climatic record of our planet. Note in the record for the past 10,000 years (lower left panel of Figure 1) that the 3 warm periods were much longer than the intervening cold periods. This seems to imply that the warm period of today, shown in the lower right panel, has a few hundred years yet to go. While this figure is admittedly an artist's rendition, it appeared under the authorship of Samuel W. Matthews (1976), who has impeccable credentials with academic and international scientific organizations. It is also consistent with most of what we think we know about our past climate.

In any case, in the past there were two warm portions of this cycle which have been relatively well documented, the "Climatic Optimum" (about 6,000 years ago) and the "Medieval Little Optimum" (about 900 years ago), when the global mean temperature is believed to have been warmer than now (Lamb, 1965; WMO/ICSU, 1975). And we currently have no reason to believe that the level of carbon dioxide in the atmosphere at those times differed from the preindustrial level of about 280 ppm (parts per million). In other words, increased carbon dioxide does not appear to be required for our mean global temperature to rise 0.5 to 2°C (0.9 to 3.6°F) above the present level (Lamb, 1965; WMO/ICSU, 1976).

What the record tells us to expect in the longer term

Now, let's take a look at the record over the longer term. Note in the upper panel of Figure 1 the approximately 90,000-year periods of cold climate—the glacials, interspersed by approximately 10,000-year periods of warm climate—the interglacials. Approximately 17 of these cycles are believed to have occurred over the past 2 to 3 million years of terrestrial history and we have no reason to believe that anything has changed to interrupt this cycle. Since our present interglacial has already lasted for 10,800 years, we presume that the next 90,000-year glacial period could begin at any time. I have been rather amazed that almost no one wants to suggest that increased greenhouse warming is just what we need to prevent, or at least delay, the onset of this next glacial period.

On even longer time scales, the Earth has been both cooling and losing atmospheric carbon dioxide. During the time of the dinosaurs about 100 million years ago, the mean global temperature is believed to have been 5 to 10°C (9 to 18°F) warmer than now and the level of carbon dioxide to have been 5 to 10 times greater than now. During the peak of the last glacial about 20,000 years ago, atmospheric carbon dioxide was down to about 200 ppm, that is only about twice the level at which many of our plants begin to undergo carbon dioxide starvation. Again we have no reasons to believe that this evolutionary decline in carbon dioxide will not continue until our atmosphere can no longer support plant life, and if the plants go, so will the animals—including humans.

The fact that man, for completely different reasons, turned to the consumption of fossil fuels, thus increasing the atmospheric content of carbon dioxide, which in turn may prevent or delay the onset of the next glacial period and apparently has already led to improvements in plant growth due to the sub-optimum preindustrial level of carbon dioxide in the atmosphere, is the type of action for which man would have claimed divine inspiration in the past! But since Rachel Carson's (1962) *Silent Spring,* we merely condemn ourselves for mucking up the planet or fouling our nest.

You may be interested to know that Sherwood Idso (1989, p. 120) has even suggested that the rising carbon dioxide content of the atmosphere "may possibly be implicated as a contributing factor to the significant worldwide downturn in circulatory heart disease experienced over the past two decades."

Development of our temperature record

Now, let us take a look at the more recent climatic record. Aside from a few individual efforts, it has been less than 30 years since people have seriously begun to try to put past temperature data together to determine what has been happening to our climate, or to the hemispheric or global mean temperature. The reason for this is quite simply that, without computers, it is very difficult to process the millions of observations that make up the record. The first curves were for the land stations of the Northern Hemisphere, as shown in the upper curve of Figure 2.

While there had been many reports of warming from single stations with long records and for larger areas around the North Atlantic in the late 1920s and early 1930s, following the so-called Arctic Warming circa 1920 (note the approximately 0.4°C [0.72°F] jump in the curve near 1920); the first generally accepted curves for the average temperature of the Northern Hemisphere appeared in 1961. These indicated that the warming had already peaked in 1938 and that since then there had been a cooling.

However, those meteorologists who were interested in climate, were already convinced that increasing carbon dioxide had to lead to greenhouse warming and instead of questioning the greenhouse the-

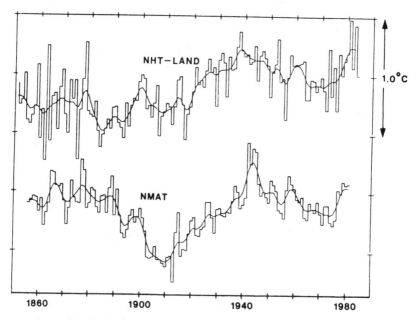

Figure 2. Annual mean surface air temperatures from Northern Hemisphere land stations 1851–1984 (from Jones et al., 1986) and Northern Hemisphere ships 1856–1981 (from Folland et al., 1984).

ory, they searched for processes that might be counteracting the expected warming.

Any of you who may have been following the subject in those days will recall; that cooling—due to reflection of sunlight by the increasing number of particles that man was putting into the atmosphere, became the prevailing explanation as to why greenhouse warming was not observed in the 1960s and early 1970s. That was my first disillusionment with science by consensus, even when that consensus included the National Academy of Sciences.

Atmospheric particulates and air pollution in general were subjects I had been following and all the data I was aware of indicated that airborne pollutants, including particles, were if anything declining. I collected all the available data I could find and wrote an extensive (35-page) review article which was belatedly published in 1975 under the title: "The Upward Trend in Airborne Particulates That Isn't (Ellsaesser, 1975)." (You will notice that I wanted to get my message across, even if the reader didn't get beyond the title.) Publication was belated for two reasons, scientific journals do not like long papers and they particularly do not like those that go against the prevailing consensus. It's a Catch-22 situation, because if I had tried to publish my conclusion without the supporting data, there would have been a ready excuse to reject it—it didn't fit the prevailing consensus.

The warming of the 1980s

Fortunately for the climate people but unfortunately for the rest of us, Mother Nature stepped in. As you can see on the Northern Hemisphere land temperature curve in Figure 2, there was an abrupt warming of 0.58°C (1.04°F) from 1976 to 1981. Everyone admits, of course, that this was much too rapid to be due to increasing carbon dioxide alone and the climate models give us no reason to believe that the warming should occur in discrete jumps. But that did not prevent the greenhouse warming theory from taking on new life.

Is the Warming We Have Seen Due to the Greenhouse Effect?

We now find that the five years 81, 83, 87, 88 and 89 all had mean Northern Hemisphere temperatures higher than that of 1938, the previous warmest year in the record. Does that mean that the greenhouse effect is here? It depends on whom you ask.

The nature of the current consensus

Dr. Solow (1989), a statistician at Woods Hole Oceanographic Institute says, no; "the bunching of high values at the end of a record is to be expected when the record has an upward trend. . . . in 1932 six of the warmest years in the record through 1932 had occurred in the previous 8." On the other hand, Dr. Jim Hansen (1990) of NASA thinks this bunching of record high values in the 1980s is evidence of greenhouse warming occurring but that the greater part of the warming is yet to come. Professor Pat Michaels (1989) of the University of Virginia, on the other hand, thinks that most of the effect of the greenhouse gases now in the atmosphere is already here and that most people will like the change. Academician Mikhail Ivanovich Budyko of Leningrad agrees with Jim Hansen on the amount of the warming and that it will take a few decades to show up but he doesn't think we should be trying to slow down the release of greenhouse gases. In fact, he recommends that we speed up the release of greenhouse gases to hurry over the possibility unfavorable transition period, with greater aridity in mid-latitudes, to the globally more favorable climate he sees for higher levels of carbon dioxide (Budyko and Sedunov, 1988).

Now I have a question for you. Do you see a consensus emerging here?

Well, let me tell you. There is a consensus. And it is, or at least has been, a strongly supported consensus. But, it is among those people who see greenhouse warming as a vehicle by which they can advance another agenda—whether it be selling news; procuring research support; inducing contributions to green organizations; guaranteeing a market for such products as gasohol, freon substitutes, or solar energy devices; increasing the share of the budget under the control of a bureaucracy or congressional committee; or building governmental and international organizations with the political muscle to tell people exactly how they are to live.

Other evidence against grreenhouse warming

If you look back at the Northern Hemisphere land temperature record (upper curve in Figure 2), you will note that most of the warming since the beginning of the record had occurred by 1938. Of man's additions of greenhouse gases to the atmosphere to date, no more than 25 to 40% could have been in the atmosphere by that time. During the addition of the remaining 60 to 75% of these gases (and remember that we are already half way to an equivalent doubling of carbon dioxide), our temperature actually dropped for 40 years. The annual mean temperature for 1965 was 0.39°C (0.7°F) below that of 1938; for 1976 it was 0.42°C (0.76°F) colder. This seems to say that the bulk of the greenhouse gases we have added to the atmosphere have produced no warming—unless we can attribute the very rapid warming of 0.58°C (1.04°F) from 1976 to 1981 to greenhouse gases. We have no other explanation for this recent rapid warming, which has persisted with little change throughout the 1980s, unless it is somehow related to the more frequent and stronger El Niño-Southern Oscillations which occurred in the eastern tropical Pacific in 1982–83 and 1986–87 (Trenberth et al., 1988).

Over the years, discussions have led to the concept of a "fingerprint" by which we can recognize greenhouse warming. For example, the models predict that the warming from increased greenhouse gases will be greatest in winter and will increase toward the poles. They also predict that the warming will decrease with altitude in polar regions and will increase with altitude in the tropics.

While it is true that the warming up to 1938 was greatest in high latitudes in winter—in fact it was first identified as the "Arctic Warming" of the 1920s—this is not true of the most recent warming since 1976. Temperatures in the Arctic have actually cooled since then and the Antarctic has shown essentially no change. Similarly, for the period since 1958 when we have had temperature data above the surface, the mid and upper tropical troposphere has been cooling—rather than warming faster than the surface, as predicted by the models.

Supporters of greenhouse warming tend to ignore all these details and simply to claim that the "model predicted warming is consistent with the warming that has been observed." And they rarely if ever point out that to make even this claim, the model predicted warming has to be restricted to the lower third of the 1 to 5°C (1.8 to 9°F) warming predicted by the models for an equivalent doubling of carbon dioxide. Can you imagine what they say when confronted with the observational evidence contradicting the greenhouse warming "fingerprint"? They say, and I quote from the IPCC (1990, p. xxix) Report; "We do not yet know what the detailed 'signal' looks like because we have limited confidence in our predictions of climate change patterns [emphases mine]." However, as you have no doubt noted, they have no lack of confidence in the predictions of warming itself.

Delay of greenhouse warming due to thermal inertia of the oceans

In an effort to explain why the observed warming has lagged further and further behind that predicted by the models, the argument has been accepted that the air (whose temperature we keep track of) cannot warm until the underlying ocean surface warms. But even this is not enough lag. So the new argument has evolved that the additional heat deposited in the surface layer of the oceans will be mixed down into the deeper ocean so that it will take decades to centuries before the ocean surface can warm and in turn warm the surface air. However, this argument is leading to other problems. First, there is the problem of, how is warm lighter surface water forced to mix down into denser colder water? And secondly, as Professor Richard Lindzen of MIT has pointed out; to be consistent with the warming observed to date, the warming for a doubling of carbon dioxide must be no more than 1°C (1.8°F) or the delay until the equilibrium warming appears must be more than a century. That is, we have lots of time to adjust to it. And to add insult to injury, the observational data seem to show, that since circa 1905 at least, the warming over the oceans has been more rapid than over the continents (see lower curve of Figure 2).

The greenhouse warming argument is over 50 years old

As recently pointed out again by Solow (1989), the greenhouse warming argument is not new; it has been going on among those interested in climate for over half a century. Most of the arguments we hear today were already brought into the literature by Callendar's (1938) paper of 1938 and the accompanying discussion. The only things that are new are the development of climate models, which were developed to study rather than to predict climate, and a new high in the temperature record.

The nature of climate models

Our best climate models—or General Circulation Models—as they are also called, were developed to study the general circulation of the atmosphere; that is, what makes the atmosphere move and behave as it does. One strong argument in justifying the relatively large budget required for them, even 30 years ago, was that they would help us to learn how the atmosphere behaved and thus be useful in improving the atmospheric models used for Numerical Weather Prediction for day-to-day weather forecasting. However, once a few such models had been developed and described in the literature, the technical journals no longer accepted such papers. The modelers soon found that the only way to get additional papers published was to make climate forecasts; that is, what happens if you double carbon dioxide? for example.

You would think that to answer such questions, all the modeler would need to do would be to run the model with the perturbed condition, doubled carbon dioxide for example, and then subtract the present climate to see what the changes were. But this assumes that the models can forecast the present climate perfectly—which, unfortunately, they do not do.

As you can see from the upper curve in Figure 2, the annual mean surface temperature of the Northern Hemisphere has varied by perhaps plus or minus 0.5°C over the past 140 years. None of the present climate models come close to reproducing the present climate to this accuracy. So the models have to be run twice, for the present or control climate and again for the perturbed climate and the two subtracted to get the predicted change.

But, if the models can't predict the present climate accurately, why should we believe that they can predict the climate change resulting from any particular perturbation accurately? This is a very good question. And insofar as I am aware, it has received almost no discussion.

References

Budyko, M.I. and Yu. S. Sedunov, Anthropogenic climatic changes, paper presented at the "Climate and Development Conference," Congress Centrum, Hamburg (FRG), 7–10 November, 1988.

Callendar, G.S., The artificial production of carbon dioxide and its influence on temperature, *Quarterly Journal of the Royal Meteorological Society* 64, 223–240, 1938.

Carson, Rachel, *Silent Spring*, Fawcett Crest, New York, 1962.

Denton, G.H. and W. Karlen, Holocene climatic variations—their pattern and possible cause, *Quaternary Research* 3, 155–205, 1973.

Ellsaesser, Hugh W., The upward trend in airborne particulates that isn't, pp. 235–269 in *The Changing Global Environment*, S.F. Singer, (ed.), D. Reidel Publishing Company, Dordrecht-Holland, 1975.

Ellsaesser, H.W., M.C. MacCracken, J.J. Walton and S.L. Grotch, Global climatic trends as revealed by the recorded data, *Reviews of Geophysics* 24(4), 745–792, 1986.

Folland, C.K., D.E. Parker and F.E. Kates, Worldwide marine temperature fluctuations 1856–1961, *Nature* 310, 670–673, 1984.

Grove, J.M., *The Little Ice Age*, Methuen, New York, 1988.

Hansen, James, Watch out! Here comes the greenhouse, *Science*, p. 549, 4 May 1990.

Idso, Sherwood B., *Carbon Dioxide and Global Change: Earth in Transition*, IBR Press, Tempe, AZ 85282, 1989.

IPCC, *Climate Change, The IPCC Scientific Assessment*, J.T. Houghton, G.J. Jenkins and J.J. Ephraums (eds.), Cambridge University Press, Cambridge, 1990.

Jones, P.D., R.S. Bradley, H.F. Diaz, P.M. Kelly and T.M.L. Wigley, Northern Hemisphere surface air temperature variations: 1851–1984, *Journal of Climate and Applied Meteorology* 25, 161–179, 1986.

Lamb, H.H. The early medieval warm epoch and its sequel, *Palaeogeography, Palaeoclimatology, Palaeoecology* 1, 13–37, 1965.

Matthews, Samuel W., What's happening to our climate? *National Geographic*, 576–620, November 1976.

Michaels, Patrick J., Crisis in politics of climate change looms on horizon, *Forum for Applied Research and Public Policy*, The University of Tennessee, Winter, 1989.

Solow, Andrew R., Is it getting stuffy in here, or is it just my imagination? *Chance: New Directions for Statistics and Computing* 2(3), 40–46, 1989.

Trenberth, K.E., G.W. Bramstator and P.A. Arkin, Origins of the 1988 North American Drought, *Science* 242, 1640–1645, 1988.

WMO/ICSU (World Meteorological Organization-International Council of Scientific Unions), The physical basis of climate and climate modeling, *GARP Publication Series*, Vol. 16, 265 pp., 1975.

Carbon Dioxide and Global Change

End of Nature or Rebirth of the Biosphere?

Sherwood B. Idso, Ph.D.

Dr. Sherwood B. Idso is a research physicist, U.S. Water Conservation Laboratory, Agricultural Research Service, U.S. Department of Agriculture.

He is a former adjunct professor in the Departments of Geology, Geography, Botany and Microbiology at the Arizona State University; and a past president of the Institute for Biospheric Research.

Dr. Idso received a B. Phys. in physics, 1964; an M.S. in soil science, 1966; and a Ph.D. in soil science, 1967 from the University of Minnesota.

He is the author of *Carbon Dixide: Friend or Foe?* published in 1982; and *Carbon Dioxide and Global Change: Earth in Transition* published in 1989 and has more than 400 journal articles published. He has been a contributor to *Science, Nature,* and the *Journal of Geophysical Research.* He was awarded the Arthur S. Flemming award in 1977.

Abstract

The rapidly rising CO_2 content of earth's atmosphere has been claimed by many environmentalists to be the greatest ecological threat currently facing mankind and all of the other life forms with which we share the planet. Diametrically opposed to this idea, it has been claimed by others that the upward trend in the air's CO_2 concentration is actually a blessing in disguise, and that a CO_2-enriched atmosphere will increase the biological carrying capacity of the earth by as much as an order of magnitude. Evidence is presented to refute the first of these hypotheses and support the second.

Introduction

In a provocative doomsday book[1] and magazine article,[2] journalist Bill McKibben suggests that the rapidly rising CO_2 content of Earth's atmosphere will shortly lead to "the end of nature." The chief culprit in this all-too-familiar scenario, of course, is the unbridled intensification of the CO_2 greenhouse effect, fueled by mankind's felling of forests and burning of ever-increasing quantities of coal, gas, and oil.[3] It is very possible, however, that in place of the mind-boggling catastrophe envisioned by McKibben, there could well be an equally dramatic beneficent outcome, and that humanity's enriching of the air with CO_2 may lead to a great "greening of the Earth," a biological stimulation of such magnitude that its amplification of the totality of Earth's life processes could only be described as a veritable rebirth of the biosphere.[4]

Sound incredible? Too good to be true? Perhaps. But it is really no more unlikely to come to pass than are the doom-and-gloom prognostications of McKibben. Indeed, there is a wealth of sound scientific data to support just such an ultra-optimistic world view. Before describing this evidence, however, it is necessary to evaluate the underpinnings of McKibben's thesis; for if the CO_2 greenhouse effect is as potent as he suggests, what I have to say becomes pretty much a moot point.

Climate Model Predictions

So how may we ascertain the climatic consequences of our energy intensive way of life? McKibben's approach, like that of all anti-CO_2 activists, is to look to the predictions of a small group of general circulation models of the atmosphere or GCMs, complex mathematical constructs designed to simulate our planet's climatic behavior within the innards of a super computer. These models characteristically predict that, for a 300 to 600 part-per-million (ppm) doubling of the air's CO_2 content, the mean surface air temperature of the globe will rise by about 4°C.[5] But expected contemporaneous increases in other trace "greenhouse gases," which are even more adept at absorbing and reradiating thermal radiation than is CO_2, are expected to be equally as significant in this regard.[6,7,8] Hence, by the time the Earth's atmospheric CO_2 content has doubled, its "equivalent" CO_2 concentration will have risen to 900 ppm and, according to the models, produced an equilibrium global warming of fully 6.5°C. In addition, equivalent atmospheric CO_2 contents are expected to ultimately rise well above this nominally doubled level of 900 ppm, perhaps tripling or even quadrupling,[9] which would push the equivalent CO_2 concentration of the atmosphere to 1500 to 2100 ppm and the mean global warming to 9.5 to 11°C. Yet other scientists suggest that "biogeochemical feedbacks" may more than double these concentration increases,[10] which would put the ultimate warming of the Earth in the 12.5 to 14.5°C range. What is more, the GCMs predict that the warming in polar regions will be two to three times greater than that of the globe as a whole,[11] or as much as 25 to 43.5°C.

Is it any wonder that McKibben and others like him are alarmed? Such a warming would actually rival the greenhouse effect that is currently provided (34°C) by the entire real-world atmosphere![12] Clearly, this observation, in and of itself, should be sufficient to totally discredit the GCM predictions; but for those who require independent empirical evidence, there are a number of other facts that may be brought to bear upon the issue.

Greenhouse Warming on Mars and Venus

Consider our nearest planetary neighbors, Mars and Venus, whose atmospheres are both within but a few percent of being totally composed of pure CO_2. As a result of a number of spacecraft missions to these distant worlds, we know that the greenhouse warming of Venus is approximately 500°C,[13,14] while that of Mars is only 5 or 6°C.[15,16] Likewise, we know that the atmospheric pressure of CO_2 on Venus is on the order of 90 bars, or a quarter of a million times more than what it is on Earth, while on Mars it oscillates over the year between only 0.007 and 0.010 bar,[17] but which still amounts to approximately 25 times more surface pressure than that exerted by the CO_2 of Earth's atmosphere.

Now if we plot the two points defined by these data on a gridwork relating greenhouse warming to the surface atmospheric pressure of CO_2, we find that there is no simple line which can be drawn through them which, when extrapolated to Earth conditions, will produce a 300 to 600 ppm CO_2-induced planetary warming in excess of 0.4°C.[18] Hence, the greenhouse warming predicted for Earth by today's state-of-the-art GCMs must be at least an order of magnitude too large. And, in fact, the palaeohistory of Earth itself confirms this estimate; for the relationship defined by the increasing luminosity of the sun[19] and the decreasing CO_2 content of Earth's atmosphere[20] over the past four billion years or so (a 25% rise in solar luminosity and a 99% drop in CO_2 content) is identical to the most simple of the relationships which may be constructed on the basis of the Mars and Venus data and which yields the maximum 0.4°C warming noted above.[18]

But how can our best climate models, which are the products of some of the finest minds at work in the atmospheric sciences today, possibly be so wrong? The answer is simple: as intricate and complex as the GCMs are, they are still no match for the complexity of nature.

Effects of Clouds

Consider the common phenomenon of clouds. Clouds possess both greenhouse and anti-greenhouse properties at one and the same time. That is, they trap and reradiate downwards a certain portion of the thermal radiation emanating from the surface of the Earth that would otherwise escape to space and be lost, thereby warming the planet, and they reflect back to space a certain portion of the incoming solar radiation from the sun, thereby cooling the globe.[21] Which effect dominates? And how would both be changed by a CO_2-induced impetus for warming?

For years the first of these questions was totally unanswerable. Now, however, as a result of information obtained from the Earth Radiation Budget Experiment (a several-year satellite study), we know that the net effect of the presence of clouds is to dramatically cool the surface of the Earth.[22,23,24] Likewise, we have learned from the construction of historical analogue climate scenarios—composites of blocks of comparatively cool and warm years extracted from the historical climate record—that cloud cover invariably expands with increasing surface air temperature, most probably as a result of increasing land and oceanic evaporation rates.[25,26,27] Hence, it is clear that our planet's finely tuned hydrologic cycle, as expressed by its ever-changing cover of clouds, possesses great negative feedback potential.

Another example of strong negative feedback, that comes from the climate models themselves, is highlighted by the work of J.F.B. Mitchell in Britain.[28,29] By merely allowing for a transformation of a realistic fraction of ice-crystal clouds into water-droplet clouds in the face of a CO_2-induced temperature rise, Mitchell found that the equilibrium warming predicted by his GCM was essentially cut in half, due to the fact that typical cloud water droplets are generally smaller than typical cloud ice crystals and are therefore depleted more slowly by precipitation processes, which results in longer-lasting clouds and, therefore, more extensive cloud cover as the climate warms (and, as noted above, is generally always observed in historical analogue climate scenario studies). Mitchell also showed that by including the reflectance-enhancing effect of the greater cloud liquid water contents that result from increased land and oceanic evaporation rates, the warming predicted by his model was reduced to only one-third of its original value.

Biological Feedbacks

Many other negative feedbacks based upon sound physical principles, all of which are treated either poorly or wrongly by the models, could be enumerated; but perhaps the greatest challenge to the predictions of the GCMs comes from the realm of biology, as the models operate almost exclusively within the confines of physics and physical chemistry, with but little attention being given to the even more complex workings of the living world.

Consider, for example, that if the world's oceans were ever to warm (an absolute prerequisite for truly global climate change), the productivity of the unicellular algae or phytoplankton that live in their surface waters would be dramatically increased, as we know from the results of several laboratory and field investigations.[30,31] Consider next that, as these phytoplankton photosynthesize more rapidly and increase in number, they produce more copious quantities of a substance called dimethylsulfoniopropionate,[32] which is believed to buffer them against the high osmotic pressure of seawater.[33] While they yet live, this substance remains within them; but when they either die or are eaten by zooplankton, it is released to the surrounding water,[34,35] where a portion of it slowly decomposes to produce dimethyl sulfide or DMS,[36] a volatile compound of sulphur that diffuses into the atmosphere where it is rapidly oxidized by OH and NO_3 radicals to produce sulfuric and methanesulfonic acid particles,[37,38,39] which subsequently function as cloud condensation nuclei or CCN.[40,41,42] And with more CCN in the atmosphere, new clouds form where before there were none, while pre-existent clouds experience an increase in their droplet number concentrations, both of which effects tend to increase the planet's reflectance of incoming solar radiation and cool the globe.[43,44,45] Just the latter of these two effects, in fact, has been estimated to be of equivalent magnitude to the (greatly inflated) CO_2 greenhouse effect and, of course, to act in opposition to it.[46]

Not to be outdone by the marine biota, life on land does its part too. DMS emission rates from soils, for example, generally double with each 5°C increase in temperature,[47] due to the increase in sulphur-emitting microbial activity associated with greater warmth.[48,49,50] In addition, DMS fluxes from soils have been observed to increase by a factor of five or more with increases in soil organic matter,[47] which, as we shall see shortly, comes about as a direct result of the aerial fertilization effect of atmospheric CO_2 enrichment.[4] Hence, it is clear that Earth's intricate web of life has great power to regulate the surface temperature of the globe and maintain it within a range suitable for its own continued existence. That it has done so for neigh unto four billion years is also evidence for the fact that it will likely continue to do so for yet a long while to come, the many onslaughts of man notwithstanding.

The Historical Climate Record

In spite of the many shortcomings that currently plague the science of climate simulation, some people point to the apparent warmth of the past decade and suggest that it presages the imminent fulfillment of the computer prophecies.[51,52,53,54] The trouble with this type of thinking is that it is extremely myopic. For one thing, the apparent warming may be nothing more than that, apparent; for the global land-surface temperature record is fraught with many problems, not the least of which is the considerable warming experienced at many of the measurement sites over the past century, due to intensification of the urban heat island effect as cities have grown in size and population.[55,56,57] Secondly, much of the world was a degree or two warmer about 6,000 and 1,000 years ago (remember the Vikings?), when the CO_2 content of the atmosphere was fully 80 ppm less than it is today.[58] And finally, there are at least three solar-modulated cycles of climate known to operate on time scales of centuries to millennia; and all of them are presently in an ascending phase indicative of warming. As a result, nothing in the historical climate record can be construed to suggest the likelihood of an imminent CO_2-induced greenhouse catastrophe, nor will any trend of the next few decades—even substantial warming—be capable of resolving the issue, as there are other reasons for expecting the Earth's temperature to be rising at the present time and it is clearly possible to have considerably warmer temperatures than those of the present with much lower atmospheric CO_2 contents.

In spite of this unsettling state of affairs, i.e., our obvious inability to validly attribute any current or near-future warming to atmospheric CO_2 enrichment, or, more truthfully, because of this very fact, many anti-CO_2 activists have felt justified in promoting their policies all the more. They say that we cannot afford to wait for unambiguous evidence of CO_2-induced warming, suggesting that by the time we can be sure of the cause-and-effect connection (if CO_2 is indeed implicated as the responsible agent), it will be too late to avert the predicted catastrophe.[60,61,62,63]

Now if this were all there were to the story, I would be the first to agree with them. But, as is usually the case in the real world, things are just not that simple.

The Human Connection

To begin with, the primary factor believed by most people to be the ultimate cause of nearly all of our environmental problems is the ever-increasing mass of humanity.[64,65] In fact, global human population growth is perhaps the best predictor we have for estimating future atmospheric CO_2 concentrations, as demonstrated by the relationship of Figure 1.[4]

Nevertheless, the true source of our real environmental problems is not human population per se, but the disparity that exists between the sum total of our needs and wants and the ability of the Earth to provide for them.

But even this acknowledgement does not go to the heart of the problem; for it is recognized by those who have carefully analyzed the

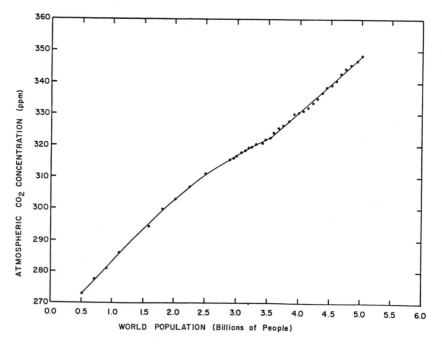

Figure 1. Global atmospheric CO_2 concentration vs. world population

situation that we presently have the ability and resources to adequately provide for several times the number of people presently inhabiting the planet and that it is "only" the political differences that exist among the peoples of the world and the criminal elements that are found in almost all societies that prevent the attainment of such a utopian state.[66,67] Because of these two aspects of the "dark side" of human nature, a sizeable fraction of Earth's human population goes hungry and undernourished each year, with some even starving to death.[68] Yet, in the face of all of the misery wrought by this sad state of affairs, the proliferation of our species continues unabated.

Clearly, it is the pressure created by this mismatch of human needs and wants and current levels of biospheric productivity which impels man to desecrate the environment in an unrelenting quest for an improved standard of living and, in some cases, for mere survival. Of course, the problem could be resolved by a basic change in human behavior; but based on everything we have learned from human history, such a radical transformation of all of humankind is just not in the offing. Likewise, the problem could perhaps be held in check by the stabilization of our numbers; but that too does not appear to be something that will soon occur. Hence, there remains but one other way of reducing the pressure we are presently placing on our global life support system and that is to make the system itself more efficient and productive everywhere.[69]

But how can we have any positive influence on the way the global biosphere conducts its business? For one thing, we can leave it alone. It's done just fine without us for billions of years; and it would probably be doing quite well today without the many assaults that we regularly make upon it. This philosophy, of course, is the foundation of the "thou shalt not" approach to dealing with environmental problems, an approach that has characterized almost all groups concerned about global change from day one. It is simple, and it has generally proven beneficial. Nevertheless, there are numerous indications that the effectiveness of this negative-action philosophy may have just about run its course in terms of our interaction with the biosphere, and that a whole new paradigm may be needed to insure the future well-being of our planet's many life forms.

CO_2 and Plant Growth

Consider the fact that CO_2 is the primary raw material used by plants in the photosynthetic production of the ultimate food source of almost all life. Consider also that, over most of the 3.8 billion years of life's existence on our planet, CO_2 has generally been present in the atmosphere in much greater quantities than those characteristic of the entire period of human history.[70,71,72] Given these two established facts, it is only logical to expect that plants would be better adapted to a higher atmospheric CO_2 concentration than that of the present, and that with more CO_2 in the air they would produce more organic matter and do it more efficiently.

Fortunately, not only is this conclusion logical, it is true as well.[4] We know from literally hundreds of field and laboratory experiments, for example, that a simple 330 to 660 ppm doubling of the air's CO_2 content will raise the productivity of all plants, in the mean, by about one-third.[73,74] Likewise, we know that such a CO_2 increase will decrease the per-unit-leaf-area transpiration or evaporative water loss rate of all plants, in the mean, by about one-third.[75] Hence, in terms of the amount of organic matter produced (P) per unit of water transpired in the process (T), plant water use efficiency (WUE) essentially doubles with a 330 to 660 ppm doubling of the air's CO_2 content, i.e., WUE = P/T = 1.333/0.666 = 2.00. And that is why biologists refer to the rapidly rising CO_2 content of Earth's atmosphere as atmospheric CO_2 enrichment, and why agriculturalists refer to it as aerial CO_2 fertilization. It is probably the single best thing that could ever happen to the biosphere.[4]

But this phenomenon is only the beginning of the good news; for as atmospheric CO_2 concentrations more than double, plant water use efficiencies more than double, with significant improvements occurring all the way out to CO_2 concentrations of a thousand ppm or more.[76,77,78]

Think of what such a biological transformation will mean to the world of the future. Grasslands will flourish where deserts now lie barren.[79,80] Shrubs will grow where only grasses grew before.[81,82] And forests will make a dramatic comeback to reclaim many areas presently sustaining only brush and scattered shrubs.[83] Increased plant cover, in turn, should greatly stabilize the world's valuable topsoil and protect it from wind and water erosion; while greater plant growth will return more organic matter to the soil, and deeper-penetrating roots will mine essential nutrients from greater soil depths to increase the fertility of upper soil horizons.[4]

Greater root production will additionally lead to greater populations of important soil microorganisms and fungi in the rhizosphere region surrounding the roots;[84] and greater rhizosphere activity should better protect host plants against the deleterious effects of soil toxins,[85,86] while enhancing their uptake of needed trace nutrients.[87,88,89] Likewise, increased soil microbial activity should enhance the rate at which polluted water moving through the soil is naturally detoxified, thereby helping to improve the quality of our strategic groundwater supplies.[90,91]

Larger and more vigorous populations of rhizosphere organisms will also accelerate weathering processes and soil formation,[92] as well as nitrogen fixation from the atmosphere.[93,94,95] And greater fungal colonization of more extensive root systems should promote an underground mutualism that fosters cooperation among species and leads to greater biological diversity and species richness,[96,97,98] characteristics of past high-CO_2 periods of Earth's history that are somewhat akin to those currently found in tropical rain forests.[99]

Biological Bootstrapping

As just one example of this multitude of growth-promoting biological feedback phenomena, consider the lowly earthworm, whose legendary soil-forming abilities were studied by Charles Darwin well over a century ago[100] and which continues to amaze plant and soil scientists.[101] By ingesting and digesting plant residues and then egesting the transformed cast material, earthworms redistribute nutrients and make them more available to plants.[102] Their burrows provide drainage for water that accumulates on the soil surface during intense rainfall, saving more of it for later utilization by deep-rooted plants, and they enable plant roots to explore deeper soil layers via channels they construct.[104] In a nutshell, these oft-neglected members of the soil ecosystem play a major role in maintaining and improving the fertility, structure, aeration, and drainage of both natural and agricultural lands.

So what do all of these beneficent by-products of earthworm activities have to do with CO_2? Simply this: higher levels of atmospheric CO_2 promote greater levels of plant productivity, which returns more organic matter to the soil. And the most important factor in maintaining good earthworm populations is that there be an adequate supply of soil organic matter.[101] Hence, atmospheric CO_2 enrichment promotes greater plant growth, which promotes greater earthworm activity, which promotes greater plant growth, and so on, until before you know it, the biosphere is pulling itself up by its own bootstraps, so to speak, as it also does via the other indirect effects mentioned previously. Consequently, the long-term biological benefits of atmospheric CO_2 enrichment are likely to be much greater than what is suggested by its already phenomenal direct effects.

This "compound interest" phenomenon is currently manifesting itself in a number of experimental studies. At the U.S. Water Conservation Laboratory in Phoenix, Arizona, for example, sour orange trees planted as seedlings just three years ago and grown in open-top chambers continually supplied with an extra 300 ppm of CO_2 have nearly tripled in biomass as compared to originally identical trees maintained in similar ambient-CO_2 chambers.[105,106] And in a similar 3-year study of a natural wetland ecosystem on a subestuary of Chesapeake Bay, yearly carbon sequestering in a number of twice-normal-ambient open-top CO_2- enrichment chambers has maintained itself at more than double the rate exhibited in the control chambers for all three years of the experiment.[107]

But even these studies cannot possibly be exhibiting the ultimate effects of the many positive feedback phenomena noted above; yet they already show productivity enhancements that are many times greater than those generally found in short-term studies of plants in greenhouse and growth chamber experiments. Hence, in consideration of these findings with respect to the CO_2-induced stimulation of in situ vegetation, plus the very real potential for vegetation to expand into new territories under such circumstances, it is not inconceivable that the rapidly rising CO_2 content of Earth's atmosphere could ultimately lead to a full order of magnitude increase in the totality of Earth's life processes.[4]

Brave New World

Is this, then, humanity's redeeming grace, the one good deed we have performed (albeit unknowingly) for the other life forms with which we share the planet? Is our returning to the atmosphere of a sizeable fraction of the life- giving carbon locked away in the bowels of the Earth so many millennia ago our great redemptive environmental act? I believe that it is, and that it will ultimately prove the salvation of our entire global life support system.

Consider, for example, that the pre-industrial CO_2 content of the air (270 ppm)[4] represented a drop of 99.6% from the 70,000 ppm value believed to have been characteristic of Earth's early atmosphere some 3.5 billion years ago at the birth of the biosphere.[108] Consider also that most of the plants presently inhabiting the planet cannot survive below CO_2 concentrations in the range of 50 to 100 ppm,[109] and that during the coldest part of the last ice age the CO_2 content of the air dropped to a value of about 180 ppm.[110,111] As a result of these two observations, even James Lovelock, the father of the Gaia hypothesis (the idea that the biosphere keeps the physical-chemical state of the atmosphere within bounds that are always conducive to its own continued existence), has gone so far as to state that the demise of the biosphere, geologically speaking, is just around the corner;[20] for the primary mechanism whereby the biosphere has modulated Earth's climate in the face of an ever-increasing solar luminosity,[112] i.e., its gradual reduction of the air's CO_2 content,[113] has just about run its course, in terms of further temperature regulation capability, while simultaneously creating such a low atmospheric CO_2 content as to be a bona fide threat to the ability of many plants to maintain essential life functions.

Clearly, the CO_2 content of Earth's atmosphere must be increased in the long-term, to prevent the extinction of life which would surely result from the disappearance of our major planetary food source; and it must also be increased in the short-term, to provide the increased global productivity needed to relieve the pressures currently being exerted by our species on the rest of the biosphere. In addition, the very small greenhouse effect that a doubling, tripling, or quadrupling of the air's CO_2 content might induce may be just what is needed to prevent our slipping into the next "scheduled" ice age.[114] Hence, guided by God, Gaia, or just plain Global Serendipity, we could well look upon man's flooding of the atmosphere with CO_2 as a blessing in disguise. Planned by a superior intelligence, orchestrated by the blind forces of nature, or but a grand piece of cosmic good luck, it appears to be just what is needed at the present time.

This is not to say, however, that all of man's activities are of benefit to the biosphere, or even benign, or that it is time to totally transform the past paradigm of environmental activism into a new form of passivism that would disavow the responsibility that our intellect confers upon us. Clearly, there are many problems arising from our position of planetary dominance that are very real, very current, and for which we do possess solutions.

We can act now, for example, to curtail the emissions of greenhouse gases other than CO_2, which are, in fact, estimated to have an aggregate climatic impact equivalent to that of CO_2,[6-8] and which have a number of other deleterious environmental effects as well.[115] We can act now to prevent the wholesale destruction of tropical forests, a regrettable enterprise that is estimated to be driving perhaps as many as 100 different species to extinction every single day,[116] and which surely must stand as a moral indictment of our own species. We can act now to become more efficient in our utilization of energy, which is but the most remedial of common sense actions to be encouraged in a world of debt-strapped nations. And we can act now to plant trees, which, like beauty, are their own excuse for being, in addition to providing a whole host of environmental and economic benefits.

Yes, there is much that the tried and true ways of the past may yet validly contribute to the future well-being of the biosphere; but we must gain a grander vision of our place in nature to truly transform the globe into the biological paradise which once it was and which again it could become. Throughout its entire history, *homo sapiens* has known nothing but a carbon-depleted world. Now, however, as a result of our own doing, the natural world is beginning to breath deeply the breath of life, as evidenced by the yearly-increasing amplitude of the seasonal CO_2 cycle,[117] which is primarily driven by the returning prowess of the planet's global photosynthetic capacity in response to mankind's flooding of the air with CO_2.[118] And as the biosphere awakens from the great lethargy of the past two millennia of CO_2 starvation, it will likely orchestrate its destiny with increasing vigor, and thereby preserve its vitality, diversity, and integrity for many millennia to come.

At this point in time we thus find ourselves confronted by two incredible futures: the end of nature or the rebirth of the biosphere. Let us not, through ignorance, do that which would revoke our newfound birthright, and prevent our passage into maturity as a living, breathing, cognitive world, functioning to its fullest potential, by disrupting the basic fabric of industrialized society in an ill-advised and futile attempt to turn back the very phenomenon that may prove our salvation. The choice is ours; and we will choose, one future or the other.

References

1. W. McKibben, *The End of Nature*, Random House, New York, NY, 1989.
2. W. McKibben, 'Reflections: the end of nature,' *The New Yorker*, Vol. 65, No. 30, 1989, pp. 47–105.
3. F.A. Koomanoff, *Atmospheric Carbon Dioxide and the Greenhouse Effect*, U.S. Department of Energy, Washington, DC, 1989.
4. S.B. Idso, *Carbon Dioxide and Global Change: Earth in Transition*. IBR Press, Tempe, Arizona, 1989.
5. S.H. Schneider, 'The changing climate,' *Scientific American*, Vol. 261, No. 3, Sept. 1989, pp. 70–79.
6. J. Hansen, A. Lacis and M. Prather, 'Greenhouse effect of chlorofluorocarbons and other trace gases,' *Journal of Geophysical Research*, Vol. 94, 1989, pp. 16, 417–16, 421.
7. D.A. Lashof and D.R. Ahuja, 'Relative contributions of greenhouse gas emissions to global warming,' *Nature*, Vol. 344, 1990, pp. 529–531.
8. H. Rodhe, 'A comparison of the contribution of various gases to the greenhouse effect,' *Science*, Vol. 248, 1990, pp. 1217–1219.
9. R.H. Gammon, E.T. Sundquist and P.J. Fraser, 'History of carbon dioxide in the atmosphere,' in J.R. Trabalka (ed), *Atmospheric Carbon Dioxide and the Global Carbon Cycle*, U.S. Department of Energy, Washington, DC, 1985.

10. D.A. Lashof, 'The dynamic greenhouse: feedback processes that may influence future concentrations of atmospheric trace gases and climate change,' *Climate Change,* Vol. 14, 1989, pp. 213–242.

11. M.E. Schlesinger and J.F.B. Mitchell, 'Climate model simulation of the equilibrium climatic response to increased carbon dioxide,' *Reviews of Geophysics,* Vol. 25, 1987, pp. 760–798.

12. S.B. Idso, 'An empirical evaluation of earth's surface air temperature response to an increase in atmospheric carbon dioxide concentration,' in R.A. Reck and J.R. Hummel (eds), *AIP Conference Proceedings No. 82: Interpretation of Climate and Photochemical Models, Ozone and Temperature Measurements,* American Institute of Physics, New York, NY, 1982, pp. 119–134.

13. Y.I. Oyama, G.C. Carle, F. Woeller and J.B. Pollack, 'Venus lower atmospheric composition: analysis by gas chromatography,' *Science,* Vol. 203, 1979, pp. 802–805.

14. J.B. Pollack, O.B. Toon and R. Boese, 'Greenhouse models of Venus' high surface temperature, as constrained by Pioneer Venus measurements,' *Journal of Geophysical Research,* Vol. 85, 1980, pp. 8223–8231.

15. J.F. Kasting, O.B. Toon and J.B. Pollack, 'How climate evolved on the terrestrial planets,' *Scientific American,* Vol. 258, No. 2, 1988, pp. 90–97.

16. J.B. Pollack, 'Climate change on the terrestrial planets,' *Icarus,* Vol. 37, 1979, pp. 479–553.

17. C. McKay, 'Section 6. Mars,' in R.E. Smith and G.S. West (eds), *Space and Planetary Environment Criteria Guidelines for Use in Space Vehicle Development, 1982 Revision (Volume 1),* National Aeronautics and Space Administration, Marshall Space Flight Center, Alabama, 1983.

18. S.B. Idso, 'The CO_2 greenhouse effect on Mars, Earth and Venus,' *The Science of the Total Environment,* Vol. 77, 1988, pp. 291–294.

19. M. Schwarzschild, R. Howard and R. Harm, 'Inhomogeneous stellar models. V. A solar model with convective envelope and inhomogeneous interior,' *Astrophysics Journal,* Vol. 125, 1957, pp. 233–241.

20. J.E. Lovelock and M. Whitfield, 'Life span of the biosphere,' *Nature,* Vol. 296, 1982, pp. 561–563.

21. Y. Fouquat, J.C. Buriez, M. Herman and R.S. Kandel, 'The influence of clouds on radiation: a climate modeling perspective,' *Reviews of Geophysics,* Vol. 28, 1990, pp. 145–166.

22. V. Ramanathan, R.D. Cess, E.F. Harrison, P. Minnis, B.R. Bankstrom, E. Ahmad and D. Hartmann, 'Cloud–radiative forcing and climate: insights from the earth radiation budget experiment,' *Science,* Vol. 243, 1989, pp. 57–63.

23. V. Ramanathan, B.R. Barkstrom and E.F. Harrison, 'Climate and the earth's radiation budget,' *Physics Today,* Vol. 42, 1989, pp. 22–32.

24. J.T. Kiehl and V. Ramanthan, 'Comparison of cloud forcing derived from the earth radiation budget experiment with that simulated by the NCAR community climate model,' *Journal of Geophysical Research,* Vol. 95, 1990, pp. 11,679–11,698.

25. A. Henderson-Sellers, 'Cloud changes in a warmer Europe,' *Climatic Change,* Vol. 8, 1986, pp. 25–52.

26. A. Henderson-Sellers, 'Increasing cloud in a warming world,' *Climatic Change,* Vol. 9, 1986, pp. 267–309.

27. K. McGuffie and A. Henderson-Sellers, 'Is Canadian cloudiness increasing?' *Atmosphere—Ocean,* Vol. 26, 1988, pp. 608–633.

28. J.F.B. Mitchell, C.A. Senior and W.J. Ingram, 'CO_2 and climate: a missing feedback?' *Nature,* Vol. 341, 1989, pp. 132–134.

29. A. Slingo, 'Wetter clouds dampen global greenhouse warming,' *Nature,* Vol. 341, 1989, p. 104.

30. R.W. Eppley, 'Temperature and phytoplankton growth in the sea,' *Fisheries Bulletin,* Vol. 70, 1972, pp. 1063–1085.

31. G.-Y. Rhea and I.J. Gotham, 'The effect of environmental factors on phytoplankton growth: temperature and the interactions of temperature with nutrient limitation,' *Limnology and Oceanography,* Vol. 26, 1981, pp. 635–648.

32. A. Vairavamurthy, M.O. Andreae and R.L. Iverson, 'Biosynthesis of dimethylsulfide and dimethylpropiothetin by *Hymenomonas carterae* in relation to sulfur source and salinity variations,' *Limnology and Oceanography,* Vol. 30, 1985, pp. 59–70.

33. K. Caldeira, 'Evolutionary pressures on planktonic production of atmospheric sulphur,' *Nature,* Vol. 337, 1989, pp. 732–734.

34. B.C. Nguyen, S. Belviso, N. Mihalopoulos, J. Gostan and P. Nival, 'Dimethyl sulfide production during natural phytoplankton blooms,' *Marine Chemistry,* Vol. 24, 1988, pp. 133–141.

35. J.W.H. Dacey and S.G. Wakeham, 'Oceanic dimethylsulfide: production during zooplankton grazing on phytoplankton,' *Science,* Vol. 233, 1988, pp. 1314–1316.

36. S.M. Turner, G. Malin, P.S. Liss, D.S. Harbour and P.M. Halligan, 'The seasonal variation of dimethyl sulfide and dimethylsulfoniopropionate concentrations in nearshore waters,' *Limnology and Oceanography,* Vol. 33, 1988, pp. 364–375.

37. A.D. Clark, N.C. Ahlquist and D.S. Covert, 'The Pacific marine aerosol: evidence for natural acid sulfates,' *Journal of Geophysical Research,* Vol. 92, 1987, pp. 4179–4190.

38. M.O. Andreae, H. Barresheim, T.W. Andreae, M.A. Kritz, T.S. Bates and J.T. Merril, 'Vertical distribution of dimethylsulfide, sulfur dioxide, aerosol ions and radon over the northeast Pacific Ocean,' *Journal of Atmospheric Chemistry,* Vol. 6, 1988, pp. 149–173.

39. S.M. Kreidenweis and J.H. Seinfeld, 'Nucleation of sulfuric acid-water and methanesulfonic acid-water solution particles: implications for the atmospheric chemistry of organosulfur species,' *Atmospheric Environment,* Vol. 22, 1988, pp. 283–296.

40. W.P. Elliott and R. Egami, 'CCN measurements over the ocean,' *Journal of the Atmospheric Sciences,* Vol. 32, 1975, pp. 371–374.

41. W.A. Hoppel, 'Measurement of the size distribution and CCN supersaturation spectrum of submicron aerosols over the ocean,' *Journal of the Atmospheric Sciences,* Vol. 36, 1979, pp. 2006–2015.

42. V.K. Saxena, 'Evidence of the biogenic nuclei involvement in Antarctic coastal clouds,' *Journal of Physical Chemistry,* Vol. 87, 1983, p. 4130.

43. J.H. Conover, 'Anomalous cloud lines,' *Journal of the Atmospheric Sciences,* Vol. 23, 1966, pp. 778–785.

44. J.A. Coakley, R.L. Bernstein and P.A. Durkee, 'Effect of shipstack effluents on cloud reflectivity,' *Science,* Vol. 237, 1987, pp. 1020–1022.

45. R.A. Scorer, 'Ship trails,' *Atmospheric Environment,* Vol. 21, 1987, pp. 1417–1425.

46. J.E. Lovelock, *The Ages of Graia: A Biography of Our Living Earth,* W.W. Norton & Co., New York, NY, 1988.

47. R. Staubes, H.-W. Georgii and G. Ockelmann, 'Flux of COS, DMS and CS2 from various soils in Germany,' *Tellus,* Vol. 41B, 1989, pp. 305–313.

48. F.B. Hill, V.P. Aneja and R.M. Felder, 'A technique for measurement of biogenic sulfur emission fluxes,' *Environmental Science and Health,* Vol. 13, 1978, pp. 199–225.

49. P.F. Adams, S.O. Farwell, E. Robinson, M.R. Pack and W.L. Barnesberger, 'Biogenic sulfur source strength,' *Environmental Science and Technology,* Vol. 15, 1981, pp. 1493–1498.

50. P.L. MacTaggart, D.F. Adams and S.O. Farwell, 'Measurement of biogenic sulfur emissions from soils and vegetation using dynamic enclosure methods: total sulfur gas emissions via MFC/FD/FPD determinations,' *Journal of Atmospheric Chemistry,* Vol. 5, 1987, pp. 417–437.

51. S. Begley, M. Miller and M. Hager, 'The endless summer?' *Newsweek,* Vol. 122, No. 2, 1988, pp. 18–20.

52. I. Anderson, 'Greenhouse warming grips the American corn belt,' *New Scientist,* Vol. 118, No. 1619, 1988, p. 35.

53. A.C. Revkin. 'Endless summer: living with the greenhouse effect,' *Discover,* Vol. 9, No. 10, 1988, pp. 50–61.

54. P. Shabecoff, 'Global warming has begun, expert tells Senate,' *The New York Times,* Vol. 137, No. 47,546, 1988, p. A1, A14.

55. F.B. Wood, Jr., 'Comment: on the need for validation of the Jones et al. temperature trends with respect to urban warming,' *Climatic Change,* Vol. 12, 1988, pp. 297–312.

56. T.R. Karl, H.F. Diaz and G. Kukla, 'Urbanization: its detection and effect in the United States climate record,' *Journal of Climate,* Vol. 1, 1988, pp. 1099–1123.

57. R.C. Balling, Jr. and S.B. Idso, 'Historical temperature trends in the United States and the effect of urban population growth,' *Journal of Geophysical Research,* Vol. 94, 1989, pp. 3359–3363.

58. S.B. Idso, 'Greenhouse warming or Little Ice Age demise: a critical problem for climatology,' *Theoretical and Applied Climatology,* Vol. 39, 1988, pp. 54–56.

59. P.E. Damon, 'The new warm epoch: solar activity vs. the greenhouse effect,' Proceedings of the NASA ERIM conference on Earth Observations and Global Change Decision Making: A National Partnership, submitted, 1990.

60. E.J. Barron, 'Earth's shrouded future: the unfinished forecast of global warming,' *The Sciences,* Vol. 29, No. 5, 1989, pp. 14–20.

61. J. Ratloff, 'Governments warm to greenhouse action,' *Science News,* Vol. 136, 1989, pp. 394–397.

62. Anonymous, 'The great sky greenhouse,' *Nature*, Vol. 345, 1990, p. 371.

63. C. Flavin, 'Slowing global warming,' *American Forests*, Vol. 96, Nos. 5 & 6, 1990, pp. 37–44.

64. R.A. Bryson, 'Environmental opportunities and limits for development,' *Environmental Conservation*, Vol. 16, 1989, pp. 299–305.

65. F.H. Borman, 'The global environment deficit,' *Bioscience*, Vol. 40, No. 2, 1990, p. 74.

66. R. Revelle, 'Food and population,' *Scientific American*, Vol. 231, No. 3, 1974, pp. 161–170.

67. P.F. Low, 'Realities of the population explosion,' *The Ensign*, May, 1971, pp. 18–27.

68. J.L. Simon, 'Resources, population, environment: an oversupply of false bad news,' *Science*, Vol. 208, 1980, pp. 1431–1437.

69. R. Dudal, 'Land degradation in a world perspective,' *Journal of Soil and Water Conservation*, Sept.–Oct., 1982, pp. 245–249.

70. M.H. Hart, 'The evolution of the atmosphere of the earth,' *Icarus*, Vol. 33, 1978, pp. 23–29.

71. H.D. Holland, *The Chemical Evolution of the Atmosphere and Oceans*, Princeton University Press, Princeton, NJ, 1984.

72. R.A. Berner, 'Atmospheric carbon dioxide levels over Phanerozoic time,' *Science*, Vol. 249, 1990, pp. 1382–1386.

73. B.A. Kimball, 'Carbon dioxide and agricultural yield: an assemblage and analysis of 430 prior observations,' *Agronomy Journal*, Vol. 75, 1983, pp. 779–788.

74. B.A. Kimball, *Carbon Dioxide and Agricultural Yield: An Assemblage and Analysis of 770 Prior Observations*, U.S. Water Conservation Laboratory, Phoenix, AZ, 1983.

75. B.A. Kimball and S.B. Idso, 'Increasing atmospheric CO_2: effects on crop yield, water use and climate,' *Agricultural Water Management*, Vol. 7, 1983, pp. 55–72.

76. H.Z. Enoch, J. Rylski and M. Spigelman, 'CO_2 enrichment of strawberry and cucumber plants grown in unheated greenhouses in Israel,' *Scientia Horticulurae*, Vol. 5, 1976, pp. 33–41.

77. H.H. Rogers. J.R. Thomas and G.E. Bingham, 'Response of agronomic and forest species to elevated atmospheric carbon dioxide,' *Science*, Vol. 220, 1983, pp. 428–429.

78. P.J. Breen, J.D. Hesketh and D.B. Peters, 'Field measurements of leaf photosynthesis of C_3 and C_4 species under high irradiance and enriched CO_2,' *Photosynthetica*, Vol. 20, 1986, pp. 281–285.

79. S.B. Idso, 'What if increases in atmospheric CO_2 have an inverse greenhouse effect? I. Energy balance considerations related to surface albedo,' *Journal of Climatology*, Vol. 4, 1984, pp. 399–409.

80. S.B. Idso, 'Industrial age leading to the greening of the Earth?' *Nature*, Vol. 320, 1986, p. 22.

81. C.H. Donaldson, *Brush Encroachment with Special Reference to the Blackthorn Problem of the Molopo Area*, Department of Agricultural and Technical Services, Pretoria, South Africa, 1969.

82. T.T. Veblen and V. Markgraf, 'Steppe expansion in Patagonia?' *Quarternary Research*, Vol. 30, 1988, pp. 331–338.
83. S.B. Idso and J.A. Quinn, *Vegetational Redistribution in Arizona and New Mexico in Response to a Doubling of the Atmospheric CO_2 Concentration*, Laboratory of Climatology, Arizona State University, Tempe, AZ, 1983.
84. M.R. Lamborg, R.W. Hardy and E.A. Paul, 'Microbial effects,' in E.R. Lemon (ed.), *CO_2 and Plants: The Response of Plants to Rising Levels of Atmospheric Carbon Dioxide*, Westview Press, Boulder, CO, 1983, pp. 131–176.
85. J.W. Donovan and M. Alexander, 'Microbial formation of volatile selenium compounds in soil,' *Soil Science Society of America Journal*, Vol. 41, 1977, pp. 70–73.
86. J.T. Koch, D.B. Rachar and B.D. Kay, 'Microbial participation in iodide removal from solution by organic soils,' *Canadian Journal of Soil Sciences*, Vol. 69, 1989, pp. 127–135.
87. R.J. Luxmoore, E.G. O'Neill, J.M. Ells and H.H. Rogers, 'Nutrient-uptake and growth responses of Virginia pine to elevated atmospheric CO_2,' *Journal of Environmental Quality*, Vol. 15, 1986, pp. 244–251.
88. R.J. Norby, E.G. O'Neill and R.J. Luxmoore, 'Effects of atmospheric CO_2 enrichment on the growth and mineral nutrition of Querous alba seedlings in nutrient-poor soil,' *Plant Physiology*, Vol. 82, 1986, pp. 83–89.
89. E.G. O'Neill, R.J. Luxmoore and R.J. Norby, 'Elevated atmospheric CO_2 effects on seedling growth, nutrient uptake, and rhizosphere bacterial populations of *Liriodendron tulipifera* L., *Plant and Soil*, Vol. 104, 1987, pp. 3–11.
90. M. Alexander, 'Biodegradation of chemicals of environmental concern,' *Science*, Vol. 211, 1981, pp. 132–138.
91. S.R. Hutchins, M.B. Tomson and C.H. Ward, 'Microbial involvement in trace organic removal during rapid infiltration recharge of ground water,' in D.E. Caldwell, J.A. Brierley and C.L. Brierley (eds.), *Planetary Ecology*, Van Nostrand Reinhold, New York, NY, 1985, pp. 370–382.
92. N.J. Rosenberg, B.L. Blad and S.B. Verma, *Microclimate: The Biological Environment*, Wiley Interscience, New York, NY, 1983.
93. E.G. Mulder and W.L. Van Veen, 'The influence of carbon dioxide on symbiotic nitrogen fixation,' *Plant and Soil*, Vol. 13, 1960, pp. 265–278.
94. R.W.F. Handy and U.D. Havelka, 'Symbiotic N2 fixation: multifold enhancement by CO_2-enrichment of field-grown soybeans,' *Plant Physiology Supplement*, Vol. 48, 1973, p. 35.
95. D.A. Phillips, K.D. Newell, S.A. Hassel and C.E. Felling, 'The effect of CO_2 enrichment on root nodule development and symbiotic N_2 reduction in *Pisum sativum* L.,' *American Journal of Botany*, Vol. 63, 1976, pp. 356–362.
96. C.P.P. Reid and F.W. Woods, 'Translocation of ^{14}C-labelled compounds in mycorrhizae and its implications in interplant nutrient cycling,' *Ecology*, Vol. 50, 1969, pp. 179–187.
97. N. Chiariello, J.C. Hickman and H.A. Mooney, 'Endomycorrhizal role for interspecific transfer of phosphorus in a community of annual plants,' *Science*, Vol. 217, 1982, pp. 941–943.

98. J.H. Warcup, 'Mycorrhizal associations and seedling development in Australian Lobelioideae (Campanulaceae),' *Australian Journal of Botany*, Vol. 36, 1988, pp. 461–472.

99. D.I. Axelrod, 'An interpretation of high montane conifers in western Tertiary floras,' *Paleobiology*, Vol. 14, 1988, pp. 301–306.

100. C. Darwin, *The Formation of Vegetable Mould through the Action of Worms, with Observations on their Habitats*, John Murray, London, UK, 1881.

101. C.A. Edwards, 'Earthworms and agriculture,' *Agronomy Abstracts*, Vol. 80, 1988, p. 274.

102. A.N. Sharpley, J.K. Syers and J. Springett, 'Earthworm effects on the cycling of organic matter and nutrients,' *Agronomy Abstracts*, Vol. 80, 1988, p. 285.

103. W.D. Kemper, 'Earthworm burrowing and effects on soil structure and transmissivity,' *Agronomy Abstracts*, Vol. 80, 1988, p. 278.

104. S.D. Logsdon and D.L. Linden, 'Earthworm effects on root growth and function, and on crop growth,' *Agronomy Abstracts*, Vol. 80, 1988, p. 280.

105. S.B. Idso, B.A. Kimball and S.G. Allen, 'CO_2 enrichment of sour orange trees: two and a half years into a long-term experiment,' *Plant, Cell and Environment*, in press.

106. S.B. Idso, B.A. Kimball and S.G. Allen, 'Net photosynthesis of sour orange trees maintained in atmospheres of ambient and elevated CO_2 concentration,' *Agricultural and Forest Meteorology*, Vol. 54, 1991, pp. 95–101.

107. B.G. Drake, W.J. Arp, L. Balduman, P.S. Curtis, J. Johnson, D. Kabara, P.W. Leadley, W.T. Pockman, M.L. Seliskar, D. Sutton, L.D. Whigham and L. Ziska, *Effects of Elevated CO_2 on Chesapeake Bay Wetlands. IV. Ecosystems and Whole Plant Responses. April–November 1988*, U.S. Department of Energy, Washington, DC, 1989.

108. M.H. Hart, 'The evolution of the atmosphere of the earth,' *Icarus*, Vol. 33, 1978, pp. 23–29.

109. F.B. Salisbury and C.W. Ross, *Plant Physiology*, Wadsworth Publishing Company, Belmont, CA, 1978.

110. R.J. Delmas, J.-M. Ascencio and M. Legrand, 'Polar ice evidence that atmospheric CO_2 20,000 yr BP was 50% of present,' *Nature*, Vol. 284, 1980, pp. 155–157.

111. J.M. Barnola, D. Raynaud, Y.S. Korotkevich and C. Lorius, 'Vostok ice core provides 160,000-year record of atmospheric CO_2,' *Nature*, Vol. 329, 1987, pp. 408–414.

112. D. Ezer and A.G.W. Cameron, 'A study of solar evolution,' *Canadian Journal of Physics*, Vol. 43, 1965, pp. 1497–1517.

113. T. Owen, R.D. Cess and V. Ramanathan, 'Enhanced CO_2 greenhouse to compensate for reduced solar luminosity on early Earth,' *Nature*, Vol. 277, 1979, pp. 640–642.

114. R.A. Bryson, 'Civilization and rapid climatic change,' *Environmental Conservation*, Vol. 15, 1988, pp. 7–15.

115. U.S. National Research Council, *Causes and Effects of Stratospheric*

Ozone Reductions: An Update, National Adademy Press, Washington, DC, 1982.

116. N. Myers, 'Threatened biotas: "Hotspots" in tropical forests,' *The Environmentalist,* Vol. 8, 1988, pp. 1–200.

117. G.H. Kohlmaier, E.-O. Sire, A. Janecek, C.D. Keeling, S.C. Piper and R. Revelle, 'Modelling the seasonal contribution of a CO_2 fertilization effect of the terrestrial vegetation to the amplitude increase in atmospheric CO_2 at Mauna Loa Observatory,' *Tellus,* Vol. 41B, 1989, pp. 487–510.

118. S.B. Idso, 'Comment on "Modelling the seasonal contribution of a CO_2 fertilization effect of the terrestrial vegetation to the amplitude increase in atmospheric CO_2 at Mauna Loa Observatory" by G.H. Kohlmaier et al.,' *Tellus,* in press.

119. Contribution from the Agricultural Research Service, U.S. Department of Agriculture.

It's The Water!

Peter E. Black, Ph.D.

Peter E. Black is Professor of Water and Related Land Resources at the SUNY College of Environmental Science and Forestry in Syracuse, New York where he has taught and conducted research since 1966. He was a Research Forester at the U.S. Forest Service' Coweeta Hydrologic Laboratory in North Carolina from 1956 to 1959, and taught at Humboldt State College in Arcata, California from 1961 to 1965. He has taught forest management and surveying, and currently offers courses in forest hydrology and watershed management, as well as related offerings in soil and water conservation policy, environmental impact analysis, and a new course in watershed hydrology and nonpoint source pollution. His research has focused on interception, runoff from watershed models, and water quality at the urban/rural interface. He was awarded the MF and MS degrees from the School of Natural Resources at the University of Michigan, and the PhD degree in Watershed Management from Colorado State University in 1961.

Black has published an educational film, numerous article on hydrology and water resources, and three books entitled *Environmental Impact Analysis* (Praeger, 1981), *Conservation of Water and Related Land Resources* (Rowman &

Littlefield, Second Edition, 1987); and *Watershed Hydrology* (Prentice-Hall, 1991). In 1974, he co-founded IMPACT CONSULTANTS, a private firm in Syracuse, for which he served as EIS project manager for twelve years. He is a Charter Member and Fellow of the American Water Resources Association, serving as its President in 1991.

I'm sure that we all can recall hearing how ubiquitous and important water is. Probably back in the early grades, and again, and again. If I fully understand a couple of recent books I've read, it's a lot more important than I thought. The books are *The Descent of Woman* by Elaine Morgan (1972), and *The Ages of Gaia* by John Lovelock (1990).

Morgan simultaneously contradicts several traditional evolutionists and fills in a major missing gap in fossil remains between the time our ancestors swung through the trees and the time they allegedly romped the savannahs and were officially declared to be homo erectus. She suggests, in contrast, that our ancestors spent a considerable period living at the waters' edge. Her detailed analysis includes explanation of salty tears, as well as the lack of cavemen's caves in the savannah, the partial web between the thumb and forefinger, the loss of body hair that was apparently and heretofore inexplicably coincident with the deposition of a layer of subcutaneous fat, how we managed to survive the lengthy Pliocene drought, develop speech, the use of tools and weapons, aggression, and ventral copulation. Her critical and entertaining presentation shows what we have in common with aquatic mammals and explains in simple concepts what the traditional evolutionists had developed complex theories for. In addition, she awakened many to the major issue which the traditional evolutionists failed to account for, the female's role in human evolution, as her book title proclaims. It makes sense: the diversified edge of the land-sea interface would have eroded away the fossil remains, and provided a variety of food, caves, safety of the water, and perhaps most important of all, intellectual stimulation and challenge. *One might have guessed at the primary evolutionary medium:* it's the water.

Lovelock presents compelling arguments for the early appearance of life forms that subsequently helped set the conditions under which that life would thrive. His application to the entire planet of the concept of homeostasis was a next and natural step in the evolution of thought concerning our view of the Earth, especially as we looked back at our own spaceship from the moon-bound Gemini and Apollo missions. I was so impressed with Lovelock's work that I assigned the book as one of the readings in a new course in hydrology and water quality this Spring. Part of the reason for this choice is because I wanted the students to be exposed to some new and convincing ideas, even if they are largely untestable; and the main reason was that fully two-thirds of the work treats the fundamental chemical relationships of our environment. That environment is, of course, dominated by water, and thus comprehension of the linkages between hydrology, land use, and water quality begins with understanding the fundamental relationships that exist all around us. In fact, I thought one day as I answered some questions concerning the homeostatic role of life in perpetuating and regulating environmental conditions, we have actually watched it happen on TV. When we scan a month's worth of whole-Earth atmospheric activity from the GEOS satellite, we are seeing a pulsating degree of cloud cover that is the self-limiting regulator: as the sun warms the earth and seas, the lower layers of the atmosphere are warmed, increasing evaporation, raising the moisture content of the atmosphere and increasing cloud cover; the greater degree of cloud cover increases the reflectivity of the planet, in turn putting a limit on the amount of solar radiation that reaches the surface, thus cooling us off. *One might have guessed at the primary regulatory mechanism of the planet: it's the water.*

Now, maybe you had been aware of both of these facts long before I slapped my forehead with my palm and said "Eureka!" If you were, I wish you would have told me about it. On the other hand, I'm glad I had the chance to discover it for myself, because it gave me some more insight into the importance of this wondrous substance to which I, and all of you, have decided to devote our careers. I hold it—and in fact, our basic circumstances—in a new awe.

The next step, it seems to me, is to ask the question: "Is there some message in the fact that the primary and all-important evolutionary and regulatory mechanisms on Earth are one in the same, water?" The answer is: "Of course." As we stress this planet with between the current 5 billion and the 20 billion people expected by the year 2020, the focus of our onslaught will be Earth's waters. The most obvious, but not necessarily the most important, ramification of that stress is pollution.

Not only was the 1972 goal of "zero discharge of pollutants" unreasonable and unattainable, it was folly. Wastes are a natural part of our existence and to eliminate them is impossible. We can, of course, be less wasteful, which is a horse of a very different color. Natural aquatic systems swell with the natural waste products of the flora and fauna that inhabit them. Flushing is a natural part of virtually all our wildland aquatic environments, indeed, our entire environment. Nutrients surge in the near-coastal zones, then dilute. That flushing action, the balance of Lovelock's daisies and sunlight, of foxes and rabbits, whales and plankton, carbon dioxide and oxygen are locked forever in a dance that was choreographed eons ago and is an inherent part of our very existence.

Of even greater importance is the intimate relationship between water and energy. Thus, global warming is the direct result of our inadvertent tampering with the fundamental regulatory mechanism water or, more specifically, the greenhouse gas water vapor (along with carbon dioxide and methane). Like flushing, global climate change is normal: however, we also know that the normal range of temperature is greater than those extremes we have thus far caused, and that our changes are so small we cannot fully evaluate their significance. Perhaps, more importantly, we cause those oscillations to occur more rapidly than normal: the rate of change may be a more serious problem owing to the inability of our Earth to buffer the extremes.

In a recent issue of *Science,* Paul E. Waggoner urgently and correctly points out that what is most important is our high degree of vulnerability to long term changes in precipitation caused by less severe changes in climate[1]. I believe the changes in temperature and precipitation we effect will result in extremes that are well within the normal, historical limits *but that far exceed the levels we have determined as necessary for our civilization's comfort and convenience.* It may well get too warm for our air conditioners to function to the desired level of effectiveness; or too cold for us to rapidly convert old Sol's prehistoric storage of energy in fossil fuels without adverse (unwanted) effects on the local and global climates. We may even decide that life is not worth living under conditions that are less than those to which we have become accustomed (if we manage to leave ourselves a choice). Life will go on. Just not the life that we might like to be a part of. Human beings will survive; just not in the comfort-controlled places of business and the homes we currently enjoy; not with a fancy car (or two) in the garage, a cold beer in the fridge, or a ready computer on the desk. Gaia will survive. One might have guessed that *Gaia's primary evolutionary and regulatory mechanism would survive: it's the water.*

What, then, can we do about this expected onslaught on the life-embracing waters of the Earth? As citizens and role models for the young, we can build into our daily lives that which we believe to be necessary in order to make the Earth more liveable. As leaders in our aquatic professions, we can set examples, conduct research, teach, and help others understand the importance, urgency, and fundamental nature of our circumstances. And, as members of professional and scientific associations[2], we can creatively exploit the interdisciplinary forums that are available for that purpose.

References

1. P. E. Waggoner, 1991. "U.S. Water Resources Versus an Announced But Uncertain Climate Change," *Science* 251:1002.
2. Prepared orginally as "The President's Page" in the Feburary, 1991 *Water Resources Bulletin*.

Landfills

Today's Landfills Are Light-Years Away from Yesterday's Dumps
Jay H. Lehr

South Dakota Is the Answer, What Is the Question?
Jay H. Lehr

An Alternative to Municipal Solid Waste Landfills
G. Fred Lee and R. Anne Jones

Today's Landfills Are Light-Years Away from Yesterday's Dumps

Jay H. Lehr, Ph.D.

Please see editor's biographical sketch on page v.

When I was a young boy growing up in New Jersey, our community carried its trash to an old dump on the outskirts of town. We were fortunate to have a convenient dump (an abandonment gravel quarry once used by the highway department) which created no visual eyesore.

As the years went by, I became interested in geology and wondered about the wisdom of this garbage dump. The townsfolk also became concerned as the pit filled with refuse, attracting vermin and pests. Soon, the community turned it into a sanitary landfill by covering it every few days with a layer of dirt to inhibit its attraction to animals and the movement of airborne bacteria from the site. But I, as a budding geologist, still wondered about the wisdom of our actions.

Two decades later, in the early '70s, I wondered aloud on behalf of the National Water Well Association (NWWA) before congressional committees of the House and the Senate debating first the Safe Drinking Water Act (1972–74) and then the Resource Conservation and Recovery Act (1974–76). NWWA played a major role in the protection of our ground water with the successful passage of both these laws. The distance the nation has traveled from the incredible ignorance that turned gravel quarries into garbage dumps to today's sophisticated landfills is equivalent, in astronomical terms, to light-years.

The members of NWWA, water well contractors, ground water scientists, manufacturers and suppliers alike, have dramatically benefited from the nation's new knowledge of the value of its ground water resources. The conscientious effort of scientists and the advanced technologies developed by consultants and instrument manufacturers are leading to non-polluting landfills. The ground water resource that supplies water well contractors with their livelihood is being protected for future generations. At the same time, the massive effort underway to design and install monitoring wells at old and new landfills and develop remedial management programs at old facilities has swelled the ranks and bank accounts of the ground water community.

441

We should spread the word that major advances in landfill containment and leachate collection, aimed at ground water protection, are now being put into place and are tremendously effective.

No one today can say that the state-of-the-art landfill resembles the garbage dump of yesteryear in either appearance or impact. It would be easier, in fact, to make a case that overly restrictive requirements and unrealistic risk assessments have limited the number of available landfills.

For the past 10 years NWWA has, with a growing army of like-minded environmental groups, succeeded in sharpening EPA's recognition and regulation of optimum landfill facilities. Through advancing knowledge, federal landfill requirements have improved to nearly fail-safe operations.

Tomorrow's criteria will dictate:

- Location standards

- Facility design and operating criteria

- Closure and post-closure care requirements

- Financial assurance standards

- Ground water monitoring and corrective action standards.

Regulations proposed and published in the *Federal Register* on August 30, 1988, are likely to be approved at the end of this year, and states will have 18 months to revise their own solid-waste disposal regulations so as to be at least as stringent as federal requirements.

While most of the landfill regulatory requirements are contained in amendments to the Resource Conservation and Recovery Act, landfills must also conform with requirements of other legislation as diverse as the National Historic Preservation Act of 1966, the Endangered Species Act, Coastal Management Act, Wild and Scenic Rivers Act, Fish and Wildlife Coordination Act, Clean Water Act, Clean Air Act, and the Toxic Substance Control Act.

Those who feel this country does not have a strong enough regulatory program with regard to municipal waste disposal in landfills, simply are not reading the many laws that positively impact public and private activities. Siting limitations alone are complex, because they take into account every imaginable use of adjoining land. Requirements run the gamut from the obvious to the incredible. Landfills cannot be within 200 feet of a fault that has displaced in the last 9000 years; landfills cannot be near wetlands, and strenuous requirements are placed upon them if they are within 100-year floodplains. If a landfill is sited near an airport, operators must prove that the site will not attract birds, which could interfere with the aircraft.

In the operation of landfills, detailed programs must be established to exclude all hazardous waste. They include random inspection of suspicious loads, maintaining records of these inspections, and the training of operating personnel to locate hazardous waste in municipal landfills. Tight restrictions govern the cover material that must be used daily over the disposed wastes to control disease, fires, odors, blowing litter, and scavengers. Public access to landfills must be controlled to protect human health and the environment. The systems have to be designed, built and maintained in a manner that will enable water from a 24-hour, once-in-25-year-storm event to run off without any significant negative consequence to the landfill. Landfills can no longer cause a discharge of pollutants into waters covered under the Clean Water Act or other related surface water control programs.

Today's landfills must have strict closure programs delineating how they will be managed when they are no longer actively receiving waste. These programs will be maintained for a minimum of 30 years to ensure that no significant recharge waters are allowed to move through the landfills, creating potentially deleterious leachate. Proof of financial capability to manage the landfill during post-closure is also required.

While everyone applauds the comprehensive, thoughtful, and thoroughly protective program that is being laid out to ensure that tomorrow's landfill is a totally innocuous waste-disposal facility, nothing is taken for granted. Extremely stringent ground water monitoring requirements are being emplaced. The ground water monitoring system required at all landfills must be able to detect the presence of any leakage from the site. The numbers, spacing, depth, and construction of the wells will vary with site-specific conditions. Part of the final design of the system must include a thorough characterization of the aquifer, its thickness, ground water flow rates and directions, as well as descriptions of saturated and unsaturated geologic units overlying the uppermost aquifer, its thickness, hydraulic conductivities, and porosities.

The no-nonsense approach to America's future landfills is already leading to the construction of model facilities that make it possible to dispose of waste in an environmentally sound manner. A case in point is in Cape May, New Jersey, the state with the most restrictive of all waste disposal ordinances. Cape May's new landfill, designed to protect the aquifer below it, consists from top to bottom of a loose layer of sand; a layer of clay; a synthetic PVC membrane; another layer of clay; a second sand layer; and a primary containment membrane made of

Hypolon, a chloro-sulfinated polyethylene synthetic rubber. The refuse is dumped on the top sand layer, which contains leachate collection pipes to draw off rain water and pipe it to collection areas for later treatment. There is a secondary collection system in the sand layer immediately above the primary Hypolon membrane that serves as a leak protection system. It picks up any leachate that may have passed through the PVC membrane before it encounters the primary membrane. In New Jersey nearly half a century ago, such a high-tech approach to garbage disposal in place of our community dump would have attracted a great deal of laughter.

The nation faces a major waste disposal crisis as a result of too few acceptable sites and too much waste, but the problem is no longer one of too little knowledge, technology or resolve to get the job done. There is now a need to let the public and the media know that significant advances are being made, and that there is light at the end of the tunnel. Last year's garbage-barge debacle made the plight of municipal waste appear hopeless and ludicrous. Yet that single load was but a fraction of what is received daily by many landfills.

The National Water Well Association has completed 20 years of effort in convincing the government and the public that ground water is a resource worthy of protection. In those 20 years, the United States has passed a variety of significant pieces of federal legislation that, while not perfect, will indeed protect our ground water for future generations.

Yesterday's dump, though still haunting us, is clearly a legacy of the past. Tomorrow's landfill will be a good neighbor to both our homes and playgrounds, as well as our ground water.

Reprinted, with permission, from
Water Well Journal, August, 1989

South Dakota Is the Answer, What Is the Question?

Jay H. Lehr, Ph.D.

Please see biographical sketch for Dr. Lehr on page v.

In 1987, a single garbage barge wandering from the East Coast to the Caribbean for a place to unload came to symbolize a nation with too much trash (D'Angelo, 1990). Having historically dumped 85 percent of our waste in landfills now reaching capacity, state and municipal landfills were forced to turn away garbage haulers.

Over a third of the nation's landfills will be full by the turn of the century. New York and Los Angeles will fill up before then, and Philadelphia is now full. However, the problem is not of a technological nature. We have the capability to create and run landfills that will pass the most stringent environmental requirements. The problem is that no one wants them in their own backyards.

A dramatic change in this attitude will be necessary for us to survive this gridlock. Reduction of waste and recycling will help significantly in the coming years, but they alone have no chance of being full solutions. At best, we can reduce our landfill requirements by 45 percent, not a percentile more. That will leave us with yet 100 million tons a year to dispose of. With costs already rising tenfold over the $10 a ton cost a decade ago, there is a lot of room for innovative management.

With the crisis looming ever larger, it is surprising that we have not even begun to close the gap toward a solution. It is especially puzzling when one weights a landfill with the relatively dirty industries that are widely sought by communities for their economic benefits. Tire factories and steel mills have little trouble finding a home. While "A landfill is not a rose garden," as former EPA Administrator and now Browning Ferris CEO William Ruckelshaus (1989) once put it, by requiring tight environmental controls and offering significant economic benefits, a community could become comfortable with this type of industry.

State government, though capable of taking a leadership role in this arena, has abdicated its position to local municipalities by offering few incentives whatsoever to face up to this ubiquitous problem.

Perhaps the difficulty in solving the problem lies with our inability

445

to find a villain. In this case, Pogo was never more correct than when he said, "We have net the enemy and it is us." It is hard to find solutions to problems wherein each shares the blame.

Quite frankly, few nations have the enormous (and enormously safe) landfill capabilities that the United States has (Rathje, 1989). Have you ever taken a flight from San Diego to Philadelphia? For 3,000 miles, you look down out of a plane at vast landscapes, not all of which have great human potential.

Obviously, Long Island with its high water, high population, and dependence on ground water offers no suitable locations; thus, each week over 1,500 20-ton trailers leave the Island loaded with garbage bound for points unknown. Even those points are getting fewer and fewer as more and more communities attempt to reject garbage from neighboring states and localities.

It is clearly time for some state to intelligently right this wrong and turn it to their economic advantage without environmental endangerment. And so I submit to you the question that should precede the answer I posed in the title: What state can solve the nation's landfill problems?

A well-designed solid waste management system located in an environmentally suitable hydrogeologic environment is nothing less than many other commercial enterprises that depend on utilization of a natural resource. If they are properly sited, designed, and managed, they are less destructive to the environment and pose fewer health hazards than many other intellectually "acceptable" commercial enterprises.

Sound, solid waste management is dependent upon the utilization of a suitable natural resource. That natural resource is a host geologic unit that fulfills the containment requirements for the system. Such a host geologic unit is in abundance in western South Dakota, says State Geologist Merlin Tipton. Figure 1 shows the area of western South Dakota where upper Cretaceous shales (referred to as the Cretaceous confining unit) are present at the land surface. This area represents many hundreds of thousands of acres of potentially ideal host geologic units for solid waste management systems. The physical characteristics that Tipton believes makes these Cretaceous shales a nearly textbook host area are listed below.

1. They cover a large geographic area.

2. Host geologic units vary from several hundred to several thousand feet in thickness (see Figure 2).

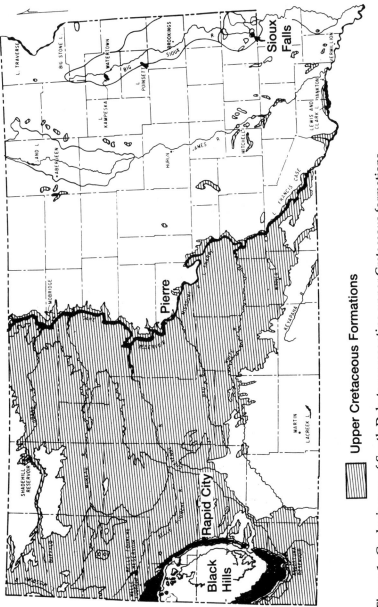

Figure 1. Geologic map of South Dakota representing upper Cretaceous formations.

	FORMATION	SECTION	THICKNESS IN FEET	
QUATERNARY	SANDS AND GRAVELS		0–50	Tertiary and Upper Cretaceous sandstone, silt, and clay.
TERTIARY — PLIOCENE	OGALLALA GROUP		0–100	
MIOCENE	ARIKAREE GROUP		0–500	
OLIGOCENE	WHITE RIVER GROUP		0–600	
PALEOCENE — FORT UNION FORMATION	TONGUE RIVER MEMBER		0–425	
	CANNONBALL MEMBER		0–225	
	LUDLOW MEMBER		0–350	
— ? —	HELL CREEK FORMATION (Lance Formation)		425	
	FOX HILLS FORMATION		25–200	
CRETACEOUS — UPPER	PIERRE SHALE		1200–2000	Upper Cretaceous confining unit
	Sharon Springs Mem.			
	NIOBRARA FORMATION		100–225	
	Turner Sand Zone			
	CARLILE FORMATION		400–750	
	Wall Creek Sands			
	GREENHORN FORMATION		(25–30) (200–350)	
GRANEROS GROUP	BELLE FOURCHE SHALE		300–550	
	MOWRY SHALE		150–250	
LOWER	NEWCASTLE SANDSTONE		20–60	
	SKULL CREEK SHALE		170–270	
INYAN KARA GROUP — LAKOTI FM	FALL RIVER [DAKOTA (?)]		10–200	
	Fuson Shale		10–188	
	Minnewaste ls.		0–25	
			25–485	

Figure 2. Host geologic rocks of upper Cretaceous confining unit and overlying Tertiary sediments.

3. The vertical and horizontal hydraulic conductivity of unweathered shale is very low. Coupled with probable low regional potentiometric surface gradients, lateral flow would be insignificant or nonexistent on a regional basis.

4. The upper portion of the host rock unit is weathered, causing increased vertical and horizontal hydraulic conductivity. Although some local horizontal movement would occur through the weathered shale, significant regional horizontal movement would be precluded by the relatively low hydraulic conductivity of the weathered host unit and the low regional gradients of the shallow groundwater regime.

5. Water in the buried aquifers underlying the entire area is under artesian pressure. This artesian pressure causes water to rise to near land surface in some areas and causes flowing wells over the rest of the

area. The combination of high hydraulic pressure in the underlying aquifer and low hydraulic conductivity of the shale precludes or substantially reduces the movement of water and chemicals form the land surface to the buried aquifers (see Figure 3).

6. Precipitation in most of the host rocks area generally ranges from 14 to 18 inches annually, resulting in semiarid conditions. These conditions minimize leachate generation and surface-water runoff.

The above statements referring to the potential desirability of shale as a host unit are based on general knowledge of shale characteristics and the assumed hydrologic similarity to till which have been under investigation by the South Dakota Geological Survey for several years. Additional research on hydrogeology of shale and site-specific characterization would need to be conducted.

Listed below are positive factors favoring the designated host geologic units shown on Figure 1.

1. Much of the land is unsuitable for anything except grazing. Even much of the cultivated land is marginal for agriculture production.

2. Land is relatively inexpensive.

3. The area is generally sparsely populated.

4. The area offers no problems relating to recent siting concerns:
   A. proximity to airports;
   B. interference with 100-year floods;
   C. interference with wetlands;
   D. proximity to fault zones;
   E. proximity to seismic impact zones.

5. South Dakota has laws, rules, and regulations which, if implemented, can effectively protect against serious environmental impacts and control solid waste management development.

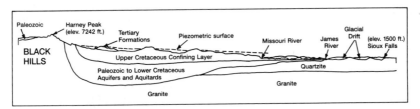

Figure 3. **Idealized cross section of South Dakota showing relationship of aquifers, water movement, and upper Cretaceous confining unit.**

Highly vocal environmentalists have already decried solid waste management proposals presented in the state with emotionally appealing references to South Dakota becoming the "garbage dump of the nation" without offering any realistic alternative, thus ignoring the problem by adhering to the NIMBY (not in my back yard) syndrome. In reality, South Dakota has already perceptually accepted solid waste management as a *positive alternative*. Those factors indicating acceptance are:

1. Some state legislators and citizenry do support solid waste management.

2. Laws, rules, and regulations governing solid waste management have been promulgated, thereby implicitly accepting receipt of solid waste by defining the conditions under which it will be accepted.

3. The legislature has not taken the initiative to render illegal the importation of out-of-state solid waste for disposal in South Dakota.

In summary, if we are to continue pursuing what Lee and Jones (1990) call the "dry tomb" approach to landfilling, we need precisely this type of South Dakota setting where there is essentially no possibility of ground-water contamination of waters that could be used for domestic water-supply purposes. This is because eventually all "dry tomb" landfills will leak when their plastic membranes degrade in a century or three with still potentially obnoxious waste residing in the fill. In tightly populated areas seeking local landfill opportunities, we may need to shift to pretreatment of wastes in fermentation and leaching cells prior to burning the relatively inert residue in a more conventional landfill cell (Lee and Jones, 1990).

I am not recommending that South Dakota open its borders to the development of a growth industry in solid waste disposal. I realize this would foster a gold rush mentality upon the discovery of the mother lode of proverbial ore in the form of the state's vast wasteland so well-suited for the "ultimate" landfill. Rather, I am recommending that the state of South Dakota become the proprietor of that landfill, joint venturing with one or another contractor to operate different aspects of the operation. This would, no doubt, include a major recycling effort which would effectively mine from the waste valuable resources made economically profitable by the sheer size of the effort. Likewise, incineration, storage, and perhaps even some fermentation and leaching opportunities for field-scale research could be a part of this operation. A veritable metropolis of waste disposal activity could be created and controlled under the direction of the state's ultimate entrepreneurs (its citizens) operating through its government. I say this as a fiscally conservative Republican who normally favors privatization of most government activities. However, major league waste disposal might rank second to national defense as a function that should remain in the control of the government. I want to further emphasize this point in light of my strong opposition to the pressure being put on Indian reservations in South Dakota and other western states to accept private waste disposal facilities because of their legal capacity to flaunt the laws of federal and state jurisdictions.

Ultimately, as Nevadans have long reaped the economic advantages of their national gambling industry and Alaskans received regular dividends from their oil, South Dakotans could watch their coffers fill as their long depressed economy floats back up to a healthy level while serving the desperate needs of a nation.

Readers can quickly point out how often government-controlled monopolies screw up the business. Examples are indeed readily available, ranging from the postal service to the printing office, but they have also built a few good roads, a damn good military machine, and some reasonably successful research laboratories, many of which are successful joint ventures and subcontracts. I think the people of South Dakota, its elected government, and dedicated bureaucrats are equal to the task. They own the ball, so they can establish all the rules of the game. Clearly, a safe and efficiently run solid waste disposal program must depend on a rigid set of standards dealing with the quantity and quality of waste delivered on well-designed schedules. This can be achieved. Surprisingly, this becomes even more possible as a result of their potential role as an entrepreneur rather than just a government body which comes under greater control of the Feds.

South Dakota Law Professor John H. Davidson (1989) has pointed out that although the Commerce clause of the United States Constitution prohibits state governments from enacting legislation that would discriminate in economic or commercial matters to the disadvantage of other states, the state of South Dakota could operate a landfill as a proprietor rather than as a government and thereby escape some aspects of the Commerce clause. Were the state as a proprietor to monopolize the field as I firmly recommend they should, existing landfill business would have to be allowed to operate under the previous rules of the game (grandfathered) or fairly compensated for any losses they may incur from a change in the rules.

It would thus be possible to place real limits on the importation of solid waste as long as the same rules and regulations are applied to in-state people and their garbage. If the state can convince the court that their restrictions and regulations placed upon the importation of out-of-state garbage and the handling of intrastate garbage as well are tied closely to legitimate health, safety, and public welfare goals, the state can run the tightest ship in the waste disposal business. The citizens of South Dakota would, thereby, be assured of good health and enhanced wealth.

A major associated advantage to the development of a state-owned national landfill would be the revitalization of America's railroads, rising like a phoenix out of the ashes. Clearly, the railroads would be the safest and most efficient form of transportation for the states on South Dakota's client list. With no end in sight for cargo opportunities, the railroads could redevelop the efficient service of yesteryear with assured black ink on the bottom line.

Indeed, South Dakota holds an answer for us all.

References

D'Angelo, P.A. 1990. Waste management industry turns to Indian reservations as states close landfills. *Environmental Reporter*. Dec. 28.

Davidson, John H. 1989. Minutes of testimony before the Solid Waste Study Committee. June 28.

Lee, G.F. and R.A. Jones. 1990. Managed fermentation and leaching: an alternative to MSW landfills. *Biocycle*. May.

Rathje, William L. 1989. Rubbish. *The Atlantic Monthly*. December.

Ruckelshaus, William P. 1989. The politics of waste disposal. *Wall Street Journal*. September 5.

Reprinted, with permission, from
Ground Water Vol. 29, No. 3, May–June 1991

An Alternative to Municipal Solid Waste Landfills

G. Fred Lee, Ph.D., PE and R. Anne Jones, Ph.D.

Dr. G. Fred Lee, PE is president of G. Fred Lee & Associates, an environmental consulting firm located in El Macero, CA. His past affiliations include: distinguished professor, Civil and Environmental Engineering, New Jersey Institute of Technology; professor of Civil and Environmental Engineering and director for the Institute of Environmental Studies, University of Texas at Dallas; and professor of Water Chemistry and director of the Water Chemistry Program, University of Wisconsin, Madison.

He received a BA in Environmental Health Sciences from San Jose State University in 1955; an M.S. in Public Health, Environmental Sciences-Water Quality, University of North Carolina, Chapel Hill in 1957; and a Ph.D. in Environmental Engineering and Environmental Sciences, Harvard University in 1960.

Dr. Lee has more than 450 professional publications in the areas of water supply, water quality, water and wastewater treatment, water pollution control, and solid and hazardous waste management. He has contributed to *Environmental Science and Technology, Journal American Water Works Association, Journal Water Pollution Control Federation,* and *Journal Water Research.*

He is a diplomate, American Academy of Environmental Engineers, and co-author with Dr. R. Anne Jones of "Is Hazardous Waste Disposal in Clay Vaults Safe?" judged by the Resources Division of the American Water Works Association as the best paper in the journal in 1984.

Dr. R. Anne Jones is vice president, G. Fred Lee & Associates. She was an associate professor of Civil and Environmental Engineering at the New Jersey Institute of Technology, and has also held professorial and research positions in the Departments of Civil and Environmental Engineering at Texas Tech University and Colorado State University.

She received a BS degree in biology from Southern Methodist University; an MS degree in environmental sciences from the University of Texas at Dallas; and a Ph.D. in environmental sciences from the University of Texas at Dallas with emphasis on water supply, water quality and water pollution control, aquatic toxicology and chemistry.

Dr. Jones has published more than 175 papers and reports on water quality and solid and hazardous waste management contributing to *Environmental Science and Technology, American Water Works Association,* and *Water Pollution Control Federation.*

The current approach for landfilling of municipal solid wastes (MSW) is to try to design and construct "tombs" consisting of liners and caps to keep buried wastes dry (US EPA, 1988). Keeping the waste dry will prevent the decomposition of the wastes and the production of leachate; prevention of leachate formation will prevent pollution of groundwaters. While such a system, if properly designed, constructed, and maintained, should, in theory, keep the wastes dry as long as the engineered system maintains its integrity, this approach is not without significant deficiencies that will ultimately threaten public health and environmental quality. First, experience has shown that theoretical integrity of liner construction, placement, and endurance is not achieved in MSW landfills. Manufacturing defects and installation imperfections breach integrity from the outset; the materials used for liners will eventually deteriorate. Liner materials are typically warranted for about 20 years. These deficiencies and failure will promote leachate generation and transport to groundwaters. Second, even if the integrity of the system could be ensured and maintained until materials failure, the contents of the "dry tomb" would, at that time, be largely the same as they were when placed; thus the leachate formation and transport from the landfill is simply being delayed until the liner and cap materials fail. Proper maintenance and remediation would involve vigilant monitoring and preparedness for exhumation forever, since the materials in the "dry tomb" landfill would represent a threat to groundwater quality forever.

Being advanced herein is a system for MSW handling that would ferment and leach the wastes prior to final disposal of residues. The leachates generated in the process would be treated as wastewaters. The stabilized leached residues would be buried. Since those residues would have already been fermented and leached, they would not represent the perpetual, significant threat to groundwater quality and public health that they would in a conventional MSW landfill. Harper and Pohland (1988) have also been critical of the US EPA's proposed approach for municipal solid waste management involving the attempt to create "dry tombs." They have also advocated a fermentation/leaching approach.

Treatment Approach

The authors feel that consideration should be given to an alternative approach for landfilling of municipal solid wastes. Rather than trying to keep the wastes dry *forever* and not succeeding, a waste disposal area/cell should be used as a treatment reactor in which the wastes are actively fermented and leached. This process would stabilize the decomposable organic matter by providing the needed moist anaerobic environment and would generate methane and carbon dioxide. By following approaches similar to those typically used for stabilization of municipal wastewater sludges, it should be possible to ferment MSW's to a near-maximum extent within a few years. This process would also be designed to leach the wastes, removing chemicals that would be expected to eventually leak out of a conventional MSW landfill.

It is well known that the moisture content in a landfill is a key to optimizing the rate of "stabilization" of decomposable organics. Much of the moisture that would be needed to stabilize wastes can, in many parts of the US, be derived from the precipitation on the landfill surface, and through the recycle of leachate through the landfill. Where necessary, those sources could be supplemented.

Since the success of this system will depend in large part on the even distribution of moisture within the wastes, several changes would have to be made in the design and operation of landfills if this rapid stabilization and leaching is to be achieved. First, the daily cover used in many sanitary landfills does not necessarily result in even distribution of moisture within a landfill. This means that some parts of the landfill stabilize at a slower rate than others. The placement of wastes, addition of water, and application of daily cover should be done in a manner to allow all parts of the landfill cell to contain an amount of moisture to bring about fermentation of the waste at an optimum rate. Second, even distribution of water will likely require that the wastes be shredded before placement in the treatment cell. Robert Ham of the University of Wisconsin, Madison has shown that shredding of municipal solid waste significantly improved its stabilization as measured by methane and carbon dioxide production. The cost of shredding municipal solid waste would be expected to add a few cents per person per day to the cost of solid waste disposal and would significantly improve waste handling, fermentation, and leaching. It may be necessary to install a header system at various depths in the waste to ensure that all parts of the landfill are receiving optimum moisture. Third, some changes will also likely have to made in the way landfill cells are constructed to optimize rates of methane generation.

Because liquids would be added to and would need to be retained within the system, attention must be given to liquid containment during treatment. For this purpose double-composite-lined cells are recommended. Each composite liner should be composed of a high-density polyethylene (HDPE) flexible membrane liner (FML) at least 100 mil thick, underlain by and in intimate contact with a 3-ft compacted clay layer having a field-measured permeability no greater than 1×10^{-7} cm/sec. It may be desirable to admix polymers or polymer/bentonite mixtures with the clay to further improve its structural integrity and reduce its permeability. There should be a leachate collection and removal system above the upper liner and a leachate detection system between the liners. It would be important not to allow the leachate head on the upper liner to build up to the point at which it would create a significant additional potential for passage of liquid through the upper composite liner into the leachate detection system. Significant leachate in the detection system would indicate that the upper composite liner system has been breached and the cell should be taken out of service for repair.

Because of the difficulties of siting MSW landfills, this system should be established at existing landfills. While with MSW landfills being designed today with liner systems there is concern about the long-term threat to public health and groundwater quality because of the eventual degeneration of the engineered systems, this is not a problem with the proposed system. This is because the cycle of fermentation and leaching would be expected to be on the order of 5 to 10 years. Thus the liner system would be accessible for inspection, and repair or replacement periodically during its expected lifetime, and, importantly, in response to detection of leachate generation.

The recommended approach will result in the leaching and removal for treatment of materials from the waste that would otherwise eventually leak into groundwaters. Leachate recycle through the system during the fermentation/leaching process will promote the leaching of those materials that would be expected to be readily leachable under the conditions that exist in a landfill. It has been known (as discussed by Lee et al. [1986]), that leachate recycle at sanitary landfills tends to improve the character of the leachate. Additional research needs to be done to understand how the composition of the recirculating leachate and water added to the landfill affects the leaching of wastes. Little is known about the conditions that would be needed to optimize leaching of wastes to maximize removal of materials that are potentially leachable under landfill conditions.

Residues

Once the waste has been stabilized and leached, it can be removed from the treatment cell. The residue can be sorted and classified by particle size/density. The "soil-like" residue can be evaluated to determine if, based on the heavy metal and other contaminant characteristics, it is suitable for use as a soil conditioner/humus. While the wastes would have been leached extensively, the non-usable residues still should be fixed with cement/silicates or other reagent and buried in a permanent, lined landfill. The amount of material that would have to be landfilled would be substantially less than that originally fermented and leached. The leachate would have to be treated as a wastewater and discharged to surface waters.

Basically this approach is similar to that being used by some municipalities for mining of existing landfills. Important differences are that the stabilization process would be greatly accelerated in the proposed system and the wastes would be extensively leached to remove solubilized contaminants.

The proposed approach is in many respects similar to composting of municipal solid wastes, except that it would be done under anaerobic rather than aerobic conditions. The purpose of composting is to produce a "stabilized" residue that will not be "offensive" to the public. That residue, however, is not leached. That is a major and significant difference between the two approaches. Waste that has been composted will likely leach contaminants that can have an adverse effect on surface and groundwater quality. Care must be exercised in using residues from MSW and wastewater sludge composting because of potential problems of this type. Epstein and Epstein (1989) recently discussed public health issues associated with composting of solid waste.

Active Management

The approach outlined herein is active management rather than the traditional, passive approach used today in which moisture addition is not controlled, or the "dry tomb" approach advocated by the US EPA (1988) and some states. By actively managing the fermentation and leaching, many of the problems being encountered today with sanitary landfills and those that threaten public health and groundwater quality forever should be greatly minimized and possibly even eliminated.

Another advantage of using a fermentation/leaching approach is the increased safety of the facility. The production of gases, methane in particular, is of concern in sanitary landfills because of the potential for explosion upon ignition when the methane exceeds a few percent of the landfill gas/air mixture. While gas production problems can be controlled at sanitary landfills, the control of gas formation and gas recovery/utilization could be more readily done with the proposed approach because fermentation would be controlled and optimized. For any given treatment cell, the duration of gas production would be significantly less than that typically encountered in a MSW landfill.

Finally, the advantages to long-term protection of public health and groundwater quality of this approach are significant. The long-term problems of groundwater contamination by MSW landfill leachate are difficult, if not impossible, to control. The proposed approach would generate residues that would pose a significantly smaller threat to public health and groundwater quality because of their treatment.

References

Epstein, E., and Epstein, J.I., "Public Health Issues and Composting," *Biocycle* 30:50–53 (1986).

Harper, S.R., and Pohland, F.G., "Landfills: Lessening Environmental Impacts," *Civil Engineering* 58:66–69 (1988).

Lee, G.F., Jones, R.A., and Ray, C., "Sanitary Landfill Leachate Recycle," *Biocycle*, 27:36–38 (1986).

US EPA, "Solid Waste Disposal Facility Criteria: Proposed Rule," *Federal Register* 53(168):33314–33422, 40 CFR Parts 257 and 258, US EPA Washington, D.C. (1988).

Published as:
Lee, G.F., and Jones, R.A., "Managed Fermentation and Leaching: An Alternative to MSW Landfills," *Biocycle 31*(5):78–80,83 (1990)

Reprinted, with permission, from
Biocycle 31(5):78–80, 83, 1990.

Media Coverage

Perspectives on the Cost Effectiveness of Life Saving
Bernard L. Cohen

Is Environmental Press Coverage Biased?
Jane S. Shaw

Good Press for Bad Science
G. Brent Dalrymple

Perspectives on the Cost Effectiveness of Life Saving

Bernard L. Cohen, Ph.D.

Bernard L. Cohen has been at the University of Pittsburgh since 1958 and is presently a professor of Physics and Radiation Health. Prior to 1958 he was affiliated with the Oak Ridge National Laboratory, 1950 to 1958.

Dr. Cohen received a B.S. in 1944 from Case-Western Reserve; an M.S. from the University of Pittsburgh in 1947; and a Ph.D. from Carnegie-Mellon in 1950.

He is the author of *The Nuclear Energy Option* published by Plenum, 1990; *Before It's Too Late* published by Plenum, 1983; and *Concepts of Nuclear Physics* published by McGraw-Hill, 1971, and three other books.

Dr. Cohen received the American Physical Society Bonner Prize for research in nuclear physics.

How much money are we willing to spend to save a life? A common answer is "Sky's the limit." But our government does not act that way. For example it is well known that divided highways have far fewer traffic deaths per mile than non-divided highways; why not make all highways divided? It would save thousands of lives every year, but we still don't do it. It costs too much money.

Methodologies and examples have been presented previously,[1] giving numerical estimates of how much our government is barely willing, or barely unwilling, to spend to save a life in various contexts. Here we consider a few such situations that powerfully illuminate some of the problems in our Society's response to technology. In particular, we calculate the cost we are marginally willing to pay to save a life for two primary examples in each of three categories—health programs in underdeveloped nations, normal health and safety measures in the United States, and nuclear technology. We then discuss the reasons for the tremendous differences we find, and their effect on the well-being of our nation.

Health Programs in Under-Developed Nations

1. Immunization[2]

About 100 million children are born each year in under-developed nations and 80 million of them are not protected by immunization. The

461

estimated number of deaths among them that could be prevented by immunization are:

Measles	2.4 million
Pertussis	1.6 million
Neonatal tetanus	0.8 million
Polio	0.4 million
Total	5.2 million

The World Health Organization has conducted studies and pilot projects on providing this immunization, and estimates of costs have been developed. These costs include sending doctors and nurses to these countries to carry out the programs, all equipment, supplies, and logistical support, etc. The results in cost per life saved are listed in Table 1. From that Table and the above list, it is clear that several million lives could be saved each year at a cost of about $200 per life saved.

2. Oral Rehydration Therapy (ORT)[2,4]

There are about 5 million deaths per year from diarrhea in under-developed countries, and it is estimated that 50–75% of these could be prevented by oral rehydration therapy (ORT). This consists of feeding a specific mixture of glucose (sugar), NaCl (salt), KCl ("salt substitute"), and $NaHCO_3$ (baking soda) mixed with a proper amount of water, on a specific schedule. The World Health Organization has conducted stud-

Table 1. Cost per life saved for immunization programs in under-developed nations[3,2,5,6]

Disease	Country	$/life saved
Several	Indonesia	210
Measles	Ivory Coast	850
		480
Tetanus	Gambia	276
Pertussis	Gambia	99
Diphtheria	Gambia	87
Measles	Gambia	50
Polio	Gambia	6600
Measles	Cameroon	50

Table 2. Cost per life saved by ORT programs for diarrhea in under-developed nations[2,4]

Country	Date	Deaths averted/ 1000 children	$/life saved
Honduras	1981	12	150
Indonesia	1984	5.7	396
Egypt	1980	7.6	550
Zaire	1982	5.0	320
Unspecified	1981	—	450
Unspecified	1981	—	580

ies and pilot programs on this, and the results are listed in Table 2. Again we see that millions of lives could be saved each year at a cost of a few hundred dollars per life saved.

3. Other Programs

Perhaps the next most cost effective method of life saving in under-developed countries is in malaria control, and this has an important additional advantage of improving the quality of life for those who do not die from the disease. Hundreds of millions of people suffer from the disease, and perhaps a million children die from it each year. Malaria can be effectively combatted by stopping mosquito breeding, by providing nets to keep mosquitos away, and by medical treatment for the disease. Costs per life saved vary from $440 with DDT to $3,320 without DDT.[7]

Other cost effective methods for life saving in under-developed countries are simply providing improved health care ($1,930)[8], improving water sanitation ($4,030),[7] and providing nutrition supplements to the basic diet ($5,300).[9]

Health and Safety Programs in the United States

It is not unreasonable for us to be willing to spend more money to save our own lives than to save the lives of very different type people in distant lands. We therefore discuss two examples of cost effective life saving programs in the United States.

1. Cancer Screening Programs

Table 3 lists the costs per life saved for various cancer screening programs that could be instituted in the United States.[1] For example, only 50% of American women who should get PAP tests for cervical cancer do get them. Some cities have utilized mailings and contacts by public health nurses to raise the percentage participation to 90%. As a result, many cancers were detected at an early stage and lives were saved. An analysis of the data gave the cost as $50,000 per life saved, as listed in Table 3.

As another example, a textile mill in North Carolina instituted a battery of cancer screening tests for its employees, and after several years of operation added up the cost of the program and the number of lives saved by early detection. The ratio of these was $52,000 per life saved as listed for "multiple screening" in Table 3.

The number for hypertension control (high blood pressure) includes not only the cost of the medicine but a generous bribe to convince people to take it (a major problem here is people not taking the medicine). Mobile intensive care units, ambulances with well trained technicians and radio contact with a physician, save lives at a cost of about $24,000 per life saved when serving 100,000 people per unit. But areas with less than 20,000 people (this estimate differs from that in

Table 3. Cost per life saved for cancer screening and medical care programs in the United States. (Except for the last entry, all costs are from Ref. 1, but since they are given there in 1975 dollars, they have been doubled.)

Item	$/life saved
Cervical cancer screening	$ 50,000
Breast cancer screening	160,000
Lung cancer screening	140,000
Colo-rectal cancer	
Fecal blood tests	20,000
Proctoscopic exams	60,000
Multiple screening	52,000
Hypertension control	150,000
Kidney dialysis	400,000
Mobile intensive care units in smaller towns	120,000
Neo-natal intensive care[10]	26,000

Ref. 1) often do not have this service because it is too expensive; this corresponds to being unwilling to spend over (5 x 24,000 =) $120,000/life saved.

2. Highway Safety Measures[11]

There are many measures that can be taken to improve highway safety and thereby save lives. Table 4 lists some of these along with the number of lives estimated to be saved per year by measures taken in 1983, and the cost per life saved.

For example, $80.5 million was spent on improving guardrails which were causing 300 highway deaths per year. This is expected to reduce the number of deaths by 40%, saving 120 lives per year. It has an estimated average service life of 6.7 years, which means that it will eventually save (6.7 x 120 =) 800 lives. Saving 800 lives with an expenditure of $80.5 million corresponds to spending $101,000 per life saved.

With the many cost-effective ways to save lives listed in Table 4, the U.S. Office of Highway Safety disdains to spend more than about $300,000 per life saved. That is why it hesitates to require air bags, which cost about $600,000 per life saved.[12]

Table 4. Evaluation of recent projects undertaken to improve highway safety[11]

Improvements	Lives saved per year	$/life saved
Improved traffic signs	79	$ 31,000
Improved lighting	13	80,000
Upgrade guard rails	119	101,000
Breakaway sign supports	2	125,000
Obstacle removal	8	160,000
Median barrier	28	163,000
Impact attenuators	6	167,000
Median strip	11	181,000
Bridge-guard rail transition	3	260,000
Channels; turn lanes	75	290,000
New flashing lights at railroad crossings	11	295,000
Permanent grooving	6	320,000

There are many very cost effective ways to save lives with improved highway safety that are not considered in Table 4 because they involve human effort that is not readily translated into a dollar cost. For example,[13] programs to improve voluntary seat belt usage could save 3300 lives per year at a cost of $3000 per life saved (strictly enforced mandatory seat belt usage could save 10,250 lives at a cost of $1000 per life saved). Motorcycle lights on day and night could save 110 lives/yr for $8000/life saved. Mandatory safety helmets for motorcycles would save 1050/yr for $18,000 per life saved, and enforcement of the 55 mile speed limit could save 3600 per year for $88,000/life saved. In all, measures of this type could save close to 15,000 lives per year in the United States at a cost of $1.4 billion per year, or $92,000 per life saved.

One case where the cost of human effort can be estimated is driver education. This was done by asking high school students what kind of a bribe it would take to make them happy about taking such a course as an addition to their schedule. The result, including the direct cost of giving the course, was $180,000 per life saved.[1]

Safety in the Nuclear Power Industry

1. Radioactive Waste Management

An example was given previously[1] for management of the high level waste from the Savannah River Plant in South Carolina where the option selected by Dept. of Energy costs $300 million per life saved, according to their estimate. A similar situation has developed in the waste storage tanks at West Valley, New York were Dept. of Energy selected an option of removing and vitrifying the waste and shipping it to a repository over the option of in-tank solidification. According to their estimates,[14] the cost is $150 million to save an estimated 0.55 lives, or $270 million per life saved. This ignores health effects on workers which, if taken into account, give more deaths for the more expensive option than for the less expensive one, converting the option chosen to infinite cost per life saved.

The most important case is our commercial waste disposal program which is estimated to cost 0.1¢/Kw-hr, or $8.8 million/GWe-yr (GWe=$10^9$ watt electric). It is estimated[15] that random burial of this waste with simple procedures would lead to 0.02 eventual deaths/ GWe-yr, so we may crudely estimate that half of this money is being spent to avert these deaths. The cost is then ($4.4 \times 10^6/0.02 =$) $220 million per life saved.

There are several strange aspects to this tremendous money expenditure on waste management. In the first place, the lives saved are those of people living many thousands of years in the future, who bear no closer relationship to us than those now living in under-developed countries whose lives we disdain to save at one-millionth of these costs. In the second place,[16] there is an excellent chance that a cure for cancer will be found in the next few thousands years, in which case these deaths will never materialize and the money will be wasted. In the third place,[16] if only a tiny fraction of this money were invested even at minimal interest, it could provide enormous benefits to these future potential victims, including the saving of tremendous numbers of lives. Equivalents of such an investment are spending the money on biomedical research, or simply using it to reduce the national debt and thereby making more money available to later generations to spend on themselves.

With any reasonable consideration of these matters, we are spending the equivalent of innumerable billions of dollars per life saved in our radioactive waste management programs.

2. Nuclear Reactor Safety Upgrade

In the 1975–1985 decade, the Nuclear Regulatory Commission (NRC) was constantly tightening requirements on reactor safety in a program that increased the cost of a nuclear power plant five-fold over and above inflation, raising the cost by over $2 billion.[17] Before this process began, the NRC's own Reactor Safety Study[18] estimated that accidents in plants of that era could be expected to cause an average of 0.8 deaths over their operating life. This program of improving reactor safety therefore corresponds to spending ($2 billion/0.8 =) $2.5 billion per life saved.

An ironic aspect of these NRC reactor safety upgrading activities is that the cost increases it has caused have forced utilities to build coal burning power plants instead of nuclear plants. A typical estimate[19] is that the air pollution from 1 GWe of coal burning plants kills 25 people per year, or about 1000 people over its operating lifetime. Considering the fact that the nuclear plant is expected to kill 0.8 according to NRC[18] (or 100 according to the anti-nuclear activist organization, Union of Concerned Scientists[20]), that means that every time a coal burning plant is built instead of a nuclear plant, something like 1000 extra Americans are condemned to an early death.[21]

As a result of this NRC program, at least 30 additional GWe of coal burning plants have been started instead of nuclear plants, causing 30,000 extra deaths. Perhaps another 50 GWe of coal burning will be started as a result of these policies, raising the total to 80,000 needless deaths.

The 60+ nuclear plants that will eventually be completed have cost an average of at least $1.6 billion extra, for a total cost of $100 billion in an effort to save these (60 x 0.8 =) 50 lives. Just one year's interest on this money alone could save 100,000 lives if spent on medical or highway safety measures ($10 billion/$100,000 per life saved = 100,000 lives saved).

There are additional indirect consequences of this NRC program. Essentially the same nuclear power plant costs 2½ times as much in United States as in France[17,22] and the Japanese are not far behind the French. Since projected costs for coal-burning electricity and nuclear electricity in the United States are about equal, this means that electricity will probably be twice as expensive in the United States as in Western Europe and Japan. This puts a direct bite on our standard of living. But more important, since many economists believe that a large part of the reason for past U.S. economic success has been our relatively low cost of energy, it is not unlikely that the reversal of that advantage will contribute substantially to our unemployment problems. It is estimated[23] that a 1% increase in unemployment in the United States causes an extra 37,000 deaths per year, including about 20,000 from cardiovascular failures, 900 suicides, 650 homicides, and 500 deaths from alcohol-related cirrhosis of the liver. In addition to the deaths, it causes 4200 admissions to mental hospitals, and 3300 admissions to state prisons.

The waste of $100 billion also has important health consequences. There is a very strong correlation among the nations of the world between per capita income and health indexes. In particular, life expectancy ranges from well over 70 years among well-to-do nations to less than 40 years in poor nations. Wealth brings health. Wasted wealth must surely, therefore, have high health consequences.

Summary and Search for Explanations

We have shown that our government is willing to spend 1000 times more to save an American life than to save a life in an under-developed country, and is requiring expenditure of 10,000 times more to save lives from the dangers of the nuclear industry than it is willing to spend to save American lives from more common risks. The first discrepancy is perhaps understandable in terms of "charity begins at home," but the discrepancy is so enormous that it should become more widely known and discussed. There may be other considerations than cost in the decision not to act, but they should be made explicit rather than implicit. Allowing millions of people to die unnecessarily is surely an important decision in which the public should be involved.

However, this is hardly a question for the scientific community. We therefore concentrate here on the second discrepancy, which is causing many thousands of needless deaths and wasting many billions of dollars in the United States every year.

Why is all this money being spent to avert nuclear hazards when it could be used thousands of times more effectively on medical and highway safety programs? That is an easy question to answer: one need only ask the NRC officials who make these decisions. They tell you that the first responsibility of a government official is to be responsive to public concern. This can also be understood as a natural selection process: a government official who is not responsive to public concern will not remain a government official for very long. Moreover, that is the way it ought to be. We certainly want our government to be responsive to public concern.

The problem here is that public concern has been misdirected. The public has literally been driven insane over fear of radiation—one definition of insanity is loss of contact with reality.[25] Its concepts of the dangers of reactor meltdown accidents and buried radioactive waste have been grossly distorted relative to the understanding of these problems by the overwhelming majority of the scientific community that deals with those subjects.

Direct responsibility for this must be attributed to the news media. They are the source of the public's information on these matters, and they greatly distort the message from the scientific community.[26] As one example of this, consider the coverage by the *New York Times* of a report of excess leukemias among workers at the Portsmouth (NH) naval shipyard where nuclear submarines are serviced. The original story, generated by a clinical physician and reporters from a Boston newspaper, was covered by 14 articles over a 3-month period in the *New York Times*, most of them on page 1. As a result, there was a three year study by the National Center for Disease Control which found that there was no excess leukemia among those workers. The *Times* reported that finding in 9 lines on page 37 of a weekday edition.

Numerous other examples of media distortion on nuclear power issues may be found by looking under "media" in the index of my book.[17] Why do they do this? Media have their own priorities and ways of doing things, developed primarily to maximize public interest, because, to a large extent they are in the entertainment business. Here, again, natural selection operates effectively. A one point increase in the Nielsen rating for a network evening news program directly generates $11 million in additional revenue from advertisers. In such a situation, if a producer decided to give higher priority to providing proper perspective on hazards than to entertaining the public, how long would his job last?

But the media cannot create stories out of nothing. They must be fed material on the dangers of nuclear power. This has been supplied directly or indirectly by a handful—perhaps a dozen—scientists who have built careers on opposition to nuclear power. These scientists do not publish in regular scientific journals and have very low credibility in the scientific community.[25] But they have been very effective because of the way the media have used them. For example, a typical story will quote one of them and, to appear unbiased, a bland-spoken utility executive on the other side, with every intimation that this is a brave scientist fighting to protect the public from an insensitive corporate exploiter. The vast majority of scientists and their leadership, elected officers of scientific organizations, members of National Academy of Sciences committees, etc. are rarely quoted.[25]

Who Is to Blame?

With many thousands of Americans dying needlessly, and many billions of dollars being wasted every year, it is only natural to ask who is to blame. From the above discussion one might blame the anti-nuclear scientists. But when many thousands of scientists are involved, there is bound to be a spectrum of opinions and we must therefore expect that one percent may have very negative opinions about nuclear power. There are also ego and personality problems involved. It is only natural for some people to covet the media coverage and numerous lecture invitations that are readily available to a scientist who opposes nuclear power. Here again there is a natural selection process. If an anti-nuclear scientist changes his opinion, he is no longer contacted by the media and invited to give lectures. Given the rewards available to any anti-nuclear scientist, it is inevitable that there will always be some of these available to the media.

The NRC is far from blameless. The public concern over the hazards of nuclear power acts through the political process—mainly via Congressmen—to put heavy pressure on NRC to tighten regulations. NRC can even argue that another accident like Three Mile Island, even though it causes no deaths, could shut down the entire nuclear industry which would have far worse consequences.

But an NRC official could stand up to a Congressional Committee and/or the Media and say "You are asking us to kill people, and that is a violation of basic moral principles." Perhaps such a message would get through, but it is more probable that he would be crucified. Still, people who believe in moral principles should take that personal risk. It is hard to understand why their consciences are not bothered by the knowledge that they are condemning tens of thousands of innocent people to an early death.

Media people must survive, but they could do much more to transmit the correct message to the American public. Especially disturbing is their arrogance. On over a dozen occasions, I have submitted papers to professional journals for journalists in an effort to explain to journalists the errors in their coverage of nuclear power issues. Each of my papers was rejected summarily without explanation. The writing could not have been too bad, because most were later published elsewhere.[26]

My book gives many other examples of media arrogance,[17] but the arrogance of power is probably an inevitable fact of life. The media wield tremendous power in this country, and it is only natural that this leads to a feeling that they don't need unsolicited advice. A *New York Times* reporter, in defense of his story implying that there was no substantial difference between containments in U.S. reactors and the Chernobyl reactor (the former would contain essentially all radioactivity in 98% of all meltdowns, while the latter would protect only against the rupture of one of the 1700 tubes in that reactor) told me, in effect, "I don't tell you how to do scientific research, so why should you tell me how to practice journalism." The problem is that his journalistic practice is targeted on selling newspapers more than on informing the public correctly.

While we can blame individual NRC officials, media people, and anti-nuclear scientists for not being guided by their consciences, we must recognize that they are the products of natural selection processes. In our political system, natural selection operates to favor government officials who respond to political pressure, to favor media people who value entertainment over providing information to the public, and to favor anti-nuclear positions among scientists.

On the other hand, our political system gives us all the opportunity to fight against the forces of irrationality. With many tens of thousands of lives at stake, there is plenty of incentive to fight very hard.

References

1. B.L. Cohen, Society's Valuation of Live Saving in Radiation Protection and Other Contexts, Health Phys. 38, 33 (1980). Costs from this paper have been doubled to account for inflation since 1975.
2. PRITECH (Technologies for Primary Health Care Project, U.S. Agency for International Development), Infant and Child Survival Technologies, Sept., 1984.
3. H.N. Barnum, D. Tarantola, and I.F. Setiady, Cost Effectiveness of an Immunization Programme in Indonesia, Bul. of World Hlth. Org. 58, 499 (1980).
4. H. Barnum, Cost Effectiveness and Cost Benefit Analyses of Programmes to Control Diarrhoeal Diseases, WHO Subcommittee on Diarrhoeal Dis-

eases of the WPACMR, Manila (April 22–28, 1981), and unpublished Report.

5. World Health Org. Wkly. Epidem. Rec. 22 (June 4, 1982), and similar pieces.

6. D.S. Shepard, L. Sanoh, and E. Coffi, Cost Effectiveness of the Expanded Programme of Immunization in the Ivory Coast, Social Science and Medicine (submitted).

7. J. Walsh and A. Warren, New Eng. Jour. Med. 301, 967 (1979).

8. J.R. Evans, K.L. Hall, and J. Warford, Health Care in the Developing World, New Eng. Jour. Med. 305, 1117 (1981).

9. J. Briscoe, Am. Jour. Pub. Hlth., Sept. 1984.

10. S.L. Kaufman and D.S. Shepard, Costs of Neonatal Intensive Care by Day of Stay, Inquiry 19, 167 (1982).

11. U.S. Dept. of Transportation, The 1984 Annual Report on Highway Safety Improvement Programs, April 1984. Figures there are given in terms of cost/fatal accident, and we assume 1.1 deaths/fatal accident.

12. Dept. of Transportation, 49CFR Part 571: Federal Motor Vehicle Safety Standard; Occupant Crash Protection; Final Rule, Federal Register 49, 138, 28962 (July 17, 1984).

13. U.S. Dept. of Transportation, National Safety Needs Study: 1981 Update of 1976 Report to Congress, DOT-HS-806 283 (Oct. 1981).

14. U.S. Dept. of Energy, "Long Term Management of Liquid High Level Radioactive Waste Stored at Western New York Service Center, West Valley," Document DOE/E1S-0081 (June 1981).

15. B.L. Cohen, "Risk Analysis of Buried Waste from Electricity Generation," Am. Jour. of Phys. 54, 38 (1986).

16. B.L. Cohen, "Discounting in Assessment of Future Radiation Effects," Health Phys. 45, 687 (1983).

17a. B.L. Cohen, "Before It's Too Late," Plenum Publ. Co. (New York), 1983; Chap. 8.

17b. B.L. Cohen, "Nuclear Power: Economics and Prospects," in S.F. Singer (Ed), "Free Market Energy," Universe Press (1984).

18. U.S. Nuclear Regulatory Com., "Reactor Safety Study," Document WASH-1400, or NUREG 75/014 (1975).

19. R. Wilson, S.D. Colome, J.D. Spengler, and D.G. Wilson, "Health Effects of Fossil Fuel Burning," Ballinger Publ. Co. (Cambridge, MA) 1980. Many other references given in Ref. 16, p. 113–115. A more recent summary is H. Ozkaynak and J.D. Spengler, "Analyses of Health Effects Resulting from Population Exposure to Acid Precipitation Precursors," Env. Hlth. Perspectives 63, 45 (1985).

20. Union of Concerned Scientists, "The Risk of Nuclear Power Reactors," Cambridge, MA (1975)

21. A list of studies that have concluded that nuclear power causes far fewer deaths than coal burning power include: National Academy of Sciences Committee on Nuclear and Alternative Energy Systems, Energy in Transition, 1985–2010 (W.H. Freeman & Company, 1980); American Medical Association Council on Scientific Affairs, "Health Evaluation of Energy

Generating Sources," Journal of the American Medical Association, 240 (1978), p. 2193; United Kingdom Health and Safety Executive, "Comparative Risks of Electricity Production Systems" (1980); Norwegian Ministry of Oil and Energy. Nuclear Power and Safety (1978); Science Advisory Office, State of Maryland, Coal and Nuclear Power (1980); Maryland Power Plant Siting Program, "Power Plant Cumulative Environmental Impact Report" (1975); Legislative Office of Science Advisor, State of Michigan, "Coal and Nuclear Power" (1980); C.L. Comar and L.A. Sagan, "Health Effects of Energy Production and Conversion," Annual Review of Energy, 1 (1976), p. 581; L.B. Lave and L.C. Freeburg, "Health Effects of Electricity Generation from Coal, Oil, and Nuclear Fuel," Nuclear Safety, 14, No. 5 (1973), p. 409; S.M. Barrager, B.R. Judd, and D.W. North, "The Economic and Social Costs of Coal and Nuclear Electric Generation," Stanford Research Institute Report (Mar. 1976); Nuclear Energy Policy Study Group, Nuclear Power—Issues and Choices (Cambridge, Mass.: Ballinger, 1977); Herbert Inhaber, "Risks of Energy Production," (Ottawa: Atomic Energy Control Board, 1978); Robert L. Gotchy, Health Effects Attributable to Coal and Nuclear Fuel Cycle Alternatives (1977); David J. Rose, P.W. Walsh, and L.L. Leskovjan, "Nuclear Power—Compared to What?" American Scientist, 64 (1976), p. 291; "Impacts on Human Health from Coal and Nuclear Fuel Cycles" (Ohio River Basin Energy Study, Environmental Protection Agency, July 1980); William Ramsay, Unpaid Costs of Electrical Energy (Johns Hopkins University Press, 1979); B.L. Cohen, "Impacts of the Nuclear Energy Industry on Human Health and Safety," American Scientist, 64 (1976), p. 550; Richard Wilson and William J. Jones, Energy, Ecology, and the Environment (New York: Academic Press, 1974); Harris Fischer et al., "Comparative Effects of Different Energy Technologies" (Upton, NY.: Brookhaven National Lab, 1981).

22. Journal Official des Debats de l'Assemblee National, May 10, 1982; U.S. Energy Information Adm. Document DOE.EIA-0356-11 (1982).

23. R. Marshall (Univ. of Texas, Austin), Health and Unemployment, Annual Mtg. of Am. Pub. Hlth. Assn., Dallas, TX, Nov. 1983 and unpublished manuscript.

24. B.L. Cohen, "Discounting in Assessment of Future Radiation Effects," Health Phys. 45, 687 (1983).

25. B.L. Cohen, "A Poll of Radiation Health Scientists," Health Phys. 50, 639 (1986).

26. B.L. Cohen, "Nuclear Journalism," Policy Review, Sept. 1983, p. 70; "The Wrong Answers to the Right Questions," Electric Perspectives, Jan. 1983, p. 23; "Most Scientists Don't Join in Radiation Phobia," The Wall Street Jour., Nov. 30, 1983; also, Ref. 25.

Is Environmental Press Coverage Biased?

Jane S. Shaw

Please see biographical sketch for Jane S. Shaw on page 267.

The January 2, 1989, issue of *Time* magazine deviated from its usual "Man of the Year" cover, heralding instead the "Planet of the Year" and describing the perils facing the earth and what might be done about them. Plant and animal species are disappearing at a rate of 1,000 times faster than they did in the past, *Time* said; the "greenhouse effect" is warming up the earth; chlorofluorocarbons (CFCs) are destroying the ozone layer; hazardous waste is befouling land and sea; the "population bomb" is endangering the world.

"Let there be no illusions," sermonized writer Thomas A. Sancton. "Taking effective action to halt the massive injury to the earth's environment will require a mobilization of political will, international cooperation and sacrifice unknown except in wartime."[1] Reading this, you had to sit up and take notice.

The "Planet of the Year" issue is already something of a classic because it was an almost perfect efflorescence of the contemporary doomsday mood, rich with unfounded claims, exaggerations, errors, and bad advice. Indeed, every claim stated above is either scientifically unsupportable (that is, opinion rather than fact), or vastly overstated.

This issue gained additional notoriety some months later when Charles Alexander, the science editor at *Time* who was primarily responsible for that issue, spoke at a conference on environmental issues. "As the science editor at *Time* I would freely admit that on this issue we have crossed the boundary from news reporting to advocacy." Andrea Mitchell of NBC News echoed the same theme. "Clearly the networks have made that decision now, where you'd have to call it advocacy."[2] David Brooks, a *Wall Street Journal* editor who attended the conference, took down these quotes and in an article asked the obvious question: Whatever happened to objective reporting?

Is it appropriate for a news reporter to be an outspoken advocate of a specific environmental agenda when it is considered unprofessional to be a proponent of a political candidate or a political policy? In other words, is environmental protection a non-controversial issue of the sort that motherhood used to be? The answer, of course, is no. Environmental issues have many ramifications. Once you start advocating environmental positions, you must take positions on public policies that can affect jobs, health, and even lives.

Time's editors came out with specific recommendations to save the Earth. For example, they urged that the government insist that auto makers improve fuel efficiency to 45 miles per gallon by the year 2000; urged a complete ban on manufacture of CFCs; said the government should set standards for recycling of waste; proposed that the government should fund family planning organizations, including the UN Fund for Population Activities; and urged ratification of the UN Convention on Law of the Sea, which would regulate mining and other commercial development of the sea. Every one of these raises important issues—issues of ethics, costs, health, and the proper role of the government.

The Costly Policies They Blithely Propose

Yet journalists ignore even rudimentary implications of their policy proposals. For example, *Time*'s recommendation for toughening the federal fuel efficiency standards would cost lives. To meet the tighter standards, automakers would have to lighten their cars even more than they have already. Lighter cars give passengers less protection in accidents. Robert W. Crandall of the Brookings Institution and John D. Graham of the Harvard School of Public Health have concluded that mandatory fuel economy standards cause hundreds of deaths a year.[3]

There are harmful effects from banning chlorofluorocarbons, too. CFCs are inert gases that are widely used in air conditioning and refrigeration (and formerly in aerosol sprays). The theory is that as these gases rise into the stratosphere, they destroy the protective ozone layer; they also are believed to contribute to global warming. The reason that CFCs are used in the first place is that they are relatively inert and nontoxic; substitutes will almost certainly be less safe and more expensive. Robert Watson of NASA, for example, says that if CFCs were banned, "probably more people would die from food poisoning as a consequence of inadequate refrigeration than would die from depleting ozone."[4]

Sometimes, of course, the results of furor about supposed dangers are merely more government boondoggles, such as the $10 billion Superfund program that supposedly cleans up hazardous waste sites. Fear that places like the Love Canal dump in New York, Times Beach in Missouri, and Stringfellow Acid Pits in California were causing birth defects and cancer created and bankrolled this program. But Superfund, a political porkbarrel, has cleaned up only about 50 sites since it was enacted in 1980.[5] Yet reputable studies have failed to confirm any serious health effects from such sites. A 1985 compilation of health studies at 21 well-publicized waste sites did not find epidemiological evidence of any long-term health effects. Researchers from the Environmental Defense Fund reviewed these and other studies and agreed that no "serious, life-threatening" diseases had turned up in statistically significant numbers, although they argued that better designed studies might have revealed "subtle effects."[6]

The perversity of media-fed scares is epitomized by the 1989 Alar scare. Alar is a growth regulator that is used on apples to keep them from falling from the tree too early and ripening too fast in storage; it is regulated by the EPA as a pesticide, though that isn't really what it is. The EPA has banned a number of pesticides and was considering a ban on Alar because it caused tumors in animal tests. Suddenly, the Natural Resources Defense Council pushed for a ban and, following a well-orchestrated public relations campaign, newspaper articles and television shows terrified mothers about giving their children apple juice made from apples that might have been treated with Alar.

Exactly how dangerous was Alar? Bruce Ames, head of the biochemistry department at the University of California at Berkeley is a well-respected researcher on environmental risks (he developed the Ames screening test for possible carcinogens). In a letter published in *Science* Magazine,[7] Ames and a colleague attempted to put the Alar scare in perspective. He noted that the potential hazard from UDMH, the carcinogenic breakdown product of Alar, from a daily lifetime glass of apple juice is less than the risk from the natural carcinogens you get from eating one mushroom daily, and less than the risk from the carcinogens you get from the aflatoxin in a daily peanut butter sandwich. Yet because Alar is "synthetic" rather than "natural," we focus on it while ignoring the other carcinogens. Ames estimates that we are ingesting about 10,000 times more natural pesticides than synthetic ones! (It may be hard to believe that pesticides are natural, but that is how plants protect themselves from predators.)

Furthermore, says Ames, the use of Alar reduces the need for pesticides on apple orchards in some places; and Alar-treated apples may be less susceptible to molds, so that juice form Alar-treated apples may have fewer toxins. By not being able to use Alar, producers may supply fewer apples, making the price higher and leading consumers to substitute less healthy foods. At the height of the scare, the New York Public School system stopped selling apples. Think of the junk food that school children probably ate instead.

The Dubious Issue of Global Warming

With global warming, the press has pulled out the stops. An Associated Press article[8] in December, 1989, reported that the "threat of an environmental cataclysm is replacing nuclear holocaust as the scariest menace to civilization." The writer cited a number of problems—clean air, ozone depletion and extinction of species. Then: "All these concerns will be secondary, however, to the one overriding issue that touches them all—global warming."

Yet global warming is far from an "overriding issue." Many scientists doubt that global warming is even a problem. Less than fifteen years ago, a book called *The Cooling*[9] predicted a new Ice Age and received respectful scientific comment. Now, largely but not entirely because we had a spate of warm years, the fear has changed direction.

Readers of *Liberty* are aware that the global warming issue is highly debated, as climatologist Patrick Michaels indicated in a recent issue.[10] Although it is an undisputed fact that carbon dioxide has been increasing in the atmosphere, by more than 35% since the Industrial Revolution, it's far from certain that CO_2 and other "greenhouse gases" are trapping more heat than they used to. There is evidence that global temperatures have increased slightly over the past century—although not in the continental United States. But the increase in temperature that has apparently occurred is far less than the computer models predict. Michaels says that "the globe has warmed up approximately one-half as much as the lower limit suggested by combinations of climate and ocean models."[11] Furthermore, these models on which scientists base their views about global warming predict greater warming at high latitudes than near the equator, but, says Michaels, "high latitude temperatures have simply not responded in the predicted fashion"[12]—and, in fact, temperatures rose rapidly before the major emissions of greenhouse gases. If computer models designed to predict temperatures on the basis of the greenhouse theory can't describe accurately what has happened so far, how can we count on them to predict the future correctly?

Michaels is not the only skeptic on global warming. Others include Reid Bryson, director of the Institute of Environmental Studies at the University of Wisconsin at Madison; Kenneth E.F. Watt, Professor of Environmental Studies at the University of California at Davis; Hugh Ellsaesser of Lawrence Livermore National Laboratory; and Andrew Solow of Woods Hole Research Center, among others. Some scientists contend that the small amount of warming that has apparently occurred represents a natural evolution from the "Little Ice Age" that ended during the 19th century.

In general, the doubting scientists have been ignored by television and the press. About 18 months after this issue first became headline news, a few skeptical articles began to appear, and some writers have begun to insert caveats in their articles. In its "Endangered Earth Update," *Time* did mention some critics of the environmental craze. However, it devoted less than a page of its seven-page "Update" to the critics, lumped them with Reagan officials James Watt and Anne Gor-such Burford as destructive naysayers, and claimed that they are on the defensive.[13] The juggernaut rolls on.

The Lurid Reporting of Chemical Dangers

Don Leal and I have conducted an informal survey of how two prominent newspapers, The *New York Times* and The *Washington Post,* report on chemical dangers. Our purpose was to compare coverage with the same newspapers' treatment of AIDS. By most standards, the advent of AIDS truly is a crisis, yet we found that the tone and language of articles about AIDS are carefully chosen to avoid alarm and clarify the risks—far different from the way chemical risks were discussed.[14]

To give you an idea of the way these newspapers treat chemical risks, consider a *New York Times* article headlined "Congress Again Confronts Hazards of Killer Chemicals."[15] It began: "'The most alarming of all man's assaults upon the environment is the contamination of air, earth, rivers, and sea with dangerous and even lethal materials,' Rachel Carson wrote a quarter of a century ago in her celebrated book *Silent Spring.* Today there is little disagreement with her warnings in regard to such broad-spectrum pesticides as DDT, then widely used, now banned."

In fact, there is and has been substantial dispute about her warnings. The author of the *Times* article notes that the book was "excoriated by the chemical industry," but never mentions that scientific opinion then and now also questions many of Ms. Carson's conclusions. For example, Norman Borlaug, who received the Nobel Peace Prize in 1970 for his agricultural research, said in a 1971 speech that *Silent Spring* presented a "very incomplete, inaccurate and oversimplified picture of the needs of the interrelated, world-wide, complex problems of health, food, fiber, wildlife, recreation and human population."[16]

Toxic waste provides a particularly attractive opportunity for fear-mongering. In a *New York Times* article, "Trying to Shut Off The toxic Spigot,"[17] Philip Shabecoff describes "a rapidly emerging belief among scholars, policy makers and environmentalists, and even within industry, that simply attempting to dispose of the increasing stream of hazardous effluents is a case of applying a Band-Aid to a hemorrhage." Yet even the Environmental Protection Agency's internal report, *Unfinished Business*, which reflects the views of career professionals, concluded that public concern about chemical waste disposal exceeds the actual health dangers[18] and, as I reported earlier in this article, no long-term health effects have been confirmed scientifically from any of the hazardous waste sites that have been extensively studied.

Perhaps the most extreme sort of coverage turns up in "comprehensive" articles about a particular supposed disaster. In late 1986,[19] The *Washington Post* reviewed the case of the Stringfellow Acid Pits, a southern California waste dump that figured in the political tangle leading to the resignation of Anne Gorsuch Burford in 1983. The front-page article by Michael Weisskopf says that the Pit "brought environmental havoc to Glen Avon [the town where the pit is located]: property loss, livestock deaths and human illnesses, including high rates of cancer and heart trouble. More ominous is the poisonous horizontal plume that is spreading underground as fast as three feet a day toward the Chino basin, which provides water for 500,000 people within a 30-mile radius."

The story is sprinkled with phrases such as "the plume of carcinogenic chemicals," "DDT-laced soil," "walking time bombs," and statements such as "[E]cological trouble was brewing beneath the surface," and "A list of the toxic substances dumped at Stringfellow looks like a chemical alphabet soup."

Yet the hazards themselves are rather different. Weisskopf reports that the level of TCE (trichloroethylene) in the water under Glen Avon was 40 parts per billion or "eight times the state's public health standard." Sounds terrible, but Bruce Ames wrote in an article published months prior to the *Post* story that this level of TCE is less hazardous than ordinary chlorinated tap water!

And about a month after the Post article appeared, the State of California issued its report on Stringfellow.[20] It concluded that "[b]ased on the present study, there is no reason to believe that the Stringfellow site has had a serious impact on the community's health." The state found no unusual incidence of cancer, miscarriages, or birth defects. The only increased incidence was reported ear infections and skin rash, symptoms not normally associated with groundwater contamination. Of course, the state of California is not an unimpeachable source, and some impacts over time still could occur. But where was Michael Weisskopf when this reassuring news came out?

The Source of the Problem

Why does the press treat environmental issues in an inflammatory way? One underlying reason is that news is entertainment, as economist Michael C. Jensen observed in an insightful lecture,[21] and certain kinds of stories entertain people better than others.

Keeping up with the news is interesting to many people. More of us probably watched the news when the earthquake hit San Francisco than would have watched the World Series that night. "60 Minutes" is consistently one of television's most popular shows. But because we rarely have a chance to affect the news, we have little incentive to develop an accurate understanding of what is happened. For the most part, we simply want to be superficially informed and therefore entertained.

Reporters, editors, and producers advance professionally to the extent that they come up with entertainment their readers and viewers want. What they want seems to consist of two chief elements: a simple, clear story-line (in Jensen's words, readers have an "intolerance of ambiguity") and a dramatic story that pits good against evil.

Even the least controversial stories must have a strong story line. A personal experience at *Business Week* illustrates how the need for a strong story line may get in the way of the truth. I was assigned the lead news story (not the cover story, but the first story in the section about what had happened in business the previous week). This was a few years after the oil crisis of 1979, and my job was to find out whether the recent drop in gasoline prices was sizable enough for Americans to "hit the road" the coming spring—that is, to plan on increasing their automobile travel. *Business Week* reporters around the country called the national parks, Disney World, travel agencies, and other such places to find out if bookings were up and if they could discern a trend. Relying on such reporting, I had to decide whether tourism and travel would be going up.

If travel was going up, I had a good lead story; if the drop in gas prices was having no discernible effect, I had either no story or a very unimportant one at the end of the news section. It was impossible to have a story that said, "On the one hand, some people are going to travel more this summer, but on the other hand, a lot of people are going to stay home." That wouldn't qualify as news.

Unfortunately, the reporting was ambiguous. In some places, it looked as though tourism was on the rise; in others, reservations were similar to what they had been for a few years. I was faced with a decision. If I decided that the travel was not going to increase, I would lose a prominent spot in the magazine that week. Certainly, my incentive was to focus on an upswing. In the end, I did.

A reporter has to go with a strong story line or "spin" if he wants the story to be published. Of course, readers want more than that, too. Jensen points out that they like stories about people and about conflicts between good and evil—that is, good and evil people. The demand for such stories is nothing new. Jensen quotes H.L. Mencken: "In so far as our public gazettes have any serious business at all, it is the business of snouting out and exhibiting new and startling horrors, atrocities, impending calamities, tyrannies, villainies, enormities, mortal perils, jeopardies, outrages, catastrophes—first snouting out and exhibiting them, and then magnificently circumventing and disposing of them."[22]

Mencken wrote that in 1920, when federal funds were not so readily available as they are today to pour money into "solving" or regulating problems like global warming, Alar, and CFCs. Today, the press doesn't have to "dispose" of the problems it "uncovers"—eager politicians will grab headlines claiming to correct them.

The Role of Ideology

To understand why the press acts as it does, one must also consider ideology. In *The Media Elite*,[23] S. Robert Lichter, Stanley Rothman, and Linda S. Lichter surveyed the elites in journalism as well as top business executives. Each person was asked to rate himself or herself on the political spectrum; the majority of journalists described themselves as liberal, only 17% as conservative more than 80% voted Democratic in presidential elections.

I believe that ideology influences journalists in their daily work. This shouldn't be surprising—many of us in journalism were attracted to the profession partly because we felt it offered a chance to correct some of the world's problems.

But when you combine ideology with the need for every story to have a "spin," and as dramatic a spin as possible, you almost inevitably will get less than purely objective stories. Rothman and Lichter documented the bias in a detailed study of journalistic treatment of the safety of nuclear power in recent years.[24] They found that the journalists they surveyed were far more opposed to nuclear power than were the scientists who actually study nuclear issues. They also found that journalists tended to interview and quote scientists who were more opposed to nuclear power than was typical of their peers; furthermore, these scientists tended to write for the public at large more than for their colleagues in peer-reviewed journals—a fact that Rothman and Lichter interpret as evidence that these outspoken scientists are not among those most respected by their peers. Undoubtedly, however, they offer the press a better story than other scientists do.

And Then There Is "Herd" Journalism

A third factor that contributes to poor reporting is the phenomenon known as "pack" or "herd" journalism (symbolized by the crowd of noisy reporters following every move of the President). Once a story reaches the front page, competing newspapers and magazines fall over themselves to report the newest development and to "be on top of" the story as it evolves. The need to stay with the pack means that a few leading publications often determine which stories get covered and what angle they take. As a reporter for *Chemical Week* magazine, I remember being told to change the spin on a story to conform to The *New York Times'* version; at *Business Week*, which would eschew such overt pressure, stories in The *New York Times* often influenced the topics the magazine chose and the way it treated them.

On some subjects, cracking the prevailing wisdom is just about impossible. I am sure that most people in the nation still think that Hooker Chemical was responsible for the leakage of chemicals from Love Canal, even though a well-documented report in *Reason* showed that the school board forced Hooker to sell the land and then ignored its warnings about the chemicals. The article (published in 1981) was picked up by ABC's *Nightline,* but, even then, it was not widely reported. It did not penetrate public consciousness and has been ignored ever since.

We are currently seeing a similar tendency in the treatment of acid rain. A ten-year government study assessing the effects of acid rain was recently completed. In the journal *Regulation,*[25] the original research director of the study wrote the following: "Extensive surveys in natural forests and commercial plantations over the eastern and northwestern states have failed to identify any regional decline that could not be attributed to natural causes, with the possible exception of red spruce trees in the high elevations of the northeastern Appalachians." Now consider the *New York Times* headline on its story about the study: "Researchers Find Acid Rain Imperils Forests Over Time."[26]

It's a mistake to put the entire blame for poor environmental reporting on journalists. Erroneous ideas and assumptions are deeply entrenched in the minds of influential people today—as anyone who has tried to defend free markets at a cocktail party surely knows. Nevertheless, a good part of the cause is inherent in the nature of journalism. Reporters seek news that tells a dramatic story; to find it, they often identify crises that don't actually exist. Furthermore, since they tend to be on the left of the political spectrum, they are willing, perhaps eager, to encourage more government involvement—and apparent crises help that along. Voices that challenge the prevailing way of looking at a problem are largely ignored, and the articles frequently lead to the adoption of policies that are costly and damaging as well as unwarranted.

References

1. *Time*, January 2, 1989, p. 30.
2. David Brooks, "Journalists and Others for Saving the Planet," *The Wall Street Journal*, October 5, 1989, p. A20.
3. Robert W. Crandall and John D. Graham, "The Effect of Fuel Economy Standards on Automobile Safety," *The Journal of Law and Economics*, Vol. XXXII, April 1989, pp. 97–118.
4. Quoted in Alston Chase, "The Ozone Precedent: We've Got a Policy, But Do We Have a Problem?" *Outside*, March 1988, pp. 37–38.
5. This estimate was reported in "Profits Are For Rape and Pillage," by Gretchen Morgenson with Gale Eisenstodt, *Forbes*, March 5, 1990, p. 94.
6. Amanda M. Phillips and Ellen K. Silbergeld, "Health Effects Studies of Exposure from Hazardous Waste Sites: Where Are We Today?" *American Journal of Industrial Medicine* (8:1–7, 1985).
7. Bruce N. Ames and Lois Swirsky Gold, "Pesticides, Risk, and Applesauce," *Science*, Vol. 244, May 19, 1989, pp. 755–757.
8. By Mitchell Landsberg, an Associated Press reporter, in *The Billings Gazette* on Dec. 27, 1989.
9. By Lowell Ponte (Englewood Cliffs, New Jersey: Prentice-Hall, 1976).
10. Michaels, Patrick, "The Greenhouse Effect: Beyond the Popular Vision of Catastrophe," *Liberty*, January 1990, pp. 27–33.
11. Michaels, p. 28.
12. Michaels, pp. 28–29.
13. *Time*, December 18, 1989, p. 68.
14. For more information on this subject, see "AIDS and Hazardous Waste Reporting," by Jane S. Shaw, in *The Washington Times*, July 3, 1989.
15. October 11, 1987.
16. Quoted in *Toxic Terror*, by Elizabeth Whelan (Ottawa, Illinois: Jameson Books, 1985), p. 67.
17. By Philip Shabecoff, September 21, 1986.
18. U.S. Environmental Protection Agency, *Unfinished Business: A Comparative Assessment of Environmental Problems* (*Volume I: Overview*), February 1987, p. 76.
19. Michael Weisskopf, "Toxic-Waste Site Awash in Misjudgment," *The Washington Post*, Nov. 16, 1986, p. 1.
20. "A Report to the Legislature on the Stringfellow Health Effects Study," State of California Health and Welfare Agency, Department of Health Services, December 1986.
21. "Toward a Theory of the Press," given at the Third Annual Interlaken Seminar on Analysis and Ideology, Switzerland, June 1976, by Michael C.

Jensen. The talk is reprinted in *Economics and Social Institutions: Insights from the Conference on Analysis and Ideology,* edited by Karl Brunner (Boston, The Hague, London: Martinus Nijhoff, 1979).

22. This quote from H.L. Mencken's, "On Journalism," in *The Smart Set,* April 1920, is reprinted in *A Gang of Pecksniffs,* ed. by Theo Lippmann, Jr. (New Rochelle, N.Y.: Arlington House, 1975), p. 65, and cited by Jensen.

23. *The Media Elite,* by S. Robert Lichter, Stanley Rothmann, and Linda S. Lichter (Bethesda, Maryland: Adler & Adler, Publishers, Inc., 1986).

24. Stanley Rothman and S. Robert Lichter, "Elite Ideology and Risk Perception in Nuclear Energy Policy," *American Political Science Review,* Vol. 81, No. 2, June 1987, pp. 384–404.

25. J. Laurence Kulp, "Acid Rain: Causes, Effects, and Control," *Regulation,* Winter 1990, p. 45.

26. By William K. Stevens, *The New York Times,* December 31, 1989, p. 1.

An abridged version
Reprinted, with permission, from
Liberty, vol. 4, no. 1, Sept. 1990

Good Press for Bad Science

G. Brent Dalrymple, Ph.D.

G. Brent Dalrymple is president, American Geophysical Union; geologist, U.S. Geological Survey; and consulting professor, Stanford University.

He received an A.B. from Occidental College in 1959, and a Ph.D. from the University of California, Berkeley, in 1963.

Dr. Dalrymple is the author of *Potassium-Argon Dating* published by Freeman in 1969, and *The Age of the Earth* published by Stanford Press in 1991.

He has more than 175 journal articles published and has contributed to the *Journal of Geophysical Research, Contributions to Mineralogy & Petrology,* and *Geophysical Research Letters.*

Dr. Dalrymple received the Meritorious Service Award, Department of Interior in 1984.

Controversy makes good press. Less obvious but no less true is that controversy can be good for the advancement of science. Press coverage of scientific controversy sounds like a perfect marriage: the press gets the good story and science gets a boost. But, in fact, the couple needs a marriage counselor. A recent event, or rather the lack of one, shows how the cultures of science and the press can sometimes conflict, with potentially serious consequences to the public.

December 3 was to have been the date of a major earthquake centered in the New Madrid, Mo., seismic zone, or so predicted an individual who believed that a particular alignment of the Sun and Moon with the Earth would trigger movement along the fault zone. Media around the country ran the story and the prediction was taken seriously. The National Guard was placed on alert in some states. Many families packed their belongings and left the region. Local governments canceled vacations for employees and stocked emergency supplies. Schools and businesses were closed to minimize the danger to a frightened public.

The fact that there was no scientific basis to the prediction got lost in the news reports. Also lost was the historical perspective a rich history of bogus earthquake predictions based on tidal effects, missing pets, psychic visions, and other zany ideas going back several decades, whose success rate is no better than, as one prominent geophysicist put it, "throwing darts at a calendar."

A panel of the nation's leading seismologists reviewed the "methodology" that led to the prediction and concluded that it should not serve as the basis of public policy. The scientists' view was reported subsequently as the counterweight to the bogus prediction, as if the two views were of comparable intellectual standing just another controversy within the scientific community. Of course, there was no major quake in the Midwest. The San Francisco area, the Middle East, and Tokyo, all of which were fingered by the same individual for a December 3 earthquake, also escaped disaster.

Skepticism and challenge to the status quo are the stuff of scientific breakthrough and new insight, that is when they are coupled with the arduous, unglamorous, often tedious process of gathering evidence and analyzing data to test a hypothesis. When an individual comes along who short-circuits the process, dodges the scrutiny of the peer-reviewed scientific literature, and takes his hypotheses directly to the press, the scientific community quite rightly views him as suspect, at best over-zealous, and at worst a charlatan. Too often, however, the press depicts such individuals as misunderstood and persecuted mavericks bucking an intolerant and self-protective establishment. That may make for a better story, but the truth is usually something quite different.

Another recent example illustrates this point. A major national magazine ran a cover story on the geologic integrity of the only site currently under study for the nation's first repository for commercial high-level radioactive waste. The site at Yucca Mountain in Nevada may or may not be suitable for waste burial: the jury is still out on the geological history of groundwater levels and other matters that bear on the question of whether waste can be kept dry and safe at the site for the next 10,000 years. But the reader of the magazine article would not know this. Instead, the reader was left with the impression that the site is totally unsuited for disposal, that the repository (if it were built) could blow up and create an environmental catastrophe beyond the magnitude of Chernobyl, that nearly all the government scientists are afraid to come to this conclusion, and that only a lone government employee has the brains and the conscience to see the hydrologic history at the site for what it is a sure sign of disaster.

The journalist reported the views of the lone employee the little guy fighting the big, dumb bureaucracy. The only problem is that the story bears little resemblance to reality. As with the purveyor of the New Madrid earthquake prediction, the lone employee at Yucca Mountain bypassed the conventional routes toward testing and gaining scientific support for his hypotheses and instead released his findings to the popular press and to those politicians eager to be rid of any waste disposal site. The particular questions he raises had been debated for years before he came along and continue to be under intense and careful study. Dozens of scientists have reviewed his work and found little merit in it. The information required to put the matter in perspective was readily available, but gathering it required more homework than the reporter apparently was willing or able to do.

These two examples make good press, but they also leave the public poorly informed about critical scientific questions that have a significant impact on important public policy decisions. While scientists have no reason to expect journalists to be defenders of the faith, we do expect that reason and soundness of method the scientific equivalent of checking the veracity of one's sources should also be prerequisites for science reporting. An elaborate, if imperfect, superstructure of peer-reviewed literature has developed in the sciences to ensure this base line of reason without suppressing rational dissenting views. Is it too much to ask of the press that similar caution be exercised in science reporting?

The marriage between science and the media can be saved if scientists and journalists will learn to communicate better with each other and to understand each other's needs and methodology. For their part, the scientific societies, the American Geophysical Union among them, have been working of late to improve their communication with the press and the public. Despite fundamental differences in purpose and style between scientists and journalists, there is hope that the two can work toward achieving a well-informed, scientifically literate public. That is in everyone's interest.

Reprinted from with permission from
January 29, 1991 issue of *EOS*,
Transactions American Geophysical Union
Washington, D.C.

Medicine

Environmental Fears and the Treating Physician
Ronald E. Gots

**Scientific Truths Versus Legal Truths
in Toxic Tort Litigation**
Ronald E. Gots

Environmental Fears and the Treating Physician

Ronald E. Gots, M.D., Ph.D.

Ronald E. Gots, M.D., Ph.D., founder and President of National Medical Advisory Service (NMAS) in Bethesda, Maryland, received his A.B. in chemistry and his M.D. degrees at the University of Pennsylvania. He received his Ph.D. in pharmacology from the University of Southern California. Since 1975, as President of NMAS, he has devoted his professional activities to medicine, science, occupational medicine and particularly to toxicology and environmental risks. He has focused on the scientific methods for assessing diseases associated with chemicals and other toxins. Dr. Gots is intensely interested in promoting effective risk communication education for the treating physician. He has been called upon to meet with workers and community groups in dozens of toxic exposure matters across the country. A number of corporations and public utilities currently engage Dr. Gots to direct their health and safety programs. He is active in numerous professional organizations—the American Public Health Association, the American Medical Association and the American Occupational Medicine Association among others. He has testified in congressional subcommittee hearings concerning victim compensation legislation He is the author of three books and 60 articles in toxicology, pharmacology, and in claims and legal literature. He is currently completing his new book entitled "Toxins: Science, Perception and Regulation" due Spring 1992.

With each episode of "60 Minutes," common perceptions of environmental health hazards are moved farther and farther away from the science that most closely approximates the truth as we know it today. The practicing physician, trained in science, but treating patients whose health education comes primarily from newspapers, magazines and television, often finds him or herself an unwitting participant in this clash between belief and knowledge.

> *"I'm having trouble breathing, Doc. It's those chemicals at the plant isn't it?"*

> *"Don't you think the cancer came from the welding fumes?"*

> *"Ever since they sprayed that pesticide around our house, we've had trouble with headaches, difficulty sleeping, and problems remembering things. It's that pesticide, isn't it?"*

A few years ago a patient would rarely ask questions like these. Today they are commonplace. Actually, the questions are rhetorical in

491

nature, because the patient has generally reached his or her final conclusion, whether you agree or not. Your professional affirmation would simply prove that his conclusion was correct. It does something else, however. It stamps the imprimatur of medical science onto this lay impression. That stamp carries the weight of medical expertise in a court of law.

Thus, agreeing with your patient when he or she asks whether an environmental exposure caused his illness may seem to be a harmless pleasantry, but the implications are quite serious. In today's climate, that pleasantry forms the expert basis of a medical claim—either a tort action (product liability) against a chemical manufacturer, or a workers' compensation claim against your patient's employer.

This leaves today's practitioners with a moral dilemma when they are asked to comment about an illness perceived to be environmental. If they don't know the answer (and in most cases neither they nor the vast majority of practitioners will) they have two choices: either admit that they don't know; or, agree with their patient anyway. If they admit uncertainty, they run the risk of dissatisfying their patient and losing him or her to a more agreeable colleague.

If, on the other hand, they certify that the perceived cause is valid when they are not certain that it is, they have violated a societal trust. You see, juries, insurers, and workers' compensation commissions believe that physicians are the only ones who know the true answers. They can be believed. Insurers (and companies that fund these claims) assume that affirmative statements indicating that certain chemical(s) caused an injury are backed by: (1) a full knowledge of the toxicology of the alleged offender(s); (2) a detailed understanding of the working environment; (3) a survey of the relevant exposure date—concentrations and duration; and (4) a careful differential diagnosis in which other causes have been systematically evaluated and excluded. In other words, a physician's word is considered by lay reviewers to be the considered statement of an expert.

If you are like most practitioners, you neither have these data bases at your fingertips, nor the time to acquire them. Thus, you are caught in an ethical and moral bind: How do you serve your patient without compromising your professional integrity?

One way is to rationalize, adopting the position that "everyone knows" that that substance is harmful and, therefore, it probably injured your patient. We, as physicians, are not immune from popular beliefs. We, too, watch "60 Minutes" and read magazines. Onslaughts of horror stories about towns decimated by aerial sprayings, or of water supplies dangerously contaminated, can make believers of even the most empirically-minded physician.

I would caution you, however, to resist that; for if physicians don't insist that good science form the basis of modern medical thought, we will be returned to the days of witch doctors and evil humors. We will be no better than the medical quacks with their magical cures against whom we so actively protest.

There are no easy answers to this dilemma. And it is problem that will increase. The science of environmental medicine lags far behind its popularization. Patients will confront you more and more. They will wonder if they are in danger because of a recent chemical spraying, wonder about risks to unborn children, wonder whether their cancer, or heart, liver, lung or kidney disease was caused by an occupational exposure or environmental pollutant. It is tempting to say, "yes," and much harder to admit that you don't know.

Reprinted with permission from
The Upstate Physician Journal
Vol. 1, No. 7, September 1983

Scientific Truths Versus Legal Truths in Toxic Tort Litigation

Ronald E. Gots, M.D., Ph.D.

Please see biographical sketch for Dr. Gots on page 491.

For most of this century the public has looked towards science to find and explore fundamental truths about diseases, their causes and their cures. Recently, toxic tort litigation has placed dramatic new pressures upon the biological sciences to find "truths" before science is ready or able to do so. This is particularly so in claims of injuries arising from chronic low-level environmental exposures. Allegations by litigants in these claims that long-term low-level exposures to chemicals in the water, the air, the soil, or just nearby, have produced unacceptable risks or current illnesses are often unprovable from the scientific standpoint and are based more upon supposition, and belief than they are upon medically established realities. This has led to serious arguments that toxic tort claims are and should be, rather than a forum for fact finding, an institution for social redress and public policy.[1]

People are convinced that environmental contaminants are responsible for serious health effects and sometimes they are. Popular belief, however, is non-discriminatory. It covers all chemicals under any circumstances. It is also highly emotional and only minimally factual. The established science of toxicology and toxicological effects is, by contrast, quite specific and quite rigorous. Facts such as the structure of the specific chemical in question, the species of animals tested with the chemical, human epidemiological data, among others, are critical to the assessment of the capability of a chemical to produce a toxic effect. For a specific individual, how much was absorbed into the system, how well other potential causes of the disorder have been considered and excluded[2] and whether a risk is real or based upon regulatory assumptions[3] are essential to the scientific evaluation of cause and effect.

In claims of illness or risk associated with environmental contaminants, these required "facts" often provide insurmountable hurdles for claimants for they ask for measurements which are unavailable and studies which do not exist. Consider claims of cancer risk as an example. Nowhere is the merging of public and social policy with common law "truths" more apparent.

The fact that an agent is regulated as a potential human carcinogen is regularly used to argue risk to a specific claimant. Quantitative risk assessment methodologies are even used to weigh that risk. The numerous assumptions and speculations underlying risk assessment have been discussed extensively in other papers.[4] A tool of public policy, risk assessment recognizes the need to protect against perceived, possible or uncertain risks. Assumptions, possibilities and speculations are acceptable grounds for regulation, but, in personal injury claims, a higher level of proof is expected.

Proving that current illnesses are due to environmental pollutants similarly often outstrips the availability of scientific facts. Nearly fifteen hundred residents of a major city are currently suing for damages because of drinking water contaminated by trichlorethylene. They have claimed several hundred disorders as actionable injuries ranging from diabetes to depression to menstrual dysfunction to cancer. The public's clamor for redress notwithstanding, modern science knows of no actual or potential harm produced by such levels of this chemical and, certainly, not those alleged impairments. Science cannot, in this case provide the facts demanded by these claims.

The public cannot mandate that science cure cancer or AIDS or multiple sclerosis tomorrow merely because it wants it to. By the same token, it cannot demand that science uncover and catalogue all of the health effects of all chemicals at all concentrations before it is ready to do so. The scientific process cannot be rushed, but the judicial process presses on by popular demand.

The claimants' response to this factual vacuum has been twofold: to argue for changes in evidentiary requirements lessening the claimant's burden of proof; and, to purchase testimony from a growing industry of "fact" manufacturers.

In scholarly treatises, it has been argued that these difficulties in proof bespeak a need to change our standards of proof.[5] This argument recognizes the difficulties of the claimants' burdens, understands the limitations of science in providing answers, and honestly asks for a reassessment of the role of the court. Whether or not one agrees with this recommendation, at least it has the virtue of honestly recognizing and separating today's scientific capabilities from today's litigation demands.

Far more disturbing is another trend, a burgeoning of science-for-hire designed to provide an imprimatur for the courtroom in the guise of scientific knowledge. A growing contingent of physicians have departed from the empirical sciences to fashion this new medicine specifically molded to the demands of toxic tort cases. Tests, without proper study or controls are being introduced at an astounding rate claiming to show neurological dysfunction, immunological disorders, neuropsychological aberrancies, metabolic derangements and other biological malfunctions. Commonly, these are solely tort tests, not used clinically to evaluate patients in actual health care, but only to provide data for claims. The blink test, sway test, certain measurements of pollutants in bodily tissues and fluids and various antibody tests fall into this category.[6] Other tests may have a place in the research laboratory or in a few special clinical settings but in the context of these claims are being used inappropriately or interpreted incorrectly. In some instances these methodologies have been formally repudiated by the peer community.[7] In many cases, however, the scholarly medical academies and respected academicians are unaware of these activities and are shocked to learn of them. In essence, this testing and testimony industry has grown to meet claimants' demands for "proof," when proof is unavailable, to satisfy a system which insists that the claimant meet his evidentiary burden.

The purveyors of this courtroom science depend upon several factors for their success. One is the law of supply and demand, which we have already discussed. The second is the popular belief in the dangers of chemicals. The third is the simplistic, easily understood nature of the claimant's allegation. "I was exposed. I am sick or will be. Therefore, the chemical is responsible." Finally, they depend upon the fact that judges, jurors and opposing counsel are scientifically and medically naive and, therefore, readily misled. To understand how this courtroom science is constructed and, how laymen are misled, a passing understanding of two disciplines is necessary. The first is the science of causation analysis. The second is the concept of controlled scientific studies.

To a nonscientist it may seem that there is no proper methodology for assessing causal relationships. Both the methodology and the conclusions may appear to be purely matters of opinion, subject to cross-examination and properly weighed by the trier of fact. While it may be true that proper methodology has not been explained to the courts, it is not true that methods for assessing causal relationships are purely matters of opinion to be determined by the preference of each expert. That would be like saying there is no proper way of diagnosing illnesses, no correct approach to differential diagnosis, and no rational methodology for deciding what is and what is not wrong with a patient. Methods for investigating causal relationships between agents and illnesses do exist and are well described in the scientific literature.[8]

To introduce the proper methodology for assessing causal relationships, let us begin by using examples from actual cases. These illustrate basic and easily understood flaws in customarily admitted testimony.

A witness testifies that a claimant's lung cancer was probably caused by exposure to toluene diisocyanate. The methodological flaw is that this chemical is not a known carcinogen. If a chemical is not known to cause cancer, should it be permissible for a witness to confuse the trier of fact by testifying that it probably did in this case? A conclusion about a cause and effect relationship must be supported by scientific evidence that such a relationship has been demonstrated and, therefore, can occur. Otherwise it is not based upon reasonable, accepted scientific methodology and it is speculative and misleading to the trier of fact.

Another common methodological error is the misleading equating of organ system disorders. Such terms, for example, include "kidney injury," "liver disease," and "nerve damage" as though there were no distinctions among specific disease entities.

Most diseases have classical clinical and morphological (seen under the microscope if specimens are available) patterns. Diabetes causes a specific kind of kidney disease; while certain chemical solvents produce quite another pattern. Carbon tetrachloride causes a characteristic liver injury; viral infections generally produce a different kind. The clinical course of viral encephalitis is generally different from that of chronic lead poisoning. Smoking-induced lung disease is different from asbestosis. The peripheral neuropathies associated with hexane are different from those associated with lead poisoning, alcoholism, or diabetes. Benzene is known to produce a certain kind of leukemia, but only that kind and no other cancer. Vinyl chloride is known to cause a rare cancer, hemangiosarcoma of the liver, not squamous cell carcinoma of the lung or carcinoma of the uterus.

Despite the fact that there exists a specificity of causal agent to disease process, testimony is routinely admitted which pretends that no such specificity exists. Formaldehyde can cause renal injury and therefore it caused this man's kidney failure opined an expert, despite the fact that the man had classical diabetic nephropathy and no features of a kidney disorder ever described in association with formaldehyde exposure.

The field of toxicology is founded upon principles of dose response. Everyone knows that drinking a fifth of whiskey can cause unconsciousness, but that a teaspoonfull will not. The same is true of many biological responses to environmental agents. It takes certain concentrations of carbon tetrachloride to injure the liver and certain levels of lead to cause a peripheral neuropathy or an encephalopathy. This is not to say that our information about safety is complete or that no one could ever be injured by exposures below levels considered to be safe. However, a toxic tort claim requires the plaintiff to prove that an exposure probably or certainly produced an illness. How can such a statement be made when the exposure was less than that known to be able to produce such an illness? In general, it cannot; yet commonly experts testify to causality despite the fact that exposures were well below levels known to be injurious. This is especially true in the environmental claims in which parts per billion are common units of exposure.

Expert testimony is commonly admitted despite other analytical errors. An oncological surgeon provided opinion testimony that his patient's lung cancer was caused by exposure to a chemical in a paint product. The chemical is not a known carcinogen and the first exposure occurred only 4 months before metastatic lung cancer was diagnosed, far too soon to satisfy known behavior of carcinogens.

The proper principles of the methodology of causation analysis can be summarized as follows:

Can: Can the agent in question produce the disease at issue?

1. Is there substantial and properly relevant animal data?

2. Is there human evidence, particularly epidemiological support?

Did: Did it cause it in this case?

1. Have other causes been properly considered and ruled out?
2. Has the exposure been confirmed?
3. Was the exposure sufficient in duration and concentration?
4. Was the clinical pattern appropriate?
5. Is the morphological pattern appropriate?
6. Is the temporal relationship appropriate?
7. Is the latency appropriate?

Just as proper medical practice requires a systematic consideration of the differential diagnoses, proper causation analysis requires the use of the methodology outlined above. The importance of this has been delineated by other authors beginning with Koch in the 19th century and continuing today among environmental pathologists and clinical scientists studying adverse drug reactions.

Once this approach to causation analysis is understood, a second question arises. What kinds of studies need to be conducted to tell us whether a chemical is capable of producing a disorder? The answer lies in scientific study design. Both studies performed in the laboratory and epidemiological studies examining populations of people share common characteristics including the need for proper controls. A simple and excellent discussion of experimental design is found in the book *Statistics Concepts and Controversies*.[9] It or others like it should be required reading for judges and attorneys embroiled in toxic tort litigation. It is quite impossible in this short paper to discuss in any detail the scientific method, but I shall try to provide a brief overview of modern experimental design and to illustrate how tort scientists deviate from that design.

A scientific study identifies relationships in a consistent, reproducible fashion. It must be sure to control other factors so as to avoid a mere appearance of a relationship, when there really is none. Assume, for example, that we gave a new antihistamine to 50 hay fever sufferers and sugar pills (a placebo) to 50 others. Assume that 40 of the treated patients got better, but none of the controls did. Those facts alone might suggest a salutary affect of the drug. Now we learn, however, that those 40 who got better were all members of the same Elk Lodge and all left hay fever country together for a three-week Pacific cruise. This confounding new variable, for which we had not controlled, leaves us uncertain as to whether the improvement was due to the drug or the change in environment (or, perhaps, to some other uncontrolled factor). When we test the effectiveness of a new narcotic, we must control for the placebo affect. Why? Because 30% of patients get relief from even postoperative pain through placebos alone.

A second essential in experimental design is in the statistical analysis, There must be a statistically significant difference between the control and the experimental population to draw meaningful conclusions. Thus, if 10 out of 30 appear to improve with a particular treatment, but only six out of 30 of the untreated controls do, there may appear to be a benefit to the therapy, but only proper mathematical analysis can tell us whether that is only a perceived or a statistically significant difference.

Next, to properly study the effect of any agent, we must know that the agent actually entered the subject's body, how much of it entered and how that differed from the controls. Unless we know this, we cannot conclude that the test agent was the cause of a disorder.

Toxic tort testing, both laboratory and epidemiological, is commonly flawed in all of these respects. Consider these examples. In a mass tort claim it was alleged that a test known as the "blink" test was abnormal in a group who drank TCE contaminated well water and that the TCE was the cause. Even if the "normal" response to the blink test were well-established, which it is not, and the residents' tests were abnormal, it would be literally impossible to point to the drinking water as the culprit.[10] Why? Because the levels actually drunk by individuals was unknown; studies had never established a relationship between low levels of ingested TCE and blink test abnormalities; and because at levels of parts per million or billion we are exposed to thousands of chemicals any of which could potentially be causal—thousands of confounding variables.

In an epidemiological study it was claimed that residents living near a site of air emissions (in the part per billion range) developed a variety of complaints due to that exposure. The "controls" had no demonstrably different exposure level (there were no comparative measurements) and, at the level measured, those chemicals are just as high routinely in indoor household air.[11] Thus, conclusions about relationships between the exposure and symptoms suffered from an invalid control group and from failure to control the confounding variables.[12]

Immunological tests are commonly used to "prove" a toxic cause of immune dysfunction. Wide day-to-day variations in "normal" permits witnesses to claim abnormalities when a repeat test would disprove that. Frequently exposures are alleged, but not defined; therefore the alleged exposure cannot be properly connected to a test result. Effects of the chemical in question upon that test are often unknown; therefore pure guesswork. Confounding variables are never controlled.

There are really only a few solutions to this growing gulf between the availability of valid science and the public's demand for answers and testimony in toxic tort cliams. One is to formalize the proposition that the resolution of these claims are matters of policy, not truths. If, however, evidentiary requirements are not changed—and I, for one, believe that they should not be—then those of us who are alarmed by pseudotruths masquerading as truths and by the distortion of science must help judges and juries to become better informed.

There is some indication that judges are becoming better informed and better able to protect juries from prejudicial or misleading testimony. There have been a number of recent cases in which "scientific" testimony has been excluded or overturned for lack of realiability. Some of those cases are documented below.[13] Recent reviews of those decisions are available.[14] The courts face growing challenges, however, as these new "sciences" become increasingly sophisticated and espoused by well-credentialed witnesses.

Convincing juries is even more challenging for the defense. Complex information has to be simplified and made interesting for lay jurors. Such concepts as: small amounts—parts per million or billion; notions of the chemical world in which we live and the context of the subject chemical in this chemical sea; concepts of risk and relative risk; the prevalence of natural carcinogens; how the immune system works; and what makes one test valid and another invalid are among the concepts which we must get across in these matters. The most basic of toxicological and scientific primers are essential if fact-finders are to have any understanding of what constitutes a fact and why. Computer graphics and videoprograms are increasingly part of the courtroom scene. They will play a key role, I believe, in helping jurors to learn what they need to know in ways that are familiar and comfortable. Ultimately, however, jurors must be protected from misleading testimony. Otherwise, as causes of cancer, immune disorders, neurological diseases and most other ills are legally determined, Westlaw will soon replace Harrison's Principles of Internal Medicine in medical school curricula.

References

1. Novey, L.B., ed. *Causation and Financial Compensation for Claims of Personal Injury from Toxic Chemical Exposure: Conference Proceedings.* Washington, D.C.: Georgetown University Institute for Health Policy analysis, 1986; Schwartzbauer, E.J. and Shindell, S. "Cancer and the adjudicative process: the interface of environmental protection and toxic tort law." *American Journal of Law & Medicine* 14:1–67 (1988).

2. Gots, R.E. "Medical causation and expert testimony." *Regulatory Toxicology and Pharmacology* 6:95–102 (1986); Evans, A.S. "Causation and disease: the Henle-Koch postulates revisited." *Yale Journal of Biology and Medicine* 49:175–95 (1976); Hackney, J.D. and Linn, W.S. "Koch's postulates updated: a potentially useful application to laboratory research and policy analysis in environmental toxicology." *American Review of Respiratory Diseases* 119:849–52 (1979); Irey N.S. *Syllabus*. Division of Experimental Pathology, Armed Forces Institute of Pathology, Bethesda, MD (1982); Ruckelshaus, W.D. "Science, risk, and public policy." Science 221:1026–8 (1983); National Research Council. *Risk Assessment in the Federal Government: Managing the Process.* Washington, D.C.: National Academy Press, 1983; Office of Technology Assessment. *Assessment of Technologies for Determining Cancer Risks from the Environment.* Washington, D.C. U.S. Government Printing Office, 1981; Wilson, R. and Crouch, E.A. "Risk assessment and comparisons: an introduction." *Science* 236:267–70 (1987).

3. Schwartzbauer and Shindell, "Cancer and the adjudicative process"; National Research Council. *Risk Assessment in the Federal Government;* Ruckelshaus, "Science, risk, and public policy"; Ruckelshaus, W.D. "Risk, science, and democracy." *Issues in Science and Technology* 1:19–38 (1985); *Risk Assessment and Management: Framework for Decision Making.* Washington, D.C.: Environmental Protection Agency, 1984; Russell, M. and Gruber, M. "Risk assessment in environmental policy-making." *Science* 236;286–90 (1987); Prera, F.P. "Quantitative risk assessment and cost-benefit analysis for carcinogens at EPA: a critique." *Journal of Public Health Policy* vol. 1987:202–21 (1987).

4. Gleeson, J.G. and Gots, R.E. "Understanding and defending claims of increased risk of contracting disease." *DRI Damages and Jury Persuasion* 6:54–69 (1987); *Risk Assessment and Management: Framework for Decision Making,* Washington, D.C.: Environmental Protection Agency, 1984; Schwartzbauer and Shindell, "Cancer and the adjudicative process"; Davis, D.L. "The shotgun wedding of science and the law: risk assessment and judicial review." *Columbia Journal of Environmental Law* 10:67–109 (1985).

5. Schwartzbauer and Shindell, "Cancer and the adjudicative process"; Rosenberg, D. "The causal connection in mass exposure cases: a public law vision of the tort system." *Harvard Law Review* 97:849–929 (1984); Trauberman, J. "Statutory reform of toxic torts: relieving legal, scientific and economic burdens on the chemical victim." *Harvard Environmental Law Review* 7:177–297 (1983).

6. Feldman, R.G., Chirico-Post, J. and Proctor, S.P. "Blink reflex latency after exposure to trichloroethylene in well water." *Archives of Environmental Health* 43:143–7 (1988).

7. Terr, A. "Clinical ecology." *Annals of Internal Medicine* 111:168–78 (1989).

8. Evans, A.S. "Causation and disease"; Hackney, J.D. and Linn, W.S. "Koch's postulates updates"; Irey, N.S., *Syllabus,* Division of Environmental Pathology; Irey, N.S. "Tissue reactions to drugs." *American Journal of Pathology* 82:617–47 (1976); Irey, N.S. "Diagnostic problems in drug-induced diseases." In: Meyer, L.P. *Drug-Induced Diseases.* Vol. 4. Amsterdam: Excerpta

Medica, 1972; Meyer, L. and Herxheimer, A. *Side Effects of Drugs.* Vol. 7. Amsterdam: Excerpta Medica, 1972; Henninger, G.R. "Drug and chemical injury." In: *Pathology,* 6th ed. St. Louis: Mosby, 1971. pp. 174–241; Gots, R.E. "Medical/Scientific Decision Making in Occupational Disease Compensation: Analytical System, Operational Approach." Presentation, Crum and Forster Corporation, November 1, 1981; Gots, "Medical Causation and expert testimony"; Gots, R.E. "The science of medical causation: its application in toxic tort litigation." *DRI Toxic Tort Litigation* 4:18–27 (1986).

9. Wade, D. *Statistics: Concepts and Controversies.* New York: W.H. Freeman, 1985.

10. Feldman, R.G., Chirico-Post, J. and Proctor, S.P. "Blink reflex latency"; Feldman, R.G., Firnhaber, W.R., Currie, N.N., et al. "Long-term follow-up after single toxic exposure to trichlorothylene." *American Journal of Industrial Medicine* 8:119–12 (1985).

11. Environmental Protection Agency. *The Total Exposure Assessment Methodology (TEAM) Study.* Springfield, VA: National Technical Information Service, 1987. (EPA-600-6-87-002)

12. Ozonoff, D., Colten, M.E., Cupples, A., et al. "Health problems reported by residents of a neighborhood contaminated by a hazardous facility." *American Journal of Industrial Medicine* 5:581–97 (1977); Ozonoff, D. "Medical aspects of the hazardous waste problem." *American Journal of Forensic Medicine and Pathology* 3:343–8 (1982).

13. Richardson V. Richardson-Merrell, 649 F. Supp. 799 (DDC, 1986); Sterling v. Velsicol Chemical Corporation, 647 F. Supp. 303 (W.D. Tenn., 1986; In re Agent Orange Product Liability Litigation, 611 F. Supp. 1223 (E.D.N.Y. 1985), aff'd 818 F.2d 145 (2nd Cir. 1987); In re Paoli Railroad Yard PCB Litigation, 706 F. Supp. 358 (E.D. Penn. 1988); Arnett v. Dow Chemical Corporation, California Superior Court no. 729586, March 21 1983.

14. Elliott, E.C. "Toward incentive-based procedure: three approaches for regulating scientific evidence." Proceedings of the Yale Law School Program in Civil Liability, New Haven, Connecticut, April 8–9, 1988; Austrian, M.L. "Significant evidentiary issues concerning expert testimony: a case study of the bendectin litigation." Defending Toxic Tort Litigation: Seventh Annual Law and Science Defense Seminar, October 27–28, 1988, San Francisco, California; Schwartzbauer, E.J. and Shindell, S. "Cancer and the adjudicative process"; Gleeson, J.G. and Shelton, H. "Reasonable scientific certainty: a proposed evidentiary standard for the disqualification of experts in toxic tort litigation." *DRI Toxic Tort Litigation* 4: 1–17 (1986); Black, B. "Causation case law—where theory and practice diverge." In: Novey, L.E. ed. *Causation and Financial Compensation for Claims of Personal Injury from Toxic Chemical Exposure: Conference Proceedings.* Washington, D.C.: Georgetown University Institute for Health Policy Analysis, 1986. pp. 233–41.

Originally presented in 1989, at a workshop "Trying Mass Toxic Court Cases," for the American Bar Association.

Nutrition

**Unproven (Questionable) Dietary and Nutritional
Methods in Cancer Prevention and Treatment**
Victor Herbert

**Perceptions of Food Safety Issues
in the United States**
Fredrick J. Stare

Nutrition
Elizabeth M. Whelan and Fredrick J. Stare

Unproven (Questionable) Dietary and Nutritional Methods in Cancer Prevention and Treatment

Victor Herbert, M.D., J.D.

Dr. Victor Herbert is a Professor of Medicine and Chairman, Committee to Strengthen Nutrition, Mount Sinai & Bronx Veterans Affairs Medical Centers. He formerly was an assistant professor at Harvard Medical School; a clinical professor at Columbia University, New York; and chairman of Medicine, Hahnemann University, Pennsylvania.

Dr. Herbert received a B.S. in chemistry from Columbia University in 1948; an M.D. from Columbia University in 1952; and a J.D. from Columbia in 1974.

He is the author of *Nutrition Cultism: Facts & Fictions,* Stickley-Lippincott, 1981; *Vitamins & Health Foods: The Great American Hustle,* co-authored with Steve Barrett, published by Stickley-Lippincott in 1985; and is editor (and author of nine of the 41 chapters) of *The Mount Sinai School of Medicine Complete Book of Nutrition,* published by St. Martin's Press in 1990. He has published over 655 scientific articles and has contributed to *New England Journal of Medicine, Journal of the American Medical Association (JAMA),* and *American Journal of Clinical Nutrition,* among others.

He is the recipient of many national academic awards.

To quote the overview of the role of diet and nutrition in cancer by Weinhouse[1]:

> . . . we are deluged with falsehoods and distortions about food and nutrition. There is a constant drumbeat of pernicious misinformation from the media, the "health food" industry, a host of misguided food faddists and from a variety of charlatans who live off the anxieties of people. This is why guidelines based on sound science are desirable to keep the public reliably informed.

It is well recognized[2] that many adverse nutrient-nutrient interactions may result from immoderately high doses of any one vitamin or mineral, or of fiber. Most of the quackery in dietary and nutritional methods in cancer prevention and treatment relates to use of exactly such immoderately high doses of vitamins and/or minerals and/or fiber and/or various herbs, pills, powders, potions, and perfusions represented as "food supplements," "alternative therapy," and/or "metabolic balancing," and/or "anti-cancer" agents. There is no such thing as "alternative therapy." The phrase is largely a euphemism for quackery. There are therapies that work, therapies that do not work, and experimental therapies.

Questions Which Make Something "Questionable": The Four Basic Canons of Therapy

"Unproven" is a euphemism for questionable. The definition of a questionable method is that it has not successfully answered the basic consumer protection questions[3-5] of efficacy and safety with respect to any proposed cancer preventive or treatment modality: Has the proposed modality been objectively and reproducibly demonstrated in humans in the peer-reviewed literature to be (1) more effective than suggestions (doing nothing) and, in addition, either: (2) as safe as doing nothing? or (3) if there is any question with respect to safety, to have a potential for benefit which exceeds its potential for harm? The above three questions are basic rules of epistemology. Only by insisting on adequate answers to these questions can we avoid being brainwashed by the promoters of health frauds.

An implicit fourth canon is that the burden of proof is on those who propose the treatment.[3,5] It is for the proponents to demonstrate efficacy and safety, and not for responsible cancer researchers to give up their cancer research in order to demonstrate lack of safety and lack of efficacy of every new lucrative quack cancer remedy that comes down the pike.

Proponents of questionable cancer remedies who say they are too busy taking care of patients to prepare for publication in the peer-reviewed literature proof of efficacy and/or proof of safety and/or proof that the benefit exceeds the harm of their proposed remedy are usually promoters of quackery and are usually really saying that they are too busy piling up profits to bother to find out whether their remedy can be objectively demonstrated to work or not, help or hurt, cure, or kill.[6-10] Their patients and other boosters are usually victims of their own blind hope, compounded by ignorance of epistemology (the science of "how do we know the things that we know" and "what are the questions for which we should insist on meaningful answers before believing something may work?"), and of the hucksters' (usually false) claim that "at least it won't hurt you." This claim is insidious and irresponsible.

This succinct definition of questionable as not successfully answering the basic questions regarding safety and efficacy gets away from semantics, such as the use of terms like "orthodox," "unorthodox," "alternative" (i.e., alternative to what works), "holistic," "nutritional and metabolic," "establishment," and "nonestablishment." There is no such thing as "orthodox versus alternative" therapy. There is simply responsible therapy (therapy that works), irresponsible therapy (therapy that does not work), and experimental therapy. Therapy that works meets the safety and efficacy basic canons. Experimental therapy does not meet the canons. It is unethical and irresponsible (1) not to inform a patient that you are going to do experiments on him or her, and (2) to charge the experimentee to be your "guinea pig." If, after experimentation, the proposed treatment proves ineffective or to produce more harm than good, then it becomes quackery to continue using it. Direct harms from quackery have ranged from minor to death. Indirect harm is often death from a cancer which responsible therapy could have cured (see testimony of Harvey Wachsman, MD, JD[10]).

Inadequate (Irresponsible) Informed Consent Identifies Quackery

I believe we should change the name of the American Cancer Society's Unproven Methods Committee to "The Committee on Questionable Methods." [1] An alternative name could be the "The Committee on Irresponsible Methods," because it is irresponsible to give a patient a remedy which has not successfully met the efficacy and safety questions without first getting an informed consent from the patient, which informed consent clearly spells out to the patient that the therapy has not been demonstrated to be more effective than suggestion or doing nothing, has not been demonstrated to be as safe as doing nothing, and there is a definite possibility that the therapy may do more harm than good.

Promoters of quack remedies frequently claim they are "ahead of their time," i.e., a euphemism that they are doing experimental therapy, but they almost never inform their patients that they are carrying out experiments on them. Their "informed consent" forms tend to be deceptive and misleading, with wholesale "deception by omission." In fact, their informed consent forms may include sufficient concealment of material facts, misstatements, misrepresentations, and false promises as to constitute actionable consumer fraud. I have never seen a consent form of a promoter of quackery which informs the patient that what they are doing is experimental, that it has not been demonstrated to be more effective than suggestion or doing nothing, and that it may do more harm than good. Their informed consent forms appear designed to protect the doctor, not the patient. This failure to get an adequate informed consent distinguishes promoters of quackery from responsible health professionals, who always submit their experimental protocols to a responsible Human Studies Committee for evaluation that the patient is being adequately protected, including submission to the Committee of the informed consent form they are going to use, so that the Human Studies Committee can make an objective determination that the informed consent form in fact adequately informs the patient. The informed consent forms used by the responsible health professionals are designed to protect the patient.

The sellers of snake-oil cures for cancer are usually advised by "hired gun" attorneys that the way to sell fraudulent cures and not pay the penalty for fraud is to get the patient to sign an "informed consent" form prepared by the hired gun stating that the patient recognizes that the therapy is "unorthodox" or "not establishment" but wants it anyway, and stating that the patient recognizes that he or she is not being promised a cure. The hired guns tell the promoters not to make promises in their offices, but to imply then through the media: radio, television, newspapers, magazines, and books. The media campaigns steer the public to the snake-oil providers. They frighten patients away from proper treatment by referring to surgery as "cutting," to radiation as "burning," and to responsible chemotherapy as "poisoning." They then tell the patient that the snake-oil remedy is "alternative therapy" which provides painless treatment, without disfigurement, and a good result. All the quack has to do is nod in sympathetic understanding when the patient says he has come for painless treatment and a good result. A case in point is Schneider v. Revici.[11] In that case, the patient had a small cancerous lump in one breast which a competent physician had recommended be removed by lumpectomy. She heard Gary Null,[4] a syndicated radio commentator and "cancer expert" for Penthouse magazine, promoting the "Revici method," one of the quack remedies for cancer listed by the Unproven Methods Committee of the American Cancer Society.[12]

Twenty years ago, it had been reported to be worthless in the *Journal of the American Medical Association* after a thorough investigation.[13] Null, of course, did not inform his audience that the method had never been objectively demonstrated to be more effective than suggestion or doing nothing. In *Schneider v. Revici,* Revici claimed in his own defense testimony that he had told the plaintiff to have a lumpectomy and pointed out that she signed an informed consent agreeing to take his treatment. The jury awarded $1 million to the plaintiff for Revici's malpractice in giving her his snake-oil remedy, but reduced the damages by 50% for the plaintiff's "contributory negligence" for not having a lumpectomy! Her failure to have a lumpectomy during the 14 months Revici was giving her his snake-oil resulted in her breast cancer spreading widely, with the result that when she left Revici and went to a competent physician, she required bilateral mastectomy, radiation, and chemotherapy. Two years before the jury brought in its verdict, Revici's license to practice medicine had been suspended for the allowable 60 days by the New York State Commissioner of Health on the grounds that he was "a danger to the health of the people of New York State." Revici has been selling snake-oil to treat cancer (a mixture of oils and selenium—a literal snake-oil) for 30 years in the state of New York, and for 20 years after his remedy was put on the American Cancer Society's Unproven Methods list, before the state of New York finally temporarily suspended his license to practice medicine. However, such suspensions are only valid for 60 days in New York. No one knows how many patients have died because they went to Revici instead of competent physicians for treatment.

The Revici case illustrates that more important than the money, quackery steals time from patients. The time to treat cancer is usually early in its course. Quackery deludes the patient into not getting early treatment, and gives the tumor time to spread, often fatally, before the patient moves from quackery to responsible therapy.[10]

Who Should Pay for Experimental Therapy?

It should be stressed again that Health Care Financing Administration (HCFA) does not approve experimental therapies for reimbursement under Medicare and Medicaid, and most insurance companies follow this rule. I believe this is as it should be. I believe developing new therapies should be paid for by government and private agencies working in the public interest, and not by the individual hopeful patients on whom they are being used, who often make their decision to participate because of desperation, and who are performing a social good by participating in responsible protocols for development of responsible therapies. [2]

As stated above, a proposed cancer prevention or treatment modality which has not successfully answered two of the above three questions is by definition questionable. It is "questionable-experimental" if it is new, and not only questionable but very probably quackery if it is not new (i.e., if it has had time to successfully answer two of the above three questions, and has not done so).

In general, society (in the form of the American Cancer Society or the National Cancer Institute or other responsible organization) bears the costs of experimental therapy. If patients are asked to pay for questionable therapy, it is likely to be quackery rather than experimental. It is important to recognize the HCFA does not (and should not) allow payment for experimental therapy, so promoters of quackery who designate their modalities as experimental are usually unable to get third party payment from federal or insurance sources, and try to force such payments by political pressure.[10]

Whom Can One Trust?

Unreliable sources include nearly all those who profit from selling or taking advertising for questionable diagnostic tests like hair analysis to falsely diagnose a need for "food supplements" and cytotoxicity testing to diagnose nonexistent clinical allergies, and questionable therapies like megadoses of vitamins and minerals, through various herbs, and overt poisons like Laetrile. Many of them are mentioned by Herbert[3,4] and in the two Pepper Committee documents, "Quackery: A $10-Billion Scandal. Report"[9] and "Quackery: A $10-Billion Scandal Hearing".[10] Other sources I would not trust include those trade associations whose names conceal rather than reveal their economic interest, such as the Council for Responsible Nutrition (which is a trade association of nutrition supplement manufacturers, with 70 corporate members).

Nature of the Problem

The Oregon conference focus and objectives were stated as follows:[14] "Health care mis-information, fraud and quackery are a national outrage."

It is estimated that each year millions of Americans are bilked out of as much as $26 billion in worthless, unproven, and very often dangerous—even deadly—advice, treatment and prescriptions. Quacks are everywhere, operating under the guise of legitimacy, reaching out to us through "medical" offices, newspaper advertisements, letters and flyers, door-to-door salespeople, and through thousands of retail outlets.

And yet, as the National Council Against Health Fraud reminds us:

Health fraud and quackery have become socially acceptable. Misguided politicians openly work to legalize Laetrile, DMSO, Gerovital, ortho-molecular psychiatry, chelation therapy for heart disease and other questionable treatments. Some health care professionals have abandoned the scientific rigor of their disciplines to promote a myriad of pseudosciences under the guise of "holistic health." Several established drug companies encourage unnecessary nutrition supplements by the public. A number of medical laboratories promote hair analysis inappropriately, as a means of determining nutritional status, cytotoxic testing for food allergies, and misrepresent health information on computer print-outs. A number of 'medical providers' continually seek to expand their scope of practice through lobbying, threatening antitrust suits and public indoctrination. At the same time, many of these same people actively oppose immunization, pasteurization, fluoridation, and other scientifically proven public health measures.

Quacks rob us of our money, our dignity, our health and our lives.

At the very best the laws of regulatory agencies which should defend us are weak and ineffectual. At worst, laws are being put on the books which allow, even encourage, quackery to flourish.[14]

How to Evaluate Claims of Cure

As noted elsewhere,[5] in our health-professional schools, we learn that health anecdotes and testimonials are worthless as evidence of either the safety or efficacy of a claimed remedy, because, by definition, health anecdotes and testimonials are allegations that have not been verified by responsible health professionals as cause and effect rather than coincidence, suggestion, misunderstanding, or falsehood. Patients frequently believe they have a disease they do not have, and promoters of quackery use a battery of worthless "diagnostic tests"[10,15] to "diagnose" cancer or other diseases in people who do not have them. We learn that before accepting a claim of cure as valid, it is necessary to ascertain that (1) The patient had the disease (i.e., the diagnosis was established by responsible criteria); (2) The claimed cure resulted from the therapy rather than merely being coincidental with the therapy; (3) The disease is in fact cured, rather than progressing silently (i.e., without symptoms); and (4) The patient is alive (i.e., survived the treatment). Claims of cancer cure with various quack "nutritional and metabolic" or "alternative" (alternative to what works) therapies that have been investigated by responsible health professionals invariably prove due to coincidence, suggestibility and/or the natural history of the disorder, and fall into one of the following five categories of "cures that never were" as follows:

1. The patient never had cancer. Women with hormone-dependent cystic mastitis, which waxes before menses, wanes after, and often goes away with no treatment, have been told by quacks that the lump

in their breast is cancer and they will cure it with Laetrile. When the lumps have disappeared, the quacks have taken the credit, and the women have sworn in court that they were cured of cancer. Competent antiquackery attorneys do not let them testify, because they have had no biopsy to demonstrate that they ever had cancer.

2. The cancer was cured or put in remission by proper therapy, but laetrile or other "nutritional and metabolic" therapy was irrelevantly given and erroneously credited for the cure (see Rutherford case[3,16]).

3. The cancer is progressing silently but represented as a cure (see the cases of Joe Hofbauer and Chad Green[3,16]). Joey Hofbauer had a Stage IA Hodgkin's disease, curable by proper therapy. Instead, he got Laetrile and immunoaugmentative therapy and was dead in 2 years. The autopsy showed his lungs largely replaced by Hodgkin's disease. Chad Green had null cell leukemia curable in 80% of cases by proper therapy. Instead, he got Laetrile and was dead in 2 years. His autopsy showed leukemia cells in many of his tissues and cyanide (presumably from the Laetrile) in his liver and spleen.

4. The patient is dead, but represented as cured (see case of *Scott v. McDonald*[13,16]).

5. The patient had a spontaneous remission, publicized as a cure, but the 100 or more patients who received the same quack remedy in the same time frame from the same promoter and died were not publicized.

How Quacks Identify Themselves

Quacks identify themselves by using anecdotes and testimonials to support their claims of efficacy and safety, and simultaneously failing to point out that they have not separated the claimed cause and effect result from coincidence, suggestibility, or the natural history of the disorder.

They further identify themselves by claiming they are doing experimental and/or avant garde medicine, but either not providing the patient an informed consent form, or asking the patient to sign one aimed at exculpating the doctor rather than protecting the patient.

Professor James Harvey Young provided a ten-point profile of nutrition cultism and quackery, which we paraphrased elsewhere[5] and repeat here: (1) Exploitation of fear. (2) Promise of painless treatment and good results. (3) Claims of miraculous scientific breakthrough. (4) Simpleton science: Disease has but one cause, and one treatment is all that is needed to fight it. (Bad nutrition causes all disease; good nutrition cures it.) (5) The Galileo ploy: (Like Galileo, we cult gurus are misunderstood by blind scientists but are destined to be heroes to future generations). (6) The conspiracy theory (also known as "the establishment is out to get us"). (7) The moving target: Shifts in theory to adjust to circumstances. (Laetrile went from drug to "vitamin," from cure to palliative, from palliative to preventive; from low to high dosage; from working alone to never working alone; from one chemical formula to another,; and so forth. "B_{15}," (pangamate), another quack remedy for cancer and whatever else ails you, is any chemical or combination of chemicals that the seller chooses to put in the bottle; there is no such substance.[3,17–19] (8) Reliance on testimonials. (9) Distortion of the idea of freedom: by distorting "freedom of informed choice" to mean "freedom of misinformed choice," snake-oil salesmen acquire freedom to defraud their victims. (There is no freedom of choice based on false information.) (10) Most significantly, large sums of money are involved. (Cancer quackery is a more than $5-billion-a-year industry in the US today[8] [see Appendix]. Laetrile was a billion-dollar-a-year industry for product alone in the United States as of June 1979,[3] and despite the multi-institution study published in 1982 demonstrating once again that Laetrile is worthless,[20] Laetrile was still a billion-dollar-a-year industry in 1984).

Young summarized the success of Laetrile and other "nutrition" quackery in cancer therapy as "fear of cancer, suspicion of government, a primitivistic retreat from complex civilization to 'natural' ways, skillful organization, adept lobbying, and a shrewdness at borrowing time-tested techniques from quackery's well-stocked past."[21]

Nutritional Methods in Prevention of Cancer

As Ames and others[3,22–26] have pointed out, the human diet contains a great variety of natural mutagens, antimutagens, carcinogens, and anticarcinogens, some of which are present in substantial quantities in fruits and vegetables, rancid fats, and charred meats, or are formed as a result of cooking reactions involving proteins or fats. To a large extent, they probably cancel each other out in sound diets incorporating the three basic rules of sound nutrition: moderation, variety, and balance. One such diet is that recommended by Health and Welfare Canada[27] and the US Department of Agriculture[28]: four daily portions of fruits and vegetables, four daily portions of grains, two portions of milk products, and two of meat products (which include meat, fish, poultry, eggs, nuts and legumes). It is important to remember that variety includes not only the variety of the four basic food groups, but also variety within each of the four groups.[29-31] Using the four basic food groups, and employing moderation and variety within each group, one has essentially the "anticancer" diet suggested by the National Academy of Sciences[25] and the American Cancer Society.[32] The anticancer diet advises intake of food from the four basic groups, with emphasis on reducing the average American's intake of fat, minimizing salt-cured, smoked (but not smoke-flavored, which may be no problem and is more usual in the US[25]) and carcinogen-contaminated (aflatoxin, etc.) foods, and moderation in alcohol consumption. We agree[33] with the position[34,35] that it is questionable whether cancer rates will be lowered by the anticancer diet, but "the changes are unlikely to be harmful and may well be beneficial in the context of other diseases,"[35] particularly if one realizes these are not changes so much as they are adherence to the rules of moderation, variety, and balance.[27-31]

Poor Science but Good Public Relations: Referring to Dietary Influences on Cancer as Dietary Causes of Cancer

Some enthusiasts of the role of diet and nutrition in cancer confuse the phrase "caused by" (i.e., primary etiology) with the phrase "influenced by" (i.e., in later pathogenic events). The cause of a human being is the union of a sperm and an ovum, but the development of the fetus is heavily influenced by diet and nutrition. The cause of cancer is a carcinogen, which may be a virus, oncogene, or other primary inducer of uncontrolled growth.[36] Diet and nutrition are factors in modulation of the speed of that uncontrolled growth.[36,37] Failure to separate cause from promotion (i.e., failure to separate etiology from subsequent pathogenetic events) leads to deceptive and misleading claims that diet and nutrition are the cause of a large percentage of cancers. At the 1985 annual meeting of the Federation of American Societies for Experimental Biology, Peto[38] indicated that the data leading Doll and Peto[39] to previously speculate that "10% to 70% of cancer was due to diet" was exceptionally soft, and that the hard figure was less than 3%.[38a,39]

An example of this confusion is the claim that prostate cancer is caused by diet. The data suggests that the incidence (the number of cancer foci in the prostate found at autopsy) is the same in vegetarians as in omnivores, but the progression to clinical disease is much greater in omnivores. Since prostate cancer grows on male sex hormones, and male sex hormone levels are higher on meat diets than in vegetarians, diet seems likely involved in the progression of this cancer, but not a likely cause.[40] What is needed is attack on cause as well as on progression.

Nutrition Enthusiasts Versus Hard-Nosed Nutrition Scientists

Too often, enthusiasts develop slogans which blind them to science.[41,42] They rely on epidemiologic studies, whose conclusions are not evidence, but are inferences (or, at best, circumstantial evidence). The watch-words of hard-nosed nutrition scientists are evidence, efficacy, and safety. The enthusiasts for the slogan "vitamins A and C against cancer" appear to ignore the data that too much vitamin A may produce deformed infants[43] and too much vitamin C may promote progression of clinical human leukemia,[44] and ignore that epidemiologic data do not show that diets containing more than the Recommended Dietary Allowance (RDA) amounts of vitamins A and C are more "protective" than RDA amounts: the data only suggest without quantitation that diets substantially below the RDA for vitamin C are less protective, as are diets low in carotenoids.[45] Only 10% of carotenoids are convertible to vitamin A and vitamin A may not be a factor at all,[46] a possibility[45] the vitamin A enthusiasts ignore.

Similarly, the slogan "fat causes cancer" ignores substantial data that too many calories (regardless of source) may promote cancer, more than too much fat in the diet per se.[47] The "enthusiast" versus "hard-nosed nutrition scientist" controversy has been inaccurately characterized as a fight between "activists" and "traditionalists."[42a] Too much fiber may actually promote cancer.[42a]

Harms From Quack Cancer "Nutritional and Metabolic" Therapy

We have examined in detail elsewhere[3,5] the destructive cancer "treatment" regimens that are represented as nutrition science and provided literature citations to the harmful effects of each facet of these diets. Such diets promote deficiencies of iron and zinc when they lack meat, fish, and fowl and/or contain too much fiber; deficiency of calcium when they lack dairy products and/or contain too much fiber; deficiency of protein and of vitamin B_{12} when they lack all animal protein.[5] They often promote poisoning with various toxins: cyanide (Laetrile),[3] toxic megadoses of vitamins A[10,48] and C,[5] and toxic megadoses of selenium.[10] As others have noted,[2] megadoses of one vitamin or mineral (or fiber) have adverse effects on other nutrients. Almost all diet and nutrition quackery in cancer therapy uses diets and nutrition pills, powders, potions, poultices, and perfusions which have adverse effects and no objective benefit against cancer. EDTA chelation therapy,[10] promoted to "cure" heart disease and even cancer, may in fact promote osteoporosis by chelating calcium from bone. Megadoses of nutrients are toxic.[48a]

The lack of safety of macrobiotic diets for cancer patients has recently been reviewed.[3,49] Cutaneous nocardiosis in cancer patients receiving immunoaugmentative therapy (IAT) in the Bahamas was reported in 1984,[50] and worse was to come from this "nutritional and metabolic" therapy. According to *Oncology Times*,[51] doctor (of zoology) Lawrence Burton and his Immunology Researching Center, Inc., pool samples of blood products from their patients, process them, and then give the product intravenously (and sells bottles of serum product to patients to take back to the US), and have been treating AIDS patients as well as cancer patients since 1983. As any competent health professional would have expected on mixing AIDS serum with non-AIDS serum, the AIDS-linked HTLV-III virus antibody was found in 8 of 18 bottles of IAT serum, all 18 of which were positive for hepatitis B,[51] both of the two patients who received the 18 bottles from Burton's clinic in the Bahamas were seropositive for hepatitis B surface antigen (HBsAg) and the antibody to HTLV-III virus, and the virus was isolated from one of the nine specimens placed in lymphocyte culture.[51] Thus, compounding the bizarre with the macabre, the alleged (by the promoters without ever passing peer review, on *60 Minutes,* one of the most popular television shows in the US) "immunoaugmentative" therapy in fact may prove immunodestructive, by spreading AIDS to desperate cancer victims exploited by the "cancer underground" of "alternative therapy." Since it was well known in 1984 that AIDS serum can spread AIDS, Mr. Burton and his Immunology Researching Centre Ltd. in Freeport, Grand Bahama Island, may have no defense should any of the many people they must have made seropositive for HBsAg sue them not only for product liability but also for knowingly and deliberately giving them a blood product from AIDS patients. Nevertheless, as with all lucrative questionable remedies, willing victims, unaware that they are victims, continue to promote the remedy, with the help of the media,[52] who join in the deception of the public by omitting to publish with each IAT story that in all the years IAT has been given, IAT has never been objectively demonstrated to be more effective than doing nothing.

Pertinent recent reviews that should be read in the original include, "Snake-oil nutrition: the scam of the century,"[53] "Food faddism, cultism and quackery,"[54] "Foods, drugs or frauds?"[55] and the book *Examining Holistic Medicine*,[55a] which separates the noble rhetoric of holistic medicine form its ignoble performance.

Promotion Pattern of Cancer Quackery

The eleven-point common promotional pattern among lucrative questionable remedies which usually prove to be quackery includes the following[7,8]: (1) The treatment is based on an unproven theory. (2) "Special" nutrition support is needed, rather than responsible nutrition support. (3) The treatment is claimed to be painless and nontoxic. (4) Claims are published only in the media, and not in responsible peer-reviewed journals. (5) Claims of benefit are compatible with placebo effect. (6) The major proponents are not recognized as responsible experts in cancer treatment. (7) The proponents decry responsible methods of treatment. (8) The proponents claim only specially trained physicians can produce results with the treatment, or claim the treatment product is a secret. (9) Responsible oncologists are attacked as "the establishment." (10) The promoters demand "freedom of choice." (11) The promoters are making millions of dollars.

Quackery thrives by creating and exploiting public fear. In nutrition quackery, it is fear of the American diet, which is falsely represented as killing us all.[38,40] In cancer quackery, it is fear of responsible therapy, with surgery, radiation and chemotherapy depicted as "cutting," "burning," and "poisoning."

Quacks are health robbers.[56] Dripping charisma, pity and charm, they rob cancer patients of their money, their health, and their lives. They are the cruelest killers.[12] Most importantly, they rob patients of time. The time patients spend with quacks gives the cancer time to spread, so proper treatment is delayed to a time when the cancer is so widespread that proper treatment is powerless to stop the inexorable course to death.

New York physicians who wish to protect their patients against questionable practices should be aware that, under New York law, it is not only actionable professional misconduct to charge for unnecessary services, tests, or procedures, but failure to report such misconduct is itself misconduct.[57]

Note: The most recent review of this subject is in "The Mount Sinai School of Medicine Complete Book of Nutrition", edited by Herbert V., Subak-Sharpe G.J., and Hammock D.A., St. Martin's Press, New York, 1990.

Footnotes

[1] In 1990, the American Cancer Society formally adopted this recommended committee name change.

[2] There is no need to go to the "cancer underground" for experimental therapy with unproved drugs. As Philip M. Boffey noted in the *New York Times* (January 7, 1986; C1): "Thousands in US receive treatment in experiments"), "there is a staggering array of responsible clinical cancer studies now under way, enough to accommodate the vast majority of patients who seek to participate. . . Access to clinical cancer trials has also been simplified considerably in recent years. The National Cancer Institute has a new computerized data bank called PDQ, for Physician's Data Query, which provides up to date information on clinical trials being offered around the country for treating various types and stages of cancer. The data bank currently lists some 1000 active studies, scattered around the country, that are accepting patients, as well as some 10,000 physicians and 2000 organizations that provide cancer care. Any doctor can check the data bank from a library or personal computer; the system receives from 300 to 500 inquiries a month. In addition, patients and their families can call (the NCI) toll-free number, 1-800-4-CANCER, to get information about cancer, including the location of clinical trials that might be suitable for the patient. Many experts say an important factor in determining whether a patient gets into an appropriate clinical trial is the aggressiveness of physicians and family members in seeking to find such a trial."

References

1. Weinhouse S. The role of diet and nutrition in cancer: An overview. *Cancer* 1986; 58: 1791–1794.
2. Levander OA, Cheng L, eds. Micronutrient interactions: Vitamins, minerals, and hazardous elements. *Ann NY Acad Sci* 1980; 355:1—372.
3. Herbert V, Nutrition Cultism: "Facts and Fictions". Philadelphia: George F. Stickley, 1981.
4. Herbert V, Barrett S. Vitamins and "Health" Foods: The Great American Hustle. Philadelphia: George F. Stickley, 1981.
5. Herbert V. Faddism and quackery in cancer nutrition. *Nutr Cancer* 1984; 6:196—206.
6. Pepper C. Executive Summary. Quackery: A $10-Billion Scandal, Washington, DC: Pepper Subcommittee on Health of the Select Committee on Aging, US House of Representatives, May 31, 1984.
7. American Society of Clinical Oncology Subcommittee on Unorthodox Therapies, Ineffective cancer therapy: A guide for the layperson. *J Clin Oncol* 1983; 1:154—163.
8. American Society of Clinical Oncology Subcommittee on Unorthodox Therapies. Ten ways to recognize ineffective cancer therapy (Brochure). Columbus, OH: Adria Laboratories, 1984.
9. Pepper Committee. Quackery: A $10-Billion Scandal. Report (by the Chairman of the Subcommittee on Health and Long-Term Care of the Select Committee on Aging, US House of Representatives, May 31, 1984. Washington, DC 20402: Superintendent of Documents, US Government Printing Office. Committee Publication no. 98-435.

10. Pepper Committee. Quackery: A $10-Billion Scandal. Hearing (before the Claude Pepper Subcommittee, May 31, 1984. Chairman of the Subcommittee on Health and Long-Term Care of the Select Committee on Aging, US House of Representatives, May 31, 1984. Washington, DC 20402: Superintendent of Documents, US Government Printing Office. Committee Publication no. 98-463.

11. Edith and Herman Schneider versus Emanuel Revici, MD, and Institute of Applied Biology, Inc. US District Court, Southern District of NY, 83 Civ 8035 (CBM), Constance Baker Motley, Chief Judge, December 1985.

12. Wood GC, Presley BM. The cruelest killers. In: Barrett S, ed. The Health Robbers. Philadelphia: George F. Stickley, 1980; 108—122.

13. Lyall D, Schwartz S, Herter FP et al. Treatment of cancer by the method of Revici. *JAMA* 1965; 194:165—166.

14. Multnomah County Medical Society. Understanding and Combatting Health Fraud and Quackery. Portland, OR: Multnomah County Medical Society, 1985.

14a. Budiansky S. Direct-mail medicine criticized. *Nature* 1984; 4:309.

15. American Cancer Society: Unproven Methods of Cancer Management. New York: The American Cancer Society, 1979.

16. Herbert V. Laetrile: the cult of cyanide. Promoting poison for profit. *Am J Clin Nutr* 1979; 32:1121—1158.

17. Herbert V. Pangamic acid ("vitamin B_{15}"). *Am J Clin Nutr* 1979; 32:1534—1540.

18. Herbert V, Herbert R. Pangamate ("vitamin B_{15}"). In: Ellenbogen L, ed. Controversies in Nutrition. New York: Churchill-Livingstone, 1981; 159—170.

19. McPherrin EW, Herbert V, Herbert R. "Vitamin B15": Anatomy of a Health Fraud. New York: American Council on Science and Health, 1981.

20. Moertel CG, Fleming TR, Rubin J et al. A clinical trial of amygdalin (Laetrile) in the treatment of human cancer. *N Engl J Med* 1982; 306:201—207.

21. Young JH. Laetrile in historical perspective. In: Markle G, Petersen JC, eds. Science and Cancer. Boulder, CO: Westview Press, Inc, 1980.

22. Gross L. Oncogenic Viruses. ed. 3. New York: Pergamon Press, 1983.

23. Committee on Food Protection of the Food and Nutrition Board. Toxicants Occurring Naturally in Foods, ed. 2. Washington, DC: National Academy of Sciences, 1973.

24. Ames B. Dietary carcinogens and anticarcinogens. *Science* 1983; 221:1256—1263.

24a. Ames B. Dietary carcinogens and anticarcinogens. *Science* 1984; 224:659—670, 757—760.

25. Committee on Diet, Nutrition and Cancer, Assembly of Life Sciences, National Research Council. Diet, Nutrition and Cancer. Washington, DC: National Academy of Sciences Press, 1982.

26. Black CA. Diet, Nutrition and Cancer: A Critique. Special publication no. 13, Ames, IA: Council for Agricultural Science and Technology, 1982.

27. Canada's Food Guide: Handbook. Ottawa: Ministry of National Health

and Welfare, Bureau of Nutritional Sciences, Health Protection Branch, (catalog no. H21-74/1977) 1977.

28. US Department of Agriculture. Food. Home and Garden Bulletin no. 228. Washington, DC: Human Nutrition Center, US Department of Agriculture, 1979.

29. Herbert V. Health foods in nutrition: Science or scam? *National Forum* 1984; 69:3–7.

30. Herbert V. No gourmet needs supplements. *Gastronome* 1985; 7:36–41.

31. Herbert V. Towards healthful diets: The role of health foods. In: van den Berg, EME, Bosman W, Baedveld BC, eds. Proceedings of the 4th European Nutrition Conference, May 24–27, 1983. Voorlichtings-bureau voor de Voeding. The Hague, The Netherlands, 1985; 75–87.

32. American Cancer Society. Nutrition and Cancer: Cause and Prevention. American Cancer Society Special Report. New York: American Cancer Society, 1984.

33. Herbert V. A book whose time has not yet come. *Am Med News* 1977; 20:22.

34. Young VR. Diet as a risk factor for cancer: A brief overview. In: Jansen GR, Anderson J, eds. Proceedings of the 1982 Lillian Fountain Smith Conference for Nutrition Educators. Fort Collins, CO: Colorado State University, 1983.

35. Willett WC, MacMahon B. Diet and Cancer: An overview. *N Engl J Med* 1984; 310:633–638, 697–703.

36. Gross L. *Oncogenic Viruses*, vols. 1 and 2, ed. 3. New York: Pergamon Press, 1984.

37. Gross L. Reduction in the incidence of radiation-induced tumors in rats after restriction of food intake. *Proc Nat Acad Sci USA* 1984; 81: 7596–7598.

38. Peto R. Epidemiological investigation on diet and cancer: What are the proper methodological approaches? Symposium: Dietary Causes of Cancer, Fed Amer Soc Exper Biol (FASEB), 69th Annual Meeting, Anaheim, CA, April 23, 1985.

38a. Peto R. Why cancer? *The Times (London) Health Supplement,* June 11, 1981.

39. Doll R, Peto R. The causes of cancer. *J Natl Cancer Inst* 1981; 66:1191–1308.

40. Herbert V. Toward healthful diets: The role of health foods. In: van den Berg EME, Bosman W, Breedveld E, eds. Proceedings of the 4th European Nutrition Conference. Voorlichtingsbureau voor de Voeding. The Hague, The Netherlands, 1985; 75–87.

41. Olson R. The decline and fall of the Food and Nutrition Board. Nutrition Today 1985; 20:17–23.

42. Marshall E. Diet advice, with a grain of salt and a large helping of pepper. *Science* 1986; 231:537–539.

42a. Herbert V. Vegetables, fruits, and oncologists. Diet and Cancer (Letters). *Science* 1986; 232:11; 233: 926.

43. Lammer EJ, Chen DT, Hoar RM et al. Retinoic acid embryopathy. *N Engl J. Med* 1985; 313:837–841.

44. Park C. Biological nature of the effect of ascorbic acids on the growth of human leukemic cells. *Cancer Res* 1985; 45:3969–3973.

45. Olson JA, Hodges RE. The scientific basis of the suggested new RDA values for vitamins A and C. *Nutr Today* 1985; 20:14–15.

46. Munoz N, Wahrendorf J, Bang LJ et al. No effect of riboflavine, retinol, and zinc on prevalence of precancerous lesions of esophagus. *Lancet* 1985; 2:111–114.

47. Pariza M, Simoupolis A, eds. International Life Sciences Symposium: Calories and energy expenditure in carcinogenesis. *Am J Clin Nutr* (Suppl) 1987; 45:149–372.

48. Herbert V. Toxicity of 25,000 IU vitamin A supplements in "health" food users. *Am J Clin Nutr* 1982; 36:185–186.

48a. Marshall CF. Vitamins and Minerals: Help or Harm? Philadelphia: George F. Stickley Co., 1985

49. Bowman BB, Kushner RF, Dawson SC, Levin B. Macrobiotic diets for cancer treatment and prevention. *J Clin Oncol* 1984; 2:702–711.

50. Centers for Disease Control. Cutaneous nocardiosis in cancer patients receiving immunotherapy injections: Bahamas. *Morbid Mortality Weekly Rep* 1984; 33:471,472.

51. Anonymous, OT Briefs. *Oncol Times* 1985; 7:26.

52. Associated Press. Cancer patients willing to risk Bahamas AIDS. *New York Daily News,* July 31, 1985:2.

53. Herbert V. Snake oil nutrition: The scam of the century. In: Currie MN, ed. Patient Education in the Primary Care Setting. Kansas City, MO: Project for Patient Education in Family Practice, St. Mary's Hospital, 1984; 15–23, 34–36.

54. Jarvis WT. Food faddism, cultism, and quackery. *Ann Rev Nutr* 1983; 3:35–52.

55. Anonymous. Foods, drugs or frauds? *Consumer Rep* 1985; 50:275–283.

55a. Stalker D, Glymour C, eds. Examining Holistic Medicine. Buffalo: Prometheus Books, 1985.

56. Barrett S. The Health Robbers. Philadelphia: George F. Stickley Company, 1980.

57. Herbert V. A proposed solution to the malpractice problem. *NY J Med* 1986; 86:394–395.

Excerpted and abridged, with permission, from
Cancer, Volume 58, number 8, October 15, 1986.

Perceptions of Food Safety Issues in the United States

Fredrick J. Stare, Ph.D., M.D.

Dr. Fredrick J. Stare is Professor of Nutrition, Emeritus, Harvard School of Public Health and founder of Harvard's Department of Nutrition.

He received a Ph.D. in biochemistry from the University of Wisconsin in 1934, and an M.D. from the University of Chicago in 1941.

Dr. Stare is the author of *Balanced Nutrition Beyond the Cholesterol Scare*, Bob Adams, Inc., 1989; *The 100% Natural, Purely Organic, Cholesterol-Free, Megavitamin, Low-Carbohydrate, Nutrition Hoax*, Atheneum 1984; and *Living Nutrition*, Wiley, 4th Edition, 1984.

He has over 450 articles published and has contributed to *Journal of Nutrition, American Journal of Clinical Nutrition, and New England Journal of Medicine*.

Dr. Stare has received an Honorary Doctor of Science from Trinity College, Dublin, Ireland; Suffolk University, Boston; and Muskingum College, Ohio. He is the recipient of the Distinguished Service Award of the University of Chicago School of Medicine.

Many people in our country have the perception that our foods and drinking waters are not safe, in part, because some of them contain residues of various pesticides and insecticides, or smallish amounts of various antioxidants, nitrites, synthetic flavorings or dyes, or other man-made chemicals added to food and beverages for good and specific reasons. Recently, even the safety of fluoride in fluoridated water has been questioned. But these are perceptions, really misperceptions, and, in my opinion, usually are not based on good science or are a misinterpretation of the findings of good science. An example of the latter is the recent flap about fluoridation. Just over a month ago, the National Toxicology Program issued a preliminary report titled "Toxicology and Carcinogenesis Study of Sodium Fluoride in Rats and Mice." The study was well done. No evidence was found of carcinogenic activity of sodium fluoride in female rats or in male and female mice receiving fluoride in drinking water at 11, 45, or 79 ppm fluoride for two years, and only equivocal evidence of carcinogenic activity, an osteosarcoma in 1 of 50 rats receiving drinking water containing 45 ppm fluoride for two years and in 4 of 50 rats receiving water with 79 ppm fluoride for two years.

Fluoride in the fluoridation of water for humans is provided at a level of 1.0 ppm.

The report stated: "Epidemiological studies from the United States, Australia, Austria, Canada, New Zealand, and the United Kingdom have failed to show an association between cancer mortality in humans and the fluoride content of drinking water." The report also stated: "The International Agency for Research on Cancer has concluded that none of the studies reported (through March 1987) had provided any evidence that an increased level of fluoride in water was associated (in man) with an increase in cancer mortality."

In my opinion, this recent report has absolutely no bearing on fluoridation nor did it question the safety or value of fluoridation. Yet, it was interpreted by the still few ardent anti-fluoridationists as new evidence that fluoridation might be a hazard to public health and should be halted. Unfortunately, some of the public will also obtain this misperception despite the fact that fluoridation began 45 years ago, that somewhat more than half of all Americans, and many people in other parts of the world, drink fluoridated water, and fluoridation has been hailed as one of the four greatest advances of all times in public health.

I don't know how many of you have seen James Harvey Young's recent book titled *Pure Food*. Harvey Young is Professor Emeritus of American Social History at Emory University and our pre-eminent scholar in health and medical quackery. His book, *The Medical Messiahs*, published 23 years ago, is a classic. In it, he goes back to colonial days in his search for clues to the quacks' persistence and success on the American scene. His recent book, published only a few months ago, is essentially a history of the enactment by Congress of the Federal Food and Drug Acts of 1906. In those days, and before, many of our foods were grossly adulterated, and filth and microbial contamination were rife.

Not so today. Our foods are safe, absolutely safe. In collaboration with Dr. Elizabeth Whelan, a former student and colleague at Harvard's Department of Nutrition and now President of the American Council on Science and Health, we emphasized this in a book for the public published about 15 years ago under the title of *Panic in the Pantry*. The gist of *Panic in the Pantry* is, there is no reason for panic in the pantry. Our foods are perfectly safe. Unfortunately, too many American pantries have too much food and regrettably some do not have enough. Dr. Whelan and I are now working on an updated version of this book. Our tentative title: "*Still—Panic in the Pantry.*"

Last fall, a book titled: *Balanced Nutrition Beyond the Cholesterol Scare* was published. It was authored by me and two former members of our Department, Drs. Robert Olson and Elizabeth Whelan. In it, we have a chapter on food safety and recount the year ago media-fueled panic about apples poisoned by trace residues of Alar which stands out as another recent overblown food-related scare. The Alar cancer scare was another of the emotionally charged issues of food safety in America. Meryl Streep, whose celebrity appearances played a major role in bringing Alar to national attention, is admittedly an excellent actress. She turns out, however, to be an inept and ill-informed toxicologist. Those who are interested in learning the facts are advised to give greater heed to the opinions of the many scientists and health professionals who have spent entire careers in the relevant fields.

There was a time about 30 years ago when many cancers were thought to have been caused by chemicals in the general environment. Although it was firmly established by the late 1950s that cigarette smoking was the leading preventable cause of cancer in the United States, the scientific community, understandably, wanted to reduce cancer further, and believed that all that was necessary was to identify certain cancer-causing chemicals and eliminate them. It was also thought that those chemicals emanated strictly from industrial sources. Accordingly, scientists tested only chemicals that were man-made.

The chosen means of testing these chemicals was through animal studies, in other words, feeding large amounts of the chemicals to laboratory animals, primarily to rats and/or mice. If the animals developed cancer at a rate higher than expected, the assumption was that people ingesting much smaller amounts might also develop cancers.

The objective, then, was a clear one: get rid of any measurable amounts of the chemicals that caused cancer in the laboratory, and the population would be healthier. Over the next few decades, however, a number of new developments changed the landscape dramatically.

First, of course, the technology advanced. Today, we can detect levels of chemicals that were completely undetectable thirty years ago. And we now realize that the mere presence of trace levels of a substance does not, by definition, demonstrate a health hazard. Specifically, such trace levels do not necessarily reflect any increased risk of cancer.

Second, and more dramatic, was the realization that cancer-causing agents occur not only from synthetic sources, but also throughout nature. In other words, 'carcinogens'—that is, chemicals that cause cancer in high does in animals—are present throughout nature and therefore in our food supply.

This does not, by any stretch of the imagination, lead to the conclusion that most foods pose hazards. These natural carcinogens, like most synthetic ones, show up at barely measurable levels. All the discovery means is that the assumptions by which toxicologists worked in the 1950s and 1960s (namely, that nature was "carcinogen-free' and only man-made chemicals suspect) were in error.

It is important to keep in mind that the naturally occurring carcinogens have little in common with the chemicals and other materials that have been linked with occupation-related cancer. These agents would include asbestos and vinyl chloride—substances to which, unlike natural carcinogens, a person would have to be subjected for long-term, high-dose exposures to assume a high risk of developing specific types of cancer. Nevertheless, in determining that cancer-causing agents are much more common than initially believed, scientists have concluded that the world abounds with carcinogens—of both natural and synthetic origin—at very low levels.

It is possible that naturally occurring carcinogens may pose a problem in less developed countries, where food quality and processing methods are inferior. Of course, living in such a country poses a great many health risks when compared with living in developed nations like the United States, England or Sweden.

The main point is that finding any carcinogen—man-made or naturally occurring—is not enough. Furthermore, the idea that most foods could somehow be produced without trace levels of carcinogens or toxins is simply wrong. The dose of whatever chemical is being discussed, not its mere presence, is the key in determining if a health risk is present.

Part of the problem is the terminology used to discuss the issue of "chemicals in foods." Most of us would probably choose a dinner that is "free of chemicals" over one that "contains chemicals." That dinner, however, would have to be an empty plate, because all food is composed of chemicals as are all plates.

Take, for instance, a potato. A potato is a complex aggregate of at least 150 different chemicals, all put there by nature. Nature also put into the natural potato such "chemical" niceties as arsenic, solanine, nitrate, and a number of other toxic chemicals—all of which could be extremely dangerous if you took them in high doses. But there is such a minute amount of these chemicals in each individual potato that ordinary quantities of potatoes are perfectly safe to eat. (For that matter, the same could be said for virtually any quantity you'd be able to consume that wasn't ordinary. In other words, you could certainly eat several dozen potatoes, or even several bushels of potatoes, without worry—though probably not without indigestion.)

Back to the simple, all American apple. It may contain near-unmeasurable amounts of a substance that can, at high doses, cause an increase of cancer in mice—namely Alar, a chemical used in the processing of low-cost, high-quality apples. Why, if we can justify eating a potato that contains arsenic, should some consumers be terrorized by apples, apple sauce, or juice that may contain comparable levels of Alar?

BHA and BHT are abbreviations for butylated hydroxyanisole and butylated hydroxytoluene, two antioxidants added to many foods including margarine, crackers and breakfast cereals. These additives are perfectly safe in the concentrations used in foods. They are used to lessen the oxidation process and hence slow down the development of "off" flavors, help retain crispness, and preserve nutrient content.

Not only are BHA and BHT safe, but some research of many years ago indicates that they may actually play an important role in the prevention of stomach cancer. When included in the diets of laboratory animals, there is a marked reduction in stomach cancer. Some cancer epidemiologists believe that the decline in the human death rate from stomach cancer since the 1940s—when the widespread use of BHA and BHT began—may actually in part be a result of the increased use of these two additives.

Because of the scare tactics of some who should know better, many manufacturers stopped using BHA and BHT.

In a recent issue of the *Journal of The American Medical Association,* Dr. Whelan and I commented that the interplay between diet and cancer is not well understood. The incidence of different cancers varies greatly from country to country. This implies that environmental factors either contribute to or protect against certain cancers. Although there is some epidemiologic evidence that diet may be a factor in the etiology of malignancies of the colon and breast, the roles played by specific components of diet are still speculative. The association of high fat diets and some forms of cancer may be related ultimately to the excess calories provided by the fat. Repeated recent attempts have failed to substantiate an epidemiologic relationship between dietary fat intake and increased breast cancer. In animal models, the only fatty acid shown to enhance carcinogenesis is linoleic acid. By contrast, a naturally-occurring derivative of this essential fatty acid actually inhibits experimental neoplasia.

As previously stated, foods are known to contain naturally occurring carcinogens but also contain potentially protective agents. A few of these have been identified but their significance remains to be established. The quest for a fuller understanding of the relationship of diet and cancer preventives has led to the investigation of the possible role of fiber, vegetables, fruits, vitamins A, C, and E, and the mineral selenium in cancer prevention. However, biomedical scientists have not yet identified specific components of human diets as causative or preventive agents for cancer.

A current trend is the fostering of the view that a single food is either a panacea or a poison. Unfortunately, a few of our major health associations have tried to popularize this approach. This "good food/bad food" dichotomy ignores the wide consensus among nutritionists that: all foods are appropriate foods for health when used in moderation as part of a balanced, varied diet of foods selected from among and within the Basic Four Food Groups.

The March 1990 issue of the *FDA Consumer* points out that more than 96 percent of the fruits, vegetables, grains, dairy and other products analyzed by FDA in 1988 either contained no residues of pesticides or the levels found were well below legally permitted limits, according to the agency's latest annual pesticide monitoring report.

The report is based on FDA's analysis of 18,114 domestic and imported food samples, a 25 percent increase over the number tested in 1987. Samples included foods from all the states and Puerto Rico, along with products from 89 foreign countries.

The report describes FDA findings for all samples analyzed during the fiscal year ending September 30, 1988, and discusses FDA's various pesticide monitoring activities.

FDA testing methods permit the agency to detect any one of 256 pesticides. Of that number, 118 actually were found last year—about the same as in 1987. In general, residues present at 0.01 parts per million or above can be measured. For some pesticides, levels in the parts per billion range can be measured.

Concerns about possible health hazards caused by food additives and pesticide residues are at an all time high. Yet, this feeling of misgiving is not scientifically justified: strict regulatory standards that are in place ensure the safety of our plentiful food supply. With the exception of specific allergens (e.g., sulfites), to which small numbers of people are sensitive, I am not aware of a single case of ill health reported in the peer-reviewed literature that has been linked with residues of food additives or pesticides on foods purchased in any grocery store in our country.

Given the exaggerated perception of risk, physicians and scientists should play a major role in educating the public and restoring confidence about food safety. A return to the classic toxicological adage of "only the dose makes the poison" would begin to make the point.

A recently released video available to schools and educational TV and narrated by Walter Cronkite, brings a breath of fresh air into the food safety debate. Mr. Cronkite's opening statement is "As surprising as it may seem, there is increasing evidence the presence of trace amounts of chemicals in the environment pose no threat to human health."

The video is titled: "Big Fears—Little Risks." It was prepared by the American Council on Science and Health and is available from their offices at 1995 Broadway, New York, NY 10023. The scientists appearing in the video include: Bruce Ames, John Higginson, Stephen Sternberg, Elizabeth Whelan, and Norman Borlaug.

In closing these introductory remarks, I want to quote a few words from one of our leaders in the food industry, Mr. D. Wayne Calloway, Chairman and Chief Executive Officer of the PepsiCo Companies. In a talk a few months ago at the National Food Processors Association's Annual Meeting in San Francisco, Mr. Calloway stated: "Industry should stand up to the fear-mongers and agents of anxiety." And, I might add, so should academics. Continued Mr. Calloway: "Safety (in foods and beverages) is something we can never and will never risk."

As we begin a new decade, the science of nutrition faces the continued challenge of sorting out hypothesis from scientific conclusions, and of flamboyant media reports from factual reports. Special emphasis should remain on the fundamentals of nutrition that have helped Americans reach their current level of good health: choose a variety of foods from among and within the Four Basic Food Groups; maintain reasonable weight to height by restricting total caloric intake and increasing physical activity; avoid fad diets, particularly those that restrict whole food categories; and be skeptical of those who try to popularize the idea that our food and water supply is unsafe, or inherently unhealthy.

An Address made to the International
Conference on Food, Safety and Toxicology,
Michigan State University, 1990

Nutrition

Elizabeth M. Whelan, M.P.H., D.Sc. and Fredrick J. Stare, Ph.D., M.D.

Please see biographical sketch for Dr. Whelan on page 223.

Please see biographical sketch for Dr. Stare on page 522.

The focus of nutrition has shifted substantially over the past few decades. In the past, concern over health hazards focused mainly on the consequences of inadequate intakes of protein, vitamins, and minerals. Today, the possible role of nutrition in the causation of chronic disease, frequently the result of excess nutrient consumption, is receiving the most attention.

Current nutritional issues include obesity; the relationship between diet and the two leading causes of death, coronary heart disease (CHD) and cancer; and the dangers of nutrition misinformation such as that featured recently regarding alleged chemical hazards from food.

According to *The Surgeon General's Report on Nutrition and Health,* there are 34 million obese adults living in the United States.[1]

The list of health risks associated with obesity includes CHD, stroke, hypertension, hypercholesterolemia, diabetes, cancer, and complications during pregnancy. In fact, almost all conditions of ill health are worsened by obesity.

Physicians should teach patients to respect the balance between the different food groups as sources of essential nutrients and between energy intake, physical activity, and overall good health. Often, obesity goes hand in hand with a sedentary life-style. Lack of physical activity also contributes to hypertension, a major risk factor for CHD and for stroke. A recent study concluded that even a modicum of physical exercise yields significant health benefits.[2]

The relationship between diet and hypercholesterolemia, one of the potentially modifiable risk factors contributing to CHD, is the most hotly debated item on the nutrition agenda today. The National Cholesterol Education Program[3] recommends that the entire US population reduce dietary intake of saturated fats and cholesterol to reduce CHD risk. The National Cholesterol Education Program defines elevated serum cholesterol as any level above 5.15 mmol/L and recommends measurement of cholesterol levels in all Americans 20 years of age and older.[3]

Serious questions remain about the efficacy and cost efficiency of such a program to prolong life.[4-7] Many health professionals challenge the mass intervention approach.[7-10] Rather than trying to change the diet of the entire population, they propose a solution that targets high-risk individuals such as those with an elevated serum cholesterol level above 6.20 mmol/L and other specific CHD risk factors, particularly cigarette smoking and high blood pressure. What is the danger of mass dietary campaigns to lower serum cholesterol levels? Such advice may lead to an overly zealous restriction of foods rich in many essential nutrients, thereby adversely affecting nutritional balance and health.[8] Many elderly, especially the economically disadvantaged, may be avoiding eggs, meat, and dairy products, and, although it is uncertain whether these individuals have anything to gain from reducing their serum cholesterol levels,[6] it is likely that they have much to gain from an adequate, well-balanced diet. The role of diet in the causation of CHD remains unresolved, and overemphasis on diet may shift attention away from more promising regimens, such as quitting smoking and controlling hypertension.

Another area of disagreement concerns the application to children of cholesterol screening and treatment recommendations that were originally developed for high-risk adults.[10] The Committee on Nutrition of the American Academy of Pediatrics has stated its opposition to such routine blood screening.[11] Less than 5% of American children have a genetic predisposition to high serum lipid levels, and therefore an increased risk for future heart disease. Healthy children exhibit fatty streaks in their intima, a normal, nonpermanent phenomenon found in all children independent of diet or other environmental factors. Fatty streaks do not predict later development of CHD.[8] Overzealous dietary restrictions for children are not without health consequences. Regarding a low-cholesterol diet for children, a report from the Committee on Nutrition of the American Academy of Pediatrics states: "In fact, such an unwarranted diet or treatment could be deleterious to growth and development."[11]

The interplay between diet and cancer is not well understood. The incidence of different cancers varies greatly from country to country, which implies that environmental factors either contribute to or protect against certain cancers. Although there is some epidemiologic evidence that diet may be a factor in the etiology of malignancies of the colon and breast, the role played by specific components of diet is still a matter of speculation.[12-14] The association of high-fat diets with some forms of cancer may be related to the excess energy intake provided by the fat.[15] Recent repeated attempts have failed to substantiate an epidemiologic relationship between dietary fat intake and increased breast cancer risk. In animal models, the only fatty acid shown to enhance carcinogenesis is linoleic acid. In contrast, a naturally occurring derivative of this essential fatty acid actually inhibits experimental neoplasia.[16]

Foods are known to contain both naturally occurring carcinogens and potentially protective agents. A few of these have been identified, but their significance remains to be established. The quest for a fuller understanding of the relationship of diet to cancer preventives has led to investigations involving the possible roles of fiber; vegetables; fruits; vitamins A, C, and E; and the mineral selenium in cancer prevention. Although the role of many identified specific components of human diets as causative or preventive agents for cancer remains to be elucidated,[17] some progress is being made in identifying anticarcinogenic elements in the vegetable and fruit portion of the diet.[18]

There is a current trend toward the view that a single food is either a panacea or a poison. Unfortunately, this approach is being gradually adopted by major health associations. This "good food/bad food" dichotomy ignores the wide consensus among nutritionists that all foods can be compatible with health when used in moderation as part of a balanced, varied diet.

Concerns about possible health hazards caused by food additives and pesticide residues are at an all-time high. Yet this worry is not scientifically justified:[19] strict regulatory standards ensure the safety of our plentiful food supply. With the exception of specific allergens (e.g., sulfites) to which small numbers of people are sensitive, there has never been a known case of ill health linked with the residues of food additives or pesticides used in the approved manner. Given the exaggerated public perception of risk, physicians and scientists can play a major role in educating the public and restoring confidence about food safety. A return to the classic toxicological adage of "only the dose makes the poison" would begin to make the point.

As we begin a new decade, the science of nutrition faces the continued challenge of sorting out hypotheses from scientific conclusions and flamboyant media reports from factual reports. Special emphasis should remain on the fundamentals of nutrition that have enabled Americans to reach their current level of good health: choose a variety of foods from among and within the four basic food groups; maintain reasonable weight in proportion to height by restricting total energy intake and increasing physical activity; avoid fad diets, particularly those that restrict whole food categories; and be skeptical of those who try to popularize the idea that our food supply is unsafe or inherently unhealthful.

References

1. *The Surgeon General's Report on Nutrition and Health.* Washington, DC: Dept. of Health and Human Services; 1988, Publication (PHS) 88 50210.

2. Blair SN, Kohl HW III, Paffenbarger RS Jr, Clark DG, Cooper KH, Gibbons LW, Physical fitness and all-cause mortality: a prospective study of healthy men and women. *JAMA.* 1989; 262:2395–2401.

3. Report of the National Cholesterol Education Program Expert Panel on detection, evaluation, and treatment of high blood cholesterol in adults. *Arch Intern Med.* 1968; 148:36–69.

4. Olson RE. A critique of the report of the National Institutes of Health Expert Panel on the detection, evaluation, and treatment of high blood cholesterol. *Arch Intern Med.* 1989; 149: 1501–1503.

5. Garber AM. Where to draw the line against cholesterol. *Arch Intern Med.* 1989; 11:625–627.

6. Garber AM, Littenberg B, Sox HC Jr, et al. *Costs and Effectiveness of Cholesterol Screening in the Elderly.* Washington, DC: Office of Technology Assessment; 1989. Series on Preventive Health Services Under Medicare, No. 3

7. Palumbo PJ. Cholesterol lowering for all: a closer look. *JAMA.* 1989; 262:91–92.

8. Stare FJ, Olson RE, Whelan EM. *Balanced Nutrition: Beyond the Cholesterol Scare.* Holbrook, Mass; Bob Adams Inc; 1989.

9. Wilson PWF, Christiansen JC, Anderson KM, Kannel WB. Impact of national guidelines for cholesterol risk factor screening: the Framingham Offspring Study. *JAMA.* 1989; 262:41–44.

10. *Report of the 95th Ross Conference on Pediatric Research: Prevention of Adult Atherosclerosis During Childhood.* Columbus, Ohio: Ross Laboratories; 1988.

11. American Academy of Pediatrics Committee on Nutrition. Indication for cholesterol testing in children. *Pediatrics.* 1989; 83:141–142.

12. Rohan TE, Bain CJ. Diet in the etiology of breast cancer. *Epidemiol Rev.* 1987; 9:120–145.

13. Schatzkin A, Greenwald P, Byar DP, Clifford CK. The dietary fat-breast cancer hypothesis is alive. *JAMA.* 1989; 261:3284–3287.

14. Seitz HK, Simanowski UA, Wright NA, eds. *Colorectal Cancer: From Pathogenesis to Prevention?* New York, NY: Springer-Verlag NY Inc; 1989:361–374.

15. Willett WC, MacMahon B. Diet and cancer—an overview. *N Engl J Med.* 1984; 310:633–638, 697–703.

16. Pariza MW. Dietary fat and cancer risk: evidence and research needs. *Annu Rev Nutri.* 1988; 8:167–183.

17. Pariza MW. A perspective on diet, nutrition, and cancer. *JAMA.* 1984; 251:1455–1458.

18. Skolnik A, Smith J. Innovative ways to fight cancer dominate 1989 AMA Houston science news conference. *JAMA.* 1989; 262:2056–2064.

19. Ames BN, Magaw R, Gold LS. Ranking possible carcinogenic hazards. *Science.* 1989; 236:271–280.

Reprinted, with permission, from *JAMA*
The Journal of the American Medical Association,
May 16, 1990, volume 263.

Ozone

My Adventures in the Ozone Layer
S. Fred Singer

An Atmosphere of Paradox: From Acid Rain to Ozone
Hugh W. Ellsaesser

My Adventures in the Ozone Layer

S. Fred Singer, Ph.D.

Please see biographical sketch for Dr. Singer on page 393.

Mrs. Thatcher's 123-nation Conference to Save the Ozone Layer, held in London this March, ended with a whimper. The developing nations, principally China and India, were quite unconvinced by the evidence and unwilling to go along with the European Community and the United States in rushing to completely phase out chlorofluorocarbons (CFCs) and other widely used chemicals. The developing nations have a point.

The London conference has been followed in the last three months by gatherings in the Hague and in Helsinki. All this after the 1985 Vienna Convention, the Montreal Protocol (September 1987), Geneva, Toronto, and who knows how many other international gabfests in between. Who can keep track of them? Norway's Prime Minister, Mrs. Gro Harlem Brundtland, a devout environmentalist, hardly spends time in Oslo any more. When do these people ever govern?

The hyperactivity this created in U.S. government agencies, mainly the State Department and EPA, has to be seen to be believed. The congressional Government Accounting Office would have done an investigation and totaled up the thousands of hours and the huge resources spent on this issue—except that Congress and its staffs are just as involved. Things are building up to a fever pitch—spurred on by lurid stories in the media. "Arctic Ozone Is Poised for a Fall," scream the headlines. "Skin Cancer Is on the Rise!" Is it all hype? Or are there real grounds for worry? As we'll see, the scientific basis for the much-touted ozone crisis may be evaporating—leaving the new breed of geo-eco-politicians high and dry.

It all started with SST, just twenty years ago. The emerging environmental movement scored its first great victory by convincing Congress to cancel the program to build two SST prototypes that would have been tested in the stratosphere. When objections concerning noise and sonic booms didn't bring down the program, the activists discovered the stratospheric ozone layer and the fact that a fleet of five hundred planes might have some effect on the ozone content of the upper

535

atmosphere. Most influential was the argument that a reduction in ozone would allow more solar ultraviolet (UV) radiation to reach the earth's surface, and thus increase the rate of skin cancers. That did it: the skin-cancer scare has been with us ever since, inextricably intertwined with the stratospheric-ozone issue.

Throughout these past two decades many truths have been uncovered by imaginative researchers, but many of these have not been revealed to the public—and quite a few things have been propagated that departed from scientific truth. Scientists, by and large, behaved honorably, although egos and ambitions sometimes collided with facts, leading to a temptation to ignore the facts. Politicians had no hesitation in manipulating science. And the media had a field day. Let me give you a personal account of this convoluted history.

I first got involved in the SST issue in 1970 while serving as a deputy assistant administrator of the EPA. I was asked to take on the additional task of chairing an interagency committee for the Department of Transportation on the environmental effects of the SST. (I had some background in atmospheric physics, having been active in the earliest rocket experiments on the ozone layer, and I even invented the instrument that later become the main ozone meter for satellites.) There were many false starts. We knew so little about the upper atmosphere. The ozone problem didn't come up until sometime in 1970, as I recall; and then only in the context of the effects of the water vapor from the burning of the SSTs' fuel. It was a year later before we came to realize that the main culprit would be, not H_2O, but the small amount of nitrogen oxides (NOX) created in any combustion process.

The first estimates suggested that some 70 percent of the ozone would be destroyed by an SST fleet; without the ozone shield, "lethal" ultraviolet radiation would stream down to sea level, and an epidemic of skin cancers would sweep the world. This scare campaign led to the cancellation of the SST project. Of course, the two prototypes—all that was authorized—wouldn't have caused any noticeable effect; but the SST opponents had succeeded in confusing the issue. England and France went on to build the Concorde—with no apparent environmental consequences.

Only later was it discovered that there were also natural sources of stratospheric NOX, and the SST effect soon fell to 10 percent. But then laboratory measurements yielded better data, and by 1978 the effect had actually turned positive: SSTs would add to the ozone! It became slightly negative again after 1980, but by then the SST had been forgotten and all attention was concentrated on the effects of CFCs.

Few outside my special field know about these wild gyrations in the theoretical predictions. But those of us who lived through them have developed a certain humility and affection toward the ozone layer. It's a matter of some irony that current theory predicts that aircraft exhaust counteracts the ozone-destroying effects of CFCs. But remember: it's only a theory, and it could change.

Science is supposed to be value-free. I learned differently when I conducted a modest survey among my colleagues during the SST controversy. I found that those who opposed SSTs for economic (or less valid) reasons also tended to believe that the environmental effects would be serious. Those who liked the idea of supersonic transportation tended to belittle the ozone effects; they turned out to be right a few years later, but how did they know?

Once Pandora's box had been opened, we all began to look for other ways to affect stratospheric ozone. During my EPA tenure I became intrigued by the idea that human-produced (or, at least, human-related) methane could affect the stratosphere. Methane is a long-lived gas in the atmosphere, very difficult to destroy. It was thought to be mainly due to natural sources, swamps and things like that. But I soon realized that many important sources are related to human activities: rice paddies, cattle, and oil and gas wells, for example.

I reached two conclusions: that about half the methane input is anthropogenic and should therefore increase as population and GNP grow; and that methane can percolate up into the stratosphere, there to be attacked by solar UV radiation, eventually adding to the water vapor in the dry stratosphere. To my surprise I found that methane's contribution to water vapor is about as large as that feared to come from a hypothetical SST fleet.

Public interest in my methane theory was mild, to say the least. The American journal *Science* turned it down, based on the recommendation of the reviewer—a good friend, who then called to tell me he did it to protect my scientific reputation! It was finally published in 1971 in the British journal *Nature*. But no one got excited about it: stopping cows from belching and emitting other gaseous exhalations didn't ignite the environmental community. Cows are so—natural and low-tech. Not a great cause. And besides, controlling their emissions could be messy.

CFCs are different. Brilliant work by a British scientist, James Lovelock, and the calculations of two Californians, Mario Molina and Sherwood Rowland, demonstrated in 1974 the possibility that long-lived and normally quite inactive CFCs would percolate up into the stratosphere, and there be decomposed and attack ozone.

The ecofreaks were ecstatic. At last, an industrial chemical—and produced by big bad Du Pont and others of that ilk. What a worthy successor to the SST, now that that issue was dead.

Regulation was not long in coming. By 1975, voluntary restraints were adopted on the use of CFCs in spray cans, an important but non-critical application. By 1978, the United States and some other Western nations had unilaterally banned CFC use in all aerosol applications.

But that was all for a while. The other applications of CFCs didn't have easy replacements. Substitutes hadn't been developed; and they might turn out to be hazardous, toxic, or expensive—perhaps all of the above. Besides, replacing refrigerators, air conditioners, plastic-foam blowers, and electronic cleaning equipment loomed as an expensive undertaking. Most Europeans and the Japanese were not interested in joining any global agreement, and further unilateral action by the U.S. wouldn't have been very effective globally.

On top of all this, the data from the labs and computers were reducing the threat. A National Academy of Sciences study in 1980 predicted an 18 per cent ozone decrease, based on a certain standard CFC scenario. By 1982 the estimate had decreased to 7 percent, and by 1984 to between 2 and 4 percent. Ironically, much of the reduction was due to the discovery of the counteracting effects of other pollutants: methane, nitrogen oxides, and carbon dioxide. So—putting these polluting gases into the atmosphere hastens the arrival of global warming by the greenhouse effect while reducing the destruction of ozone.

Then along came the Antarctic "ozone hole" (AOH).

In 1985, a British group operating an ozone observing station at Halley Bay, Antarctica, published a result that came out of the blue. Beginning around 1975, every October, they observed a short-lived decline in the amount of stratospheric ozone. The amplitude of the decrease had grown steadily, reaching nearly 50 percent of the total ozone. The finding was quickly confirmed by satellite instruments, which also indicated that the phenomenon covered a large geographic region.

The "smoking gun" had been found—so it seemed. CFCs were immediately suspected; and indeed, chlorine compounds were observed in the region of ozone destruction. The process itself was a new one and had not been studied before; it involved the presence of ice clouds that formed in the polar winter in the coldest region of the earth's atmosphere. The growth of the hole was "obviously" connected to the rise in the atmospheric CFC concentration; and it seemed only a matter of time before the hole would expand and "swallow us all"—or at least all the world's ozone.

The AOH put new life into the anti-CFC crowd. Dropping their earlier opposition, the industry rolled over and played dead, finally joining the environmental activists. It may have dawned on businessmen that with demand rising and supply limited or even declining, prices and profits could grow nicely. Those with safe substitutes might even gain market share and keep out competitors. Within the government, the strong push came from EPA and State, where mid-level bureaucrats fashioned a steamroller that pushed the White House to propose, as a compromise, a CFC production freeze, followed by a rollback to 50 percent. That was the upshot of the 1987 Montreal Protocol.

But some things didn't quite fit. I was puzzled by the sudden onset of the AOH in 1975. It suggested some kind of trigger, unlikely to come from the steady increase of the atmospheric CFC content. What could it be? A climate fluctuation that cooled the stratosphere enough for the ice crystals to form? But if the cause was a cooling fluctuation, then the hole could disappear if the fluctuation went the other way. Or—the AOH might have existed before. I sent a letter to the editor of *Science* suggested this—no luck. So, in November 1988, I finally published a short note in *Eos*, the house journal of the American Geophysical Union, the major professional society in this field.

Meanwhile, the hype was deafening. I remember one congressional hearing in 1987—there were so many—where the witness was a noted dermatologist. He explained that since 1975, malignant melanoma has increased nearly 100 percent—a frightening but true statistic. He simply did not explain three other facts to the Congress or to the media:

- An Antarctic hole should have no effect whatsoever on cancer rates in the United States.

- In any case, melanomas have not been related directly to increased UV exposure.

- And finally, melanoma rates have been increasing by about 800 percent since statistics were first collected in 1935. There has been no corresponding change in the ozone layer or in the UV reaching sea level. To the contrary, measurements of UV-B (the biologically active component) have shown a pronounced and steady decline at every location; UV intensities in America cities are lower today than in 1974. The cause of melanoma must include more than UV exposure.

There does exist a correlation between UV-B intensity and benign, non-melanoma skin tumors. Their frequency clearly increases as one approaches the equator, where the sun and the UV are both stronger, with tumor incidence more than doubling between Minnesota and south Texas. But we should not assume that all the increase is due to higher UV intensities. Lifestyles in warmer climates are conducive to longer exposures and may therefore contribute at least as much to skin tumors as the UV values themselves.

One other fact they don't much talk about: a 5 percent decrease in the ozone layer, as calculated by some of the more pessimistic scenarios, would increase UV exposure to the same extent as moving about sixty miles south, the distance from Palm Beach to Miami, or from Seattle to Tacoma. An increase in altitude of one thousand feet would produce the same result.

The latest phase in the war against CFCs began in March 1988 when the NASA Ozone Trends Panel (OTP) announced its findings, after a massive re-analysis of data from ground stations and satellites. After subtracting all the natural variations they could think of—some of them as large as 50 percent within a few months, at a given station—they extracted a statistical decrease of 0.2 percent per year over the last 17 years. Making these corrections is very difficult and very technical and very uncertain—especially when the natural variations are a hundred times larger than the alleged steady change.

Furthermore, there is the matter of choosing the time period of study. When people ask me whether the climate is getting warmer or colder, I generally just answer "yes." It all depends on over what time scale we average. If the time scale is a few months, then the answer in the spring would of course be "warmer" and in the fall "colder." Now, 17 years is only one and a half solar cycles; and solar cycles have a very strong influence on ozone content. Another letter to *Science*—not accepted.

The Panel announced its result with great fanfare, an "executive summary," and a press conference—but no publication (as yet) that would allow an independent check of its analysis. One item stands out from its announcement: the trend found is greater than calculated from the theory. Now this could mean that the theory is wrong, or the trend is spurious, or both. But the Panel's conclusion was different: the trend is "worse than expected." and therefore CFCs must be phased out completely and quickly. The logic of this conclusion escapes me; but this has now become the U.S. position. Can you blame the Chinese and Indians for not going along?

It's not difficult too understand some of the motivations behind the drive to regulate CFCs out of existence. For scientists: prestige, more grants for research, press conferences, and newspaper stories. Also the feeling that maybe they are saving the world for future generations. For bureaucrats the rewards are obvious. For diplomats there are negotiations, initialing of agreements, and—the ultimate—ratification of treaties. It doesn't really much matter what the treaty is about, but it helps if it supports "good things." For all involved there is of course travel to pleasant places, good hotels, international fellowship, and more. It's certainly not a zero-sum game.

I have left environmental activists till last. There are well-intentioned individuals who are sincerely concerned about what they perceive as a critical danger to humanity. But many of the professionals share the same incentives as government bureaucrats: status, salaries, perks, and power. And then there are probably those with hidden agendas of their own—not just to "save the environment" but to change our economic system. The telltale signs are the attack on the corporation, the profit motive, and the new technologies.

Some of these "coercive utopians" are socialists, some are technology-hating Luddites; most have a great desire to regulate—on as large a scale as possible. That's what makes the CFC/ozone issue so attractive to them. And it showed tellingly at the Hague conference this March—to which the U.S. was not invited. You can perhaps guess why. These geo-eco-politicians actually proposed a new UN agency, aptly named "Globe." Globe was supposed to invoke and enforce sanctions on nations that did not knuckle under to the environmental dictates of those who knew better, Wow!

Globe didn't fly—this time round. Here is David Doniger, senior attorney for the activist Natural Resources Defense Council, writing in the National Academy's *Issues in Science and Technology* in 1988: "[The CFC protocol] serves as a precedent for . . . [protocols on] carbon dioxide and a dozen other trace gases." So that's what they are headed for. Doniger fairly chortles when he recounts how "hard-liners" and "anti-regulatory elements" in the White House fought a losing battle against tough control on CFCs because they "seemed either to disbelieve the scientific evidence of ozone depletion or to belittle its consequences."

As one of those hard-liners, I need to explain where I stand and why I am unrepentant in considering any extreme controls on CFCs to be premature. I tried to explain all this in a letter to the editor of *Issues,* but he turned it down. Twice, in fact. So much for open discussion of important scientific and public-policy issues.

I am not against CFC control at all; but look at the poor state of the scientific evidence. The case against CFCs is based on a theory of ozone depletion, plausible but quite incomplete—and certainly not reliable in its quantitative predictions. Doniger himself does a good job of undermining the credibility of the theory—his only "witness for the prosecution." In his own words:

- "Current models for predicting ozone depletion are inadequate."
- "A National Academy of Sciences [NAS] report . . . quickly became outdated because of new scientific information."

He neglects to inform us that during the past decade the NAS results have varied all over the place. To make matters worse for Doniger's case, evidence is firming up that volcanoes, and perhaps salt spray and bio-chemical emissions from the oceans, contribute substantially to stratospheric chlorine, and thus dilute the effects of CFCs. And new scientific results, from the laboratory and the stratosphere, are pouring in constantly; the theory has been in a state of flux and is bound too change.

Having impugned the CFC/ozone theory—the only basis for making predictions—Doniger nevertheless insists on immediate draconian measures to control CFC production. Not content with a temporary freeze or a rollback, he argues for a complete phase-out of CFCs—without waiting for better scientific data.

The standard CFC/ozone theory did not predict the ozone hole, nor can it account for its future course. According to recent reports, an ozone hole is just about to open in the Arctic—and, by implication, all over the globe. That's a scary thought—and it has made a great impact on the public as well as on governments. It probably was the main impetus for the Montreal Protocol.

This sudden growth of the AOH may, however, as I mentioned before, simply signal the presence of a triggering mechanism that has nothing to do with the steady increase in CFC concentration. Under this hypothesis, the AOH would not continue to grow as CFCs build up, and could even be ephemeral.

In reaction to my suggestion published in *Eos,* Professor Marcel Nicolet, a distinguished Belgian atmospheric physicist, has reminded us in a note to the same journal of a long-forgotten publication by G.M.B. Dobson, the Oxford professor who started modern ozone observations. Dobson recounts that when the Halley Bay Antarctic station was first set up in 1956, the monthly telegrams showed that "the values in September and October 1956 were about 150 [Dobson] units [50 percent] lower than expected . . . In November the ozone values suddenly jumped up to those expected . . . It was not until a year later, when the same type of annual variation was repeated, that we realized that the early results were indeed correct and that Halley Bay showed a most interesting difference from other parts of the world."

As noted earlier, the Ozone Trends Panel of NASA has not yet released its full report for general review. Yet much political action has already been initiated on the basis just of the announcement. For example, Western nations, principally the UK, are pushing to tighten the Montreal Protocol by completely phasing out most CFCs, instead of just freezing and gradually rolling back CFC production to 50 percent as agreed to in the protocol.

While the OTP report itself is not available, a parallel report from the Center for Applied Mathematics of Allied-Signal, Inc., was distributed at a UN Ozone Science Meeting at the Hague in October 1988. The Allied study deals with many of the corrections necessary to establish a believable trend. The estimated change in total ozone over the 17 years 1970–86 is somewhat less than the OTP result. But the change shows a surprisingly strong dependence on the choice of time period. A simple explanation may be that the 1970–86 period covers only one and a half solar cycles and includes two solar flux decreases versus only one increase, i.e., two periods of decreasing ozone trends and one increase. Thus the reported global ozone decline may just be an artifact of the analysis procedure.

On the other hand, if the ozone trend is real, then there are several possible explanations:

- CFCs are indeed lowering the average concentration.

- Human-related factors other than CFCs are decreasing ozone levels. One such factor could be methane, as mentioned earlier. Another could be commercial jet aircraft, which are increasingly penetrating the lower stratosphere. But current theory does not envision ozone destruction from this source.

- Natural effects related to solar-cycle variability may be responsible for an observed ozone decline. For example the decline in strength of the solar cycle after 1958 could account for an ozone decline. This hypothesis lead to an interesting aside. Solar cycles have varied greatly. In recent times, sunspot numbers at the peak of the cycle have been as low as 40 (in 1817) and as high as 190 (in 1958). During the Maunder Minimum (1645–1715) sunspots were essentially absent. This suggests that there could have been substantial changes in average ozone levels in the past, approximating those feared to result from the release of CFCs. It would be interesting, therefore, to search the historical records for any biological consequences to humans, agricultural crops, or marine life caused by low ozone levels around 1700.

The current situation can fairly be summarized as follows: The CFC/ozone theory is quite incomplete and cannot as yet be relied on to make predictions. The natural sources of stratospheric ozone have not yet been delineated, theoretically or experimentally. The Antarctic ozone hole may be ephemeral; it may be controlled by climate factors rather than by CFCs. The reported decline in global ozone may just be an artifact of the analysis. Even if real, its cause may be related to the declining strength of solar activity rather than to CFCs. The steady increase in malignant melanomas has been going on for at least fifty years and has nothing to do with ozone or CFCs. And the incidence of ordinary skin tumors has been greatly overstated.

So—the basis for all of the control efforts, the negotiations, the protocol, and the international conferences is pretty shaky. Now may be the time to reflect on the decisive words of the immortal Comrade Lenin: "*Shto dyelat'?*—What to do?"

The regulatory regime for CFCs adopted in the Montreal Protocol is immensely complicated. Enforcing it will be a nightmare, involving the use of trade barriers and sanctions applied not only to CFCs but to products manufactured with CFCs—such items as foam plastics and electronic circuit boards that go into computers and TV sets. It will prove to be a contentious issue, particularly since special concessions were given to Third World nations and the USSR. The Common Market operates under a European production cap, which further complicates the situation.

The stakes involved are high. Recent newspaper accounts have the EPA predicting a six-fold increase in price, as growing demand for CFCs presses against a limited supply. You can see the struggle for market share and profits reflected in Doniger's choice of words. When Du Pont was fighting the protocol, it was said to be concerned about "price" (read: profits); but once it decided to manufacture substitutes, he talked only about the "right market incentives." An interesting and contentious question: Should Du Pont and other chemical manufacturers keep the profits that will be created as a result of government regulation?

Any finding substitutes for CFCs is no simple matter. A *New York Times* report of March 7, 1989, talks about the disadvantages of the CFC substitutes. They may be toxic, flammable, and corrosive; and they certainly won't work as well. They'll reduce the energy efficiency of appliances such as refrigerators, and they'll deteriorate, requiring frequent replenishment. Nor is this all. About $135 billion worth of equipment uses CFCs in the U.S. alone, and much of this equipment will have to be replaced or modified to work well with the CFC substitutes. Eventually, that will involve one hundred million home refrigerators, the air conditioners in ninety million cars, and the central air-conditioning plants in 100,000 large buildings. Good luck!

These are some of the costs we are now trying to impose on developing countries, which can ill afford them. Sanctions through a new UN agency seem to be out—at least for the time being. But trade barriers can accomplish the same results and won't make us beloved. Third World countries are already heard accusing the West of protecting its pocketbooks as well as its fair skins. (Keep in mind that olive-skinned and dark-skinned people are not very susceptible to skin tumors caused by solar ultraviolet rays.)

Of course, if we in the west should be inclined to pay the bill for this major industrial *perestroika*, then others might just go along. So it's environmental blackmail versus environmental imperialism.

Governments have yet to address what I regard as the real policy issue: how to make decisions about controls on CFC production, and the timing of these controls, in the light of incomplete and often conflicting scientific information. What is needed, it seems to me, is a more complete analysis that weights the risks to society stemming from a delay in instituting production controls against the possibility of substantially improving both observations and the theory so that the predictions can be relied upon. At least, when George Bush decided to go along with born-again environmentalist Maggie Thatcher, he qualified his support for a complete CFC phase-out depending on the availability of safe substitutes. He should have added careful science as another condition.

Reprinted, with permission, from
the *National Review*
June 30, 1989, pp. 34–38

An Atmosphere of Paradox: From Acid Rain to Ozone

Hugh W. Ellsaesser, Ph.D.

Please see biographical sketch of Dr. Ellsaesser on page 404.

Introduction

In the early 1960s I became quite conscious of a sea change in our society. Since we had just moved to the Bay Area in California, one of the more obvious and puzzling signs of the change was the free-speech movement at Berkeley; another was the reaction to the Vietnam War, and a third was the constant haranguing for "clean air" and "quality of life." To me, it was as though the inmates had taken over the asylum and the previous administrators had lost their tongues. And the most baffling and worrisome aspect of the whole thing was; why were the media going along with the inmates and refusing to tell it like it was?

After more than a quarter of a century of puzzling over the matter and reading what others have said about it, I have come to believe that it can most simply be explained as a shift in paradigm.

A Shift in Paradigm

My desk Webster dictionary says paradigm is an example, or a pattern. In science it is used to mean a theory, usually a currently applied but not necessarily proven theory such as Newton's theory of gravity, or the theory of greenhouse warming. And, as in these cases, paradigms become names through which one can easily convey very complex and involved ideas, frequently by a single word. That's what makes them both so useful—and so dangerous when misused to push invalid policies, as they are so frequently used today.

I am using the word paradigm more in the sense of a pattern or template. Templates, or stencils, are what road crews use to speed the writing of STOP and PED XING on new pavements. In using the word in this sense I am also trying to convey the concept of the guided mindset or instantaneous brain-washing that recent inventors and popularizers of current paradigms hoped to accomplish by their introduction.

The Switch from Competing with Nature to Competing with Each Other.

When I was in grammar school we read *The Little Red Hen*—this installed in our minds the paradigm: "He who doesn't work doesn't eat." When you are wresting your livelihood from nature, as by farming, as most of us were in those days, this is a useful paradigm. However, as we have progressed . . . (I use this term advisedly because I know there are those who are trying to convince us that we haven't advanced in the interim.) As we have progressed—to our present state in which about 2% of our population grows all of our food, most of the rest of us compete—not with the natural world—but with each other. And that is what gives rise to many of our current problems. In dealing with the real world, it is crucial to arrive at correct diagnoses. This puts a high value on accuracy and truth. In dealing with other humans, perception becomes more important than reality. We humans have repeatedly demonstrated that we play the rules and not the game. In dealing with other people, many find it advantageous to use misleading information; that is, the trickery and swindle of the con artist. Unfortunately, once one has departed from the path of truth and honesty, more and more reasons appear to crop up for continuing to do so. And falsehoods, repeated often enough, seem to acquire the ring of truth—even to those uttering them.

Most of us are now involved in selling something. The media sell subscriptions and advertising space; politicians try to buy votes by selling ideas, programs and policies; the advertising industry has evolved to a quite sophisticated level to help anyone sell anything. I should stress that over the past 2 to 3 decades even scientists have had to compete in selling their research efforts to Congress and to the bureaucracies Congress has created. And, as you are all aware, salesmen are not noted for acknowledging truths which might hamper their business.

The Multiplication of Pressure Groups

This same period has seen the development of a host of pressure groups; desegregationists, equal-rightists, environmentalists, consumer-advocates, anti-abortionists, freedom-of-choicers, pacifists, antinuks, defenders-of-the-homeless, gay-rightists, Earth First!ers and eco-terrorists. These groups have caused not just a shift in paradigm—they have given the construction of paradigms a whole new meaning with the introduction of such terms as pollution, acid rain, climate change, Love Canal, Three Mile Island, Chernobyl, The Snail Darter,

The Spotted Owl, the Ozone Hole, the *Exxon Valdez* and now Greenhouse Warming.

The mere recitation of a few of these paradigms in an ominous manner puts large numbers of the public in a mood to lynch the nearest operator of a power plant, or of a paper or saw mill, the driller of an oil well, the captain of an oil tanker, or an automobile manufacturer. Or, as an alternative, to assuage their anger and frustration—if not their conscience, many people mail off checks to such organizations as the Sierra Club, Greenpeace or the Union of Concerned Scientists.

The Studied Use of Pejorative Words.

However, if you dig into the details of these emotion-provoking and mind-polluting paradigms, you find that they are very much of a mixed bag. It's as with the Queen in *Alice in Wonderland*—these paradigms mean exactly what the user wants them to mean; which may vary from use to use. The word "pollution" itself, other than condemning that to which it refers, conveys even less logical meaning than the word "weed;" it doesn't even tell whether what is being discussed is animal, vegetable or mineral; or—solid, liquid or gas.

As you well know, we have come to accept that a "weed" is "a plant out of place." Pollutants, particularly the original six that EPA was formed to combat (particles, hydrocarbons, ozone, carbon monoxide, oxides of nitrogen and sulfur dioxide), are not out of place. They are simply augmenting natural levels of these same substances but—significantly, apparently—they were put there by man! That alone is presumed to be enough to condemn them. And by consistently and repeatedly calling them pollutants, the public has been brainwashed into believing they have to be removed at any cost.

Most of these so-called pollutants, with far greater logical justification, could be called fertilizers. You may be interested to know that one of the most recent concerns expressed with respect to the burning of tropical forests, is that this represents a rate of loss in biologically fixed nitrogen equal to "~9–20% of the estimated annual, global, terrestrial nitrogen fixation rate (Lobert et al., 1990)." In other words, buring of biomass can be thought of as a significant application of negative fertilizers—applied by man. One would have thought that this should have been one of our earliest concerns!

You are no doubt aware that great concern has been expressed concerning Alaska's salmon industry following the rupture of the *Exxon Valdez* in Prince William Sound last year. However, I doubt that many of you have heard that this year's salmon catch is the biggest in history: over 40 million salmon caught in 1990 versus 29 million in

1987, the previous biggest catch! (*Wall Street Journal*, p. A10, 4 Sept. 1990). This is probably no accident. There are organisms in the sea which thrive on oil, starting a food chain which is eventually available to fish such as salmon.

Paracelsus' Dictum: "The Dose Makes the Poison."

I am sure that most of you know what happens when you apply an excess of fertilizer. The results can be rather drastic; but merely illustrate again Paracelsus' dictum: "The dose makes the poison." That is, even medicines and substances essential to life become poisons at sufficiently high doeses. Even oxygen, water and food can kill at sufficiently high doses. The recent concern with dieting, jogging, and weight loss support my argument. Calling an excess of fertilizer a case of pollution that requires digging up and disposing of tons of topsoil, at consumer and/or taxpayer expense, is not a particularly helpful solution. But that is essentially what is now coming to be standard practice as a result of the legislative monsters initiated by NEPA (The National Environmental Protection Act), which in turn was left to us primarily as the legacy of Senator Muskie's bid to become President of the United States in 1972.

Love Canal

In January 1981 I was at the annual meeting of the AAAS (American Association for the Advancement of Science) in Toronto, Canada. This was some months after Love Canal became a new paradigm. The published program appropriately listed a session devoted to Love Canal with several speakers scheduled, mainly from EPA. When I arrived at the session I found that not one of the scheduled speakers showed up; the session had to be cancelled. What happened? In the few months since the schedule was put together, the speakers apparently decided that they no longer wanted to defend the proposition that Love Canal was the simple toxic disaster the head lines had implanted in the public mind. But the paradigm lives on and it is still regularly invoked.

Acid Rain

After 10 years of study, costing over half a billion dollars, those who are interested now know that acid rain is a perfectly natural phenomenon. Lakes, which have become more acid as forests have been allowed to regrow in their drainage areas, are still less acid today than they were before those areas were first cleared of trees. Many of these same lakes would never have been able to support fish if man had not intervened and cleared the surrounding forests in the first place, thereby reducing their natural acidity (Krug, 1990). If man is making any contribution to acid rain and acid lakes through air pollution, it comes primarily from our screening the acid neutralizing particles from smoke stack plumes before allowing the acidic gases to escape into the atmosphere.

The Ozone Hole

The Ozone Hole, if due to man, reflects a process which can occur only in those portions of the atmosphere which are maintained at temperatures below about minus 80°C (-112°F) for 2 to 3 months, during at least the latter half of which, they must also be exposed to sunlight. Such temperatures occur only in restricted vertical layers, roughly 12 to 20 km, within the polar vortices which develop due to radiative cooling when sunlight is absent over the pole in winter, and at the tropical tropopause.

A self-limiting process

In 1987 the level of ozone within this cold layer over Antarctica fell essentially to zero—less than 5% of normal. In other words, the maximum possible ozone hole occurred in 1987. The phenomenon does not occur to any appreciable extent over the N. Pole because the north polar vortex breaks up and rewarms about the same time as the sun comes up there. Also, ozone loss is unlikely to ever be detected at the tropical tropopause both because there is little ozone there to be destroyed and because the air there is constantly being flushed out by a slow updraft through the tropical tropopause. Thus, unless there are changes other than the simple addition of more chlorine to the stratosphere, the ozone hole does not appear likely to become any more important than it was in 1987. It should also be noted, that the ozone hole merely causes ultra violet fluxes at the surface over Antarctica in spring, comparable to what is experienced there every summer.

Global trend in ozone.

While it is true that the globally averaged depth of the ozone layer appears to have declined some 3 to 5% over the past 15 to 20 years, the level today appears to be higher than it was back in 1962. There are at least five papers in the scientific literature reporting 4.3 to 11% increases in the depth of the ozone layer from 1962 until the early 1970s (Komhyr et al., 1971; Christie, 1973; Johnston et al., 1973; London and Kelley, 1974 and Angell and Korshover, 1976). These numbers are up to twice as large as the declines that have been reported in recent years.

The Equivalence of Ozone Reduction to Equatorward Displacement in Causing Skin Cancer

There are additional aspects of the so-called ozone destruction issue—for which the big hazard is supposed to be increased ultra violet radiation leading to increased incidence of skin cancer. The NAS (National Academy of Sciences, 1975) found that over the U.S. the doubling distance for skin cancer incidence is 8 to 11° of latitude, or roughly 600 miles. This means that skin cancer incidence increases roughly 1% for each 6 miles of displacement towards the equator. The NAS (1975) also concluded that a 1% decrease in the ozone layer was equivalent to a 2% increase in skin cancer incidence; that is, to a displacement of 12 miles toward the equator.

For approximately 15 years now the ozone modelers have computed that the continued use of freons, or hydrochlorofluorocarbons, even at the unconstrained 1976 level, until they came into their maximum effect at equilibrium (after 75 to a 100 years), would produce a decrease in the ozone layer of less than 10%. In other words, the maximum computed effect would be comparable to that resulting from a displacement of 50 to 100 miles to the south. I doubt that many of you consider that to be a very serious hazard.

I was called foolish for pointing out this equivalence at a conference on stratospheric ozone back in 1976 (Dotto and Schiff, 1978—footnote p. 283). I have repeated it in several papers published since then. But I'll bet none of you have ever heard it before. Am I wrong? Will anyone, who has heard that a one percent decrease in ozone is equivalent—in terms of skin cancer incidence—to moving 12 miles south, please raise his or her hand.

The One-Way Filter: Ignoring the Benefits

But even this is not the whole of "The Ozone Hole" story. From the very beginning, the fact that all terrestrial vertebrates, including man, have to have ultra violet light to generate the so-called vitamin D required to metabolize calcium into skeleton, has been studiously ignored. Shortage of vitamin D or ultra violet leads to rickets in the young (Loomis, 1970) and to osteomalacia (bone deterioration) in the elderly (Dantsig et al., 1967). In the United States today, the number of bone fractures per year from osteomalacia, typically of the femur, is about twice the incidence of new skin cancers. And these bone fractures, particularly if of the femur, constitute far more serious medical problems than ordinary skin cancers.

Since increased ultra violet exposure would presumably alleviate these conditions in future generations, there is good reason to believe that a decrease in the ozone layer would actually result in a net benefit to human health, particularly since our bodies are much better able to tell us when we are getting too much ultra violet than they are at telling us when we are getting too little.

You are no doubt aware that ultra violet light is a biocide; that is, that it kills germs, viruses, bacteria, microbes. Something you may not know is, that since EPA is now trying to eliminate the use of chlorine in the treatment of drinking water, ozone and ultra violet irradiation may have to be instituted as part of the replacement treatment for drinking water. Here is another benefit of increased ultra violet which the do-gooders are spending our money to protect us from.

It is worth noting, that in essentially all environmental issues that have been built into paradigms, just as in the paradigm of the "Ozone Hole," a one-way filter (Ellsaesser, 1974) has been applied in looking at the effects cascade; that is, possible beneficial effects have been studiously ignored and all possible detrimental effects have been carefully searched out and emphasized. It is obvious, of course, that by using this technique, you can easily convert any activity by man- or womankind into an apparent intolerable crime.

And, to a lesser extent, the converse is also occurring. From the observational data presently available, one can calculate that if EPA is successful in removing smog from Los Angeles, then the local population there will be subjected to an increase in ultra violet of about 30%, which equates to a skin cancer increase of about 60%. Have you ever heard anyone mention this? Do you have any doubts as to why you haven't?

References

Angell, J.K. and J. Korshover, Global analysis of recent total ozone fluctuations, *Monthly Weather Review* 104, 63–75, 1976.

Christie, A.D., Secular or cyclic change in ozone, *Pure and Applied Geophysics* 106–108, 1000–1009, 1973

Dantsig, N.M., D.N. Lazarev and M.V. Soklolov, Ultraviolet installations of beneficial action, *Applied Optics* 6(11), 1872–1876, 1967.

Dotto, L., and H. Schiff, *The Ozone War*, Doubleday, Garden City, NY, 1978.

Ellsaesser, Hugh W., The dangers of one-way filters, *Bulletin of the American Meteorological Society* 55(11), 1362–1363, 1974.

Johnston, H.S., G. Whitten and J. Birks, Effects of nuclear explosions on stratospheric nitric oxide and ozone, *Journal of Geophysical Research* 78, 6107–6135, 1973.

Komhyr, W.D., E.W. Barrett, G. Slocum and H.K. Weickmann, Atmospheric total ozone increase during the 1960s, *Nature* 232, 390–391, 1971.

Krug, Edward C., Fish Story, *Policy Review*, No. 52, 44–48, Spring 1990.

Lobert, J.M., D.H. Scharffe, W.M. Hao and P.J. Crutzen, Importance of biomass burning in the atmospheric budgets of nitrogen-containing gases, *Nature* 346, 552–554, 1990.

Loomis, W.F., Rickets. *Scientific American* 223, 77–91, 1970.

London, J. and J. Kelley, Global trends in total atmospheric ozone, *Science* 184, 987–989, 1974.

NAS (National Academy of Sciences), *Environmental Impact of Stratospheric Flight,* National Academy of Sciences, Washington, D.C. 1975.

Population

**Scientific Establishment Now Agrees: Population
Growth is Not Bad for Humanity**
Julian L. Simon

**Vision 2020: World Food Abundance, Trade, and the
Natural Advantage of America**
William J. Hudson

Scientific Establishment Now Agrees: Population Growth Is Not Bad for Humanity

Julian L. Simon, Ph.D.

Dr. Julian L. Simon is affiliated with the College of Business and Management, University of Maryland.

He received a B.A. in experimental psychology in 1953; an M.B.A. from the University of Chicago in 1959; and a Ph.D. in business economics from the University of Chicago in 1961.

He has written several books and has published numerous articles and reports.

Any sample of the television news or the newspapers shows notables from Andrei Sakharov to Dan Rather repeating that more people on earth mean poorer lives now and worse prospects for the future. In a typical sample from the few months prior to this writing, World Bank president Barber Conable calls for population control because "poverty and rapid population growth reinforce each other" (*Washington Post,* July 16, 1990, p. A13). Prince Philip advises us that "It must be obvious by now that further population growth in any country is undesirable" (*Washington Post,* May 8, 1990, p. A26). 37 Senators write President Bush in support of funding for population control (*Washington Post,* April 1, 1990, p. H1). The Trilateral Commission and the American Assembly call for reduction in population growth (*U. S. News and World Report,* May 7, 1990). *Newsweek*'s year-ending cover story concluded that "Foremost of the new realities is the world's population problem" (December 25, 1990, p.44). The president of NOW warns that continued population growth would be a "catastrophe" (Nat Hentoff in the *Washington Post,* July 29, 1989, p. A17). The quotes could be multiplied indefinitely.

Paul Ehrlich had an entire series, Assignment Earth, on the Today show in January, 1990, fingering overpopulation for "destroying the entire ecological system" and for every ill of humanity, without even token counter-comment. He was on the same Today show for five full minutes on each of three days in May, 1989, about the "problem" of "overpopulation," with nary a whiff of the "balance" that journalists pride themselves on. But this is an old story. Ehrlich has been on

Johnny Carson's show for an unprecedented full hour—more than once—all the while complaining that the danger of population growth is being ignored by the media.

The editorial and opinion writers chime in. Ellen Goodman laments "People Pollution" (*Washington Post,* March 3, 1990, p. A25). Herblock cartoons the U. S. neglecting the "world population explosion" (*Washington Post,* July 19, 1990, p. A22) Hobart Rowen likens population growth to "the pond weed [which] grows in huge leaps" (*Washington Post,* April 1, 1990, p. H8). A *Newsweek* "My Turn" suggests giving every teen-age girl a check for up to $1200 each year that she does not have a baby "in order to stop the relentless increase of humanity" (Noel Perrin. "A Nonbearing Account," April 2, 1990, p. 9). A typical editorial in the *Washington Post* (June 3, 1989, p. A14) says that "in the developing world . . . fertility rates impede advances in economic growth, health, and educational opportunities."

These ideas affect public events. In 1973, Supreme Court Justice Potter Stewart's vote in *Roe v. Wade* was influenced by this idea, according to Bob Woodward and Scott Armstrong: "As Stewart saw it, abortion was becoming one reasonable solution to population control" (quoted in *Newsweek* of September 14, 1987, p. 33). In 1989, when hearing the Webster case, Justice Sandra Day O'Connor again brought the idea of overpopulation into a hypo thetical question she asked of Charles Fried, former solicitor-general, "Do you think that the state has the right to, if in a future century we had a serious overpopulation problem, has a right to require women to have abortions after so many children?" The decision said: "If the secular analysis were based on a strict balancing of fiscal costs and benefits, the economic costs of unlimited childbearing would outweigh those of abortion" (whatever that means). And in a series of state supreme court decisions, unfounded demographic-economic notions have been adduced to support anti-abortion arguments, too (Judith Simon, 1989).

Erroneous belief about population growth has cost dearly in material terms. It has directed attention away from the factor that we now know is central in a country's economic development, its economic and political system. For a quarter century our "helping" institutions misanalysed such world development problems as starving children, illiteracy, pollution, supplies of natural resources, and slow growth. The World Bank, the State Department's Aid to International Development (AID), The United Nations Fund for Population Activities (UNFPA), and the environmental organizations have asserted that the cause is population growth—the population "explosion," or "bomb," the population plague." Economic reforms away from totalitarianism and central economic planning in poor countries probably would have been faster and more widespread if slow growth was not explained by recourse to population growth.

And in rich countries, misdirected attention to population growth and the supposed consequence of natural resource shortage has caused waste through such programs as synthetic fuel promotion and the development of airplanes that would be appropriate for an age of greater scarcity.

Our anti-natalist foreign policy also is dangerous politically because it risks being labeled racist, as happened to us when Indira Ghandi was overthrown because of her sterilization program. Furthermore, misplaced belief that population growth slows economic development provides support for inhumane programs of coercion and the denial of personal liberty in one of the most sacred and valued choices a family can make—the number of children that it wishes to bear and raise—in such countries as China, Indonesia, Vietnam.

Given the nearly uniform blanket of assertion and belief, it must come as a surpise to many that for almost as long as this idea has been the core of U. S. theory about foreign aid, there has been a solid body of statistical evidence that contradicts this conventional wisdom about the effects of population growth—evidence which falsifies the ideas which support U. S. population policy toward less-developed countries. And it must also come as a great surprise that scientists who study these matters have recently reversed their views from the earlier consensus that population growth is a key force holding back economic development. Yet all this is indeed the case.

In this paper, I shall first review key data. Then I shall discuss the amazing U-turn in scientific opinion that occurred in the 1980s with respect to the economics of population growth.

The Key Facts About Population Growth, Resources, and Environment

There now exist perhaps two dozen competent statistical studies covering the few countries for which data are available over the past century, and also of the many countries for which data are available since World War II. The basic method is to gather data on each country's rate of population growth and its rate of economic growth, and then to examine whether—looking at all the data in the sample together—the countries with high population growth rates have economic growth rates lower than average, and countries with low population growth rates have economic growth rates higher than average.

The clearcut consensus of this body of work is that faster population growth is not associated with slower economic growth. On average, countries whose populations grew faster did not grow slower economically. That is, there is no basis in the statistics for the belief that faster population growth causes slower economic growth.

Additional powerful evidence comes from pairs of countries that have the same culture and history, and had much the same standard of living when they split apart after World War II—East and West Germany, North and South Korea, and China and Taiwan. In each case the centrally-planned communist country began with less population "pressure," as measured by density per square kilometer, than did the market-directed non-communist country. And the communist and non-communist countries in each pair also started with much the same birth rates and population growth rates.

The market-directed economies have performed much better economically than the centrally-planned countries. Income per person is higher. Wages have grown faster. Key indicators of infra-structure such as telephones per person show a much higher level of development. And indicators of individual wealth and personal consumption, such as autos and newsprint, show enormous advantages for the market-directed enterprise economies compared to the centrally-planned, centrally-controlled economies. Furthermore, birth rates fell at least as early and as fast in the market-directed countries as in the centrally-planned countries.

These data provide solid evidence that an enterprise system works better than does a planned economy. This powerful explanation of economic development cuts the ground from under population growth as a likely explanation. And under conditions of freedom, population growth poses less of a problem in the short run, and brings many more benefits in the long run, than under conditions of government planning of the economy.

One inevitably wonders: How can the persuasive common sense embodied in the Malthusian theory be wrong? To be sure, in the short run an additional person—baby or immigrant—inevitably means a lower standard of living for everyone; every parent knows that. More consumers mean less of the fixed available stock of goods to be divided among more people. And more workers laboring with the same fixed current stock of capital means that there will be less output per worker. The latter effect, known as "the law of diminishing returns," is the essence of Malthus's theory as he first set it out.

But if the resources with which people work are not fixed over the period being analyzed, then the Malthusian logic of diminishing returns does not apply. And the plain fact is that, given some time to adjust to shortages, the resource base does not remain fixed. People create more resources of all kinds. When horse-powered transportation became a major problem, the railroad and the motor car were developed. When schoolhouses become crowded, we build new schools— more modern and better than the old ones.

As with human-made production capital, so it is with natural resources. When a shortage of elephant tusks for ivory billiard bills threatened in the last century, and a prize was offered for a substitute, celluloid was invented, followed by the rest of our plastics. Englishmen learned to to use coal when trees became scarce in the sixteenth century. Satellites, and fiber-optics derived from sand replace expensive copper for telephone transmission. And the new resources wind up cheaper than the old ones were. Such has been the entire course of civilization.

Extraordinary as it seems, natural-resource scarcity—that is, the cost of raw materials, which is the relevant economic measure of scarcity—has tended to decrease rather than to increase over the entire sweep of history. This trend is at least as reliable as any other trend observed in human history; the prices of all natural resources, measured in the wages necessary to pay for given quantities of them, have been falling as far back as data exist. A pound of copper—typical of all metals and other natural resources—now costs an American only a twentieth of what it cost in hourly wages two centuries ago, and perhaps a thousandth of what it cost 3,000 years ago; the history since 1800 in the U. S. in shown in Figure 1. And the price of natural resources has fallen even relative to consumer goods, as Figure 2 shows.

The most extraordinary part of the resource-creation process is that temporary or expected shortages—whether due to population growth, income growth, or other causes—tend to leave us even better off than if the shortages had never arisen, because of the continuing benefit of the intellectual and physical capital created to meet the shortage. It has been true in the past, and therefore it is likely to be true in the future, that we not only need to solve our problems, but we need the problems imposed upon us by the growth of population and income.

The idea that scarcity is diminishing is mind-boggling because it defies the commonsense reasoning that when one starts with a fixed stock of resources and uses some up, there is less left. But for all practical purposes there are no resources until we find them, identify their possible uses, and develop ways to obtain and process them. We perform these tasks with increasing skill as technology develops. Hence, scarcity diminishes.

The general trend is toward natural resources becoming less and less important with economic development. Extractive industries are

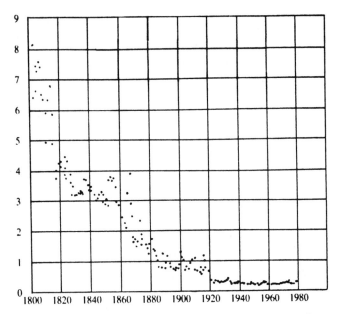

Figure 1. The Scarcity of Copper as Measured by Its Price Relative to Wages

only a very small part of a modern economy, say a twentieth or less, whereas they constitute the lion's share of poor economies. Japan and Hong Kong are not at all troubled by the lack of natural resources, whereas such independence was impossible in earlier centuries. And though agriculture is thought to be a very important part of the American economy, if all of our agricultural land passed out of our ownership tomorrow, we would be the poorer by only about a ninth of one year's Gross National Product. This is additional evidence that natural resources are less of a brake upon economic development with the passage of time, rather than an increasing constraint.

There is, however, one crucial "natural" resource which is becoming more scarce—human beings. Yes, there are more people on earth now than in the past. But if we measure the scarcity of people the same way we measure the scarcity of economic goods—by the market price—then people are indeed becoming more scarce, because the price of labor time has been rising almost everywhere in the world. Agricultural wages in Egypt have soared, for example, and people complain of a labor shortage, because of the demand for labor in the Persian Gulf, just a few years after there was said to be a labor surplus in Egypt.

Nor does it make sense to reduce population growth because of the supposedly-increasing pollution of our air and water. In fact, our air and water are becoming cleaner rather than dirtier, wholly the opposite of conventional belief.

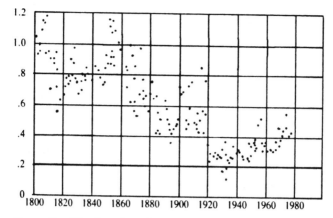

**Figure 2.   The Scarcity of Copper as Measured by Its
Price Relative to the Consumer Price Index**

The most important and amazing demographic fact—the greatest human achievement in history, in my view—is the "recent" decrease in the world's death rate. It took thousands of years to increase life expectancy at birth from just over 20 years to the high '20's. Then in just the last two centuries, life expectancy at birth in the advanced countries jumped from less than 30 years to perhaps 75 years. What greater event has humanity witnessed?

Then starting well after World War II, life expectancy in the poor countries has leaped upwards by perhaps fifteen or even twenty years since the 1950s, caused by advances in agriculture, sanitation, and medicine. Is this not an astounding triumph for humankind? It is this decrease in the death rate is the cause of there being a larger world population nowadays than in former times.

Let's put it differently. In the 19th century the planet Earth could sustain only one billion people. Ten thousand years ago, only 4 million could keep themselves alive. Now, 5 billion people are living longer and more healthily than ever before, on average. The increase in the world's population represents our victory over death.

One would expect lovers of humanity to jump with joy at this triumph of human mind and organization over the raw forces of nature. Instead, many lament that there are so many people alive to enjoy the gift of life because they worry that population growth creates difficulties for development. And it is this misplaced concern, that leads them to approve the inhumane programs of coercion and denial of personal liberty in one of the most precious choices a family can make—the number of children that it wishes to bear and raise.

Then there is the war-and-violence bugaboo. A typical recent headline is "Excessive Population Growth a Security Threat to U. S.," invoking the fear of "wars that have their roots in the unrestrained growth of population." This is reminiscent of the Hitlerian cry for "lebensraum" and the Japanese belief before World War II that their population density demanded additional land.

There is little scientific literature on the relation of population to war. But to the extent that there has been systematic analysis—notably the great study of war through the ages by Quincy Wright (1968), the work on recent wars by Nazli Choucri (1974), and a study of Europe between 1870 and 1913 by Gary Zuk (1985)—the data do not show a connection between population growth and political instability due to the struggle for economic resources. The purported connection is another of those notions that everyone (especially the CIA and the Defense Department) "knows" is true, and that seems quite logical, but has no basis in factual evidence.

The most important benefit of population size and growth is the increase it brings to the stock of useful knowledge. Minds matter economically as much as, or more than, hands or mouths. Progress is limited largely by the availability of trained workers. The main fuel to speed the world's progress is the stock of human knowledge. And the ultimate resource is skilled, spirited, hopeful people, exerting their wills and imaginations to provide for themselves and their families, thereby inevitably contributing to the benefit of everyone.

Even the most skilled persons require, however, an appropriate social and economic framework that provides incentives for working hard and taking risks, enabling their talents to flower and come to fruition. The key elements of such a framework are respect for property, fair and sensible rules of the market that are enforced equally for all, and the personal liberty that is particularly compatible with economic freedom.

The Amazing Recent Shift in the Scientific Consensus

In the 1980's a revolution occurred in scientific views toward the role of population growth in economic development. Revolution is usually the most newsworthy of events. Yet this revolution has gone unreported in the popular press, and conventional ideas therefore continue as before the revolution.

By 1990, the economics profession has turned almost completely away from the previous view that population growth is a crucial negative factor in economic development. There is still controversy about whether population growth is even a minor negative factor in some cases, or whether it is beneficial in the long run. But there is no longer any support for the earlier view which was the basis for the U. S. policy and then the population policy of other countries.

The "official" turning point came in 1986 with the publication of a report by the National Research Council and the National Academy of Sciences, entitled Population Growth and Economic Development, which almost completely reversed an earlier report on the same subject from the same institution. The 1971 report said at the beginning of its "Overview":

> "[A] reduction in present rates of population growth is highly desirable from many points of view, because high fertility and rapid population growth have seriously adverse social and economic effects (p. 1) . . . Rapid population growth slows down the growth of per capital incomes in less developed countries (p. 2)."

And it then proceeded to list the supposed ill effects upon savings, investment, food supplies, unemployment, modernization, technological change, industrialization, social effects, education, health and child development, and the environment.

It is worth noting, however, that—in a pattern now familiar in such reports—many of the background chapters exhibit far less alarm, and state many more qualifications, than does the summary. For example, the 1971 report even notes:

> "Providing for the required resources seems overwhelmingly difficult when we look into the future, and yet, looking backward, we learn that man seems not to have failed . . . Specifically, mankind has not 'run out' of any critical resources" (pp. 16–17).

While asserting that population growth is an obstacle to economic growth, it noted that "Unprecedented and still accelerating population growth has not prevented very rapid economic advance. Population growth, though not a negligible block to development and modernization, has also not been an overriding factor . . . Among individual less developed countries, moreover, the relation between income . . . and rates of populatin growth has been low during the 1950's and again in the 1960's. The statistical relations are, if anything, positive in each case." (p. 27; interestingly, the volume did not mention Kuznets's path-breaking investigations showing much the same thing, published four years earlier). And it goes on even more surprisingly: "Much the same patterns have held for the developed economies during the postwar decades. Here, however, one would expect population growth to have a positive impact on economic growth" (p. 28).

Furthermore, the 1971 report makes clear that

"We have limited outselves to relatively short-term and clear-cut issues ... We have endeavored to examine the population problem as it affects us now—and for the next 5 to 30 years," a period too short for any of the beneficial effects of babies currently being born to come into play (p. vi).

Only now do I notice this moderation and balance from the 1971 volume. Undoubtedly typically, I took away an impression of only the alarming notes sounded in the summary, which referred only to speculations about the future, and ignored the evidence cited from the past that contradicted the speculations. The biased summary is an example of how science has been perverted in the service of zealous belief that population growth is a demon to humanity, and must be suppressed at all costs.

Compare the 1986 report. On the specific issue of raw materials that has been the subject of so much alarm, 1986 NRC- NAS concluded: "The scarcity of exhaustible resources is at most a minor constraint on economic growth ... the concern about the impact of rapid population growth on resource exhaustion has often been exaggerated." And it was quite unworried about most of the other effects that caused alarm in the 1971 report. The general conclusion goes only as far as "On balance, we reach the qualitative conclusion that slower population growth would be beneficial to economic development for most developing countries ..." That is, 1986 NRC-NAS found forces operating in both positive and negative directions, its conclusion does not apply to all countries, and the size of the effect is not known even where it is believed to be present. This is a major break from the past monolithic characterization of additional people as a major drag upon development across the board.

The background paper on macroeconomic-demographic models for the NAS-NRC report by Dennis Ahlburg concluded:

It is widely believed that population growth has an adverse effect on economic growth in developing nations ... However, a group of scholars have recently argued that the effects of population growth are neutral or may even be positive (no date, p. 2) ...

The early models found a very large negative impact of population growth on economic development. Subsequent models have found that while the short-run impact is negative it may not be as large as previously thought (Barlow-Davies, Bachue-Kenya, Simon, Kelley-Williamson) and may even be positive in the very long-run (Simon, Mohan). Other models have shown that demographic effects can vary widely across countries (Wheeler) and that population change has had little impact on the degree of urbanization in developing countries (Kelley-Williamson, Mohan).

On the basis of this review of economic-demographic models, we concur with Preston that "population growth is not so overwhelmingly negative a factor for economic advance as to swamp the impact of all other influences. That is a worthwhile lesson that bears repeating, but it is no argument for faster demographic growth" (p. 47).

Even earlier "establishment" recognition came in the 1984 report of the World Bank. That institution for many years has been the strongest and shrillest voice calling for reduction in the rate of population growth on the grounds that the world is running out of natural resources. In its 1984 World Development Report, however, the World Bank did an about-face and said that natural resources are not a reason to be concerned about population growth.

Still earlier, in his Presidential Address to the Population Association of America (PAA) John Kantner (not an economist) took notice of the new thinking, but also denounced it. He recognized the excesses until then: "In this war against rapid population growth, truth often fell before expediency." And he asserted that though many demographers understood the "full complexity" of the "problem of rapid population growth . . . few of us made more than mild protest as the profession was borne to prominence on a Malthusian tide of alarm" (1982, p. 430). And he speculated why this was so:

> Heady times those, and something in it for everyone—the activist, the scholar, the foundation officer, the globe circling consultant, the wait-listed government official. World conferences, a Population Year, commissions, select committees, new centers for research and training, a growing supply of experts, pronouncements by world leaders and, most of all, money—lots of it (p. 430).

Though he recognized that "There has been a sea change" in thinking, he lambasted it:

> The recent spate of revisionist writing . . . asserts that the effect of population growth on development . . . is at best indeterminate and, as Colin Clark has argued for years, can sometimes be beneficial. The most prevalent form of analysis employed by the revisionist school is to selectively cite negative cases . . .

The conclusions derived from selective citing of negative evidence have been reinforced by appeal to that arcanum of modern policy analysis, the computerized simulation—a useful policy tool which, as in this case, is often abused. One can select among economic-demo graphic models, as among facts, to get the desired results—a familiar problem to demographers. A current model (Simon, 1981) much admired by the revisionists and used to give further solidity to their argument.

> It is unarguably not legitimate (p. 436). "In sum, the revisionist case is a weak one, flawed in logic, method and empirical understanding (p. 437).

Kantner treated the new thinking as dangerous. With Ismail Sirageldin he wrote: "This is not an area for frivolous approaches [they were referring to my work] or one where academics may contend confusedly with no great harm to anyone. It is an area where an effective mobilization of public will and commitment based on understanding of issues is essential" (1982, p. 173).

The 1986 PAA Presidential Address by Paul Demeny also took notice that

> "Since the early 1980s a substantial shift has occurred in the balance of views" (p. 473). "The various lines of attack on the orthodoxy converge in a newly optimistic assessment of the population problem ... In the extreme formulations, the problem is disposed of entirely. The more typical revisionist views, however, merely put the problem in its presumed deserved place: several drawers below its former niche" (p. 474).

In a later article Demeny quoted with approval Colin Clark's assessment of the current situation as "Malthusianism in retreat" (1989, p. 38). (I do not quote Clark directly because the views of that great scientist are discredited for many by his being a Catholic, and even worse, a convert to that church.)

Demeny is respectful of the "revisionist" view, but he says early on that "I will find it wanting" (p. 474) because the revisionist view does not call for population-policy rearrangements of society to allow for the negative externalities of children (though the present revisionist, at least, calls for the rearrangement of internalizing the externalities by privatising such services as housing and transportation in China).

The revolution in economic-demographic thought is evident in an outpouring of scholarly "reviews." Though the reviews vary in whether they consider the shift to be important but not revolutionary, or completely revolutionary, they all agree that what has come to be called the "revisionist" viewpoint cannot be ignored. To all, the view that population growth is either neutral or favorable to development in the long run is at least a controversial pole in a legitimate debate. And there is consensus that if population growth has a negative effect in any given country, it is not a factor of overwhelming importance.

From the Introduction to a United Nations Expert Group Meeting on "Consequences of Rapid Population Growth in Developing Countries," August, 1988 (pre-published June, 1989):

> [T]he majority of popular writings during the 1970's portrayed population growth as a major obstacle to achieving economic development in the Third World. A major exception was the more balanced view found in the second edition of the Determinants and Consequences of Population Growth," published by the United Nations in 1974. During the 1980's, this more balanced view emphasizing the complexity of demographic-economic interrelations has re-emerged. The negative effects of population growth are now portrayed as less important than had been asserted during the previous decade, and some scholars have even indicated that potentially favourable effects may be non-negligible" (p. 1)

From Geoffrey McNicoll of the Population Council in that organization's Population and Development Review:

> [T]here was a casual assumption by many that early efforts to model economic-demographic relationships had wrapped up the subject, demonstrating to general satisfaction the net adverse results of rapid population growth for the development effort . . . The assumption that the various studies of the consequences problem have cumulatively settled the matter might be plausible were there a reasonable consensus on where the balance of growth consequences lies. Such a consensus probably did exist in the 1960's, but is much less evident today. In the last decade a revisionist stream of thought has emerged that seems to cast doubt on the previous orthodoxy; rapid population growth, according to scholars of this persuasion, is often a neutral and can even be a positive factor in development. Hence the odd current situation of fundamental disagreement about the net impact of one of the most profound changes in social circumstances in the modern world—a disagreement found, moreover, not in variant political or philosophical premises but in economic modeling and in readings of the empirical record (1984, pp. 177–178).

From a recent article by T. N. Srinivasan entitled "Population Growth and Economic Development":

> Much of the concern about the deleterious effect of a rapid growth of population on economic development is based largely on the view that either household fertility decisions are exogenous or if endogenous, pervasive and significant externalities distort them.
>
> It is argued that this view is mistaken and that many of the alleged deleterious consequences result more from inappropriate policies and institutions than from rapid population growth. Thus policy reform and institutional change are called for, rather than policy interventions in private fertility decisions to counter these effects . . . (1987, abstract)

To conclude, most of the arguments for a policy intervention in private household fertility decisions appear to be based either on an inappropriate association of undesirable social consequence due to other distortions in the society with individual fertility choices, or on associations that cannot be ruled out in theory but are empirically weak, if not exaggerated (p. 25).

From "Population Growth Versus Economic Growth" by David E., Horlacher and F. Landis MacKellar:

> Contrary to the [earlier views of Coale and Hoover], this paper will advance the thesis that population growth has both beneficial and adverse effects and that we remain relatively uncertain concerning its net effect on development (1987, p. 1).

The principal author of the 1986 NAS-NRC report, Samuel Preston, recently summarized the situation as follows:

> "Discussions of the role of population growth in economic progress have become markedly less alarmist in the past decade" (1897, p. 1987).

Allen Kelley recently reviewed the reviews, as well as the literature as a whole, for the American Economic Association's "official" *Journal of Economic Literature.* He finds that the recent work converges at a position much like that of the 1986 NRC-NAS, and he adds the interpretation that "in a number of countries the impact of population was probably negligible, and in some it may have been positive" (1988, p. 1715). Here as elsewhere he endorses the view that there has been a "revision" in thinking about the consequences of population. (In fact, he may be the person who coined the term.) Kelley himself has broadly shared this general view since at least 1970, one of the very few demographers who has done so.

Until I re-examined the above statements when writing this article, I did not realize how massively the professional opinion has shifted on this matter. And the statements cited above are only those that arrived spontaneously in my mail, without my attempting to systematically survey the literature.

The fallback position of some of the population-control enthusiasts says that the "revisionist" shift is due to the shift of larger political forces rather than to new discoveries of facts and theory—"just politics." For example, Barbara Crane writes that "There is no doubt that the relatively sudden reversals we have seen in recent years reflect the influence of the New Right with the Reagan White House as well as in Congress" (1989, p. 126). But then why should one not think that previous positions were also "just" politics, especially given the huge political organizations maintained by the population and environmental organizations in Washington and elsewhere? What Crane et. al. are saying is: When government agencies agreed with us, that was because what we said was the scientific truth.

But when the government no longer agreed with us, that is due to the other side's political machinations. Indeed, Hodgson (1988) extends this sort of analysis to the rise of the population-control orthodoxy, too.

Orthodoxy's emergence is attributed to two sets of factors: first, the inability of demographic transition theory to explain several postwar demographic trends; second, the manner in which the Cold War, decolonization, and the influx of funds for fertility control changed American demographers' approach to the study of population trends. The recent rise of revisionism is attributed both to orthodoxy's difficulty in digesting the favorable economic and demographic trends of the 1970s and to the changes in the political and funding environments within which American demographers work (Abstract, p. 759).

Please note that "revisionism" is treated in the above quote as simply a fait accompli. This indicates, I think, that it is not necessary to further multiply the quotations to prove that this shift in thinking has surely occurred.

The issues that have been subject to revision include all the purported negative effects with which the doomsaying authors and organizations have taxed population growth—income growth slowing, reduction in saving and investment, natural resource exhaustion, worsening of the food supply, hindering the supply of education—the works. In each case, examination of the evidence leads to complete or near-complete exoneration of population growth and size. There has even been some "revisionist" research rebutting the charge that population growth promotes violence and war (Zuk, 1985; Simon, 1989, Cuhsan, forthcoming).

This is not the place to review the intellectual history of the ideas underlying this "revision," or to discuss credit for the revolution, especially because it is difficult for me to either understand my own role clearly or to assess it without bias. Suffice it to say that William Petty in 1682 stated the key idea that human minds are the key element in human progress, and more minds on balance lead to more progress. Friedrich Engels grasped this idea in the context of modern science. Colin Clark was the first economist in the twentieth century to work systematically on these ideas, both theoretically and empirically, but he was dismissed as being "just" a Catholic and, even worse, a convert. Simon Kuznets provided the first statistical evidence that nations' economic growth is not negatively affected by their population growth. Ester Boserup surveyed anthropological and historical evidence to show that increased population density has led to an adoption of already- known but more arduous agricultural practices that use land more intensively, a process that points toward modern economic development. Harold Barnett provided theory and data proving that natural resources become more plentiful rather than more scarce as demand increases due to the growth in income and population. My 1980 article and 1981 book brought these ideas as well as my own and others' earlier research on this subject to the public at large, and by good luck they received sufficient attention that the scientific public was forced to take notice, which catalyzed the revision in thinking. (This technique of reaching the scientists by way of inquiries to them by non-scientists—a technique which John K. Galbraith designed for use with his 1952 book American Capitalism—probably was crucial. Until that time my 1977 book which provided the technical research underlying my 1981 book had received no attention at all, and almost surely would have continued in obscurity.)

The scientific U-turn does not seem to matter to makers of policy and public opinion, however. The president of the same World Bank, Barber Conable, was still saying at the 1988 annual meeting of the World Bank that "curbing excessive population growth" is one of the key elements of the Bank's strategy with respect to world poverty. He said that it is "imperative that developing countries renew and expand efforts to limit population growth" (1988, p. 753).

The commitment of the official organizations to the old beliefs and policy is evidenced in this comment by the then-head of AID's population program quoted in a Science article on the 1986 National Academy of Science work. Steven Sinding "said he felt enormous 'relief' at the [NAS] committee's conclusions."

That is, AID was pleased that the report was sufficiently ambiguous that it could be interpreted as a warrant for business as usual.

So what's to be done? I'm afraid the answer is: Nothing.

Efforts to change the beliefs of the public and the assertions of journalists are likely to be a waste of time. A few columnists protest. Editorials appear in *The Wall Street Journal* and in a few other newspapers across the country, mostly in small cities. The *Economist* carries some right-thinking discussion, but even it still retains a sense of limits: "[I]t is surely inconceivable that the earth can support 14 billion people—the eventual total now predicted by the United Nations—at anything like the standard of living enjoyed today in middle-income countries such as Mexico and Malaysia, let alone that of the rich West" (Sept 12, 1989, p. 16). Some Catholic leaders speak out. A think tank such as Cato may hold a meeting. But the overall state of thought will remain as it is because this revolution in thought will not penetrate the fortress of belief.

Typically, in the service of maintaining its treasured belief and urging U. S. funding for population programs, the *Washington Post* editorial of June 3, 1989 says that "The Chinese vehemently deny" forced abortions. This is despite the fact that the columns of the same newspaper not many months earlier contained a physician's account of forced abortions in Tibet, a couple of years earlier the paper's own staff reported on page 1 about coercive population control in China, and not long before *The Wall Street Journal* had carried quotations from official Chinese statements showing unmistakably that what the Chinese deny to the *Washington Post* they affirm as policy at home. Yet the editorial staff of the Post proceeds as if nothing more were known about the topic that what people thought they knew as fact two decades ago.

The public often assumes that the very absence of contrary information means that the prevailing wisdom must be sound. Would that it were always so. But every school child in the Soviet Union has been taught a pack of lies about Soviet history. And there have been many episodes in the West when monolithic public belief has been all wrong on the facts. Near-unanimity is no guide to truth.

Malcolm Muggeridge was the only foreign reporter in the famine area of the U. S. S. R. unaccompanied by Soviet officials in 1933 when Stalin's program of collectivization of agriculture killed perhaps 16 million human beings. Muggeridge described the horrors of the famine, and wrote of Stalin as "a bloodthirsty tyrant of unusual ferocity even by Russian standards." In response he was simply accused of being a liar.

Meanwhile, the prevailing newspaper coverage—typified by the *New York Times*—talked about "plump cows and apple-cheeked dairy maids and plump, contented cows." Muggeridge concludes that "People, after all, believe lies, not because they are plausibly presented, but because they want to believe them. So their credulity is unshakable."

Similarly, almost no one in American journalism seems to care about the tens of millions of children the Chinese government prevents couples from bringing into life with the one-child policy, with loss to the potential human beings as well as to the potential parents. Nor do discussions of human rights mention the scores of millions of women in China whose bodies are violated by Chinese policies which include forced insertion of IUD's, followed by subsequent X-ray surveillance to ensure that the IUD's have not been removed, and banning childbearing among those said to be "psychotic." A recent review of the politics of the matter noted that even the Reagan administration has paid no attention: "From the President and the Secretary of State on down, US diplomats dealing with China and speaking in public on overall US-China relations have scarcely even mentioned, much less condemned, China's population policy." (Crane and Finkle, 1989, p. 25).

The intellectual arena is again in the hands of the yahoos, the same people with the same ideas that have been discredited by events and by scientific research. The educated public and the journalists seem even more credulous than ever, and there are large numbers of professors on campuses who have been raised in this movement and now teach it as professionals, in comparison to the amateur zealots who powered the movement two decades ago without professional stake in carrying on. An incredible 80 percent of the U. S. public now agree that "protecting the environment is so important that requirements and standards cannot be too high, and continuing environmental improvements must be made regardless of cost," up from about 40% in 1981. And in Britain, the proportion who "say the environment is one of the most important issues facing Britain" jumped from about 5 percent to about 35 percent from December, 1988 to July, 1989, extraordinary testimony to the power of modern communications. (The Economist, September 2, 1989, p. 4). The weight of these numbers is overwhelming.

True, there has been a shift in the declared policy of the U. S. From the days of President Lyndon Johnson in the early 1960's until 1984, it was the policy of the U.S. to urge and even coerce developing countries to reduce their population growth rates. (See Phyllis Piotrow (1973) for an excellent history.) At the World Population Conference in Bucharest in 1974, the U.S. voice was the loudest in calling for population control. At the next World Population Conference in Mexico City in 1984, the official U.S. position was exactly the opposite, and said that population growth is "neutral" with respect to economic development. This adjustment of policy to scientific fact certainly is gratifying. Unfortunately, however, the policy statement is interpreted by both the pro-abortion and anti-abortion camps as pertaining solely to the abortion issue, though the entire statement contained just a few sentences about abortion. This partisan attention from both sides of that issue has caused the rest of the message to be forgotten except by a few foreign demographers. And its impact on larger issues gets lost in the welter of the argument about abortion.

Planned Parenthood responded to the 1984 Mexico City policy with a continuing public-relations campaign of great breadth and expense. They have run a long series of full-page advertisements in such newspapers as the *Washington Post* and the *New York Times*, and large signs on buses and subways, attacking the U. S. position. Rather than considering the scientific argument on its merits, they simply refer to a putative conspiracy of "right-wing extremists" said to be "in the White House" who are supposed to lack compassion for the poor women of the world.

Implicated in both cause and effect of this state of affairs is that there is not a single pro-people organizational address in journalists' rolodexes for them to turn to, in contrast to the enormous roster of organizations that work against population growth. The everlasting abortion wars absorb the energy that might otherwise be attracted to the more general issue of the value of human life. The abstraction of human life cannot compete in drama with hacked-up fetuses and back-alley operations.

This situation inevitably worsens because the few people who might try to publicize the truth are scared off to start with, or give up in despair. At some point even a not-so-prudent person sees the hand-writing on the wall, and stops devoting chunks of life to a losing cause.

People who hear or read the central message expressed by "our side" often say that we are "just" optimists. We are indeed optimistic about the long-run future of humanity, though I reply with a line swiped from Herman Kahn: I'm not an optimist, I'm a realist. With respect, however, to the short-run likelihood that people in the Western countries will have an accurate assessment of the issues discussed here, and the probability of avoiding great losses of life and wealth as a result of faulty assessments, I am extremely pessimistic.

In the long run, the inevitable forces of progress will roll over these intellectual obstacles. Population will grow, knowledge will increase, economies will develop, liberty will flourish. But the meantime there will be innumerable avoidable tragedies because the good news goes unreported. How sad that is.

References

Ahlburg, Dennis A. "The Impact of Population Growth on Economic Growth in Developing Nations: the Evidence from Macroeconomic-Demographic Models," no date.

Bachrach, Peter, and Elihu Bergman, Power and Choice: The Formulation of American Population Policy (Lexington, Mass: Lexington Books, 1973).

Choucri, N. (1974) Population Dynamics and International Violence. Lexington, MA: Lexington Books.

Coale, Ansley, and Edgar M. Hoover, Population Growth and Economic Development in Low-Income Countries (Princeton: PUP, 1958).

Conable, Barber, extracted in *Population and Development Review*, vol 14, December, 1988, pp. 753–755.

Crane, Barbara, "Policy Responses to Population Growth: A Global Perspective," in Population and U. S. Policy, Principia College, 20–22 April, 1989

———and Jason L. Finele, "The United States, China, and the United Nations Population Fund," *Population and Development Review*, vol 15, March, 1989, pp 23–60.

Cuhsan, Alfred G., "Demographic Correlates of Political Instability in Latin America: The Impact of Population Growth, Density, and Urbanization," forthcoming in *Review of Latin American Studies*.

Day, Lincoln H. and Alice Day, Too Many Americans (New York: Houghton Mifflin, 1964).

Demeny, Paul, "Population and the Invisible Hand," *Demography*, vol 23, November, 1986, 473–488

———, "Demography and the Limits to Growth," *Population and Development Review*, forthcoming, 1989.

Ehrlich, Paul R., Anne H. Ehrlich, and John P. Holdren, Ecoscience: Population, Resources, Environment, 2nd ed., (San Francisco: W. H. Freeman, 1977).

———, review of The Resourceful Earth, in *Bulletin of the Atomic Scientists*, February, 1985, page 44.)

Everett, Alexander H., New Ideas in Population (New York: 1826/1970, A. M. Kelley).

Fornos, Werner, "A Time For Action," *Populi*, Vol. 11, No. 1, 1984, pp 32–35.

Hamburg, David A., "Population Growth and Development," *Science*, p 73 Volume 226, Number 4676, November 16, 1984.

Hardin, Garrett, *The New Republic*, October 28, 1981, pp. 31–34.

Hazlitt, Henry, Economics in One Lesson (New York: Arlington House, second edition, 1962)

Hirshleifer, Jack, "The Expanding Domain of Economics," *The American Economic Review*, Vol. 75, December, 1985.

Hodgson, Dennis, "Orthodoxy and Revisionism in American Demography," *Population and Development Review*, 14, December, 1988, 541–570.

Holden, Constance, "A Revisionist Look at Population and Growth," Science, Vol. 231, March 28, 1986, pp 1493–1494.

Horlacher, David E., and F. Landis MacKellar, "Population *Growth Versus Economic Growth (?)," xerox, 1987

"Implications for U.S. Policy of the NAS Report on Population Growth and Economic Development: Policy Questions," Xerox, March 6, 1986.

Jaffe, Frederick S., cited in Simon, 1981, taken from Elliott, Robin; Lynn C. Landman; Richard Lincoln; and Theodore Tsuruoka. 1970. U.S. population growth and family planning: a review of the literature. In Family Planning Perspectives, vol. 2, repr. in Daniel Callahan, ed., The American population debate (New York: Anchor Bks., 1971), p. 206.

Kaminskaya, Dina, Final Judgment: My Life as a Soviet Defense Attorney (New York: Simon and Schuster, 1982).

Kantner, John F., "Population Policy and Political Atavism," *Demography*, volume 19, November, 1982, pp. 429–438

Kelley, Allen C., "The National Academy of Sciences Report on Population Growth and Economic Development," Xerox, April 10, 1986.

———, "Economic Consequences of Population Change in the Third World," *Journal of Economic Literature*, Dec, 1988, 1685–1728)

Lee, Ronald, "Economic Consequences of Population Size, Structure and Growth," International Union for the Scientific Study of Population Newsletter No. 17, January–April 1983, pp 43–59.

———, review of World Development Report 1984, *Population and Development Review*, 11, March, 1985, pp. 127–130.

McGreevey, William Paul, "World Development Report 1984—Two Years Later," Xerox, April 1986.

McNicholl, Geoffrey, "Consequences of Rapid Population Growth: An Overview and Assessment," *Population and Development Review*, Volume 10, June, 1984, 177–240.

Muggeridge, Malcolm, Chronicles of Wasted Time, Vol I: The Green Stick, New York: William Morrow, 1972, pp. 258, 274.

National Academy of Sciences, Rapid Population Growth: Consequences and Policy Implictions (Baltimore: The Johns Hopkins Univ. Press, 1971).

National Research Council, Committee on Population, and Working Group on Population Growth and Economic Development, Population Growth and Economic Development: Policy Questions (Washington, D.C.: National Academy Press, 1986)

Oldenburg, Don, "Whistle Blower's Anguish," The *Washington Post*, March 31, 1987, p. C5.

Piel, Gerard, "Let Them Eat Cake," *Science*, Volume 226, Number 4673, October 26, 1984.

Piotrow, Phyllis, World Population Crisis (New York: Praeger, 1973).

Population and Family Planning Section American Public Health Association, Memo to "All US Citizens, Members and Attendees at the IUSSP Conference," undated.

Preston, Samuel H., "The Social Sciences and the Population Problem," Sociological Forum, Vol 2, Fall, 1987, pp. 619–644.

Repetto, Robert, "Why Doesn't Julian Simon Believe His Own Research?," Letter to the editor, The *Washington Post*, Nov. 2, 1985, p. A21).

Schultz, Theodore W., Investing in People. (Chicago: University of Chicago Press, 1981.)

Simon, Judith, memo concerning superior court decisions using economic-demographic arguments, October 3, 1989.

Simon, Julian L., "The Concept of Causality in Economics," Kyklos, Vol. 23, Fasc. 2, 1970, pp. 226–254.

——— The Economics of Population Growth (Princeton: PUP, 1977).

———, "Resources, Population, Environment: An Over-supply of False Bad News," *Science*, 208, June 27, 1980, pp. 1431–1437.

———, The Ultimate Resource (Princeton: PUP, 1981).

———, "Disappearing Species, Deforestation, and Data," New Scientist, 15 May, 1986, pp. 60–63.

———, "Lebensraum: Paradoxically, Population Growth May Eventually End Wars," *Journal of Conflict Resolution*, Vol. 33, No. 1, March 1989, pp, 164–180.

———, and Paul Burstein, Basic Research Methods in Social Science (New York: Random House, third edition, 1985)

Sirageldin, Ismail and John F. Kantner, "Review," Population and Development Review, March, 1982, pp. 169–173.

Srinivasan, T. N., "Population Growth and Economic Development," xerox, August, 1987.

United Nations, Expert Group, "Consequences of Rapid Population Growth in Developing Countries," xerox, June, 1989.

Wolfson, Margaret, Profiles in Population Assistance (Development Centre of the Organization for Economic Co-operation and Development, 1983).

Wright, Q. (1968) "War: the study of war," in D. Sills (ed.) International Encyclopedia of the Social Sciences. New York: Macmillan-Free Press.

Zuk, Gary (1985) "National Growth and International Conflict: A Reevaluation of Choucri and North's Thesis," in *The Journal of Politics*, Vol. 47, pp. 269–281.

Vision 2020: World Food Abundance, Trade, and the Natural Advantage of America

William J. Hudson

William J. Hudson is an independent consultant in Maumee, Ohio. He was formerly a senior vice president of research at The Andersons.

Hudson received a B.A. in mathematics, English and history; and his masters in English.

Hudson is a past member of the National Users Advisory Board for Research and Extension to the U.S. Department of Agriculture, and speaks widely on grain markets, fertilizer, and other agricultural topics.

He is the author of *Business Without Economists*, Amacom; *Vision 2020*, The Andersons; and *Intellectual Capital*, Wiley (forthcoming).

What works best for me, contrary to what one always hears about business, is *long-term* series of data. The further back in time I can take something, the more illuminating and helpful it is to business persons. As an example, the price of wheat is given later in this book, back to the year 1800. At that time, a bushel of wheat was worth the wages from 30 hours of work. By 1900, the wage required to buy a bushel had fallen to about five hours. By 1940, two hours. And today, less than 1 hour's wages will buy a bushel of wheat. Lesson for business: Food price always trend down over the long haul. When prices are up for a year or two, maybe because of unusual weather or politics, keep the profits for the next downturn. Don't build a business which depends on high prices. Commodities always get cheaper.

Few subjects can be dealt with in depth without quantification.

What we want, for our understanding to reach the highest order, is a minimum of numbers, but the right ones, cast in the right light.

In my studies of grain data I have found that the bushel is the best unit to use, not only in domestic markets but internationally. I hasten to add that by no means is the bushel completely familiar to all Americans today. With fewer and fewer Americans occupied in farming, the cultural sense of how much a bushel represents has faded. I find, however, that the bushel is intrinsically better related to human scale than metrics, and that an awareness of how much is in a bushel can be quickly re-established with a lay audience.

The bushel is a unit of volume, not of weight.

$$1 \text{ bushel} = 64 \text{ pints}$$

In the case of wheat, a basic and primary foodstuff, one pint weighs about one pound. These are very "human" dimensions. An old saying goes, "A pint's a pound the world around."

Now consider further the size of an average human stomach. Picture one. Some stomachs are certainly larger than others, but the average stomach will be found to hold about one pint.

The question now becomes, "How often must we be able to fill our stomachs?" We hear the saying in this country that we each want to have "three square meals a day." If one meal fills a stomach which holds a pint, then three square meals will be about three pints.

If we eat three pints a day for a month, how much is this, in bushels? Three pints multiplied by 30 days would be 90 pints, which is more than 1 bushel. Since a bushel is 64 pints, our monthly intake would be 90 divided by 64, or about 1½ bushels.

How much then is "three square meals a day" for an entire year? The answer is 1½ bushels per month times 12 months, which gives 18 bushels per year. If everyone in the world is to have his or her stomach full three times a day, then they each need about 18 bushels per year.

Taking the bushel method globally, we find that if three square meals a day means 18 bushels per year, then the grain needs of the people of the world are:

18 bushels each year x 5 billion people = 90 billion bushels

One benefit of this simple arithmetic is that it allows us immediately to ask, "How well are we doing?"

According to official data, we have nearly 60 billion bushels of world grain production in the face of 90 billion bushels of "demand"—or at least of "desire."

Economists argue that perfect abundance is not possible. As evidence, they cite the sheer existence of price. If things were perfectly abundant, the price for them should be zero. Everything would be free.

Although prices are not zero, I think there are many grounds for optimism. Here is a quick history of international grain price since World War II:

Year	Price of Wheat (in 1986 dollars)
1950	8.10/bushel
1955	7.10
1960	5.75
1965	4.10
1970	3.45
1975	6.00
1980	4.30
1985	3.35
1987	2.80

The price rise of commodities in the 1970s had to do directly with the success—which turned out to be temporary—of a cartel. Cartels arise not from shortage, but from abundance. Likewise, the appearance of government price supports, for food or other commodities, is not a sign of scarcity but of overabundance.

But over any period of time longer than a few years, the price of commodities always falls. The above data on wheat show this in the post-war period, and here is a longer term series of prices from Julian Simon:

Year	Price of Wheat (hrs of labor req'd to buy a bu)
1800	30 hours/bushel
1850	15
1900	5
1950	2
1960	1
1970	0.5
1975	1.2
1980	0.7
1985	0.4

There are other even more tangible indications of food abundance. In the Spring of 1988, the United States government, at the direction of Congress, paid U.S. farmers not to produce grain on about thirty percent of all available acres. The magnitude of this "paid non-production" was roughly three billion bushels.

We learned that world grain production was 60 billion bushels, which, divided among a world with a population of five billion people, means an average consumption of 12 bushels each.

The American "paid non-production" of three billion bushels is five percent of the world's production of 60 billion bushels. Can it be said, in other words, that food is so abundant we need only employ 95 percent of our capacity? Or is something amiss? What about hunger?

The American "paid non-production" of three billion bushels, if distributed at 12 bushels apiece, would feed 250 million people, completely, for an entire year. Experts estimate that there are no more than 50 million acutely starving people.

Alternatively, the three billion bushes could be distributed, let's say at three bushels each, to the one billion people most in need. Experts say there are (at most) about a billion malnourished people in the world.

Is something amiss? Do American taxpayers actually realize that a monetary expenditure of $15 billion is going to their farmers *not* to produce an amount of grain that would potentially solve world hunger?

Another set of mind-boggling numbers is associated with annual growth of world population. Some 80 million people are apparently added to the earth every year. If each is to receive the average 12 bushels, then the yearly increase in grain production must reach 1 billion bushels.

One billion bushels is a large number, certainly, but only about one-third of the quantity we will pay American farmers *not* to produce in 1988.

To the student of history, the existence of a permanent surplus of grain in North America comes as no surprise. The continent was settled late in the story of mankind. Today it contains only five percent of the world's population but nearly 25 percent of the world's food production capacity.

Additional production depends more on yield than acreage. In the case of corn, the national average yield is about 120 bushels per acre; however, in experiment stations the yield can be doubled and tripled. A farmer in Champaign County, Illinois, has achieved a corn yield of 385 bushels/acre under contest conditions. The same genetic potential exists with wheat, sorghum, and the other grains. In fact, in advanced experiments such as those conducted by NASA for food production in highly controlled, extra-terrestrial conditions, the genetic potential of grains appears to be as much as *ten times present yields*.

Suppose the world wants to move from 60 billion bushels of grain production closer towards the ideal of 90 billion, which we calculated as "three square meals a day" for all humanity. If the U.S. doubled its yields, the result would be an additional 12 billion bushels. This would move the world total to 72 billion bushels. If the U.S. tripled its yields, the new total would be 84 billion bushels. If all the other grain exporting nations—Canada, Brazil, Argentina, Western Europe, South Africa, Thailand, Australia, et al.—did the same thing, then the total would be above 90 billion bushels.

My question is this: Why not at least consider such a vision, as opposed to the insistence on perpetual scarcity? In terms purely of technology, the job looks possible; crops are genetically capable of vastly greater yields.

But, in advocating the positive vision, is there some other "limit" we are not properly considering, perhaps trade, or the environment?

Some peoples have a much greater natural advantage in producing food than others. Those who produce in relative abundance ship food to those in relative scarcity, in trade for other items of value. Those in rural areas ship food to those in urban areas. In recent centuries, as the price of food has gradually fallen, and as the wealth from manufacturing has gradually increased, the amount of trade has grown.

If American grain production were tripled from 12 billion bushels to 36 billion, and if other major grain producers added 6 billion bushels to their shipments, world trade could be increased from 7 to 37 billion bushels. World usage could go from 60 to 90 billion bushels, the "ideal" of three square meals a day for everyone. In such a potential world, grain trade would be 40 percent of usage. Is there something inherently wrong with this much trade?

The prevailing opinion today, so far as I can tell, is that it is somehow undesirable for a nation to have to "resort" to trade, especially in something so basic as food. This opinion is interwoven with basic nationalism, and an apparently moral imperative to be self-sufficient.

If this attitude is locked in place, trade will not grow, or it will grow only slowly and begrudgingly, as the price of export supplies becomes cheaper and ever more attractive. We will thus continue to witness the spectacle of surplus coexistent with severe shortage. We will see the most fertile farms in the world, let's say in central Illinois, paid by their government not to produce, while in other parts of the world millions of people starve to death.

Self-Reliance Versus Self-Sufficiency

There are many prominent cases of self-reliance among nations, without food self-sufficiency. Japan, Hong Kong, Singapore, Taiwan, and Singapore lead the list.

Perhaps the most important of these is Japan—an island country whose population markedly exceeds its natural advantages in food production. Japan is the leading importer of grain in the world. Japan is able to do this because of its enormous success at manufacturing, and because its own exports of automobiles and other such goods are well in excess of the "food bill" it runs on the import side.

An extreme case of seeking self-sufficiency in an arid land would surely be Saudi Arabia. In 1987, using irrigation water from underground deposits, the Saudis harvested 80 million bushels of wheat—more than the states of Ohio and Michigan—whereas Saudi domestic consumption was less than 35 million bushels. In December of 1987, the Saudis joined other grain exporters and sold a two million bushel vessel of wheat to the Soviet Union, presumably at the prevailing world price, which would have been about $3.00 per bushel. This is one-fifth of the $15 per bushel price paid by the Saudi government to produce the wheat.

So Saudi Arabia is demonstrating that if money is no object, food security in arid lands is no problem. The example is extreme, but in concept I find it not so much different than other investments to enable food production from inherently unpromising soils and climates. It can be done, but at what cost? What did the Saudis invest—perhaps $10,000 per acre? If we did this, let's say in our arid lands of Utah or Nevada, when prime Illinois land is vacant at only $1,000 per acre, wouldn't we declare it bad business?

Farmers ask me, "How many World Bank and United Nations projects amount to the same thing, an expensive investment in poor land which could be better served by full production at home?" Why invest additional billions in African land, and set aside more in Illinois and Iowa?

Is the claim true that "self-sufficient people are less dependent on other cultures" and free to pursue their own agendas? Is not genuine "freedom from dependency" a matter of overall economic success rather than being tied, day by day, to the production of one's own bare minimum of food? In Europe of the middle ages, peasant farmers were virtual prisoners of the land; they remained that way until markets and trade flourished and freed them. Why apply capital today to create conditions barely beyond those we have sought so desperately to outgrow? Why not enlist money and technology today in bigger leaps?

From a purely technology point of view, the major exporting countries of the world, led by the United States, could produce and ship another 30 or 40 billion bushels of grain around the world, raising the world's total food consumption to 90 billion bushels or more—or, alternatively, providing for more meat in the average world diet, or providing for a much greater population than five billion people. To do this, grain trade might rise to 30 or 40 percent of usage, which is a new record, but not an unusual percentage compared with other essential raw materials.

From a political and cultural point of view, however, we are presently prevented from this by a goal of country by country food self-sufficiency.

Environment

The World Commission on Environment and Development, in a 1987 report sponsored by the United Nations, said, "The increasingly desperate efforts of poorer people to increase agricultural output are damaging the land, perhaps irreparably." Overuse of poor soils in hot countries encourages deserts to expand. The felling of trees for fuel and for timber export also helps destroy arable land, and "Much of this forest is converted to low-grade farm land unable to support the farmers who settle it." My reaction, as you will certainly guess, is that if the struggle to help others achieve subsistence or self-sufficiency on marginal land is seen to endanger environment and global climate, why not consider with fresh vision the role of trade in supplying food, from lands which have a better record in care of the environment?

I submit that the United States is such a land. I believe we are in fact the world's leader in environmental programs, even if these are as yet thought to be inadequate. A major conservation club held their world conference in the U.S. in 1988, because they said, "There is no doubt the U.S. is literally light years ahead of the rest of the world in general conservation."

Consider erosion. The United States, virtually alone among nations, has conducted some five major surveys of erosion in this country over the past five decades—in 1934, 1958, 1967, 1977, and 1982. The surveys are difficult to compare, because with advancing knowledge, different tests were made and different language was used to describe the results. My study of this data shows the following trend:

	Portion of Land Considered "Satisfactory"
1934	56%
1958	38
1967	36
1977	66
1982	66
1987	75

A useful distinction in considering the environmental impact of farming, worldwide, is its degree of "intensification." All farming can be put on a scale of intensification, from high to low. Highly intensified farming involves the cultivation of row crops with large amounts of chemical fertilizers and substantial amounts of pesticides. At the other end of this scale is the so-called "organic farming," which uses only natural fertilizers, no pesticides, but lots of management—as in the practice of "integrated pest management." At this end of the scale, farming is called "low intensity"—or sometimes extensified rather than intensified.

Every continental land mass, and islands as well, can be classified into subregions on the scale of intensification-extensification. The combination of soil, climate, and terrain (hilliness) divides the world into an infinite set of very different, natural "farms." The American cornbelt, for instance, has a large number of subareas which can be used intensively, without adverse environmental consequences, such as heavy erosion. Not all parts of the cornbelt are this way, however, and certainly not all parts of North America share this capability.

My argument is that globally those regions with a natural tolerance to intensification should be so used, and those regions with less tolerance should be managed with extensification—all in appropriate, varying degrees. The surplus production from intensified areas should be exported to the other areas, and paid for by the peoples of those areas by means of adding value to their labor in ways other than agricultural—as is the case now in Japan, Singapore, Hong Kong, and many other countries.

Climate is another enormous problem—perhaps there is none any larger. A scientific dispute is raging over whether Man is changing the earth's climate. I'm not sure if he is or not, but I think the more important fact of which we can be completely sure is this: Climate is changing. Climate always has changed, for "natural" reasons, associated with fluctuations in the solar furnace, the impingement of asteroids on the earth, etc. The question is not whether or not climate will change, but rather how do we adapt to it as it does? My answer is that we must become evermore comfortable with Trade, and substitute self-reliance for self-sufficiency. If the climate over Siberia becomes mild, while that over Iowa becomes arid, is not the answer for both peoples still to trade?

One aspect of pollution is of special importance to world food, namely the so-called "non-point-source pollution" associated with the leaching of agri-chemicals into ground water. When such chemicals are delivered at the right time and the right place, pollution does not occur; the chemicals are almost fully absorbed by the target plant, and little is wasted. It is unprofitable for such waste to occur; technologies and management practices are being designed today to recapture the lost profit now called "ground water pollution." I expect substantial progress in this concern within the next five years.

Radiation and Nuclear Energy

Radiation Around Us
Dixy Lee Ray

The Radioactive Waste Disposal Problem
Bernard L. Cohen

The Truth About Three-Mile Island
Bernard L. Cohen

Radiation Around Us

Dixy Lee Ray, Ph.D.

Dr. Dixy Lee Ray is retired and lives on the family farm in Washington. She has been an Associate Professor of Zoology at the University of Washington; past chairman of the Atomic Energy Commission; and past director of the Pacific Science Center.

She received her Ph.D. in Marine Biology from Stanford University in 1945.

Dr. Ray is the author of *Trashing the Planet*, Regnery Gateway, 1990; and *Marine Boring and Fouling Organisms*, University of Washington Press, 1957.

She has more than 200 published articles and has contributed to *International Journal of Science and Technology, Journal of Nuclear Materials Mgmt.*, and various zoological journals.

Her many honors and awards include 21 Honorary Doctoral Degrees, the Woman of Achievement in Energy Award, 1988; the Susan B. Anthony Award, Women's Caucus, Washington State Legislature, 1987; and the Centennial Medal, Institute of Electrical Engineers, 1984. She was the governor of the state of Washington from 1977–1981.

For all those who do not like radioactivity, the Earth is no place to live.[1]

The simple fact is we inhabit a radioactive world, always have, and always will. Our bodies receive the impact of about 15,000 radioactive particles per second—that's 500 billion per year and 40 trillion in a lifetime. We don't feel them or suffer any apparent ill effect from this constant bombardment.

We have no human sense to detect radioactivity. No sight, sound, or smell reveals it. Radiation has been something like magnetism or gravity or molecules-unknown or at least not understood until instruments were developed that measure the phenomenon with incredible accuracy.

Indeed, one of the difficult aspects of the radiation phobia is that our ability to measure radioactivity has become so accurate and precise that it is now possible to detect the scintillation of a single atom. Unbelievably small amounts are measurable; for example, one part per billion is easily counted.

How much, or rather how little, is that? How can we visualize one part per billion? One way is by analogy: One part per billion (called a nanocurie) is equivalent to one part of vermouth in five carloads of gin. Now that is a very dry martini.

Or, look at it another way: There are now about five billion people living on this planet. Therefore, one family of five persons represents one part per billion of the entire human population.

What about radioactivity at the level of one part per trillion? This unit is referred to as a picocurie (pCi) and is 1,000 times less than a nanocurie (nCi). It would be analogous to one drop of vermouth in 5,000 carloads of gin!

When clouds containing radioactivity from the Chernobyl accident in the U.S.S.R. in April 1986 reached the West Coast of the United States, the popular press was full of dire warnings about possible fallout, even reporting how many picocuries had been measured in the clouds. But nowhere did reporters explain that a pCi is one part in a trillion. A person would have to drink 63,000 gallons of rainwater, all at once, to receive as much radioactive iodine from the fallout as a patient receives in a diagnostic test for thyroid problems. A formidable task.[2]

There is radioactivity everywhere, in the ground, in sand and stone and clay. In the words of Walter Marshall (Lord Marshall of Coring), who served for years as head of Great Britain's Atomic Energy Council, and is now director of the Central Electricity Generating Board:[3]

> In my own country , the United Kingdom, I like to point out that the average Englishman's garden occupies one tenth of an acre. By digging down one metre, we can extract six kilograms of thorium, two kilograms of uranium, and 7,000 kilograms of potassium, all of them radioactive. In a sense, all of that is radioactive waste, not man-made, but the residue left over when God created the planet.

There is plenty of radioactivity in the upper levels of the earth's crust, but there's more inside. It is the heat of radioactive decay that helps to keep the earth's core molten and provides warmth from inside that contributes to Earth's habitability. It is the heat of radioactive decay that provides much of the driving force for movement of the Earth's tectonic plates. This in turn accounts for the size and shape of continents and contributes to mountain building, to earthquakes, and to volcanic eruptions.

There is radioactivity in water and in the atmosphere. In all, about 70 different radioactive elements occur in our natural environment. The energy that reaches Earth from the Sun includes cosmic rays that strike molecules in the upper atmosphere and produce the isotopes Carbon-14 and Potassium-40, both of them radioactive and found in the bodies of all living things.

Of the total radiation that is received on average by each American every year, 82 percent comes from natural sources. Only 18 percent is man-made.[4] Of the natural radioactivity, a little more than half, 55 percent, is from radon; 8 percent is from cosmic sources and solar flares; another 8 percent is from terrestrial sources, mainly uranium and thorium, and 11 percent comes from internal Potassium-40. The remaining

18 percent that is man-made consists of medical X-rays, which account for 11 percent; nuclear medicine, 4 percent; consumer products, such as smoke detectors, tobacco, and ceramics, etc., 3 percent, and all other sources about 1 percent. The latter figure, 1 percent, includes the entire nuclear energy industry, which contributes no more than 0.1 percent of our entire radiation exposure.

Two points should be emphasized. First, radioactivity is radioactivity, whether made by humans or found in nature. Alpha particles, for example, consist of two protons and two neutrons. They are positively charged and strongly Ionizing—regardless of their origin. Beta rays are electrons, no matter where or what they come from. Gamma rays are short electromagnetic rays; at low energy, they behave like X-rays. Alpha particles can be easily blocked. They do not penetrate through a sheet of paper, whereas gamma radiation penetrates tissues readily.

I emphasize the nature of alpha, beta, and gamma radiation, *regardless of source*, because some anti-nuclear activists have recently adopted the curious attitude—a position amounting to a strange mental ambivalence—that somehow man-made radiation is different from and, therefore, more dangerous than the benign radiation of nature. They are wrong. Natural radiation also includes cosmic rays, which are a mixture of high energy photons and particles, mainly protons and electrons, with a smaller number of helium nuclei and metallic nuclei.[5] Cosmic rays originate both in space and in the Sun. They are abundant in solar flares and can cause brilliant auroras when they collide with oxygen or nitrogen in the upper atmosphere. Collision with oxygen produces red or green colors and interaction with nitrogen causes blue or violet auroras. Finally, ultraviolet rays have wave lengths shorter than visible light. They are a component of sunlight and are also produced in tanning parlors. The ultraviolet rays from these two sources are indistinguishable and the effects on human skin are the same.

The second point to emphasize is this: With all this radiation bombarding us from every direction, why aren't we, all of us, stricken with cancer? The reason is that, although exposure to very high levels of radiation can result in the type of cellular damage that leads to growth of a cancer, the situation is by no means clear with respect to low level exposures, even when they are chronic. The risk appears to be very small, indeed, and evidence is accumulating that low levels of radiation may be either harmless or a positive benefit! So radical a statement requires thorough examination.

For one thing, cells, which evolved from Day One in a radiation-rich environment, have an incredible ability to repair themselves from moderate radiation damage. This has been especially well studied in plant tissues and lower organisms. Additionally, most direct strikes from radioactive particles result in dead cells, not injured ones, and dead cells do not grow to produce cancers. It takes a very special kind of non-lethal damage for a cell to be genetically damaged or to become cancerous. The exact nature of this damage is unknown, as is the exact amount and kind of radioactivity that may cause it in humans or other species.

For another thing, the risk is very small. Let's compare the odds to those of a lottery. Of course, it's possible that any person who buys a lottery ticket may win the jackpot, but the reality is that most do not. In the case of natural radioactivity, the chance that any one of the 40 trillion particles that strike every person over a lifetime will cause a cancer or a genetic effect is one in 50 quadrillion. That's one in 50 million billion, or one in 50,000,000,000,000,000.[6] Pretty good odds—for no effect at all. And most people do not worry about natural radioactivity.

Even so, we know that a high enough exposure can cause cancer. So, how much is high enough? We generally speak in average exposures, but the amount of natural, background radiation varies greatly from place to place. In the United States, the average annual natural or background radioactivity is about 300 to 350 millirems, but the range is great; it may be as low as 60 millirems or as high as 600. A millirem is one-thousandth of a rem, which stands for Roentgen Equivalent Man. This unit represents the actual Ionizing effect on the human body. One millirem would be the dose from one year of watching color TV for a few hours per day. For each one millirem of radiation we receive, it is calculated that our risk of dying from cancer is increased by about one chance in eight to ten million. Put another way: One millirem is equivalent to being struck by about seven billion particles of radiation. A Curie equates to 37 billion disintegrations per second, which is the activity of one gram of radium.

Extensive research has established that it takes exposures well in excess of 100,000 millirems for a detectable effect. Cancer will result in half of the cases at exposures of 400,000 millirems. As a result of the Three Mile Island nuclear power plant accident, the surrounding population received an additional radiation exposure of 1.2 millirems. Compare this to the natural background of:

- 30 millirems per year from cosmic rays.

- 20 millirems per year from the ground.

- 10 millirems per year from building materials (except for such buildings as Grand Central Station in New York or the Capitol building in Washington, D.C., where the granite or marble gives exposures in excess of 100 millirems annually).

- 25 millirems per year from internal Potassium-40.

- 80 millirems per year from medical procedures.
- 180 millirems or more from radon.

The totals are about 360 millirems per year for every American. In the Rocky Mountain states, where there is a higher uranium-thorium concentration in the soil and the altitude means more exposure to cosmic rays, the natural radiation is about twice the average, and in Florida it is 15 percent below.

In some places, the natural background radiation reaches abnormally high concentrations. Hot springs and mineral water resorts usually have elevated amounts of radioactivity. For example, the "wondrous waters" of the English city of Bath[7] have a radon content of 1,730 pCi per liter; recall that the EPA has set a level for homes of 4 pCi per liter, above which remedial action should be taken. The radon in natural gas at Bath is 33,650 pCi per liter. The Romans built a temple there in 43 A.D. and dedicated it to the Goddess of Wisdom and Health, and in 1742, the Royal National Hospital for rheumatic diseases was established there. It is still a premier research center for the study of rheumatism. People flock to Bath and to other hot spring spas, all of them with elevated radioactivity, for their reputed beneficial health effects. There is no proof that the radiation makes people feel better, but there is also no evidence that it has any ill effect on the visitors who both bathe or soak in the springs and drink the waters. It is unfortunate that most of the information about the health benefits of hot springs is anecdotal, for here is a wonderful opportunity to study both the patients (visitors) and the resident population, not only at Bath but also at Baden Baden; Warm Springs, Georgia; White Springs, Virginia; and multitudes of other resorts.

Even more, one wonders why good, thorough, modern studies have not been conducted on the resident populations in Cochin-Ernakulum in the state Kerala on the southwestern coast of India, where the thorium-bearing soils give annual exposures of up to 16,000 millrems per year, and on people living in several areas of Brazil. Many seacoast beaches in northeastern Brazil are composed of black mineral sands called Monagite, noted for its high radiation levels. Visitors flock to these beaches by the thousands for their reputed health benefits. Exposures on the black sands there are estimated to exceed 500 millirems per year. Yet radiation experts in this country consider these populations to be "too small to permit meaningful epidemiologic investigations. . . ."[8]

And then there are the two extraordinary regions of high natural radiation, one in Africa and the other in Brazil. In Gabon, West Africa, near a place called Oklo, where there is now a uranium mine, the concentration of the fissionable isotope U-235 was once so high that 1.8 billion years ago a natural chain reaction set in. Nature produced a nuclear fission "reactor" that "operated" or sustained criticality for a period of one million years. Tons of uranium were burned and both plutonium and other typical fission reactor isotopes (transuranics) were produced. Even though the reactor area has been subject to rainfall and other weathering agents , the plutonium and fission isotopes have migrated only a few meters from the place of their production. This natural nuclear reactor has been thoroughly studied.

Not so with the Brazilian area called Morro do Ferro, a weathered mound 250 meters tall that is formed of an ore body containing an estimated 30,000 metric tons of thorium and 100,000 metric tons of rare earths. The radiation level is one to two mRoentgens per hour over an area of 30,000 square meters. The mound supports both animal and plant life. So high is the absorbed radioactivity in the vegetation that photographs can be taken—autoradiographs—showing the plants truly glowing in the dark.

A colony of rats occupies burrows in the mound. Measurements show that they breathe an atmosphere containing radon at levels up to 100 pCi per milliliter! (100,000 pCi/liter). The radiation dose to the rats' bronchial epithelium is estimated to be between 3,000 and 30,000 rems per year, roughly three times the concentration that should produce tumors or other radiation effects. Fourteen rats were trapped and autopsied, and no abnormalities were found. Yet the investigator who made these measurements commented, "This is of little significance, since the Morro do Ferro is a relatively small area."[9]

It also seems likely that radiation scientists are loath to believe that life exists—plant, animal, and human—in regions where the background natural radiation is abnormally high and yet suffer no apparent ill-effect, even with exposures exceeding 10,000 millirems per year, the generally accepted cutoff between low-level and high-level radiation. I'll have more on this interesting topic after we consider two well known radioactive elements, radium and radon.

Of course radiation is dangerous, and that includes natural radioacctivity. The degree of danger depends on the dose—the amount, kind, and duration of the exposure—and on the knowledge of how to handle radioactivity. The latter is of great importance. It can be said with confidence that, because of extensive research programs pursued since the 1940s, more is known about the effects of Ionizing radiation than about the consequences of exposure to any of the many toxic substances that exist in nature or have been introduced into the environment.

From its discovery up to the 1940s, radium was applied to a number of commercial products. Using radioluminescent paint on the numbers on the faces of watches, clocks, and other instruments so they could be read in the dark, was a widespread practice. Less well known was a similar application—on the lid and handles of the chamber pot, before the days of indoor plumbing and electric lights in the home! About two pounds of radium were extracted from the ground and put in commercial or medical use during the first 50 years of its availability. Worldwide, at least 100 people died from its improper use or mishandling. Since 1942, various atomic energy programs have produced the radioactive equivalent of many tons of radium. But because of greater knowledge and understanding, not one human death has resulted from exposure or internal deposition of a wide variety of artificially produced radio-nuclides.[10]

Of all the naturally occurring radioactive materials, none has received more public attention recently than radon. It is believed that this widespread element in nature is responsible for between 5,000 to 30,000 deaths from lung cancer annually in the United States. Is this so, and is it a new threat? In a sense, yes. I'll return to the question shortly.

The history of our knowledge of radon started in the Erz Mountains (between Germany and Czechoslovakia) in the sixteenth century. In a book about mining, it was reported that most men who worked the metal mines eventually died from "mountain disease," a respiratory illness. It was not until 1879 that it was recognized as lung cancer.[11]

After the discovery of radioactivity and the identification of radon as a breakdown product of radium, it was found that the Erz Mountain mines had exceptionally high levels of radon. A connection between radon and lung cancer was suspected but not established until the 1950s, when studies were undertaken of U.S. uranium miners. The U.S. mines provided a far healthier work environment than those in the Erz Mountains, but the levels of radon were similar. The study results showed conclusively that where radon concentrations were greatest, so also were the number of cases of lung cancer—and where radon levels were lowest, so was the rate of lung cancers. Further studies of the Czech miners confirmed the U.S. results, so that by 1976, radon-induced lung cancer became the best quantified of all health effects of radiation.

Improved ventilation to remove the radon from mines has essentially eliminated the problem for miners, but now it has become clear that radon alone is not the culprit. Smoking, particularly cigarette smoking, now appears to be an important triggering agent. Moreover, at the Austrian mine at Joachinosthal, A. Pirchman reported in the *American Journal of Cancer*, 1932, that there were no lung cancers in any of the working miners—only among the "pensioners," two-thirds of whom died of lung cancer. The average life expectancy then was 55 years; the miners lived longer and retired at 65![12] The moral is, if you worked the mines and breathed radon, you got lung cancer—*but only after you outlived the general population.*

Now, let's take a closer look at radon.[13]

Remember, radon is a gas. Inert, chemically neutral, and radioactive, it appears in the natural decay chain of uranium. There are 14 different steps in this chain, and the important immediate precursor to radon is the radioactive element, radium, itself a decay product of uranium.

Some atoms of radium undergo a radioactive process called "alpha decay," meaning that the nucleus of these atoms spontaneously eject a cluster of nuclear particles consisting of two neutrons and two protons bound together. This is the alpha particle. It is the loss of the two protons that converts radium (Atomic Number 88) to radon (Atomic Number 86). Alpha-emitting substances are not dangerous unless they get inside the body, and they are not "penetrating"—that is, they can be blocked by barriers. Even a sheet of paper is sufficient.

So, the process that produces radon starts with uranium. Uranium is ubiquitous, and since it is widespread throughout the earth, radon gas can appear anywhere. Typically, about six atoms per second from every square inch of soil come bubbling up through the ground, where they dissipate into the atmosphere. At least, that is what happens in nature.

When people build homes and other structures, these natural radioactive processes in the ground beneath are in no way affected. The only difference is that when radon gas seeps to the surface under a home, it moves into and through the construction. Radon enters buildings through the foundation, through cracks, crevices, and fissures, and through the freshwater supply. It is then trapped indoors for some time.

In an average home, the normal radon level may be as much as ten times what it is outdoors—sometimes 100 times. How does the trapped radon gas get out of houses and other buildings? Essentially, the same way it came in—through cracks and fissures, except that this time the openings are above ground, around windows and doors, and through walls and roofs. In old-fashioned homes, minute openings and spaces around the doors, windows, and elsewhere permit a ventilation rate, or exchange of atmosphere, once or twice per hour, and radon accumulation is usually not a problem. But in a modern, tightly sealed, energy-conserving home, this ventilation rate is reduced to about one exchange of the atmosphere per day.

The rate of radon entering the energy-efficient home from the ground beneath is not changed, but the rate of leaving is. Therefore, the radon is trapped longer and builds up inside. *Sealing up a home for the purpose of energy conservation inevitably leads to higher levels of indoor radon.* Is this bad? Yes. That, too, needs explanation.

The focus of concern about the hazard of indoor radiation is on radon, but it is not the gas itself that is the problem. Rather, it is the further radioactive breakdown of radon leading to the formation of what are called "radon daughters" that causes trouble. Radon itself radiates only weakly, and its half-life is 3.82 days. Since it is inert, it does not react chemically in the body, and when inhaled, radon gas will be exhaled again. But if radon daughters are inhaled, the situation is quite different. They are solid particles, have short half-lives, emit alpha particles or beta particles, and are intensely radioactive. Moreover, radon daughters tend to adhere to dust particles, which easily become airborne and thus inhaled. They also adhere to smoke particles. That's one reason "passive" smoking—that is, breathing smoke in a closed atmosphere—is so dangerous. Carried to the lungs, the radon daughters will likely stick to the mucous membrane and bombard the sensitive tissue with intense radioactivity. This is what may cause lung cancer.

Radon daughters can pile up in rooms that are not well ventilated. Some typical radon daughters are:

- Polonium-218, half-life 3.05 minutes, an alpha emitter.
- Bismuth-214, half-life 19.7 minutes, a beta emitter (electron).
- Lead-214, half-life 26.8 minutes, beta emitter.

Let's use Bismuth-214 as an example. In indoor air that has accumulated radon daughters for three minutes (three-minute air), Bismuth-214 will contribute 0. 5 percent of the total radioactive particle energy; it will contribute four percent in ten-minute air and 38 percent in non-circulating air.

At a ventilation rate of one or two times per hour (in most American homes), the radon accumulation is approximately one picocurie per liter of air (to repeat: one picocurie is one part in a trillion). This translates to an exposure of roughly 100 millirems per year, about one-third of the average annual radiation exposure and well within the 500-millirem maximum recommended for the population as a whole by the Environmental Protection Agency.

Two things, however, can increase this average exposure to radon: living in an energy-efficient home and living in an area where the natural concentration of uranium is high. Most granitic soils contain larger than normal amounts of uranium; in parts of New England, for example, radon exposures tend to be elevated. The geological formation best known for its high uranium-radon levels is the Reading Prong, which extends roughly from Reading, Allentown, and Easton, Pennsylvania, through Morristown, New Jersey, and into New York State. Some homes built on this formation have been found to have 1,000 times the average radon level, and dozens have been found with levels hundreds of times the average. To date, the highest radon level measure is in a home in Boyertown, Pennsylvania—2,500 times the average![14]

The naturally occurring outdoor radon, to which every American is exposed to some degree, might lead to lung cancer. In a population the size of the United States, approximately 10,000 cases per year are attributed to this source. Energy conservation, as urged by the U.S. government, will approximately double this number. Tobacco smoke further increases the cancer danger of this exposure. Perhaps the fuel-saving brought about by tightly sealing up one's domicile is worth the slight additional risk of lung cancer, but, if so, the decision should be made by the homeowner, by individual citizens knowledgeable about the consequences of their acts—not by a government enforcing strict energy codes or by utilities offering attractive incentives to "leakproof" one's residence. The choice of taking a risk should be left to the person involved, but he should have sufficient and accurate information to make an informed decision.

Although the facts about the increased risks of radon/radon daughter buildup in energy-efficient homes have been known since the late 1970s, the EPA issued no warnings and didn't release its long expected guide on indoor radon concentration until August 1986. What took the agency so long? Is energy conservation so overwhelmingly important that it transcends public health problems?

The EPA guide says nothing about the dangers of sealed up buildings. Instead, it calls for remedial action to be taken whenever the indoor radon level rises above four picocuries per liter of air (equivalent to roughly 400 millirems exposure per year). According to EPA estimates, this level is already exceeded by some 11 million homes, involving more than 26 million people in the United States. Ironically, the four picocuries per liter of air is 40 times higher than the very strict limit that applies to the nuclear industry! Why should stricter limits apply to the workplace, where perhaps an adult family member spends, say, one-third of each day, than to the domestic residence, where the entire family, including infants and children, spends two-thirds or more of each day?

One explanation for this regulatory absurdity is that whereas the EPA regulates radiation hazards in homes, the Nuclear Regulatory Commission (NRC) regulates radiation hazards in nuclear facilities. The NRC sets very strict and very low limits for the nuclear industry; indeed, if the *de minimus* rule adopted were to be enforced by the NRC, as urged by environmental activists, most natural, outside air would be illegal.

Two recent episodes involving radiation exposure should be remarked upon. First, there was the tragic accident at the Chernobyl plant in Russia. *Outside* the evacuation zone, the radiation exposure to

Soviet citizens was equivalent to the exposure New England citizens get every year from living in their radon-rich homes for about eight months.

Second, in 1984, a federal judge in Utah awarded $2.6 million damages to ten plaintiffs whose radiation exposures from atmospheric bomb tests between 1961 and 1962 did not exceed 500 millirems per year. This is far less than the exposures in many energy-efficient homes; some in New England exceed it by factors of ten or more. In delivering his verdict, Federal judge Bruce S. Jenkins made a statement that should command the attention of both governmental agencies and the news media.15 He said the government was negligent in its "failure to adequately warn plaintiffs . . . of known or foreseen long-range biological consequences" of its actions, and, in addition, of failure to "adequately and continuously inform individuals of well-known and inexpensive methods to prevent or mitigate the effects of inhaled radioactive particles."

Where are the government-issued warnings about radon in energy-efficient homes? Our government has actively promoted energy-efficient homes with everything from do-it-yourself literature to tax breaks for insulating your home. Where has the press been? Aiding and abetting the government.

Perhaps the time has come to let in some fresh air, to improve home ventilation by opening up the windows, even at the cost of higher energy consumption. Perhaps it's not too much to expect that the cause of common sense and public health could be served by a truly free press—freed from the silly notion that energy conservation is a good that transcends all and freed from the mindless assumption that radon trapped in a sealed house won't hurt you while radon seeping from uranium mine tailings (where nobody lives) will. It's all the same radiation. We should open the windows and let the fresh air in—and the radon out.

Or, better yet, let's stop building air-tight homes and buildings that not only develop radon accumulation problems but also suffer from another very modern phenomenon, "indoor air pollution." So bad is the poorly ventilated indoor air in some newly constructed energy-efficient buildings that an entirely new problem has emerged, the Sick Building Syndrome.

Increasingly, the occupants and workers in newly built energy-conserving buildings are complaining of adverse health effects. Respiratory illnesses, allergies, headaches, and skin problems appear to be common, and adequate ventilation, reduced by energy-conserving construction is becoming recognized as a serious problem. Few conclusive studies have been conducted, but one is instructive. It involves the health of Army recruits living in groups of 100 to 250 men in new, tightly sealed barracks with forced ventilation, compared to other groups of men assigned to old barracks.[16]

Four Army training centers, located in the southeastern and south-central United States, were included. Recently constructed , well-insulated, mechanically ventilated quarters were paired with old, World War II-vintage barracks. The study covered different groups of trainees over a 47-month period from October 1, 1982, to September 1, 1986. It was found that for the entire period the risk of respiratory illness was increased in the modern barracks by 45 percent or more at each of the four centers. During epidemic periods, the relative risk of respiratory disease in the modern barracks reached 100 percent. The conclusion is that "In tight buildings with closed ventilation systems, airborne pathogens are not only re-circulated and concentrated but also efficiently dispersed through indoor living spaces, while in 'leaky' buildings, airborne-transmitted agents are diluted by fresh outdoor air and relatively quickly exhausted from indoor spaces." So much for energy-efficient homes and buildings.

These paired modern problems of indoor radon and indoor air pollution, amounting now to a public health issue for all Americans, have been brought about by environmental zealots who insist that conservation is the preferred, if not the only, way to deal with assuring sufficient energy. If the same amount of devotion and effort had been directed toward building new electricity generating plants as has been wasted on tightly sealing our buildings, we would have a healthier population, as well as enough electricity to heat (and cool) structures in a clean, safe, and healthy manner—without respiratory problems. Energy conservation, however noble it may seem, is no more a "source" of energy than money under the mattress is a "source" of income.

Now it is time to return to the question: How harmful is exposure to low levels of radiation; that is, levels below 10,000 millirems per year? We've seen examples of apparently healthy populations living in areas of exceedingly high natural radiation. What about living with elevated levels of radon? How hazardous is it? Fortunately, there are now available extensive data from very recent investigations, which may cause some fundamental change in our attitude toward and understanding of chronic exposure to low-level radioactivity. Here we must introduce the concept of Hormesis.

Hormesis

A very long time ago, in the sixteenth century, a German physician known as Paracelsus (Theophrastus Bombastus von Hohenheim) developed a proposition that has become a fundamental principle of toxicology.

"What is it that is not poison?" he wrote. "All things are poison and none without poison. Only the dose determines that a thing is not poison."[17]

This is the concept of Hormesis. The validity of this principle has been confirmed over and over again. Even seemingly benign or necessary substances, such as water, can be toxic if taken in large enough doses. In 1979, a man in Germany died because he drank 17 liters of water within a very short time. The immediate cause of death was cerebral edema and electrolytic disturbance due to excess water.

This simple, common-sense principle, "the dose is the poison," also holds true for radiation. Clearly high doses of radiation can cause injury or death; conversely, low doses (as those from natural radioactivity) are, for all practical purposes, harmless. At what level of exposure does radioactivity become a "poison"?

Extensive research has established that exposures above 10,000 rems (10 million millirems) are lethal. Exposures of 300 rems (300,000 millirems) are fatal for roughly half of those who are exposed. Radiation injury is likely at exposures from 100 rems to 300 rems, but no deleterious reactions have been found at levels below 100 rems (100,000 millirems).[18]

Recall that the average exposure to natural background radiation in the U.S. is 350 millirems. Empirically, therefore, on the basis of many observations, measurements, and experiments, it would seem that there is a threshold for damaging radiation below which no adverse effects are found, and that threshold is about 100,000 millirems. But there is great argument about this, with most scientists accepting the "linear hypothesis," which holds that some effect occurs, decreasing toward zero, that is perhaps not detected nor measurable but can be calculated by extrapolating downward from effects at high level exposure.

A few scientists (and many laymen) go further and believe that *no* level of radiation is safe. We might note, in passing, that the linear hypothesis does not explain the absence of any detectable effect from living in a radioactive world. Since the radiation science community has not settled on the threshold or linear hypothesis, let us consider the evidence.

Early in the Manhattan District days (1943), radiation scientists were concerned about the toxicity of uranium. They exposed a colony of rats to an atmosphere laden with uranium dust with another colony breathing fresh air as a control. Although the researchers expected the high level of uranium to be fatal, the experimental rats lived out their normal life span and outlived the control rats. Moreover, they appeared healthier than the control rats and had more offspring. They developed no tumors. Most radiation scientists today consider this experiment flawed and dismiss the result as an anomaly.[19]

But similar results keep accumulating. In 1980, Professor T.D. Luckey published the conclusions from 1,239 separate studies of many investigators involving living things from cell cultures and bacteria to plants (800 references) and animals (200 references) of many different species exposed to varying amounts of Ionizing radiation of all types. He reports that the results are consistent: There is a threshold or cutoff point below which Ionizing radiation is either harmless or beneficial. Luckey concluded that ionizing radiation is generally stimulating in low doses; that low doses give accelerated development, increased resistance to disease, greater reproductivity, and longer life span; that low doses do not give proportionate harmful effects; that radiation is less dangerous in low doses than usually believed; and that chronic irradiation in doses slightly above ambient may be beneficial for both animals and plants.[20]

Still, many radiation scientists are reluctant to accept these findings as applicable to humans. Experiments on humans are out of the question, but the extensive data relating radon exposure to human lung cancer provide a unique opportunity to compare lung cancer rates among populations living in areas with different amounts of radon naturally present. The results are interesting:

- In Cornwall, England, radon exposures are 100 times larger than the average British exposure, but there is no increased incidence of lung cancer.

- Cumberland County, Pennsylvania, has nine times the average U.S. radon exposure but is well below the average U.S. lung cancer rate.

- In Finland, where indoor radon levels average 2.5 picocuries per liter (2.5 times the world average), the lung cancer incidence among Finnish women is only 70 percent of that of other industrialized countries.

Similar data are available for Colorado and other high mountain states with more than average natural radioactivity. Recently, in 1987, a group of Austrian scientists presented the results of a carefully documented epidemiological study of various types of cancer in the United States (at the Fourth International Symposium on Natural Radiation Environment).[21] They concluded that when the background radiation was between 350 and 500 millirems per year, the smallest number of cancers were found.

Professor B. L. Cohen of the University of Pittsburgh, analyzing 39,000 measurements of radon exposures versus lung cancer in 411 U.S. counties, has found the correlation to be negative at low levels—

the more radon, the less lung cancer! His data are corrected for housing differences in cities versus rural areas and for cigarette smoking.[22]

Additional evidence for Hormesis or for a threshold below which adverse effects are lacking comes from many careful modern studies that fall to find the results predicted if one extrapolates downward from high level exposures. These include studies of chromosome damage by radiation in human white blood cells, malignancy due to radiation in mouse embryo cells, cancer incidence in mice exposed to various doses of gamma rays, cancer incidence in mice injected with radioactive material, leukemia rates in survivors of Hiroshima, and data on watch dial painters exposed to radium.

Whether one chooses to believe that exposure to low level radiation can be beneficial, we can say with confidence that assuming a linear relationship (rather than a threshold) will overestimate the possible damage—and the public prefers to believe the worst. With exposure to low levels of radiation—below 10,000 millirems—we have a situation in which belief can override evidence. This is a mental phenomenon that is not unknown. For centuries, even well educated persons believed that, among humans, men had one less rib than women because of the biblical story that God created Eve from Adam's rib. This belief persisted, despite the availability of proof to the contrary: Count the ribs in skeletons. There is nothing in our philosophy, Horatio, that is true or false—but thinking makes it so.

So it is with radiation.

References

1. Cobb, Charles E., Jr., 1989, "Living With Radiation," *National Geographic*, April 1989, pp. 403–437.

    Eisenbud, Merrill, 1987, *Environmental Radioactivity From Natural Industrial and Military Sources*, Academic Press, 1250 Sixth Avenue, San Diego, CA 97101.

    Wagner, Henry N., Jr. and Linda E. Ketchum, *Living With Radiation—The Risk, The Promise*, Johns Hopkins University Press, 701 West 40th Street, Baltimore, MD 21211.

    Moghissi, A. Alan, editor, 1978, *Radioactivity in Consumer Products*, USNRC, NUREG/CP-003.

    Cohen, Bernard L., 1981, "How Dangerous Is Radiation?," *AECL Ascent*, Vol. 2, No. 4, 1981, pp. 8–12.

    Cohen, Bernard L., 1982, "The Genetic Effects of Natural Radiation," *AECL Ascent*, Vol. 3, No. 3, 1982, pp. 8–13.

2. Beckmann, Petr, 1986, "Iodine 131 and Chernobyl," *The American Spectator*, July 1986.

3. Marshall, Walter (Lord Marshall of Goring), 1986, "Nuclear Power: Energy of Today and Tomorrow," ENC International Conference, 2 June 1986.

4. Young, Alvin L. and George P. Dix, 1988, "The Federal Approach to Radiation Issues," *Environmental Science and Technology,* Vol. 22, No. 7, pp. 733–739.

    Grant, R.W., 1988, "Radiation Exposure by Source," in *Trashing Nuclear Power,* p. 33ff, Quandary House, Box 773, Manhattan Beach, CA 90266.

5. Luckey, T.D., 1980, *Hormesis and Ionizing Radiation,* p. 16, CRC Press, Inc., 2000 NW 24th Street, Boca Raton, FL 33431.

    Eisenbud, Merrill, 1987, op. cit., p. 160.

    Lapp, Ralph E., 1979, *The Radiation Controversy,* Reddy Communications, Inc., 537 Steamboat Road, Greenwich, CT 06830.

    Cohen, Bernard L., 1983, *Before It's Too Late,* see especially Chapter 2, "How Dangerous Is Radiation?," Plenum Press, New York and London.

    Beckmann, Petr, 1990, "Death From Outer Space," *Access to Energy,* Vol. 17, No. 8, 1990.

6. Cohen, Bernard L., *Before It's Too Late,* 1983, op. cit.

    Luckey, T.D., 1980, op. cit.

7. Beckmann, Petr, 1982, *Access to Energy,* Vol. 9, No. 5, Box 2298, Boulder, CO 80306.

8. Eisenbud, Merrill, 1987, op. cit.

    Beckmann, Petr, 1985, *The Health Hazards of NOT Going Nuclear,* Golem Press, Box 1342, Boulder, CO 80306.

    Cohen, Bernard L., 1988, *Health Effects of Low Level Radiation,* report from American Council on Science and Health, 47 Maple Street, Summitt, NJ 07901.

9. Eisenbud, Merrill, 1987, op. cit.

10. Ibid.

11. Cohen, Bernard L., 1983, op. cit.

12. Pirchman, A, 1932, "Working Miners and Lung Cancer at Joachinosthal," *American Journal of Cancer,* 1932.

13. Ray, D.L., 1986, "Who Is Radon and Why Are His Daughters So Bad?," *World Media Report,* Winter 1986.

    Thomas, Ron, 1989, "Radon's Troublesome Daughers Stir Up Controversy," *AECL Ascent,* Vol. 8, No. 2, summer 1989.

    Brookes, Warren T., 1989, "Radon Terrorism Unleashed by EPA," *The Washington Times,* 29 June 1989.

    Brookes, Warren T., 1990, "Radon, Anatomy of Risk-Hype," *The Detroit News,* 5 March 1990.

14. Nero, A.V. et al. 1986, "Distribution of Radon 222: Concentrations in U.S. Homes," *Science,* 21 November 1986, pp. 992–997.

    Nero, A.V., 1988, "Controlling Indoor Pollution," *Scientific American,* May 1988.

    Lapp, R.E., 1989, *Radon Health Effects?,* radon panel, Health Physics Society meeting, Albuquerque, NM, 29 June 1989.

15. Jenkins, Judge Bruce S., 1984, *Radiation Expsoures in Utah From Bomb Tests,* 1951–62, decision in Federal Court.

16. Brundage, J.F. et al, 1988, "Building-Associated Risk of Febrile Acute Respiratory Illness in Army Trainees," *Journal of the American Medical Association*, 8 April 1988, pp. 2108–2112.

Marcus, Amy Dockser, 1989, "In Some Workplaces, Ill Winds Blow," *The Wall Street Journal*, 9 October 1989.

Lawrence, Henry J., 1989, "Is Your Office Out to Kill You?," *Seattle Post-Intelligencer*, 14 August 1989.

Holzman, David, 1989, "Elusive Culprits in Workplace Ills," *Insight*, 26 June 1989.

17. Efron, Edith, 1984, *The Apocalyptics*, Chapter 12, "The Case of the Missing Thresholds," p. 344, Simon & Schuster, Inc., Rockefeller Center, 1230 Avenue of the Americas, NY 10020.

Luckey, T.D., 1980, *Radiation Hormesis*, op. cit.

18. Ibid.

Sagan, Leonard A., 1987, "What Is Hormesis and Why Haven't We Heard About It Before?," *Health Physics*, guest editorial, Vol. 52, No. 5, pp. 521–525, May 1987.

Cohen, Bernard L., 1987, "Tests of the Linear No Threshold Dose Response Relationship for High LET Radiation," *Health Physics*, Vol. 52, No. 5, pp. 629–636, May 1989.

Fremlin, J., 1989, "Radiation Hormesis," *Atom*, London, April 1989.

Luckey, T.D., 1988, "Hormesis and Nurture With Ionizing Radiation," in *Global 2000 Revisited*, Hugh Ellsaesser, editor, Paragon House publication, 1988.

19. Letter from Marshall Brucer to *Time* magazine, quoted in *Access to Energy*, Vol. 16, No. 7, March 1989.

20. Luckey, T.D., 1988, op. cit.

21. Fleck, C.M., H. Oberhummer and W. Hofmann, 1987, *Inference of Chemically and Radiologically Induced Cancer at Environmental Doses*, Fourth International Symposium on the Natural Radiation Environment, Lisbon, Portugal, 7–11 December 1987.

22. Cohen, Bernard L., 1989, "Lung Cancer and Radon: Hermesis at Low Levels of Exposure in American Homes," *Access to Energy*, Vol. 16, No. 9, 1989.

Cohen, Bernard L., 1989, "Expected Indoor Radon-222 Levels in Counties With Very High and Very Low Lung Cancer Rates," *Health Physics*, Vol. 57, No. 6, December 1989, pp. 897–906.

Lou Guzzo, a TV/radio commentator in Seattle, is a former managing editor of *The Seattle Post-Intelligencer* and reporter for *The Cleveland Plain Dealer* and *The Seattle Times*. He is the author of several books, including a biography of Dixy Lee Ray.

The Radioactive Waste Disposal Problem

Bernard L. Cohen, Ph.D.

Please see biographical sketch for Dr. Cohen on page 461.

High-Level Waste[1]

When the fuel from a nuclear reactor has been mostly burned up, it is removed from the reactor. Ideally, it should be shipped to a reprocessing plant where it would be put through chemical procedures to remove the valuable components. The residual material, which contains nearly all of the radioactivity produced in the reactor, is called high-level waste. Concern has been raised about its disposal.

One important aspect of the high-level waste disposal question is the quantities involved: the waste generated by one large nuclear power plant in 1 year is about 6 cubic yards. This waste is 2 million times smaller by weight, and billions of times smaller by volume, than wastes from a coal-burning plant. The electricity generated by a nuclear plant in a year sells for about $400 million, so if only 1 percent of the sales price were diverted to waste disposal, $4 million might be spent to bury this waste. Obviously, some very elaborate protective measures can be afforded.

Once the radioactive waste is buried, the principal concern is that it will be contacted by groundwater, dissolved into solution, and moved with the groundwater to the surface where it can get into food and drinking water supplies. How dangerous is this material to eat or drink? To explain this, we will take the quantity that would have to be ingested to give a person a 50 percent chance of death. When the waste is first buried it is highly toxic and a fatal dose is only 0.01 ounce. However, the radioactivity decays with time, so that after 600 years a fatal dose is about 1 ounce, making it no more toxic than some things kept in homes. After 10,000 years a lethal dose is 10 ounces.

When some people hear that nuclear waste must be carefully isolated for a few hundred years, they react with alarm. They point out that very few man-made structures, and few of our political, economic, and social institutions can be expected to last for hundreds of years. Such worries stem from our experience on the surface of the earth, where most things are short-lived. However, 2,000 feet below the surface the environment is quite different. Things remain essentially unchanged for millions of years.

In order to understand the very long-term (millions of years) hazard, the natural radioactivity in the ground is a good comparison. The ground is full of naturally radioactive materials, so that by adding nuclear waste to it the total radioactivity in the top 2,000 feet of U.S. soil would increase by only one part in 10 million per plant-year. Moreover, the radioactivity in the ground (except that very near the surface) does virtually no harm.

Waste burial plans would delay the release of the waste to the environment for a very long time, thus giving near-perfect protection from the short-term problem. Under these plans the rock formation chosen for burial will be well isolated from groundwater and expected to remain isolated for at least 1,000 years. If water did enter that rock formation it would have to dissolve a reasonable fraction of the surrounding rock before reaching the waste. The least favorable situation for this factor would be if the waste were buried in a salt formation, because salt is readily dissolved in water. However, in the New Mexico area being considered for an experimental repository, if all the water now flowing through the ground were diverted through the salt formation, the quantities of salt are so vast and the amount of water so meager that it would take 100,000 years to dissolve the salt around the buried waste from 1 year of all nuclear electricity in the U.S.

A third protection is the specific backfill material surrounding the waste package. Clays selected for this purpose swell up to seal very tightly when wet, thereby keeping out any appreciable amount of water. These materials are also highly efficient filters; if groundwater did get to the waste and dissolve some of it, these clays would filter the radioactive material out of solution before it could escape with the water.

Another safeguard is that the waste will be sealed in a corrosion-resistant casing. Casing materials are available that would not be dissolved even if soaked in groundwater for many thousands of years. Also the waste itself will be a rock-like material that would require thousands of years of soaking in water before dissolving. Groundwater is more like a "dampness" than a "soaking," thus dissolving things hundreds of times more slowly.

There is also a time delay. Groundwater moves quite slowly, usually only inches per day, and ordinarily must travel many miles before reaching the surface from 2,000 feet underground. Hence, even if the dissolved radioactive material moved with the groundwater, it would take about 1,000 years to reach the surface. But there are processes by which the rock constantly filters the radioactive materials out of the groundwater, causing it to migrate about a thousand times slower than the water itself. It would therefore take most of the radioactive materials a million years to reach the surface even if they were already dissolved in groundwater. Most of the radioactive materials are highly insoluble under geological conditions; thus, if they were in solution when the water encountered these conditions (chemically reducing, alkaline), they would precipitate out and form new rock material.

Finally, if radioactivity did reach surface waters, it would be detected easily—one millionth of the amounts that can be harmful are readily detected—and measures could be taken to prevent it from getting into drinking water or food.

With all these safeguards it seems almost impossible for much harm to result during the first few hundred years while the waste is highly toxic, and there is substantial protection over the long term.

One way of estimating the distant effects is to assume that an atom of buried waste has the same chance of escaping and of getting into a person as an atom of average rock. It can be shown that an atom of average rock submerged in flowing groundwater has about a one in 100 million chance per year of escaping into surface waters. Once in surface waters its chance of getting into a human body is about one in 10,000. If these probabilities are combined and applied to buried radioactive waste, the result indicates that the waste would eventually cause 0.018 fatalities per plant-year. Note that this is still 1,000 times less than the health effects of air pollution from coal burning.

If there is a problem in the above arguments, it would be in how buried radioactive waste differs from average rock. There are basically three differences. First, a shaft must be dug to bury the waste, giving a connection to the surface not usually present for rock; second, the radioactive waste emits heat, which is not a normal property of rock; and third, the waste is a foreign material, not in chemical equilibrium with the rock-groundwater regime. Solving the first problem depends on our ability to seal the shaft, and the technical community seems highly confident that this can be done to make the area as secure as if the shaft had never been dug.

The heat radiated from buried waste is enough to raise the temperature of the surrounding rock by about 200 degrees Fahrenheit. There has been concern that this might crack the rock, producing new pathways by which groundwater can reach the buried waste and through which the dissolved waste might escape. This problem has been studied intensively for over a decade, and the conclusion seems to be that there are no serious problems of this type. These studies are continuing, however.

If it is decided that the temperature must not be allowed to rise so high, there are two easy remedies; the waste can be distributed over a wider area to dilute the heating effect, or burial can be delayed to allow some of the radioactivity to decay. The latter option is especially effective since the rate of heat emission is decreased 10-fold after 100 years and 100-fold after 200 years. Also, the protective casings in which the waste will be enclosed are highly resistant to high-temperature groundwater.

The chemical equilibrium between rock and groundwater is a surface phenomenon. If a foreign rock, such as radioactive waste converted to a rock-like material, is introduced, the groundwater begins to dissolve it, but in the process precipitates out a highly insoluble material on its surface. Further dissolution of the foreign material can then only take place by diffusion through this surface layer, and that process thickens the latter which slows down the diffusion process. After a short time, chemical equilibrium is reached, with only a tiny quantity of the waste having been dissolved.

Since we have mentioned the ways in which buried waste is less secure than most rock, the ways in which it is more secure should be pointed out. The geological environment for the waste will be carefully selected and will be much more favorable than for average rock. The waste will be buried in a region with little or no groundwater, whereas our average rock is submerged in groundwater. Finally, the buried waste will be sealed in a leach-resistant casing that provides a complete and independent safety system which should avert danger even if all other protections fail.

Since most of the health impact of radioactive waste is expected to occur millions of years in the future, it is instructive to compare this with the cancer-causing solid wastes released in coal burning. Some of these, like arsenic, beryllium, cadmium, chromium, and nickel, are very long-lasting and their effects can therefore be calculated in a similar way as for radioactive wastes. When this is done they can be expected to cause about 70 eventual fatalities per plant-year, an effect thousands of times larger than the effects of nuclear waste. Also, solar electricity technologies require vast amounts of materials, and deriving these requires the burning of larger quantities of coal—about 3 percent as much coal as would be used to produce the same amount of energy by direct coal burning. Consequently, the wastes from solar technologies are many times more harmful than nuclear wastes. In addition, some solar technologies use large quantities of cadmium, which increases the health consequences considerably.

Reference:

1. This discussion is based on a group of papers reviewed in B.L. Cohen, *Long Term Waste Problems from Electricity Production, Nuclear and Chemical Waste Management* 4,219 (1984).

The Truth About Three Mile Island

Bernard L. Cohen, Ph.D.

Please see biographical sketch for Dr. Cohen on page 461.

Introduction

If our democracy is to function properly in public decision-making situations, the public must have the required basic information, and it is the responsibility of journalism to provide it. The purpose of this paper is to point out a situation in which this system has failed miserably, with far-reaching adverse consequences for the author's nation, the U.S.A.

The decision is on the acceptability of nuclear power for generation of electricity. It may seem that this is an extremely complex question, but we have boiled it down to three very simple non-controversial questions that have been crucial in the public's decision-making, but to which the public's answers are wrong according to all but the most far-out fringes of the scientific community. This last statement may be easily checked, for example, by presenting this material to professors of relevant scientific disciplines in universities of one's choice.

The three questions, stated over-briefly, are:

(1) How does the radiation expected from the nuclear power industry compare with the natural radiation to which mankind has always been exposed?

(2) Was the Three Mile Island accident a "close call" on a public health disaster?

(3) How do the risks of nuclear power compare with other common risks in our Society?

The public's answers to these questions were determined by mailing questionnaires to names and addresses randomly selected from the telephone directories for 12 areas representing all major sections of the country. None of the names or addresses meant anything to us; from the 300 questionnaires sent (12x25), 75 were returned.

Questions and Results

The questions asked, and the results, were as follows:

Question 1: How can we expect the average human exposure to radiation from the nuclear industry (if it flourishes), including accidents (taking into account how frequently they may occur), wastes (including their exposures to future generations), transporting radioactive materials, etc. to compare with the natural radiation to which mankind has always been exposed? The response was as follows:

35%   Nuclear power will give much more exposure.

31%   Nuclear power will give somewhat more

14%   They are about equal

 7%   Nuclear power will give somewhat less

14%   Nuclear power will give much less.

We see that 80% of the responders believes that nuclear power will give as much or more radiation than natural sources.

Since different sources of radiation expose different organs of our bodies, the scientists' answer is most simply given in terms of the extra number of cancer deaths expected to result from the radiation. From all nuclear industry sources other than reactor accidents, typical estimates are about 10 extra deaths per year in the United States, including those projected for the future.[1] For reactor accidents, based on probabilistic risk analyses, government studies estimate an average of 5 extra deaths per year,[2] while the principal antinuclear activist organization, Union of Concerned Scientists, estimates an average of 600 extra deaths per year.[3] These estimates are based on assuming twice as many nuclear plants as are now in operation and under construction.

Natural radiation, according to the same estimating procedures, causes 2500 deaths per year in the United States,[4] not counting the several times larger effects of the naturally radioactive gas, radon, whose importance has only been fully recognized in the past few years.[5] Clearly, the correct answer is that nuclear power will give much less radiation than natural sources; 86% of the public has a less favorable wrong understanding.

Question 2: Was the Three Mile Island Accident a "close call" on a disaster causing many dozens of deaths or more? (By "close call" we mean if 2 or 3 relatively minor things had happened differently, there would have been a disaster.) The response was as follows:

65%   Yes, it was a "close call."

35%   No, it was not a "close call."

So, nearly two-thirds of the responders believe that it was a close call.

The accident[6] was terminated by closing a valve to stop the escape of water. This was done at the suggestion of one man. What if he had failed to make the suggestion? Within one minute after the valve was closed, a call was received from an expert analyzing the situation from his home, suggesting that the crucial valve be closed. What if he also had failed to understand the problem? Escape of water would have continued for 30 to 60 minutes before a meltdown would have become inevitable, and during this time new symptoms would have developed that would have made the situation and its cure much more evident, so that the proper action very probably would have been taken.

But what if the valve had never been closed and a meltdown had occurred? All reactors are sealed inside a powerfully constructed building called the "containment" which is designed to hold the radioactivity inside in the event of a meltdown. Only if it is somehow broken open during the course of the accident can appreciable quantities of radioactivity escape. But all post-accident reports on the Three Mile Island accident conclude that the containment was never in danger there. At least two independent further major system failures would have been required to compromise its security by "major," we here mean something more than a pump failing or a valve sticking, because these systems have redundant pumps and valves to protect against such minor failures. Thus, even if there had been a meltdown in the Three Mile Island accident, there would not have been a public health disaster.

The correct answer to question No. 2 is that the Three Mile Island accident was not a close call on a disaster. Nearly two-thirds of the public is badly misinformed on that matter.

Question 3. How does the average person's risk of death resulting from nuclear power operations (whether or not it is recognizable as such) compare with that person's risk of death from some other dangers many of us face? Please check which of the following risks is greater than the risk of nuclear power (assuming that the nuclear industry flourishes). For example, if you think our risks from nuclear power are very great, few of any of these should be checked; if you think our risks from nuclear power are very slight, all or nearly all of these should be checked.

76%   smoking cigarettes (6.5 years for 1 pack per day)

28%   being 15 lb overweight (1.3 years)

85%   automobile accidents (200 days)

20%   being poor (about 5 years)

31%   drowning (40 days)

62%   fires (27 days, including burns)

31%   gas leaks (7.5 days from asphyxiation)

41%   being murdered (90 days)

34%   being killed in a fall (39 days)

The figures in parentheses following each item, which were not included in the questionnaire, feature the amount of life expectancy lost due to these risks;[7] for smoking cigarettes, overweight, and being poor they refer to those who take those risks, and for all other cases they refer to the total U.S. population.

The loss of life expectancy due to the risks from a flourishing national nuclear power program can be derived mathematically from the effects described above following question 1. It is 0.05 days (about one hour) according to government sponsored studies, or 1.5 days according to the estimates of Union of Concerned Scientists. Even if we use the latter, it is clear that each of the risks listed in question 3 is much greater than the risk of nuclear power. We see that only three of the nine common risks are recognized by the majority of the public as being larger than those from nuclear power, and most of them were so recognized by only one-third of the public. The risks of overweight and of poverty are at least several hundred times, and probably over 10,000 times higher than those of nuclear power, but 3/4 of the public believes that they are lower.

Discussion

Of course the public cannot be well informed on all subjects, and it is only important for the functioning of democracy that it be well informed on questions vital for public decision making. Do our three questions fit into that category? I believe that they are the most vital questions, and that if the public would just understand the answers to these three simple questions, the great majority of opposition to nuclear power would disappear.

Perhaps the most important source of this opposition is fear of radiation, which is addressed by question 1. How could people be so fearful of this radiation if they realized that it is only a tiny fraction of what they receive from natural sources, a similarly tiny fraction of the extra radiation received by residents of Colorado and neighboring states due to the fact that natural radiation in that area is nearly double the national average, and less than the radiation they receive (due to radon) from staying home one extra day per year? The public's fear of radiation is constantly fanned by media coverage of accidents involving radiation, ranging from a package containing radioactive material falling off a truck to releases from nuclear power plants. Wouldn't it defuse the fear generated by these stories if it were stated that the radiation doses were comparable to what we all receive every day from natural sources? Scientists always use these comparisons in explaining radiation to the public, but journalists hardly ever do.

Question 2 addresses what are probably the most important specific fears about nuclear power, the danger from a meltdown accident which is widely (and incorrectly) viewed as a horrible public health disaster, and disbelief of government and industry assurances that such a disaster is highly improbable based on the idea that it nearly happened at Three Mile Island. If the answer to question 2 became widely known, the great increase in public opposition to nuclear power generated by the Three Mile Island accident would be largely counteracted. Journalists have many opportunities to let the truth be known on this matter, such as in the reviews they present on each anniversary of the accident, but they never give the public this very simple bit of vital information.

Question 3 addresses a much more general problem, the public's failure to quantify and understand risk and keep it in perspective, but this problem has reached new heights on the nuclear power issue. If people understood that their risk from nuclear power was equal to that of an overweight person eating one extra slice of bread and butter every 10 years, and very much less than many other risks that they face ever day and regard as negligible, how could they possibly be very fearful of the nuclear power risk?

If a person knew the correct answers to our three questions, he could not be fearful of nuclear power. And if this fear were removed, public opposition would all but vanish. These three questions are therefore vital for public decision-making.

Regulatory Ratcheting

What have been the effects of the public misunderstanding, represented by its overwhelmingly wrong answers to our three questions? Public fear of nuclear power has materialized as ever-tightening government regulations, called "regulatoring ratcheting"[1] which involve new requirements at nuclear plants for equipment and procedures designed to improve safety. As a result, the cost of nuclear power plants has increased five-fold over and above inflation since the early 1970s. Some would have us believe that the skyrocketing costs are due to incompetence, but these plants are being built by the same utilities, with the same architects, engineers, and constructors that had so much success and so little difficulty before the public opposition began driving the government's nuclear regulatory process. If this regulatory ratcheting has actually improved safety, which is considered doubtful in many technical circles, the cost based on the government's own figures has been US$2 billion per life saved. By comparison, there are many ways of saving lives through biomedical research, medical screening, and highway safety programs for less than $100,000 per life saved. The money spent to save one life by nuclear regulatory ratcheting could thus save 20,000 lives if spent more wisely.

But the most important practical effect of this regulatory ratcheting has not been to save lives from nuclear dangers, but rather to force utilities to build coal burning plants rather than nuclear plants. It is very widely recognized that the former have far greater impacts on human health through their air pollution. In fact, every time a coal burning plant is built instead of a nuclear plant, something like a thousand extra people are condemned to an early death—this is true even if we accept the estimate by the Union of Concerned Scientists. Due to regulatory ratcheting, several new coal burning plants have been constructed each year, which means that the process is killing several thousand Americans per year rather than saving the few lives for which it was intended.

But the harm done goes far beyond that. As a result of regulatory ratcheting, U.S. power plants now cost more than twice as much as nuclear plants being built in Western Europe and Japan. Historically, United States has always had cheaper energy than those countries, and economists consider this to have been an important ingredient in the economic success of our country. Thus, the economic effects of our electricity becoming twice as expensive as that of our competitors may well have dire future consequences for our unemployment problems and for our standard of living.

Conclusions

We have shown that the public has been grossly misinformed about three questions that are vital to its decision-making on nuclear power. We have shown that this misinformation is unnecessarily killing thousands of Americans and wasting billions of dollars every year, and that it is jeopardizing our nation's economic future. Surely, journalists have a sacred duty to correct this misinformation. Their failure to do so represents a horrible breakdown in the workings of our democracy.

References

1. B.L. Cohen, Before It's Too Late. Plenum Press, New York, NY, 1983.
2. Reactor Safety Study, U.S. Nuclear Regulatory Commission Document WASH-1400, NUREG 75/014, 1975.
3. Union of Concerned Scientists, The Risks of Nuclear Power Reactors, Cambridge, MA, 1977.
4. U.S. National Academy of Sciences, Committee on Biological Effects of Ionizing Radiation (BEIR-III). The effects on Population Exposure to Low Levels of Ionizing Radiation, Washington, DC, 1980.
5. U.S. National Academy of Sciences, Committee on Biological Effects of Ionizing Radiation (BEIR-IV), Health Effects of Radon and Other Internally Deposited Alpha Emitters, Washington, DC, 1989.
6. J.B. Kemeny (Chairman), Report of the President's Commission on the Accident at Three Mile Island, Washington, DC, 1979; M. Rogovin (Director), Three Mile Island: A Report to the Commission and to the Public, U.S. Nuclear Regulatory Commission, Washington, DC, 1980.
7. B.L. Cohen and I.S. Lee, A catalog of risks, Health Phys., 36 (1979) 707; B.L. Cohen, Catalog of risks updated and extended, Health Phys. (submitted).

This article is printed with permission of Elsevier Science Publishers BV, publisher of the *Journal of Hazardous Material*, in which this article appeared under the title "Journalism's Failure on Nuclear Power Information" in Volume 21, pages 255–260, 1989.

Radon

Radon—Risk and Reason
Keith J. Schiager

Radon: A Way to Make a Little Money
Jane M. Orient

Radon—Risk and Reason

Keith J. Schiager, Ph.D.

Dr. Keith J. Schiager is director, Radiological Health, University of Utah. Formerly he was a professor of Health Physics at the University of Pittsburgh; and an associate professor in radiology and radiation biology, at Colorado State University.

He received a B.S. in physics from Colorado State University in 1956; an M.P.H. in radiological health from the University of Michigan in 1962; and a Ph.D. in environmental health from the University of Michigan in 1964.

He has published more than 70 journal articles and major research reports. He was the president of the American Academy of Health Physics in 1990; and is president-elect of the Health Physics Society, 1991–92.

What Is the Controversy?

For several decades, encompassing the professional lifetime of most present health physicists, we have endorsed and followed basic principles of radiation protection that have evolved through the conscientious efforts of many competent scientists working on national and international advisory committees. In the few instances where radiation protection has been inadequate, the reason has usually been the failure to follow these principles. We have also worked diligently to replace irrational fear of radiation with understanding based on factual information. It is the deliberate rejection of these principles by some federal bureaucrats that is a major source of the controversy in radiation protection.

One of the basic tenets of our profession is that radiation protection should be optimized, i.e., that radiation exposures should be kept as low as reasonably achievable (ALARA), taking into account social and economic considerations. Controversy arises primarily between those who believe strongly in this principle (mostly within the health physics profession) and those who believe in some other basis for radiation protection (mostly in special interest groups and governmental agencies).

Because of its high visibility and universal impact, radon has become a focal point of the controversy. Among the points of contention are:

- the magnitude of the radon risk,
- the reliability of measurements,
- the costs versus the benefits of mitigation,

- the public's right to full disclosure, and
- the methods of communicating with the public.

As scientists and technologists, we tend to focus on accuracy and precision of numerical values and calculations, but the basic controversy stems from the fact that politicians and regulatory agencies don't follow the same principles as we do, i.e., balancing costs and benefits of radiation protection, and communicating by informing rather than inflaming the public.

The fact that radon produces larger radiation doses than most other natural or manmade sources of radiation does not necessarily mean that it produces an exceptionally large health risk nor that it requires extraordinary remedial actions. It may simply demonstrate that other radiation sources are very well controlled. Conversely, the fact that radon is ubiquitous in the natural environment is not, in itself, a sufficient reason to consider the exposure acceptable and to make no effort to control it.

Because it is impossible to define an absolutely "safe" amount of radiation exposure, it is always necessary to ask "How safe is safe enough?" The easy answer to that question, "as safe as possible, using the best available technology, without regard to costs," would be recognized immediately as a completely self-serving response from those who earn their livelihood by providing radiation protection. Instead, our ethics demand that we encourage responsible decisions, based on balancing the costs and benefits of reducing radiation exposures.

Exposure to moderate levels of radon may be an unavoidable byproduct of living in durable, affordable, comfortable houses. On the other hand, excessive exposure to radon may be avoided at relatively low cost, especially in new construction. We should encourage reduction of indoor radon to the level at which the benefits of further reduction are too small to justify the additional costs.

Is Radon Really Harmful?

Radon gas (and its decay products) occurs naturally in indoor air and is generally recognized as the source of the largest effective radiation dose to the average person and, therefore, to the population as a whole. For some people, this newly acquired awareness has led to unnecessary anxiety over radiation exposures.

The evidence that radon is a causative agent for lung cancer is based on epidemiological studies of miners. Extrapolation of risks to the general population from the data on miners increases the uncertainties. However, most scientists agree that the estimated risk is sufficient to warrant careful evaluation and, in some cases, protective actions. The Environmental Protection Agency (EPA) bases its risk estimates on a relative risk model derived from models published by the International Commission on Radiological Protection (ICRP, 1987, *Lung Cancer from Indoor Exposures to Radon Daughters*) and by the National Academy of Sciences Committee on the Biological Effects of Ionizing Radiation (BEIR, 1988, *Health Risks of Radon and Other Internally Deposited Alpha-Emitters*). Although EPA's risk estimates are higher than those of some scientific organizations, it should be noted that most of the risk estimates are within approximately a factor of two of each other.

A major uncertainty in calculating the number of lung cancers that may be attributable to radon is due to lack of data on national average exposures. The EPA assumes a higher indoor radon concentration than is generally accepted as representative of living spaces in single-family residences. This higher average is then applied to the entire U.S. population, including those living in high-rise buildings, trailers, mobile homes, house boats, etc.

On the basis of these assumptions—assumptions with which many scientists disagree—the EPA calculates that almost 1 percent of all deaths are attributable to radon. However, whether deaths attributable to radon are 0.1 percent or 1 percent of the total, the questions remain: What to do, how to do it, and who should do it?

Is Everyone's Risk the Same?

Based on the relative risk model used by the EPA, the risk from exposure to other carcinogens, such as cigarette smoke, multiplies the risk of lung cancer from radon exposure. For equal indoor radon exposures, the risk of lung cancer to persons who have never smoked is approximately 1/4 of the risk to the average members of the population and 1/10 of the risk to smokers. The increased life expectancy for each picocurie per liter (pCi/L) of radon reduction and for each year of reduced exposure is estimated to be 0.1 day for a non-smoker, and 1 day for a smoker. For ex-smokers and those exposed to second-hand smoke, the population average value of 0.4 day provides a reasonable estimate. Since the risks are not uniform, then it is obvious that the balancing of costs and benefits cannot be the same for everyone.

Should Exposure Be Reduced?

Indoor radon concentrations can be reduced in several ways. Some methods require no maintenance and remain effective for the lifetime of the building. Other methods, e.g., sub-slab ventilation, are effective only as long as they are properly operated and maintained. A reasonable estimate of the annualized cost for a typical installation, including operation, maintenance and amortization of equipment, would be $200 per year for the life of the house. Methods that cost less might be used effectively in new construction, but may require changes in building codes. Although $200 per year may not seem like a large expense, it should be justified on the basis of the benefit derived.

The risk of radon-induced lung cancer is presumed to be proportional to the total lifetime exposure to radon, not simply to the instantaneous concentration of radon present at the time of measurement. At present, there is an unwarranted emphasis on a short-term measurement as the basis for taking action; this erroneously implies a sharp dividing line between "safe" and "unsafe" radon concentrations. However, when benefits and costs of radon reduction are properly balanced, a concentration that is unacceptable in some cases may be acceptable in others. The benefit of radon reduction in a given building is determined by the decrease in the long-term average concentration, the number of people affected, and time spent in the reduced concentration.

A reduction from 40 to 20 pCi/L is only 50 percent effective, but eliminates exposure to 20 pCi/L; a reduction from 5 to 1 pCi/L is 80 percent effective, but provides only one-fifth as much real benefit. Radon reduction in a single-family residence might benefit only a few people, but for a school building the number would be in the tens, hundreds or even thousands.

For exposure in homes, the EPA assumes that the average individual spends 75 percent of the time indoors at home. Based on occupancy times, the benefit to an individual child from radon reduction in the school is about one-fifth of the benefit from the same reduction in the home. However, in the case of a school or other public building, the benefit to the community that bears the costs depends on the number of years the reduction remains effective, whereas in the case of a home, the benefit to the owner who pays for the reduction depends only on the duration of occupancy by that family. Rarely, if ever, would the mitigation last as long as a human lifetime.

The relative benefit of radon reduction depends heavily on the smoking habits of those exposed. For any level of radon exposure, the risk to those living with one or more smokers is much greater than the risk to those living in a smoke-free environment. Therefore, the relative risk model implies that smokers obtain much more benefit than non-smokers from radon reductions, even though the remaining risk due to smoking would remain high.

How Much Benefit Can We Afford?

The EPA justifies its radon mitigation recommendations by calculating "lives saved" and comparing the cost per life saved with costs for other EPA programs. For any reduced cancer risk, "lives saved" is a euphemism for "cause of death changed" or "death delayed." The benefit of risk reduction can be more accurately expressed as increased statistical life expectancy. For each lung cancer prevented, the statistical life expectancy gained is about 15 years, or 5,000 days. Given an average personal income in the U.S. of $16,444 per year (1988), the maximum statistical benefit of a lung cancer avoided is about $250,000.

As a practical matter, we must ask, "How much of current income can the average person spend for each day of increased statistical life expectancy?" Since some of our productivity must go into maintaining life, health and happiness in the present, we can't spend it all on the future. The U.S. population actually spends a little more than 10 percent of its gross national product, or approximately $5 per person per day of life, on health-related products and programs. At about 5,000 days of life expectancy gained per lung cancer avoided, the expenditure consistent with current public actions would be approximately $25,000 per lung cancer avoided.

It is not the place of health physicists to decide how much home owners or taxpayers ought to spend for a given health benefit. The amount we are willing to spend to prolong life depends on the immediacy of the health threat, but since radon exposure represents a statistically small threat far in the future, the range from the average daily health expenditure of $5 to the average daily income of $45 provides a suitable frame of reference.

To illustrate how factors in addition to the radon concentration affect the benefits of radon reduction, consider three situations, each with initial radon concentrations of 8 pCi/L that could be reduced to 2 pCi/L, or a decrease of 6 pCi/L. First assume a non-smoking, middle-aged couple, who expect to move to another residence within five years. The calculated benefit of radon mitigation would be six days of added life expectancy (6 pCi/L × 10 person-years × 0.1 day per person-year-pCi/L). This statistical benefit would justify an expenditure of at least $30, but no more than $300, over five years.

Second, assume a family of two adult smokers and three children, all of whom expect to live in the same house for at least 10 years. Since this family involves several ages of smokers and non-smokers, the reference population risk may be more applicable, giving a calculated benefit to the family of 120 days of added life expectancy (6 pCi/L × 50 person-years × 0.4 day per person-year-pCi/L). This statistical benefit would justify an expenditure of at least $600, but no more than $6,000, over 10 years.

Finally, consider a school with elevated radon in rooms occupied by 500 children, each of whom attend only a few hours per day for a few years; however, assume that the mitigation will be effective for 20 years. The ultimate benefit to the community depends on how many of these children will eventually become smokers. If the risk factor for the reference population is assumed to apply, the total benefit of radon reduction is calculated to be 4,800 days of added life expectancy (6 pCi/L × 10,000 person-years × 0.2 occupancy × 0.4 day per person-year-pCi/L) and an expenditure of at least $24,000 would be justified.

These examples illustrate a method of calculating the statistical benefit of radon reduction and appropriate expenditures. It must be remembered that any benefit of current reductions will only appear in the population after at least ten years or more. One thing that such calculations clearly demonstrate is that reducing radon in schools will generally be much more beneficial to those bearing the costs than will mitigation in private residences.

What About the "National Goal"?

The Radon Abatement Act of 1988 states that the "national goal" is to reduce radon indoors to the same level as radon in the air outdoors. Because this "national goal" does not address the tremendous costs that would be incurred in trying to reach this very elusive goal, most health physicists cannot support it. The costs per "life saved" calculated by the EPA, range from approximately $200,000 for an action level of 20 pCi/L to an incremental cost of $1,963,000 for the additional "lives saved" if the action level is reduced from 4 to 2 pCi/L. Many health physicists believe that the actual costs would be much higher. The EPA considers that remedial action at any level down to 2 pCi/L would be "cost effective," in spite of the fact that the statistical years of life expectancy gained would cost up to eight times more than the U.S. annual average personal income! This blatant disregard for costs is a major source of controversy between health physicists and the EPA.

What Should Be Done?

• Appeal to reason instead of emotion!

Emotional scare tactics should not be used to overcome apathy. The enhancement of radiation phobia among the public may easily have undesirable repercussions, e.g., reluctance to undergo medical examinations involving radiation or to have a home tested for radon.

- Give the public realistic expectations!

Even if a national radon abatement program is extremely effective, the reduction in lung cancer incidence would occur only after a decade or more, and the actual reduction would probably be too small to be statistically detectable.

The case of indoor radon is quite different from most radiation sources for which professional organizations or public agencies have attempted to reduce exposures. In this case, there is no "villain" to take the rap, no "deep pocket" to tap for remediation expenses and, in most cases, no legal requirement to do anything. Because measurement and mitigation must be initiated and paid for by individual home owners or by local school boards, and because risks and benefits are not the same for everyone, balancing benefits against costs of radiation protection must be done on an individual, case-by-case basis, similar to decisions for purchasing smoke detectors or fire extinguishers.

It is clear that reducing indoor radon concentrations by tens of pico-curies per liter provides sufficient benefit to justify the cost to most individual home owners. Reductions of less than 10 pCi/L are likely to be justified only in households with many children or in schools because a larger number of individuals would benefit. In short, specific remediation actions should be recommended only on the basis of an analysis of expected benefits and costs.

- Give home owners and elected officials realistic information on the benefits and costs of radon reduction!

The public should be clearly informed of the synergistic effects of smoking and radon exposure and the fact that the total benefit of radon reduction depends on the number of people affected and the total duration of the exposure reduction. Although some health physicists accept short-term, screening measurements for indicating a potential need for mitigation, decisions on mitigation should be based on measurements that are representative of normal, long-term occupancy.

- Base priorities on likely exposure avoidance!

The EPA should reconsider its recommendation to use a single action level, e.g., 4 pCi/L, for all situations. Emphasis should be placed on (1) radon resistant design for new construction, (2) identification and mitigation of homes and schools with radon concentrations in the tens of pCi/L, and (3) assistance to the real estate industry and mortgage lenders in developing programs for radon testing and abatement at the time of property transfer.

Emphasis should be placed on the number of people affected and the lifetime exposure avoided, i.e., person-years-pCi/L avoided, rather than on any arbitrary starting or ending radon concentrations. Since it is the reduction in lifetime exposure that determines the benefit, it can be calculated only from before and after measurements that truly represent long-term, average radon concentrations in actual living areas and conditions.

Long-term reduction of exposure to the population could be achieved in a more cost-effective manner by treating indoor radon like many other geological risks faced by home owners. Prospective buyers have a right to inspect a house and detect any potential hazards. For existing houses, a good time to conduct a thorough evaluation of geological risks, including radon, is at the time of purchase. Mitigation costs could be guaranteed by buyer's insurance or by escrow accounts, based on long-term measurements in living areas after the transfer of the property. Very high risk situations should be corrected as soon as possible. Situations involving low to moderate risks should not be the subject of regulation other than to assure disclosure and an opportunity for buyer and seller to negotiate knowledgeably.

Reprinted, with permission,
from the *Health Physics Society's Newsletter*,
Volume XVIII, number 10, October 1990

Radon: A Way To Make A Little Money

Jane M. Orient, M.D.

Please see biographical sketch for Dr. Orient on page 186.

What should people do if they live in a house that has a high radon level?

One possibility is to charge admission. Another is to sue somebody for damages.

People flock to famous European spas, notably Bath in England and Badgastein in Austria, where the level of natural radioactivity is high. And many have even paid money to sit in abandoned mines, taking in the salubrious "rays"—of radon and its daughters.

Your house might not have the ambiance of an exclusive spa or an abandoned mine shaft. But the dose within might be higher than that now permitted for uranium miners by Environmental Protection Agency (EPA) standards.

The maximum level allowed for miners is four picocuries per liter. The mean concentration in American homes ranges from 0.45 picocuries per liter in San Francisco to 7.6 picocuries per liter in eastern Pennsylvania. An occasional home has a level exceeding 50 or even 100 picocuries per liter. One solar home in Maine had a concentration of 200 picocuries per liter.

What is the effect of these rays, long prized by health-seekers?

Some authorities estimate that radon causes about 10,000 lung cancers per year in the United States. The evidence comes from studies of uranium miners, who were previously exposed to much higher levels of radon than today's, possibly along with other carcinogens, for example, arsenic. Because of these studies, mines are now well ventilated.

Theoretically, the risk of getting lung cancer should be proportional to the amount of radon inhaled: doubling the dose should double the cancer rate. Although this "linear hypothesis" is widely used to predict cancer incidence, it might not be true at low doses. A study of counties with relatively high average radon levels showed a lower than average incidence of lung cancer (*Health Physics*, May 1987).

Some people (besides those who have mine shafts for rent) actually believe that low dose radiation may be good for you. (Everybody agrees that high doses do increase the risk of cancer.) There is a lot of

627

evidence for this "hormesis" effect; it is collected in a book called *Hormesis with Ionizing Radiation* by T.D. Luckey, CRC Press, 1980. To give just one other example, workers who inhaled plutonium at Los Alamos, Rocky Flats, and Mound Laboratory have had a death rate only 70 percent as high as the normal population. Their incidence of cancer, especially lung cancer, is also unusually low.

But wait, with radon aren't we talking about high dose radiation? "Unidentified experts" consulting with plaintiff's attorneys in Tucson have stated that homes in a radon-plagued neighborhood are exposed to more radiation than most victims of a nuclear plant meltdown would receive. This is quite true. To be precise, a house that is considered safe by EPA standards gives its inhabitants, over their lifetimes, about eight times as much radiation exposure as a hypothetical power plant accident that is supposed to require evacuation. Antinuclear activists rant about the second potential exposure, but until recently almost everyone had yawned over the first.

If you own a private, radon-rich environment, you could advertise the potential hormesis effect and charge by the hour. But all things considered, it might not make financial sense in the long run. It would tend to undermine your lawsuit, and we all know where the money is these days.

Reprinted from *The Phoenix Gazette*
Used with permission. Permission does not imply endorsement.

Recycling

**The Growing Abundance of Natural Resources
and the Wastefulness of Recycling**
George G. Reisman

Dumping: Less Wasteful than Recycling
Clark Wiseman

The Growing Abundance of Natural Resources and the Wastefulness of Recycling

George G. Reisman, Ph.D.

Dr. George G. Reisman has been a professor of Economics at Pepperdine Universitys' School of Business and Management, Los Angeles, California since 1986, and was an associate professor 1979–1986. He was an associate professor of Economics at St. John's University, Jamaica, New York, 1966–1979, and was an assistant professor 1964–1966.

Dr. Reisman received an A.B. (cum laude) from Columbia College in 1957; an M.B.A. from NYU-GBA in 1959; and a Ph.D. from New York University, Graduate School of Business Administration in 1963. His mentor was Ludwig Von Mises.

He is the author of *The Government Against the Economy,* published by Jameson Books in 1979; the translator of Von Mises' *Epistemological Problems of Economics,* published by Van Nostrand in 1960; and the translator of Heinrich Rickert's *Science and History,* published by Van Nostrand in 1962. Currently, he is completing a comprehensive defense of the free market titled *Capitalism: A Treatise on Economics.*

The picture I have painted of a free economy is one of continuous progress and improvement. And so it has been in the United States over the last two hundred years, during most of which time we had a substantially free economy. As the free economy has come to be steadily undermined and the transition to a form of socialism drawn even closer, however, the foundations of economic progress have been eroded. For reasons that should become progressively clearer from now on, a controlled or socialist economy cannot have economic progress. I believe that the advocates of socialism know this, or at least that they sense it, and that, as a result, they have launched a widespread campaign to try to deny the very possibility of continuous economic progress. The nature of their attempt is summed up in the phrase "The Limits to Growth." The motivation of the supporters of that phrase, I believe, is to be able to blame the end of economic progress not on the end of capitalism, but on the fundamental nature of the world.

Therefore, let us consider the basic facts that underlie the possibility of continuous economic progress.

As far as man himself is concerned, the basic fact is that knowledge can be transmitted from generation to generation and that each generation has the ability to add to the total of what it has received. The only limit to this process would be the attainment of omniscience.

Let us consider the physical world in which man lives. Is there a limit to the supply of natural resources on earth? [1]

Yes, there is. But the limit is utterly irrelevant to human action. For practical purposes it is infinite, because the limit is *the entire mass of the earth*. The entire earth, from the uppermost limits of its atmosphere to its very center, four thousand miles down, consists exclusively of natural resources, of *solidly packed natural resources*. For what is the earth made out of? It is made exclusively out of chemical elements found in different combinations and in different proportions in different places. For example, the earth's core is composed mainly of iron and nickel—millions of cubic miles of iron and nickel. Aluminum is found practically everywhere. Even the soil of the Sahara desert is composed of nothing but various compounds of silicon, carbon, oxygen, hydrogen, aluminum, iron, and so on, all of them having who knows what potential uses that science may someday unlock. Nor is there a single element that does not exist in the earth in millions of times larger quantities than has ever been mined. [The amount of energy supplied by nature is equally immense. In a single thunderstorm nature releases more energy than is generated by mankind in an entire year. And heat from the sun every year provides a constantly renewed supply of energy that is millions of times greater than the energy consumed by man.] . . .

Now because the world is composed entirely of natural resources and possesses a virtually irreducible and practically infinite supply of energy, the problem of natural resources is simply one of being able to obtain *access* to them, of being able *to obtain command over the resources*, that is, of being in a position to direct them to the service of human well-being. This is strictly a problem of science, technology, and the productivity of labor. Its solution depends merely on learning how to break down and then put together various chemical compounds in ways that are useful to man, and having the equipment available to do it without requiring an inordinate amount of labor. Human intelligence certainly has the potential for discovering all the knowledge that is required, and in a free, rational society, the incentive of profit virtually guarantees that this knowledge will both be discovered and provided with the necessary equipment to be put to use.

The record of the last centuries, certainly, demonstrates that such a society has no problem of a scarcity of accessible natural resources. While the total volume of chemical elements in the world has remained the same, the volume of *useful* elements and compounds *at the disposal of man* has been enormously *increased.* Today, for example, because of improved knowledge and equipment, it is probable that man can more easily extract minerals from a depth of a thousand feet than he could a century ago from a depth of fifty feet. In the same way, he has learned how to use elements and compounds he previously did not know how to use—such as aluminum and petroleum, which have only been in use for approximately a century, and, more recently, uranium. There is no reason why, under the continued existence of a free and rational society, the supply of accessible natural resources should not go on growing as rapidly as in the past or even more rapidly. Further advances in mining technology, for example, that would make it possible to mine economically at a depth of, say, ten thousand feet, instead of the present limited depths, would so increase the portion of the earth's mass accessible to man, that all previous supplies of accessible minerals would appear insignificant by comparison. And even at ten thousand feet, man would still, quite literally, just be scratching the surface, because the radius of the earth extends to a depth of four thousand *miles.* In the same way, dramatic advances are possible in the field of energy, such as may occur through the use of atomic energy, hydrogen fusion, solar power, tidal power, or thermal power from the earth's core, or still other processes as yet unknown.

Because the earth is literally nothing but an immense solid ball of useful elements and because man's intelligence and initiative in the last two centuries were relatively free to operate and had the incentive to operate, it should not be surprising that the supply of accessible minerals today vastly exceeds the supply that man is economically capable of exploiting. In virtually every case, there are vast *known* deposits of minerals which are not worked, because it is not necessary to work them. Indeed, if they were worked, there would be a relative overproduction of minerals and a relative underproduction of other goods—i.e., a waste of capital and labor. In virtually every case, it is necessary to choose *which* deposits to exploit—namely, those which by virtue of their location, amount of digging required, the degree of concentration and purity of the ore, and so forth, can be exploited at the lowest costs. Today, enormous mineral deposits lie untouched which could be exploited with far less labor per unit of output than was true of the very best deposits exploited perhaps as recently as a generation or two ago—thanks to advances in the state of mining technology and in the quantity and quality of mining equipment available.

As just one example, and a very important one, consider the fact that there are petroleum deposits in shale rock and tar sands in our own Rocky Mountain states and in Canada of a size far exceeding the petroleum deposits of the Arab countries. Until now, these deposits have not been exploited, because it has been cheaper to obtain petroleum from liquid deposits. Even though oil obtained in these ways would be more expensive than oil obtained in its liquid state, still, it is undoubtedly cheaper—in terms of the labor required to produce it—to obtain oil in these ways today than it was to obtain liquid petroleum a century ago and probably even a generation or two ago. There is no reason why further advances in mining technology and in the availability of mining equipment would not enable oil obtained in these ways in the future to be less expensive than oil obtained in its liquid state today. Similarly, there are vast untapped known coal fields in the United States containing enough coal to supply present rates of consumption for many centuries.

In some important respects, these coal fields must be considered not merely a substitute, but the full equivalent of petroleum deposits. For it is possible to produce some of the identical products from coal as from oil—for example, gasoline. This too has not been done commercially until now, because it has been cheaper to produce gasoline from petroleum. But there is no reason why, with the further progress of technology and the availability of equipment, gasoline produced from coal in the future should not be cheaper than gasoline produced from oil today, just as gasoline produced from coal today would undoubtedly be cheaper than was gasoline produced from oil in the past. If it were necessary, a free American economy could respond to a loss of foreign supplies by turning to such other sources of oil and gasoline as these, and, in not very much time, both through reducing their costs of production and by developing other, newer sources of fuel, would enjoy lower costs and more abundant supplies of energy than ever before. In a free American economy, it would not matter in the long run if the Arabian peninsula and its oil simply did not exist. As a free economy, we would not need Arab oil. Neither our survival nor our long-run progressive prosperity would depend on it.

The growing threat to the supply of natural resources that people are beginning to complain about is not the result of anything physical—no more than it was the result of anything physical in the days when these terrible words of despair were written:

"You must know that the world has grown old, and does not remain in its former vigour. It bears witness to its own decline. The rainfall and the sun's warmth are both diminishing; the metals are nearly exhausted; the husbandman is failing in the fields, the sailor on the seas, the soldier in the camp, honesty in the market, justice in the courts, concord in friendships, skill in the arts, discipline in morals. This is the sentence passed upon the world, that everything which has a beginning should perish, that things which have reached maturity should grow old, the strong weak, the great small, and that after weakness and shrinkage should come dissolution." [2]

That passage is not a quotation from some contemporary ecologist or conservationist. It was written in the *third century*—ages before the first chunk of coal, drop of oil, ounce of aluminum, or any significant quantity of any mineral whatever had been taken from the earth. Then as now, the problem was not physical, but philosophical and political. Then as now, men were turning away from reason and toward mysticism. Then as now, they were growing less free and falling ever more under the rule of physical force. That is why they believed, and that is why people in our culture are beginning to believe, that man is helpless before physical nature. There is no helplessness in fact. To men who use reason and are free to act, nature gives more and more. To those who turn away from reason or are not free, it gives less and less. Nothing more is involved.

There are no significant scarcities of accessible raw materials as yet. But the enemies of reason and capitalism sense the consequences of the social system that they hope to impose, and they project them on to the present. Thus they admonish us to save every little tin can and every scrap of paper. Their world, if it ever comes, will have to live like that. But we, who are capable of producing in abundance—we do not have to regard bits of garbage as priceless treasures. To us, used tin cans, paper wrappings, and the like, which cost us hardly any labor to produce or to replace, are generally not worth the trouble of saving or reusing. In fact, it is usually wasteful for us to do so: it wastes *our labor and our time,* which are the only things in life we should be concerned about not wasting. For if we can produce new tin cans easily, by scooping iron ore out of the earth in ten or twenty-ton loads, it is simply ludicrous to take the trouble to gather up each little tin can and carry it off to some recycling center, because in doing so we spend far more labor than we save.

Nor is it "wasteful" or uneconomic in any way that we use so many tin cans or so many paper wrappings. If we consider how little labor it costs us—in terms of the time it takes us to earn the money we spend for it—to have things brought to us clean and fresh and new, in new containers and new packaging, and what the alternatives are for the spending of that money or the use of that time, it becomes clear that the expenditure is well made. For consider the alternatives: We could have our food and other goods wrapped in old newspapers and put in jars, bags, or boxes that we would have to carry along with us whenever we went shopping, or which we would have to make a special trip to go and fetch whenever we came on something unexpectedly that we wanted to buy. We could then use the money we saved in that way to buy a handful of other goods. Conceivably, we could use the money we saved to work a few minutes less at our jobs each day, and earn correspondingly less. But these alternatives would simply be bizarre, because neither a handful of extra goods nor working a few minutes less at our jobs each day would compensate us for the loss of cleanliness, convenience, aesthetic satisfaction, and also time saved in shopping that is provided by modern packaging.

Let the ecologists adopt the poverty-stricken life-style of Eastern Europe if they choose. Let them go about like old Russian grandmothers in Moscow, with an ever-present shopping bag and herring jar, if that is what they like. Let them pick through garbage pails while pretending that they live in a spaceship—"spaceship Earth," they call it— rather than in the richest country of the planet earth. But there is absolutely no sane reason why anyone should or needs to live this way, and certainly not in modern America. Above all, let them keep their peculiar values to themselves and not seek to impose them on the rest of us by the enactment of laws.

Footnotes

[1] I limit the discussion to the resources available on earth. Actually, advances in space technology are making it clear that this restriction is far too narrow.

[2] The quotation appears in W.T. Jones, *The Medieval Mind*, Volume II of *A History of Western Philosophy*, Second Edition (New York: Harcourt, Brace, and World, 1969), p. 6.

Dumping: Less Wasteful than Recycling

Clark Wiseman, Ph.D.

Dr. Clark Wiseman is an associate professor of economics at Gonzaga University and former visiting fellow at Resources for the Future.

He received an M.A. in economics from Washington State University in 1964, and a Ph.D. in economics from the University of Washington in 1967.

Dr. Wiseman has published more than 21 journal articles and has contributed to *Review of Economics and Statistics, American Journal of Agricultural Economics,* and *Public Choice.*

The proposal by the hard-pressed government of New York City to suspend its recycling program for a year is a direct result of the high cost of recycling. At around $300 per ton, the cost has proven to be well in excess of the $65 per ton figure that was originally estimated. True, the program has been plagued by labor problems and a low level of citizen participation, but it is wishful thinking to believe that either more cooperation from sanitation unions or the achievement of greater civic support and a higher recycling rate will bring the costs of recycling down to an acceptable level.

Curbside recycling programs across the U.S. typically cost far more than landfilling, frequently twice as much, even when sales revues and avoided waste disposal costs are included in the calculation. On a strictly economic basis, large-scale recycling is simply wasteful, leaving taxpayers and users of solid waste disposal services paying a larger bill. The frenzied national push for recycling is largely the result of grossly mistaken beliefs about landfilling and the magnitude of the disposal problem, together with a seriously flawed decision-making process in the siting of landfills.

What most people don't know about landfills could fill a landfill. At the current rate, if all the nation's solid waste for the next 500 years were piled or buried in a single landfill to a depth of 100 yards—about half the eventual height of Staten Island's Fresh Kills landfill—this "national landfill" would require a square site less than 20 miles on a side. With compaction, even this volume could be halved.

Most people also don't know that the amount of solid waste generated nationally has grown at only a 2% average rate over the past 30 years, considerably less than the growth of GNP. This means that our "throw away society" is actually throwing out a progressively smaller share of its output. There are indications that this rate of growth is declining as the economy becomes more service-oriented.

The view is widely held that landfilling should be minimized because of the great environmental risks. But landfills are constantly becoming less obnoxious. New federal and state performance standards are comprehensive and stringent, with environmental considerations entering into all relevant aspects of landfill construction and operation, including location; fencing; groundwater and gas monitoring and control; frequency of earth covering for rodent, bird, and odor control; closure; and post-closure gas and groundwater monitoring. Many landfills designed and operated with this degree of environmental control already exist; some have already filled and closed, and the land has been converted to other (often recreational) uses.

If our landfills are to be environmental Cadillacs, the issue then becomes one of sticker price. As might be expected, this will vary according to differences in land prices. A new landfill can cost up to five times as much as a standard 1975 landfill. Even so, landfill costs account for only about 25 cents of the cost of disposing of the garbage in a standard 32-gallon can.

The remainder of what one pays is the relatively high cost of collection, hauling and perhaps hidden and explicit taxes. Even where land is expensive it is seldom more than a small fraction of the landfilling portion of waste disposal charges. Even with the sky-high land prices and the long hauls that are necessary in most metropolitan area, landfilling is a bargain.

The solid waste problem is not only of space, ecology or even cost. The problem is a political one—that of siting new landfills. Anticipating the loss of amenities or property values, potentially affected property owners unite into a group capable of bending government to its will. The special interest nature of the resulting policies is not different in nature from farm subsidies, protective tariffs and unnecessary military installations, all of which confer losses upon citizens at large.

The landfill siting problem is directly related to population densities. In some of the more sparsely populated areas of the Western states that are virtually no siting difficulties. By contrast, in the East, permitting new landfills is political suicide.

Fortunately, a decision-making procedure is available that helps the creation of new landfills, while still preserving control over the environmental consequences of landfills. The state of Wisconsin has since 1982 legally required municipal and county governments to establish local negotiating committees in response to applications for the creation of a landfill by a private landowner. The committees, which must include a prescribed number of private citizens as well as elected officials, are empowered to negotiate the financial and other contractual relations between the landfill owner and local governments. Environmental and technical matters are not negotiable but are handled by a separate process at the state level. Although—or perhaps because—failure to reach an agreement can result in outside mediation and possibly arbitration by a state agency, agreements have been negotiated by committee in almost all cases.

The workability of a system along these lines results from the explicit recognition of a prescribed set of rules. Although such rules constrain their powers, local elected officials do not complain, since their longevity in office can only be enhanced by the inability to make "unpopular" decisions.

The choking off of a viable alternative like low cost and environmentally sound landfills is wasteful of society's resources. Before continuing to run headlong toward politically popular but costlier alternatives—including recycling—it would be wise to give increased attention to the real cause of the so-called solid waste "crisis."

Reprinted, with permission, from
The Wall Street Journal, July 22, 1991

Reverse Effects

"Toxins" Can Act as Nutrients;
"Nutrients" Can Act as Toxins
Walter A. Heiby

Two Theses in Radiobiology
T.D. Luckey

A Critique of the Statistical Methodology
in Radiation and Human Health
Jane M. Orient

"Toxins" Can Act as Nutrients; "Nutrients" Can Act as Toxins[1]

Walter A. Heiby

Walter A. Heiby graduated from Wright Jr. College in Chicago where he majored in mathematics, physics and chemistry. Heiby then worked days in design engineering while he "moon-lighted" as a math instructor. Heiby developed an interest in philosophy and his book *Live Your Life* was published in the United States by Harper and Row and in Germany by Vulkan Verlag. This book is widely quoted and has been used as a college text. Subsequently, he added geology and oil exploration to his areas of interest. As a consequence, he helped find and develop several midcontinent fields.

About 20 years ago, Heiby became interested in nutrition and years ago discovered, in hundreds of scientific and clinical studies, the prevalence of a phenomenon which he named "the reverse effect." His current book, a 1216-page volume, is entitled *The Reverse Effect: How Vitamins and Minerals Promote Health and CAUSE Disease.* The 4821 references of *The Reverse Effect* make it the most highly documented, single-author book ever published.

Currently, Heiby teaches literature research seminars in nutrition, medicine, and dentistry at the University of Illinois.

The ex-body environment can impact the body's internal environment with often-unexpected reverse effects. There is a good probability that the activity of any entity (toxins, drugs, vitamins, minerals, X rays, exercise, etc.) that is health promoting or health inhibiting at a given concentration may reverse its action and become respectively health inhibiting or health promoting at a different concentration. I call this reversal from health promotion to health inhibition and vice versa the "reverse effect."

Suppose a set of experimental data involving the varying concentrations of a test entity (shown on the X-axis) and the values of a dependent physiological variable (shown on the Y-axis) are properly graphed as shown in Figure 1. Often a scientist will improperly extrapolate and will presume that the function would graph as shown in Figure 2. Unknown to the scientist, reality might involve a reverse effect at a lower concentration of the test entity as indicated in Figure 3. Or a reverse effect might exist at a higher concentration as shown in Figure 4. On the other hand, reverse effects might be present at both higher

Figure 1.

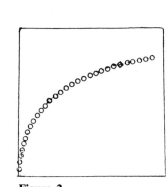

Figure 2.

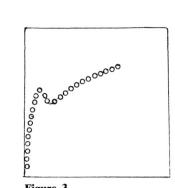

Figure 3.

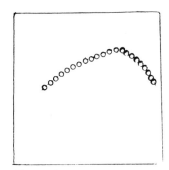

Figure 4.

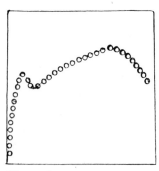

Figure 5.

Figures 1–5 are hypothetical plots of experimental data involving the varying concentrations of a test entity shown on the x-axis (vertical) and the values of a dependent physiological variable shown on the y-axis (horizontal).

and lower concentrations as shown in Figure 5. If the scientist is unaware of one or more reverse effects in the functional relationship he is studying, he could easily draw false conclusions.

I'd like to start illustrating reverse effects with some examples that involve radiation. When Roy E. Albert et al.[1] exposed rats to beta rays they found that small dosages of 230 to 1,900 rads produced a few tumors. When the Albert group employed higher concentrations of beta rays, the tumor incidence increased. As shown in Figure 6, the tumor count maximized at about 4,000 rads. Then, a reverse effect set in with fewer tumors being produced at 5,000 rads and fewer tumors still at 10,000 rads.

Another example of the reverse effect occurred in data relating to the incidence of death that followed the atomic bomb explosions ending World War II. The *Encyclopaedia Britannica*[2] states, "A sample of nearly 100,000 survivors of Hiroshima and Nagasaki yielded the anomalous result that groups exposed to doses between 11 and 120 rads actually had a lower death rate in the ensuing 15 years than those receiving a lesser dose." (The result is anomalous only in the sense that there is little appreciation for how common the reverse effect really is.) A more recent analysis of these events by G. W. Beebe et al.[3] showed a death rate during the 25-year span from 1950–1974 of 23.3% for persons who received no radiation, then rising death rates up to 25.5% at a dose level of 99 rads. At a level of 100–199 rads a reverse effect set in and death rates declined to 23.8%, declined further to 22.6% at 200–299 rads, and further yet to 21.6% at 300–399 rads. Then a second reverse effect occurred and at dosages of about 400 rads the death rate was highest at 28.4% (as expected). It is interesting to note that the death rates for those persons exposed to levels between 200 and 399 rads were lower than controls that received no radiation.

A book by T.D. Luckey[4] entitled *Hormesis With Ionizing Radiation* illustrates the work of various investigators who found that when wheat seeds and also strawberry plantlets were exposed to X rays their yields increased up to a dosage of 1,000 to 10,000 rads and then showed a reverse effect by declining as the dosage was increased further. This phenomenon is shown in Figure 7. (Those concerned with increasing of crop yields and/or with problems of world hunger may find this especially thought provoking.)

Many examples of the reverse effect can be found in the physiological actions of vitamins. At extremely low dosages, vitamin C can act detrimentally as a pro-oxidant (i.e., it may help foster harmful reactions with oxygen).[5–17] Then, at higher dosages vitamin C shows a reverse effect and acts beneficially as an antioxidant.[7,8] This is the well known favorable effect of vitamin C. However, the work of Linda Chen[18] has

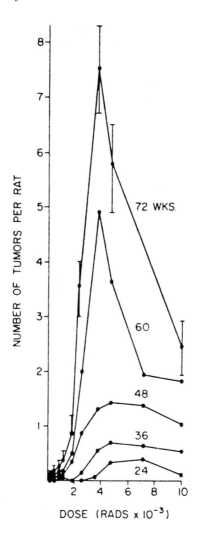

Figure 6. Effect of beta rays on the number of tumors in rats. Reproduced from Roy E. Albert et al., *Radiation Research* 15: 410–430 (1961) with permission of Academic Press. Note the "reverse effect" at about 4,000 rads.

shown that at still higher dosages vitamin C may at times again show a reverse effect and act detrimentally as a pro-oxidant.

The action of vitamin C on cancer cells is especially interesting. The research of N. Bishun et al.[19,20] showed that vitamin C killed two types of cancer cells but then, at higher dosages, there was a reduction in this killing power. This reverse effect (as I call it) is shown in Figure 8.

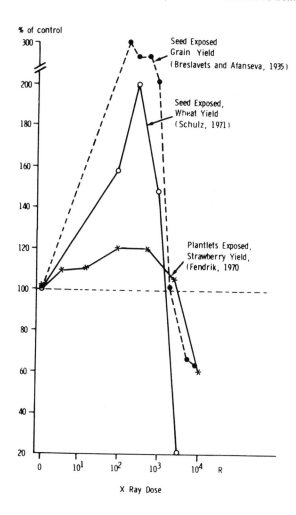

Figure 7. Effect of X rays on yields of grain and of strawberry plantlets. Reproduced from T.D. Luckey, *Hormosis With Ionizing Radiation* (Boca Raton, Florida: CRC Press, 1980) with permission of CRC Press. Note the "reverse effect" at about 100 rads.

It has been known for many years that retinyl acetate (a vitamin A compound) is very effective in preventing breast cancer induced in animals by carcinogens. Clifford W. Welsh et al.,[21] on the other hand, found that massive administration of retinyl acetate to estrone- and progesterone-treated mice showed a reverse effect by resulting in a substantial increase in mammary carcinomas. The concentration of retinyl acetate the Welsh group used was so high, however, that the study should not cause concern among women using modest vitamin A supplementation.

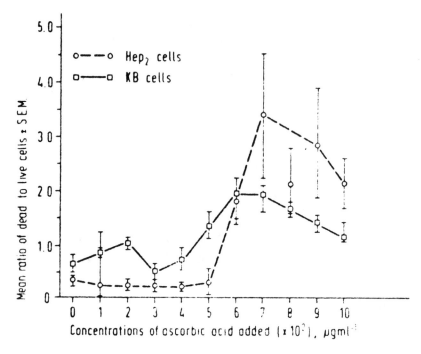

Figure 8. Cancer cells being killed by vitamin C. Note the "reverse effect" at high dosages. Reproduced from N. Bishun et al., *Oncology* (1978) 35: 160–162 with permission of S. Karger AG, Basel, Switzerland.

These examples of reverse effects suggest that we avoid rapid conclusions regarding the benefits of "nutrients" and the threats of "toxins." M. Alice Ottoboni[22] says:

> The phenomenon of beneficial effects from trace exposures to foreign chemicals, although often a subject of conversation among toxicologists, particularly with regard to why such effects occur, is rarely mentioned in the scientific literature. If the phenomenon does occur in a chronic toxicity experiment, the text of the paper reporting the results will seldom mention the fact. It is only by careful perusal of the data tables and figures presented in the body of the text that the phenomenon is revealed. Such subtleties are lost on people who read only the abstracts of scientific papers. Unfortunately, there are some scientists who may be counted among the abstract-only readers.

The doctrine of "sufficient challenge" of H. F. Smyth, Jr.[23] is relevant to our discussion. Muscles, brains and sex organs are made stronger through challenges and perhaps other organs may also be benefitted. Smyth reported that rats inhaling small amounts of the poison carbon tetrachloride grew better and were more fertile than controls.

Ottoboni[24] quotes Smith as saying: "I think that most of the small non-specific responses which we measure in chronic toxicity studies at low dosages are readjustments or adaptations to sufficient challenge. I interpret them as manifestations of the well-being of our animals, healthy enough to maintain homeostasis. They are beneficial in that they exercise a function of the animal. Only when challenge becomes overwhelming does injury result." The doctrine of "sufficient challenge" obviously casts doubt on the attitude expressed by the Delaney Clause. The Delaney Clause is based on the idea that if a substance is carcinogenic in large dosages it may also be dangerous in small dosages and it should therefore not be permitted as a food additive. *The Federal Register*[25] perpetuates the line of reasoning behind the Delaney Clause by saying, "The exposure of experimental animals to toxic agents in high dosages is a necessary and valid method of discovering possible carcinogenic hazards in man." Those favoring the Delaney Clause seem to ignore the possibility that such "carcinogens" might, at certain dosages, constitute a "sufficient challenge" and offer protection against cancer. [2]

There is much research suggesting that large amounts of vitamins and minerals may at times be harmful. Conversely, I speculate that small quantities of toxins may beneficially stimulate the liver to be more efficient just as muscles become stronger when they are occasionally exercised. (The rationale for this proposed action may involve stimulation of the liver's production of microenzyme cytochrome P-450.)[32-34] The advice to eat a well-balanced diet may be good information not only in order to achieve a broad spectrum of needed nutrients but because it introduces the body to a well-balanced but minor intake of toxins, at least some of which in small amounts may act as nutrients. Nutrition and pharmacology must become sciences of dose-response relationships in fact as well as in theory.

Thus, I am suggesting that we do not always know which entities are good and which are bad for the body. **That which is good may be bad; that which is bad may be good.** It depends on the circumstances and one of the circumstances is the quantity of the presumed nutrient or presumed toxin that is involved. W. W. Duke,[35] three-quarters of a century ago, tested the effects on blood platelets of many different agents such as toxins, bacteria, chemical poisons and X rays. He observed that agents (diphtheria toxin and benzol) causing rapid and large rises in the platelet count were the ones producing the most rapid and extreme falls in the count. It was an early recognition of the presence of a reverse effect.

Several decades later, Lawrence P. Garrod[36] gave examples of bacterial growth being stimulated when small dosages of chemotherapeutic agents were used. Among those examples he cited was a study of G. E. Foley and W. D. Winter[37] which showed that penicillin increased the mortality of chick embryos inoculated with *Candida albicans*. Garrod noted that "superinfections" sometimes occur clinically during penicillin treatment. On the other hand, T. D. Luckey[38] has reported that antibiotics such as sulfa drugs, succinylsulfathiazole, streptomycin and 3-nitro-4-hydroxyphenyl-arsonic acid can be used in very small quantities to support life and to promote growth in animals. [3]

About a quarter-century ago, N.V. Medunitsyn[43] found that various painful stimuli given rabbits had an influence on immunological reactivity. A weak stimulus increased phagocytic activity of the leucocytes and increased the antibody titer in the blood while strong painful stimuli showed a reverse effect and suppressed these processes. Alcohol in moderation can add to the joy of life and may possibly bring improved health and greater longevity. In excess, alcohol can produce liver cirrhosis and death. A high dose of not only alcohol, but also of sodium pentobarbital, acts as a behavioral depressant, while a low dose of either drug produces behavioral activation.[44] Small amounts of vitamin D are good—they help put calcium in the bones. Large amounts are bad—they pull calcium out of the bones and dump it in the soft tissues, including those of the kidney and of the joints.

Sometimes antivitamins can fulfill some of the functions of the vitamins they ordinarily oppose. Pantoyl taurine can, according to Robert E. Hodges et al.,[45] both cause a pantothenic acid deficiency or reverse that effect and mimic some of the properties of the vitamin. Hodges also reports that the antagonist omega methyl pantothenic acid may possibly act as a substitute for pantothenic acid in aiding production of antibodies, thus reversing its generally antagonistic nature. The two-faced nutritional and antinutritional properties of vitamins and their antagonists make the science of nutrition far more subtle than it was once thought to be.

Reverse effects are very common, albeit often unrecognized. Prolactin stimulates milk secretion in lactating women. Prolactin appears also to facilitate processes associated with sperm capacitation (the process, occurring in the vagina, by which sperm become capable of fertilizing an ovum).[46,47] A deficiency of prolactin may be associated with benign prostatic hypertrophy and nominal amounts seem to be required for prostatic health.[48] Rubin et al.[49] reported that prolactin seems to be capable of increasing plasma testosterone in adult men. Furthermore, according to S. P. Ghosh et al.,[50] rat experiments suggest that prolactin acts with testosterone to regulate acid phosphatase activity in the male accessory sex organs. [4] On the other hand, excessive prolactin, hyperprolactinemia, shows a reverse effect compared with normal amounts of prolactin and (like a deficiency of prolactin) is often associated with impotence, hypogonadism, decreased semen volume and reduced spermatic density in men and lessened orgasmic capacity in women.[54–57] Thus, too little or too much prolactin may lead to dire sexual effects.

John Bancroft[58] reported that L. Lidberg[59,60] and, later, Lidberg and Sternthal[61] found that oxytocin may enhance sexual responsiveness in men. Bancroft noted that two of the three studies showed the effect was greater at lower dosages and he refers to the results as "these surprising findings." It is another example of the reverse effect.

Pharmacists preparing anticancer drugs may become victims themselves according to a study at the M. D. Anderson Hospital, Houston, Texas.[62,63] The urine of nine pharmacists preparing such drugs became highly mutagenic showing that somehow they absorbed substances that mutate cells, leading possibly to cancer and thus exemplifying the reverse effect. A number of other studies[64–67] have reported that urine of nurses and other hospital personnel handling cytotoxic drugs showed positive mutagenicity. Smoking may potentiate the hazard.[66] Marja Sorsa et al.[67] published additional evidence of the hazards of occupational exposure to anticancer drugs. Many others have commented on the problem.[69–72] [5] Then too, a reverse effect of the cytotoxic drugs may be shown in the cancer patients themselves after they have been "cured." Robert Hoover and Joseph F. Fraumeni, Jr.[75] observe that various alkylating agents (including not only cyclophosphamide but also melphalan and chlorambucil) used in treating malignant neoplasms can all cause cancer (perhaps, in part, because they break chromosomes). *Lancet*[71] comments, "In patients who have been cured of malignant disease, clinicians gloomily accept that further neoplasia may arise as a late effect of treatment." Would this occur if the therapies took full cognizance of dose-related reverse effects?

Small amounts of an allergen may stimulate Immunoglobulin E (IgE) antibody production, whereas larger amounts of that allergen may suppress such production.[76] E. R. Stiehm et al.[77] reported that the alpha, beta and gamma families of interferons, which are produced when mammalian cells are stimulated in the inflammatory process, can augment the immune function or can show a reverse effect (my language) by suppressing immunity. Interferon has the ability to inhibit growth of some malignant cells. On the other hand, a study by Gene P. Siegal, et al.[78] showed that at least three types of interferon can increase the ability of Ewing sarcoma cells to invade healthy tissue, which also exemplifies the theory of the reverse effect. And so it is in the case of many other drugs, vitamins and minerals.

Foods can also display reverse effects. Frederick Hoelzel and Esther DaCosta,[79] in reporting that a protein-deficient diet caused ulcers to form in the stomach and duodenum of mice, observed that an excess of protein also tended to produce gastric lesions. Then too, water sustains life and also displays the reverse effect of causing death by drowning. [6]

The fact that many drugs have side effects corresponding with the same condition for which they are curative provides many examples of the reverse effect. This phenomenon also serves as a warning that drugs must be carefully studied for dose responses and possible reverse effects that can be exacerbative rather than curative. If you study several drugs in the *Physicians' Desk Reference*[81] you will probably discover that, for at least a few of them, the list of side effects contains a symptom for which the drug is sometimes curative. The diazepam Valium® is one such drug. [7] Among the indications for its use is included "management of anxiety disorders or for the short-term relief of the symptoms of anxiety." Among its adverse reactions are included "hyperexcited states" and "anxiety."[83] The barbiturate, Nembutal® Sodium is used as a "sedative hypnotic." Among its side effects are "agitation," "nightmares," "nervousness," and "anxiety."[84] Quinidine sulphate (Quinora®) is used for "paroxysmal atrial tachycardia" and "paroxysmal atrial fibrillation" and "paroxysmal ventricular tachycardia when not associated with complete heartblock." Among its adverse effects are "ventricular tachycardia and fibrillation" and "paradoxical tachycardia."[85] The rauwolfia product, Harmonyl® is employed for its "antihypertensive effects" and it also displays "sedative and tranquilizing properties." Among its adverse effects are "nervousness," "paradoxical anxiety" and "nightmares."[86] Asellacrin®, a growth hormone (somatropin or somatotrophin) which is used in treating short children in the attempt to increase their height, may sometimes be subject to a reverse effect. The *Physicians' Desk Reference* relates that "bone age must be monitored annually during Asellacrin® administration, especially in patients who are pubertal and/or receiving concomitant thyroid replacement therapy. Under these circumstances, epiphyseal maturation may progress rapidly to closure."[87] In other words, sometimes there is a reverse effect and, instead of the child growing taller, his epiphyses close and he completely stops growing.

In each of these drugs the normal action shows a reverse effect at some concentrations in some persons. At a given dosage a drug may cause in one person the same condition which the same dosage cures in another. A different dosage will often show a reverse effect and be curative to the first but detrimental to the second. The side effects of many of the drugs listed in the *Physicians' Desk Reference* give testimony to the reality of the reverse effect. Medicine, like nutrition, must study dose-response relationships and, by administering individually determined dosages in therapeutic applications, strive to eliminate iatrogenic reverse effects. [8] (The research-minded reader will discover other reverse effects in the *Physicians' Desk Reference* and may also desire to search for additional examples in *AMA Drug Evaluations*[88] or in *Drug Facts and Comparisons*.[89] If your library does not have *Drug Facts and Comparisons* and its monthly supplements, your pharmacist may let you look at his.)

Let me caution against quickly concluding that small doses of a given substance may often be beneficial with large doses of that substance being dangerous. Studies show that estrogen can promote carcinogenesis. On the other hand, Joseph Meites et al.[90] have shown that a large dose of estrogen (20 mg. of estradiol benzoate) effectively inhibited growth of DMBA-induced mammary tumors in rats. Meites cites the work of C. Huggins[91] and of R.I. Dorfman[92] that also showed large doses of estrogen resulted in tumor regression in rats. In addition, Meites cites a study of J. Hayward[93] in which large doses of estrogen (or of androgen) induced remission of human breast cancer in more than 20% of treated patients. In further regard to estrogen, Hoover and Fraumeni[75] cite three studies that suggest oral contraceptives may offer some protection against ovarian cancer. Apparently, more (in the case of estrogen) may sometimes induce a beneficial reverse effect.

Large amounts of dihomo-gamma-linolenic acid (DGLA), a substance which is present in human milk and which is metabolized in the body (especially if evening primrose oil is consumed), can inhibit the growth of malignant cells *in vitro*. However, Robinson and Botha[94] found that a small amount (50 micrograms per ml) of DGLA, on the other hand, can produce a reverse effect (my language) and promote cancer. Here again, a large amount of a substance may work for our benefit while a small dosage may be dangerous. (Since evening primrose oil is widely sold, it is important that any possible reverse effect be precisely delineated.)

Physicians must increasingly come to realize that the answer to ineffectiveness of a therapy may be to increase the dose or it may be to *decrease* the dose. Similarly, scientists must learn to construct experimental protocols that make it possible to discover reverse effects. Often experiments are done with dosage increments that make it impossible to discover what, if any, phenomena might exist at intermediate levels.

Could the theory of the reverse effect have implications as a research tool, e.g., in the search for anticancer agents? Podophyllotoxin from the herb *American mandrake* (mayapple) is reported to cause cancer. It is reported to also actively inhibit carcinoma and sarcoma in animals. Perhaps it might prove rewarding to look for cancer-treating substances among factors known to be mitogenic, mutagenic or carcinogenic. Such a search might be a valuable application of the theory of the reverse effect. Mary E. Caldwell and Willis R. Brewer[95] published a list of 50 plant extracts from 42 species that have been shown to significantly enhance tumor growth. A total of 300 extracts were found to exhibit some enhancement of tumor growth. The work was well done in terms of its objectives and tests were performed at various toxic and nontoxic levels. However, I think the entire group of 300 extracts should be studied for reverse effects and possible antitumor action at some yet-to-be discovered dosages. [9]

The Ex-body Environment vs. *In vivo* Actions

Let's emphasize that it is difficult to draw conclusions regarding the impact of so-called "toxins" and of so-called "nutrients" on the internal environment of the body. It is challenging to attempt to relate *in vitro* results to possible *in vivo* actions. The *in vivo* mileau, however, is almost infinitely complex and variable compared with the relative simplicity of *in vitro* conditions. Furthermore, the *in vivo* concentration of each toxin, vitamin, mineral, drug, enzyme, hormone or other body chemical may vary from organ to organ. Vitamin C, for example, concentrates in the adrenal gland and zinc in the prostate. Thus, a certain nutrient might show antioxidant action (as vitamins C and E generally do) in most areas of the body. However, in organs concentrating that nutrient a detrimental pro-oxidant action might occur and I speculate that this could be an initiating cause of disease and of side effects. Lipid peroxidation due to vitamin C might conceivably occur even in an organ that does not concentrate this vitamin but instead accumulates iron. An *in vitro* study by David Blake et al.[98] showed that vitamin C in the presence of iron may cause lipid peroxidtion in phospholipid liposomes. Could this phenomenon occur *in vivo* in iron-rich organs? Furthermore, vitamin C can concentrate iron in the liver and spleen (of mice) while having no such effect on the heart.[99]

Future physicians will continually face these difficult therapeutic questions: Is it wise to set the dosage of a nutrient or a drug at a level that will promote beneficial actions in most areas of the body while risking a detrimental reverse effect in an organ that concentrates that nutrient or drug? How can I set the dosage level to achieve the desired action in my patient while minimizing the risk of side effects caused by a reverse effect in a concentrating organ? Much research for generations to come will be required to correlate *in vitro* reverse effects with estimates, organ by organ, of the probable *in vivo* actions of various nutrients and drugs. In the more distant future, reverse effects will be controlled by sending nutrients and drugs to specific organs, thereby reducing their entrance into organs in which reverse effects might occur. Directed delivery of nutrients and drugs will be achieved by attaching them to carrier molecules with a chemical bond that will be broken only by enzymes in the target organ, thereby avoiding reverse effects elsewhere in the body.

Dramatic new breakthroughs regarding the impact of the external on the internal environment can occur only through the process of overthrowing the well-established practice of data extrapolation. The doctrine that extrapolation can predict effects in physiology must be viewed as being obsolete. A reverse effect might be present and produce a reaction quite the opposite from what data extrapolation would suggest—perhaps even a cure from a known disease-causer. I propose that the theory of the reverse effect as a tool for developing new therapies is a revolutionary paradigm.

In general, scientists find only that for which they are searching, serendipitous discoveries being rather rare. In attempting to set an appropriate dosage for a clinical trial of retinyl acetate for use in treating histopathologic lesions of the cervix, Seymour L. Romney et al.[100] wrote: "The 3 mg dose was eliminated as part of the empiric effort to establish the maximally tolerated dosage." It seems obvious they gave no thought to the possibility that the 3 mg dose could be more effective than the 9 mg dosage they finally decided to use. (That is, no thought was given the possibility that a reversal could exist somewhere between a dose of zero and one of 9 mg.) The scientific literature contains innumerable other examples indicating that the possibility of a reverse effect was simply not considered. Toxicologists and other scientists should design their experiments using wide dosage parameters so that reverse effects, instead of remaining hidden, can be discovered. The possibility that "toxins" can act as nutrients and that "nutrients" can act as toxins must no longer be ignored.

References

1. Roy E. Albert et al, *Radiation Research* (1961) 15: 410–430.
2. *Encyclopaedia Britannica*, 15th ed., vol. 15, "Radiation, Effects of". (Chicago: Encyclopaedia Britannica, 1979) pp. 378–392 at 386.
3. G.W. Beebe et al., cited by Alice M. Stewart, *Jrl. of Epidemiology and Community Health* (1982) 36: 80–86.
4. T.D. Luckey, *Hormesis with Ionizing Radiation* (Boca Raton, Florida: CRC Press, 1980).
5. H. Abramson, *Jrl. of Biological Chemistry* (1949) 178: 178–183.
6. Gottfried Haase and W.L. Dunkley, *Jrl. of Lipid Research* (1969) 10: 561–567.
7. E.D. Wills, *Biochemical Jrl.* (1969) 113: 315–324.

8. Subal Bishayee and A. S. Balasubramanian, *Jrl. of Neurochemistry* (1971) 18: 909–920.

9. O.P. Sharma and C.R. Krishna Murti, *Jrl. of Neurochemistry* (1976) 27: 299–301, at 300.

10. A.A. Barber, *Lipids* (1966) 1: 146–151.

11. A. Seregi et al., *Experientia* (1978) 34: 1056–1057.

12. G.B. Kovachich and O.P. Mishra, *Experimental Brain Research* (1983) 50: 62–68.

13. Ikuo Abe et al., *Sci. Rep. Res. Inst. Tokoku Univ.* (1979) 26, nos. 3 and 4: 39–45.

14. Walter C. Brogan, III et al., *Environmental Health Perspectives* (1981) 38: 105–110.

15. Albert W. Girotti et al., *Photochemistry and Photobiology* (1985) 41, no. 3: 267–276.

16. Peter J. Hornsby and Joseph F. Crivello, *Molecular and Cellular Endocrinology* (1983) 30: 1–20.

17. E. D. Wills in *Oxidative Stress*, ed. Helmut Sães (New York: Academic Press, 1985) pp. 197–218, at 210.

18. Linda H. Chen, *Amer. Jrl. of Clinical Nutrition* (June, 1981) 34: 1036–1041.

19. N. Bishun et al., *Oncology* (1978) 35: 160–162.

20. N. Bishun et al., *Cytobios* (1979) 25: 29–36.

21. Clifford W. Welsh et al., *Jrl. of the National Cancer Institute* (1981) 67: 935–938.

22. M. Alice Ottoboni, *The Dose Makes the Poison* (Berkeley, CA: Vincente Books, 1984) pp. 91–92.

23. H.F. Smyth, Jr., *Food Cosmetic Toxicology* (1967) 5: 51–58.

24. M. Alice Ottoboni op. cit. p. 94.

25. *Federal Register* (Oct. 5, 1983) 48, no. 194: 45508.

26. House of Representatives no. 2284, 85th Congress, Second Session (1958).

27. *Chicago Tribune* (June 28, 1985).

28. Marjorie Sun, *Science* (1984) 223: 667–668.

29. Marjorie Sun, *Science* (1985) 229: 739–741.

30. David A. Kessler, *Science* (1984) 223: 1034–1040.

31. W. Gary Flamm in *Carcinogens and Mutagens in the Environment*, vol. 1, ed. Hans F. Stich (Boca Raton, Florida: CRC Press, 1982) pp. 275–281.

32. R.F. Heller et al., *Atherosclerosis* (1983) 48: 185–192.

33. P.N. Durrington, *Clinical Science* (1979) 56: 501. Cited in ref. 32.

34. A.H. Conney et al., *Clinical Pharmacology and Therapeutics* (1976) 20: 633. Cited in ref. 32.

35. W.W. Duke, *JAMA* (Nov. 6, 1915) 65, no. 19: 1600–1610.

36. Lawrence P. Garrod, *British Medical Jrl.* (Feb. 3, 1951): 205–210.

37. G.E. Foley and W.D. Winter, *Jrl. of Infectious Diseases* (1949) 85: 268. Cited in ref. 36.

38. T.D. Luckey, *Federation Proceedings* (Feb. 1978) 37 no. 2: 107–109.

39. Committee to Study the Human Health Effects of Subtherapeutic Antibiotic Use in Animal Feeds, National Research Council, *The Effects on Human Health of Subtherapeutic Use of Antimicrobials in Animal Feeds* (Washington, D.C.: National Academy of Sciences, 1980) p. XIII.

40. Ibid, p. 7.
41. Ibid, pp. 22–23.
42. Scott D. Holmberg et al., *New England Jrl. of Medicine* (1984) 311, no. 10: 617–622.
43. N.V. Medunitsyn, *Zh. Mikrobiol. Epidemiol. Immunobiol.* (1960) 31: 742–743.
44. Frank R. George et al., *Pharmacology, Biochemistry and Behavior* (1983) 19: 131–136.
45. Robert E. Hodges et al., *American Jrl. of Clinical Nutrition* (Aug., 1962) 11, no. 2: 85–93.
46. M.L. Smith and W.A. Loqman, *Archives of Andrology* (1982) 9: 105–113.
47. M.C. Chang, *Jrl. of Andrology* (1984) 5: 45–50.
48. C. Selli et al., *European Urology* (1983) 9: 109–112.
49. Robert T. Rubin et al., *Jrl. of Clinical Endocrinology and Metabolism* (1978) 47, no. 2: 447–452.
50. S.P. Ghosh et al., *Jrl. of Reproduction and Fertility* (1983) 67: 235–238.
51. P. Marrama et al., *Maturitas* (1982) 4: 131–138.
52. Herbert Y. Meltzer, *Jrl. of Pharmacology and Experimental Therapeutics* (1983) 224, no. 1: 21–27.
53. John T. Clark et al., *Science* (Aug. 24, 1984) 225: 847–849.
54. Mark F. Schwartz et al., *Biological Psychiatry* (1982) 17, no. 8: 861–876.
55. A. Rocco et al., *Archives of Andrology* (1983) 10: 179–183.
56. L.C. Garcia Diez and J.M. Gonzalez Buitrago, *Archives of Andrology* (1982) 9: 311–317.
57. D. Ayalon et al., *Int. Jrl. of Gynaecology and Obstetrics* (1982) 20: 481–485.
58. John Bancroft in *Biological Determinants of Sexual Behavior*, ed. J. B. Hutchison (New York: John Wiley & Sons, 1978) pp. 493–519, at 511.
59. L. Lidberg, *Pharmakopopsychiat.* (1972) 5: 187. Cited in ref. 58.
60. L. Lidberg, *Hormones* (1972), 5: 273. Cited in ref. 58.
61. L. Lidberg and V. Sternthal, *Pharmakopsychiatr, Neoropsychopharmakol.* (1977) 10, no. 10:21–25. Cited in ref. 58.
62. Anon., *Science Digest* (July, 1982) 9: 94.
63. J.C. Theiss, *International Colloquium on Cancer*, Houston: M.D. Anderson Hospital and Tumor Institutes, 1981. Cited by *Lancet* (Dec. 11, 1982): 1317–1318.
64. Michael L. Kleinberg and Michael J. Quinn, *Amer. Jrl. Hosp. Pharm.* (1981) 38: 1301–1303.
65. Hanna Norppa et al., *Scand. Jrl. Work Envir. Health* (1980) 6: 299–301.
66. R.P. Bos, *Int. Arch. Occup. Environ. Health* (1982) 50: 359–369.
67. Marja Sorsa et al., *Mutation Research* (1985) 154: 135–149.
68. S. Venitt et al., *Lancet* (Jan. 14, 1984): 74–76.
69. *Lancet* (Dec. 11, 1982): 1317–1318.
70. J.F. Gibson et al., *Lancet* (Jan. 14, 1984): 100–101.
71. *Lancet* (Jan. 28, 1984): 203.
72. Luci A. Power and Michael H. Stolar, *Lancet* (March 10, 1984): 569–570.
73. Sherry G. Selevan et al., *New England Jrl. of Medicine* (Nov. 7, 1985) 313, no. 19: 1173–1178.
74. Eula Bingham, *New England Jrl. of Medicine* (Nov. 7, 1985) 313, no. 19: 1220–1221.

75. Robert Hoover and Joseph F. Fraumeni, Jr., *Cancer* (1981) 47, no. 5: 1071–1080.
76. M.A. Firer et al., *British Medical Jrl.* (1981) 283: 693–696.
77. E.R. Stiehm et al., *Annals of Internal Medicine* (1982) 98: 80–93. Cited by Richard D. De Shazo and John E. Salvaggio, JAMA (Oct. 26, 1984) 252, no. 16: 2198–2201.
78. Gene P. Siegal et al., *Proceedings of the National Academy of Sciences* (1982) 79: 4064–4068.
79. Frederick Hoelzel and Esther DaCosta, *Amer. Jrl. of Digestive Diseases and Nutrition* (1937) 4: 325–331.
80. James C. White, *New England Jrl. of Medicine* (1985) 312, no. 4: 246–247.
81. *Physicians' Desk Reference*, 39th ed. (Oradell, N. J.: Medical Economics Co., 1985).
82. David F. Horrobin, *Medical Hypotheses* (1981) 7: 115–125.
83. *Physicians' Desk Reference*, 39th ed. (Oradell, N. J.: Medical Economics Co., 1985) pp. 1723–1724.
84. Ibid., pp. 538 and 540.
85. Ibid., p. 1051.
86. Ibid., p. 532.
87. Ibid., p. 1940.
88. *AMA Drug Evaluations*, 5th ed. (Chicago: American Medical Assn., 1983).
89. *Drug Facts and Comparisons* (St. Louis, Mo.: Facts and Comparison, Inc. 1985).
90. Joseph Meites et al., *Proceedings of the Society for Experimental Biology and Medicine* (1971) 137: 1225–1227.
91. C. Huggins, *Cancer Research* (1965) 25: 1163. Cited in ref. 90.
92. R.I. Dorfman, *Methods in Hormone Research* (1965) 4: 165. Cited in ref. 90.
93. J. Hayward, Recent *Results in Cancer Research* (Berlin: Springer-Verlag 1970) p. 69.
94. K.M. Robinson and J.H. Botha, *Prostaglandins, Leukotrienes and Medicine* (1985) 20: 209–221.
95. Mary E. Caldwell and Willis R. Brewer, *Cancer Research* (Dec., 1983) 43: 5775–5777.
96. D.R. Stoltz et al., *Environmental Mutagenesis* (1984) 6: 343–354.
97. S. Morris Kupchan and E. Bauerschmidt, *Photochemistry* (1971) 10: 664–666.
98. David R. Blake, et. al., *Annals of the Rheumatic Diseases* (1983) 42: 89–93, at 92.
99. L.H. Chen and R.R. Thacker, *International Jrl. of Vitamin and Nutrition Research* (1986) 56: 253–258.
100. Seymour L. Romney et al., *Gynecologic Oncology* (1985) 20: 109–114.

Footnotes

[1] Some of this material was previously published in *The Reverse Effect: How Vitamins and Minerals Promote Health and CAUSE Disease* (1988) by Walter A. Heiby, Deerfield, Illinois: MediScience Publishers.

[2] Since the Delaney Clause was enacted in 1958, it has been interpreted to mean that no substance known to cause cancer in animals may be added to food. (It might be naturally present, but may not be added.)[26] Saccharin was subsequently given a legal exemption. In June 1985, Dr. Frank Young, FDA commissioner, stated that his reading of the law permits use of the legal concept known as de minimus, meaning "the law does not concern itself with trifles."[27] Young maintains that a *de minimus* risk does not mean the substance must be banned. Two articles by Marjorie Sun,[28,29] one by Kessler[30] and one by Flamm[31] discuss some of the problems occasioned by the Delaney Clause.

[3] However, antibiotics stimulate growth not only of the animals, but of the biota they are meant to oppose. In 1978, 48% of the antibiotics produced were designated for use in animal feeds.[39] They are used not only for growth promotion but for better feed conversion, prophylaxis and the treatment of certain diseases.[40] However, there is a danger in the practice. It appears that antimicrobial-resistant bacteria from animals can cause infection in humans and there have been a number of investigations of epidemics that may have been caused by these bacteria.[41] Scott D. Holmberg et al.[42] have identified 18 persons in four midwestern states that have been infected by *Salmonella newport*, a strain resistant to antibiotics. The onset of the illness was often triggered by the taking of amoxicillin or penicillin. They concluded that "antimicrobial-resistant bacteria of animal origin can cause serious human disease, especially in persons taking antimicrobials, and that the emergence and selection of such organisms are complications of subtherapeutic antimicrobial use in animals. We advocate more prudent use of antimicrobials in both people and animals."

[4] Intrestingly, unless prolactin secretion increases in young (but not in older men) during sleep, their testosterone production is unlikely to follow its normal tendency to peak the following morning.[51] The herb yohimbine from bark of the evergreen tree yohimbe, often used as an aphrodisiac, may stimulate prolactin secretion.[52] (Scientists for many years have declared aphrodisiacs to be nonsense. Experiments at Stanford University by John T. Clark et al.[53] have shown, on the contrary, that yohimbine has both immediate and more lasting effects in increasing the sexual appetite of male rats. Clinical studies showing yohimbine's sexual-arousing effect in men will be discussed in detail in a sex book I have in preparation.)

[5] The antineoplastic drugs, cyclophosphamide, diethylstilbestrol, fluorouracil, methotrexate, and vincristine show teratogenic and mutagenic *in vivo* effects in animals. Selevan et al.,[73] in a case-control study involving 17 hospitals in Finland, found that pregnant nurses exposed to antineoplastic drugs had a statistically significant increase in fetal loss. Nurses who lost their fetuses were twice as likely to have been exposed to antineoplastic drugs during the first trimester as those who gave birth. In a related article, Bingham[74] has observed that, "Institutions such as hospitals and university laboratories have not been in the forefront of occupational-disease prevention."

[6] Even excessive amounts of drinking water may be dangerous to epileptic patients. White[80] observes that hydration has long been known to induce seizures. He calls attention to a diet plan which advocates drinking eight to twelve 8-ounce glasses of water per day and cautions that epileptic patients should be aware of the possible dangers.

[7] The patent on Valium® has recently run out, so there are likely to be many more diazepam drugs to offer it competition. Horrobin[82] speculates that diazepams and possibly the benzodiazepines, including chlordiazepoxide hydrochloride (Librium®), may promote tumor growth. (Tumor promoters do not usually cause cancer when used alone but can enhance the effect of a cancer initiator.) Is it possible that diazepams and/or benzodiazepines at some dosage might show a reverse effect and act to cure cancer?

[8] Studying the *Physicians' Desk Reference (PDR)* at intervals of a decade or more, I get the impression that drugs reported by the *PDR* as having a "paradoxical effect" are associated with a tendency to be retired. However, a phenomenon is paradoxical only until the paradox is resolved. Would it be beneficial to keep using some of these retired drugs but more clearly define dose-response relationships?

[9] D.R. Stoltz et al.[96] found that 22 varieties of fruits and vegetables had mutagenic activity. Grapes, onions, peaches, raisins, raspberries and strawberries showed the most potent mutagenicity. Could components (perhaps flavonoids) from these six be concentrated to act as anticancer agents, as the reverse effect suggests? Many flavonols found in foods are mutagenic and references for this finding are found in *The Reverse Effect*. I have not seen studies indicating that 3,4'-dimethoxy-3'5,7-trihydroxyflavone and centaureidin are mutagenic and/or carcinogenic. However, S. Morris Kupchan and E. Bauerschmidt[97] found that these flavonols from the plant *Baccharis sarothroides* showed significant inhibitory activity against cells derived from human carcinoma of the nasopharynx carried in cell culture.

Two Theses in Radiobiology

T.D. Luckey, Ph.D.

Dr. T.D. Luckey is Professor Emeritus of The University of Missouri, Columbia, Missouri; and president of ORALU Corp., 1009 Sitka Ct., Loveland, CO 80538.

Dr. Luckey received a B.S. degree in chemistry from Colorado State University (1941) and M.S. and Ph.D. degrees from the University of Wisconsin. He was Assistant, 1946–52, and Associate, 1952–1954, Professor of Research at LOBUND, the University of Notre Dame. From 1954 to 1968 he was Chairman of Biochemistry, School of Medicine, University of Missouri, Columbia. He evaluated science teaching at the University of Qatar in 1983. In 1984 he retired from teaching, was named Honorary Professor of Olde Herborn Free University, Herborn, Germany, and was knighted, *Ritter von Greifenstein*.

Dr. Luckey has lectured extensively abroad, is a member of numerous professional societies, and has published more than 200 scientific papers and chapters in books. He is author of *Germfree Life and Gnotobiology*, Academic Press, New York, 1963; *Hormesis with Ionizing Radiation*, CRC Press, Inc., Boca Raton, 1980; and *Radiation Hormesis*, CRC Press, Inc., Boca Raton, 1991. The Japanese translation of *Hormesis with Ionizing Radiation* was published by Soft Science, Inc., Chiba, 1990. He is coauthor of *Heavy Metal Toxicity, Safety and Hormology*, Georg Thieme Publishers, Stuttgart, 1975 and *Metal Toxicity in Mammals*, vols. 1 and 2, Plenum Press, New York, 1977 and 1978.

Dr. Luckey was nutrition consultant with NASA Apollo Space flights 12–17. He has consulted with industry in the United States, Germany and Japan for more than 40 years. He is presently Consultant in Life Sciences at his home in Loveland, Colorado.

We are confronted with two opposing theses for chronic, whole-body exposures to ionizing radiation: the "zero" thesis argues that "all radiation is harmful; the "hormesis" thesis argues that "small and large doses produce opposite effects." Although both agree that large doses are harmful, their positions on small doses are quite contradictory. Where the zero thesis predicts real harm (e.g., mutations and cancer), the hormesis thesis predicts benefit from small doses.

The zero thesis spawned several linear models that interpolate between data points from populations exposed to large doses of ionizing radiation and controls receiving about 2 mGy y^{-1} of whole-body radiation. Given the abundant data from exposures to low doses of ionizing radiation, it is inexcusable to make linear interpolation from results with large doses to controls. The results from large doses of ionizing radiation are applicable to those exposed to medical treatment, radiochemical accidents, or nuclear explosions. These traumatic occasions are not, however, pertinent to everyday life. Such results are of value only for subjects having no specialized cells, no hormones, no neurologic system, no immune cells, and almost no communication between differentiated cells. The zero thesis lacks evidence from whole-body exposure to low doses of ionizing radiation in vertebrates. Critical review of the vertebrate literature provides no substantial support for the zero thesis or any linear model (*Radiation Hormesis*, T.D. Luckey, CRC Press, Inc., in press).

Despite the scarcity of data, linear models are promulgated on the basis that they provide a safe basis for limiting ionizing radiation to the lowest possible or reasonable dose. In spite of this unscientific support, the zero thesis and its linear models are invalid and counter-productive.

The hormesis thesis, on the other hand, is the basis for the radiation hormesis model. Stimulation by low doses of potentially harmful agents has been accepted for many centuries in toxicology and pharmacology. The inclusion of ionizing radiation broadens the base of this general thesis and, at the same time, adds validity to the radiation hormesis model. In like manner, data from vertebrates is well supported by data from invertebrates, plants, and micro-organisms (*Hormesis with Ionizing Radiation*, T.D. Luckey, CRC Press, Inc., 1980).

Information of everyday concern involves chronic, whole-body exposures of vertebrates to doses less than 1,000 times background radiation levels. The literature from whole-body exposures to low doses of ionizing radiation provides overwhelming support of the hormesis thesis with adequate statistically significant results to invalidate the zero thesis.

The conclusion is that we need more ionizing radiation, not less! In other words, we live in a partial deficiency of ionizing radiation. Added radiation would improve the quality of life as measured by growth, neurologic development, reproduction, immune competence, resistance to cancer, and longer average lifespan. Depriving populations of adequate amounts of this natural, beneficial agent is unreasonable. Safe supplementation of ionizing radiation to populations should be considered. We should discard the invalid linear models and accept radiation hormesis as the basis for changing current recommendations and regulations.

Future health physicists should be concerned less about probing for minimum exposures and become active in promulgating ways to provide safe supplementation, 20 to 100 mGy y^{-1}. Except for persons with genetic inability to repair DNA, this is well below harmful effects of chronic, whole-body exposures, estimated to be over 1,000 mGy y^{-1} for low-LET radiation. The basic challenge is acceptance of the hormesis thesis as a practical basis for a new plateau of health.

Reprinted, with permission, from
the *Health Physics Society's Newsletter*,
Volume XVIII, Number 12, December 1990.

A Critique of the Statistical Methodology in Radiation and Human Health

Jane M. Orient, M.D.

Please see biographical sketch for Dr. Orient on page 186.

An innovative method for calculating the risk of delayed health effects, primarily cancer, from low dose radiation is presented by John Gofman in his book *Radiation and Human Health*. Intended for a diverse audience—worried mothers, lawyers settling compensation claims, policymakers setting standards for industry, and physicians making decisions about diagnostic procedures—it claims to "demystify a whole field of science." His "methodological breakthrough", by extirpating all mathematics more complex than percentages and by solving relatively simple algebraic equations, makes matters previously in the realm of elite specialists easily comprehended by lay people, even those with math anxiety.

Gofman's startling estimates, for example, that 950,000 worldwide deaths will eventually result from plutonium released by atmospheric weapons testing, are likely to alarm the public and to discredit scientists. Could his results be correct, and those of the vast majority of scientists wrong by orders of magnitude? In evaluating his method, two questions will be posed: (1) are his assumptions reasonable and (2) are his predictions compatible with the epidemiologic evidence currently available?

Basic Assumptions

Gofman assumes several broad principles. First, he assumes that the carcinogenic potential of a given dose of radiation depends upon the age of the subject. In fact, he considers age to be essentially the only important variable. While some studies support his assertion that the risk is much higher for the young, some have come to the opposite conclusion (UNSCEAR77). His second assumption is that all tissues have the same latency period and the same dose-response curve for radiation damage, except the bone marrow. Thus, the distribution of specific types of cancer should be the same among radiation-induced and spontaneous tumors. This assumption is clearly not correct (Be78; BEIR80). The third assumption is that the linear hypothesis, or proportionality principle, holds under all circumstances. That is, 100 rad administered to 1000 people will have the same effect as 10 rad given to each of 10,000 or 1 rad to each of 100,000. Actually, Gofman argues for the supralinear hypothesis, which states that lower doses induce more cancers per rad. Since all his calculations are based on linearity, he considers them to underestimate cancer risk. The concept of a threshold is called "fraudulent," and work on DNA repair mechanisms is dismissed as a search for "gremlins."

Calculation Methods

Although questionable, these assumptions will be accepted for purposes of the discussion below, since Dr. Gofman invites those who differ with him to substitute their own parameters. The method of calculation itself will be examined.

The book graphically illustrates the belief that the carcinogenic effect of radiation increases for 40 yr after exposure, and declines symmetrically thereafter. Dr. Gofman does acknowledge that "we do not know this to be the exact curve for every type of human cancer.... For now, this curve is an exceedingly reasonable approximation derived from valid human evidence that we have in hand right now...." The descending limb of the curve is based on sheer speculation. The particular evidence for the first 45 yr seems to come from studies with the longest available follow-up periods (Be78; Bo77; He75; Sh77), which he refers to in a later letter (Go81b). A review of these studies shows that, whereas radiation risk may still be present as late as 40 yr, it is not apparent that the risk accelerates up to that time. Twenty-nine yr after the atomic bombs, the relative risk of lung cancer in survivors was still increasing, but that of breast and stomach cancer was declining (Be78). In women subjected to repeated fluoroscopy for tuberculosis, the relative risk ratio for breast cancer increased in the periods from 35 to 45 yr after exposure, but the calculation is based on only six cases of cancer. The confidence interval is very wide, overlapping substantially that for excess cancers at 15–34 yr (Bo77). Irradiation in infancy has been associated with thyroid neoplasms; between 35 and 40 yr later, two malignant and four benign ones have been reported (He75). Elsewhere in the book, Gofman himself cautions readers about the uncertainties of small numbers.

Based on a handful of cases, Gofman's curve is used to extrapolate to populations of millions. Reading figures from his diagram, he calculates to three significant figures tables of conversion factors, which allows one to convert the excess percent per rad observed in a study with limited follow-up to a new universal constant, the peak excess percent cancers per rad, which presumably would be observed at 40 yr post-exposure. The excess percent per rad is equal to 100 (O − E)/(E multiplied by mean dose in rads), where O is the number of cancers observed, and E the number expected, based on a control group or standard mortality tables. Since this quantity is not defined in the same way as a standard relative risk ratio, established statistical methods for calculating variance do not apply, and no attempt is made to determine confidence intervals. Though the peak excess percent per rad hypothetically should be the same for all studies involving subjects of the same age group, in practice it is not, because of the "small numbers problem". In fact, it varies from 0 to 366.7. The solution is to combine all the studies in the world literature to arrive at an optimum value.

Pooling data from different studies is not unprecedented; for example, trials of anticoagulants in myocardial infarction patients were subjected to this type of analysis (Ch77), in an effort to obviate the need for still more experiments. Among the pitfalls of the method is to average percentages, rather than to return to the raw data (Go79). To illustrate: if a cancers are observed in m subjects in Study 1, and b cancers in n subjects in Study 2, the mean proportion of subjects developing cancer is $(a + b)/(n + m)$, not $((a/m) + (b/n))/2$. Gofman violates this basic rule of arithmetic. Furthermore, he weights his means of percentages by the number of cancers observed. In this way, he ablates all information about the number of subjects in the studies, and neatly eliminates all negative results by multiplying them by zero. These "best values" are then applied to many practical situations. He demonstrates how a lawyer might use his tables to proclaim in court that a client's cancer has precisely a 14.7% probability of causation by occupational radiation received a given number of years previously.

Comparing Gofman's Predictions with Available Data

Since Gofman projects the occurrence of cancer so far into the future, up to 72 yr after exposure, many of his predictions cannot be verified yet. Half the lung cancer deaths from plutonium in fallout are not scheduled to appear until after the year 2000. Nevertheless, it is possible to test how well his model predicts epidemiologic results which have already occurred.

For studies of a given number of subjects who received a known dose of radiation, the number of cancers predicted by Gofman's model for any desired follow-up year can be calculated. Of course, because of random variation, the actual number observed in the study will not be exactly the same. We can determine, however, how likely it is that an experiment would give a result deviating from Gofman's by at least the amount observed, if Gofman's model is assumed to be correct. Because the outcome (cancer or no cancer) is a dichotomous variable, the results of a large series of identical experiments would form a binomial distribution. The expected number of cancers, based on the Gofman hypothesis, $E(r)$, is calculated from his assumptions. The probability of cancer, π, follows from the properties of the distribution and is equal to $E(r)/n$, where n is the number of subjects. The probability of an experimental result of r or fewer cases could be calculated exactly from the binomial distribution, but for a large number of cases, the normal approximation can be used, and this probability is given by the normal tail area beyond a standardized normal deviate (Ar71):

$$u = \frac{|r - n\pi| - 0.5}{\sqrt{n\,\pi\,(1 - \pi)}} \tag{1}$$

Gofman's procedure for estimating the results of long term studies from those of shorter studies can be tested using data for women treated with radiation for post-partum mastitis (Sh77) because the authors have plotted the incidence of cancer in experimental and control groups by year after exposure. Using age 30 as the approximate mean age at treatment, and inferring from the discussion that follow-up began approximately seven years after radiation, the Gofman conversion factors can be found. Reading from Shore's text-figure 2, the cumulative number of breast cancer cases in 571 subjects would be about 28.5 in the 22nd yr, compared with about 14.8 spontaneous cancers expected, when an adjustment is made for the age distribution in irradiated and control subjects by the Hankey-Myers method. Since the mean dose was 247 rad, the excess percent cancers per rad would be 0.375%. Using Gofman's conversion factor 4.37, the peak excess percent per rad should be 1.64%. This value is hypothesized to be the same at 34 yr. With the appropriate conversion factor for that interval, the excess percent per rad expected in 34-yr study should be 1.64%/1.76, or 0.930%. The cumulative spontaneous incidence at that time, from the morbidity curve for the control group, should be 29.1. From the equation,

$$\frac{(O - E)\,100}{E \times 247} = 0.930, \tag{2}$$

substituting $E = 29.1$, O is found to be 95.9. This is taken to be the mean of the binomial distribution, $E(r)$, by the Gofman hypothesis. The

$\pi = 0.168$. In the study, 36 cases of cancer occurred in irradiated subjects, but the figure adjusted for age gives 54.2, which will be taken as the value of r. Thus,

$$u = \frac{\left| 54.2 - 95.9 \right| - 0.5}{\sqrt{571(0.168)(0.832)}} = 4.61. \tag{3}$$

From a standard normal table, the probability of this experimental observation, or one still more deviant, is less than 3×10^{-5}.

A similar argument can be applied to data for survivors of the Hiroshima and Nagasaki atomic bombs. Beebe et al. have tabulated extensive data about the composition of the sample and the mean T65 dose estimates (Be78). The mean dose to persons in the 50–99 and 100–99 rad ranges was about 100 rad. Predicted numbers of cancer cases between 1950 and 1974 (5 and 29 yr of follow-up) were calculated by Gofman's method for this subset. Spontaneous cancer incidence was estimated from figures in the zero rad category. The number of persons in each cohort was assumed to be proportional to the number of person-years of follow-up. Thus, the approximate number of persons in each category of age at the time of bombing could be calculated from the data in Tables I and II (Be78). Observed numbers, and Gofman's predictions, are given in Table I. His extremely improbable predictions become slightly less improbable if we assume that persons in the zero-rad group actually received about 10 rad, and lowered the estimate of spontaneous cancers accordingly.

A further example is a 29-yr follow-up of 6560 army radiology technologists, compared with 6826 medical, laboratory, and pharmacy technologists (Ja78). While Gofman refers the reader to this "possibly negative" study, he does not use the results because dosimetry was not done. The authors of the paper, however, give reason to assume that the radiology technologists must have received at least 10–20 rad. That group experienced 133 cancer deaths (other than from leukemia), actually 12 fewer than the number predicted from the incidence in the control group. Gofman's model predicts 172 deaths if 10 rad were delivered, and 199 if 20 were given. The probability of an experimental result this far deviant would be 0.00149 in the former case, and less than 3×10^{-5} in the latter.

Conclusions

Gofman asserts that the risk of cancer induced by low-dose radiation, such as that resulting from the commercial use of nuclear power, is

Table 1. Gofman Method Applied to Hiroshima-Nagasaki Data:

(a) Estimated spontaneous cancer rate

Age ATB*	Persons Receiving O rads a	Cancer Deathsγ (except leukemia) b	Cancer Death Rate c=b/a	Persons Receiving Mean 100 Rads d	Expected Spontaneous Cancers e=cd
0-9	7505	8	.00107	1242	1.33
10-19	7817	42	.00537	1906	10.24
20-34	7429	169	.02275	1618	36.81
35-49	8092	675	.08342	1675	139.7

* ATB means at the time of bombing
γ from Table VI (Be78)

(b) Gofman's predictions for persons receiving 100 rad

Age ATB	Peak excess % per rad* f	Conversion factorγ g	Excess % per rad at 29 yrs h=f/g	Gofman's predicted number of cancers $e + \dfrac{eh(100\ rads)}{100}$
0-9	54.8	2.89	19.0	26.6
10-19	4.41	2.46	1.79	28.6
20-34	3.71	2.32	1.60	95.7
35-49	3.57 ‡	2.55	1.40	335.3

* from Table 17, Combination of Miscellaneous Cancer Analyses with Breast-Cancer Analyses to Yield Best Estimates, Overall, for Peak Percent in Cancer, Per Rad (Go81a)
γ From Table 5 (Go81a), with interpolation as directed
‡ Estimated from Figure 5 (Go81a)

(c) Comparison of Gofman's predictions with observed results

Age ATB	Predicted Number of Cancers	Observed Number of Cancers	Standardized Normal Deviate	P
0-9	26.6	2	4.72	<.00003
10-19	28.6	12	3.03	<.0013
20-34	95.7	47	5.077	$< 3 \times 10^{-7}$
35-49	335.3	154	11.06	$<< 10^{-9}$

"genocidal" in magnitude. Prophetically, he projects most of the cases decades into the future. However, following his advice about reality testing, one can determine the likelihood of observing the present epidemiological data to be on the order of $(3 \times 10^{-5})(1.3 \times 10^{-3}(1.49 \times 10^{-3})$ $= 5.8 \times 10^{-11}$, using the most favorable probabilities for each the above independent experiments. Are his hypotheses unlikely enough to be rejected? He himself considers the null hypothesis to be suspect when the data have a 10% probability of being observed under that assumption, lest we disregard important results because of rigidity about "magical" levels of statistical significance.

Gofman's frighteningly high estimates of radiation risks are derived from questionable assumptions about tumor biology, and questionable statistical analysis. They are unsupported by goodness of fit tests and unqualified by calculations of confidence intervals. They should be regarded with great skepticism.

References

Ar71 Armitage P., 1971, *Statistical Methods in Medical Research* Oxford: Blackwell Scientific Publications).

Be78 Beebe G.W., Kato H. and Land C.E., 1978, "Studies of the Mortality of A-Bomb Survivors: 6. Mortality and Radiation Dose, 1950–1974", *Rad. Res.* 75, 138–201.

BEIR80 Report of the Committee on the Biological. Effects of Ionizing Radiation (BEIR III), 1980, *The Effects on Population of Exposure to Low Levels of Ionizing Radiation* (Washington, DC: National Academy Press).

Bo77 Boice J.D., Jr. and Monson R.R., 1977, "Breast Cancer in Women After Repeated Fluoroscopic Examination of the Chest", *J. Natl. Cancer Instit.* 58, 823–832.

Ch77 Chalmers T.C., Matta R.J., Smith H., Jr. and Kunzler A., 1977, "Evidence Favoring the Use of Anti-coagulants in the Hospital Phase of Acute Myocardial Infarction", *New Engl. J. Med.* 297, 1091–1096.

Go81a Gofman J.W., 1981, *Radiation and Human Health* San Francisco, CA: Sierra Club).

Go81b Gofman J.W., 1981, "Response to H. David Maillie's Letter", *Health Phys.* 41, 204–208.

Go79 Goldman L. and Feinstein A.R., 1979, "Anticoagulants and Myocardial Infarction: The Problems of Pooling, Drowning, and Floating", *Ann Intern. Med.* 90, 92–94.

He75 Hemplemann L.H., Hall W.J., Phillips M., Cooper R.A. and Ames W.R., 1975, "Neoplasms in Persons Treated with X-rays in Infancy: Fourth Survey in 20 Years", *J. Natl. Cancer Instit.* 55, 519–530.

Ja78 Jablon S. and Miller R.W., 1978, "Army Technologists: 29-Year Follow Up for Cause of Death", *Radiology* 126, 677–679.

Sh77 Shore R.E., Hempelmann L.H., Kowaluk E., Mansur P.S., Pasternack B.S., Albert R.E. and Haughie G.E., 1977, "Breast Neoplasms in Women

Treated with X-rays for Acute Postpartum Mastitis", *J. Natl. Cancer Instit.* 59, 813–822.

UNSCEAR 77   United Nations Scientific Committee on the Effects of Atomic Radiation, 1977, *Sources and Effects of Ionizing Radiation,* 1977 Report to the General Assembly, with Annexes (New York: UN).

Risk

Toxicological Risk Assessment Distortions
Jay H. Lehr

A New Measure of Risk
Jay H. Lehr

Toxicological Risk Assessment Distortions

Jay H. Lehr, Ph.D.

Please see biographical sketch for Dr. Lehr on page v.

Without Bias

Before starting on risk assessment, let me establish credibility for my subject. Over the years, I have become reasonably competent in toxicology as a result of my efforts to promote ground-water protection legislation. I was among the first to stand up in Washington and demand federal regulation of ground water. In 1968, my colleagues and I pointed out that half the water we drink is ground water, and that it was totally unprotected. We succeeded in getting only a limited amount of ground-water protection language into the amended Water pollution Control Act that passed in 1972, but we immediately went to work on a new law, which became the Safe Drinking Water Act of 1974. Soon thereafter, I joined another group who were working on regulating surface disposal of waste and helped them develop the Resource Conservation and Recovery Act, which closed many loopholes involving activities that were polluting ground water. Later, I worked on the Surface Mining Control and Reclamation Act, the Toxic Substance Control Act, and finally, spent many years assisting in the development of Superfund. This background is important because it will soon become apparent to you that I think we have gone too far. Nearly everything that is driving the Association of Ground Water Scientists and Engineers today deals with ground-water pollution. I do not intend to undermine your careers, but I do intend to reduce the crisis mentality that is driving this system beyond any reason, common sense, or scientific knowledge.

Common Science/Media Distortions

I recently found an excellent example of today's distortion problems on the front page of a *Des Moines Register* (October 3, 1989). The headline read "Underweight Babies, Rural Drinking Water Linked in University's Study." In the next day's paper (October 4, 1989, on the fourth

673

page of section 5), there appeared a rebuttal from the people who had been wounded by the aforementioned unreviewed study at the University of Iowa. Now, there isn't anything wrong with studying the relationship between birth weight and environmental factors such as water supply as long as many potential influences are considered. In this case, if you read the full article, you learned that a correlation was made using a small statistical sample with a very small effect and then correlating that effect to the water that was drunk. Dozens of other factors that might affect the birth weight of babies were ignored.

The way science is being practiced by some people today is all but scandalous. There are scientists who like to parade in front of the press with glib comments that elicit much-desired publicity, and they like to avoid the peer review of their more conservative colleagues. Their studies easily find the light of day in publications of the Natural Resource Defense Council (for example, the ALAR study); The National Wildlife Federation (e.g., the Drinking Water study); and many other overzealous environmental group publications.

What is even more wrong is using the media for peer review of science. Research should first stand the test of peer review in scientific journals. There is, unfortunately, an attitude at some universities that praises people who get publicity for science done on the university campus regardless of how spurious it may be.

Many studies of ground-water pollution show that the percentage of wells exceeding maximum contaminant level is small, if not insignificant. But the news media that report these studies focus not on what can be considered good news, but rather on the percentage of wells in which a particular chemical is detected.

A 1987 Iowa Department of Natural Resource Public Water Survey that tested 853 water systems is a case in point (Fleetwood and Martin, 1989). The fact that 14.7 percent of the systems had detectable pesticides got all the publicity; the fact that only 1 percent exceeded maximum contaminant levels did not.

That we find so many chemicals in our water supply is less related to an expanded use and abuse of chemicals than it is to an ever-improving capability to perceive smaller and smaller quantities of chemicals in laboratory tests. In many cases, the presence of pesticides in ground water has evolved into a public concern that borders on outrage (Spaulding, 1989). This "chemophobia" tends to result in a loss of objectivity.

Much of this concern has evolved from quantification of compounds which, because of their constituents (many contain halogens or nitrogen), can be detected in sub-part per billion amounts. Once the part per million was the visible limit, then we improved three orders of magnitude to parts per billion, and now we commonly measure things 1/1000 of that quantity or a part per trillion basis. We will achieve common recognition of 1/1000 of that in a part per quadrillion in the next decade. One day we may recognize that there is something of everything in everything else and that a glass of water likely contains a molecule of every compound on earth.

Eventually, it may dawn on the public that they are being hoodwinked by environmental zealots who aggressively promote a nondetection limit for all undesirable chemicals. But analytical technology changes. Yesterdays zero is no longer zero, and today's zero will not be zero tomorrow. Driving toward nondetection is unreasonable and may even be unethical. Although many chemicals are undesirable, safe standards for all chemicals based on reasonable toxicological studies can be established, and the nation's paranoia or chemophobia can be diminished.

Reality

Environmental scares are not all bad. They often create new industries. The ground-water monitoring industry is certainly a case in point. The radon industry is another. Environmental groups would quickly go out of business were they not able to send out mailings describing the latest threat and asking for money to fight it. In other words, environmentalism has created a whole group of vested interests who fare better when there are many problems than when there are few. That tends to tilt the public debate toward solutions even when knowledgeable scientists are skeptical about the seriousness of the threats and the insistence of urgency.

There is an element of make-work involved in all of this. And it's not just our industry that is plagued with it. Scares about the food on the shelves of your own markets have shaken the food industry to its foundation. The driving force in the food scare movement consists of hundreds of organizations that bill themselves as environmentalists. In a way, they have become an industry as well. Their ability to attract foundation grants and individual contributions rests on their ability to make news, and they often have allies in the press who suspend their skepticism when these groups make claims—even wild ones. They say they represent the public interest, but they don't do so badly for their own interests either.

Do we need an environmental industry? Maybe we do but the answer would be more certain in my mind if this industry exercised more quality control over its science. And indeed there is a very bad aspect to it all, namely an unconscionable effort to distort scientific reality in order to provide a constant sense of menace for a population that is terribly deficient in scientific aptitude. I don't intend to eliminate environmental concern among my readers, but I do hope to arm you with information that will allow you to stand toe-to-toe and argue over the seriousness of many environmental threats.

Risk

We all accept risks every day. When we accept these voluntarily, we don't worry. Problems arise when the risks are forced upon us. Many of you smoke in spite of the proven risk of lung cancer. But that is your choice. It's the involuntary risks that really scare us. We are willing to accept some involuntary risks like being struck by lightning. The chance of being struck is about one in a million, and we consider that an acceptable risk. In the legal profession, that is called a "deminimus" risk—a negligible risk.

The chance of your contracting cancer in today's society is about one in four. Yet we regulate chemicals based on a lifetime ingestion of a maximum contaminant level that would increase our risk of getting cancer by one in a million. This is conservative, to say the least. And while that calculation is commonly based on so many parts per billion of a particular chemical, many environmental groups and some at U.S. EPA are dissatisfied. They prefer zero quantities of unwanted chemicals in our water supply.

Risk Assessment and Management

Risk assessment has always been with us. When cavemen recognized that animals could be a source of food, they had to weigh the hazards of being mauled by that animal versus courting starvation. We actually have writings about risk assessment that date back about 3,000 years, yet our present level of concern began only in 1960 (Paustenbach, 1989a). We became aware then of radiation in our environment and decided to quantify how much we would be willing to put up with from a variety of sources.

Risk management deals with the need for risk reduction (Rodricks, 1987). Are there significant risks that can significantly reduce the length of our lives and, if so, what should we do to eliminate them? We are now reaching a point where acceptable risk is considered somewhere in the order of a one-in-a-hundred-thousand chance of dying from exposure over a lifetime. People, in fact, are willing to accept much greater risk than that, but this is the risk below which government says it will not try to regulate. After you see how it reaches those numbers, you will realize that it is likely already regulating risks that are much smaller.

The goal of risk management is to select the options that balance the benefits of an action against a real or perceived risk with the costs of eliminating that risk. Risk assessment yields critical data to risk managers and has a number of parts, including hazard identification, dose-response assessment, exposure assessment, and risk characterization.

Hazard identification is defined here as the process of determining whether human exposure to an agent could cause an increase in the incidence of illness. It involves characterizing the nature and strength of the evidence of causation. Although the question of whether a substance causes cancer or other adverse health effects in humans is theoretically a yes-no question, there are few chemicals or physical agents on which the human data are definitive.

Dose-response assessment is the process of characterizing the relation between the dose of an agent administered or received and the incidence of an adverse health effect in exposed populations.

Exposure assessment is the process of measuring or estimating the intensity, frequency, and duration of human or animal exposure to an agent currently present in the environment or of estimating hypothetical exposures that might arise from the release of new chemicals into the environment.

Risk characterization is the process of estimating the incidence of a health effect under the various conditions of human or animal exposure described in the exposure assessment.

So far, we have studied about 600 chemicals that are carcinogenic to animals. We only know of about 25 that are definitely carcinogenic to man. The studies take many years. We can't use man as a guinea pig, so we make certain assumptions between animal studies and that of the human.

Carcinogenicity

A carcinogen is a chemical that will cause genes and cells to duplicate themselves in an imprecise manner so that we get a repetition of incomplete cells in the body. An unhealthy environment is created in some organs or parts of the body until the whole body ceases to function properly. A mutagen is a substance that will alter genes, but not necessarily make them reproduce. A teratogen is a substance that creates birth defects. There are six factors that we consider in determining carcinogenicity of a chemical. We look at the number of animal species to which a chemical is carcinogenic. We study the number of animals affected by the chemical; the number and types of tumors that the different animals get from the chemical; what the incidence of cancer is in the animal population that is subject to it; what kind of a dose is required for the chemical to be carcinogenic; what the dose-response relationship is to those given such and such a dose and exactly what happens. In order to come out with some kind of numbers, we take an animal population and feed it x milligrams of a chemical per kilogram of weight every day. We then determine at what dose does the whole population get a tumor? We then drop down and determine at what dose does half the population get a tumor and then, perhaps, at what dose does five percent of the population get a tumor, and at what dose does none of the population get a tumor?

We use those dose-response figures to plot a curve and interpolate with mathematical models a calculation of what kind of an exposure man could withstand without having a serious risk of getting cancer. We usually define that risk as one chance in a million—very conservative—and the extrapolation is very difficult. Normally, we are feeding the animals between 100 and 10,000 times the amount that would have an impact on the human species. This kind of research was never intended to be used to define numbers that man could live with. It was intended only to determine whether a chemical was a hazard at any level at all.

We now end up with interpolations sometimes as small as a part per trillion as being the upper limit for some chemicals. A part per trillion is the same as the ratio of 1/16th of an inch to the distance to the moon. In other words, 1/16th of an inch is one-trillionth of the distance to the moon. We're chasing things that small while 400,000 people die every year from smoking and another 100,000 die from drinking alcohol. Also, the risks are based on unrealistic ingestion rates for a lifetime of 70 years. With water, we assume you drink two liters a day (approximately one-half gallon) for 70 years. You can't leave your house, go on vacation, or leave town for any reason; you must keep drinking the same water.

Human Data

Many of the studies intended to scare us cannot be duplicated, but are never discarded. We do rat studies, feeding them tons of chemicals, then extrapolate back from a dose 10,000 times what a human would ever consume while simultaneously disregarding real human data. An example in point; EDB, a fumigant for agriculture, was determined to be a carcinogen through a rat study. Based on the maximum contaminant level established for EDB, an individual who worked eight hours a day for 40 years in a workplace exposed to 20 parts per million of EDB would have a 99.9 percent chance of getting cancer. Yet we have data on people who have worked in exactly those conditions for decades, working to manufacture EDB, and we don't have a single incident of cancer. Has the MCL for EDB been changed? Not at all.

We have similar data on dioxin. There is no proven evidence as yet that dioxin has done anything more than cause a skin rash. Yet we do not change our acceptable risk levels. In addition, the government always chooses the most conservative mathematical model to extrapolate from the animal studies. We are unnecessarily wreaking havoc on our economic system and creating anxiety-related problems for the public in general. People do not understand what risk is really about or that this 70-year exposure yields only a one-in-a-million chance of cancer.

While a few chemicals with which we deal allow a risk of one in 100,000 or even one in 10,000, one in a million is the norm. What does this really mean? Your chances of eventually contracting cancer are currently about one in four. Supposedly if you drink a water supply just above a particular chemical's maximum contaminant level for 70 years, you increase your risk of getting cancer by one in a million. This means that in a population of a million, instead of 250,000 people dying of cancer, 250,001 might die of cancer. Or, your chances of getting cancer are increased from 25 percent to 25.001 percent, an incredibly small increase in risk. To prove this statistically would require a sample larger than the earth's entire population (Paustenbach, 1989b).

Prudence in all matters of health is certainly desirable, but we need a balance between the chemophobia that has overtaken much of our population and the high cost. We need sensible management of industrial chemicals. We don't need crazed obsession.

If you want to look at the necessity for introducing economics into the health arena, you only have to look at the debate that is going on in the medical profession regarding how much medical effort should be expended to save the life of a very elderly person. We are able to keep people alive long after the quality of life ceases to be desirable. The health care system is running out of money and it is having to ethically debate this issue. The "health care" system's argument is tangible because it's not a mathematical model that determines whether a person is allowed to die or allowed to live with the expenditure of money. It's a sure thing. We're spending similar amounts of money where it's as far from a sure thing as anything imaginable.

Thresholds

Until the recent chemophobia scare, the main rule in toxicology has been that the dose makes the poison. We don't assume in toxicology that something is a poison, because everything is a poison given a certain dose. And yet when it comes to carcinogenicity in a water supply, we treat matters entirely differently. We have adopted the no-threshold idea that if a lot of something is bad, then a tiny amount of it is bad, also. This doesn't make sense in medical science.

We know that the carcinogenic effect is the square of the dose (Higginson, 1988). You double the dose, you square its impact, But we also know that it rises to the fourth power of time; that is to say if you double the time of exposure, that time interval has a fourth power exponential impact. If we take a 60-year-old person who has been smoking a pack of cigarettes a day for 20 years, that person actually has one-tenth the chance of getting lung cancer compared to a 60-year-old person smoking one-half pack a day for 40 years. We should also be capable, therefore, of extrapolating back to reasonable No Observable Effect Limits (NOEL).

Cancer and Carcinogens

The population is getting older, and it was as our population aged that we really saw cancer as a major death threat. Diet looks like the key factor in cancer. The Japanese have a lot of stomach cancer (Higginson, 1983). We have little. Japanese get little colon cancer. We have a considerable amount. We are looking at what our dietary differences are. We are learning that industrial chemicals are not nearly as important as we once thought, and that natural things are more important in the whole evolution of cancer and people's health (OTA, 1989). In America, we equate nature with benevolence. Historically, nature has not been benevolent. Nature produces more poisons than man ever thought of.

Overall, in the decade between 1974 and 1983, cancer declined. Stomach cancer was down 20 percent, cervical cancer was down 30 percent, and ovarian cancer was down 8 percent (Ames, 1986a). Only lung cancer increased, 15 percent in men and 70 percent in women. Our life expectancy has risen from 45 years in 1900 to 65 years in 1940, and 76 years in 1982. We anticipate it will be 82 at the turn of the century.

This is true because chemicals are not killing us. In fact, natural pesticides make up 5 to 10 percent of dry plant weight. We are exposed to 10,000 times as many natural carcinogens as we are to man-made carcinogens (Ames, 1986b), and thus 99.99 percent of our carcinogen-intake is natural rather than man-made. A cup of coffee has 4,000 parts per billion (ppb) of hydrogen peroxide. Our chlorinated tap water averages 83 ppb of chloroform. A cola drink has 7900 ppb of formaldehyde. Beer has 7700 ppb of formaldehyde, and 50 million ppb of the well-known carcinogen, alcohol. A peanut-butter sandwich has 75 times the risk of the maximum contaminant level of EDB. A mushroom has 200 times that risk, and yet we outlawed EDB without looking at what the long-term effects of radiation as a substitute to sterilize food will be. We also replaced TCE. TCE came along because earlier solvents were flammable and dry-cleaning plants went up in flames. So we developed TCE, but it is a carcinogen. However, the limits we are requiring on TCE use are ridiculous. You've all read about Silicon Valley, where officials shut off 35 water-supply wells. Only two of the wells had significant health risks (Ames, 1987), and it was determined that the two worst wells in Silicon Valley were 1,000 times safer than drinking two glasses of wine, and 15 times safer than breathing most indoor air (Tierney, 1988).

The wells in Woburn, Massachusetts have fewer carcinogenic substances than normal tap water, and yet they were all closed. The reasons for cancer in America are not determined scientifically. Doll and Peto (1981) explain that between 25 and 40 percent of our cancer is a result of tobacco; between 10 and 70 percent is a result of our diet; occupational cancer causes are between 2 and 8 percent; alcohol causes between 2 and 4 percent; viruses are estimated to cause between 1 and 10 percent; and pollution causes between 1 and 5 percent.

Improving the Atmosphere

None of this changes the law, which all of you must abide by. You have to respond to the EPA maximum contaminant levels. We have no choice. Thus you may say that everything I've written up to this point is irrelevant to your performing your job, whether it's determining if someone can use a water supply, or determining if something needs to be cleaned up. I recognize that, and I don't expect to change that. What I'm striving for is an atmosphere in which the public can sleep better, where the public recognizes the validity of protecting our water supply and the importance of cleaning up some of the water supply without getting far sicker from anxiety-related diseases than they will ever get from environmental pollution.

I don't regret anything I've ever done in contributing to the passage of environmental protection laws. I resent, however, the ways in which those laws and imprecise science are used to rile the public into an atmosphere of unnecessary fear. Now some will say that if people don't stay scared, they won't do anything. I have a lot more confidence in my fellow man. I think we can tell people the truth, tell them what the real risk is, and be confident they are not going to walk away and forget about health risks. The public can help us make wise decisions.

Cleaning up to the public's expectation is impossible. The risk has already been distorted, and analytical techniques are too good. The public wants everything out. This was fine when we could only see a part per million. Now we are seeing a part per trillion, and a part per quadrillion is right around the corner.

We need to let the public know that we have been doing a magnificent job in the area of prevention. Our ground-water pollution threats are landfills, agricultural chemicals and industrial waste lagoons, underground storage tanks and leaking pipelines, septic tanks and highway salting, and on and on. There isn't one activity threatening ground water that we have not addressed competently in the last five years. We've probably already reduced by 80 percent emissions into the ground water from activities that are initiated today. That will not make your job any less important, because we've been making a mess for 40 years in many different ways. We have to at least assess the mess. We have to monitor it. We have to clean up some of it. And we will. But in the area of prevention, we are doing a good job. What's driving the crisis mentality now are largely environmental organizations that have been overzealous in describing the magnitude of the problem. They are made up of two kinds of people: those who want to save the world, and those who simply hate capitalism. Barry Commoner, one of the latter, said that MCLs of any kind are no good because basically they create a channel through which industry can pollute. He says you have to go to zero. What do you do when you go up into the North Woods, a great, beautiful unspoiled area where there is no industry for miles? You inhale the pine odor and you find out that pine odor is made up of polycyclic aromatics, carcinogens, in the cleanest air we have in the country.

You've read about one scientist who helped create the scare and is now trying to undo what he did—Bruce Ames. He's the fellow who developed the test for carcinogenicity with bacteria which indicated that so many things are carcinogenic, and he was the one who first alerted people to not have anything to do with these things. Then he began to realize that the levels we were ingesting were not a problem, and he started to look at nature and came up with many of the numbers that I've quoted in this article.

These two men have two different views toward society and water pollution. Barry Commoner concludes that there are natural alternatives to all man-made chemicals and, therefore, he wants to dismantle the chemical industry. Bruce Ames sees no need for organically grown food, but he has no interest in shutting down the health food markets, because that's the American way. We don't need to drink bottled water in most places in this nation, but there's nothing wrong with an entrepreneur's convincing people that bottled water is a good drink.

The environmental movement is the third wave in our redemptive struggle in western society. The first was Christianity, the second was

Socialism, and the third is Environmentalism. All, obviously, have good ideas to offer. But carried to extreme, they become negative. I believe that we, as ground-water scientists or engineers, health officers, or regulators involved in the environment, owe it to the people we serve to be honest and straightforward in dealing with risk assessment.

In "Toxicological Risk Assessment Distortions: Part II," which will be in the March–April 1990 issue, I will deal in greater length about the no-threshold controversy and the still relevant concept that the dose does, indeed, make the poison.

References

Ames, Bruce N., Renae Magaw, and Lois Swirsky Gold. 1987. Ranking possible carcinogenic hazards. *Science*, v. 236, April, pp. 271–280.

Ames, Bruce N. 1986a. Six common errors relating to environmental pollution. *Water*, v. 27, no. 4, Winter, pp. 20–22.

Ames, Bruce N. 1986b. Water pollution, pesticide residues, and cancer. November 11, 1985, Testimony to California Senate Committee on Toxics and Public Safety Management. *Water*, v. 27, no. 2, Summer.

Doll, Richard and Richard Peto. 1981. *The Causes of Cancer*. Oxford Press, New York, p. 1249.

Fleetwood, Scott and Paula Martin. 1989. Sensible ground water policy: an industry perspective. *Ground Water Monitoring Review*. Fall, pp. 71–72.

Higginson, John. 1988. Changing concepts in cancer prevention: limitations and implications for future research in environmental carcinogenesis. *Cancer Research*, v. 48, March 15, pp. 1381–1389.

Higginson, John. 1983. The face of cancer worldwide. *Hospital Practice*. November, pp. 145–156.

Office of Technology Assessment, Congress of the United States. 1989. Identifying and Regulating Carcinogens. Marcel Dekker, Inc., 270 Madison Ave., New York, NY 10016.

Paustenbach, Dennis J. 1989a. Health risk assessments: opportunities and pitfalls. *Columbia Journal of Environmental Law*, v. 14, no.2, pp. 379–410.

Paustenbach, Dennis J. (editor). 1989b. The Risk Assessment of Environmental Hazards. Wiley Interscience, New York. 1155 pp.

Rodricks, Joseph V., Susan M. Brett, and Grover C. Wrenn. 1987. Significant risk decisions in federal regulatory agencies. *Regulatory Toxicology and Pharmacology*, v. 7, pp. 307–320.

Spaulding, Roy F. 1989. Complexities associated with the interpretation of trace pesticide levels in ground water. *Ground Water Monitoring Review*, Fall, pp. 79–80.

Tierney, John. 1988. Not to worry. *Hippocrates*. January/February, pp. 29–38.

Reprinted, with permission, from
Journal of *Ground Water*,
Vol. 28. No. 1, January–February 1990

A New Measure of Risk

Jay H. Lehr, Ph.D.

Please see biographical sketch for Dr. Lehr on page v.

Misconceptions abound today in the field of risk assessment. Let me make an effort to clear up some of the confusion by simplifying the process through some innovative though basic techniques. They should provide you with the necessary tools to educate others in your sphere of influence.

My concern for risk assessment stems from my being a bleeding heart liberal, concerned with the inequities dealt various groups and individuals in life. I want to see public and private funds used in a manner that will help improve the human condition.

While I have been earning a paycheck as a hydrogeologist for 37 years, I only began studying risk assessment five years ago. My effort culminated in a series of editorials I wrote for the Journal of *Ground Water* called "Toxicological Risk Assessment Distortions," which also appear in *Rational Readings on Environmental Concerns*. This paper deals with a new concept for which I drew heavily from the work of Dr. Bernard Cohen at the University of Pittsburgh.

Costs Must Be Considered

Risk management attempts to balance the benefits of action against risk with the cost of risk reduction. Some people think that costs of risk shouldn't even be considered. They believe we should just eliminate risk wherever we find it. That is an incredibly naive attitude and one that would not only bankrupt our nation but, in fact, also bring excessive premature death to the population.

The easiest way to talk about the economics of health-related decisions is in terms of hospital administrators who make daily decisions with regard to life. They may have to make a decision to cut off life-support systems when continuing costs cannot be justified in light of poor chances of recovery. This is a life and death decision that involves money.

When you chase risks too far, you not only do not eliminate further risk, you actually increase other risks. This happens when money is taken out of the community that could be used to eliminate other, more important risks instead of trying to reduce a negligible risk to zero.

All of the readers of *Ground Water Monitoring Review*, to some extent, earn a living chasing risk and eliminating contamination. On the other hand, most of us know that we are all better off to play it straight by helping the public understand what the real risks are. We don't need to expand or distort the risks of ground water contamination. There's enough work to be done in this field that is very real. For some of you, what I say here might reduce the importance of the work we do. While this is arguable you certainly will not be out of a job. There's more than enough work to employ us, our children, and our grandchildren with ground water contamination problems in a realistic and undistorted way.

The Diagnosis

The reason we have many problems with risk assessment and toxicology is that toxicology is a very immature science, which bases itself on one-time experiments and linear extrapolation models. The data developed can at times have little bearing on fact. Risk assessments are based on maximum contaminant levels that are developed from rodent bioassay studies. Few rodent bioassay studies will hold up under scrutiny during the next few years. We are finding out that the megadoses we are giving rats are causing cancer, not the molecular construction of the chemicals to which they are being exposed. There have been a number of papers in *Science* recently recognizing that "mitogenesis," the splitting of cells caused by high-dose inoculation, is actually causing cells to mutate. We are, in essence, creating a kind of artificial cancer, which is not related to real cancer risks.

The Simple Arithmetic of Risk

Risk assessment is grocery-store arithmetic. You determine how many people die vs. how many people are exposed to a risk. Guesses as to exposure can throw things off, but generally we can come up with good numbers. The easiest example is the automobile. There are 250 million people in America. 50,000 die every year in automobile accidents, thus your chance of getting killed in an automobile this year is one in 5000. Assuming a 70-year life-span, you have that chance of one in 5000 for each of your 70 years. Therefore, your actual risk of getting killed in an auto over a lifetime is about one in 70.

The new term I am going to introduce to you is the LLE, or Lost Life Expectancy. You can translate all risk into lost life expectancy, which will make sense to most of us. If 1000 people at birth have a life expectancy of 70 years, 1000 people should live a combined total of 70,000 years; but we know that of that 1000, 14 will die in an automobile accident. The 14 people who will die, will die at an average age of 35. Some will die as infants, some at 5 years, some at 10 years, some at 65. The people who die will thus lose an average of 35 years of their life. When you multiply 35 × 14, you get 490 years lost by that 1000-person population due to the automobile. By subtracting the 490 years from 70,000 years and then dividing back by the 1000 population you find that each person now has a life expectancy of 69½ years. The existence of the automobile gives you an LLE of six months. This figure indicates that if we eliminate automobiles, everyone would live six months longer. That isn't true, of course, because you would then have to ride horseback and riding horseback also has an LLE. If you walk, that also has an LLE, so you have to look at all the alternatives.

Every risk can be calculated in terms of the Lost Life Expectancy of a large population, in order to prioritize that risk to the population. This data is readily available because we know how people die. These LLEs are accurate statistics as compared to the tremendous inaccuracy of determining toxicological MCLs (maximum contaminant levels).

There is a danger in everything. There is a danger in traveling. There is also a danger in staying home. Twenty-five percent of all accidents occur in the home. There is a danger in eating food, because food has carcinogens; there is also danger in not eating food—malnutrition. There is a danger in working. There are 120,000 job-related deaths each year. There is a danger in exercising. There is a danger in not exercising. These inherent dangers do not mean we shouldn't try to reduce risks. The question is deciding how to effectively reduce risk.

Let's focus our attention on risks that have been well publicized by the media. The media tells scare stories and the public reacts. If we could teach people risks in terms of LLEs, they would be better able to discern how we should be spending our money. If you do something that has a 1 percent risk, that means the chances of losing whatever life you have left is one chance in a hundred. So if you're 40 years old and you are going to do some dangerous act with a chance of getting killed of one in a 100, you have one-hundredth of a chance of losing your remaining 30 years. With a group of 1000 people who have 30 years left allotting a 1 percent risk opportunity for the group computes to an LLE of .3 years.

There are many ways we can extend lives in America because we know there are certain things that have been proven to reduce life. Probably the best way we could save lives would be to have more **mar-**

riage brokers because single people don't live as long as married people. The life expectancy of a white single male in America is six years shorter than a white married male. For women, the discrepancy is a little less. A single woman has three years fewer to live than a married woman. A caring relationship leads to better medical attention; less loneliness; and lower numbers of alcoholism, suicide, and heart disease. I'm not saying married people are always happier, but they do, on average, live longer.

The LLE for **smoking** one pack of cigarettes a day computes to 6.4 years for men and 2.3 years for women. Now you might want to look at that and say if you like smoking so much why not live six fewer years and enjoy the 64 years that you will live. But you have to consider the quality of life too—the coughing, the short-windedness, the limited athletic ability, and so on. Still if you smoke, the odds are you are going to reduce your life expectancy.

You've heard a lot about **passive smoking** and the calculations that have been done on this. We've studied men and women who are smokers' spouses. If you live with a smoker, the LLE is 50 days. You are going to lose 50 days of your life by inhaling that smoke. That's not nearly as risky as we have been led to believe.

We weigh dead people and we know that for every pound that you are over the proper **weight** for your size, you die, on average, one month prematurely. Really, that's not all that bad considering how much fun eating is. If you're 30 pounds overweight on the average, you are going to live 2½ fewer years than someone who is the ideal weight. But then again, just like smoking, being heavy causes other problems that affect the quality of life.

What do you think is the largest cause of reduction in life span?

The answer is **poverty.** The difference in economic scale in the United States between the highest economic scale and the lowest is a seven-year life expectancy. We find that well-to-do, well-educated people with good jobs tend to live seven years longer than those in the lowest economic level. In Chicago, the difference is nine years. In Canada, the spread is 10.8 years between the highest economic level and the lowest. In Finland, it's seven years. Here are some real life expectancies that relate to the poverty problem: The life expectancy in the developed world (the United States, Japan, and Western Europe) is about 75 years. In Poland, Czechoslovakia, and Romania, it's 72 years. In Mexico and Central America, it's 67 years. In Brazil and Turkey, it's 64 years. In India, Egypt, and Iran, it's 59. In Central Africa, it's 45. In Afghanistan, the life expectancy is 38 years.

Lack of skills to hold a job, and lack of education are a deadly combination; they contribute significantly to premature death. A college-educated person will live 2.6 more years than the American average, and a boy who drops out of grammar school will live 1.7 fewer years than the average. Add 2.6 years to 1.7 and you have 4.3 years of lost life expectancy for a kid who drops out of school vs. one who goes through college. It's clear that if we want to save lives, we should educate people.

The LLE for combat soldiers is lower than you would imagine. If you served in the Army in Vietnam, the LLE was two years. If you were a Marine, it was 2.8 years. If you were in the Navy, it was .5 years. If you were in the Air Force, it was .28 years. In the Persian Gulf the LLE for coalition forces was measurable in days.

All accidents on the average have an LLE of just slightly over a year. Driving an automobile every mile, your LLE is .4 minutes, the same as walking across the street. A coal miner has a reduced life expectancy of three years. **One year of unemployment reduces life expectancy by 500 days.** This is forced unemployment, not retirement. The rates of smoking, heart attacks, stress, suicide, and homicide go up among the unemployed. These latter facts result in the following grim statistic: If the unemployment rate goes up 1 percent, 37,000 people will die a premature death. With 1 additional percent of unemployment, we will also see 4200 more hospital admissions and 3300 more people sentenced to prisons. Obviously, any action that is taken that increases unemployment does a great deal to harm the health and life expectancy of the public.

Hurricanes and tornadoes have an LLE of one day; airline crashes $\frac{1}{10}$ of a day; major fires $\frac{7}{10}$ of a day; chemical releases $\frac{1}{10}$ of a day. You hear that if you broil your steaks over charcoal you produce nitrosamines. The LLE for eating $\frac{1}{2}$ pound of broiled steak a week is four hours. Over a lifetime, a teaspoonful of peanut butter a day has an LLE of 1.1 days; chlorination of our drinking water, .6 days; a diet soda every day is two days. Alcohol abuse has an LLE of 230 days; other addictive substances, 100 days.

Remember when the Skylab was falling down? The LLE for the world population for the fall of Skylab was .002 seconds. Radioactive waste burial has an LLE of 10 seconds. If you lived in Harrisburg, Pennsylvania, during the Three Mile Island accident, the LLE was 2 minutes. Lightning is 20 minutes. High school football is $\frac{3}{10}$ of a day. College football, $\frac{5}{10}$ of a day. Boxing is eight days. Hang-gliding is 25 days. Parachuting is 25 days. (I'm a skydiver and will be happy to give up a month of my life for it, but of course I know in reality I will give up all or nothing at all.)

Averages don't mean much if you're killed in one of these ways. But in terms of spending public money, we've got to deal with averages. So let's look at the costs vs. the risks and see where we can most effectively spend public money. Let's use the LLE to determine how to get the best bang for the buck. If we used government funds to install air bags in all our automobiles, we would save 15,000 lives at a cost of $300,000 per life. If we all had better tires, we would save 1800 lives from blowouts, but that would cost $6 million per life. Smoke alarms would save 2000 lives at a cost of about $120,000 per life. General cancer screening for women could save one life at a cost of $90,000. There are a variety of cancer checks that we could require of the population that would save lives at the rate of $20,000 to $120,000 per life. Blood pressure checks could save one life for every $150,000. Mobile intensive care units that are dispatched quickly to heart attack victims can save a life for $24,000 in big cities and up to $60,000 in smaller cities.

But now look at this money and see what we can do worldwide. The World Health Organization can save a life for 50 bucks in Gambia; for $20, in Indonesia; a $550 investment saves a life from malaria in the Third World; $2000 in general health care saves a life; $4000 worth of work on water sanitation saves a life; and $5000 worth of nutritional effort all save a life in the Third World.

But look where the EPA is spending our money. A sulfur-scrubber on a smoke stack saves a life for $500,000. The removal of radium from our water supply (now we are getting down into parts per billion that affect all of you) can save a life for every $2,500,000 spent. We used to save a life for every $8 million spent on a nuclear reactor. Now with the added cost for nuclear reactors because of safety, we've added $2 billion to the cost of every nuclear reactor which saves .8 lives. This computes to about $2.5 billion per life saved.

We are not spending our money wisely in saving lives. Is it a result of bad government? No, there are a lot of good people in government who know these facts. It's the result of a scientifically illiterate public that is led around by the media.

We live in a democracy; the government must be responsive to the public. The unfortunate fact is when the public doesn't understand the situation, democracy leads you in the wrong direction.

I'm certainly not recommending a change in government. What I am recommending is that educated people who are making a living working on these problems should try to explain the realistic level of risk to others. The problem is that the people don't understand the scientific process. They don't know how to evaluate data. They don't know the right questions to ask and, therefore, the people who end up in control are members of the Green Movement.

I have helped to write every existing piece of major federal legislation that deals with ground water in this country. I think it was an important effort. It has now been taken too far, because we are chasing risks that are too small. The people behind it are misguided environmentalists who want to save the planet but are not trying to save human life. They are directing public resources into an area that does not save significant life and are taking those resources away from areas where great numbers of lives could be saved. Lack of skill, lack of education, and poverty are America's problems and our money should go to wisely address these problems. Begin to think in terms of the Lost Life Expectancy. Try to analyze, evaluate, and compare risks correctly. Share this information with your neighbors, your colleagues, and everybody in your sphere of influence. If we all do this, maybe some day we will raise the scientific literacy quotient of the nation to a point where our resources can be spent more judiciously.

Reprinted, with permission, from
Ground Water Monitoring Review,
Vol. XI, No. 2, Spring 1991.

Scientific Processes

The Credibility Gap Between Science and the Environment

Hugh W. Ellsaesser, Ph.D.

Please see biographical sketch for Dr. Ellsaesser on page 404.

Whom Do You Believe?

Now we come to the heart of the matter. Here I am, saying that almost everything you hear, including that put out by the National Academy of Sciences and many international scientific organizations, is wrong. Whom do you believe?

That's a very good question and one I have been pondering for some time.

Real Problems Versus Imaginary Problems

Until relatively recently, man- and womankind have had sufficient real problems in wresting a livelihood for themselves and their families that they did not need to dream up imaginary ones to occupy their time.

Ben Wattenberg (1974), in *The Real America*, analyzed this problem 16 years ago. He found then, that during the previous 14 years in which we had been "visited by so many curiously ephemeral crises, national failures and other such rhetorical luridities," that conditions had "actually improved more rapidly than at any time in recent American history, in fact, improving to such a degree as to create a new human situation, a society whose massive majority is 'middle class.'" But this advance brought something else; "... there is an entirely new element in American life, because the career reformers have learned the usefulness of extravagant rhetoric; they have organized as never before; ... And most disturbing of all—they have captured not just the pulpits, the cowtown editorial offices and the enthusiasm of the Women's Christian Temperance Union; *but they have now arrogated 'the media,' the reputed public conscience of America, a weapon that never even existed in the 20s* [italics mine]."

Ben Wattenberg's diagnosis of our predicament was arrived at two years earlier by John Maddox (1972), the long-time editor of the prestigious scientific journal Nature, in a volume he titled *The Doomsday Syndrome*. As one reviewer expressed it; "he dared defy the accusers, the prophets of ecological disaster whose writings are today's best-sellers."

Syndrome, you will recall, is a group of concurrent symptoms characterizing a disease. The "disease" Maddox addressed, in the words of Arnold Beichman (1972), was "doomsday prediction—inflammation of the statistic marked by a swelling of the hubris." Hubris, in case you're wondering means arrogance.

Ben Wattenberg's numbers give the beginning of this new era in human evolution as 1960. Most authors cite as the milestone of its beginning the publication of Rachel Carson's *Silent Spring* in 1962. At any rate, this new "disease" has now been rampant for nearly 30 years and as yet shows no sign of even having crested. Others have made the same diagnosis and sought its cause. The answers have all been some combination of the following: over-affluence, too long a removal from real problem, and a growing body of reformers who, in their efforts to gain sovereignty over us, concoct elaborate horror stories about the present condition of our institutions, our values, our very planet. An outstanding testimonial to the accuracy of these diagnoses is the consistency with which they have all been studiously ignored.

It is worth noting that there are now more than 3,000 national, regional and local environmental organizations in this country and that the 22 largest together receive nearly half a billion dollars a year and, not counting overlaps in their membership, represent over 13 million people, almost 5% of the U.S. population (Ron Arnold, 1990).

Look for the Bodies Before You Panic

You should also recall, that until relatively recently, we didn't pay too much attention to problems until they brought themselves to our attention, i.e., until we could count the cadavers or the invalids. Perhaps you remember all the fuss about health effects of air pollution back in the 1960s and 70s. So far as I am aware, the only evidence that air pollution has ever been responsible for a death in our smog capital, Los Angeles, is one suicide note. Even after the "Killer Smog" in London in December 1952, supposedly responsible for 4,000 "excess deaths" (the excess for the weeks ending Dec. 13th and 20th over that for the week ending Dec. 6th, or over the average for the corresponding weeks of 1947 through 1951), not one of the approximately 8,000 total cadavers found during these two weeks could be singled out as having died from "air pollution" or from a cause different from that normally responsible for deaths at that time of year. Even more significant, those interested in influenza wonder why there was no "flu" epidemic in early December 1952 in London since there were epidemics in other parts of Britain and across the channel in Europe at that time.

Since the so-called "Air Pollution Episodes" of the Meuse Valley Belgium in 1930, Donora Pennsylvania in 1948 and London England in

1952, it has become standard practice to compute mortality and morbidity rates statistically and not to worry about the fact that the actual cadavers and invalids can never be identified.

Never mind that the toxicologists are still puzzling over what actually killed those people. After half a century of trying, no toxin or air pollutant has been found to have been present in concentrations demonstrated to be hazardous to health. Those events simply became another argument, and a major one, in support of the growing presumption that theory is superior to data. In keeping with this practice, our present EPA Administrator, William Reilly, on April 1, 1990 estimated the premature deaths in this country due to air pollution at 50,000 per year (Brookes, 1990). It would be nice if he had only been conforming with the spirit of April Fool's Day, but he was reporting to Congress.

Known deaths from Chernobyl are 31, and from Three Mile Island—zero. Those statistically predicted to result from Chernobyl are settling on 28,000 (Loefstedt and White, 1990). However, Professor Don Luckey (1980), retired from the University of Missouri, has collected literally hundreds of studies which appear to show that we would all be better off—i.e., be healthier, live longer, have fewer genetic defects—if we had more exposure to radioactivity than we now do, up to 10 times more. In Figure 1, I have reproduced one of these studies. This one shows the total cancer mortality against background radioactivity by states. You will note that cancer mortality tends to be less in those states, such as Colorado, having the highest background radioactivity. Professor Luckey has predicted that exposure to the fallout from Chernobyl will actually prove to be a net health benefit (Luckey, 1991).

The Strangely Unquestioned Official Standards

Professor Luckey's arguments highlight the crucial element—the unrecognized Achilles Heel of the bureaucracies created to enforce government regulations—and that is the official standards driving the regulatory programs.

Numerous people, including myself, have claimed that the present National Ambient Air Quality Standard of 0.12 ppm for ozone could not be met in Los Angeles—even if man and his works were removed completely from the area. Yet EPA is mandated by current law to meet the standard regardless of cost. Why is there no public outcry to have this Catch-22 situation straightened out—to have this standard reconsidered?

Since this ozone standard has been exceeded on approximately half the days of every year since observations were first begun in Los Angeles over 40 years ago, it seems rather obvious to me either; that

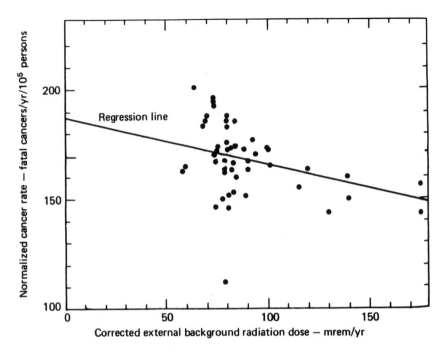

Figure 1. Total fatal cancer incidence against background radiation exposure, averaged over individual states in the United States. (Adapted from Cohen, 1980)

ozone ought to have produced some rather easily detectable health effects over this period, or that such a low standard isn't really required to protect human health. Doesn't this look like a reasonable question for public discussion before we ban the automobile?

The same kinds of arguments can be made on almost every government standard in existence. And yet the standards are never questioned. Whence comes this immunity to questioning? Where are the bloodhounds of the media and of the consumer advocate groups who so love to uncover skeletons in the closet of the executive branch of our government?

One firm conclusion that I, at least, draw from this situation is that all those pushing for blanket enforcement of our current environmental regulations, and that includes most of our media and consumer advocats as well as environmentalists, are not primarily concerned with overall public welfare.

I don't want to leave you with the impression that the paradigms into which "The Ozone Hole," "Greenhouse Warming," etc., have been constructed are due solely to the scientific specialists in these fields; although I do believe that many of them could have, and should have, done more to defuse these issues before they reached their present emotional states. Most scientists in these areas would prefer to be left alone to pursue their research in peace and quiet, particularly now that they see possible far-reaching and costly preventive programs impending.

These paradigms were built into headline issues by the hangers-on; the applied and social scientists who hoped to get paid for telling you what the impacts are going to be, the national and international scientific and political organizations who see these issues as vehicles for increasing their growth and power, the news media who know that excitement and danger best sell their product, the legislators and bureaucrats who find these ideal issues for increasing the fraction of the budget they can pull under their control and thus increase their power and prestige. Now we even have international lawyers and heads of state who hope to go down in history for their role in the creation of a Global Environmental Protection Agency or a Global Climate Control Agency.

And what about the environmental groups? It should be remembered that if environmental problems get solved, they are out of business! Their concern is not to solve environmental problems, but to find and publicize environmental issues which capture the attention of a fickle public and induce them to support environmental action with their votes and contributions.

At every level at which you look at these paradigms, essentially all of the incentives guiding the behavior of individuals is such as to reward those who can make the public believe that the problems are even bigger and more serious than previously believed and to punish those who dare to claim othrwise. Warren Brookes is about the only person I can think of who has been able to build an outstanding career out of trying to set the record straight. It's like "The Emperor's New Clothes" all over again—if you can't see and admire his non-existent new suit, it simply means that you yourself are not fit to occupy even your present status.

One Last Thought

Let me leave you with one last thought.

Even today, no one, and least of all the general public, has a clear understanding of the total economic cost of the NEPA (National Environmental Protection Act) or any of the many other environmental protection laws that have followed. If we are going to require EnIRs (Environmental Impact Reports) on every new program or construction project, isn't it about time that we also require an EcIR (Economic Impact Report) on every new environmental law or program?

Any laws or programs to eliminate the use of chlorofluorocarbons (freons) or to reduce the emissions of carbon dioxide to the atmosphere are bound to have far reaching and significant economic impacts. In my opinion, these ought to be carefully thought out *before* we make any firm decisions in these areas. Making a purchase before we determine the price is something only children and billionaires can afford.

References

Arnold, Ron, *Defending Free Enterprise,* Free Enterprise Press, 1990—see *The Wise Use Memo* 2(2), Center for the Defense of Free Enterprise, August 1990).

Beichman, Arnold, Doomsday has been canceled, *Christian Science Monitor,* 29 November 1972.

Brookes, Warren, 50,000 premature deaths?, *Washington Times,* 25 April 1990.

Carson, Rachel, *Silent Spring,* Fawcett Crest, New York, 1962.

Cohen, J.J., Natural background as an indicator of radiation-induced cancer, pp. 801–804 in *Proceedings of the 5th International Radiation Protection Association Congress, Jerusalem, March 1980,* Pergamon Press, New York, 1980.

Loefstedt, R.E. and A.L. White, Chernobyl: Four years later, the repercussions continue, *Environment* 32(3), 2–5, 1990.

Luckey, T. (Don), Hormesis and nurture with ionizing radiation, in *Global 2000 Revisited; or Reassessing Man's Impact on Spaceship Earth,* H.W. Ellsaesser (ed.), Paragon House, New York (an ICUS book, forthcoming, 1991).

Luckey, T. (Don), *Hormeis with Ionizing Radiation,* CRC Press, Boca Raton, FL, 1980.

Maddox, John, *The Doomsday Syndrome,* McGraw-Hill, New York, 1972.

Wattenberger, Ben, *The Real America,* American Enterprise Institute, Washington D.C., 1974.

The Serious Error Inherent in a Doomsday View

Max Singer, J.D.

Please see biographical sketch of Max Singer on page 363.

(Draft of talk given at annual meeting of Global Tomorrow Coalition, June 2, 1983, Washington, D.C.)

Introduction

I want to begin by complimenting those responsible for inviting me to speak here knowing that my message is that your organization and this meeting is built on a profound and harmful error, a basic misunderstanding of what is happening in the world. The extreme commitment to fairness and responsibility your officers displayed in making room on this program for my message stands out, even in this country which believes in fair and open debate as virtually no other in history. You have my thanks and my admiration.

What I owe in return is to not just lecture to you about the truth as I see it, but to make a desperate effort to reach your minds with a presentation of what I see as the fundamental issue so that you will perceive the choice that exists, or else find a resolution of the dilemma that gives full weight to both sides. I hope you will forgive me for talking to you about how I believe you think about these issues. Normally that isn't a nice thing to do; but I don't know any other way to really get my ideas before you.

Before I say why such a polarizing and portentous introduction is necessary, let me say just a few words about where I am coming from in terms of values and methods, and where I point to as far as policy.

The basic value that lights the view I will be sharing with you is humanism, a belief in the value of human life, a sense that human beings are in some way sacred—although I don't profess to know what that means. Although I do think that I know that neither belief in the sacredness of man nor pride in our success in changing the world is a challenge to or denial of God in any way, nor a claim of man's perfection or even innocence.

To descend to more practical issues, I want to emphasize that I am not speaking as an optimist. As you know there are three different kinds of optimist: First "the predicting optimist" who says that the

699

Giants are going to win the pennant this year. Second, the valuing optimist, like the little boy who says "life will be wonderful when I have a puppy." And finally the "recognizing optimist" like the little girl lost in the woods at night who reassures her little brother by saying that soon it will be morning and it will be light again. While I have to plead guilty to the third form of optimism I think I can assure you that just about all the facts or calculations I will use here today are technically—not politically—conservative. I used no hopeful or daring assumptions or calculations to reach any of my conclusions. I don't think I am depending on any long shots, or even on any even money bets.

The differences between us are not about policy. While many environmentalists and conservationists and their organizations have put forward the perspective that I will be challenging, I am not in any way challenging their values or policies in this talk. In fact I think that the alternative perspective I am going to present is a strategically sounder basis for fighting for conservation and protecting environmental values.

Therefore this talk will be humanist, technically conservative, and pro-environment and pro-conservation.

Perspective

Many of you must still be wondering how I could be so foolish as to say that this organization is based on an error when you must be very conscious of your sense of uncertainty, of complexity, or how many different issues you are concerned with, and of the many differences of opinion among yourselves. How could I think that you could all be wrong when you have such a large and diverse quantity of ideas? Since you are not monolithic and don't have a "party line" of any kind how could I say that you are in error together?

What I am going to try to show is that you share and have been applying an often unrecognized and unexpressed perspective. I want to try to show you how countless statements of facts, projections of the future, and urgent policy concerns that you accept are based on a specific perspective that I want to bring out for examination.

My impression is that many of you don't think that you ever chose a particular perspective. You feel that your perspective is only common sense, or prudence and a recognition of every generation's responsibility to future generations. How could there be an alternative to that kind of perspective?

My message today is that you do have a particular perspective—and that perspective is by no means natural or automatic; in fact it is wrong. There is an alternative perspective, one that is factually correct, that you could use in dealing with the facts you know and the concerns you wish to address.

The principal points of this talk are (i) to emphasize that two quite different perspectives are possible on the matters you are concerned with; (ii) to show how much of the discussion of Global Tomorrow has been dominated by one of these perspectives; and (iii) to show you that an alternative perspective is at least equally compatible with your values, policy programs, and many of your perceptions of the facts. Of course I will also point out that one perspective is false and the other true, but I don't care whether I convince you about that today. I will be satisfied if I can show you what the alternatives are.

In other words, while you can't all be wrong about hundreds of facts and dozens of issues, about which you have diverse opinions, to the extent that they are all informed by a single perspective, that perspective could be wrong, which is the idea I will try to present in these few minutes.

The key is not to argue about any particular facts, but to show you that there are two alternative perspectives that can be taken.

The reason I am taking this approach is illustrated by a quotation in Monday's *Post* article about the Kahn-Simon meetings in Detroit last weekend: "a consultant to Earthwatch acknowledged that the original Global 2000 projections may turn out to be wrong, but said that 'is not the point. The idea was to call attention to the fact that we are destroying our habitat. That is still true.'"

I do not want to get in an argument about the facts that are not the point. I want to challenge the fundamental perspective that the gentleman from Earthwatch felt was more important than the particular projections in Global 2000 or Global 2000 Revised.

Present Trends

Before I begin I want to tell you about a Gary Cooper movie that I vaguely remember from the days we used to see three movies every Saturday afternoon for 25 cents. It was a piece of froth in which the humor was based on Cooper, playing Hollywood's idea of a typical masculine engineer in those days, having to take care of a baby without any woman to help him. He was doing fine applying his engineering training and masculine logic to the tasks of diapering, feeding, burping, etc. But one thing troubled him. He reported to his friend that the baby's weight had already increased by 20% and that at that rate it would weigh over a hundred pounds before it was a year old, and he didn't know what to do about it. (Chart 1)

One of the supporting features of the perspective I wish to challenge is the matter of how to deal with "trends."

Nobody would project a baby's growth by saying that it grew 5% last month so the trend is that it will grow 5% this month. Everybody

Chart 1. Gary Cooper Rides Again

"At present and projected growth rates, the world's population would reach 10 billion by 2030 and would approach 30 billion by the end of the twenty-first century."

From the "Conclusions" of
The Global 2000 Report to the President

Chart 2. "Present Trends"

1. Gary Cooper (Baby grows 2% per month) — only for a while
2. G2000 (World population grows 1.7% per year) — at one phase of the demographic transition —
3. 10 year old boys and 15 year old boys
   (If they keep getting more difficult at this rate, what will they be like when they are 30?)
   — but fortunately they change regardless of what we do —
4. A ball thrown up (Slows then reverses)
5. A tree growing taller (Stops gradually)
6. A fish growing bigger (Keeps going)
7. A stone dropped from the Eiffel Tower (Accelerates then stops suddenly)
8. Pollution from increasing wealth (Gets worse then gets better)
   — at least in some cases —

knows that a baby's growth keeps getting slower. The only real trend is a comparison between the growth rate of a particular baby and some normal growth rate for that age.

On screen 1 are a list of some trends. (Chart 2) In each case the data will show something going in one direction, but being able to say anything about the continuation of the trend—whether it will reverse, like a ball thrown in the air, or accelerate and then stop suddenly, like a rock dropped from the Eiffel Tower—depends on knowing what kind of process is going on.

Using numbers from the past to think about the future requires making judgments about patterns. You have to evaluate the process and say what its nature is before you can say what the past implies about the future.

Of course we must treat the uncertainty of the future with respect. But, if we are to say anything at all about it, we have to try to understand it, not to project numbers but to talk about the future development of the patterns we think we see. It is not always hard.

For example, I will predict with confidence that the sun will rise in the East on May 12, in the year 4033. I make this prediction not on the basis of a statistical analysis of how often the sun has risen in the East in the past but on the basis of a view of the forces that move the Earth and the Sun.

Specifically, about world population, the "present trend" is not a 1.7% growth rate, which would get us to almost 30 billion by the end of the next century. The present trend of world population is a demographic transition in which falling death rates are followed by falling birth rates for a number of very practical reasons and the statistical evidence for which can be seen in imperfectly known but clearly declining fertility rates almost everywhere. Whether this "trend" will lead to a "leveling off" at a world population of 8 billion or 16 billion nobody knows. Nor does a recognition of the real trend, rather than the Gary Cooper trend, lead to any necessary conclusion about population policy. But which kind of trend you talk about says something about one's competence and perspective.

One of the reasons that perspective is so critical for thinking about the future is that the only realistic way to evaluate trends is to ask "what is happening?", that is, to have a perspective.

This talk is about two views about "what is happening" in the world, and now we can begin.

Two Perspectives

Perhaps it is unfair but I will refer to one view as the "doomsday perspective" and the other one as the "passage from a natural to a human world." I will first just present the tone of the "doomsday view." Then I will describe the passage from a natural to a human world—which is the basic thing that is happening in human history today. Then I will put the doomsday view more fairly before you in terms of some fundamental propositions, which can be said to be either true or false, and which are in fact false.

Before looking at the Doomsday View on Screen 2, (Chart 3) glance over the True Statements listed on Screen 1. (Chart 4) They are there to show you that we are living on the same world. I know as well as you about these basic facts that we cannot escape.

For example I am not oblivious to the fact that the amount of iron or cobalt or copper in the world is finite. More important, the sun's lifetime is finite, and when its fuel is used up, in a finite number of years, we do not know how human life on Earth can continue. The short-sighted among us will take pleasure from fact that sun's finite life seems very likely to last at least 100 million and 3 years from next Thursday. And

Chart 3. The Doomsday Perspective (Tone)

The world will be:
- More crowded
- More polluted
- Less stable
- More vulnerable

People will be poorer
Life will be more precarious
Threats to the future of mankind
Population approaching carrying capacity of the Earth
Better projections would show more intensifying stress

Chart 4. Some True Statements

1. There are only finite amounts of our essential raw materials, and we will need to use increasing amounts of them.
2. Population is increasing, people are getting richer, more raw materials will be needed; therefore the task of avoiding or coping with pollution will become much greater.
3. Changing technology and the effects of increased population and wealth will require major changes in the way we do things — economically and technically.
4. The multiplication of food production which is needed is a very large and complex task. Bad agricultural practices will cause much waste; there will be many harmful mistakes.
5. World population is growing rapidly and will almost certainly double in the next century.
6. Those countries whose populations are growing faster than 1% per year might become wealthy sooner if their fertility rate were to slow down faster than it is slowing down now (but the evidence for this is weak at best).

there is no reason to believe that anything else that we depend on is "more finite" than the sun. That is, none of the arguments about finiteness contain any basis for believing that the limits will bite us before the demise of the sun puts other issues on the back page. So "finite" is true but uninformative. It does not imply troublesome scarcity.

Similarly for each of the other true statements. They do not imply anything frightening or dangerous.

For example: #2 says pollution is getting to be a bigger problem. There is an old saying: "small children small problems, big children big problems." But the implication is not that it is bad that children get big. And that while pollution will undoubtedly be a bigger problem 50 years from now, by far the most careful and substantial study of the subject, by RFF, shows that the evidence is that we will have a cleaner environment in the US in 50 years while spending a smaller share of our economy on preventing and coping with major pollutants.[1]

A key point in contrasting the two perspectives is to distinguish between tasks and dangers. If you say something is a big problem it can mean either that it will be big job to take care of it (and maybe we don't even know exactly how to do it), or else it can mean that it is a threat to our lives or well-being. Those are very different kinds of things. Basically the correct perspective is that we have many problems in the sense of big challenging tasks, but practically no physical problems in the sense of major threats to human well-being.

The doomsday view talks about real problems as if they threaten the future of the human race when in fact the problems actually affect just how well we do, not whether we survive.

For example, Lester Brown, head of Worldwatch, a leading doomsayer, has frightened people with the specter of a disastrous erosion of the topsoil essential to the growth of our food. One who has heard about Lester Brown's works might think that the statements below about the feasibility of feeding vastly larger populations are much too optimistic because they fail to take into account the problem of topsoil loss. But in fact Brown's pessimistic estimates of topsoil loss don't contradict the confidence expressed here about having plenty of food in the future. Brown himself agrees that, even if his predictions come true, soil erosion would not be enough to noticeably increase the long-term cost of food. (Actually the biggest cost of soil erosion probably comes not from the harm to the land the soil leaves but from the harm caused where the soil goes. But these costs, while worth doing something to prevent, are not dangerous.)

The third true statement suggests a false dilemma (Chart 5) that is often used to defend the Global 2,000 Report and other examples of the Doomsday Perspective. There is no choice between the present and change. The present *is* change.

The fact that there needs to be change does not necessarily mean that we have to get excited about making sure that the change will come.

I have the Doomsday Perspective up at the same time as the list of true statements in order to emphasize that the true statements do not give a basis for accepting the doomsday perspective. But perhaps some people are a little careless and take these facts as implying the doomsday perspective.

Chart 5. A False Dilemma

"How can you believe that things will be alright if we fail to make changes? Sticking with present policies would produce disaster."

But the present is like a river, it is continuously changing. "Present Policies" include the dynamics of continuing change.

The choice between:
> 1. Current practice, and
> 2. Change

is a false dilemma;
> The first alternative is the same as
> the second; current practice is change

The real choice concealed by the false dilemma is between:
> 1. Continuing rapid evolutionary change, and
> 2. Radical change.

"Life ... will be more precarious ... unless the nations of the world act decisively to alter current trends." "Efforts now underway around the world fall far short" "An era of unprecedented global cooperation and commitment is essential." From *The Global 2000 Report to the President*

We don't need to go into detail about the doomsday view at this point. It is nicely summed up in the phrase from the *Washington Post,* "we are destroying our habitat." Let's turn to the alternative perspective which should replace it.

Passage Perspective

The basic situation is very clear—and exciting. (Charts 6 and 7) From man's beginnings on Earth until 100 or so years ago all societies and nations were poor. Recently man learned how to create wealth. Already some one quarter of us live in wealthy societies, and within another 100 years or so the passage will be essentially completed and most of us will live in wealthy societies.

It is an illusion of short-term perspective to see the world as divided between poor countries and rich countries. All countries were poor before. All countries will be rich in the future. What is seen today is a situation in the middle of a common passage from poor to rich in which some are ahead of others, breaking the path.

Chart 6. Where We Are in History: The Passage from a Natural to a Human World

Before the Passage the Natural World 100,000 Years	The Passage About 1800–2166	After the Passage Human Worlds 100,000 Years
People die young		People live full human life span
	Change	
Nature — mainly dirt, disease and death — occupies most of human attention.	and	People spend most of their time concerned with human creations: science, art, politics, commerce, and war.
	differences	
No human society is wealthy.*	among countries	Most people live in wealthy countries

*A society or nation is "wealthy" if the great majority of its people live in healthy conditions, are educated, have "human" working conditions, can make some choices and know many people and places. (Financial indicator: About $3,000 per capita (1980 $1)

Chart 7. Where We Are in History: The Passage from a Natural to a Human World (continued)
(All numbers are approximate.)

Before the Passage the Natural World 100,000 Years	The Passage About 1800–2166	After the Passage Human Worlds 100,000 Years
Population growth rate:		
Very Slow (.1% per year)	**Fast** (.6% per year)	**Very Slow** (.02% per year)
Increase in human life expectancy:		
Almost None (from 20 to 30)	**Much** (from 30 to 75)	**Almost None** (from 75 to 85)
Increase in average income:		
Very Slow ($100 to $250)	**Fast** ($250 to $25,000)	**Very Slow** ($25,000 to ?)

Chart 8. Why We Don't Have to Worry About Raw Material Scarcity (Key Points)

1. Practically the only raw materials we use are food and fuel. (e.g., U.S. spends only $60 per person for all the metals we use.)
2. Conventional agriculture can provide enough food for 30 billion people — without scientific breakthroughs.
   - There are another 2,000 million hectares of agricultural land available at reasonable cost   (Now use 1200 mil. h.)
   - Over 200 years, we can get at least 5 times as much food per acre as we do now   (World-wide average)

   (Including off-farm methods for avoiding waste)

   (Unconventional sources of food, or new science, could probably replace a significant share of conventional food crops if needed.)

3. If necessary solar energy can — after 200 years — provide all the energy we want at affordable costs.

   (In an all-solar world, 200 years from now, energy would not cost more than 30% of GWP, even if the unit cost of solar energy were more than 4 times higher than the average unit cost of the energy we use today.)

It is like going to a family with a 5 year old and a 10 year old and saying I see you have a large son and a small son—how unfair it is for the small son.

I use the words "wealth" and "poverty" without moral connotations to refer only to the extent of control over resources that can be used to provide things that enable people to live what we would regard as human lives.

Past societies were poor because they didn't have enough resources to take good enough care of the mass of their people so that they could live a full life and have some opportunity to choose among human creations. A society is wealthy when the great majority of its citizens live a human life—which is distinctly different from the lives that virtually all people lived until this century.

There are two questions that make people doubt whether the passage to a human world is a realistic picture of what is happening. They are: roughly: "Is there enough stuff? and why should we expect wealth to continue to spread? I can only give you a sentence or two on the answers to each of these here. (Charts 8, 9, and 10)

Chart 9. Why We Should Expect That the "Whole World" Will Be Wealthy Soon (In 100 — 200 Years)

1. Wealth comes from productivity.
2. Productivity comes from the society learning how (plus the accumulation of capital)
3. Many kinds of society have learned how

Economic growth is primarily a social learning process.

4. The more wealth and wealthy societies there are in the world, the easier it is for a country to learn how to become wealthy.
5. Learning and capital accumulation take time:

Effects of Different Possible Long-Term Growth Rates

	Annual Rate of Growth	Amount Income Will Be Multiplied in a Century
Par	2%	8 times
Good	3%	16 times
Very good	4%	32 times
"Miracle"	5%	64 times

Conclusion

Let me say before finishing why I care so much about this question of which perspective should be used. (Chart 11) Your feeling may be that even if the doomsday perspective is a bit frayed at the edges it can't hurt; better to be too cautious than too hopeful. Here's why you shouldn't imbibe that nearly invulnerable cop-out.

We don't have time for all the reasons but I do want to say something about #2. The necessary implication of the Doomsday Perspective (Chart 12) is that we are sailing together on spaceship Earth and that eventually babies will be dying of thirst. Meanwhile we Americans are using the drinking water for long leisurely showers. Who can live with himself if he has that picture of what he is doing? There is no moral justification for the United States that is consistent with the Doomsday Perspective. The Doomsday perspective is a direct attack at the moral foundations of this country. Why should we tell slanderous falsehoods that serve no useful purpose about ourselves? That is carrying "an excess of caution" too far.

Chart 10. Thinking Poor and Thinking Rich

(In relation to raw materials, pollution, etc.)

1. The medium term perspective — the next 100 or 200 years
   - Getting to the rich, long term world
   - Problems caused by shortage of money in poorer countries
     (Bad short-cuts often used where there isn't much money.)
   - One big issue is how fast countries that are now poor learn how
     to increase productivity,* and how fast they accumulate capital.

2. The long term perspective — after the next 100 or 200 years
   - Potential problems of a **rich world** for a long time
   - Over 10 billion people
   - World-wide average per capita income over $25,000 (1980 $)
   - 90% of people in countries with per capita income over $3,000.
   - Tens of thousands of years

Conclusion: While there is still room for "thinking poor," it is a dying
perspective; for the long term we must "think rich."

*This is not just an economic or technical matter. It is a social, psychological,
cultural, process of great subtlety and wide variation.

Chart 11. Why It Makes a Difference Which View You Have

Both views are compatible with an urgent concern about real problems
and practical tasks.

1. The realistic view is a better long run basis for supporting environmental and conservationist values.

2. The doomsday view inevitably and falsely implies that the US is
   fundamentally immoral and harmful to the world.

3. The doomsday view diverts attention rom real problems.

4. Truth is to be preferred.

   It is good for children to see the world as it is.

I have put up on Chart 13 a way of integrating not really the two
perspectives, but your perceptions with the passage to a human world
perspective. The basic point is just that before getting upset at some
horrible news about our habitat—maybe that 1/4 of our soil is about to

Chart 12. "The Doomsday View" (Propositions)

1. The overall **physical condition of the world is getting worse** as more and more people use it more and more intensely.
   - Raw material scarcity
   - Pollution
   - Dangers to health (chemicals, nuclear wastes, etc.)

2. As raw materials are consumed at ever-increasing rates, the world is eventually **doomed to have great raw material scarcity** problems (perhaps combined with worsening pollution).

3. Current world population trends **threaten to exceed the earth's carrying capacity** in the next few centuries. If they continue there would be ten times the current population in only a few centuries and there would be a shortage of living space.

4. Already the amount of **arable land is shrinking** and losing quality as farmland is converted to urban use and to roads, and as topsoil is eroded.

5. Therefore, the basic long term problem for mankind will be **how to adjust to increasing scarcity.**

Chart 13. Integrating the Two Perspectives

Dilemma

1. We are rapidly changing from a natural to a human world.

2. But all these terrible problems and dangers exist.

Resolution

1. For each "terrible problem or danger" ask:
   "Does it really threaten the passage to a human world?"
   (It almost certainly won't.)

2. If not:
   Work on the problem, but put it in the correct perspective.
   - Physically we can expect to be OK
   - Work and wisdom can speed the passage and reduce the costs
   - Overreaction and radical change usually make things worse
   - The US is part of the solution not the problem

Chart 14.

Some Things I Am Not Saying

1. We're safe. There is no danger.

2. Nothing needs to be done.

3. We know the future.

4. We now know exactly how to solve all our problems.

5. "Progress" is inevitable. (i.e., "The world keeps getting better.")

6. The world will be better in the future.

What I Am Saying

1. The threats to mankind from scarcity, pollution, or any limits imposed by the physical condition of the earth, are small.

2. Our basic expectation should be that almost all countries will be wealthy within a century or possibly two.

3. Mankind's real worries are human — not physical
   - Decadence (Too much wealth, power, freedom, and knowledge)
   - War
   - Tyranny

float out to sea or something which sounds terrible—ask whether if true it matters enough so that it would jeopardize the passage to a human world. If not work at the job of fixing the problem but don't use it to frighten the kids.

But I want to close by making sure that no one goes home thinking that I am saying any of those things listed on the top part of Chart 14, the lower half of which summarizes much of my conclusions.

Reference

1. *To Choose A Future,* R.G. Ridher and W.D. Watson, Johns Hopkins U. Press, Baltimore, 1980.

The Scientific Process—Part I: Do We Understand It?

Jay H. Lehr, Ph.D.

Please see biographical sketch for Dr. Lehr on page v.

From January to May of this year, I harangued my readers over the public's and the scientific community's inability to sort out reasonable assessments of toxicological risk. At the root of the problem is, I believe, an inadequate understanding of the philosophical and intellectual principles upon which scientific thought is based.

When I was young, the world held only a fraction of the information and knowledge available today. It was possible to be "a man for all seasons," a Renaissance man, a person with both breadth and depth of knowledge. In a real sense, that is no longer possible because the depth of available knowledge is so great that one with broad interests can only scratch the surface. The result is that we barely try to do even that and, instead, focus narrowly on our specialty attempting to absorb all of the information in that field. We lose sight of what ties all of science together: the scientific thought process. We become so busy learning facts, procedures, equations, regulations, instrumentation, and legal comportment that we are less and less scientists and more and more human computers dedicated to the recognition and utilization of technical information. We no longer think like scientists or act like scientists; we find ourselves buried in data.

Every one of us who fits this description, and I see few denying the legitimacy of my indictment, can continue on our present path to a prosperous, productive, and useful career. But we will not concurrently advance the cause of science or improve the scientific literacy of either our clients or the public. We can, however, accomplish all of the above with only a little extra dedication to becoming a real scientist rather than a ground-water technologist.

What Science Is

The humanities, which include such subjects as philosophy, religion, and the arts, deal with ideas about human nature and the meaning of life. Although important, none of the ideas expressed in these subjects

713

can be scientifically proven. There are no tests to determine the correctness of a philosophical system; no one can quantify the feelings expressed in a painting, nor can we find errors in a symphony or poem.

A theory developed by a scientist, however, cannot be accepted as part of scientific knowledge unless it is verified by the studies of other researchers. This characterization of science sets it apart from other branches of knowledge. Joseph Dauben, writing for the *World Book Encyclopedia* (1988), describes six methods commonly found in the development of scientific discoveries: (1) observing nature, (2) classifying data, (3) using logic, (4) forming a hypothesis, (5) conducting experiments, and (6) expressing findings mathematically.

Society today appears to be watering down the components of the scientific process and beginning to treat science as loosely in its relevance and acceptance as it does the humanities.

Daniel Koshland, Jr., in a humorous piece in *Science* ("Two Plus Two Equals Five," 1990), emphasizes this point in a tongue-in-cheek interview in which a scientist defends his mandate to tell people that two plus two equals four. "That's exactly where your views are wrong," a public relations expert tells the scientist. "A recent poll shows that 50 percent of the people think two plus two equals five and almost every TV network agrees with them. These people have rights; they believe, sincerely, that two plus two equals five and you take no account of their wishes and desires. Simply imposing two plus two equals four on them is not democracy." And, indeed, it is not. But it is science, and it is not amusing to recognize that the public fails to understand what science is.

Science is much more than a body of knowledge. It is a way of thinking. Science invites us to let the facts in even when they don't conform to our preconceptions. It counsels us to carry alternative hypotheses in our heads and see which ones best match the facts. It urges us to maintain a fine balance between a sincere openness to new ideas, however heretical, and the most rigorous skeptical scrutiny of everything. No one has said it more eloquently than Thomas C. Chamberlin, one of the greatest geologists who ever lived. Writing for the *Journal of Geology* in 1897, Chamberlin said,

> "In developing the multiple hypothesis, the effort is to bring up into view every rational explanation of the phenomenon in hand and to develop every tenable hypothesis relative to its nature, cause or origin, and to give all of these as impartially as possible a working form and a due place in the investigation."

One way to view science is as an intellectual approach to thinking with the primary goal of minimizing the chance of being misled by an observation. In other words, it is a set of rules to avoid being fooled. The scientific method teaches us to make many observations and examine them as a group. Bad science develops rules based on single observations. The poor scientist, the pseudoscientist, and the public suffer a high probability of being fooled in this manner. As we approach the 21st century, there is still a great need for reliable observation on what is happening in the real world, for testable hypotheses and for well-designed experiments. Concurrently, the field of statistics can be a valuable servant of science by offering rules for determining the need for multiple observations required to yield a high probability of truth.

It is also important to differentiate science from technology. Although they are intimately related, they are different. Science is the systematic search for new knowledge; technology is the practical application of that knowledge. Science discovers that electricity can be changed into light; technology engineers a practical light bulb. A technologist can tell you how light bulbs are made; a scientist can explain how they give light.

Science in History

Science is a never-ending search for truth. We may not always reach our goal, but the search is satisfying nevertheless. Ignorant people may turn on self-assured scientists with the oft-quoted expression of Socrates: "If I am the wisest man, it is because I alone know that I know nothing." In reality, Socrates pretended ignorance to lure others into propounding their views, then by a series of innocent questions, forced them into an admission of ignorance. From this clever technique, we get the Socratic system of examination in law schools and in our courts, which is why scientists almost always lose to lawyers in courtroom battles. (Actually, Socrates was such an arrogant know-it-all that it is a marvel that the Athenians waited until he was 70 before giving him a poison cocktail!)

As smart as the Greeks were thought to be in Aristotle's time, they were known for untestable hypotheses. A case in point was his explanation for gravity: "Objects fall downward because the ground is their rightful place." While Aristotle in his day, and we in our day, can observe a phenomenon closely and think we know what's going on, if we don't make careful measurements, we are likely to be fooled. This may be a common enough scientific principle now, but it took a great deal of effort for Galileo to introduce it in the 16th century and Lavoisier to finally establish it as recently as the 18th century.

Over the long history of Western civilization, probably no other person has been cited more often as an authority on every subject than Aristotle. Despite his enormous influence, Aristotle was wrong so often, it has been argued that he did more to delay the progress of science than to advance it. Roger Bacon, a 13th century scientist striving to encourage a new approach to knowledge, suggested burning all of Aristotle's books. "The study of them," he said, "leads to loss of time, production, fear and increased ignorance" (Asimov, 1989).

Reason is what was lacking throughout most of Man's history when it came to science. Reason is the faculty which perceives, identifies, and integrates the material provided by man's senses. It raises our knowledge from the perceptual level of the animal kingdom to the conceptual level which we alone can reach. Reason employs logic which is the art of noncontradictory identification.

On the opposite side from science we find what can be called mysticism. Today, to be more diplomatic, we might define it as pseudoscience: a collection of unsubstantiated claims. It is the acceptance of allegations without evidence or proof either apart from or against the evidence of one's senses and one's reason. It is a claim to some nonsensory, nonrational, nondefinable, nonidentifiable means of knowledge such as instinct, intuition, or revelation. Reason is the only objective means of communication and understanding that allows reality as its standard frame of reference. When people claim to possess some special means of knowledge, no persuasion, communication, or understanding are possible. And if others subordinate their minds to the illogical or unreasonable beliefs of charlatans, they are sowing the seeds of the destruction of the technological society as we know it. No one should be allowed to diminish your natural curiosity or exploratory drive. We must continue to seek answers and not necessarily those offered by the latest authority or even our former teachers. We must give up only those ideas that have proven false or irrelevant.

Imagination is an important scientific tool. Without imagination, a scientist's discoveries will be limited. Nobel laureate Linus Pauling has said, "The essence of scientific discovery requires that you look at what all people see but notice something nobody has seen yet" (Osiatynski, 1984). To ultimately break new ground, one must add to that 10 percent imagination and inspiration Edison's well-known 90 percent of perspiration. Thus, most discoveries have resulted from arduous experiments testing numerous hypotheses containing all kinds of suppositions or assumptions. The very moment of discovery may not be expected, but it is rarely an accident. Pauling also said, "There is room for neither mystery nor mysticism." A true scientist must believe that the world in all its manifestations is totally rational and comprehensible. All phenomena can be reduced by the human mind to some basic and physical mechanism.

Science Questioned

Scientists are not always as objective and dispassionate in their work as we would like to think. We often get our ideas through hunches and wild guesses rather than rigorous, logical processes. As individuals, we often come to believe something to be true long before we assemble the hard evidence that will convince others. Motivated by faith in our ideas and a desire for acceptance by our peers, some scientists labor for years believing a theory to be correct and vainly devising experiment after experiment in hopes of obtaining results which may support that position.

Important findings should not be accepted until other scientists have repeated the experiment or application in their own laboratories or field tests with their own searches for extraneous influences. The cold fusion experiments of our colleagues in physics are certainly the best case in point. Yet in our brushes with toxicology with regard to animal dose response tests, due to lack of time and money, we accept single tests for the incessant carcinogenic hypothesis.

The degree of rigor common to any specialized field of science varies considerably, usually as a function of that field's history and ability to do experiments. As a field develops, it accumulates enough observations to permit astute thinkers to perceive regular patterns in the phenomena observed. Chemistry and physics are two of the oldest sciences and perhaps the most developed. Toxicology is one of the youngest sciences and one of the least developed, and thus it is fraught with weak observational evidence.

Science need not, however, be complex; in fact, an innocuous, oft-forgotten principle of science is "parsimony" which states that one should choose the simplest explanation for a phenomenon from among more complex alternatives. It may not always be correct, but the laws of statistics are on your side. The greatest advances in scientific theory are usually based on the adoption of simpler sets of events to account for a phenomena. This particular principle was coined by the English philosopher, William of Occam, and is commonly known as "Occam's razor." My favorite example of it is "if upon a walk in your favorite pasture, you hear hoofbeats behind you, assume them to be those of a bull rather than a zebra." Another might be not to assume that a few parts per billion of anything will bring the human body to its knees.

What Do We Know for Sure?

There is a movement afoot in our ever-expanding technological world to discredit science by virtue of the ease with which one can find individuals with some form of scientific credential to dramatically disagree on a subject (e.g., consultants on either side of a court case). Those who wish to pursue a path to a preconceived objective, rather than to pursue truth or science, add fuel to the situation by offering up age-old bromides like "we don't know anything for sure;" or its inverse parallel, "nothing is impossible."

Bertrand Russell (1948) clarified this dilemma when he wrote "although every part of what we should like to consider knowledge may be in some degree doubtful, it is clear that some things are almost certain while others are matters of hazardous conjecture."

Rothman (1989) makes the point that scientists occasionally err on the side of caution in judging the certainty of their knowledge. In so doing, they feed the myth that "nothing is known for sure." Jacob Bronowski (1973) fell into this trap when he said most eloquently, "there is no absolute knowledge. And those who claim it, whether they are scientists or dogmatists open the door to tragedy. All information is imperfect. We have to treat it with humility."

In spite of a distinct need for humility, science today does know a great deal with a high degree of certainty. The person who prefaces all arguments with the statement that we don't know anything for sure, is exhibiting an extreme form of skepticism. Actually, his doubt is usually directed against scientists. The scientist, on the other hand, must be a pragmatic skeptic. Understanding the difference between dogmatic and pragmatic skepticism is important in applying the principles of science to the analysis of pseudoscience. The pragmatist is practical, even-handed, and objective, while to be dogmatic is to be prejudiced and one-sided.

If we start with the premise that nothing is known for sure, then it becomes impossible for anyone to say that a proposed action is impossible. However, if we agree that at least some of our knowledge is certain, or verified to a high degree of certainty, then we are in a position to say that some events are not permitted by nature. Our best computers have a hard time predicting next week's weather. On the other hand, what we know about the physics of meteorology allows us to be accurate in predicting the kinds of things that cannot happen. Perhaps one day this type of reasoning will assuage the public's daily fears toward the latest industrial or agricultural chemical.

Another myth: "Whatever we think we know is likely to change radically in the future," is a gross exaggeration. Actually, while our level of knowledge does continually change, it is less commonly the result of overturning a concept than the evolution of theories from less correct to more correct, commonly as a result of improved instrumentation and experimentation.

In many ways, environmental terrorism is thrusting society back into a Dark Ages mentality of ignorance and fear. John Galt, Ayn Rand's protagonist in *Atlas Shrugged* (1957) said, "to a savage, the world is a place of unintelligible miracles where anything is possible to inanimate matter and nothing is possible to him. His world is not the unknown, but the irrational horror: the unknowable. He believes that physical objects are endowed with a mysterious volition, moved by causeless, unpredictable whims while he is a helpless pawn at the mercy of forces beyond his control. He believes that nature is ruled by demons who possess an omnipotent power and that reality is their fluid plaything."

Can you see a parallel here with many of today's attitudes in spite of the fact that civilization is supposed to have advanced light years from the jungle?

Today, the voices proclaiming disaster are so fashionable that people are battered into apathy by their monotonous existence. Still, the anxiety under that apathy is real. What you see among modern intellectuals is the terrible spectacle of militant uncertainty and crusading cynicism. Why would people want to cling to the conviction that doom, darkness, and ultimate disaster are inevitable? Psychologists will tell you that when a man suffers from neurotic anxiety, he seizes upon any available rationalization to explain his fear to himself. He clings to that rationalization in defiance of logic, reason, reality, or any argument assuring him that the danger can be averted. What we are seeing today is "the neurotic anxiety of an entire culture" (Rand, 1960).

Critical Thinking

In our haste to train scientists to deal with ground-water problems, or any other technical discipline, we pay too little attention to the rules of critical thinking.

Since you're not likely to encounter it elsewhere, let us embark on a brief discourse on critical thinking. James Lett (1990) has distilled the subject down to six rules which are worthy of consideration. They are: (1) sufficiency, (2) replicability, (3) comprehensiveness, (4) honesty, (5) logic, and (6) falsifiability.

The evidence offered in support of any scientific claim must be **sufficient** to establish the truth. The absence of evidence to the contrary is never sufficient, and the credentials, knowledge, or experience of a proponent must not influence one's attitude toward an unsupported claim. No amount of expertise in any field is a guarantee against human fallibility, and even the sincerest human perception is notoriously inaccurate and thus inadequate. Unfortunately, because so much of ground-water science cannot be precisely proven, many issues must be decided on our judgment as consultants. In these cases, we should make every effort to adequately prove things to ourselves before we attempt to influence the beliefs of a judge or jury.

The rule of **replicability** provides a safeguard against the possibility of error, fraud, or coincidence. A single experimental result is never adequate in and of itself, whether the experiment concerns the production of nuclear fusion or the existence of animal carcinomas. Every experiment, no matter how carefully designed and executed, is subject to the possibility of explicit bias or undetected error. The rule of replicability, which requires independent observers to follow the same procedure and to achieve the same results, is an effective way of correcting bias or error. In ground-water science, we don't often do what would be described as experiments, but we do aquifer tests to determine hydraulic characteristics, and we commonly repeat the tests and/or analyze the data with different methods. This is an important part of replicability to us.

The evidence offered in support of any claim must be **comprehensive**—that is, all of the available evidence must be considered. It is never reasonable to consider only the evidence that supports the theory and to discard the evidence that contradicts it. For example, the proponents of biorhythm theory are fond of pointing to airplane crashes that occurred on days that the pilot, copilot, and/or navigator were experiencing low points in their intellectual, emotional and/or physical cycles. The evidence considered by the biorhythm proponents, however, does not include the even larger number of airplane crashes that occurred while the crews were experiencing higher or neutral points in their biorhythm cycles (Hines, 1988). Quite obviously, we will forever be besieged by the lunacy of the water witching phenomenon due to this same omission of the full body of comprehensive data.

The evidence offered in support of any scientific claim must be evaluated without self-deception. This rule of **honesty,** like the rule of comprehensiveness, is frequently violated. The rule of honesty means that we must accept the obligation to come to a rational conclusion once we have examined all the evidence. Denial, avoidance, rationalization, and all other familiar mechanisms of self-deception would constitute violations of the rule of honesty. This rule alone invalidates the entire discipline of parapsychology. After more than a century of systematic scholarly research, extrasensory perception remains wholly unsubstantiated and unsupportable (Hyman, 1985). From all indications, the number of parapsychologists who observe the rule of honesty pales in comparison with the number who delude themselves.

An argument offered as evidence in support of a scientific claim must be sound. By rules of **logic,** an argument is sound if its conclusions follow unavoidably from true premises. If you can conceive of a single instance whereby the conclusion would not necessarily follow from the premises, even if the premises were true, then the argument is invalid.

It may sound paradoxical, but in order for any scientific claim to be true, it must be **falsifiable.** That is to say, if it isn't true, you could prove it to be false. This means that the evidence must matter. it is a fundamental rule of evidential reasoning. If nothing conceivable could ever disprove a claim, the claim is invulnerable to any possible evidence. This would not mean, however, that the claim is true; instead, it would mean that the claim is meaningless. Every true claim is falsifiable. The claim that ground water moves at velocities proportional to its hydraulic gradient is falsifiable. The claim that water freezes at 32 degrees F is falsifiable. Any claim that is not falsifiable is not a factual assertion but an emotional statement. Nonfalsifiable statements can be true but only when vague and insignificant such as those appearing in one's horoscope.

Because we are often motivated to rationalize and fool ourselves, because we can make mistakes, and because perception and memory are problematic, we must demand that the evidence for any scientific claim be evaluated without self-deception, be carefully screened for error, fraud, and appropriateness, and that it be substantial and unequivocal. Passing all six tests, of course, does not guarantee that the claim is true, but it does yield good reasons for believing the claim.

This relates to my assertion that we must at least convince ourselves of the validity of our judgment about a scenario that describes a set of causes and effects. Do we always do that or do we become biased toward the position which favors our client? This is advocacy, and as scientists, we must not be advocates.

The Real Scientist

"The real gulf is not the gap between the arts and the sciences but the canyon between those who practice genuine scientific thinking (whether or not they have a scientific background) and everyone else—including many scientists and engineers highly trained in narrow specialties" (Jones, 1989).

Science is the very best method of problem-solving we know. Scientific thinking, in essence, is the method of guess and test. It is beholden to no particular subject matter. All is grist for its mill. Science in this sense is basic to all rational inquiry. Confusion between general scientific thinking and knowledge of a restricted field of expertise is still very much in evidence. A scientific thinker may know next to nothing of physics, but will be used to asking, "How would I test that? What is the evidence? How likely is it that this was no more than a coincidence? Is there an alternative explanation? Was there a control group?" They look across the canyon described by Jones in puzzlement at the behavior of educated individuals who either no not think to ask these questions or simply divine answers, much the same as Aristotle often did. Most of the latter never take heed of their proper, pragmatically skeptical colleagues because as Jones relates, they are too busy examining their horoscopes, cultivating their alpharhythms, and having their ailments diagnosed by charlatans.

To the layman, science is assumed to mean physics and chemistry with perhaps a little biology thrown in. This bundle of knowledge has been referred to as "Science Two" by Jones (1989) based partially on Snow's (1959) original work in which he defined true scientific thinking as Science One. There are scientific thinkers whose daily work does not deal with Science Two subjects just as there are professional scientists who seem unable to handle Science One.

Francis Crick, the famous British biophysicist who codiscovered the double helix structure of DNA for which he won a Nobel prize, was once asked what characterized a man who understood science. Crick's response, "Knowing the difference between an atom and a molecule" (Jones, 1989) was disappointing because it was tied to a specific knowledge of physics rather than the scientific thought process.

One of the hardest things for a science expert to do is freely acknowledge his status as a layman in matters outside his specialty. This accounts for why so many nongeologic scientists accept the validity of water witching. In my own defense as a self-proclaimed expert in areas only tangential to ground water, I spend incredible amounts of time and effort studying these disciplines and even more time working up the nerve to address them.

Laypersons seldom realize that Science Two specialists may not even be competent to investigate one another's disciplines. Richard Nisbett and Lee Ross (1980) have drawn together a vast amount of material in support of this position. Getting to the moon, they say, "was a joint project if not of 'idiot savants' at least scientists whose individual areas of expertise were extremely limited." One scientist knew a great deal about the propellant properties of solid fuels but little about the guidance capabilities of small computers, another scientist knew a great deal about the guidance capability of small computers but virtually nothing about gravitational effects on moving objects, etc. Finally, those scientists included people who believed that redheads are hottempered, who bought their last car on the cocktail party advice of an acquaintance, and whose mastery of the formal rules of scientific inference did not spare them the personal foibles of the rest of society.

If this seems a little harsh, some of my readers may recall that Werhner von Braun, formerly Deputy Associate Administrator for Planning at NASA until 1972 and a major figure in the American space program, wrote the foreword to a book for young people entitled *From*

Goo to You by Way of the Zoo Hill, 1985), which was intended to entirely discredit Darwinian evolution. He also authored letters to the California Board of Education advocating the teaching of creationism in our schools. Far removed from the narrow, scientific perspective of von Braun, lies the mind of Isaac Asimov, the well-known science fiction author. When queried by his father about where he learned all the wealth of information in his books, he responded, "From YOU, Papa," His Dad looked up in disbelief and said, "I don't know one word about these things." "You taught me the value of learning," he said. "That's all that counts. All other things are just details" (Asimov, 1989).

Society's Inclination

One reason that may explain why individuals hold on to rational unscientific beliefs deals with their desperate need to find predictability in our uncertain world. Gustav Jahoda (1970) offers samples showing that no matter how often they are disappointed, human beings will cling to various schemes of divination. The temporary feeling of security offered by these techniques continues to make them attractive. While a given prediction ritual may give success only 50 percent of the time, it can give temporary relief from the feeling of helplessness 100 percent of the time. In support of this premise, there is an old joke among psychiatrists concerning the patient who is convinced that he is dead. "Do corpses bleed?" the psychiatrist asked the patient. "No, they don't," he replied. Whereupon the psychiatrist cut the patient's finger and pointed to the blood. "My God," said the patient, "they do bleed!" It's a classic case of a deficiency of Science One training or understanding.

In a world growing ever more deficient in citizens capable of the scientific thought process, there has been a continued growth of pseudoscience and the resulting misrepresentation of knowledge about scientific information. An alarming amount of damage is done by outspoken individuals of great reputation who appear as the darlings of the environmental media. Like Aristotle, their reliance on the powers of the mind alone to offer answers does a great disservice to science. When the scientific revolution came, it emphasized the need for observation. Too often today, we substitute mathematical models for observations. Observed facts always take precedence over theory no matter how beautiful the logic. Aristotle and many of today's oracles, however, make no distinction between science and the purely mental practice of philosophy. Truth cannot be perceived by the exercise of the developed powers of the mind, yet many famous people with too great an ego are attempting it today. They are succeeding in part, due to our feeble effort to educate our children about science, and also as a result of the media's proclivity to reinforce our leanings toward unsupportable claims of the pseudoscientists. These subjects will be the focus of the concluding part of this editorial series in the next issue of *Ground Water*.

References

Aristotle, 1952. *The Works of Aristotle, Volume I: Great Books of the Western World.* Encyclopedia Britannica Inc.
Asimov, Isaac. 1989. *The Relativity of Wrong.* Doubleday/Pinnacle Books.
Bronowski, J. 1973. *The Ascent of Man.* Little Brown, Boston.
Chamberlin, Thomas C. 1897. *Journal of Geology,* v. 5, pp. 837–848.
Dauben, Joseph. 1988. *World Book Encyclopedia,* v. 17.
Hill, Harold, Mary Elizabeth Rogers, and Irene Burk Harrell. 1985. *From Goo to You by Way of the Zoo.* F.H. Revell Co., Old Tappan, New Jersey
Hines, Terence. 1988. *Pseudoscience and the Paranormal.* Prometheus Books, Buffalo, New York.
Hyman, Ray. 1985. A critical historical overview of parapsychology. *A Skeptic's Handbook of Parapsychology,* by Paul Kurtz. Prometheus Books, Buffalo, New York.
Jahoda, Gustav. 1970. *The Psychology of Superstition.* Penguin Books, Baltimore.
Jones, Lewis. 1989. The two cultures: a resurrection. *The Skeptical Inquirer.* Fall.
Koshland, Daniel, Jr. 1990. Two plus two equals five. *Science,* v. 247, no. 4949, March.
Lett, James. 1990. A field guide to critical thinking. *The Skeptical Inquirer.* Winter.
Nisbett, Richard and Lee Ross. 1980. *Human Inference: Strategies and Shortcomings of Social Judgment.* Prentice-Hall.
Osiatynski, Wiktor. 1984. *Contrasts: Soviet and American Thinkers Discuss the Future.* MacMillan Publishing Co., New York.
Rand, Ayn. 1957. *Atlas Shrugged.* Random House, New York.
Rand, Ayn. 1960. "Faith and Force: The Destroyers of the Modern World," Lecture at Yale University.
Rothman, Milton. 1989. Myths about science . . . and belief in the paranormal. *The Skeptical Inquirer.* Fall.
Russell, B. 1948. *Human Knowledge, Its Slopes and Limits.* Simon & Schuster, New York.
Snow, C.P. 1959. *The Two Cultures and the Scientific Revolution.* Cambridge Univ. Press.
Socrates. 1909. The apology of Socrates. *The Harvard Classics,* v. 2. P.F. Collier & Son, New York.

Reprinted, with permission, from
Ground Water, Vol. 28, No. 5, 1990.

The Scientific Process—Part II: Can We Learn It?

Jay H. Lehr, Ph.D.

Please see biographical sketch for Dr. Lehr on page v.

Science and the School Systems

Every year brings a fresh set of revelations about America's seemingly boundless superstition and ignorance. Fully 40 percent of the nation's adults think alien creatures have visited earth, according to studies by John Miller, Director of Northern Illinois University's Public Opinion Laboratory (Miller, 1987). Only 45 percent of us know the planet revolves annually around the sun, and just 46 percent have accepted that humans evolved from an earlier species. A 1989 report by the National Research Council estimates that three-quarters of the nation's graduating high school seniors leave school without the skills to survive a college-level math or engineering course.

Bruce L. Edwards, an associate professor at Bowling Green State University speaking to the Heritage Foundation last spring, put things in an even more dramatic perspective (Edward, 1990).

"Since Rudolph Flesch published his book, *Why Johnny Can't Read* in 1955, we have heard a lot about Johnny . . . Not only can Johnny not read, he also can't write. Johnny can't spell, can't do arithmetic, can't tell you when the Civil War was fought, can't distinguish the words of Stalin from Churchill, and can't identify Central America or his own state, for that matter, on a map. There's one other thing Johnny can't do. Johnny can't fail."

No matter how poorly has performed in school, our benevolent, paternalistic educational system—the perfect analog to our welfare system—has, by-and-large, found a way to promote Johnny to the next grade, found a way to keep his self-esteem intact, lest he be disgruntled, discouraged, or guilt-ridden.

If Edwards is correct (and it would appear he is), science, one of our more rigorous subjects, is one of our most substantial educational casualties. This is largely because so many view science as a complex quagmire of confusing facts. As I pointed out in Part I of this editorial series, that estimation of our discipline misses the point by a wide mark. Science is simply a way of looking at the world. It consists of asking questions, proposing answers, and testing them against available evidence. Science invites us to open our minds to all sides of an issue, whether or not they coincide with our preliminary opinions. science is truly a course in analytical thought.

Unfortunately, American science education serves not to nurture children's natural curiosity but to extinguish it with catalogs of dreary facts and definitions. At the lower levels, we teach our youth to handle test tubes without breaking them and at the upper level, to solve differential equations without understanding them. Most schools still teach science by lecture, textbook, and memorization. These three elements often combine to produce a familiar chemical reaction in the brain; we call it boredom. This style of teaching doesn't benefit any subject, but it's particularly deadly in a field where understanding the proper thought process is central to all learning. Nothing is more contradictory to science than rote memorization. Students should observe, measure, collect, categorize, record, and interpret data. It matters not whether a student is observing plant root growth in moist soil, measuring the temperature of water below an ice-covered surface, or studying how fast a hockey puck moves across ice.

Science education must emphasize ideas and thinking at the expense of jargon and cookbook procedures. Our goal should not be to make kids scientists or engineers but to hammer home the idea that individuals should create hypotheses, gather evidence, and weigh the correctness of their ideas. That this is not commonly being done is likely the result of the fact that the U.S. education industry, basically government-owned and run, is displaying the same classic symptoms of socialism that have ultimately destroyed the Soviet economy. These are (1) the political allocation of resources; (2) expanding bureaucratic overhead; (3) mismatching of supply and demand; (4) top/down decision-making; and (5) a replacement of reason by rote.

It can be argued in science that there is no absolute right or wrong but rather that some things are more right than others. This concept wreaks havoc with today's teaching and testing methods. Having exact answers and having absolute rights and wrongs minimizes the necessity of thinking and, sadly, that pleases both teachers and students (Asimov, 1989). Students and teachers alike prefer short-answers tests to essay tests; multiple-choice over fill-in-the-blank short-answer tests; and true/false tests over multiple-choice. Such tests are relatively useless as a measure of a student's understanding of a subject; they merely test one's ability to memorize.

In a spelling test where students are asked to spell pollution, one student may write, "pollushun," and another "plzrn." While they are both wrong, clearly the first is more correct. In a math quiz if students are asked to add 4.5 and 5.9 and one answer is 10.2 and another is 17, both are wrong but one is more correct. Doesn't the requirement of an exact answer fail to distinguish between various wrongs and thus set an unnecessary limit to understanding? This is part of the reason that so much of our science education fails.

Occasionally, I have an opportunity to teach an elementary school class a little about ground water. The children are surprisingly curious and intellectually alert. They ask provocative and insightful questions with amazing enthusiasm. On the other hand, when I have an opportunity to teach high school science, I find a very different atmosphere. These students rarely exhibit the joy of discovery and while they have a grasp of some factual information, they show no inclination to build knowledge upon these facts. They fear asking dumb questions. They easily accept inadequate answers. They rarely follow up on questions and constantly give sidelong glances to judge how their peers are approving of them when they venture into oral comment. As with all things in the life of an adolescent, peer pressure takes an exacting toll.

The problem, I believe, lies with our high school education system. I find far less fault at the college level but, unfortunately, it is too late then for most of our children.

Science and the Media

While primary blame for America's scientific illiteracy must be placed on our school system, the second reason is surely the irresponsibility of the mass media who exploit the public taste for nonsense. Throughout America, there are intelligent people who have a passion for science, but the passion is often unrequited because of what they are fed from their primary information source, the media. More information on astrology, biorhythms, channeling, and water witching is printed than on physics, chemistry, and biology. Sadly, society has rarely been taught how to distinguish real science from the cheap imitation. Virtually every newspaper in America has a daily astrology column. How many have a daily science column? Commercial TV is filled with off-beat pseudoscience shows. Prestigious companies such as *Time/Life Books* continuously promote a book series entitled "Mystic Places." Only PBSs small audience is treated to real science education through shows such as "NOVA." Carl Sagan (1990) said it best: "We live in a society exquisitely dependent upon science and technology, in which hardly anyone knows anything about science and technology." Thus, our task is not just to train more scientists but to increase public understanding of science. But scientists have been unwilling or unable to come to the aid of society by interacting with the media.

The solution to the problem of scientific illiteracy is education. It seems reasonable to conclude that once people understand science and technology, they will see that it is generally environmentally benign. Unfortunately, calm reason and environmental alarmism do not coexist. The public is far more likely to believe the opponents of science and technology than to believe its supporters.

When science and the environment are debated evenhandedly in public forums, the opposition to science almost always wins. An opponent need make only a single preposterous charge against science. The burden of proof falls upon the scientist, and he must refute the claim by scientific methods in just a few minutes—a difficult task for most, an impossible task for many. Good science and reacting glibly under pressure do not often coexist.

If we want people better educated in science, and thus competent to make decisions on technical matters that affect them, we must realize that most of the education comes from the media which serves as the scientific interpreter, not from scientists directly. The scientists and the media must join together for public purposes, if the public is ever going to get the necessary knowledge for proper decision-making. So far, there are not signs that such a union will ever take place.

Scientists work in a narrow field where quality far outweighs quantity, where tight deadlines do not control the place of work, and judgment is only by one's peers. A journalist, on the other hand, interprets a broad range of subjects under considerable time constraints, nearly always focusing on quantity rather than quality. They are primarily judged only by their editor whose objectives are the ultimate number of viewers, readers, or listeners.

A good scientist desires to be precise and thus speaks in a deliberate manner. A good journalist desires a fast response that is compact, reasonably accurate, and will make the greatest impact on an audience. This creates three problems (Gastel, 1983):

1. Conflict between technology and social interest makes good press; bad news is far more easily accepted than good.

2. The constant repetition of false, exaggerated or misleading statements become as facts in the minds of most people.

3. Since good scientists limit their remarks within disciplinary boundaries and good reporters extrapolate into a broad or common context, the result is often misinterpretation. "I was misquoted," says the scientist and vows never to talk to a reporter again. This reaction is a mistake, because it leaves the responsibility of communicating with the media to those scientists who avoid peer review of their work, have a mission or cause, or are charlatans and quacks. Science has its quota of the latter, as does every profession.

Defining Accuracy

"Whereas science is accurate to ten decimal points," Arthur J. Snyder, Science Editor of the *Chicago Daily News* observed, "newspapers like to settle for round figures" (Gastell, 1983). So while both scientists and reporters may strive for accuracy, they define the concept quite differently. A journalist may consider a story that captures the gist of a scientist's message as being accurate, while the scientist wants precision and comprehensiveness. Communicating with the public does, indeed, entail simplifying terms and omitting details and often soft-pedaling exceptions. To do otherwise, journalists will lose their audience of readers.

The skillful science writer can usually make science understandable to the public without grossly distorting the main ideas. But sadly, few newspapers, TV stations, or magazines can afford skillful science writers.

The Scientist's Dilemma

When to release scientific findings to the public is an unsettled question. Most scientists desire peer review before risking their reputation in the public press. In fact, some journals will not publish findings that have already been released publicly. Such a policy ensures that research results are adequately reviewed.

Readers of this journal who have attempted publication in it know full well the scientific value of peer review as proven by the difficult obstacle course placed before them. The process of peer review which leads only 45 percent of all papers submitted to *Ground Water* to ultimate publication is so thorough and challenging as to make clear to any author just what is expected of him or her in terms of the scientific process. The same may be said for the more than 150 men and women who serve as peer reviewers and see the process operate and, in fact, their own review efforts held up to comparison with the very best scientific minds in our fields.

Frequently, however, pressure exists not to wait for publication. Journalists want to know now, and scientists can be coerced into believing that public health or welfare could benefit by premature release. Discretion, as with all things, is the better part of valor and poses not a simple dilemma for many a scientist.

Standards of news judgment need revision. Journalism has finite resources and a limited claim on the public's attention. The TV networks in their manic and comic narcissism hurl their anchors hither and yon for no discernible journalistic reason, adding nothing to America's understanding of what is important. We need to strike a new balance when deciding which news matters most. Is a proper portion of the American public's precious and scarce resource, attention, being allocated properly? I think not.

Dixy Lee Ray, former Governor of Washington State and Chairman of the U.S. Atomic Energy Commission, summed it up best (Ray, 1990):

> "We should be very jealous of who speaks for science, particularly in our age of rapidly expanding technology. How can the public be educated? I do not know the specifics, but of this I am certain: The public will remain uninformed and uneducated in the sciences until the media professionals decide otherwise. Until they stop quoting charlatans and quacks and until respected scientists speak up."

Ground Water's Shared Problem: Models

Models are both the lifeblood of today's science and its most deadly poison. Ground-water hydrology offers as valid a support of this supposition as any other scientific discipline. I began defining models to the readers of this journal more than a dozen years ago. Still, I am not convinced that all of today's readers, as bright as they may be, fully understand what is intended by the use of models or even how they can be accurately defined. So, at the risk of insulting a few of you, I'll restate what may or may not be so obvious.

If you're trying to study some phenomenon that you can't entirely see or comprehend, modeling is a useful approach. This could be a physical model of a car, a molecule, or the solar system. It could be a verbal model, a simile, a metaphor, or analogy of what you think a system might be. Today, we are more familiar with models that consist of mathematical formulas, often found in a computer program. If our mathematical model is good, it may predict what should happen in a real situation on the basis of numerical information which we feed the computer.

Every area of science uses models as intellectual devices for making natural processes easier to understand. The model that reliably predicts the outcome of real events, or that continues to fit new data, is essentially a kind of theory, a broad statement of how nature works.

Commonly, however, models are so seductively attractive that they live way beyond the time when empirical findings refute them. Some models capture popular attention and hold it until long after scientists have discarded them. The model of an atom as a little solar system is one such example. This model maintains that electrons orbit the atomic nucleus like planets orbiting the sun. It was, however, discarded more than 50 years ago when physicists realized that electrons move every which way inside shells enclosing atomic nuclei.

The model of the nuclear winter and a Draconian global warming fall under the same category. In ground-water hydrology, we really have no models that have fallen widely from grace because we have none that ever achieved the pinnacle of precision. Sadly, a few among us, and many who direct us in government and industry, are forever seduced by the three-dimensional, color-coordinated graphics which so artfully turn garbage to gospel.

The Science Trap

Scientific knowledge is only one criterion among others that affect decisions, so hydrology must accept its opportunity to play a more complex role in the drama of policy-making. Hydrogeologic information is accepted or rejected in highly political ways. Michael Bradley (1986) said it well, writing in this journal: "Regardless of how strongly hydrologists want their information to be used in a strictly objective and neutral manner, the democratic decision making process will demand its due, and information will be structured by political criteria and constraints."

While all of this comes as no surprise to a public policy specialist, it seems to be something of a shock to many ground-water scientists. Decision-makers listen most carefully to information which supports their programs and projects while outside information or criticism is buffered and ignored. Whether you serve as a scientific advisor, a water resource administrator, or an expert witness in court, an increased appreciation of how scientific knowledge interacts with public policy is a clear advantage. Unfortunately as we get caught up in this process, our desire to see the most correct truth emerge as the most widely accepted position often causes us to fall into the advocacy trap.

A lawyer properly acting as an advocate has the task of presenting a case with persuasive force. However, a lawyer does not have to vouch for the evidence submitted in a cause, the value of which a judicial body is responsible for assessing. This is never the case for a scientist. He or she must be reasonably certain of all the evidence presented and must present all the evidence. In the course of conducting a consulting assignment, scientists are too often affected by the chance of losing a client and thus increase the chances of losing their objectivity. Ignoring data and softening interpretations can often result. Surprisingly, however, lawyers for whom the advocacy concept was developed have a considerable burden of additional responsibility often unrecognized by the scientist playing advocate. An advocate does, indeed, have a duty to disclose directly any adverse data to the authority in control which may not have been disclosed by opposing parties.

"A lawyer shall not knowingly fail to disclose a material fact to a tribunal when disclosure is necessary to avoid assisting a criminal or fraudulent act by the client; not fail to disclose to the tribunal legal authority in the controlling jurisdiction information known to the lawyer to be directly adverse to the position of the client and not disclosed by the opposing counsel. In an ex parte proceeding, a lawyer shall inform the tribunal of all relative facts known to the lawyer that should be disclosed to permit the tribunal to make an informed decision, whether or not the facts are adverse" (Nemeth, 1986).

Clearly, then, a lawyer is obliged to ensure that his or her arguments are correct and viable recitations of law and fact. How many consultants realize that even when they err in becoming an advocate in the first place, they are commonly not even following the rules laid down for lawyers who are intended to be advocates?

Advancing Our Capacity to Implement the Scientific Process

In an editorial I wrote entitled, "Monitoring and the Information Explosion" (*Ground Water Monitoring Review,* Spring, 1989), I made an argument for widening one's reading horizons regardless of how difficult it is to keep up-to-date in our own field of ground-water hydrology. I recommended an outstanding book entitled, *The 100,* which is Michael Hart's personal analysis of his choice for the 100 most influential people in history. At this time, I wish to make a parallel case for broadening one's horizons in other areas of science and thereby enhancing one's understanding of the scientific process. We will then be better able to avoid either accepting or rejecting aspects of other fields of science for reasons less of a lack of knowledge than for a lack of understanding of the processes involved.

Surprisingly, many areas of science outside our obvious realm relate significantly to what we do. Today, it is impossible to avoid recognition of the value of biotransformation as a process in many ground-water remediation techniques. It behooves us, therefore, to learn something of biochemistry. If we follow that path, it will lead us to a basic study of life itself which, perhaps better than any other field of science, will break down the barriers of our minds that previously allowed any mysteries to exist or be explained by nonscientific means.

While it is not my intention to eliminate one's propensity toward unsupported faith, such holidays in science can only lead to errors in judgment. My own favorite case in point deals with the initiation of life itself at a level not very far from the bacteria that helped break down many of today's ground-water contaminants.

Microbiologists and geneticists readily accept that many of the complex molecules that are part of the chemistry of life can arise through entirely nonliving processes. At least 13 inorganic compounds have been detected drifting in outer space, apparently formed there by spontaneous chemical reactions (Rensberger, 1986). Many more can be formed in the laboratory. In 1953, Stanley Miller and Harold Urey took three gases common to the earth's early atmosphere—hydrogen, ammonia, and methane—sealed them in a flask with water, and zapped it with electric sparks simulating lightning in the primordial atmosphere. After a week, the water turned brown and was found to contain several different amino acids, fatty acids, sugars, and urea. A decade later, Sidney Fox found that gently heating a mixture of amino acids caused them to link up, forming long chains that would be called proteins if a living cell had made them. Fox called them "proteinoids" and found that, when placed in water, they assembled into spheres the size of bacteria and that with time they would reproduce themselves. These laboratory experiments point toward the processes that formed life initially on earth. As we know, before life can truly exist, genes must be synthesized to control reproduction. Though DNA, the primary component of genes, has been synthesized in the laboratory, it has only been done using enzymes derived through living cells. The point of all this is that while none of these experiments proves how life really arose, it shows that a great many steps toward life result from ordinary chemical reactions that work every time. Whether they are the reactions that helped give rise to life we may never know, but that they could have is inescapable.

Advocates of unscientific origins of life point to the complexity of a modern cell and the odds of a bunch of random molecules coming together in precisely the right combination. They would be correct that the billions of combinations and permutations would take more time to assemble than the universe has been in existence. Where they go wrong is in assuming that the entire process must be done randomly. The Miller-Urey experiment showed that amino acids form in an orderly, not random, manner as is also the case for Fox's experiment (Rensberger, 1986). The oddsmakers also overlook the fact that the first living cell could have been a very primitive organism and still survive very well in a world with no living competitors.

An understanding of such processes removes more and more mystery through the process of observation and evaluation which is the backbone of good science. A broader knowledge of what goes on in associated fields of science will insulate us from the gullibility and naivete that can exist outside the boundaries of our own expertise.

For this purpose, I recommend an unusual book entitled, *How the World Works*, by Boyce Rensberger (1986). This inexpensive, 378-page paper back published by William Morrow, describes the discoveries and the discoverers responsible for two dozen of the most significant fields of scientific endeavor. While it is written in dictionary form, it is difficult to put down regardless of the entry at which one begins reading.

The Public Problem

If I am correct in thinking that we have difficulty coping with science outside our own profession, consider the dilemma of the poorly educated public. Not long ago, the enormity of this problem came home to me while watching the master magician, David Copperfield, take an audience to the edge of illusional capability. In another day at another time, a man who could so openly make people and large objects disappear, walk through a steel-plated mirror directly in front of us and seemingly disappear and reappear at will would have been considered a sorcerer and revered, or a witch and burned. We know today that he is merely one of the world's greatest illusionists—not really a magician, not a charlatan, but in fact a scientist, engineer, and showman. What he does for the intelligent person is bring full recognition that our eyes cannot and must not be believed. They are not proper judges of reality but only of perception. Sadly, we act on our perceptions as though they were reality, scientific fact, or engineering proofs. But they are not, and we must not. Yet the acceptance of Copperfield's incredible skills can only lead us to sympathize, empathize, and understand the situation that leads lesser-trained people to believe in flying saucers because they saw a moving light; water witching because they saw a stick point down; and mental control over objects because they saw a spoon bend.

Without the knowledge of the scientific method to insulate us from false perceptions, we are so obviously subject to the influence of charlatans who, in command of but a small fraction of Copperfield's skills, can nevertheless play with our minds due to the weakness of the perceptive capability of our eyes.

If you ever have the chance to see David Copperfield in person, run, do not walk to the ticket office. Not for the entertainment value, though indeed he has that, but rather because his show will dramatically advance your understanding of our collective limitations. Seeing him on TV is of no value. Our minds will never seriously accept what we see through a remote camera whether it is real or not. You must see him in person to experience this scientific revelation.

We Can All, Teacher and Student, Be

"Do not surrender your mind to the minds of others. When your understanding of reality clashes with the assertions of others, do not fear independence of thought and renounce rational faculty" (Rand, 1957). When at the crossroads of "I know" and "They say," reject the authority of others. Choose to understand rather than to believe. Never surrender your power to perceive. If you do, you'll be changing your standards from the object to the collective.

For some, this treatise on science may seem sophomoric and simplistic. For others, I hope it will open minds to a clear and more precise path toward a greater professional contribution to your chosen field of ground-water science. If you find it enlightening, I will be so bold in giving credit to others by quoting the greatest scientist in the history of our still young world, Isaac Newton (1967): "If I have seen farther than other men, it is by standing on the shoulders of giants."

References

Asimov, Isaac. 1989. *The Relativity of Wrong.* Doubleday/Pinnacle Books.

Board of Mathematic Sciences (and) Mathematical Sciences Education Board, National Research Council, 1989. Everybody Counts: A Report to the Nation on the Future of Mathematics Education. National Academy Press, Washington, DC.

Bradley, M.D. 1986. An introduction on how to slant a ground-water investigation for political purposes. *Ground Water*, v. 24. no. 3

Edwards, Bruce L. 1990. Notable and quotable. The *Wall Street Journal*, Monday, Aug. 26.

Gastel, Barbara. 1983. *Presenting Science to the Public.* ISI Press, Philadelphia, PA.

Miller, John D. 1987. The scientifically illiterate. *American Demographics.* v. 9, no. 6, pp. 26–31.

Nemeth, Charles P. 1986. *Para Legal* Handbook—Theory, Practice and Materials. American Institute for Para Legal Studies Inc., Prentice Hall.

Newton, Isaac. 1687. Principia.

Rand, Ayn. 1957. Atlas Shrugged. Random House, New York.

Ray, Dixy Lee. 1990. Who speaks for science? *Chemical Times* and *Trends.* January.

Rensberger, Boyce. 1986. *How the World Works.* William Morrow Publishing Co., New York, New York.

Sagan, Carl. 1990. Why we need to understand science. *The Skeptical Inquirer.* Winter.

Reprinted, with permission, from
Ground Water, Volume 28, number 6, 1990

The Green Lobby's Dirty Tricks

Gerald Sirkin, Ph.D.

Please see biographical sketch for Dr. Sirkin on page 119.

The abusers of universities in the interest of "politically correct" causes have struck again.

The Natural Resource Defense Council (NRDC), a powerful lobbying organization, undertook in the summer of 1990 to delay a study on the economics of pesticide regulation by two agricultural economists at the University of California, Berkeley. The NRDC feared that the study would provide California voters in the November election with information adverse to Proposition 128, "Big Green," which NRDC had helped to draft.

While the study was eventually published and Proposition 128 defeated, this story still remains important—as further evidence that environmental regulation is a scientific and economic issue that is being guided not by science and economics but by the skills of street fighters in the back alleys of politics, the courts and the media.

The *San Francisco Chronicle* broke the story on Oct. 31 but without mentioning the NRDC by name. The *Chronicle* revealed that the two state legislators who chair the subcommittees that control University of California budget requests, Sen. Nicholas Petris and Assemblyman Robert Campbell, had written to Berkeley President David Gardner on Aug. 9 warning that the research project might jeopardize the university's funding:

"We strongly urge to you to reconsider the research project that is now planned for publication this fall. . . . It is evident that we are entering a period of austerity in which various institutions will be asked to eliminate programs. . . . We urge you not to risk the university's standing by publishing research involving such highly charged political issues."

Behind the legislators' implicit threat lay a letter from the NRDC's Albert Meyerhoff to President Gardner dated June 25 with copies to Vice President Kenneth Farrell, the 31 members of the Board of Regents, U.S. Sens. Pat Leahy and Richard Lugar, and State Assemblyman Campbell. The NRDC objected to the study and its publication "on an expedited basis . . . prior to the election" as a misuse of public funds.

736

Mr. Meyerhoff then met with Mr. Farrell, head of the division of agriculture and natural resources. Mr. Meyerhoff followed up with an Aug. 6 letter to Mr. Farrell summarizing their meeting. In that letter, Mr. Meyerhoff said that the NRDC still believed "it entirely inappropriate for the University to utilize public resources, on an unprecedented and expedited schedule, in order to generate this report prior to the election date." But he added that "should you decide to the contrary and go forward with this effort, we wish to confirm our understanding of the 'ground rules' you proposed." These, according to Mr. Meyerhoff, included enlarging the study to "include representatives of disciplines beyond economics who are expert in pest control issues" and their "health or environmental effects."

Mr. Farrell's interpretation was different. In a letter to Mr. Meyerhoff dated Aug. 27, he wrote:

> To characterize our research as being conducted on an "unprecedented and expedited schedule in order to generate this report prior to the election date" is incorrect . . . The research we have undertaken with reference to pesticide regulation is not being timed to influence votes with respect to Proposition 128 but has been undertaken as a legitimate University function intended to be useful to all parties interested in and concerned with long-term alternatives to pesticides.

He added: "University research is conducted by its faculty and other highly trained professionals. We do not instruct these individuals on matters bearing upon their professional competency or on research methodology to be employed. Nor do we direct them to time the publication of their research in consideration of anyone's political interests or objectives."

Yet on that very same day, Mr. Farrell wrote a letter to the chairman of the department of agriculture and research economics warning that "It is imperative that you comply with university policy" to remain neutral on political issues. Prof. David Zilberman, who headed the study, called it a violation of academic freedom, according to a Nov. 7 article in the student newspaper, the *Daily Californian*. "If you do something that has anything to do with public policy, it is your duty to publish it before election time," he said. Colleague Jerome Siebert, in the Oct. 31 *San Francisco Chronicle* article, called the warning a "gag order." In the end, the authors published their study as a "working paper" and distributed 300 copies before the election, Mr. Siebert stated in a telephone interview.

The impropriety of university research that might have a bearing on voting is a principle newly discovered by the NRDC. It has not objected to university activities promoting environmental regulation, nor did it object when Berkeley officials campaigned for Proposition 111 to increase the gasoline tax and bring more state funds to the university.

When the *Chronicle* exposed the legislators' tactics, Berkeley biochemist Thomas Jukes, armed with the additional information that the NRDC was the initiator, wrote to President Gardner that Proposition 128's attack on pesticides "was triggered" by the NRDC's campaign against Alar, which he characterized as a case of "irresponsible scaremongering by NRDC."

The NRDC's "scare-mongering" is about to be examined in the courts, an interesting reversal of the NRDC's usual position in the legal system.

For years, the NRDC has been engaged in legal assaults under the Clean Water Act. It goes through business's records of their violations of the act, which they are required to report, and threatens them with "citizen damage" suits. Companies settle rather than go through expensive court battles. (While the NRDC ordinarily designates third-party recipients for the settlements, the procedure adds to the organization's clout in the environmental movement.) Though Congress in permitting such suits intended to enhance enforcement, NRDC sues on violations sometimes long past that have frequently been corrected.

Now, however, the suer is being sued. On Nov. 28, the apple industry filed a multimillion-dollar class-action suit against NRDC, CBS, and two CBS affiliates in the state of Washington that broadcast the "60 Minutes" Alar program; and Fenton Communications, NRDC's public relations firm that orchestrated the apple scare. Exposure of NRDC through this suit could do wonders for cleansing the environment.

Reprinted, with permission, from
The Wall Street Journal
January 2, 1991

Species and Forest Reduction

Disappearing Species, Deforestation and Data
Julian L. Simon

Disappearing Species, Deforestation and Data

Julian L. Simon, Ph.D.

Please see biographical sketch for Dr. Simon on page 557.

The issue of species loss is heating up. In 1983, the United States Congress set up a task force to develop a strategy for the conservation of biological diversity, and the task force produced a report in 1985. The Congressional Office of Technology Assessment (OTA) commissioned 50 papers and drafted a study document, *Technologies to Maintain Biological Diversity*, which culminates in a discussion of "policy issues and options for Congressional action." People are clearly worrying about the implications of extinction, but I believe that they are calling for action before they know what action is required.

The OTA's study grows out of the 1980 *Global 2000 Report* to the President, which expressed concern over the possible loss of species between now and the year 2000. The "major findings and conclusions" section said: "Extinctions of plant and animal species will increase dramatically. Hundreds of thousands of species—perhaps as many as 20 percent of all species on Earth—will be irretrievably lost as their habitats vanish, especially in tropical forests." *Global 2000* also expressed concern about deforestation, especially in the tropics, and its effect upon species loss. "The projections indicate that by 2000 some 40 percent of the remaining forest cover in LDCs (less developed countries) will be gone."

Concern is now widespread. An article dealing with *Global 2000* in *Science* ends like this: "We cannot afford the extinction of '15 to 20 percent of all species on Earth' by the year 2000, as predicted in *Global 2000.*" And U.S. Agency for International Development, citing the relationship of humid tropical forests is one of the most important environmental issues for the remainder of this century." Both are typical responses.

The available facts, however, are not consistent with the level of concern. I do not suggest that our society, and humanity at large, should not attend to possible dangers to species. Species constitute a valuable endowment, and we should guard their survival just as we guard our other physical and social assets. But we should strive for a

741

clear and unbiased view of this set of assets in order to make the best possible judgments about how much time and money to spend in guarding them.

I will take *Global 2000* as my text. It said: "Efforts to meet basic human needs and rising expectations are likely to lead to the extinction of between one-fifth and one-seventh of all species over the next two decades." That projection is based on a statement by Thomas Lovejoy of the World Wildlife Fund that "of the 3–10 million species now present on the Earth, at least 500,000–600,000 will be extinguished during the next two decades." This estimate is just guesswork, as I shall show. And it is at variance with the existing evidence.

The basis of any useful projection for the future must be a body of experience collected in situations that encompass the expected conditions, or that can reasonably be extrapolated to the expected conditions. However, none of Lovejoy's references contains any scientifically impressive body of experience. The only published source given for his key table is Norman Myer's book *The Sinking Ark*, written under the auspices of a committee of which Lovejoy was one of three members. The writings of Myers and Lovejoy, which are not independent, appear to be the basic source of all the widely discussed forecasts of species extinction.

Myers sums up the argument like this: "At least 90 percent of all species that have existed have disappeared. But almost all of them have gone under by virtue of natural processes. Only in the recent past, perhaps from around 50,000 years ago, has man exerted much influence . . . from the year AD 1600, he became able, through advancing technology, to over-hunt animals to extinction in just a few years, and to disrupt extensive environments just as rapidly. Between the years 1600 and 1900, man eliminated around 75 known species, almost all of them mammals and birds . . . Since 1900 man has eliminated around another 75 known species—again, almost all of them mammals and birds, with hardly anything known about how many other creatures have faded from the scene . . .

"Since 1960, however, when growth in human numbers and human aspirations began to exert greater impact on natural environments, vast territories in several major regions of the world have become so modified as to be cleared of much of their main wildlife. The result is that the extinction rate has certainly soared, though details mostly remain undocumented. In 1974 a gathering of scientists concerned with the problem hazarded a guess that the overall extinction rate among all species, whether known to science or not, could now have reached 100 species per year." (Here Myers refers to *Science* 1974, pp 646–647.)

This, at any rate, is a source. But it is only a consensus "guess" among scientists of an upper limit to the rate of extinction. And it refers to all species, not just birds or mammals, Myers goes on:

"A single ecological zone, the tropical moist forests, is believed to contain between two and five million species. If present patterns of exploitations persist in tropical moist forests, much virgin forest is likely to have disappeared by the end of the century, and much of the remainder will have been severely degraded. This will cause huge numbers of species to be wiped out. . . .

"Let us suppose that . . . the final one-quarter of this century witnesses the elimination of one million species—a far from unlikely prospect. This would work out . . . an average extinction rate of . . . over 100 species per day . . . Already the disruptive processes are well under way, and it is not unrealistic to suppose that, right now, at least one species is disappearing each day. By the late 1980s we could be facing a situation where one species becomes extinct each hour."

I will restate the key points from the above quotation:

(1) The estimated extinction rate was one known species every four years between the years from 1600 to 1900.

(2) The estimated rate is about one species a year from 1900 to the present.

(No sources are given for these two estimates, either on the page from which the quote is taken or on the pages of Myers' book where these estimates are discussed.)

(3) Some scientists have (in Myers' words) "hazarded a guess" that the extinction rate "could now have reached" 100 species per year. That is, the estimate is simply conjecture and is not even a point estimate but rather an upper bound. The source given by Myers for the "some scientists" statement is a report written by a member of staff of a news magazine. Note also that the subject of this guess is different from the subject of the estimates in (1) and (2); the former includes mainly or exclusively birds or mammals whereas the latter includes all species. While this difference implies that (1) and (2) may constitute too low a basis for estimating the present extinction rate of all species, it also implies that there is even less statistical basis than there might otherwise seem for estimating the extinction rate for species other than birds and mammals.

(4) The guessed upper limit in (3) is then increased and used by Myers, and then by Lovejoy, as the basis for the "projections" quoted above. In *Global 2000* "could now have reached" has become "are likely to lead" to the extinction of between 14 and 20 percent of all species

before the year 2000. That is, an upper limit for the present—which is pure guesswork—has become the basis of a forecast for the future. That forecast has in turn been published in newspapers to be read by tens or hundreds of millions of people, and is presumably understood by them as a scientific statement.

Rates of Extinction		
Records of	1600–1900	One per four years
birds and mammals	1900–1980	One per year
Guess	1980	100 per year
Extrapolated from guess	1980–2000	40,000 per year

Given the two historical rates provided by Myers—the OTA's documents of 1986 and other recent statements cite no additional evidence on the matter—one could extrapolate almost any rate for the year 2000. Lovejoy's extrapolation has no better claim to belief than a rate, say, one-hundredth as large. Considering the two historical points alone, many forecasters would project a rate much closer to the past rate than to Lovejoy's. The common wisdom is that in the absence of additional information, the best first approximation for a variable tomorrow is its value today, and the best second approximation is that the variable will change at the same rate in the future that it has in the past. Lovejoy uses an accelerating rate of change to project from two points.

Underlying the huge jump from the observed rates of the past to the projected rates of the future is an assumed relationship between deforestation and species loss, together with an estimate of future change in forested area. I will, therefore, now examine both of these elements.

To connect an assumed rate of deforestation to a rate of species loss we need systematic evidence on the relationship. But no such empirical evidence is given by Lovejoy, Myers, *Global 2000,* the OTA, or any other reference I have checked. Nor does the 1986 OTA draft document refer to any empirical evidence on the matter. Rather, *Global 2000* relied entirely upon a hypothetical graph drawn by Lovejoy connecting the zero points for proportions of deforestation and of species lost, with the 100 percent points for both variables, in a slightly bowed curve. Other students of the subject see the matter differently, denying that (say) 10 percent deforestation would result in almost 10 percent loss in species, as Lovejoy's diagram would have it. So this supposed relationship is a frail logical link for an argument about loss of species in the future.

It might well be reasonable without further ado to adjudge as quite unproven the case that large numbers of species are in danger of being lost as a result of deforestation in the next two decades. Nevertheless, let us give every benefit of the doubt to those who warn of rapid species loss, and move on to deforestation itself. *Global 2000* says "Significant losses of world forests will continue over the next 20 years . . ." But *Global 2000* presents no time-series data on such losses.

Only a historical series of comparable observations can establish the existence of a trend scientifically and statistically. Observation at one moment can convey only impressions about a trend. The impressions may be sound if they are made on the basis of first-hand contact, previous wide experience and wise judgment. But such impressions provide a different basis for policy decisions than well-grounded statistical estimates of trends.

The Food and Agricultural Organisation (FAO) of the United Nations has published estimates of total world forest area since the late 1940s, and these show two things. First, the data are too crude and irregular to show any trend reliably. Secondly, there is no obvious recent downward trend in world forest—no obvious "losses" at all, and certainly no "near catastrophic" loss. Surveys by *World Wood* since 1965 of the countries with the largest forest show much the same. At resources for the future, a long-established research institution, Roger Sedjo and Marion Clawson have also studied deforestation. Clawson has the pre-eminent student of forest economics for half a century, and Sedjo and Clawson conclude that "there is certainly nothing in the data to suggest that the world is experiencing significant net deforestation."

Global 2000, Lovejoy, Myers and others focus on *tropical* forests (rather than total world forests) as particularly liable to deforestation. *Global Future*, a supplement to *Global 2000*, opens its section on tropical forests "The world's tropical forests are disappearing at alarming rates . . . tropical deforestation is an urgent problem . . . "

Lovejoy estimated species loss with two assumed deforestation rates for tropical forests. The "low" projection assumed a 50 percent deforestation rate between 1980 and 2000 for Latin America, a 20 percent rate for Africa (Africa is the least important in terms of total species), and 60 percent for South and Southeast Asia. The "high" projection assumed 67 percent deforestation for all regions between 1980 and 2000. Lovejoy cites Chapter 8 of *Global 2000* as his source for the alternative assumptions. But no trend data are given there to support any such estimate. The main sources in that section of *Global 2000* are studies by Reidar Persson and by R.H. Whittaker and G.E. Likens. The former provides one-time inventory of world forests as a whole. The latter contains no original survey data. Neither supports Lovejoy's assumptions about tropical deforestation.

Sedjo and Clawson dug into the available evidence and compared it with the estimate found in *Global 2000*. They write: "Information about the tropical moist forests is relatively scant. What information we do have comes more from anecdotal evidence—provided by isolated investigations at single times and places—than from systematic studies conducted over large areas and lengths of time . . . A hard look at the available data supports the view that some regions are experiencing rapid deforestation. However, the view that this is a pervasive phenomenon on a global level is questionable."

Working with Sedjo and Clawson at Resources for the Future, Julia Allen and Douglas Barnes compared, country by country, the estimates for recent deforestation from three studies. They used the FAO's report of 1979, Myers' report in 1980 to the National Academy of Sciences and an FAO-UN Environmental Programme study by J.P. Lanly published in 1982. Lanly is Forestry Coordinator for the UNEP/FAO Tropical Resources Assessment Project, and Sedjo and Clawson consider the Lanly report "the most thorough." That study "estimated the rate of deforestation of the closed tropical broadleaf forests at 7–1 million hectares per annum (0–60 percent) for 1976–80 or 30–50 percent of Myers' estimates of 20–24 million hectares per annum." Lovejoy's assumptions about deforestation rates are too high.

Sedjo and Clawson make another key point: "Importantly and somewhat surprisingly, the [UN-Lanly] study indicates that the undisturbed or 'virgin' broadleaved closed forests have a far lower rate of deforestation than the total, being only 0.27 percent annually as compared with 2.06 percent annually for logged over secondary forest. This figure indicates that deforestation pressure on the more pristine and generally more genetically diverse tropical forests is quite low." Sedjo and Clawson note that "these findings are in sharp contrast to the conventional view that the tropical forests 'are disappearing at alarming rates', and suggest that concerns over the imminent loss of some of the most important residences of the world's diverse genetic base, based on rates of tropical deforestation, are probably grossly exaggerated." Early on, Sandra Brown, professor of forestry at the University of Illinois, and Ariel Lugo, project leader at the U.S. Forest Service's Institute of Tropical Forestry, also studied the available data. Brown and Lugo concluded that "dangerous" misinterpretation and exaggeration of the rate of deforestation has become common.

Brazil and its Amazon region attract special concern. But the materials used by *Global 2000* to support its assertions that Brazil is being rapidly deforested seem quite incompetent to support such evidence. The main source for Lovejoy's report was a single set of satellite photographs taken in 1978 and reported in the *Washington Post* to show that "as much as one-tenth of the Brazilian Amazon forest has been razed." Robert Buschbacher is an ecologist currently working as a Fulbright scholar in Brazil. According to a more recent description of that Landsat study that he relates, it "concluded that 1.55 percent of the Brazilian portion of the Amazon has been deforested." On the basis of this and other evidence, Buschbacher says: "Because of a relatively low percentage of forest clearing and the remarkable capacity of the forest to recover its structure . . . the threat of turning the Amazon into a wasteland is exaggerated." Even more importantly, at best such photographs constitute a single observation which provides no trend data.

Another of the sources frequently referred to is Adrian Sommer. He predicts future deforestation in the Amazon region of Brazil by simply comparing the road networks built in the 1960s and 1970s and planned for the future, and then assuming that the forests will be cleared around the roads. This is a flimsy basis for any prediction.

Nor is there need to rely on such evidence. The two UN studies and even Myers agree closely in their estimates of the Brazilian deforestation rate—between 0.0025 and 0.004 percent per year. At such a rate, it would be a great many years before many species are threatened, it would seem. If this rate were to continue unchanged, even Lovejoy's "low" assumption would be perhaps 10 times too high.

Each year *World Wood* publishes an assessment of the commercial forest situation for Brazil (and for each other's country). Those assessments differ from accounts such as those of Lovejoy and Myers. For example, M.K. Muthoo, leader of a FAO/UNDP/IBDF Project Team, wrote in 1978: "Brazil has abundant natural forest. It holds the world's biggest tropical forest reserve, in the Amazon, which can be continuously used and improved at the same time, but has hardly been tapped."

The famous Jari project of Daniel Ludwig had planted only 100,000 hectares and planned another 300,000 before it folded, an insignificant area by any measure. Furthermore, the Brazilians are acutely aware of the importance of their forests, and the government has established incentive programmes to encourage sensible forest plantations.

Why is there such a large difference in perceptions of Brazil's and the Amazon's forests? Perhaps different people look at different aspects of the matter. The environmentalists may have in mind specific areas of the Amazon, perhaps those about which Muthoo writes: "Since over-exploited forests in the Atlantic coastal belt and the Parana area in the south and elsewhere have been eroded phenomenally, they cannot sustain supplies to suffice local requirements, much less to meet growing national and international raw material requirements." The environmentalists may also be influenced by the impressive and even frightening visages of the huge machines that now can cut down and saw up trees like a lawnmower goes through grass. The commercial foresters focus instead on the large untouched areas of forests, on the smallness of Brazilian and world demand for wood relative to the potential supply, on the lack of Brazilian demand for new agriculture lands relative to the area available, and on the possibilities for improved yields of food and tree crops.

Overexploitation of local areas near transportation networks may or may not be ugly, painful, and a serious problem in Brazil or other countries. But such local exploitation would not seem to threaten species just as long as there are isolated similar habitats nearby. There may be some species that can live in only one small area. But such species are likely to be less important to humankind and the rest of nature for that very reason.

Biologists with whom I have discussed this material agree that the numbers in question are most uncertain. But they say the numbers do not matter scientifically. The conclusion would be the same, they say, if the numbers were different even by several orders of magnitude. If that is so, why mention any numbers at all? The answer, quite clearly, is that these numbers do matter in one important way: they have the power to frighten in a fashion that numbers much smaller would not. The OTA 1986 document says: "Conveying the importance of biological diversity will require a formulation of the issue in terms that are easily understandable and convincing." These frightening numbers meet that test. I can find no scientific justification for such use of numbers.

Some have said that Rachel Carson's *Silent Spring* was an important force for good, even though it exaggerated. Maybe so, but the account is not yet closed on the indirect and long-run consequences of ill-founded concerns about environmental dangers. It seems to me that, without some very special justification, there should be a presumption in favour of statements that lead to the facts as best we know them, especially in a scientific context.

Still, the question remains: How should decisions be made, and sound policies formulated, with respect to the danger of species extinction? I do not offer a full answer. It does seem clear that we cannot simply propose saving all species at any cost, any more than we can responsibly propose a policy of saving all human lives at any cost. Certainly we must try to establish some informed estimates about the social value, present and future, of species that might be lost. In the same way we find that we must estimate the value of human life in order to make rational policies about public health care services such as hospitals and surgery, and about indemnities to survivors of accidents. Just as with human life, valuing species relative to other social assets will not be easy, especially because we must value some species that we do not know about, but the job must be done somehow.

We must also try to get more reliable information about the number of species that might be lost with various changes in the forests.

This is a tough task too, and one that might exercise the best faculties of many a statistician and designer of experiments.

Lastly, any policy analysis concerning species loss must explicitly evaluate the total cost of the safeguarding activity, for example cessation of foresting in an area. Such an estimate of total cost must include the long-run indirect costs of reduction of economic growth to a community's education and general advancement, as well as the short-run costs of foregone wood or agricultural sales. To ignore such indirect costs because they are hard to estimate would be no more reasonable, and in fact probably considerably less so, than to ignore the loss of species that we have not as yet identified.

To summarize: There is now no prima-facie case for any expensive policy of safeguarding species without more extensive analysis than has so far been done. But the warnings that have been sounded should persuade us of the need for deeper thought, and more careful and wide ranging analysis, than has been done until now.

Reprinted, with permission, from
New Scientist, May 15, 1986

Toxicology

**The Dose Makes the Poison:
Some Common Misconceptions**
M. Alice Ottoboni

Chemical Toxicity: A Matter of Massive Miscalculation
Jay H. Lehr

The Dose Makes the Poison: Some Common Misconceptions

M. Alice Ottoboni, Ph.D.

Dr. M. Alice Ottoboni has devoted more than 20 years of her professional life to public service as staff toxicologist in the California Department of Health Services. During these years she has had considerable contact with people concerned about the effects of exposure to environmental chemicals upon their health and well-being.

She received her B.A. with highest honors from the University of Texas in 1954, and her Ph.D. in comparative biochemistry from the University of California, Davis, in 1959.

Dr. Ottoboni is the author of *The Dose Makes the Poison,* 2nd Edition published by Van Nostrand Reinhold, 1991. She is the author of numerous scientific papers reporting the results of her research into the adverse effects of environmental chemicals.

She is a member of Phi Beta Kappa, the American Conference of Governmental Industrial Hygienists, the New York Academy of Sciences, and the Society of Toxicology, and she is listed in *Who's Who of American Women* and *American Men and Women of Science.*

A review of the numerous factors that govern whether a chemical will or will not produce adverse effects justifies the conclusion that the toxicity of chemicals is a very complicated subject. This complexity, quite understandably, is perplexing to many people. As a result, a number of misconceptions relating to the toxic actions of chemicals and the science that studies them have arisen. Some of these misconceptions have acquired a semblance of fact by dint of repetition and lack of challenge. Three in particular pose serious obstacles to an understanding of the toxic behavior of chemicals. They must be dispelled if informed and rational public participation in the making of government policy relating to the many aspects of chemicals in our environment is to be achieved. These misconceptions are: (1) the effects of chemicals normally considered to be toxicants are always detrimental; (2) the science of toxicology is capable of determining whether chronic exposure to trace amounts of a chemical is absolutely safe; and (3) some chemicals that enter our bodies lodge permanently inside us and build up, indefinitely, to higher and higher concentrations. These misconceptions will be discussed under the headings of "Sufficient Challenge," "Trans-Science," and "Bioaccumulation."

Sufficient Challenge

Every toxicologist who has been engaged for any period of time in research into chronic toxic effects of chemicals has observed, more often than not, that animals in the group with the lowest exposure to the test chemical grew more rapidly, had better general appearance and coat quality, had fewer tumors, and lived longer than the control animals. I know from personal experience that novice toxicologists usually consider such observations as aberrations in their data or the result of some flaw in their experimental design or conduct. They are usually loath to call attention to such findings, perhaps because to do so might bring their competence into question, or because they are unable to explain the reason for such findings. It is only with the confidence that comes with experience that the research toxicologist can comfortably acknowledge the occurrence of such results in his own experiments and broach the subject with his colleagues. The reaction from his fellow toxicologists is usually one of, "You, too?"

The phenomenon of beneficial effects from trace exposures to foreign chemicals, although often a subject of conversation among toxicologists, particularly with regard to why such effects occur, is rarely mentioned in the scientific literature. If the phenomenon does occur in a chronic toxicity experiment, the text of the paper reporting the results will seldom mention the fact. It is only by careful perusal of the data tables and figures presented in the body of the text that the phenomenon is revealed. Such subtleties are lost on people who read only the abstracts of scientific papers. Unfortunately, there are some scientists who may be counted among the abstract-only readers.

The reluctance on the part of toxicologists to acknowledge and discuss freely the observation that trace quantities of foreign chemicals can produce beneficial effects may be founded partly on the fact that they have not, as yet, formulated a unified theory for the phenomenon. Or, it may be due to the reality that we live in a time when it is not politic to make favorable statements about synthetic chemicals. Since toxicologists recognize that the phenomenon has little practical significance (unless it could be put to some therapeutic use), they have no compelling reason to emphasize it; they might be misunderstood and, as a result, lose a reputation for objectivity. (A scientist who is looked upon as being not objective is not credible.) In the absence of any public benefit, why open oneself to attack from people ready to seize upon any statement that sounds prochemical and be labeled by them as an industry apologist? Dr. Smyth learned firsthand how easy it is to be misunderstood, as he mentions in his article entitled "Sufficient Challenge" (Smyth, H.F., Jr. "Sufficient Challenge." In *Food and Cosmetics Toxicology*, Vol. 5, 1967, p. 51); his first use of the term sufficient challenge earned him the epithet, "Dr. Smyth and his fellow poisoners."

Dr. Smyth's hypothesis was developed from the familiar old proverb that tells us we must each eat a peck of dirt before we die, and from personal observations in the conduct of toxicology studies. Though not stated in his article, the evolution of Dr. Smyth's hypothesis was obviously accompanied by considerable reading and reflection during and after the developmental stage. Dr. Smyth describes the manifestations of the peck-of-dirt maxim in medical, physical, psychological, and social sciences. It is to the noted historian Arnold Toynbee that he gives credit for providing the name "Sufficient [but not overwhelming] Challenge" to his peck-of-dirt hypothesis for toxicology. Dr. Smyth writes that ". . . Toynbee concluded that no civilization has risen to importance without a 'sufficient but not overwhelming challenge' from its physical environment or its neighbors. He cites abundant examples of civilizations successful because of sufficient challenges he identifies, and unsuccessful because of overwhelming challenge or absence of challenge."

Figure 1 represents what Dr. Smyth proposes, and what considerable toxicologic data confirm, to be the complete dose-response curve. With small doses, no effect on the health of the experimental animals is seen (points a to b). Animals receiving slightly higher doses have better than normal health (points b to c, the range of sufficient challenge). At point c, the curve passes back through normal health and, from that point on, deleterious effects occur.

Dr. Smyth gives numerous toxicologic examples of benefit from small doses, but perhaps his more important contribution is his theory for why such effects occur: "A rationale to explain sufficient challenge as a general phenomenon is based on the axiom, established in so many instances, that an unused function atrophies. This is the parable of the talents; 'To him that hath shall be given, to him that hath not shall be taken away even that which he hath.' One of the functions of an organism is homeostasis. Homeostatic mechanisms must be kept active if health is to be maintained.

"I think that most of the small non-specific responses which we measure in chronic toxicity studies at low dosages are readjustments or adaptations to sufficient challenge. I interpret them as manifestations of the well-being of our animals, healthy enough to maintain homeostasis.

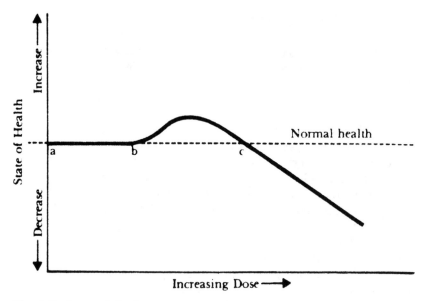

Figure 1. A complete dose-response curve.

They are beneficial in that they exercise a function of the animal. Only when challenge becomes overwhelming does injury result."

Some people reject the concept of sufficient challenge, claiming that data demonstrating beneficial effect from low doses are artifacts. However, the fact that such results occur frequently, regularly, and are reproducible speaks against the claim. It profits no one to deny truth, but there are cynics who claim it is necessary to deny sufficient challenge lest it be used to absolve environmental polluters from past transgressions and give them license to delay or disallow abatement procedures. Such fears have little foundation in reality.

A general acceptance of the theory of sufficient challenge would have no impact on chronic exposure standards or the regulatory procedures that govern them. The prohibitive cost of experimental determination of sufficient challenge dose ranges, the impossibility of determining individual responses to such doses, and public concern about environmental contamination all operate against any practical application of the concept. But if the public were aware of the existence of sufficient challenge, it could help to lessen the personal damage done by an unreasonable fear of chemicals. It could also help bring a very necessary public objectivity to the subject of environmental chemicals. In Dr. Smyth's words, "General acceptance of the concept of sufficient challenge would do much to alleviate the emotional revulsion which the thought of chemicals in daily life so often evokes."

Trans-science

The term "trans-science" was proposed by Alvin M. Weinberg to describe wisdom that cannot be achieved through scientific methodology. In his discussion of the relation between scientific knowledge and societal decisions (Weinberg, Alvin M. "Science and Trans-science." In *Minerva*, Vol. 10, 1972, p. 209), he notes, "Many of the issues which arise in the course of the interaction between science or technology and society . . . hang on the answers to questions which can be asked of science and yet *which cannot be answered by science.* I propose the term *trans-scientific* for these questions since, though they are, epistemologically speaking, questions of fact and can be stated in the language of science, they are unanswerable by science; they transcend science."

Dr. Weinberg cites three causes for the inability of science to answer trans-scientific questions: (1) "science is inadequate simply because to get answers would be impractically expensive"; (2) "science is inadequate because the subject-matter is too variable to allow rationalisation according to the strict scientific canons established within the natural sciences"; and (3) "science is inadequate simply because the issues themselves involve moral and esthetic judgments: they deal not with what is true but rather with what is valuable." The great majority of trans-scientific questions asked of toxicology can be placed in the first category, which, for our purposes, will also include questions which science does not yet have sufficient knowledge or techniques to answer.

The questions uppermost in the minds of individuals relate to whether or not exposure to some chemical or chemicals will be harmful to their health or that of their loved ones. Often these are the very questions that toxicology cannot answer with a definite "yes" or "no." Science has no way of knowing the exact biochemical makeup of any individual person or exactly what quantity of chemical would be just below that person's threshold for the most subtle adverse effect of which the chemical is capable. An answer based on judgment can be given, but science does not as yet (and may very well never) have the methodology to respond to these concerns with direct evidence.

The questions uppermost in the minds of regulatory officials relate to the nature and incidence of adverse effects that might result from exposure of large populations of humans to very small quantities of environmental contaminants. Science does not have the resources— money, trained personnel, laboratory facilities, experimental animals, etc.—to provide such information for even a few, much less all, of the many chemicals we may encounter in our daily lives.

Dr. Weinberg's example illustrating the impracticability of such an experiment deals with radiation effects, but it could as well apply to any chemical exposure: "Let us consider the biological effects of low-level radiation insults to the environment, in particular the genetic effects of low levels of radiation on mice. Experiments performed at high radiation levels show that the dose required to double the spontaneous mutation rate in mice is 30 roentgens of X-rays. Thus, if the genetic response to X-radiation is linear, then a dose of 150 millirems would increase the spontaneous mutation rate in mice by ½ per cent .... Now, to determine at the 95 per cent. confidence level by direct experiment whether 150 millirems will increase the mutation on rate by ½ per cent. requires about 8,000,000,000 mice! Of course this number falls if one reduces the confidence level: at 60 per cent. confidence level, the number is 195,000,000. Nevertheless, the number is so staggeringly large that, as a practical matter, the question is unanswerable by direct scientific investigation."

Dr. Weinberg's example tells us that, in essence, an effect that has a very low incidence of occurrence would almost certainly not be seen in an experiment using only a few animals; a study using 10 animals will not reveal an effect that occurs in only 1 out of 100,000 animals. In order to demonstrate such an effect, many times 100,000 animals would be required. Experiments of that magnitude are beyond the capability of existing resources and so are in the realm of trans-science.

Another toxicologic question that falls within the realm of trans-science relates to the shape of dose-response curves for carcinogens as they approach zero dose. The inability of toxicology to answer this question by experiment has given rise to a scientific controversy concerning whether or not there is a threshold (no-effect level) for carcinogenic effects. If there is no threshold, extension of the experimentally derived dose-response curve to zero effect would yield a line that would go through the origin (zero dose). If there is a threshold, the extended line would meet the abscissa at some point greater than zero dose (see Figure 2). Federal regulatory agencies adopted the no-threshold theory of carcinogenesis many years ago when faced with the requirement to make regulatory decisions about chemicals classed as carcinogens, proven or suspected. They still employ the no-threshold concept in their regulatory deliberations. The development of carcinogenicity risk assessment methods resulted from this regulatory need, and revision and refinement of risk assessment methodology continues to the present time.

In 1971, the National Center for Toxicological Research (NCTR) was created to provide federal regulatory agencies, such as the FDA and EPA, with the scientific data required by them for the performance of their regulatory duties. One of the first missions of NCTR was to determine how far they could penetrate into the realm of dose-response

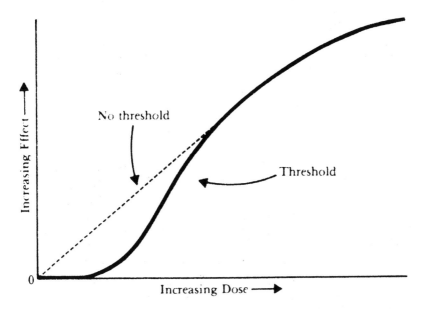

Figure 2. Extrapolation of the dose-response curve to zero dose.

curve trans-science by means of an animal carcinogenicity feeding study of hitherto unimagined size—a "Megamouse" study. In 1972, the NCTR proposed that the large-scale study be designed to answer the question of whether they could accurately describe an ED_{001}, for a known carcinogen whose effects had been extensively investigated and were relatively well understood. ED means "effective dose" (in this case the effect is cancer), and the 001 means for 0.1 percent of the animals in the study.

Such a study, it was hoped, would provide experimentally derived points lower down on the dose-response curve than had ever been obtained before, and also, perhaps, shed some light on the controversy over whether or not carcinogenic effects have thresholds. The physical resources of NCTR, however, could not accommodate the number of animals required by an ED_{001} study, so the plans were scaled down to an ED_{01}, study (effective dose for 1 percent of the animals in the study). The carcinogen selected for the study was 2-acetylaminofluorene (AAF), a chemical that fit the criteria established by NCTR better than any other known carcinogen. The study required 18 months for the planning stage, 9 months for production of the more than 24,000 mice employed in the study, and another 3 to 4 years for conduct of the study and evaluation of the data, and it cost somewhere between $6 and $7 million.

The results of this heroic effort were published in a special issue of the *Journal of Environmental Pathology and Toxicology* (Vol. 3, No. 3, 1980), and were reviewed by a special Committee of the Society of Toxicology. The review by the Society of Toxicology Committee, a few excerpts from which are quoted below, was published in *Fundamental and Applied Toxicology* (Vol. 1, No. 1, 1981, pp. 27–128). Evaluation of the massive quantities of data produced by the ED_{01}, study is still being conducted and will continue into the foreseeable future. The study has produced a great deal of information that is of extreme importance to the design and conduct of future studies of the carcinogenicity of chemicals, and will contribute immensely to the development of appropriate models and formulae for carcinogenic risk assessment. But the question of whether or not a threshold exists for the carcinogenic effects of AAF was not answered. Thus, the question still remains, and probably always will remain, in the realm of trans-science.

One finding made by the Society of Toxicology in its review of the NCTR report brings to mind Dr. Smyth's theory of sufficient challenge. The Society's review points out that the statistical model used for extrapolating effects from very low doses "provides statistically significant evidence that low doses of a carcinogen are beneficial" (p. 77). "If the time-dependent low-dose extrapolation models are correct," the Society states, "then we must conclude that low doses of AAF protected the animals from bladder tumors." However, the Society also asserts, ". . . not only is the simple model used by NCTR statisticians inappropriate to the data, but most of the models that have been proposed in the statistical literature are also inappropriate." They urge a "profound rethinking of the entire problem of chemical carcinogenesis and low-dose extrapolation" (p. 80).

Dr. Weinberg, in his article on trans-science, mentions another point that is important to an understanding of the limitations of science: " . . . no matter how large the experiment, even if *no* effect is observed, one can still only say there is a certain probability that there is in fact no effect. One can never, with any finite experiment, prove that any environmental factor is totally harmless. This elementary point has unfortunately been lost in much of the public discussion of environmental hazards."

This brings us to one of the most frustrating matters with which toxicologists must cope, namely, the subject of proving absolute safety of exposure to chemicals. It is most natural for people to demand assurance that the chemical exposures they experience are absolutely safe; it is very difficult for them to grasp the fact that this is an assurance that no one is capable of giving. Absolute safety is the complete absence of harm—the nonexistence of harm. How does anyone prove the nonexistence of anything? Consider the parent whose child is afraid of goblins. The child calls out that there is a goblin in the darkened bedroom. The parent turns on the light and says, "See, no goblin!" The child says that the goblin ran into the closet. The parent opens the closet door and says, "See, no goblin!" The child says that the goblin jumped out the window when the closet door was opened. How does the parent prove that the goblin does not exist? A "See, no goblin!" response is not proof.

Or consider the case of a friend who says he wants to borrow your book on how to get rich in the stock market, a book you have never owned. You tell your friend that you do not own such a book. Your friend may, or may not accept your word, or may be convinced that you are telling the truth because you are, in fact, not rich. But again, that is not proof. How would you prove that you do not possess the book in question?

For the toxicologist, every negative response to a question can always be followed by another question. If a toxicity experiment shows a certain level of exposure to be a no-effect level, the question can be asked, "What if you had used more animals?" If more animals yield the same results: "What if you had continued the study through one or two generations?" If a two-generation study yields the same results, then: "What might happen after five generations?"—"ten generations?"— "twenty generations?" There is no such thing as a diminishing supply of questions.

There is also no such thing as a diminishing supply of skepticism. A number of years ago, during a panel discussion on the use of herbicides in forestry, I was challenged by a man in the audience who said that the concept of the no-effect level was just not logical: there must be a chemical somewhere in the world that would be harmful no matter how small a dose was given. Obviously, there is no way to prove that such a chemical does not exist. For any chemical that does produce adverse effects down to the smallest dose that can be administered practically, there is no direct way to prove that some much smaller dose would be harmless. But, by the same token, there is also no way to prove that smaller doses would be harmful. People who hold opposing views of what the correct answer is to a trans-scientific question are on equal ground!

The fact that toxicology cannot provide absolute answers to many of the questions that are of great concern to people should not be cause for alarm. Toxicologists can make judgments about the possibilities and probabilities of harm resulting from exposure to chemicals. These judgments are based upon scientific data obtained from chronic toxicity testing, knowledge of the behavior of the chemical in animal systems, and application of appropriate margins of safety.

Bioaccumulation

Bioaccumulation is a term that is commonly used, but apparently has different connotations for different people. Thus, in discussing the subject, it is important to define the sense in which it is being used in order to avoid total confusion. Bioaccumulation is considered by some people to mean the accumulation of cellular or tissue injuries as a result of exposure to chemicals. It is more popularly used to refer to the storage of chemicals in living organisms. Toxicologists take bioaccumulation to mean an increase in the concentration of a chemical in living organisms over that which is present in the environment. This is the meaning that will be used in the following discussion. It should be noted here that "bioaccumulation" and "storage" are not synonymous. Storage means deposition of a chemical in some anatomic site. Storage results in bioaccumulation, but bioaccumulation can occur by means other than storage.

There is probably no concept in toxicology that is less understood, even in the scientific community, or more frightening than this phenomenon referred to as bioaccumulation. The public became aware of the concept several decades ago when environmental concerns about chlorinated hydrocarbon pesticides, particularly DDT, were being widely publicized. But, to date, people are still confused and concerned about what bioaccumulation is and what it means for their health and well-being.

A brief description of the nature of the chemical balances that exist in all living organisms is essential to even an elementary understanding of bioaccumulation. All living cells possess a certain specific composition, which, within normal limits, remains constant so long as they maintain their customary good health. Despite the fact that this relatively stable composition makes it appear that living organisms are static organizations, nothing could be further from the truth. All living creatures exist in what is known as a state of dynamic equilibrium. Dynamic means "moving" and equilibrium refers to the balance, or equalness, that is maintained within living organisms.

A simple illustration of a state of dynamic equilibrium would be a box that contains a specific number of marbles, where every time some new marbles are added to the box, an equal number of old marbles are removed. The total number of marbles in the box remains constant, but the individual marbles present in the box at any one time are not necessarily the same marbles that were in the box at some earlier time or that will be in the box at some future time. If the input of marbles increases, the output will also increase by the same number, either immediately (thereby keeping the total number in the box the same), or after a delay of some specific time (thereby increasing the total number in the box to a new equilibrium level, determined by the length of the delay). A decreased input will have the opposite effect. So it is with living organisms: there is a steady drive toward the achievement of equilibrium—constant change, but little or no net change. Even the bones in our bodies, which people think of as completely inert structures, contain atoms and molecules that are in a constant state of flux.

With the concept of dynamic equilibrium in mind, let us now look at bioaccumulation. Whether or not a chemical accumulates in an organism depends upon how fast it is eliminated (metabolized or excreted) relative to how fast it is absorbed into the body. The time between absorption and elimination will be referred to in the following discussion as the residence time of the chemical. "Residence time" is not a scientific term; scientists prefer the concept of half-life because it is quantifiable. The biological half-life of a chemical is the length of time required for the concentration of the chemical in the body, determined at a given point in time, to be reduced by half. The length of time required for the concentration of a chemical to come into equilibrium with the exposure concentration after onset of exposure is referred to as the time to equilibrium. For purposes of this discussion, consideration of the total time from entry to exit is more appropriate.

For chemicals with the same rate of absorption, those with longer residence times have a greater potential for bioaccumulation than those with shorter residence times. If the residence time is very short, little or no bioaccumulation will occur (at least not of the chemical, itself; a metabolite may accumulate, but that is an unnecessary complication at this point). Residence time is dependent upon such factors as pathway through the body, rate of metabolism or excretion, and the tendency of the chemical either to be held for a period of time in some metabolic pool, or to be deposited in some storage site (depot), such as fat or bone. The actual concentration that a chemical achieves in a living organism is regulated by the same kinds of forces that drive the organism's internal biochemical environment toward equilibrium.

The storage of foreign chemicals in various depot sites can involve some very complex physiologic processes and equilibria relationships. Although, for the sake of simplification, the following discussion will consider only the overall equilibrium between absorption, storage, and elimination, it must be remembered that the blood, which serves as the transport medium for chemicals in the body, participates in the process and that it has its own equilibrium relationships at the sites of entrance, storage, and elimination of chemicals.

The simplest case involving storage of chemicals occurs when the concentration of a chemical to which an organism is exposed remains constant over a prolonged period of time and, after entering the body, the chemical is not metabolized. Before the onset of exposure, there are no molecules of the chemical in the body's storage site. When exposure starts, molecules of the chemical begin entering the depot. Some of these molecules also exit the depot, but the number entering is greater than the number exiting, with the result that the concentration of chemical in the depot site gradually increases. With continued exposure, the concentration of chemical in the storage depot continues to increase until finally it comes into equilibrium with the exposure concentration. At this point, the number of molecules leaving the storage site equals the number entering, and the storage concentration remains constant (molecules of chemicals are moving constantly in and out, but there is no net change—just like the marbles in the box). If the exposure concentration increases, the storage concentration will gradually increase until it comes into a new equilibrium with the higher exposure. If the exposure concentration decreases, the storage concentration will gradually decrease until it comes into a new equilibrium with the lower exposure. If exposure ceases, the stored chemical will gradually be eliminated from the body.

The quantity of chemical that can be stored in any body depot can never exceed that which would be in equilibrium with the exposure, and the chemical cannot remain in the storage depot without being replenished continually from the outside. Thus, the popular notion that foreign chemicals stored in a depot become immobilized and permanently fixed in the body, with additional exposures increasing the quantity stored ad *infinitum,* has no basis in fact.

The quantity of storage and the length of time required to reach an equilibrium state vary from chemical to chemical and depend upon the magnitude of the exposure concentration and the nature of the chemical, i.e., its physical, chemical, and biologic properties. Many chemicals have no affinity for any storage depot in the body and, thus, do not store. A few have a large propensity for storage and, with high exposures, can build up to high storage concentrations. The remainder display varying degrees of storage between the two extremes. If exposure to a chemical varies continually in concentration and duration, interspersed with periods of no exposure—as is the case with the great majority of our exposures to foreign chemicals in the environment—an equilibrium state is never achieved. In such cases, the quantity stored in the body is also continually changing, but would be considerably less than that which would be in equilibrium with the highest exposure level.

The relationship between magnitude of exposure and storage concentration at equilibrium is not known for most chemicals. One exception is the organochlorine pesticide DDT, which has been the subject of a tremendous amount of investigation, including study of its storage and excretion patterns. The U.S. population, on average, received a daily oral exposure of approximately 0.2 ppm DDT during the several decades of heavy use of the pesticide. This level in the diet produced an average DDT concentration of 7 ppm in body fat of the American people. The increase in storage concentration relative to exposure concentration at this low level of exposure was approximately 35-fold. In studies with rats and mice, a diet containing 20 ppm DDT produced a body fat content of 200 ppm, a 10-fold increase. A diet of 200 ppm produced body fat levels of 600 ppm, only a 3-fold increase. Thus, for DDT, the relationship between exposure and storage concentrations is not a linear one. The total quantity of storage increases with increasing exposure, but low levels of exposure produce **relatively** larger storage levels than high levels of exposure. From what is known about storage and excretion of foreign chemicals, it appears that DDT can serve as a model for chemicals that store in body fat.

Another piece of misinformation that has been circulated about bioaccumulation, or more specifically, storage, is that it has put every one of us at risk of serious or fatal poisoning by chemicals suddenly released from fat depots in the event of severe weight loss, such as could occur in the case of debilitating illness or starvation. We have also been told that the chemicals stored within us have made us all walking "time-bombs." These claims are examples of the kind of tactics employed to exploit the fact that, unfortunately, the most effective method for engendering public concern about chemicals in the environment is to make people fear that those chemicals are endangering their health.

Poisoning from rapid mobilization of a chemical from fat stores is theoretically possible in laboratory animals under rigidly controlled conditions of exposure and food intake, but is a practical impossibility in the human population. The question of whether DDT could be released from body fat with sufficient rapidity and in sufficient quantity to produce symptoms of poisoning was of interest to FDA scientists during World War II, when heavy applications of DDT were being made to prevent pandemic outbreaks of certain vector-borne diseases. In the FDA studies, rats were fed 200, 400, or 600 ppm DDT in their diets for sufficient periods to build up high levels of DDT in their body fat, followed by total withdrawal of feed. Rats receiving 600 ppm suffered marked tremors typical of acute DDT intoxication. Rats receiving 200 or 400 ppm DDT suffered increased irritability, but no tremors.

The practical impossibility of such a circumstance occurring in the human population can be demonstrated also using DDT as an example. Let us assume a worst-case situation of an obese person who is 25 percent fat (a normal, well-nourished person is approximately 10 percent fat). A 140 kg (308 lb) obese person would contain 35 kg (77 lbs) of fat. The average concentration of DDT in the body fat of Americans during the years of its peak use was 7 ppm. At that concentration, the 35 kg of fat in our obese subject would contain a total of 245 mg DDT. If our obese person had an extremely high exposure to DDT, comparable to rats receiving 20 ppm DDT per day in their feed, he might have a body fat concentration of 200 ppm, for a total quantity of 7000 mg DDT. Now, let us assume that our obese person stops eating in an effort to lose weight, and that he is highly successful and manages to lose all 35 kg overnight (I can hear sighs of envy). He would suddenly release 245 mg DDT or 7000 mg, depending upon which of the above concentrations applied.

How do these quantities relate to acutely toxic doses for humans? It is known from controlled studies in human volunteers that ingestion of 35 mg DDT/kg of body weight per day, for a period of 4 to 5 years, produced no adverse effects, acute or chronic, in any of the subjects. Assuming that ingested DDT can be equated with DDT released from fat stores, 35 mg DDT/kg of body weight would be equivalent to 4900 mg DDT per day for our obese subject. It is also known from accidents and volunteer experiments that ingestion of somewhere between 50 and 70 mg/kg is less than an acutely toxic dose. For our obese person, that would be equivalent to between 7000 and 9800 mg of DDT. Our hypothetical obese person would contain less total DDT in his entire body than would be required to produce symptoms of acute illness!

Furthermore, anyone who has ever tried to lose weight knows the absolute impossibility of losing so much weight so rapidly. Human volunteers placed on regimes of total starvation can lose only about 0.6 lb of fat in a day (about 0.25 kg). The quantity of DDT released from 0.25 kg fat containing 7 ppm would be less than 2 mg, a quantity that data from human volunteers strongly suggest is a nontoxic chronic dose. Thus, people who are concerned about the potential of acute or chronic poisoning from chemicals stored in their body fat are subjecting themselves to the trauma of needless worry.

A property of chemicals that is associated with bioaccumulation and equally maligned is persistence. To be more precise, the chemicals of concern are not persistent chemicals as a class but persistent pesticides in particular. The chlorinated hydrocarbon pesticides, such as DDT, are resistant to metabolism and environmental degradation and, therefore, remain in their unaltered states for relatively long periods of time.

Persistence and bioaccumulation are not inherently good or bad, but in the public mind they are considered, almost universally, to be the latter. In actual fact, whether they are beneficial or detrimental depends upon the context in which they are viewed. The ability of an organism to store a chemical in some anatomic depot can actually be beneficial because it functions as a mechanism whereby the organism is protected from the toxic action of the chemical. The chemical does no harm while it resides in the storage site. For example, lead is a very toxic element for animals. It stores in bone where, like DDT in fat, it is in equilibrium with blood levels, which, in turn, are in, are in equilibrium with exposure and excretion. Lead isolated in bone does no harm, but lead in nerve tissue causes very serious damage. The removal of lead from blood by bone prevents the blood lead level from becoming sufficiently high to damage nerve tissue. If lead exposure is prolonged or excessive, blood lead levels do reach levels toxic to nerve tissue, despite the storage of lead in bone tissue.

A storage depot may be considered to have a buffer function: during periods of increased exposure, the chemical is deposited in the storage site, where it is prevented from exerting a harmful effect, rather than building up in the blood to a level where it will do damage to a sensitive organ. When exposure decreases, the chemical is mobilized from the depot and eliminated from the body, thereby freeing the site for some future time when exposure may again increase and storage be required to protect against adverse effects somewhere else in the body. With exposure levels that are sufficiently high to saturate the depot, there is no more storage space for the chemical and the protective effect of the storage site is abolished.

Probably the only detrimental effect of the ability of the body to isolate chemicals in storage depots is a psychological one. In the absence of knowledge about the phenomenon, the thought that potentially toxic chemicals are residing in one's body can be very distressing.

Persistence is a desirable quality in pesticides from the viewpoint of effectiveness and efficiency. A pesticide that retains its ability to kill pests for prolonged periods needs to be applied less often than a pesticide that degrades rapidly. Thus, the total quantity of pesticide required to do the job is considerably less, which reduces the cost of crop protection and production.

The undesirable aspects of persistence in pesticides relate to their continued pesticidal action after such need ceases to exist, and to the fact that they remain in the environment for prolonged periods. The persistence of DDT and its extremely heavy use during World War II resulted in its gradual distribution, in trace quantities, throughout the world. The tremendous improvements in analytic techniques during the past two decades, which enabled chemists to detect infinitesimally small quantities of chemicals, led to the discovery that DDT had become virtually omnipresent in the environment. The statistical association of these small quantities of DDT with declines in certain wildlife species spurred the campaign to ban the use of all persistent pesticides. The campaign has been eminently successful in eliminating DDT and many of the other persistent pesticides, but it has done relatively little to alleviate the problems of endangered wildlife.

The rationale that substitution of nonpersistent pesticides for persistent ones will solve all the environmental problems attributed to the latter is an example of the fuzzy thinking that permeates so many decisions relating to environmental protection. The rationale seems to be based upon the notion that a nonpersistent pesticide does its job and then immediately, in a puff, dematerializes into nothingness.

On the contrary, all nonpersistent pesticides merely degrade to other chemicals! The only difference is that these new chemicals do not have the same pesticidal action as their parent chemicals. But about the new chemicals: they may not kill pests, but what is their toxicity to other organisms? What is their fate in the environment? Do they persist? Do they accumulate?

There are a great deal of data to indicate that some degradation products of nonpersistent pesticides have at least as much potential for nontarget damage as DDT. The identities of many of these degradation products are known because one of the requirements for registration of pesticides is study of their environmental fate. But, there is absolutely no program for environmental monitoring of persistent products of nonpersistent pesticides. There is no demand from the groups who lobbied so hard to ban persistent pesticides to investigate the potential environmental damage from nonpersistent pesticides. Why? This is a philosophic question worthy of pursuit for anyone truly concerned about protection of the environment.

A nonpersistent pesticide must be applied much more frequently than a persistent one because of the very fact of its nonpersistence. This need for increased numbers of applications would greatly increase the environmental burden of persistent degradation products. By forcing a ban on persistent pesticides, environmentalists may very well have created a much larger problem than the problem they perceived as requiring the ban. Time may tell, if someone asks the right questions!

Time has already told us that the switch from persistent to nonpersisent pesticides has greatly increased the number of acute poisonings among farm workers. Cases of acute illnesses from chlorinated hydrocarbon insecticides were virtually nonexistent prior to their ban. The worst problems were cases of skin irritation. With the introduction of the organophosphate pesticides, the major class of nonpersistent insecticides, it became evident that these highly toxic chemicals would require elaborate precautions for safe use. When DDT was banned, the use of organophosphate insecticides increased greatly. This increase was accompanied by a large increase in worker poisonings, some so severe as to be lethal. The efforts to protect wildlife had the ultimate effect of producing acute health problems among workers in the agricultural industry. Despite elaborate programs of worker protection— medical surveillance, protective clothing and cleanup, automatic measuring and mixing devices to avoid human contact, restrictions on reentry into treated fields, etc.—poisonings of farm workers by nonpersistent pesticides still occur.

Adapted from *THE DOSE MAKES THE POISON,*
A Plain Language Guide to Toxicology, Second Edition,
by M. Alice Ottoboni, Ph.D.,
published by Van Nostrand Reinhold

Chemical Toxicity: A Matter of Massive Miscalculation
Jay H. Lehr, Ph.D.

Please see biographical sketch for Dr. Lehr on page v.

The Toxicological Problem

The toxicology testing laboratory dates back into the 1930s, but the science of toxicology in the United States can be traced realistically to the formation of the Society of Toxicology in 1961 (Kamrin, 1988). This Society institutionalized the growing number of scientists involved in testing the effects of environmental pollutants, food additives, drugs, and other chemicals. In the late 1970s and the 1980s, academic departments of toxicology began at many universities. Thus, toxicology is a very young discipline. Many scientists we call toxicologists have become such through experience rather than formal training. Differing levels of toxicological expertise have contributed to the conflict and uncertainty in the scientific community regarding the toxic effects of a number of chemicals.

Toxicology is a discipline with basically one goal: to understand how chemicals can adversely affect living organisms. However, due to the multiplicity of chemicals and our limited understanding of how the human organism works, there are often questions as to whether or not an adverse effect has occurred, even if some biological change can be detected. Toxicology can provide answers to many questions we wish to address, but these answers must be looked at cautiously. We should not use these data to develop a false sense of what we know and unrealistic expectations for what our actions can accomplish.

We have been taught that science produces certainty. As a result, the public is impatient with scientists who express uncertainty and tend to believe scientists who express their views without reservations.

In toxicology, the certainty that most of us seek is that a particular chemical is safe. Unfortunately, there is no such thing as an absolutely safe chemical, since all chemicals can cause toxic effects in large enough amounts. When faced with this reality, most people look for a different certainty—a "safe" amount. They want to know the exact level at which a chemical changes from nontoxic to toxic. Again, this is not a scientifically realistic goal. People vary tremendously in their responses to their environment, including the chemicals in it, so what is "safe" for one person may not be "safe" for another.

While the government sets levels for many chemicals, the rationale behind these standards is not entirely scientific. Most people would like to entirely eliminate chemicals which have been labeled as highly toxic, but the ubiquitous distribution of many such chemicals makes their elimination unrealistic. In the minds of the public, however, a single number becomes a dividing line between "safe" and "unsafe." The amount and quality of scientific evidence behind this number vary from case to case and change over time much to the consternation and almost total lack of understanding on the part of the public.

Toxicity

The *acute toxicity* of a chemical refers to its ability to do harm as a result of a one-time exposure to the chemical. This exposure is sudden and commonly produces a health emergency. *Chronic toxicity* refers to the ability of a chemical to do systematic damage as a result of repeated exposures to small quantities of low concentrations of a chemical over long periods of time. Reactions produced by these two different types of exposure bear no resemblance to one another. Chronic toxic effects are not predictable from knowledge of acute exposure effects of the same chemical.

Some chemicals have a high acute toxicity but no chronic toxicity. That is to say, small quantities over a long period of time are harmless and, in fact in some cases, are beneficial. Vitamin D would serve as one such example; sodium fluoride is another. We require small quantities of vitamin D daily for good health, and we know that fluoride is essential for good dental health. The same can be said of sodium chloride, common table salt.

In contrast to chemicals that are acutely toxic and chronically nontoxic, there are some that are just the reverse. Metallic mercury is one example. A large ingestion of a single dose of metallic mercury will pass through the body without causing significant damage, but a buildup of mercury in small amounts over a lifetime can be lethal. Although there is little correlation between acute and chronic toxicity, both in their own way are dose-related. The greater the dose, either in small continuous quantities or a single large quantity, the greater the effect will be.

We ingest many "lethal" doses of a wide variety of compounds which have no effect on us because we spread the dose out over a lifetime. Caffeine in coffee, oxalic acid in spinach, ethanol in scotch, and acetylsalicylic acid in aspirin are just a few examples. In spite of this, the concept that exposure to trace quantities of foreign chemicals may actually produce beneficial effects is unacceptable to many people who have an antichemical bias.

The poor health of people who worked at certain trades was noted by early Greek and Roman physicians. The first monograph on occupational diseases was published in 1567, 26 years after the death of its author, the Swiss physician, Paracelus (Ottoboni, 1984). He set forth one of the basic tenets of modern toxicology when he wrote: "What is it that is not poison? All things are poison and nothing is without poison. It is the dose only that makes a thing not a poison."

Thresholds

The term "threshold" is used in toxicology to describe the dividing line between no-effect and effect levels of exposure. It may be considered as a maximum quantity of a chemical that produces no effect or the minimum quantity that does produce an effect. It is common for the threshold to vary with the species involved and even with individuals within each species. For purposes of extrapolating animal data to humans, the highest level of exposure that produces no detectable adverse effect of any kind in any test animal is used by toxicologists as the threshold.

A margin of safety is an arbitrarily established separation between the threshold of a chemical found by animal experimentation and the level of exposure estimated to be safe for humans. The FDA adopted the convention of a hundredfold margin of safety years ago when it began setting standards for acceptable quantities of food additives. The assumptions behind the hundredfold margin are that humans are ten times more sensitive to adverse effects of chemicals than the test animals are and that the weak in the human population are ten times more sensitive than the healthy and thus compounding the tens, we get a hundredfold margin (Ottoboni, 1984). So, we end up extrapolating the no-effect level for humans, then reducing the quantity by two orders of magnitude for additional safety. In recent years, there has even been a demand to adopt a thousandfold margin, although no more scientific justification for that exists than for the hundredfold margin.

No matter how large the experiment or the margin of safety, one can never prove that any environmental factor is totally harmless. Absolute safety is the complete absence of harm. We can never achieve this. We can only offer probabilities that there will, in fact, be no harm.

Cancer

While there are many diseases as life-threatening as cancer, there are few as widely dreaded. Cancer-causing chemicals are set apart from other chemicals and made the subject of special regulations. Cancer is now recognized as a generic term for a whole host of malignant growths in the body that ultimately destroy functions upon which they are imposed. Their causes are not yet well-known, but variations among different populations of the world are beginning to focus attention on dramatic dietary differences. The hypothesis that differences in cancer incidence are due to environmental factors was first annunciated by Dr. John Higginson in the early 1950s (Maugh, 1979). Dr. Higginson's hypothesis was based upon an extensive study of cancer among African black populations as compared with black populations in other parts of the world. He concluded that 80–90 percent of all cancers were caused by environmental factors. With the growing concern about environmental contamination during the 1960s, it was only a short time before the term "environmental factors" became transformed into "environmental chemicals." Thus, by the mid-70s, statements to the effect that a majority of human cancers could be attributed to carcinogenic chemicals in the environment became commonplace even though that was not at all the point of view held then or now by Dr. Higginson (1988).

Occupational Risk

The people at greatest risk of developing cancer from exposure to chemical carcinogens are those exposed to the highest concentrations —generally in the workplace. Unfortunately, far more public attention has been given to trace compounds in our environment than to carcinogenic chemicals in the workplace. Distinct patterns of cancer do arise among workers in specific occupational groups.

The incidence of cancer is clearly dose-related: the higher the dose of a chemical carcinogen, the greater the number of individuals who will develop the cancer. Thus, a more moderate approach to carcinogenicity testing would be the use of doses comparable to those received during occupational exposure to chemicals. Working populations, as a rule, do not include young children, senior citizens, or people who are ill or debilitated. But their lack of representation in occupational groups is compensated for by the fact that occupational exposures to chemicals are usually several thousands to millions of times greater than those encountered by the general public. Occupational exposures are sufficiently high to provide valid data using numbers of animals that reasonably can be accommodated in a toxicology laboratory. That is to say, we could actually work backwards from exposures in the workplace to extrapolating those exposures to animals and testing them above and below those ranges. The use of exposure levels that reflect those found in occupational settings is most appropriate; a chemical that does not cause cancer in occupationally exposed people is not likely to cause cancer in the general public, particularly at dose levels that are thousands of times lower than those found in occupational settings.

M. Alice Ottoboni, in her outstanding book *The Dose Makes the Poison,* offers some wonderful analogies for the distortions offered by using high doses in animal tests. In one, she describes a team of sports physicians interested in investigating the adverse effects that may occur to the ankles, knees, and hip joints of pole vaulters. The problem is that it would take thousands of pole vaulters making daily jumps for years to obtain the necessary data to produce statistically significant results. Thus, the physicians decide that instead of a thousand people making a 20-foot jump each day for many years, they could study a hundred people making a 200-foot jump ten times a day for one year. But since no athlete could vault 200 feet, and since the trip up is of no importance to the experiment, a nearby 200-foot cliff is suggested as a jumping-off place.

This analogy grossly exaggerates the indifference to the obvious biological limitations of the test organism but, nevertheless, the effect of indifference is the same as it is in high-dose carcinogenicity testing.

Large Dose/Any Dose

Supporters of the use of large doses justify their position by claiming that if a chemical will cause cancer in high doses, it will also cause cancer in low doses. While this can neither be proved nor disproved, knowledge of biochemical mechanisms and data provided by studying the metabolism of carcinogens belie its accuracy. Clearly, the use of high doses in animal studies does provide valuable information that should not be ignored, but the acceptance of results from high-dosage exposures while rejecting data from moderate dose studies and studies of mechanisms of action, metabolic fate, etc., represents an attitude that is foreign to objective scientific inquiry.

There is no controversy about the existence of threshold doses for chronic toxic effect, but there is a relatively large segment of the scientific community that denies the existence of thresholds for chemical carcinogens. It can be argued that one molecule of a carcinogen can cause a change in a nucleic acid which may result in an altered gene, which in the right environment and with the right cofactors, will result in a cancer (Crone, 1986). Therefore, no concentration level of a carcinogen can be entirely safe. This is true. But one chemical change which occurs among all the other changes caused by natural and artificial means is not going to add much to the risk of the individual getting a cancer.

There are no absolutes in the study of toxicity; all risks are statistical relating to the dose of chemical received. If we are to understand the significance of chemicals in our world, we must stop seeing toxicity as an absolute and appreciate the shades of grey. We must also cast aside any tendency to superficial generalizations and examine each question of toxicity freshly and with reference to all the information available.

Risk Assessment

Most laboratory experiments are performed at dose levels of the chemicals that produce clear, easily measured responses in the test animals (U.S. EPA, 1987). However, exposure to chemicals in the environment often occurs at dose levels low enough that adverse effects are not immediate or obvious. Thus, it is necessary to extrapolate results obtained in high doses to results expected at low doses.

This is especially difficult in the case of carcinogens. Extrapolation must span a very large change in dose levels (four to six orders of magnitude). Carcinogenic risk assessments are generally done by exposing the laboratory animals to the same dose every day for a lifetime. This type of exposure rarely occurs in humans. Even in the workplace, exposure is frequently interrupted by vacations, sick days, and job changes.

Regardless of the threshold debate, the public should be aware that there are at least practical thresholds for all carcinogens. The incidence of carcinogenic effects and the lengths of their induction periods are definitely dose-related. If exposure to a carcinogen is sufficiently small enough as to reduce its cancer incidence to one in a billion or less or to increase its induction times to 200 years, of what practical significance would that be to the human population (Ottoboni, 1984)? The chances of winning your state's lottery would be far greater!

So, having first assumed that animal studies directly translate to humans, the cancer risk models used by regulatory agencies then assume that one molecule of a carcinogen is capable of initiating a cancer process, and that a constant exposure to the carcinogen will occur over a 70-year lifetime. Estimates derived from these models are extremely conservative as a result of these assumptions. While it is appropriate for our government to err on the side of safety in matters of public health, if the assumptions used are incorrect, the errors can be too safe by several orders of magnitude.

Society has a right to know the cost of excess safety and decide if it is worth that cost. In order for the public to make valid judgments, they should be informed of the uncertainties inherent in the risk assessment process. They should know that risk estimates are only an expression of how great or small the chances are that exposure to a given carcinogen will cause cancer and that current methods for assessing risk are extremely conservative. No one questions that those charged with protecting the public need methods to make estimates for safe exposure levels. In the process of estimating risk, however, mathematical models should be an adjunct to and not a substitute for scientific judgment. In addition, as Miss Ottoboni states clearly, "Regulatory agencies must be free from special interest pressures so that the standard they promulgate are based on concern for the public good, not political expediency."

Mathematical Models

It is important to recognize that since direct estimates of risk at low levels of exposure would require the expensive testing of prohibitively large numbers of animals (Neal, 1983), models must be used to help account for this shortcoming. One of the major problems with risk assessment today, however, is the extreme oversimplification resulting from the use of quantitative mathematical models for extrapolating from high doses to low doses. The purpose of so-called cancer models is to estimate a safe dose, more recently termed "a risk specific dose," (Paustenbach, 1989) based on the extrapolation of experimental results well outside the dose range used in animal tests; usually three to four orders of magnitude below the no-observable effect level (NOEL). Statistical procedures for estimating the low-dose response involve a mathematical model relating the probability of the specific response at a very low dose. Because of the statistical and biological problems inherent in the identification of true no-effect levels, most mathematical models for carcinogens have eliminated the concept of threshold dose where no response would be expected. The major failing of the widespread use of low-dose extrapolation models is that they often reduce the decision process that requires thoughtful and critical analysis of complex, and often conflicting data, to one driven by a computer analysis of four or more data points on a dose-response curve.

Although every regulatory agency that deals with carcinogens advocates the necessity for careful evaluation of all available data, most regulatory decisions have been overly responsive to the results of these models at the expense of the biological information. The greatest problem in relying so heavily on modeling rodent bioassay data is that it gives decision-makers the mistaken impression that not only is the analysis routine, but that a high degree of certainty is achieved in the analysis.

Human risk assessment is a very inexact exercise based largely upon theoretical assumptions concerning interspecies extrapolations. The uncertainties involved should be fully recognized by the scientific community and society. The mathematical models used attempt to predict how many test animals would respond at low-exposure levels based upon observed responses of a few animals at high-dose levels. The models tell us little about predicted human responses at any exposure level.

Upper-Bound Estimates

Mathematical models give a range of risk estimates. The upper-bound estimate is the worst case estimate of risk. By its derivation, it is unlikely that the number represents the true risk. Dr. Fred Hoerger, at a seminar in 1985, illustrated the upper-bound estimate and its inherent distortion in the following manner.

> "It can be said that the upper-bound estimate of rainfall for the United States is 15,000 inches per year. Since yearly rainfall in the United States averages from a few inches to perhaps 50 or 60 inches per year in Miami, Florida, my estimate of 15,000 inches per year sounds outlandish. For a moment, let me justify my estimate on the basis of 'prudent' predicting principles. Historical record shows the highest single-day rainfall was 43 inches in Alvin, Texas in 1989. Simply multiplying this number by the number of days in a year and extending it to the entire United States gives my estimate of 15,000 inches."

Of course, such extreme rain occurs only rarely in the United States. But in risk assessment terminology, it is just such a worst case assumption that once linearized and extrapolated can result in equally outlandish answers. The faults in the aforementioned logic are that 43 inches of rainfall is obviously an extreme case, and that dramatic climate differences across the United States preclude its being considered a homogeneous area for extrapolation purposes.

Conditions of risk assessments with such extreme conservative biases do not provide decision makers with the information they need to formulate an efficient and cost-effective regulatory strategy. A perverse and unfortunate outcome of using absurd, upper-bound estimates based on compounded conservative assumptions is that it leads us to regulate insignificant risks while ignoring more serious ones.

The increasing tendency of assessors in regulatory agencies to adopt conservative assumptions in their calculations has become a serious problem which few have dared to address. The rationale, of course, is to ensure that the true risk to everyone will be less than that predicted. While this may look like an admirable goal, it has a number of shortcomings. The repeated use of exceedingly unlikely exposure scenarios makes it difficult to compare assessments by different scientists because they incorporate widely varying levels of conservatism in their assumptions. Furthermore, many exposure scenarios are called "worst cases," implying that they are feasible when, in fact, they are not.

The Bottom Line

There are many scientists from a host of disciplines who, along with toxicologists, expound on the toxic properties of chemicals. Molecular biochemists and microbiologists, for example, may well have great expertise in the effects of chemicals at the molecular or cellular levels. But unless they understand the principles that govern the more complex organisms such as man, they have no basis for making judgments, much less public statements, on the significance for man of the effects they find in their test systems. In addition to miscellaneous scientists and physicians who render toxicological opinions, recent years have seen the emergence of a new breed of toxicologist, the environmentalist lawyer who knows the vocabulary but not the substance of toxicology. Finally, there are the high priests—scientists who have taken up the cause of protecting the environment and all God's lesser creatures against the machinations of our industrial society. They believe we should eliminate nearly all synthetic chemicals. Their gospel includes stories of the damaging effects of synthetic chemicals, and they preach it with religious zeal and passion. Unfortunately, the public has no way of distinguishing between self-proclaimed and legitimate experts. Attempts by anyone, no matter how objective, to indicate who is expert and who is not only causes further controversy and worsens the problem. The public must eventually rely on their own good judgment in deciding whom to heed.

Dismantling the Chemical Industry

It has become common for environmental consumer groups to demand that chemicals they consider detrimental to their special interests be banned. The banning of chemicals in the words of M. Alice Ottoboni (1984), is " . . . a simplistic solution to a very complex problem and often produces greater problems than those sought to be remedied by the ban. In some cases, it requires the elimination of a chemical whose toxic properties and hazards are quite well-known and the substitution of a chemical or chemicals about which there is much less information." Furthermore, the banning of chemicals denies man's ingenuity to develop methods of use that will be protective of the health of the public and of the environment. Proponents of bans either do not accept that the chemicals in question can be used safely or contend that the malevolence and greed of the petrochemical industry will not permit safe use.

References

Crone, Hugh D. 1986. Chemicals and Society. Cambridge University Press, p. 41.

Higginson, John. 1988. Changing concepts in cancer prevention: limitations and implications for future research in environmental carcinogenesis. *Cancer Research*, v. 48, March 15, pp. 1381–1389.

Hoerger, F. 1985. Some current views on risk assessment. Presented at a seminar entitled, "Understanding Environmental Risk," sponsored by the Public Service Research and Dissemination Program, University of California at Davis, Sacramento, California.

Kamrin, Michael A. 1988. *Toxicology*. Lewis Publishers, Chelsea, Michigan, p. 2.

Maugh, Thomas H. II. 1979. Research news: cancer and environment: Higginson speaks out. *Science*, v. 205, p. 1363.

Neal, Robert A. 1983. Health and risk assessment: toxicology. American Water Works Association First Atlantic Workshop Proceedings. December 8–10, 1982, Nashville, Tennessee.

Ottoboni, M. Alice. 1984. *The Dose Makes the Poison*. Vicente Books, Berkeley, California, 222 pp.

Paustenbach, Dennis J. (editor). 1989. *The Risk Assessment of Environmental Hazards*. Wiley Interscience, New York. 1155 pp.

United States Environmental Protection Agency. 1987. Toxicology Handbook PES-8, Government Institutes, Inc.

Reprinted, with permission, from
the Journal of *Ground Water*,
Vol. 28, No. 2, March-April 1990
where it appeared under the title "Toxicological Risk Assessment Distortion
Part II: The Dose Makes the Poison."

Wetlands

The Health Risks of Wetlands:
The Western U.S. Perspectives
William Hazeltine

The Swamp Thing
Rick Henderson

Wetlands: A Threatening Issue
Jay H. Lehr

The Health Risks of Wetlands: The Western U.S. Perspectives

William Hazeltine, Ph.D.

Dr. William Hazeltine is manager-environmentalist, Butte County (California) Mosquito Abatement District. His past experience includes field research with Chemagro Corp. in Kansas City, Missouri, and former manager of Lake County Mosquito Abatement District. Prior to this he was self employed as a crop production consultant in Arizona.

He received an AB degree from San Jose State College in 1950 and a Ph.D. in entomology from Purdue University in 1962.

Dr. Hazeltine is the author of *Legislative History & Meaning*, Federal Insecticide; Fungicide & Rodenticide Act as amended 1972, 1975. He has published articles in *J. Econ. Entomology, Mosquito News*, and *Proc. Calif. Mosquito & Vector Control Association*.

There is a dark side to the wetlands issue. That dark side is the high risk of human diseases which can occur as a direct result of the creation or maintenance of wetlands. In order to reduce this risk, intensive management is required in most circumstances, and there are times and places were creating or maintaining a wetland is simply ill-advised.

I want to consider the magnitude of the health risks and how these risks relate to the responsibility and liability of the landowners, creators or managers of wetlands.

I begin with the assumptions that (1) man is an integral part of any natural system, and human needs must be considered in decision making, and (2) that humans are pre-eminent to any other species of organism, in any final judgement on priorities. These assumptions are consistent with the National Environmental Policy Act or NEPA.

While there are other disease vectoring species of animals associated with wetlands, mosquitoes are the most common. A vector is an animal capable of transmitting a disease causing agent from one host to another. Mosquitoes and wetlands usually go together, and one biological definition of a functional wetland is "a place where water stands long enough to produce two consecutive broods of mosquitoes."

Not all mosquitoes produced in a wetland vector diseases. Those that can and do transmit disease causing organisms are of greater concern to those of us who are charged with protecting people's health, even though we consider the disease caused by nuisance species as important.

The major types of mosquito vectored diseases, which are of concern in California, are malaria and viral encephalitis. Both of these diseases require an efficient mosquito vector, presence of the disease causing organisms and suitable, susceptible hosts. At the risk of oversimplifying for some of you, it is necessary to understand these disease cycles, in order to understand why we consider wetlands as a major risk to human health.

Disease Cycles

Malaria is caused by a protozoan parasite which requires alternate passage through a mosquito, and then through a vertebrate host species, except when transmitted by blood transfusion. There are specific stages which occur in each host, but once present in a person or other primate, the parasites can remain infective for prolonged periods, or gradually decline and die out. Having malaria confers no immunity, and this disease can be contracted repeatedly.

Control of malaria is usually achieved by control of the mosquito vectors, or by anti-malarial drugs in humans, or both. The present estimate of malaria, worldwide, is about 400 million cases a year with a 1% or more death rate or 4 million or more deaths a year. Most of these cases are in third world Countries, and it is the worst of the child killing diseases.

Western Equine Encephalitis (WEE) and St. Louis Encephalitis (SLE) are virus diseases occasionally found in the Western U.S. The normal cycle is begun by a capable vector mosquito feeding on a viremic (actually sick) host. As the blood meal is digested, the virus multiplies in the mosquito's tissues, and moves to the salivary glands where it can be injected with the anti-coagulant saliva during the next blood meal. In a susceptible host, there may be an acute disease phase, and then antibodies are typically produced which confer a measure of protection from future exposure to that specific virus type. Migratory birds are considered to be a common vehicle for virus transport into an area, and birds are a major endemic or natural cycle host for WEE and SLE. Migratory birds are a natural transport and endemic cycle host, because they come into and then live around the wetland environment where the host mosquitoes are produced.

I will not burden you with the scientific names of particular kinds of mosquitoes, unless it seems appropriate, but recognize that more than one species of mosquito and many kinds of migratory birds can be involved in encephalitis cycles. Jack Rabbits and a flood water mosquito create a second endemic disease cycle. However, Bird-feeding mosquito species are more likely to be the vector to carry the virus to man or horses, in an epidemic condition.

The usual cycle for WEE and SLE is between birds or rabbits and mosquitoes, with man and other animals considered to be "incidental" hosts. As the numbers of viremic mosquitoes increase, the risk of feeding on and transmitting the virus to people goes up. The proximity of the infected mosquitoes to humans is also a factor in the increased risk of transmission.

The migratory bird and mosquito relationship occurs with other species of mosquitoes and other viruses, in other areas of the Country. One example is Eastern Equine Encephalitis (EEE) along the Northeastern coastal tide lands, and salt-water marshes and forests.

History of Malaria
in the Sacramento Valley

In the days before gold was discovered in California, the Central Valley as well as what is now the San Francisco Bay Delta repeatedly flooded, and large areas were wetlands.

Historians set the date of Spring, 1830 as the time when fur trappers from Oregon came into California with malaria parasites in their blood. The Native Americans were hunting where the winter floods had forced game onto the Sutter Buttes near Marysville. The trappers moved south, continuing to spread the disease after the prevalent local mosquitoes had picked up the parasites. When the trappers returned going north in the Fall, estimates of 25% to 90% are given for the numbers of Native Americans in this area that had died from malaria.

The epidemic conditions existed because the parasite, the right vector and a susceptible population of people were all in the same place at the same time.

During the gold rush days of California, locally transmitted malaria was a major cause of illness and death. The occurrence and increase of malaria mosquitoes was a result of natural wetlands in the valleys of California, as well as the breeding areas created by water diversions and ditches in the foothills, which were part of the California gold rush scene. Miners worked and lived with the Anopheles mosquitoes, which thrive in the kinds of environment the miners created. The "ague" was a local name for malaria.

"Fevers of unknown origin" were also present, and were probably due in part to viral encephalitis.

With the draining of wetlands, creation of flood control systems, good mosquito control and modern drugs, malaria has been largely eliminated in most of California. Occasional outbreaks have occurred, such as in 1952, when a veteran of the Korean war had a malaria relapse during a July 4th weekend campout in an area near a Girl Scout camp in Nevada County. The right mosquitoes were also present, probably having come from Beaver ponds and shaded pools in a nearby stream. Thirty-two cases were documented from this one parasite source.

San Diego County has reported cases of malaria occurring repeatedly over the past few years, as workers from Mexico bring in the parasites and live in areas where the right kind of mosquitoes are present to transmit the disease agent to other workers living in the same area. Most of these cases occur in streambeds where these workers live in make shift shelters, and where effective vector mosquitoes can breed.

There have also been occasional, recent, locally transmitted cases of malaria in the Central Valley and Foothills of Northern California which remind health care professionals of the continued risks. Shaded woodland pools such as beaver ponds are increasing in our area, and the better vectors of malaria thrive in this kind of habitat.

The last locally transmitted malaria case reported in our County was very near an area where the California Department of Water Resources people were trying to clean a flood control channel, but they actually created a worse mosquito breeding problem when they followed the State Fish and Game requirements on making potholes in this channel.

So the risk of malaria goes up as wetlands such as shaded woodland pools are allowed and even encouraged, as part of planned urban development in and around many cities in California.

Encephalitis

Western and St. Louis encephalitis used to be common in California, before the start of modern mosquito control. Efficient control has depended on chemicals, as well as water source management which includes natural controls or even water elimination. From a health view point, it makes more sense to control or eliminate mosquito breeding than to try to kill the flying mosquito adults after they have emerged and left these water sources.

The California incidence of human mosquito transmitted encephalitis shows a decline and then a recent increase.

1950–1959	938 human cases, 27 deaths
1960–1969	143 human cases, 2 deaths

1970–1979   28 human cases, no deaths

1980–1989   72 human cases, 1 death

During the last decade alone (1980–1989) there were 26 and 28 cases of human disease confirmed in just two of these years. One year the cases were in the Long Beach area, and the most recent episode was in the Southern San Joaquin Valley. The Long Beach cases were thought to be associated with urban mosquito sources and the presence of a large non-immune population which had no previous low-level disease exposure. The virus source was believed to be migratory wild birds. The 1989 cases were associated with the increased flooding of non-agricultural areas, primarily along drainage systems in Kern County. The implications are quite clear about more wetlands, which attract migratory birds and higher numbers of mosquitoes, causing increased risks of human disease.

The worst recorded epidemic of encephalitis in California was in 1952, where there were 420 human cases and 10 deaths recorded. Three hundred and seventy-five of these cases were WEE, and 104 of this number or 28% were infants under 1 year of age.

Mosquito control was recognized and supported as the only way these diseases could be controlled. Lots of money was made available for mosquito control, for a while, after this epidemic. Remember there is no cure, once the viral encephalitis starts, and there is no generally available human vaccination to prevent it.

Survivors of the 1952 epidemic showed that Western Encephalitis can cause extensive residual brain damage, as well as death, with the most severe brain damaged cases occurring in the very young. About 40% of the 104 infants with WEE under 1 year of age had brain damage. Another observation from this epidemic was that the disease in a pregnant mother can be passed on to her unborn offspring, as shown in identical twins, both of which had severe motor function damage. Some of these cases are graphically recorded in a film made after this epidemic.

St. Louis encephalitis, which is caused by a different virus, is particularly severe on older people, with the death rate in the neighborhood of twice that of WEE. Eastern Encephalitis on the east coast and California Encephalitis in the Midwest are reported to have an even higher death rate in humans.

Organized mosquito control agencies now collect mosquitoes and bleed and test domestic and wild birds and other wildlife, to see when any of the viruses which cause encephalitis are active. Experts believe that encephalitis viruses may produce hundreds of low level undiagnosed flu-like illnesses for each diagnosed human case. Delayed neurological problems are suspected from encephalitis infections in young people, which were not diagnosed because the disease was too mild. These delayed problems are thought to occur as the brain is developing and growing, with the risk ending when the brain reaches its maturity.

History shows that early season encephalitis virus activity in California and Arizona is associated with drainage systems, such as the Colorado River and in the San Joaquin and Sacramento Valleys. In Northern California, the Sacramento River and nearby extensive wetlands and refuges are the places where we usually first find the virus. Infected mosquitoes are usually found in urban areas 2 to 4 weeks after virus activity is seen in the wetland habitat. Migratory birds are the most likely source of the virus.

Encephalitis in Other Areas

Other areas besides California and its neighboring States have experienced recent periodic epidemics of Western and St. Louis encephalitis, along with other mosquito vectored, viral diseases.

In 1975, for example, an epidemic which began in Mississippi spread north along the major river systems into the upper Mid-west. Before this epidemic was over there were over 4000 confirmed human cases, with 95 confirmed deaths. This was a combined epidemic, with Western starting first, followed by St. Louis. Ohio experts alone reported 419 cases, and estimate another unreported 1100 human cases. In one Ohio City, one mosquito in 120, on average, was infected.

An SLE epidemic in Houston Texas in 1964 resulted in 34 deaths, and was the reason a control district was established there. In the same decade, Dallas was also sprayed by military aircraft, to stop an epidemic there.

In 1971, 10½ million acres in Texas and the Gulf States were sprayed by air in just 35 days and over 3 million acres were sprayed by ground rigs to control mosquitoes and other biting flies during an epidemic of Venezuelan Equine Encephalitis (VEE). While the concern in the U.S. seemed focused more on horses than people, the record in Venezuela showed there were about 32,000 human cases and 190 fatalities between 1961 and 1964. In 1967 and 1968, the human cases were estimated at a quarter to a half million, and 50 to 100 thousand horses had died. VEE seemed to have some of the same endemic migratory bird host patterns as WEE and SLE.

This past year (1990) Eastern Equine Encephalitis was active in Massachusetts and New Jersey. A reported 700,000 acres in 4 Counties near Boston were sprayed by air for this outbreak. This problem was associated with tidal wetland mosquitoes and migratory birds which frequent these salt marshes.

Florida and the Gulf States including Houston, Texas also experienced many cases of human SLE this past year, and many sentinel chicken flocks in Florida showed that 100% of the birds had antibodies. While the numbers are not complete, there were 96 confirmed and 42 suspect human cases, with 5 deaths, as of November 1990. Halloween trick or treat activities and evening athletic events were cancelled in many areas, because of the disease risk.

Encouraging Disease Risks

The anti-pesticide movement has caused disastrous consequences for the protection of humans from diseases. Public health pesticide uses cover small areas compared to agricultural uses and are usually applied at much smaller dosage rates. As a result, low volume uses for health protection have been the first uses dropped by the manufactures. This is simple economics. It has been our experience that the Federal EPA does not recognize the low volume, high benefit uses of pesticides for health protection as meriting special consideration.

In California, we have lost our best pasture mosquito control material because this chemical is only used for mosquitoes and household pests. The State allowed the use on pastures in the past for health protection, but new perceptions of risk and new rules on residue tolerances now even prevent this use. Ironically these pasture mosquitoes and Jack Rabbits are one important part of the endemic encephalitis virus cycle. A popular alternative pesticide for protection of people from this vicious mosquito is pyrethrin, which is a natural extract of a particular Chrysanthemum flower, and it is in very short supply. Besides costing about $300.00 per pound, we are on allocation and have no assurance we will even be able to buy or afford enough to kill these mosquitoes and the other vectors of encephalitis in dense urban areas, let alone the pasture environments outside of town. Other States are having similar problems with pesticide availability.

People want to have a decent place to live, and increased population pressures have caused people to invade mosquito areas and even to try to recreate natural wetlands. There are some in our society that have even used the wetlands issue as a way to discourage or stop urban growth. They have invoked the Federal Army Corps of Engineer's dredge and fill permits, and more recently the Endangered Species Act as ways to prevent land reclamation. The consequences of wetland and wildlife preservation workers fighting against the use of pesticides at the same time they advocate more wetlands could cast these people in the roll of promoters of human diseases, whether they realize it or not.

Liability of Creating or Maintaining Wetlands

By now, I hope you can appreciate some of the poorly comprehended social risks which go along with creating or maintaining wetlands.

In addition, I am really bothered to see conflicting data that suggests that the numbers given to support the argument about past wetland losses in the Central Valleys of California is not accurate, or even purposefully misleading. These differences may be due to different definitions of wetlands.

Numbers compiled by the USDA, Soil Conservation Service shows that in all of the Pacific States, between 1982 and 1987, the average loss of privately owned wetlands was about 8,000 acres a year. This does not square with the catastrophic loss numbers being used to achieve a political objective.

Control of the Problem

Continuing on the theme of risks associated with wetlands, let me point out some options open to public health workers, to protect people from unreasonable adverse effects on their lives and their living place (= their environment). First, reasonable people should want to have progress, as well as what one might think of as amenities, such as wetlands. The key to having a reasonable mix of both wetlands and health protection lies in a committment to wetland management. This is not the usual flood and hunt idea, or just keeping an area wet, which many people think of as wetland management. True management requires being responsible for a reasonable balance of benefits, with minimum risks. Wildlifers are going to have to get real, and recognize that pesticides and physical control are necessary parts of their control options. Said another way, pesticides have benefits and their use may be the only means for having socially acceptable wetlands.

Keeping useful pesticides available is in the best interest of true conversationists.

Whose Risk and Whose Benefit?

I have tried to understand the reasons why society is so set on increasing wetlands, at all costs. Let me list some reasons which I have heard.

1. To replace losses due to civilization—this spiritual kind of idea involves a return to primeval times, for reasons that are not clear.
2. To solve a social guilt complex such as "replacing the plunder."

3. To make a place for "sportsmen" to play.

4. Some vague sense that wetlands are the primordial birth place of land animals, and therefore worthy of preservation.

Using these possibilities or others you may want to consider as benefits, ask yourself what are the offsetting factors or problems in increasing wetlands. Obviously, I hope you now see health risks as a vital consideration.

Another question is where do people fit in this "creating wetland" equation, and who pays the hidden costs for wetlands? Under most modern thinking, the direct beneficiary should pay the cost, if that beneficiary can be identified.

Ask yourself, how many human disease cases are an acceptable amount, in order to have an incremental increase in wetlands? Realize also that statistics are fine before an incident occurs, but not when someone in your own family is that one case in ten thousand.

With this social preamble about the beneficiary paying the costs, I would like to tell you some of the alternative strategies I consider to be viable ways to protect the public from the unreasonable health risks such as mosquitoes and diseases, associated with wetlands.

1. Convince landowners to manage their land so it does not become a public nuisance. Under California law, a breeding place for mosquitoes is a public nuisance. Other States have general nuisance laws, and many have specific mosquito control laws. If wetland banking is required, it must be adequately managed and appropriately located.

2. If land is in a vector control district or the County uses the California vector control laws, the nuisance can possibly be treated with pesticides and biological control agents paid for with tax dollars, or the landowner, which may be a Local or State Agency, can be ordered to abate the nuisance and pay the costs. They can even be charged civil penalties. Federal Agency have immunity from local laws.

3. If a landowner in California buys a property with an easement, or sells or gives an easement which concerns wetland preservation on land he continues to own, the new or continuing land owner can be made to abate the nuisance, or pay the costs. In this case, civil penalties on the landowner may be more appropriate for use to fight the mosquitoes when they fly from his property.

Conservation easements can have costs which may not be recognized. Again, the Federal land immunity is an exception, but a private landowner is really setting himself up for continual grief when he sells or gives a conservation easement. Under the Federal easement program, the Federal Government can flood such land, and the private landowner can be held liable for the nuisance which the Feds have created.

4. We may be able to require a statement in the public report to any buyer of a property with a wetland easement, telling him of the liability he is accepting when he buys such land. We are also trying to get the Soil Conservation Service or other Federal Agencies such as the U.S. Fish and Wildlife Service to disclose the consequences and liability which will be incurred, in the language of their contracts with landowners which cover easements that require wetland preservation. Full disclosure of the consequences of an easement should be a requirement.

5. Any State or Local Agency in California which provides or controls the delivery of water that results in a mosquito breeding public nuisance can be cited, charged costs to abate the nuisance that water creates, and assessed civil penalties. Under this theory, even the owner of a water right could be issued an abatement order for a nuisance created with water appropriated under that right.

Federal immunities are not absolute in this case. If Federal wetlands get water from State or Local Agencies, or through their delivery facilities, or are subject to State water rights laws, this is a potential way to get compliance from Federal Agencies for wetland management. We could put pressure on the supplier or transporter of the water, and provide a disincentive to supply water to Federal landowners who allow that delivered water to become a public nuisance.

6. Finally, we can get more active with the media, where stories trickle down or up to the legislative decision makers. Consider the reaction to a press release concerning potential encephalitis from a proposed wetland, with T.V. graphics showing brain damaged infants. With work, I believe we can achieve public awareness of the risks of improperly managed wetlands.

People have a right to good health, which public health people should support with as much vigor as the wildlife people use to support their wildlife constituency.

Nostalgia

Those "good old days" which are shown on late night T.V. tend to overlook the hardships. Early settlers and miners had hardships I do not envy. Glorifying wetlands for purposes of political perceptions without equal time to explain the risks, is not ethically acceptable. I consider it to be socially dishonest. If the wetlands issue was a scientific debate, such one-sidedness (or adversary science) would be grounds for censure. Perhaps this is why the wetlands issue has been taken out of the realm of science, so the accountability will not diminish the success.

Saul Alinsky, the radical organizer has written: "If the ends don't justify the means, what the hell does?" Said another way, "don't let conflicting facts interfere with winning the game."

I frankly enjoy our present standard of living.

I can assure you that the health issues will not let me sit by and watch society taken back to those "good old days" where food was scarce, disease was prevalent, and an early grave was the expected.

The Swamp Thing

Rick Henderson

Rick Henderson has been the assistant managing editor of *REASON* since October 1990. Prior to this position, he was employed by *REASON* as both a researcher and a reporter. He was an intern at the National Journalism Center, Washington, D.C., summer of 1989.

Henderson received an A.B. in political science from the University of North Carolina, Chapel Hill, in 1979.

Last December, the CBS "Sunday Morning" program visited a beautiful marshland in the Louisiana bayou. To protect the marsh from commercial development, the state government had just made it a public wildlife preserve. As cameras panned the hardwoods and swamp grasses, host Charles Kuralt gravely intoned, "Who would want to pave this land? . . . Now it's a gift to ourselves."

To lovers of the outdoors, wetlands evoke powerful emotions and seemingly simple policy decisions: preserving nesting grounds for the whooping crane instead of building a shopping mall; protecting a habitat for shellfish rather than letting Jack Nicklaus construct a golf course. But the struggle to preserve wetlands doesn't always involve such straightforward choices. And not all of the areas threatened by development are as easy to designate as our Louisiana marshes.

Wetlands were once defined by the functions land performs: Now officials focus on technical factors. A landowner has to prove his property isn't a wetland. And the parameters are extremely elastic.

Environmental officials have begun to apply a broad and vague new definition to wetlands that adds to the nation's wetlands inventory property that would hardly qualify as swamps, marshes, or bogs: most of the eastern United States and perhaps as much as 40 percent of drought-stricken California. If strictly followed, the new definition will make millions of acres of private property unusable and require huge tax-dollar payouts to compensate property owners.

Consider the following examples, all deemed "wetlands" under the new policy:

- A North Dakota corn field where pools of water collect for a week each year during normal spring run-off.

- A muddy patch between railroad tracks in the center of the main street in an Idaho town.

- Irrigation ditches dug by farmers in Nevada, California, and other western states—some of which have been in use since the turn of the century.

Once federal or state officials designate property as a wetland, if the owner wants to do anything with it he must first apply for a permit from the Army Corps of Engineers. Corps regulators decide whether they will issue the permit, and, if so, whether the owner will have to set aside other property to make up for ("mitigate") the lost wetland. The corps gained this authority from Section 404 of the 1972 Clean Water Act, which allows it to regulate "the discharge of dredged or fill material" into "the navigable waters of the United States."

People who fill wetlands without obtaining a permit face fines and prison sentences. Perhaps the most famous case involves Hungarian immigrant John Pozsgai. A mechanic in Morrisville, Pennsylvania, Pozsgai purchased a piece of property to build a garage. After removing tires and automobile parts that had been dumped on the land three decades earlier and hauling in soil to level the property before construction, he was convicted of 41 violations of the Clean Water Act, sentenced to three years in prison, and fined $200,000 for illegally filling a wetland. Neither federal nor state regulators had listed his property as a wetland before he started to clear it. As a July '90 *Audubon* article stated, "Dumping dirty old tires into a creekbed does not seem to violate the Clean Water Act so conveniently as dumping new fill into an old wetland."

Many other wetlands cases involve farmers and ranchers. In Missouri, when corn farmer Rick McGown repaired a sunken levee on his property, he was accused of illegally filling a wetland after an Army Corps official found a "cattail" growing on the land. McGown claims the plant is a strand of sorghum he planted. If the corps wins its suit, the farmer will have to give the government one-third of his farm and pay a $7,500 fine.

After a normal spring thaw, the Idaho transportation department wanted to get rid of the mud-and-gravel mixture that collects on the sides of snowplowed dirt roads. Farmer Bud Koster allowed the department to dump this muck onto a plot of pasture. The corps later ruled that Koster had illegally filled a wetland and told him to either convert other property to a wetland, remove the dirt, or pay a fine. The case is still pending.

In Nevada, a rancher who repaired irrigation ditches dug 75 years ago has been accused of "redirecting streams." Farmers in North Dakota have been charged with illegally destroying habitats for migrating birds when they drained potholes in their fields. In California and Maryland, regulators halted construction of low- and moderate-income housing projects after charging that the construction sites were functioning wetlands. A recent Army Corps ruling suggests that when owners pull tree stumps from their land, if any chunks of dirt fall from the stumps, that may constitute filling a wetland. As we'll see, expansive government wetlands policy not only violates the rights of property owners but defies common sense.

In 1977, the Carter Administration sought to reduce wetland destruction; it began by asking federal agencies to define uniformly what they planned to protect. The Department of the Interior, the Department of Agriculture, and the Environmental Protection Agency jointly defined *wetlands* as areas flooded or saturated with ground water often enough that, under normal circumstances, they would support "vegetation typically adapted for life in saturated soil conditions." The definition emphasized that wetlands were limited "to only aquatic areas"—in other words, swamps, marshes, or bogs.

But by the mid-1980s, the government definition started to expand. The Corps of Engineers developed a new set of guidelines to help distinguish between plants that grow in wet soils and dry soils. The guidelines, which set up five classifications of plants, were intended as a measuring device, not a basis for policy. But corps officials noted that this checklist could provide ambitious regulators with an expansive new definition of wetlands.

That's what happened. The guidelines evolved into the *Federal Manual for Identifying and Delineating Jurisdictional Wetlands.* Wetlands, which were previously delineated by the functions land performed, are now defined by technical factors: the wetness of the soil (its hydrology), its chemical properties (whether the soil is "hydric"), and the varieties of plants that grow there (hydrophytic vegetation). Theoretically, land is supposed to meet all three criteria before it's declared a wetland, but the burden of proof is on the property owner. And the parameters are extremely elastic.

Bernard Goode, who headed the Army Corps's 404 regulatory office in Washington, D.C., from 1981 to 1989, helped develop the delineation manual. He says that when the government was developing its uniform definition of wetlands, each agency expanded the definition by drawing the parameters as broadly as possible. For instance, the Soil Conservation Service included hydric soils, which are moist enough to impede crop growth but not necessarily saturated or flooded, as *wetland* soils. The manual also says that soil that is inundated for as little as one week each year (the typical run-off period for farmland in a river basin or valley) qualifies as wetlands.

And the EPA insisted that facultative vegetation—plant life which *by definition* appears in uplands as often as in wetlands—be included as a wetland-defining parameter. Robert Pierce, a former Corps of Engineers regulator who helped write the enforcement guidelines for Section 404, notes that in some regions of the country. Kentucky bluegrass is facultative. In Pennsylvania, regulators consider ash trees and dogwoods to be wetland plants. The most common facultative plant, says Pierce, is the red maple tree, which can grow in standing water or on top of a mountain.

As a result of the new definition, the Soil Conservation Service estimates that as many as 70 million acres of existing farmland (privately owned, of course) could be considered jurisdictional wetlands, subject to federal control. Wildlife advocates and property developers, the *Los Angeles Times* reports, estimate that most of the eastern half of the United States fits the new guidelines. One environmental policy analyst I talked to cited an estimate that defined 40 percent of drought-stricken California as jurisdictional wetlands; San Francisco attorney Mark Pollot, who defends property owners in wetlands cases, contends "That may even be an underestimate."

Pollot notes that the technical definition simultaneously expands the authority of environmental regulators and makes their jobs easier. Before the manual guided regulators, he says, delineating wetlands was neither a simple nor a speedy process.

"Figuring out the hydrology of land is the most difficult thing to do," he says. So regulators took a short cut. Now any land with a preponderance of wetland plants—even facultative plants—can be designated as wetlands "without ever looking at the hydrology," Pollot explains. "You can dig around and find hydric soils even if that land hasn't been wet for ages. The manual will allow you to assume that the land is actually wet."

The expanded definition has contributed to already-inflated estimates of wetland losses. No one seems to know how many acres of wetlands disappear each year, although guesses abound. The Audubon Society claims we lose between 300,000 and 500,000 acres yearly; the Office of Technology Assessment says 275,000 acres. The American Farm Bureau Federation says those estimates are based on data collected in the 1950s and 1960s (when Army Corps flood-control projects were at their peak) and don't reflect current wetland losses; the Farm Bureau estimates losses of around 100,000 acres annually. When asked for a current figure, John Meagher, the EPA's deputy director of wetlands policy, admits that the most valid data was collected decades ago, but "we probably lose from 300,000 to 400,000 acres a year."

A recent EPA study casts doubt on the higher estimates. A 1989 agency study found that a 1985 law cutting off subsidies from farmers who drain and till wetlands ended almost all wetland conversion in six of the seven states studied. The Farm Bureau also estimates that around half of current annual losses come from the erosion caused by flood-control projects that prevent cities like New Orleans from being washed away every few years.

Inflated loss estimates bolster the Bush administration's policy of "no net loss" of the nation's wetland base—a long-time goal of EPA chief William Reilly. In 1986, while head of the Conservation Foundation, Reilly spearheaded the National Wetlands Policy Forum, an advisory group to the EPA composed of elected officials, government regulators, environmentalists, and agricultural interests. In 1988, the forum issued a report urging an initial policy of "no net overall loss" and eventually an increase in the quantity and quality of wetlands.

Candidates George Bush and Michael Dukakis signed onto the forum's agenda. President Bush subsequently endorsed no net loss in his first State of the Union address, and set up a task force to implement the goal. The no-net-loss policy means that whenever a property owner fills a wetland, he must convert an equal amount of property into wetlands. It doesn't matter what functions the original wetland performed: The property owner has to make an acre-for-acre trade.

Although the Bush-appointed task force hasn't yet completed its job, applications of the no-net-loss policy already exist. To preserve 600 acres of marginal wetlands in Staten Island, New York, taxpayers may have to reimburse property owners $400 million for their land and install $1 billion in storm drains.

And when a government project alters wetlands on public property, the wetlands have to be replaced—no matter how the restoration of the wetlands affects the rest of the local ecology. A riverside road-construction project near Savannah, Georgia, filled four acres of tidal marsh. To replace the filled wetland, officials cleared an adjacent mature pine- and palm-tree forest, excavating the forest floor to the mud line. They also clearcut vegetation from an island on the other side of the river. When he called for no net loss, says former regulator Pierce, "George Bush didn't know what he was getting into."

From providing wildlife habitats to controlling floods, wetlands can play important—sometimes crucial—ecological roles. Wetlands advocates, however, say wetlands perform functions that people—or other natural processes—can't replace. Preservationists end up engaging in a form of bait-and-switch: They argue that wetlands provide unique, essential ecological benefits, but we don't know enough about the functions to quantify them and set priorities. With imprecise knowledge, the "safe" answer, ecologically speaking, is to define wetlands so broadly that everything's included.

Says the EPA's Meagher, "It's really hard to [set] priorities from a functional perspective, to say that fish are more important than ducks, more than mammals, more than water quality."

Meagher says that priorities must depend on "the science of landscape ecology," based on functions that are determined at each site. Before considering priorities, regulators must look beyond the immediate area. "You can't make these decisions without considering what else is going on in the ecological system," he says. Officials would like to decide based on whether property serves wetland functions, he says, but "functions are complex," and regulators may not know which ones are most valuable. So *any* human activity that affects wetlands becomes questionable.

But there is no pure "science of landscape ecology," nor a "scientific" definition of wetlands, responds Bernard Goode. He says that "wetlands are and will be whatever combination of soils and water the government decides." The definition will always be determined politically.

As a result, in Pierce's words, "What is being called a wetland is functionally not different from uplands." The new definition fails even to distinguish between natural and man-made wetlands. A rice levee falls under the same jurisdiction as a swamp.

There are alternative policies that can preserve truly functional wetlands, as well as good sense. Goode suggests basing the delineation of wetlands on one parameter: the presence of enough surface water for a sufficient amount of time to support typical wetland plants. To meet this criterion, the water *table*—not merely a random puddle—must be at or above the surface for 30 days during the growing season. He says this degree of saturation will once again limit wetlands to truly aquatic areas—swamps, marshes, and bogs.

Now a private consultant who helps property owners and government regulators identify wetlands, Goode has attempted to sway officials in the regulatory agencies and with the wetlands task force. He says some members of the task force seem sympathetic, but he sees them as "the last hope in the executive branch to bring some reason and fairness" to the problem. Since leaving the corps, Goode says he's becoming something of a "crusader for landowners."

Attorney Pollot suggests his own alternative to acre-for-acre mitigation. "Is it necessarily true," he asks, "that a 15,000-acre wetland you use for a flood-control project needs to be replaced by another 15,000-acre wetland if a 100-acre or a 500-acre or a 1,000-acre wetland will do the same function?"

If the EPA or the Department of the Interior obtained authorization from Congress to swap publicly held parcels of land for privately owned wetlands of comparable value—instead of taking the property outright—environmental regulators could preserve the valuable functions of wetlands without strapping property owners and other taxpayers. Land swaps would require federal officials to quantitatively assess wetland functions and set priorities. But "wetlands restoration," Pollot concludes, "doesn't have to be expensive."

So far, federal agencies won't endorse these less-complicated proposals. Corps spokesperson Edward Greene says his department defers any priority-setting or other policy changes to the White House task force. The EPA's Meagher says his agency doesn't have enough information to set priorities or set aside comparable plots of land. So it will proceed with the existing policies.

Undoubtedly, restricting wetland regulations would require government policymakers to limit their own regulatory power—bad news for the Army Corps. The corps needs a new bailiwick, now that building dams is out of style, and environmental activism looks like just the thing. In February 1990, Lt. Gen. Henry Hatch, commander of the corps, officially endorsed the no-net-loss policy, declaring that his agency should become "the nation's environmental corps." *Audubon* writer John Madson notes, "for the Corps, [environmental projects] mean job security."

Meanwhile, the Washington rumor mill holds that the EPA's Reilly wants to replace embattled Interior Secretary Manuel Lujan. When the White House task force finally settles on a definition of wetlands, and on which agencies will regulate them, the EPA could lose much of its authority. The Department of the Interior could become the logical home for most wetlands regulation. And at Interior, Reilly would supervise the National Parks Service, the Bureau of Land Management, the Fish and Wildlife Service, and become chief enforcer of the Endangered Species Act—a sizable fiefdom for an ambitious regulator.

Still, Pollot believes that rationality will return to wetlands policy once taxpayers start routinely paying large sums to compensate for taking private property. Declaring property a jurisdictional wetland can eradicate almost all its market value—entitling the property owner to compensation under the Takings Clause of the Fifth Amendment. Since an estimated 80 percent of jurisdictional wetlands are privately owned, he says, payouts "could run into staggering amounts."

Last year, the U.S. Claims Court awarded a real-estate developer $2.6 million and a mine owner $1 million to compensate them after the Corps of Engineers defined their properties as wetlands. In each case, corps actions reduced the market value of the properties by more than 95 percent. In a surface mining reclamation case, another claims court ruling awarded more than $100 million to a plaintiff.

"If you go to Staten Island or Manhattan, and [force people] to shell out a couple of billion bucks to buy up marginal wetlands," Pollot says, "you are much more likely to ask the most fundamental question of all: Is this really worth it?"

Reprinted, with permission, from the April 1991 issue
of *Reason* magazine.
Copyright 1991 by the Reason Foundation,
2716 Ocean Park Blvd., Suite 1062, Santa Monica, CA 90405.

Wetlands:
A Threatening Issue

Jay H. Lehr, Ph.D.

Please see biographical sketch for Dr. Lehr on page v.

A war is being waged around us but we, in the community of groundwater scientists and engineers, are abdicating our responsibility to society to become involved. This is particularly unfortunate in light of the significant impact our considerable expertise could achieve. The subject at issue is the rapidly expanding definition of a wetland and society's irrational desire to protect them beyond all levels of scientific reason and environmental benefit. The forces at hand and the players at the table are complex, sometimes illogical, and often underhanded.

It is time for all of us to abandon our reticence, choose a side, and get involved. If I can't convince you of my point-of-view, then join the opposition. The credibility of our profession will still be raised by your active contribution.

On the following pages, I shall present the problem, its rules, the players, their rewards, the unfortunate impacts on society, and a call to action by those persuaded by my rhetoric.

Ground Water Ignored Again

One totally overlooked impact of the drive to protect wetlands that aren't even wet should be of grave interest to all of us in groundwater science. I am speaking of the significant quantity of the nation's ground water that can never be put to use. It will logically follow that the protection of a nonwet wetland will include not lowering the water table by which it is declared a wetland in order to gain benefit from its water supply. We will have to leave in place billions of gallons of valuable ground water in order to preserve some vague species of hydrophytic vegetation growing out of some ill-defined hydric soil. It makes no sense at all. We already have a largely underused groundwater resource throughout a nation that uses three times more surface water. Now, we are legislating against our utilization of one of the few simple answers to our growing water supply problems. Locking up these valuable ground waters for posterity for no discernible benefit to either man, beast, fowl, or plant (where no real wetland, i.e., swamp or marsh, exists) makes no sense.

799

It reminds one of efforts in San Antonio, Texas to leave undisturbed the 550-foot thick Edwards Aquifer, full to the brim with nearly a 400-foot artesian pressure head upon it, rather than reduce the head a few feet and in so doing, eliminate a spring which supports a variety of beneficial uses. Little question exists as to the desirability of keeping water flowing at the spring, but not at the expense of leaving as much as 50,000,000 acre-feet of water untapped in the vast Swiss-cheese-like pores and channels of the Edwards limestone.

It is likely that the ground water we will leave behind under spurious wetlands will dwarf that endangered in San Antonio by those trying to promote nearly nonexistent surface water at the expense of ground water.

A Silly Sound Bite

In 1988, candidate George Bush delighted the nation's environmentalists by pledging that during his term in office there would be "no net loss of wetlands."

Today, that pledge is mired in a swamp of controversy over the definition of a wetland which, indeed, no longer needs look at all like a swamp. The President now faces the wrath of farmers, landowners, and small communities disturbed by the original 1989 *Wetland Manual* that condemns as much as 300 million acres of mostly private property to a useless future in spite of the fact that it may appear high and dry to the "untrained" eye.

Now, the administration is being derided by environmental zealots enraged by efforts to rewrite the *Wetland Manual* with a modicum of reason, though this writer could find no such modicum in the new May, 1991 draft. Environmental groups are charging that the new directive could eliminate between 3 and 10 million acres from federal protection, a small improvement indeed from the original 300 millionacre estimate.

Bush's no netloss pledge could only be achieved by empowering the Army Corps of Engineers through Section 404 of the 1972 Water Pollution Control Act (Clean Water Act) to literally "take" the property of thousands of Americans without compensation, in spite of the Constitution's Fifth Amendment. In fact, the Clean Water Act never made mention of wetlands but was aimed at dumping (now deemed filling) in navigable waters. Only the 1985 Farm Act, sometimes referred to as the swampbuster provision, regulates wetlands that were under federal subsidy programs.

In the Old Days

Few people likely remember that up until 15 years ago, the U.S. Department of Agriculture offered direct financial assistance to farmers who would convert wetlands into arable land. Even after these subsidies ended in 1977, indirect benefits through other farm programs and income tax deductions continued until the 1985 Farm Security Act stopped price supports on converted wetlands and the 1986 tax reform ended tax deductions.

In the days of old, wetland controversy pitted preservers of whooping cranes against shopping mall moguls, and shellfish supporters against another Nicklaus-conceived golf course. Today, such clear distinctions are blurred by the strange idea that land which has a high water table saturating the surface a few days a year during the growing season and containing hydric soil which supports hydrophytic plants, is to be considered wetlands. (Hydric soils are defined as soils that are saturated, flooded, or ponded long enough during the growing season to develop anaerobic conditions in the upper part. Hydrophytic vegetation describes plants that live in conditions of excess wetness.)

Simply put, we used to believe wetlands were areas where land and water met. Now, where they don't quite meet to qualify as a wetland, it is only necessary for the vegetation to have adapted to sometimes saturated conditions.

While we can technically understand that a wetland need not be a marshy area with lush vegetation, we are sure the rank-and-file citizenry, which has often been willing to stand up for the rights of swamp critters, is being ambushed by the broad new definition of a wetland being fostered by environmental zealots. It is aimed far more at limiting the rights of the individual in favor of the higher causes of "society" rather than actually giving two hoots for the native flora and fauna of our strange new dry, and often barren, "wetlands."

Regulatory Ragu

Upon reading the 1991 update of the federal *Wetlands Manual,* many will be amazed at how little is changed from the 1989 version. The wetlands definition, for one, has changed only by doubling from 7 to 14 the required number of days of high water table during the growing season. Four agencies collaborated on this disappointing effort which included U.S. EPA, Army Corps of Engineers, the Department of Interior's Fish and Wildlife Service, and the Department of Agriculture's Soil Conservation Service. Public comment was received at hearings across the country but one wonders who showed up.

In the words of the manual, the proposed revisions:

a. make it easier for Federal or State agency staff to explain to landowners how wetlands are being delineated;

b. clarify that, except in specified circumstances such as prairie potholes which provide valuable waterfowl habitat, demonstration of all three parameters (hydrology, soils, and vegetation) is required for delineating vegetated wetlands;

c. explicitly acknowledge that there are wetlands which are difficult to identify due to seasonal dry periods, droughts and other circumstances (these areas do not have all three wetland parameters observable at all times, although they clearly function as wetlands);

d. clarify that the presence of mapped hydric soils alone cannot be used to delineate an area as a wetland;

e. require stronger demonstration of wetland hydrology by requiring direct evidence of greater than 14 days of soil saturation and/or inundation;

f. incorporate localized criteria in wetlands delineation; for example, in the growing season and soil phases; and

g. limit the use of facultative plants (that is, plants that have about the same frequency of occurrence in uplands as in wetlands) to determine wetland hydrology.

One of the most interesting sections of the new wetlands manual is labeled "Difficult-to-Identify Wetlands." It seems logical to me if land is difficult to identify as a wetland, it is probably because it isn't.

Winners and Losers

The Army Corps of Engineers is clearly enjoying its newly found status as "keeper of the nation's wetlands." Having been recently run out of the dam-building business by an improvement in the nation's collective common sense and a decline in available funds for future mega projects, work was getting scarce.

The Soil Conservation Service, likewise, could pick up as much as 70 million acres of farmland which may be deemed "Jurisdictional Wetlands," subject to federal control.

It is amazing that while the demand for water continues unabated, especially in the western United States, not only have changing societal and national priorities heralded the end of major surface water supply projects, now wetland restrictions on ground water withdrawals will further increase the size of the problem.

The Bureau of Reclamation, which was previously the architect of many of the admittedly wrong headed dam projects that did bring water online, is now focusing attention on maintaining wetlands. While we will not argue that maintenance of some wetlands will help address important concerns of sediment runoff, water quality, recharge, and flood control, "dry" wetlands will do none of the above.

I am delighted to see the Bureau of Reclamation devote itself to the more efficient management of existing projects and to address resource management issues and certainly water conservation, but they should not jump on a popular bandwagon at the expense of their prior water supply commitments to society, whether or not the public believes they were successful in the past.

Wetlands injustice has now been chronicled in the strange story of John Posygai, a mechanic from Morrisville, Pennsylvania. He was sentenced to three years in prison and fined $200,000 upon being convicted of 41 violations of the Clean Water Act. He purchased a piece of property upon which to build a garage and proceeded to "criminally" remove tires and auto parts that had been dumped on the property, and then leveled the land with fill soil.

Similarly strange stories can be related to North Dakota cornfields containing pools of standing water on rare occasions, muddy patches of dirt between railroad tracks in Idaho, and irrigation ditches dug at the turn of the century in numerous western states.

Voices in Revolt

A few individuals have gathered the nerve to voice skepticism about the strange ground swell of emotions over the nation's damper land. Senator Wyche Fowler of Georgia expressed concern that the current wetlands' definition can be applied to nearly 80 percent of Florida's, South Carolina's, and Georgia's coastal lands. Senator Jake Garn of Utah commented that "We are getting to the point where you won't be able to spit on the ground without the Army Corps of Engineers coming up behind you and declaring it a wetland." Even Representative Tim Valentine of North Carolina, a staunch environmentalist, worried publicly that "the way the wetlands issue has been handled may chill the efforts of environmental legislation in the future." The rather broad, if not ridiculous, interpretation of Section 404 of the Clean Water Act has added considerable complexity and cost to making the highest and best use of much of the nation's real estate.

While the Mosquito Awaits Its Prey

There is one more side to the wetlands issue that has been almost entirely overlooked, and it is indeed the darkest side. It is the high risk of human disease which can occur as a direct result of the creation or maintenance of real wetlands.

Mosquitoes are the most common disease-causing species in a wetland. In the vernacular of public health, they are "vectors" or animals, capable of transmitting a disease-causing agent from one host to another. Mosquitoes and wetlands commonly go together, so much so that one biological definition of a functional wetland is "a place where water stands long enough to produce two consecutive broods of mosquitoes."

While not all mosquitoes present in a wetland vector diseases, those that do frequently carry malaria and encephalitis. These diseases require, in addition to the mosquito, a disease-carrying organism and a susceptible host. Control of malaria which is caused by a protozoan parasite is usually achieved by control of mosquito vectors and anti-malaria drugs. There are 400 million cases of malaria worldwide each year resulting in 4 million deaths. It is most deadly among children. While malaria has been largely eliminated in the United States, occasional outbreaks have all been associated with wetlands. Encephalitis, which has no cure or vaccine, declined in this country until the mid-70s, but has been on the rise ever since. Combining the anti-pesticide movement with the end of wetland reclamation will cause disastrous consequences for the protection of humans from disease.

Can Common Sense Prevail?

Where and when did it become the accepted policy of this nation that man ranks on the bottom rung of the ladder of life when it comes to protection and preservation? What single environmental protection issue is saving more human lives than would be saved were the money spent to have been directed squarely at a comparable human health issue? To my knowledge, there are none.

Wildlifers are going to have to recognize that pesticides are a necessary part of a desire to maintain wetlands. Balancing benefits and risks of our new objects of religious zeal will require much more than keeping land wet or damp. In the long run, the price may be more than the public will want to pay.

Rationality may return to wetlands policy once taxpayers are forced by the constitution to pay big bucks to property owners to adequately compensate them for the taking of their land. A bill is presently in the congressional hopper which would require compensating the owners of high priority wetlands which could push the country far down the road toward bankruptcy. If one day you and I have to shell out a few billion dollars to buy up marginal wetlands on New York's Manhattan or Staten Islands, we may collectively begin to ask if this is really worth it.

Why?

Environmental zealots want to maintain wetlands for the same old reasons that are always present but never spoken: to control society's use of land and form of government. The general public is probably attracted by other reasons. They are best defined by William Hazeltine writing in *Rational Readings on Environmental Concerns* (Lehr, 1992, in press):

1. To replace losses due to civilization—this spiritual kind of idea involves a return to primeval times, for reasons that are not clear.

2. To solve a social guilt complex such as "replacing the plunder."

3. To make a place for "sportsmen" to play.

4. Some vague sense that wetlands are the primordial birthplace of land animals and, therefore, worthy of preservation.

Regardless, glorifying wetlands for purposes of political perceptions without explaining the risks and costs is unacceptable. If the wetlands issue was a scientific debate, the proponents would be censured for presenting incomplete data. Perhaps this is why the wetlands issue has been removed from the realm of science, so that accountability will not diminish its success.

A Call to Action

You, as a groundwater scientist or engineer, should carefully consider the long-range impact this issue will have on your profession. Most of us, until now, have viewed it as just another one of those environmental bandwagons that the ill-informed public was riding in pursuit of the pied pipers of the green revolution. But this issue is different. You have an innate understanding of these new wetlands which could benefit your friends and neighbors in dealing with their increasing confusion.

Real wetlands are areas saturated by surface or ground water at a frequency and duration that will support vegetation typically adapted for life in saturated soil. Wetlands are called by many names such as swamps, marshes, bogs, and even fens. Real wetlands provide a haven for rare and endangered species which depend on these areas for survival. Many are important fish spawning and nursery areas as well as nesting and feeding areas for waterfowl. These real wetlands should be largely preserved. These wetlands, however, are not what we are talking about. Society's ever-present ignorance of science has once again allowed power-grabbing regulators, headline-grabbing politicians, and environmental zealots to turn John Q. Citizen every way but loose in riding this issue down a primrose path laden with selfish benefits to a few, and enormous losses to the overall population.

Currently, congressional legislative proposals HR-2400 and HR-1330 offer to carve in stone a national wetland policy. In my opinion, the latter is better than the former, but both are bad. Write your congressman for a copy and offer him or her your expert opinion from a ground-water perspective.

Respond to the weekly, if not daily, articles on the issue in your local newspaper. No one is addressing the ground-water issue so you are sure to be heard and quoted. We, as ground-water professionals, have allowed the environmental activists (not scientists, certainly) to usurp our responsibilities in this area. To preserve these areas simply because they are natural does not make sense. In the long term (short term, geologically), nature does not even preserve them.

Reference

Lehr, Jay H. 1992. *Rational Readings on Environmental Concerns*. Van Nostrand Reinhold, New York.

Editor's Note: The author is happy to acknowledge the assistance of Editorial Board member, Peter Riordan, in the preparation of this article.

Reprinted, with permission, from
the Journal of *Ground Water*,
Vol. 29, No. 5, September–October 1991

Wilderness

Wilderness—What Can We Expect for the Future?
Jo Ann Kwong

Wilderness—What Can We Expect for the Future?

Jo Ann Kwong, Ph.D.

Please see biographical sketch for Dr. Kwong on page 277.

September 1989 marked the 25th anniversary of one of the seminal acts in environmental history—the Wilderness Act of 1964. The Act sought "to secure for the American people of present and future generations the benefits of an enduring resource of wilderness" by establishing the National Wilderness Preservation System. Under Congressional directive, appropriate wild lands from federal holdings were to be "folded" into the National System as Wilderness lands and protected from development—no roads, dams or other permanent structures; no timber cutting; no new mining claims or mining leases; no motorized vehicles or equipment; and no other uses that would indicate man's presence.

The Wilderness Act has touched off a stubborn battle between environmentalists who want to see more land set aside as Wilderness, and others who enjoy more developed uses of the public lands. Because both are fighting to achieve conflicting uses, the Act, will undoubtedly continue to generate hostilities and competition far into the future.

Is there hope for a resolution in which the various competitors for public land can peacefully coexist?

National Wilderness Preservation System

The Wilderness Act mandates that Wilderness areas be "administered for the use and enjoyment of the American people in such a manner as will leave them unimpaired for future use and enjoyment as Wilderness." In 1964, the closest cousin to modern day Wilderness was the "wild," "Wilderness," or "canoe" lands of the National Forest Service.

Under the Act, these lands, comprising 9.1 million acres, became the initial contribution to the National Wilderness Preservation System. The Forest Service was directed to study its remaining "primitive" areas to determine additional contributions to the system.

The review process has touched off a series of legislative acts, court rulings and various procedures that have become cornerstones in the history of Wilderness development in the United States.

In 1971 the Forest Service conducted a Roadless Area Review and Evaluation (RARE) to launch its mandated assessment. The Forest Service standards for Wilderness determination were challenged in the courts, eventually leading to a 1978 reevaluation.

The RARE II study, completed in 1979, recommended that 15.4 million acres of land be designated as Wilderness, 10.6 million acres be set aside for further study, and 36 million acres be open for development. These studies, despite their controversial nature, touched off the Congressional process of designating Wilderness lands.

During the RARE studies, the Bureau of Land Management received its mandate to contribute to the Wilderness system with the passage of the Federal Land Policy and Management Act of 1976. This act directed the BLM to assess its roadless lands in the lower 48 states to determine suitable additions to the system.

The agency determined about 25 million acres should be studied. With an eye toward BLM's 270 million acre domain, this decision has generated tremendous hostility from environmentalists who saw these lands as their best chance for expanding the Wilderness Preservation System.

The national parks and wildlife refuges have also been reviewed as part of the Wilderness process. In the contiguous United States, about 6 million acres of national park land and almost 660,000 acres within National Wildlife Refuges have been designated Wilderness. Other park and refuge areas have been recommended for inclusion in the system but are awaiting Congressional action.

In addition, 56.5 million acres of Alaskan lands were added to the Wilderness system under the Alaska National Interest Lands Conservation Act of 1980—5.4 million acres in forests, 32.4 million acres in national parks and 18.7 million acres in national wildlife refuges.

The battle lines have been drawn between environmentalists who want more Wilderness lands, and a range of other interests who fear that Wilderness designations preclude alternative uses of the lands.

While many of the diverse uses of public lands can be compatible, the conflicting uses are many. Wilderness buffs, for example, prefer to keep their distance from dirt bike riders; nature photographers stay out of range of big game hunters; and cross country skiers shy away from snowmobilers.

These incongruous uses are to be expected in a land as heterogeneous as ours. But the debate is being framed in terms of right and wrong—that some users have more rights to the public lands than others. Each side continues to attack the land use preferences of the other.

For example, environmentalists have broadly attacked commodity users of the lands they want withdrawn: "Voices of the extractive industries and ranchers, often backed by powerful federal officials, want the little remaining BLM wildland to be mined, logged and grazed as well—damaging it irreparably for future generations." (The Wilderness Society, "A Forgotten Legacy: BLM Lands of the American West," undated.)

Such emotional rhetoric has been characteristic of the Wilderness battle.

Environmentalists claim that non-Wilderness users will forever mar the landscape. Their opponents, however, argue that other critical considerations are masked by such sweeping emotionalism—that economic issues are also at stake precluding important development uses.

Unfortunately, there are no easy answers.

Cultural and aesthetic concerns must be balanced with social and economic ones. One way to pursue this balance is to adopt a system that forces environmentalists and other prospective users of the public lands to make responsible, accountable claims to the land. How is this possible?

The Tragedy of The Commons

The Wilderness problem essentially boils down to property rights. As long as Wilderness is reserved as public lands, owned by everyone, we can expect a chorus of voices fighting to determine "proper" uses. Unlike private lands where the owner decides what to do with his property, we have millions of owners of public lands, each with a different idea of the best use. The problems arising from commonly owned property are often referred to as the "tragedy of the commons."

In 1968 Garrett Hardin formulated his insightful theory of the commons. To illustrate, Hardin described a cattle pasture that is open to all the herdsmen in a community. With free access, more and more cattle are added to the pasture until it reaches its carrying capacity. Beyond this point, the pasture becomes overgrazed to the detriment of everyone. But no single herdsman has the incentive to concern himself with the cost of overgrazing. Instead, each will continue to add yet another cow, knowing that he will receive the benefits from the animal while the costs of overgrazing are spread among all the herdsmen.

In our public lands, we have, in essence, a commons that is owned by everyone. And each user, seeking to advance his own best interest, will fight to see his favored use dominate.

Like the herdsman who knows that the burden of his one additional cow will be borne by everyone, the users of public lands know that the cost of more Wilderness, or more dirt bike trails, will be paid by the overall taxpayer. Therein lies the tragedy of the public lands.

When the costs are borne by the nation as a whole, we can expect each user group to overstate its interests.

Wilderness buffs will argue that they must have millions of acres to preserve the pristine environment they so treasure. The timber and mining industries will argue that they need access to so many acres of forests and mineral lands in order to manage worthwhile ventures. And snowmobilers and motorcyclists will similarly request as much land as possible for their motorized enjoyment.

None of these groups have to substantiate their claims for how much they "need." But if each of these groups had to pay the full costs of their requests, we would likely see a different pattern of land allocations.

When people have to pay for things, they use them much more sparingly than when they are free. When faced with costs, people have to make trade-offs in how to spend their money. For example, although many outdoorsmen love to downhill ski, the $25 to $30 price of the lift ticket typically tempers the number of ski days. Although a bigger timber harvest may be more desirable than a smaller one, the costs of purchasing more timber land affects the choice of optimal harvest size.

Without some sort of gauge to determine the worth of alternative uses to various interest groups, we can expect to see conflicts on public lands endure forever as each group vies to achieve its benefits. In order to impart some level of accountability on the part of public land users, it would be prudent to ask the user groups to take financial responsibility for their "need" claims. Only then can we tell how much each interest group really values its proposed use.

Drive to Preserve Wilderness

Although the literature of environmental groups is full of magnanimous reasons for preserving Wilderness, there is another side to their mission that is far less altruistic. These groups, like every other special interest, are trying to obtain benefits for their constituents. With regard to Wilderness in particular, environmental leaders are appealing to those who enjoy the "untrammeled" outdoors—people who enjoy backpacking and hiking, often to escape from the noise and bustle of everyday urban life.

This constituency, however, is very limited, The environmental leaders boast of tremendous memberships, suggesting widespread support for the Wilderness movement. Although many people appreciate the Wilderness concept, few actually exercise it.

Supporting evidence can be found in a comparison between the number of visitor trips to Wilderness and non-Wilderness national parks. Visits to the Bob Marshall Wilderness Area in northern Montana pale in comparison to visits to its southern counterpart, Yellowstone National Park. And within Yellowstone, there is no doubt that the majority of people prefer to drive to their vacation spots, and hike or camp just a short distance from their cars. As a result, roads in the park are lined with vistas so people can pull out of traffic to photograph or observe wildlife and scenery. People enjoy the environment from their cars. Although there are thousands of miles of hiking trails, those closest to the roads are most heavily used. The Wilderness areas are used by only a select few of the professed environmentalists.

The misleading claims of widespread support by environmental groups stem largely from the fact that many people consider themselves to be environmentalists. Most people prefer cleaner air to dirtier air, drinkable and swimmable waters to those that are not so, and pristine landscapes over heavily congested urban landscapes. To this extent, almost everyone is an environmentalist.

But to argue that most people support the idea of Wilderness grossly exaggerates the interests of a handful of people active in the movement.

The drive to withdraw land into Wilderness for a handful of hardy outdoorsmen is extremely costly. In addition, since Wilderness lands are supported by taxpayers, it poses an undue burden on the rest of the nation, most of which actually opposes the idea on the scale envisioned by environmental leaders.

If these leaders are intent on increasing Wilderness designations, there are ways they could do so that would be fair to the taxpaying public and help resolve some of the issues confronting Congress with regard to Wilderness designations.

User Fees

One way to get Wilderness buffs to reveal their true valuation of Wilderness lands is to charge user fees to contribute to the costs of maintaining the lands. Congress could determine what percentage of the maintenance costs must be covered by users in order to justify continued Wilderness designation.

Clearly, user fees are likely to come up short, but environmental groups would also be free to make financial donations toward maintenance costs.

If one Wilderness area is not being supported with user fees, the Park Service or other administering agency could be required to present public notification of that area's financial situation. If sufficiently concerned, the environmental groups either could launch a fund-raising drive to raise the necessary money, or could make contributions from their general funds. This way, the rhetoric of environmentalism is cut out and the groups are asked to indicate their true commitment to the Wilderness concept.

If they are only willing to accept Wilderness when its cost is shouldered entirely by taxpayers, we should question their true valuation of it. Otherwise, what sense does it make to coerce taxpayers to fund designations that even the staunchest proponents are not willing to support?

Management by Environmental Groups

The maintenance fee proposal is sure to raise the ire of environmental groups. They may begin to scrutinize the government agencies' fiscal management of the parks and claim that their money is being squandered. In this case, another option could offer alternative benefits. A program could be devised that offers environmental groups the chance to bid on the management and administration of certain Wilderness areas.

If the Wilderness Society, for example, challenges the Park Service's handling of Wilderness areas in Glacier National Park, it could bid for the right to do so more efficiently themselves. This plan would impart a measure of responsibility and accountability to environmental groups by essentially giving the Society property rights to the Wilderness areas. Congress could determine the specific bounds within which the groups could operate and for which they would be responsible.

If given broad autonomy, for example, we could imagine a plan in which environmentalists decide to charge higher user fees in order to ease their maintenance costs, or in which they agree to allow a small white-water rafting company to rent rafts within their bounds, provided they collect rental fees from the proprietor.

In the absence of any sort of ownership responsibility, environmentalists have opposed almost every imaginable use of Wilderness. But when faced with a financial incentive and responsibility, they may be more willing to compromise and make beneficial trade-offs.

Selling Off Some Wilderness Lands

A third option would take the property rights concept one step further. Environmental groups could be offered the option of buying Wilderness lands outright. This way, they would truly be demonstrating their willingness to pay for the right to maintain Wilderness.

Such a plan could be launched in pilot stages by selling off very small parcels of Wilderness. The environmentalists could be given the right of first refusal. If they could not raise the funds (again, an indication that their constituents do not support the idea) or if they decide not to exercise the option, the land could either be sold to other interested buyers, or could revert to non-Wilderness lands in the national public lands system.

Environmental groups have demonstrated remarkable abilities to negotiate and compromise on land that they actually own. The National Audubon Society, for example, has permitted oil drilling on some of its lands, including the Rainey Preserve in Louisiana.

The Michigan Audubon Society has allowed the Michigan Petroleum Exploration Inc. to drill on the Bernard N. Baker Sanctuary in southern Michigan. This does not reflect callous disregard for nature on the part of the Audubon Society. Instead, it reflects the trade-offs that property owners are willing to make when faced with financial realities.

In both cases, the Audubon Society stipulated extremely restrictive conditions under which drilling could occur. The oil companies were willing to comply, given the potential benefits they faced.

If the groups independently own Wilderness areas, they could make the choices of what will be permitted and what will not, knowing that there may indeed be financial costs to prohibiting certain uses. This type of accountability is not reflected in the public land system that currently prevails.

What Hope for the Future?

The environmental groups were selected out in this article to illustrate the divergence in actions that occur between private and public land users. These organizations have been among the most vocal advocates of Wilderness use of public lands, even though the benefits of such uses go to a very small handful of people.

Because of the tremendous costs of bringing these benefits to a small group, it is argued here that we need to look at proposals to alter the stakes involved for competing users of the public lands.

If environmentalists want more Wilderness lands, they should be willing to pay the costs of the withdrawals.

This principle, however, should not apply only to environmental groups. We should consider other revisions that would similarly foster responsible decision making in the management of public lands by other competing users. Grazing lands, for example, could be leased at market value rather than at federally subsidized rates; mining laws that permit the purchase of BLM lands for $2.50 an acre could be revised to market prices; multinational mining companies could pay royalties similar to those paid on private lands, and other measures could be instituted to bring market forces to bear on land use decisions.

Undoubtedly, the environmentalists will argue that it is unfair to expect them to pay for the Wilderness benefits that they want. After all, they have argued that they are fighting not only for present benefits, but also for benefits to future generations who are not currently available to contribute to the cost.

The willingness to pay response once again provides a satisfactory answer: If the environmentalists indeed believe future generations want Wilderness, they should be free to make this generous gift—provided they pay the price. But don't coerce others, who may have different ideas about the wants and desires of future generations, to do so. After all, a reasonable case can be made that future generations would much prefer an energy secure environment rather than a pristine one, and so would approve of energy productions from the public lands.

Who are we to speak for the unborn? How many of us look back on what our forefathers did and argue that they should have left more for us?

The designation of Wilderness on public lands is a sensitive subject. But without bringing responsibility to bear, we are likely to see a continued pattern of reckless claims and demands with little true support. Unless we are willing to let taxpayers carry the burden of designating another 100 million acres of public lands in Wilderness, we should give serious thought to alternative approaches.

Reprinted, with permission, from
Our Land

In Summary

The Toxicity of Environmentalism
George Reisman

The Toxicity of Environmentalism

George G. Reisman, Ph.D.

Please see biographical sketch for Dr. Reisman on page 631.

Recently a popular imported mineral water was removed from the market because tests showed that samples of it contained thirty-five parts per billion of benzene. Although this was an amount so small that only fifteen years ago it would have been impossible even to detect, it was assumed that considerations of public health required withdrawal of the product.

Such a case, of course, is not unusual nowadays. The presence of parts per billion of a toxic substance is routinely extrapolated into being regarded as a cause of human deaths. And whenever the number of projected deaths exceeds one in a million (or less) , environmentalists demand that the government remove the offending pesticide, preservative, or other alleged bearer of toxic pollution from the market. They do so, even though a level of risk of one in a million is one-third as great as that of an airplane falling from the sky on one's home.

While it is not necessary to question the good intentions and sincerity of the overwhelming majority of the members of the environmental or ecology movement, it is vital that the public realize that *in this seemingly lofty and noble movement itself can be found more than a little evidence of the most profound toxicity.* Consider, for example, the following quotation from David M. Graber, a research biologist with the National Park Service, in his prominently featured *Los Angeles Times* book review of Bill McKibben's *The End of Nature:*

> This [man's "remaking the earth by degrees"] makes what is happening no less tragic for those of us who value wildness for its own sake, not for what value it confers upon mankind. I, for one, cannot wish upon either my children or the rest of Earth's biota a tame planet, be it monstrous or—however unlikely—benign. McKibben is a biocentrist, and so am I. We are not interested in the utility of a particular species or free-flowing river, or ecosystem, to mankind. They have intrinsic value, more value—to me—than another human body, or a billion of them.
>
> Human happiness, and certainly human fecundity, are not as important as a wild and healthy planet. I know social scientists who remind me

819

that people are part of nature, but it isn't true. Somewhere along the line—at about a billion years ago, maybe half that—we quit the contract and became a cancer. We have become a plague upon ourselves and upon the Earth.

It is cosmically unlikely that the developed world will choose to end its orgy of fossil-energy consumption, and the Third World its suicidal consumption of landscape. Until such time as Homo sapiens should decide to rejoin nature, some of us can only hope for the right virus to come along.

While Mr. Graber openly wishes for the death of a billion people, Mr. McKibben, the author he reviewed, quotes with approval John Muir's benediction to alligators, describing it as a "good epigram" for his own, "humble approach": "'Honorable representatives of the great saurians of older creation, may you long enjoy your lilies and rushes, and be blessed now and then with a mouthful of terror-stricken man by way of a dainty!'"

Such statements represent pure, unadulterated poison. They express ideas and wishes which, if acted upon, would mean terror and death for enormous numbers of human beings.

These statements, and others like them, are made by prominent members of the environmental movement. The significance of such statements cannot be diminished by ascribing them only to a small fringe of the environmental movement. Indeed, even if such views were indicative of the thinking only of 5 or 10 percent of the members of the environmental movement—the "deep ecology," Earth First! wing—they would represent toxicity in the environmental movement as a whole not at the level of parts per billion or even parts per million, but at the level of *parts per hundred,* which, of course, is an enormously higher level of toxicity than is deemed to constitute a danger to human life in virtually every other case in which deadly poison is present.

But the toxicity level of the environmental movement as a whole is substantially greater even than parts per hundred. It is certainly at least at the level of *several parts per ten.* This is obvious from the fact that the mainstream of the environmental movement makes no fundamental or significant criticisms of the likes of Messrs. Graber and McKibben. Indeed, John Muir, whose wish for alligators to "be blessed now and then with a mouthful of terror-stricken man by way of a dainty" McKibben approvingly quotes, *was the founder of the Sierra Club,* which is proud to acknowledge that fact. The Sierra Club, of course, is the leading environmental organization and is supposedly the most respectable of them.

There is something much more important than the Sierra Club's genealogy, however—something which provides an explanation in terms of *basic principle* of why the mainstream of the ecology movement does not attack what might be thought to be merely its fringe. This is a fundamental philosophical premise which the mainstream of the movement shares with the alleged fringe and which logically implies hatred for man and his achievements. Namely, the premise that *nature possesses intrinsic value*—i.e., that nature is valuable in and of itself, apart from all contribution to human life and well-being.

The antihuman premise of nature's intrinsic value goes back, in the Western world, as far as St. Francis of Assisi, who believed in the equality of all living creatures: man, cattle, birds, fish, and reptiles. Indeed, precisely on the basis of this philosophical affinity, and at the wish of the mainstream of the ecology movement, St. Francis of Assisi has been officially declared the patron saint of ecology by the Roman Catholic Church.

The premise of nature's intrinsic value extends to an alleged intrinsic value of forests, rivers, canyons, and hillsides—to everything and anything that is not man. Its influence is present in the Congress of the United States, in such statements as that recently made by Representative Morris Udall of Arizona that a frozen, barren desert in Northern Alaska, where substantial oil deposits appear to exist, is "a sacred place" that should never be given over to oil rigs and pipelines. It is present in the supporting statement of a representative of the Wilderness Society that "There is a need to protect the land not just for wildlife and human recreation, but just to have it there." It has, of course, also been present in the sacrifice of the interests of human beings for the sake of snail darters and spotted owls.

The idea of nature's intrinsic value inexorably implies a desire to destroy man and his works because it implies a perception of man *as the systematic destroyer of the good, and thus as the systematic doer of evil.* Just as man perceives coyotes, wolves, and rattlesnakes as evil because they regularly destroy the cattle and sheep he values as sources of food and clothing, so on the premise of nature's intrinsic value, the environmentalists view man as evil, because, in the pursuit of his well-being, man systematically destroys the wildlife, jungles, and rock formations that the environmentalists hold to be intrinsically valuable. Indeed, from the perspective of such alleged intrinsic values of nature, the degree of man's alleged destructiveness and evil is directly in proportion to his loyalty to his essential nature. Man is the rational being. It is his application of his reason in the form of science, technology, and an industrial civilization that enables him to act on nature on the enormous scale on which he now does. Thus, it is his possession and use of reason—manifested in his technology and industry—for which he is hated.

The doctrine of intrinsic value is itself only a rationalization for a preexisting hatred of man. It is invoked not because one attaches any actual value to what is alleged to have intrinsic value, but simply to serve as a pretext for denying values to man. For example, caribou feed upon vegetation, wolves eat caribou, and microbes attack wolves. Each of these, the vegetation, the caribou, the wolves, and the microbes, is alleged by the environmentalists to possess intrinsic value. Yet absolutely no course of action is indicated for man. Should man act to protect the intrinsic value of the vegetation from destruction by the caribou? Should he act to protect the intrinsic value of the caribou from destruction by the wolves? Should he act to protect the intrinsic value of the wolves from destruction by the microbes? Even though each of these alleged intrinsic values is at stake, man is not called upon to do anything. When does the doctrine of intrinsic value serve as a guide to what man should do? Only when *man* comes to attach value to something. Then it is invoked to deny him the value he seeks. For example, the intrinsic value of the vegetation et al. is invoked as a guide to man's action only when there is something man wants, such as oil, and then, as in the case of Northern Alaska, its invocation serves to stop him from having it. In other words, *the doctrine of intrinsic value is nothing but a doctrine of the negation of human values.* It is pure nihilism.

It should be realized that it is logically implicit in what has just been said that to establish a public office such as that recently proposed in California, of "enviromnental advocate," would be tantamount to establishing an office of Negator of Human Valuation. The work of such an office would be to stop man from achieving his values for no other reason than that he was man and wanted to achieve them.

Of course, the environmental movement is not pure poison. Very few people would listen to it if it were. As I have said, it is poisonous only at the level of several parts per ten. Mixed in with the poison and overlaying it as a kind of sugar coating is the advocacy of many measures which have the avowed purpose of promoting human life and well-being, and among these, some that, considered in isolation, might actually achieve that purpose. The problem is that the mixture is poisonous. And thus, when one swallows environnmentalism, one inescapably swallows poison.

Given the underlying nihilism of the movement, it is certainly not possible to accept at face value any of the claims it makes of seeking to improve human life and well-being, especially when following its recommendations would impose on people great deprivation or cost. Indeed, nothing could be more absurd or dangerous than to take advice on how to improve one's life and well-being from those who wish one dead and whose satisfaction comes from human terror, which, of course, as I have shown, is precisely what is wished in the environmental movement—openly and on principle. This conclusion, it must be stressed, applies irrespective of the scientific or academic credentials of an individual. If an alleged scientific expert believes in the intrinsic value of nature, then to seek his advice is equivalent to seeking the advice of a medical doctor who was on the side of the germs rather than the patient, if such a thing can be imagined. Obviously, Congressional committees taking testimony from alleged expert witnesses on the subject of proposed environmental legislation need to be aware of this fact and never to forget it.

Not surprisingly, in virtually every case, the claims made by the environmentalists have turned out to be false or simply absurd. Consider, for example, the recent case of Alar, a chemical spray used for many years on apples in order to preserve their color and freshness. Here, it turned out that even if the environmentalists' claims had actually been true, and the use of Alar would result in 4.2 deaths per million over a seventy-year lifetime, all that would have been signified was that eating apples sprayed with Alar would then have been *less dangerous than driving to the supermarket to buy the apples!* (Consider: 4.2 deaths per million over a seventy year period means that in any one year in the United States, with its population of roughly two hundred and fifty million people, approximately *fifteen* deaths would be attributable to Alar! This is the result obtained by multiplying 4.2 per million times 250 million and then dividing by 70. In the same one-year period of time, approximately fifty thousand deaths occur in motor vehicle accidents in the United States, most of them within a few miles of the victims' homes, and undoubtedly far more than fifteen of them on trips to or from supermarkets.) Nevertheless, a panic ensued, followed by a plunge in the sale of apples, the financial ruin of an untold number of apple growers, and the virtual disappearance of Alar.

Before the panic over Alar, there was the panic over asbestos. According to *Forbes* magazine, it turns out that in the forms in which it is normally used in the United States, asbestos is *one-third as likely lo be the cause of death as being struck by lightning.*

Then there is the alleged damage to lakes caused by acid rain. According to *Policy Review*, it turns out that the acidification of the lakes has not been the result of acid rain, but of the cessation of logging operations in the affected areas and thus the absence of the alkaline run-off produced by such operations. This run-off had made naturally acidic lakes non-acidic for a few generations.

Besides these cases, there were the hysterias over dioxin in the ground at Times Beach, Missouri, TCE in the drinking water of Woburn, Massachusetts, the chemicals in Love Canal, and radiation at Three Mile Island. According to Prof. Bruce Ames, one of the world's leading experts on cancer, it turned out that the amount of dioxin that anyone would have absorbed in Times Beach was far less than the amount required to do any harm and that, indeed, the actual harm to Times Beach residents from dioxin was less than that of drinking a glass of beer. (The Environmental Protection Agency itself subsequently reduced its estimate of the danger from dioxin by a factor of fifteen-sixteenths.) In the case of Woburn, according to Ames, it turned out the cluster of leukemia cases which occurred there was statistically random and that the drinking water there was actually above the national average in safety, and not, as had been claimed, the cause of the leukemia cases. In the case of Love Canal, Ames reports, it turned out upon investigation that the cancer rate among the former residents has been no higher than average. (It is necessary to use the phrase "former residents" because the town lost most of its population in the panic and forced evacuation caused by the environmentalists' claims.) In the case of Three Mile Island, not a single resident has died, nor even received an additional exposure to radiation, as the result of the accident there. In addition, according to studies reported in *The New York Times*, the cancer rate among residents there is no higher than normal and has not risen.

Before these hysterias, there were claims alleging the death of Lake Erie and mercury poisoning in tuna fish. All along, Lake Erie has been very much alive and was even producing near record quantities of fish at the very time the claims of its death were being made. The mercury in the tuna fish was the result of the natural presence of mercury in sea water, and evidence provided by museums showed that similar levels of mercury had been present in tuna fish since prehistoric times.

And now, in yet another overthrow of the environmentalists' claims, a noted climatologist, Prof. Robert Pease, has shown that it is impossible for chlorofluorocarbons (CFCs) to destroy large quantities of ozone in the stratosphere because relatively few of them are even capable of reaching the stratosphere in the first place. He also shows that the celebrated ozone "hole" over Antarctica every fall is a phenomenon of nature, in existence since long before CFCs were invented, and results largely from the fact that during the long Antarctic night ultraviolet sunlight is not present to create fresh ozone.

The reason that one after another of the environmentalists'claims turn out to be proven wrong is that they are made without any regard for truth in the first place. In making their claims, the environmentalists reach for whatever is at hand that will serve to frighten people, make them lose confidence in science and technology, and, ultimately, lead them to deliver themselves up to the environmentalists' tender mercies. The claims rest on unsupported conjectures and wild leaps of imagination from scintillas of fact to arbitrary conclusions, by means of evasion and the drawing of invalid inferences. It is out and out evasion and invalid inference to leap from findings about the effects of feeding rats or mice dosages the equivalent of a hundred or more times what any human being would ever ingest, and then draw inferences about the effects on people of consuming normal quantities. Fears of parts per billion of this or that chemical causing single-digit deaths per million do not rest on science, but on imagination. Such claims have nothing to do either with actual experimentation or with the concept of causality.

No one ever has, can, or will observe such a thing as two groups of a million people identical in all respects except that over a seventy-year period the members of one of the groups consume apples sprayed with Alar, while the members of the other group do not, and then 4.2 members of the first group die. The process by which such a conclusion is reached, and its degree of actual scientific seriousness, is essentially the same as that of a college students' bull session, which consists of practically nothing but arbitrary assumptions, manipulations, guesses, and plain hot air. In such a session, one might start with the known consequences of a quarter-ton safe falling ten stories on the head of an unfortunate passerby below, and from there go on to speculate about the conceivable effects in a million cases of other passersby happening to drop from their hand or mouth an M&M or a peanut on their shoe, and come to the conclusion that 4.2 of them will die.

Furthermore, as indicated in contrast to the procedures of a bull session, reason and actual science establish causes, which, in their nature, are *universal*. When, for example, genuine causes of death, such as arsenic, strychnine, or bullets, attack vital organs of the human body, death is absolutely certain to result in *all but* a handful of cases per million. When something is in fact the cause of some effect, it is so *in each and every case* in which specified conditions prevail, and fails to be so only in cases in which the specified conditions are not present, such as a person's having built up a tolerance to poison or wearing a bulletproof vest. Such claims as a thousand different things each causing cancer in a handful of cases are proof of nothing but that the actual causes are not yet known—and, beyond that, an indication of the breakdown of the epistemology of contemporary science. (This epistemological breakdown, I might add, radically accelerated starring practically on the very day in the 1960s when the government took over most of the scientific research in the United States and began the large scale financing of statistical studies as a substitute for the discovery of causes.)

In making their claims, the environmentalists willfully ignore such facts as that carcinogens, poisons, and radiation exist in nature. Fully half of the chemicals found in nature are carcinogenic when fed to animals in massive quantities—the same proportion as applies to man-made chemicals when fed in massive quantities. (The cause of the resulting cancers, according to Prof. Ames, is actually not the chemicals, either natural or man-made, but the repeated destruction of tissue caused by the massively excessive doses in which the chemicals are fed, such as saccharin being fed to rats in a quantity comparable to humans drinking eight hundred cans of diet soda a day.) Arsenic, one of the deadliest poisons, is a naturally occurring chemical element. Oleander, one of the most beautiful plants, is also a deadly poison, as are many other plants and herbs. Radium and uranium, with all their radioactivity, are found in nature. Indeed, all of nature is radioactive to some degree. If the environmentalists did not close their eyes to what exists in nature, if they did not associate every negative exclusively with man, if they applied to nature the standards of safety they claim to be necessary in the case of man's activities, *they would have to run in terror from nature.* They would have to use one-half of the world to construct protective containers or barriers against all the allegedly deadly carcinogens, toxins, and radioactive material that constitutes the other half of the world.

It would be a profound mistake to dismiss the repeatedly false claims of the environmentalists merely as a case of the little boy who cried wolf. They are a case of the *wolf* crying again and again about alleged dangers to the little boy. The only real danger is to listen to the wolf.

Direct evidence of the wilful dishonesty of the environmental movement comes from one of its leading representatives Stephen Schneider, who is well-known for his predictions of global catastrophe. In the October 1989 issue of *Discover* magazine, he is quoted (with approval) as follows:

> . . . To do this, we need to get some broad-based support, to capture the public's imagination. That, of course, entails getting loads of media coverage. So we have to offer up scary scenarios, make simplified, dramatic statements, and make little mention of any doubts we may have. This "double ethical bind" we frequently find ourselves in cannot be solved by any formula. Fach of us has to decide what the right balance is between being effective and being honest.

Thus, in the absence of verification by sources totally independent of the environmental movement and free of its taint, all of its claims of seeking to improve human life and well-being in this or that specific way must be regarded simply as lies, having the actual purpose of inflicting needless deprivation or suffering. In the category of malicious lies fall all of the environmental movement's claims about our having to abandon industrial civilization or any significant part of it in order to cope with the dangers of alleged global warming, ozone depletion, or exhaustion of natural resources. Indeed, all claims constituting denunciations of science, technology, or industrial civilization which are advanced in the name of service to human life and well-being are tantamount to claiming that our survival and well-being depend on our abandonment of reason. (Science, technology, and industry are leading products of reason and are inseparable from it.) All such claims should be taken as nothing but further proof of the environmental movement's hatred of man's nature and man's life, certainly not of any actual significant danger to human life and well-being.

It is important to realize that when the environmentalists talk about destruction of the "environment" as the result of economic activity, their claims are permeated by the doctrine of intrinsic value. Thus, what they actually mean to a very great extent is merely the destruction of alleged intrinsic values in nature such as jungles, deserts, rock formations, and animal species which are either of no value to man or which are hostile to man. That is their concept of the "environment." If, in contrast to the environmentalists, one means by "environment" *the surroundings of man*—the external material conditions of human life—then it becomes clear that all of man's productive activities have the inherent tendency to *improve* his environment—indeed, that that is their essential purpose.

This becomes obvious if one realizes that the entire world physically consists of nothing but chemical elements. These elements are never destroyed. They simply reappear in different combinations, in different proportions, in different places. Apart from what has been lost in a few rockets, the quantity of every chemical element in the world today is the same as it was before the Industrial Revolution. The only difference is that, because of the Industrial Revolution, instead of lying dormant, out of man's control, the chemical elements have been moved about, as never before, in such a way as to improve human life and well-being. For instance, some part of the world's iron and copper has been moved from the interior of the earth, where it was useless, to now constitute buildings, bridges, automobiles, and a million and one other things of benefit to human life. Some part of the world's carbon, oxygen, and hydrogen has been separated from certain compounds and recombined in others, in the process releasing energy to heat and light homes, power industrial machinery, automobiles, airplanes, ships, and railroad trains, and in countless other ways serve human life. It follows that insofar as man's environment consists of the chemical elements iron, copper, carbon, oxygen, and hydrogen, and his productive activity makes them useful to himself in these ways, his environment is correspondingly improved.

All that *all* of man's productive activities fundamentally consist of is the rearrangement of nature-given chemical elements for the purpose of making them stand in a more useful relationship to himself—that is, for the purpose of improving his environment.

Consider further examples. To live, man needs to be able to move his person and his goods from place to place. If an untamed forest stands in his way, such movement is difficult or impossible. It represents an improvement in his environment, therefore, when man moves the chemical elements that constitute some of the trees of the forest somewhere else and lays down the chemical elements brought from somewhere else to constitute a road. It is an improvement in his environment when man builds bridges, digs canals, opens mines, clears land, constructs factories and houses, or does anything else that represents an improvement in the external, material conditions of his life. All of these things represent an improvement in man's material surroundings—his environment. All of them represent the rearrangement of nature's elements in a way that makes them stand in a more useful relationship to human life and well-being.

Thus, all of economic activity has as its sole purpose the improvement of the environment—it aims exclusively at the improvement of the external, material conditions of human life. Production and economic activity are precisely the means by which man adapts his environment to himself and thereby improves it.

So much for the environmentalists' claims about man's destruction of the environment. Only from the perspective of the alleged intrinsic value of nature and the nonvalue of man, can man's improvement of his environment be termed destruction of the environment.

The environmentalists' recent claims about the impending destruction of the "planet" are entirely the result of the influence of the intrinsic value doctrine. What the environmentalists are actually afraid of is not that the planet or its ability to support human life will be destroyed, but that *the increase in its ability to support human life* will destroy its still extensively existing *"wildness."* They cannot bear the thought of the earth's becoming fully subject to man's control, with its jungles and deserts replaced by farms, pastures, and forests planted by man, as man wills. They cannot bear the thought of the earth's becoming man's garden. In the words of McKibben, "The problem is that nature, the independent force that has surrounded us since our earliest days, cannot coexist with our numbers and our habits. *We may well be able to create a world that can support our numbers and our habits,* but it will be an artificial world . . . ." (Italics supplied.)

The toxic character of the environmental movement implies the observance of a vital principle in connection with any measures which the movement advocates and which might actually promote human life and well-being, such as those calling for the reduction of smog, the cleaning up of rivers, lakes, and beaches, and so forth. The principle is that even here one must not make common cause with the environmental movement in any way. One must be scrupulously careful not to advocate even anything that is genuinely good, under its auspices or banner. To do so is to promote its evil—to become contaminated with its poison and to spread its poison. In the hands of the environmentalists, concern even with such genuine problems as smog and polluted rivers serves as a weapon with which to attack industrial civilization. The environmentalists proceed as though problems of filth emanated from industrial civilization, as though filth were not the all-pervasive condition of human life in pre-industrial societies, and as though industrial civilization represented a decline from more healthful conditions of the past.

The principle of noncooperation with the environmental movement, of the most radical differentiation from it, must be followed in order to avoid the kind of disastrous consequences brought about earlier in this century by people in Russia and Germany who began as basically innocent and with good intentions. Even though the actual goals and programs of the Communists and Nazis were no secret, many people did not realize that such pronouncements and their underlying philosophy must be taken seriously. As a result they joined with the Communists or Nazis in efforts to achieve what they believed were worthy specific goals, above all, goals falling under the head of the alleviation of poverty. But working side by side with the likes of Lenin and Stalin or Hitler and Himmler, did not achieve the kind of life these people had hoped to achieve. It did, however, serve to achieve the bloody goals of those monsters. And along the way, those who may have started out innocently enough very quickly lost their innocence and to varying degrees ended up simply as accomplices of the monsters.

Evil needs the cooperation of the good to disguise its nature and to gain numbers and influence it could never achieve on its own. Thus, the doctrine of intrinsic value needs to be mixed as much as possible with alleged concern for man's life and well-being. In allowing themselves to participate in advancing the cause of the mixture, otherwise good people serve to promote the doctrine of intrinsic value and thus the destruction of human values.

Already large numbers of otherwise good people have been enlisted in the environmentalists' campaign to throttle the production of energy. This is a campaign which, to the degree that it succeeds, can only cause human deprivation and the substitution of man's limited muscle power for the power of motors and engines. It is actually a campaign which seeks nothing less than *the undoing of the Industrial Revolution,* and the return of the poverty, filth, and misery of earlier centuries.

The essential feature of the Industrial Revolution is the use of *man-made power.* To the relatively feeble muscles of draft animals and the still more feeble muscles of human beings, and to the relatively small amounts of useable power available from nature in the form of wind and falling water, the Industrial Revolution added man-made power. It did so first in the form of steam generated from the combustion of coal, and later in the form of internal combustion based on petroleum, and electric power based on the burning of any fossil fuel or on atomic energy.

This man-made power is the essential basis of all of the economic improvements achieved over the last two hundred years. Its application is what enables us human beings to accomplish with our arms and hands the amazing productive results we do accomplish. To the feeble powers of our arms and hands is added the enormously greater power released by these sources of energy. Energy use, the productivity of labor, and the standard of living are inseparably connected, with the two last entirely dependent on the first.

Thus, it is not surprising, for example, that the United States enjoys the world's highest standard of living. This is a direct result of the fact that the United States has the world's highest energy consumption per capita. The United States, more than any other country, is the country where intelligent human beings have arranged for motor-driven machinery to accomplish results for them. All further substantial increases in the productivity of labor and standard of living, both here in the United States and across the world, will be equally dependent on man-made power and the growing consumption of energy it makes possible. Our ability to accomplish more and more with the same limited muscular powers of our limbs will depend entirely on our ability to augment them further and further with the aid of still more such energy.

In total opposition to the Industrial Revolution and all the marvelous results it has accomplished, the essential goal of environmentalism is to block the increase in one source of man-made power after another and ultimately to roll back the production of man-made power to the point of virtual nonexistence, thereby undoing the Industrial Revolution and returning the world to the economic Dark Ages. There is to be no atomic power. According to the environmentalists, it represents the death ray. There is also to be no power based on fossil fuels. According to the environmentalists, it causes "pollution," and now global warming, and must therefore be given up. There is not even to be significant hydro-power. According to the environmentalists, the building of the necessary dams destroys intrinsically valuable wildlife habitat.

Only three things are to be permitted as sources of energy, according to the environmentalists. Two of them, "solar power" and power from windmills, are, as far as can be seen, utterly impracticable as significant sources of energy. If somehow, they became practicable, the environmentalists would undoubtedly find grounds for attacking them. The third allowable source of energy, "conservation," is a contradiction in terms. "Conservation" is *not* a source of energy. Its actual meaning is simply using less. Conservation is a source of energy for one use only at the price of deprivation of energy use somewhere else.

The environmentalists' campaign against energy calls to mind the image of a boa constrictor entwining itself about the body of its victim and slowly squeezing the life out of him. There can be no other result for the economic system of the industrialized world but enfeeblement and ultimately death if its supplies of energy are progressively choked off.

Large numbers of people have been enlisted in the campaign against energy out of fear that the average mean temperature of the world may rise a few degrees in the next century, mainly as the result of the burning of fossil fuels. If this were really to be so, the only appropriate response would be to be sure that more and better air conditioners were available. (Similarly, if there were in fact to be some reduction in the ozone layer, the appropriate response, to avoid the additional cases of skin cancer that would allegedly occur from exposure, to more intense sunlight, would be to be sure that there were more sunglasses, hats, and sun-tan lotion available.) It would *not* be to seek to throttle and destroy industrial civilization.

If one did not understand its underlying motivation, the environmental movement's resort to the fear of global warming might appear astonishing in view of all the previous fears the movement has professed. These fears, in case anyone has forgotten, have concerned the alleged onset of *a new ice age* as the result of the same industrial development that is now supposed to result in global warming, and the alleged creation of a "nuclear winter" as the result of man's use of atomic explosives.

The words of Paul Ehrlich and his incredible claims in connection with the "greenhouse effect" should be recalled. In the first wave of ecological hysteria, this "scientist" declared:

> At the moment we cannot predict what the overall climatic results will be of our using the atmosphere as a garbage dump. We do know that very small changes in either direction in the average temperature of the Earth could be very serious. With a few degrees of cooling, a new ice age might be upon us, with rapid and drastic effects on the agricultural productivity of the temperate regions. With a few degrees of heating, the polar ice caps would melt, perhaps raising ocean levels 250 feet. Gondola to the Empire State Building, anyone?

The 250-foot rise in the sea level projected by Ehrlich as the result of global warming has been scaled back somewhat. According to McKibben, the "worst case scenario" is now supposed to be eleven feet, by the year 2100, with something less than seven feet considered more likely. According to a United Nations panel of alleged scientists, it is supposed to be 25.6 inches. (Even this still more limited projected rise did not stop the UN panel from calling for an immediate 60 percent reduction in carbon-dioxide emissions to try to prevent it.)

Perhaps of even greater significance is the continuous and profound distrust of science and technology that the environmental movement displays. The environmental movement maintains that science and technology cannot be relied upon to build a safe atomic power plant, to produce a pesticide that is safe, or even to bake a loaf of bread that is safe, if that loaf of bread contains chemical preservatives. When it comes to global warning, however, it turns out that there is one area in which the environmental movement displays the most breathtaking confidence in the reliability of science and technology, an area in which, until recently, no one—not even the staunchest supporters of science and technology—had ever thought to assert very much confidence at all. The one thing, the environmental movement holds, that science and technology can do so well that we are entitled to have unlimited confidence in them, is *forecast the weather*—for the next one hundred years!

It is, after all, supposedly on the basis of a weather forecast that we are being asked to abandon the Industrial Revolution, or, as it is euphemistically put "to radically and profoundly change the way in which we live"—to our enormous material detriment.

Very closely connected with this is something else that might appear amazing. This concerns prudence and caution. No matter what the assurances of scientists and engineers, based in every detail on the best established laws of physics—about backup systems, fail-safe systems, containment buildings as strong as U-boat pens, defenses in depth, and so on—when it comes to atomic power, the environmental movement is unwilling to gamble on the unborn children of fifty generations hence being exposed to harmful radiation. But on the strength of a weather forecast, it is willing to wreck the economic system of the modern world—to literally throw away industrial civilization. (The 60 percent reduction in carbon dioxide emissions urged by that United Nations panel would be utterly devastating in itself, totally apart from all the further such measures that would surely follow it.)

The meaning of this insanity is that industrial civilization is to be abandoned because this is what must be done to *avoid bad weather*. All right, very bad weather. If we destroy the energy base needed to produce and operate the construction equipment required to build strong, well-made, comfortable houses for hundreds of millions of people, we shall be safer from the wind and rain, the environmental movement alleges, than if we retain and enlarge that energy base. If we destroy our capacity to produce and operate refrigerators and air conditioners, we shall be better protected from hot weather than if we retain and enlarge that capacity, the environmental movement asserts.

There is actually a remarkable new principle implied here, concerning how man can cope with his environment. Instead of our taking action upon nature, as we have always believed we must do, we shall henceforth control the forces of nature more to our advantage by means of our *inaction*. Indeed, if we do not act, no significant threatening forces of nature will arise! The threatening forces of nature are not the product of nature, but of *us!* Thus speaks the environmental movement.

All of the insanities of the environmental movement become intelligible when one grasps the nature of the destructive motivation behind them. They are not uttered in the interest of man's life and well-being, but for the purpose of leading him to self-destruction.

It must be stressed that even if global warming turned out to be a fact, the free citizens of an industrial civilization would have no great difficulty in coping with it—that is, of course, if their ability to use energy and to produce is not crippled by the environmental movement and by government controls otherwise inspired. The seeming difficulties of coping with global warming, or any other large-scale change, arise only when the problem is viewed from the perspective of government central planners.

It would be too great a problem for government bureaucrats to handle (as is the production even of an adequate supply of wheat or nails—as the experience of the whole socialist world has so eloquently shown). But it would certainly not be too great a problem for tens and hundreds of millions of free, thinking individuals living under capitalism to solve. It would be solved by means of each individual being free to decide how best to cope with the particular aspects of global warming that affected him. Individuals would decide, on the basis of profit and loss calculations, what changes they needed to make in their businesses and in their personal lives, in order best to adjust to the situation. They would decide where it was now relatively more desirable to own land, locate farms and businesses, and live and work, and where it was relatively less desirable, and what new comparative advantages each location had for the production of which goods. The essential thing they would require is the freedom to serve their self-interests by buying land and moving their businesses to the areas rendered relatively more attractive, and the freedom to seek employment and buy or rent housing in those areas.

Given this freedom, the totality of the problem would be overcome. This is because, under capitalism, the actions of the individuals, and the thinking and planning behind those actions, are coordinated and harmonized by the price system (as many former central planners of Eastern Europe and the Soviet Union have come to learn). As a result the problem would be solved in exactly the same way that tens and hundreds of millions of free individuals have solved much greater problems, such as redesigning the economic system to deal with the replacement of the horse by the automobile, the settlement of the American West, and the release of the far greater part of the labor of the economic system from agriculture to industry.

Indeed, it would probably turn out that if the necessary adjustments were allowed to be made, global warming, if it actually came, would prove highly beneficial to mankind on net balance. For example, there is evidence suggesting that it would postpone the onset of the next ice age by a thousand years or more and that the higher level of carbon dioxide in the atmosphere, which is supposed to cause the warming process, would be highly beneficial to agriculture.

Whether global warming comes or not, it is certain that nature itself will sooner or later produce major changes in the climate. To deal with those changes and virtually all other changes arising from whatever cause, man absolutely requires individual freedom, science, and technology. In a word, he requires the industrial civilization constituted by capitalism.

This brings me back to the possibly truly good objectives that have been mixed in with environmentalism, such as the desire for greater cleanliness and health. If one wants to advocate such objectives without aiding the potential mass murderers in the environmental movement in achieving their goals, one must first of all accept unreservedly the values of human reason, science, technology, and industrial civilization, and never attack those values. They are the indispensable foundation for achieving greater cleanliness and health and longer life.

In the last two centuries, loyalty to these values has enabled man in the Western world to put an end to famines and plagues, and to eliminate the once dread diseases of cholera, diphtheria, smallpox, tuberculosis, and typhoid fever, among others. Famine has been ended, because the industrial civilization so hated by the environmentalists has produced the greatest abundance and variety of food in the history of the world, and created the transportation system required to bring it to everyone. This same hated civilization has produced the iron and steel pipe, and the chemical purification and pumping systems, that enable everyone to have instant access to safe drinking water, hot or cold, every minute of the day. It has produced the sewage systems and the automobiles that have removed the filth of human and animal waste from the streets of cities and towns.

Such improvements, together with the enormous reduction in fatigue and exhaustion made possible by the use of labor-saving machinery, have resulted in a radical reduction in mortality and increase in life expectancy, from less than thirty years before the beginning of the Industrial Revolution to more than seventy-five years currently. By the same token, the average newborn American child today has a greater chance of living to age sixty-five than the average newborn child of a nonindustrial society has of living to age five.

In the earlier years of the Industrial Revolution, the process of improvement was accompanied by the presence of coal dust in towns and cities, which people willingly accepted as the by-product of not having to freeze and of being able to have all the other advantages of an industrial society. Subsequent advances, in the form of electricity and natural gas, have radically reduced this problem. Those who seek further advances along these lines, should advocate the freedom of development of atomic power, which emits no particulate matter of any kind into the atmosphere. Atomic power, however, is the form of power most hated by the environmentalists.

Also essential for further improvements in cleanliness and health, and for the long-term availability of natural resources, is the extension of private ownership of the means of production, especially of land and natural resources. The incentive of private owners is to use their property in ways that maximize its long-term value and, wherever possible, to improve their property. Consistent with this fact one should seek ways for extending the principle of private ownership to lakes, rivers, beaches, and even to portions of the ocean. Privately owned lakes, rivers, and beaches, would almost certainly be clean lakes, rivers, and beaches. Privately owned, electronically fenced ocean ranches would guarantee abundant supplies of almost everything useful that is found in or beneath the sea. Certainly, the vast land holdings of the United States government in the western states and in Alaska should be privatized.

But what is most important in the present context, in which the environmental movement is operating almost unopposed, is that anyone who is afraid of becoming physically contaminated by exposure to one or another alleged toxic chemical should take heed that he does not place an indelible stain on his very existence through his exposure to the deadly poison of the environmental movement. This is what one is in danger of doing by ingesting the propaganda of the environmental movement and being guided by it. I do not know of anything worse that anyone can do than, having been born into the greatest material civilization in the history of the world, now take part in its destruction by cooperation with the environmental movement, and thus be a party to untold misery and death in the decades and generations to come.

By the same token, there are few things better that one can do than, having become aware of what is involved, take one's stand with the values on which human life and well-being depend. This is something which, unfortunately, one must be prepared to do with few companions in today's world. The great majority of those who should be fighting for human values—the professional intellectuals—either do not know enough to do so, have become afraid to do so, or, still worse, have themselves become the enemies of human values and are actively working on the side of environmentalism.

It is important to explain why there are so few intellectuals prepared to fight environmentalism and why there are so many who are on its side.

I believe that to an important extent the hatred of man and distrust of reason displayed by the environmental movement is a psychological projection of many contemporary intellectuals' self-hatred and distrust of their own minds arising as the result of their having been responsible for the destruction wrought by socialism. As the parties responsible for socialism, they have certainly been "a plague upon the world," and if socialism had in fact represented reason and science, as they continue to choose to believe, there would be grounds to distrust reason and science.

In my judgment, the "green" movement of the environmentalists is merely the old "red" movement of the communists and socialists shorn of its veneer of science. The only difference I see between the greens and the reds is the superficial one of the specific reasons for which they want to violate individual liberty and the pursuit of happiness. The reds claimed that the individual could not be left free because the result would be such things as "exploitation" and "monopoly." The greens claim that the individual cannot be left free, because the result will be such things as destruction of the ozone layer and global warming. Both claim that centralized government control over economic activity is essential. The reds wanted it for the alleged sake of achieving human prosperity. The greens want it for the alleged sake of avoiding environmental damage. In my view, environmentalism and ecology are nothing but the intellectual death rattle of socialism in the West, the final convulsion of a movement that only a few decades ago eagerly looked forward to the results of paralyzing the actions of individuals by means of "social engineering" and now seeks to paralyze the actions of individuals by means of prohibiting engineering of any kind. The greens, I think, may be a cut below the reds, if that is possible.

While the collapse of socialism is an important precipitating factor in the rise of environmentalism, there are other, more fundamental causes as well.

Environmentalism is the leading manifestation of the rising tide of irrationalism that is engulfing our culture. Over the last two centuries, the reliability of reason as a means of knowledge has been under a constant attack led by a series of philosophers from Immanuel Kant to Bertrand Russell. As a result, a growing loss of confidence in reason has taken place. As a further result the philosophical status of man, as the being who is distinguished by the possession of reason, has been in decline. In the last two generations, as the effects of this process have more and more reached the general public, confidence in the reliability of reason, and the philosophical status of man, have declined so far that now virtually no basis is any longer recognized for a radical differentiation between man and animals. This is the explanation of the fact that the doctrine of St. Francis of Assisi and the environmentalists concerning the equality between man and animals is now accepted with virtually no opposition.

The readiness of people to accept the closely connected doctrine of intrinsic values is also a consequence of the growing irrationalism. An "intrinsic value" is a value that one accepts without any reason, without asking questions. It is a "value" designed for people who do what they are told and who do not think. A rational value, in contrast is a value one accepts only on the basis of understanding how it serves the self-evidently desirable ultimate end that is constituted by one's own life and happiness.

The cultural decline of reason has created the growing hatred and hostility on which environmentalism feeds, as well as the unreasoning fears of its leaders and followers. To the degree that people abandon reason, they must feel terror before reality, because they have no way of dealing with it other than reason. By the same token, their frustrations mount, since reason is their only means of solving problems and achieving the results they want to achieve. In addition, the abandonment of reason leads to more and more suffering as the result of others' irrationality, including their use of physical force. Thus, in the conditions of a collapse of rationality, frustrations and feelings of hatred and hostility rapidly multiply, while cool judgment, rational standards, and civilized behavior vanish. In such a cultural environment, monstrous ideologies appear and monsters in human form emerge alongside them, ready to put them into practice. The environmental movement of course, is just such a movement.

But if, because of these reasons, there are no longer many intellectuals ready to take up the fight for human values—in essence, for the value of the intellect, for man the rational being and for the industrial civilization he has created and requires—then all the greater is the credit for whoever is willing to stand up for these values now and, in so doing, don the mantle of intellectual.

There is certainly ample work for such "new intellectuals" to do.

At one level, the work directly concerns the issue of environmentalism.

The American people must be made aware of what environmentalism actually stands for and of what they stand to lose, and have already lost, as the result of its growing influence. They must be made aware of the environmental movement's responsibility for the energy crisis and the accompanying high price of oil and oil products, which is the result of its systematic and highly successful campaign against additional energy supplies. They must be made aware of its consequent responsibility for the enrichment of Arab sheiks at the expense of the impoverishment of hundreds of millions of people around the world, including many millions here in the United States. They must be made aware of its responsibility for the vastly increased wealth, power, and influence of terrorist governments in the Middle East, stemming from the high price of oil it has caused, and for the resulting need to fight a war in the region.

The American people must be made aware of how the environmental movement has steadily made life more difficult for them. They must be shown how, as the result of its existence, people have been prevented from taking one necessary and relatively simple action after another, such as building power plants and roads, extending airport runways, and even establishing new garbage dumps. They must be shown how the history of the environmental movement is a history of destruction: of the atomic power industry, of the Johns Manville Company, of cranberry growers and apple growers, of sawmills and logging companies, of paper mills, of metal smelters, of coal mines, of steel mines, of tuna fishermen, of oil fields and oil refineries—to name only those which come readily to mind. They must be shown how the environmental movement has been the cause of the wanton violation of private property rights and thereby of untold thousands of acres of land not being developed for the benefit of human beings, and thus of countless homes and factories not being built They must be shown how as the result of all the necessary actions it prohibits or makes more expensive, the environmental movement has been a major cause of the marked deterioration in the conditions in which most people now must live their lives in the United States—that it is the cause of families earning less and having to pay more, and, as a result, being deprived of the ability to own their own home or even to get by at all without having to work a good deal harder than used to be necessary.

In sum, the American people need to be shown how the actual nature of the environmental movement is that of a *virulent pest*, consistently coming between man and the work he must do to sustain and improve his life.

If and when such understanding develops on the part of the American people, it will be possible to accomplish the appropriate remedy. This would include the repeal of every law and regulation in any way tainted by the doctrine of intrinsic value, such as the endangered species act. It would also include repeal of all legislation requiring the banning of man-made chemicals merely because a statistical correlation with cancer in laboratory animals can be established when the chemicals are fed to the animals in massive, inherently destructive doses. The overriding purpose and nature of the remedy would be to break the constricting grip of environmentalism and make it possible for man to resume the increase in his productive powers in the United States in the remaining years of this century and in the new century ahead.

In addition to all of this vital work, there is a second and even more important level on which the new intellectuals must work. This, ironically enough, entails a form of cleaning up of the environment—*the philosophical, intellectual, and cultural environment.*

What the cultural acceptance of a doctrine as irrational as environmentalism makes clear is that the real problem of the industrialized world is not "environmental pollution" but *philosophical corruption.* The so-called *intellectual mainstream* of the Western world has been fouled with a whole array of intellectual toxins resulting from the undermining of reason and the status of man, and which further contribute, to this deadly process. Among them, besides environmentalism, are collectivism in its various forms of Marxism, racism, nationalism, and feminism; and cultural relativism, determinism, logical positivism, existentialism, linguistic analysis, behaviorism, Freudianism, Keynesianism, and more.

These doctrines are intellectual toxins because they constitute a systematic attack on one or more major aspects of the requirements of human life and well-being. Marxism results in the kind of disastrous conditions now prevailing in Eastern Europe and the Soviet Union. All the varieties of collectivism deny the free will and rationality of the individual and attribute his ideas, character, and vital interests to his membership in a collective: namely, his membership in an economic class, racial group, nationality, or sex, as the case may be, depending on the specific variety of collectivism. Because they view ideas as determined by group membership, these doctrines deny the very possibility of knowledge. Their effect is the creation of conflict between members of different groups: for example, between businessmen and wage earners, blacks and whites, English speakers and French speakers, men and women.

Determinism, the doctrine that man's actions are controlled by forces beyond his power of choice, and existentialism, the philosophy that man is trapped in a "human condition" of inescapable misery, lead people not to make choices they could have made and which would have improved their lives. Cultural relativism denies the objective value of modern civilization and thus undercuts both people's valuation of modern civilization and their willingness to work hard to achieve personal values in the context of it. The doctrine blinds people to the objective value of such marvelous advances as automobiles and electric light, and thus prepares the ground for the sacrifice of modern civilization to such nebulous and, by comparison, utterly trivial values as "unpolluted air."

Logical positivism denies the possibility of knowing anything with certainty about the real world. Linguistic analysis regards the search for truth as a trivial word game. Behaviorism denies the existence of consciousness. Freudianism regards the conscious mind (the "Ego") as surrounded by the warring forces of the unconscious mind in the form of the "Id" and the "Superego," and thus as being incapable of exercising substantial influence on the individual's behavior. Keynesianism regards wars, earthquakes, and pyramid building as sources of prosperity. It looks to peacetime government budget deficits and inflation of the money supply as a good substitute for these allegedly beneficial phenomena. Its effects, as the present-day economy of the United States bears witness, are the erosion of the buying power of money, of saving and capital accumulation, and of credit.

These intellectual toxins can be seen bobbing up and down in the "intellectual mainstream," just as raw sewage can be seen floating in a dirty river. Indeed, they fill the intellectual mainstream. Virtually, every college and university in the Western world is a philosophical cesspool of these doctrines, in which intellectually helpless students are immersed for several years and then turned loose to contaminate the rest of society. These irrationalist doctrines, and others like them, *are the philosophical substance of contemporary liberal arts education.*

Clearly, the most urgent task confronting the Western world, and the new intellectuals who must lead it, is *a philosophical and intellectual cleanup.* Without it, Western civilization simply cannot survive. It will be killed by the poison of environmentalism.

To accomplish this cleanup, only the most powerful, industrial-strength, philosophical and intellectual cleansing agents will do. These cleansing agents are, above all, the writings of Ayn Rand and Ludwig Von Mises. These two towering intellects are, respectively, the leading advocates of reason and capitalism in the twentieth century. A philosophical-intellectual cleanup requires that all or most of their writings be introduced into colleges and universities as an essential part of the core curriculum, and that what is not included in the core curriculum be included in the more advanced programs. The incorporation of the writings of Ayn Rand and Ludwig Von Mises into a prominent place in the educational curriculum is the central goal that everyone should work for who is concerned about his cultural environment and the impact of that environment on his life and well-being. Only after this goal is accomplished, will there by any possibility that colleges and universities will cease to be centers of civilization-destroying intellectual disease. Only after it is accomplished on a large scale, at the leading colleges and universities, can there be any possibility of the intellectual mainstream someday being clean enough for rational people to drink from its waters.

The 21st Century should be the century when man begins the colonization of the solar system, not a return to the Dark Ages. Which it will be, will depend on the extent to which new intellectuals can succeed in restoring to the cultural environment the values of reason and capitalism.

Individual copies of this essay are available in pamphlet form from The Jefferson School of Philosophy, Economics, and Psychology, P.O. Box 2934, Laguna Hills, CA 92654. Single copy price: $4.25. Two to ten copies: $2.75 each. Fifty or more copies sent to the same address and used for distribution only (that is, not for resale) $1 each. Prices subject to change without notice. California residents, please add sales tax. Postage charges will be added to Canadian and foreign orders of fifty copies or more.